	III A 13	IV A 14	V A 15	VI A 16	VII A 17	He 4.003
	5 B 10.81	6 C 12.011	7 N 14.007	8 O 15.999	9 F 18.998	10 Ne 20.179
	13 Al 26.982	14 Si 28.086	15 P 30.974	16 S 32.06	17 Cl 35.453	18 Ar 39.948

10	I B 11	II B 12						
Ni 58.71	29 Cu 63.546	30 Zn 65.37	31 Ga 69.72	32 Ge 72.59	33 As 74.922	34 Se 78.96	35 Br 79.904	36 Kr 83.80
Pd 106.4	47 Ag 107.868	48 Cd 112.40	49 In 114.82	50 Sn 118.69	51 Sb 121.75	52 Te 127.60	53 I 126.905	54 Xe 131.30
Pt 195.09	79 Au 196.966	80 Hg 200.59	81 Tl 204.37	82 Pb 207.19	83 Bi 208.2	84 Po (210)	85 At (210)	86 Rn (222)

	65 Tb 158.925	66 Dy 162.50	67 Ho 164.930	68 Er 167.26	69 Tm 168.934	70 Yb 173.04	71 Lu 174.97
Gd 157.25							
Cm (247)	97 Bk (247)	98 Cf (251)	99 Es (254)	100 Fm (257)	101 Md (256)	102 No (254)	103 Lr (257)

Inorganic Chemistry

Inorganic Chemistry

PRINCIPLES AND APPLICATIONS

Ian S. Butler
John F. Harrod
McGill University

The Benjamin/Cummings Publishing Company, Inc.

Redwood City, California • Fort Collins, Colorado
Menlo Park, California • Reading, Massachusetts • New York
Don Mills, Ontario • Wokingham, U.K. • Amsterdam • Bonn
Sydney • Singapore • Tokyo • Madrid • San Juan

Sponsoring Editor: Diane Bowen
Production Coordinator: Bruce Lundquist
Copy Editor: Betty Duncan-Todd
Designer: Michael Rogondino
Art Coordinator: Pat Rogondino
Artists: Mary Burkhardt, Mark Hall, Pat Rogondino, Carol Verbeeck
Composition: Polyglot Pty. Ltd.

About the cover: This computer-generated graphic depicts a superconducting compound from the crystallographic group, the "perovskites." This compound is composed of yttrium (gray), barium (green), copper (blue), and oxygen (red). The image is reproduced with permission from The I.B.M. Corporation.

Library of Congress Cataloging-in-Publication Data
Butler, Ian S.
Inorganic chemistry.
Includes bibliographies and index.
1. Chemistry, Inorganic. I. Harrod, John F. (John Frank) II. Title.
QD151. 2. B88   1989   546   88-34400
ISBN 0-8053-0247-6

ABCDEFGHIJ-XX-89321098

The Benjamin/Cummings Publishing Company, Inc.
390 Bridge Parkway
Redwood City, California 94065

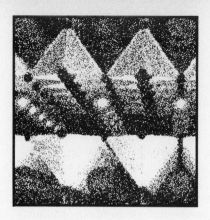

Preface

'To the writing of many books there is no end and much study is a weariness of the flesh'—Ecclesiastes 12:12

These ancient phrases echo some oft-heard laments of teachers and students down through the ages. On the other hand, if you ask any teacher or student if the perfect text has been written in his or her discipline, the answer is always negative. There are so many different approaches to the material and so many different opinions as to which subject matter is it necessary to cover and which is it not, that the best an author can aim for is to satisfy a large proportion of the potential audience.

This book is designed for use in the standard undergraduate specialized courses in inorganic chemistry and our decision to write it was predicated on dissatisfaction with the existing canon of texts. We believe that the most basic requirement of a modern inorganic chemistry text is a good balance between theoretical development, descriptive chemistry, and applications. Some of the available texts have encyclopedic coverage of descriptive chemistry and strong theoretical development, but almost no mention of applications. Others have a bias towards theory and physical inorganic chemistry, but little descriptive chemistry or applications.

The subject matter of inorganic chemistry has greatly expanded in the past decade, particularly into the realm of materials science. Solid state chemistry, especially the synthesis and characterization of new multi-component compounds, has experienced a new surge of growth, and the dominance of transition metal chemistry, which was evident in the 1960s and 70s, has begun to be challenged by a renaissance of interest in the chemistry of the compounds of the later main group elements and the bioinorganic chemistry of the main group elements. It is clearly no longer useful to write an encyclopedic text aimed at the undergraduate student which, at the same time, gives adequate introduction to all major topics and serves the interests of the beginning student in terms of clarity of focus and economy of words and dollars. We have tried to achieve a breadth of coverage sufficient to give instructors some topical choice, but not so extensive that the student buys largely unused reference material. A special emphasis is placed on the fact that inorganic chemistry has played, and continues to play, an important role in technology and that the developments in the subject at both the fundamental and applied levels are strongly coupled to each other.

Organization

The book is divided into eight parts: I, Theoretical Basis; II, Molecular Structure and Analytical Methods; III, Periodic Trends for the Elements and Simple Compounds; IV, Complex Compounds: Coordination Chemistry; V, Complex Compounds: Rings, Chains, Cages, and Clusters; VI, Solid-State Chemistry; VII, Solution Chemistry; VIII, Advanced Topics in Coordination Chemistry.

In the first part atomic and molecular theory is treated in a conventional review manner. Although all of the major empirical and semi-empirical bonding approaches are mentioned, emphasis is given to the qualitative MO view and most interpretation of experimental observations is achieved using the MO approach.

The second part reviews some of the more important experimental methods used in the structural characterization of inorganic and organometallic compounds. The inclusion of a major section on methods of structure elucidation is unusual for this type of text, but given the importance of these analytical techniques to the development of modern inorganic chemistry it is essential that the student have the opportunity to learn at least a little of the instrumental methods used to arrive at the theoretical and structural conclusions. The section is introduced with a description of symmetry and the manner in which symmetry ideas link theory and structural methods and provide a concise language for discussing structure and bonding.

The evolution of materials science in recent years has made the treatment of binary and ternary compounds an essential part of a modern inorganic course. To this end we have reviewed in Part III the chemical properties of the binary compounds of the elements in terms of the two archetypal classes (i) the hydrides and (ii) the oxides and halides. The descriptive chemistries of these classes are presented and related to the fundamental principles developed in the theoretical introduction. As far as possible it is intended that the relationships outlined in these descriptive chemistry sections will allow the student to deduce the behavior of other classes of binary compounds. This approach attempts to compromise between an encyclopedic coverage of descriptive chemistry, which inevitably would lead to an unbalanced or unduly long text, and the neglect of descriptive chemistry, which leaves the student with an undesirably impoverished knowledge of the patterns of chemical combination. We have also made every effort to emphasize the widespread importance of binary inorganic compounds in technological applications.

In the fourth part the classical coordination theory of Werner and crystal field theory are introduced. These materials are fairly conventional, reflecting the durability of the theoretical underpinnings.

The section on large molecules in Part V introduces the student to inorganic polymer chemistry, a subject of lively activity at the present time. Connections are made between bonding theories, molecular architecture, and properties. In addition, this section serves to further elaborate the student's understanding of the periodic properties of the covalent binary compounds of the p-group elements.

Part VI on solid-state chemistry is innovative in that it makes a strong connection between the basic structural chemistry of solids, treated universally in inorganic chemistry texts, and many physical and mechanical properties such as lubricity, corrosion resistance, mechanical integrity, permeability etc., which are hardly, if, ever touched upon in conventional texts. The storage battery is extensively used to illustrate the degree to which simple structural phenomena at the atomic level can be used to explain the properties of classical batteries and how they can be applied to the development of new battery systems. The material treated in Chapters 4 (The Ionic Bond), 13 (Crystal Field Theory), and 18 (Ordered Solids) permit an introductory treatment of the structure of high temperature superconductors. The solid state section also treats the glassy state in some detail and gives one of the rare accounts, to the authors' knowledge, of the chemical and physical principles of electrophotography (xerography) to be found in a chemistry textbook.

The section on solution chemistry (Part VII) covers a wide range of topics, many of which are treated in all inorganic texts, but a number of others emphasize areas in which basic inorganic chemistry has been coopted by interdisciplinary areas such as earth and atmospheric sciences and metallurgy. This section also gathers together introductory material on the mechanisms of inorganic and organometallic reactions in solution.

The remaining section (Part VIII) gives the student an introduction to three advanced topics in coordination chemistry. These areas, organometallic chemistry, catalysis, and bioinorganic chemistry, are generally recognized as having had a dominant influence on the development of inorganic chemistry over the past two or three decades.

Theory and Applications

There is a strong emphasis throughout the book on drawing the connections between the basic theoretical principles and the numerous technological applications of inorganic chemistry, including corrosion, lubrication, image reproduction, storage batteries, electronics, etc. This philosophy is summarized by the following quotation from Chapter 2:

"Yet chemistry remains very much a practical science, closely linked to old and new technologies. It is our intention to try to draw together the threads that connect the most fundamental principles of inorganic chemistry with the technological applications of the discipline. These two domains, *principles* and *applications*, are like the roots and leaves of a tree. The roots grow, but they change little from year to year; old leaves are shed and replaced by new, but each generation of leaves is indispensable to the sustenance and survival of the tree as a whole. Let us begin with the roots."

Nomenclature

Shortly after we started on this project, the Nomenclature Committee of the International Union of Pure and Applied Chemistry proposed a new group numbering system (1–18) for the periodic table. Since most chemical literature is in the older form (IA, IIA, etc.) and since the new system is presently far from universally accepted, we have opted to use both sytems throughout, e.g., IA (1), VIB (6), etc.

Study Aids

While not intending to detract from the undoubted importance of research, we recognize that the majority of students at the level for which this book is intended will not go on to become researchers. For this reason we have opted to provide a few lead literature references at the end of each chapter, rather than providing an exhaustive citation of the literature in the form of footnoted references. A Solutions Manual features solutions to a portion of the book's problems.

Acknowledgements

Our sincere thanks are due to many: the giants who have gone before us and from whose shoulders we have presumed to glimpse the wonder and beauty of chemistry; the reviewers who read various sections (Christopher Alan, University of Vermont; Alan Balch, University of California, Davis; Tristram Chivers, University of Calgary; William Evans, University of California, Irvine; Herbert Kaesz, University of California, Los Angeles; John Nelson, University of Nevada, Reno; Cortlandt Pierpoint, University of Colorado; Anthony Poë, University of Toronto; Philip Power, University of California, Davis; Robert Scott, University of Illinois, Champaign-Urbana; Sedig Wasfi, Delaware State College; Stanley Williamson, University of California, Santa Cruz) and particularly John Cooper, University of Pittsburgh, who read the entire manuscript. We also acknowledge the skillful management of the project by Diane Bowen, the impressive editorial and book production work by Evelyn Dahlgren and Bruce Lundquist and the creative marketing efforts of Rajeev Samantrai. Our thanks to our families, Pamela, Wendy, Jeffrey, Kate, and Kimberly (ISB) and Leanore, Jean-Pierre, Rachel and Ariel (JFH) for their support and understanding throughout the project. Lastly, to our departmental colleagues and students for advice and patience.

Ian Sydney Butler and John Frank Harrod,

Montreal, September, 1988

Contents

1

What Is Inorganic Chemistry?

For many years, chemistry has been conveniently divided into the subdisciplines of organic, inorganic, physical, and analytical chemistries. Such divisions are convenient for systematically separating the chemistry curriculum, for crudely identifying the interests and training of an individual, and for giving practicing chemists a tribal identity. These classical subdivisions do, however, present problems as the subject of chemistry evolves. For example, we can now identify the subdivisions of biochemistry and theoretical chemistry. Biochemistry has evolved to the point where many universities give it the stature of a separate discipline, but theoretical chemistry is still in the process of establishing its autonomy as a subdivision.

Organic and inorganic chemistries are old-established subdivisions, and at one time it could be claimed that the former was the study of the compounds of carbon, whereas the latter was the study of the compounds of the remaining elements. The tremendous growth of organometallic chemistry in the past fifty years has now blurred that elementary distinction, but it can still serve as a rough guideline.

Because inorganic chemistry deals with essentially the whole periodic table, the inorganic curriculum has become one of the main foci at which the student is introduced to qualitative atomic and molecular theory. This introduction is usually conducted at a level that is the minimum for an appreciation of the phenomena giving rise to periodicity in chemical properties and to chemical bonding. These subjects are usually treated more rigorously, but in a much narrower context, in physical and theoretical chemistry courses.

A thorough understanding of periodic properties is the cornerstone of inorganic chemistry. We gain insight into the behavior of the elements both from the ways in which they conform to elementary periodic principles and from the ways in which they deviate. Introductory theory conveniently expresses and teaches elementary principles, but a true appreciation of how real systems conform to or deviate from these elementary principles can only be gained from a thorough knowledge of the descriptive chemistry of the elements.

In the first half of the nineteenth century, inorganic chemistry was mainly concerned with identifying new elements and with systematically studying the chemistry of simple chemical compounds. This knowledge culminated in the formulation of the periodic table by Mendeleev and in a number of important new commercial applications of inorganic chemistry. Two of the main technological advances that provided impetus for the Industrial Revolution were the Bessemer process for production of steel and the invention of Portland cement, both of which are examples of complex inorganic chemistry.

In the second half of the nineteenth century, chemists became fascinated with what we today call coordination compounds. These compounds did not comply with the ideas of valence that had been developed up to that time, and by the turn of the century Alfred Werner had begun to develop his new coordination theory to explain their behavior. Werner's structural theory was a giant stride in the development of inorganic chemistry, but the weaknesses in the more general theories of valence retarded full development of coordination chemistry until the 1950s.

Figure 1.1 shows the sequence of Nobel Prizes to scientists who made their contributions in the area of inorganic chemistry. From this figure, it is clear that for the first half of the twentieth century inorganic chemistry continued to be preoccupied with filling in the periodic table, initially by filling in the gaps, but later by adding synthetic elements such as technetium, astatine, and promethium. A complete understanding of the table also required development of the concepts of isotopes, radioactive decay, nuclear fission, and nuclear fusion. Scientists working in these areas garnered nine of the eleven Nobel Chemistry Prizes awarded in the area of

Figure 1.1
A historical summary of Nobel Prizes awarded in areas strongly related to inorganic chemistry. The entries marked with an asterisk were awarded for physics, the remainder were for chemistry.

1990	
1988*	A. Müller and G. Bednorz: Mixed–metal oxide superconductors
1983	H. Taube: Mechanisms of electron-transfer reactions
1980	
1976	W. N. Lipscomb: Structure and bonding in boranes
1973	E. O. Fischer and G. Wilkinson: Organometallic sandwich compounds
1970	
1963	K. Ziegler and G. Natta: Catalytic polymerization of alkenes
1960	
1951	E. M. McMillan and G. Seaborg: Transuranic elements
1950	
1944	O. Hahn: Nuclear fission
1940	
1935	F. Joliot and I. Joliot-Curie: Synthesis of new radioelements
1934	H. C. Urey: Isotopes of hydrogen
1930	
1921	F. Soddy: Radioactivity and isotopes
1920	
1918	F. Haber: Catalytic synthesis of ammonia
1915*	W. H. Bragg and W. L. Bragg: X-ray crystallography
1913	A. Werner: Theory of the structure of coordination compounds
1911	M. Curie: Isolation and study of radium and polonium
1910	
1908	E. Rutherford: Radioactivity and radiochemistry
1906	H. Moissan: Isolation and study of fluorine
1904	W. Ramsay: Inert gases and their place in periodic table
1900	

inorganic chemistry between 1904–1951. The prizes awarded to Werner and Haber stand as lonely harbingers of the great surge of interest in coordination chemistry and catalysis in the second half of the twentieth century.

The culmination of the process of discovery and synthesis of the elements was the development of sophisticated theories of nuclear synthesis, which give a detailed description of the origins of the elements in terms of stellar evolution.

The current model of cosmic origins and evolution is the so-called big bang theory. An approximate version of this theory starts with the universe as a singularity with infinite density and temperature. Under such conditions, only radiation can exist. Sudden expansion of this singularity leads to a rapid drop in temperature and density; after a few milliseconds, it is cool enough for matter to begin forming. A few hours later, a series of nuclear particle transformations has converted most of the universe's energy to protons and deuterons. The universe today still appears to be composed mostly of hydrogen, and the ratio of protons to deuterons has changed little in the approximately 10^{10} years since the big bang was presumed to have occurred.

The initial great clouds of turbulent hydrogen gas developed local density fluctuations, and the gravity of regions of higher density began the processes of galaxy and star formation. The protostars attracted more and more hydrogen, and their ever-increasing mass and density caused the temperature to rise. Eventually, it reached about 10^7 K, the temperature at which hydrogen begins to undergo nuclear fusion according to Equation 1.1:

$$4^1\text{H} \rightarrow {}^4\text{He} + 2\beta^+ + 2 \text{ neutrinos} + 2.5 \times 10^6 \text{ MJ mol}^{-1} \qquad \textbf{(1.1)}$$

This fusion provides energy to replace that lost by radiation and to stop the further contraction of the stellar nucleus. The star then enters a period of stable equilibrium. After about one-tenth of the hydrogen has fused to helium, the temperature of the core has risen by an order of magnitude, and the fusion of helium to heavier nuclei begins.

The most important direct product of helium burning is carbon (^{12}C). The ^{12}C can then initiate a series of helium reactions of the kind shown in Equations 1.2–1.4:

$$^{12}\text{C} + {}^4\text{He} \rightarrow {}^{16}\text{O} \qquad \textbf{(1.2)}$$

$$^{16}\text{O} + {}^4\text{He} \rightarrow {}^{20}\text{Ne} \qquad \textbf{(1.3)}$$

$$^{20}\text{Ne} + {}^4\text{He} \rightarrow {}^{24}\text{Mg} \qquad \textbf{(1.4)}$$

When the helium is depleted, the core temperature rises again to about 5×10^8 K, and fusion of carbon atoms can occur:

$$2^{12}\text{C} \rightarrow {}^{23}\text{Na} + {}^1\text{H} \qquad \textbf{(1.5)}$$

$$2^{12}\text{C} \rightarrow {}^{20}\text{Ne} + {}^4\text{He} \qquad \textbf{(1.6)}$$

Processes such as those described above can account accurately for the cosmic abundances of the elements up to Sc (element 21). The abundances of the elements from Ti to Cu (with the exception of Fe) are explained by a variety of nuclear reactions, which can occur rapidly within exploding, unstable stars called supernovas. The exceptionally high abundance of iron is because it has the maximum nuclear binding energy. All elements lighter than iron can be made exothermically by fusion, but the synthesis of elements beyond iron by fusion is endothermic. Thus, synthesis of the heavier elements requires processes other than fusion. The most important of these processes involve neutron capture followed by radioactive decay, and they are believed to occur in the interiors of large, relatively cool stars known as red giants. Examples of neutron capture reactions are shown in Equations 1.7 and 1.8:

$$^{68}Zn + {}^1n \rightarrow {}^{69}Zn \rightarrow {}^{69}Ga + \beta^- \tag{1.7}$$

$$^{80}Br + {}^1n \rightarrow {}^{81}Br \rightarrow {}^{81}Kr + \beta^- \tag{1.8}$$

The abundances of elements produced by such reactions are determined by the ease with which the product isotope captures a neutron (this is quantified as the neutron capture cross section). Those isotopes that have a low-capture cross section tend to have unusually large cosmic abundances.

The great allure of nuclear and radiochemistry was tarnished by the broadening recognition of the dangers associated with nuclear weapons and with the use of nuclear fission for power generation. Whatever the rights and wrongs of the various positions taken on these issues, there is absolutely no doubt that interest in nuclear and radiochemistry research has greatly declined from its peak in the 1960s. A measure of the intense emotion surrounding issues of nuclear chemistry and physics is the fact that nuclear magnetic resonance (which takes advantage of the absolutely benign phenomenon of nuclear spin and not of nuclear transmutation) is referred to in the health-care world as "magnetic resonance imaging." The word *nuclear* is deliberately not used to avoid causing clients unnecessary anxiety! Despite these problems, radiochemistry still finds wide application in medical diagnosis.

An important spinoff of the strategic importance of nuclear chemical research during and after World War II was the enormous amount of research in inorganic chemistry that was only indirectly linked to nuclear fission. Many of the major contributors to the development of inorganic chemistry in the second half of this century began their careers in research related to nuclear chemistry and physics. The fission process gives rise to complicated mixtures of exotic and dangerously radioactive isotopes. The handling of these fission products necessitated the development of new ways of separating the elements. This in turn required intense study of the solution properties of metal complexes, mechanisms of their reactions, selective complexation of metal ions, and the development of new analytical and separation methods.

Another area of inorganic chemistry that gained some impetus from the nuclear energy field was the exploration of new materials with exceptional corrosion resistance, high-temperature resistance, and great mechanical strength. The dangers of radioactive releases placed new demands on the requirements for structural materials in the nuclear industry. The metal zirconium has many desirable properties that lend themselves to applications in nuclear reactors, including high resistance to damage by neutron fluxes, corrosion resistance, and good high-temperature mechanical performance. A great deal had to be learned about the chemistry of this relatively rare element before it could be safely used in the construction of reactors.

Despite the great growth in inorganic chemistry in the early nuclear era, the subject fell into the doldrums during the 1940s and 1950s because weaknesses in the bonding theories of the day left many of the empirical facts unexplained and disorganized. Much of inorganic chemistry resisted satisfactory explanation by the theoretical models that were serving organic chemistry so well. The transition metal complexes, many possessing unpaired electrons, presented serious problems. Electron-deficient compounds, best exemplified by the boron hydrides, strained theory to its limits. The early 1950s saw the beginning of a dramatic reversal of this situation.

The discovery of ferrocene in 1951 and the discoveries in the area of organometallic transition metal–catalyzed alkene polymerization in the mid-1950s brought matters to a head. Main-group organometallic compounds—compounds containing a main-group metal–carbon bond—had been known for many years, and some of them showed considerable chemical stability. Numerous attempts at preparation of compounds with transition metal–carbon bonds had failed, and the few that succeeded gave products with bizarre compositions that defied explanation. Therefore, it came to be generally believed that the transition elements could not form stable bonds to carbon. Imagine then the shock when Pauson and Kealey (1951) and Miller, Tebboth, and Tremaine (1952) presented the world with ferrocene, a compound containing only iron, carbon, and hydrogen. This orange crystalline solid is indefinitely stable in air, sublimable in air at temperatures greater than 100°C, and unattacked by water, dilute acids, or bases.

Ferrocene was discovered at a time when several powerful new structural analytical tools were becoming widely available for the first time. Infrared spectroscopy and nuclear magnetic resonance spectroscopy played an important part in the initial assignment of the sandwich structure of ferrocene by Wilkinson and his associates (1952), and this structure was confirmed shortly afterward by X-ray crystallography. It is interesting to note that bis(arene)chromium compounds, which contain all the structural and bonding phenomena that make ferrocene so interesting, were first made by Hein in 1919. However, the much greater reactivity of these compounds, the absence of structure determination methods, and the absence of adequate bonding theory condemned them to obscurity until the structure of bis(benzene)chromium was finally elucidated by Zeiss, Wheatley, and Winkler in 1954. The structures of ferrocene and bis-(benzene)chromium are shown in Figure 1.2.

The greatest significance of ferrocene was its impact on the theory of bonding in transition metal compounds and that it was discovered at a time when other developments made the realization of its full impact possible. It

Figure 1.2
The structures of
ferrocene and
bis(benzene)
chromium.

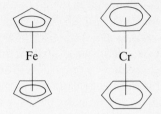

is interesting that no significant application for ferrocene has yet been found! The metal carbonyl compounds, some discovered much earlier than ferrocene, had the potential for a similar impact on bonding theories, but the chemical world was not equipped to deal with them at the time of their discovery. The metal carbonyls were nevertheless put to use in several catalytic organic syntheses before their structures and bonding were understood.

The discovery of organometallic transition metal catalysts for the polymerization of alkenes, pioneered by Ziegler and Natta, complemented the discovery of ferrocene in two important ways. First, the initial impact of Ziegler–Natta catalysis was largely commercial. At the time of Ziegler's discovery of low-pressure polymerization of ethylene, polyethylene was already a major commercial commodity. However, its manufacture required using extremely high pressures in a batch process, and the behavior of the reaction was extremely erratic. Ziegler polyethylene can be made on the laboratory bench by bubbling ethylene through a solution of a mixture of triethylaluminum and titanium tetrachloride (both the reactants and the catalyst must be carefully protected from air and water, however). Even today, the full details of the chemistry occurring during the formation of the catalyst are not understood. A few of the many likely reactions are shown in Equations 1.9–1.11:

$$TiCl_4 + Et_3Al \rightarrow EtTiCl_3 + Et_2AlCl \tag{1.9}$$

$$2EtTiCl_3 \rightarrow 2TiCl_3 + C_2H_4 + C_2H_6 \tag{1.10}$$

$$EtTiCl_3 + Et_2AlCl \rightarrow \underset{Cl}{\overset{Et}{\underset{\diagdown}{\diagup}}}Ti\underset{\diagdown Cl}{\overset{\diagup Cl \diagdown}{}}Al\underset{\diagdown Et}{\overset{Et}{}} \tag{1.11}$$

The full commercialization of Ziegler polymerization took much longer than might be expected for such a simple procedure, but it was evident from the start that it was a commercial winner. This was also true of Natta's remarkable series of discoveries of Ziegler-type catalysts that exerted a high level of control over the stereochemistry of chiral carbon atoms in polyalkenes and the geometry about the double bonds in the chains of polydienes. Again, the historical moment was right for the commercial exploitation of these discoveries: The chemical industry was rapidly moving from a coal base to a petroleum base, and vast quantities of alkenic compounds were being produced as by-products in oil refineries.

The second impact of the Ziegler–Natta work was the opening up of a vast unexplored wilderness of organometallic chemistry. It was immediately clear that the many possible variations on the theme of mixing of a main-group alkylating agent with a transition metal compound were likely to lead to exciting new organometallic compounds. Also, there was a good possibility that these compounds would catalyze commercially interesting reactions. This expectation has been amply met over the past four decades.

The pathways of discovery, which can be traced back to those exciting events of the 1950s, include the revelation of unusual new modes of bonding: nitrogen molecules bonded to metals, exotic unstable organic species such as cyclobutadiene bonded to metals, even the hydrogen molecule bonded to metals. Transition (and main-group) organometallics have provided the organic chemist with a bewildering arsenal of synthetic methods, and it is almost unthinkable that a current total synthesis of a complex natural product will not include several steps that involve organometallic reagents. Recently, there have been a number of exciting developments in the transformation of organometallic compounds into useful electronic materials and inorganic ceramics under unusually mild conditions. Organo-polysilanes have been pyrolyzed to give silicon carbide fibers of extraordinary strength and chemical and heat resistance. Unstable, volatile organotitanium compounds have been used to lay down films of titanium carbide under very mild conditions, as shown in Equation 1.12:

$$Ti[CH_2C(CH_3)_3]_4 \xrightarrow{150°C} TiC + hydrocarbons \qquad (1.12)$$

Semiconductor films can be prepared by chemical vapor deposition using volatile organometallic compounds, as shown in Equation 1.13.

$$Me_3Ga(g) + Me_3As(g) \rightarrow GaAs(s) + hydrocarbons \qquad (1.13)$$

Controlled hydrolysis of alkoxysilanes can be used to create glass films or objects at temperatures much lower than required in conventional glass-making processes.

Contemporaneous with the explosive growth of organometallic chemistry, a number of other technologies were undergoing exponential growth. Perhaps the most impressive of all was what we may loosely call electronics: the useful manipulation of electrons. The first era of electronics was dominated by the movement of charge through metals and gases (or a vacuum). The primary components of devices were made of metals and glass. The invention of the transistor changed all that, and electronics quickly moved over to the transportation of charge through solid semiconductors. A wide range of simple inorganic compounds and elements exhibit the property of semiconduction. Silicon has been, and will probably continue to be for the foreseeable future, the workhorse electronic semiconductor, but many other materials such as cadmium sulfide (CdS) and gallium arsenide (GaAs) are widely used. As the electronics of communications, information processing, and computing get more advanced, the search for materials with unusual conducting, magnetic, and photophysical/photochemical properties intensifies. A major leap in this direction was the recent discovery of high-temperature superconductivity in mixed-metal oxides (for example, $YBa_2Cu_3O_7$). These materials may have a revolutionary impact on the production, transmission, and use of electricity as well as important uses in electronics.

The pace of development in the areas of inorganic chemistry described would have been impossible without the parallel development of theory and structure determination methods. Both of these have profited from the rapid growth of computing power per dollar over the past forty years. Qualitative molecular orbital theory now allows us to systematically describe the bonding in electron-deficient and electron-excess molecules and to account for many of the subtle details of molecular structure encountered in inorganic compounds. Fast, low-cost data processing has made crystal-structure determination routine for many inorganic compounds, and many thousands of structures have now been determined.

One last branch of inorganic chemistry is bioinorganic chemistry: the study of the structure and function of biological compounds containing metals or semimetals. This subject has blossomed under the same influences as other areas of inorganic chemistry. The elucidation of the structure of extremely complex molecules lies at the heart of much bioinorganic chemistry. Crystallography plays an important role, but metalloproteins frequently cannot be obtained as crystals suitable for X-ray work. In such cases, the whole armory of spectroscopic methods is brought to bear sometimes with evident success, other times with less. Another powerful tool in bioinorganic chemistry (and, incidentally, in catalysis) is the synthesis of model compounds in which the coordination of the metal mimics that present in the natural compound.

In this introductory chapter, we have tried to give you a flavor of the history and present status of inorganic chemistry. This is summarized in the flowchart shown in Figure 1.3. The subject has had its ups and downs, but it is our conviction that for the past three decades inorganic chemistry has been experiencing and continues to experience a strong and exciting growth in fundamental knowledge, which is continually paying off in major applications.

Figure 1.3
A historical overview of the development of inorganic chemistry in the twentieth century.

1869	Mendeleev: The Periodic Table
1890s	Werner: The Coordination Theory
1916	Lewis and Sidgwick: The Coordinate Bond

1930s	Bethe and Van Vleck: Crystal Field Theory	Pauling: Valence Bond Theory Mulliken: Molecular Orbital Theory

1940s

Synthesis of Elements
Isotope Separation:
• Chromatography
• Thermodynamics
• Chelation
• Solvent Extraction
• Polarography

1950s	Ligand Field Theory	Reaction Mechanisms

1960s	Structure Determination	Homogeneous Catalysis	Organometallic Synthesis

1970s	Bioinorganic Chemistry	C_1 Chemistry[1] Selective Oxidation Photo- and Electrocatalysis	Metal–Metal Bonds

1980s	Multiple Bonds Between Heavier p-Group Atoms Electron-Rich Molecules	Inorganic Polymers Unusual Materials	Clusters

1990s	?

[1] C_1 chemistry involves study of catalytic processes for converting C_1 compounds—for example, CO, CH_4—into molecules containing more than one carbon atom. This interest is strongly coupled to fossil-fuel economics because effective use of coal (source of cheap CO) and natural gas (mainly CH_4) requires good catalytic processes for homologation of C chains.

BIBLIOGRAPHY

Suggested Reading

Basolo, F. and R. Pearson. *Mechanisms of Inorganic Reactions in Solution*, 2nd ed. New York: Wiley, 1967.

Bethe, H. *Ann. Physik*. 5, no. 3 (1929) (crystal field theory).

Burbidge, E. M., G. R. Burbidge, W. A. Fowler, and F. Hoyle. "Synthesis of the Elements in Stars." *Rev. Mod. Phys.* 29 (1957): 547.

J. Chem. Ed. 63 (1986): 188 (proceedings of an ACS symposium on industrial applications of organometallic chemistry).

Miller, S. A., T. A. Tebboth, and J. F. Tremaine. *J. Chem. Soc.* (1952): 632.

Natta, G. *J. Polymer Sci.* 16 (1955): 143 (sterospecific polymerization).

Pauling, L. *The Nature of the Chemical Bond*. Ithaca, N.Y.: Cornell University Press, 1967.

Pauson, P. L. and T. L. Kealy. *Nature* 168 (1951): 1039 (ferrocene).

"Photovoltaic Cells." Special Report *Chem. & Eng. News* (7 July 1986): 34.

Scientific American. "Materials for Economic Growth" (Oct. 1986).

Seaborg, G. T., ed. *Transuranium Elements: Products of Modern Alchemy*. Stroudsburg, Penn.: Dowden, Hutchinson and Ross, 1978.

Taube, H. *Adv. Inorg. Chem. Radiochem*. 1 (1959): 1 (mechanisms of oxidation/reduction reactions).

Transformation of Organometallics into Common and Exotic Materials: Design and Activation, edited by R. M. Laine. Dordrecht, Holland: NATO Advanced Science Institute Series, Martinus Nijhoff, 1988.

Van Vleck, J. H. *J. Chem. Phys.* 3 (1935): 803 (molecular orbital theory for metal complexes).

Wells, A. F. *Structural Inorganic Chemistry*. London: Oxford University Press, 1984.

Werner Centennial. *Advances in Chemistry Series*, Vol. 62. Washington, D.C.: American Chemical Society, 1967.

Wilkinson, G., M. Rosenblum, M. C. Whiting, and R. B. Woodward. *J. Am. Chem. Soc.* 74 (1952): 2125 (structure of ferrocene).

Ziegler, K. *Angew. Chem.* 16 (1955): 424 (low-pressure ethylene polymerization).

PART ONE

Theoretical Basis

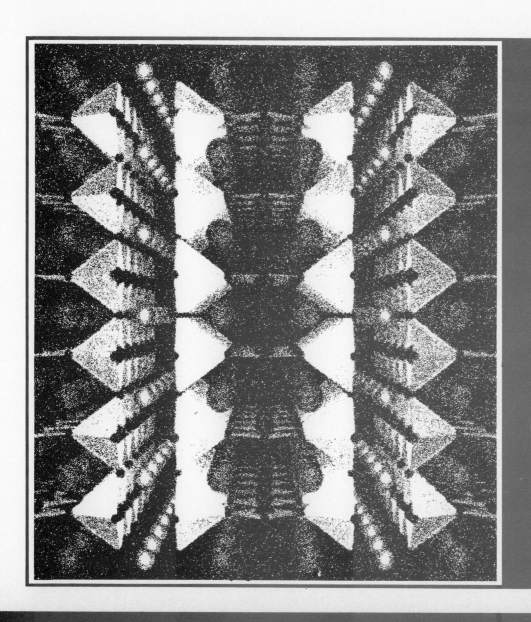

2

Review of Atomic Theory and Atomic Properties

2.1   A Historical Perspective

Since the beginning of recorded history, people have carried out activities that we would today identify as chemical arts. Understanding has been pursued on two broad fronts: inventing practical arts for the transformation of substances and developing a theoretical understanding of the fundamental composition of matter. For most of this long historical process, the theoretical framework was largely unscientific and was based only to a small extent on experimental evidence. Nevertheless, by the time the scientific revolution of the eighteenth and nineteenth centuries began to give birth to the physical sciences, the notions that matter is composed of tiny invisible particles (atoms), that these particles exist in a relatively restricted number of different types (elements), and that different combinations of these elements give rise to the innumerable forms of matter had been around, if not universally accepted, since the time of Democritus in the fourth century B.C.

The advances of the past 100 years have given us a rather clear picture of the detailed nature of atoms. We can be reasonably sure that we have identified all elements that presently exist on earth, together with some that may yet be made artificially or that may even exist in exotic parts of the universe.

Given this powerful theoretical framework, is it not surprising that we continually uncover new chemistry and new physicochemical phenomena by accident? In fact, it is not. For reasons we will develop later, the quantitative aspects of our theories are subject to some fundamental limitations that mar their precision. Because small differences in physical parameters can make significant differences in chemical behavior, lack of precision can render theoretical predictions useless. For the synthetic research chemist, the slogan will always be *vive la différence* because the real thrills in synthesis are usually the result of the unexpected. The theoretician, on the other hand, has a vested interest in imposing order and predictability on the discipline.

Before leaving this question of the differing sentiments of the exploratory practitioner and the theoretician, it is worth noting some of the differences between theoretical physicists and theoretical chemists. The theoretical physicist is conventionally perceived as being concerned with "fundamental" questions. One hundred years ago, the fundamental questions related to the nature of matter were largely focused on the nature of atoms, and those questions were being dealt with by physicists. This continued to be the case until the nucleus and fundamental particles superceded the atom as the main preoccupation of physicists; the atom was then left in the hands of theoretical chemists. The main interest of chemists was to extend the understanding of atoms into the domain of collections of bound atoms (molecules). More recently, the interests of chemists and physicists in the nature of chemical bonding have reconverged in the disciplines of solid-state chemistry and physics. The concepts of localized and delocalized orbitals, developed by chemists, are now indispensable partners to the delocalized band theory, developed by physicists to explain the electronic properties of solids.

The end of the nineteenth century and the first decades of the twentieth century saw physics in a state of profound revolution, a state brought about by the inability of classical theory to explain the interactions between matter and light and the apparent structure of the atom. The classical theories of optics and the elegant, successful equations of Maxwell (1873), based on a model of light as an electromagnetic wave, created an overwhelming prejudice in favor of light being a wave phenomenon. Similarly, the vast body of knowledge accumulated on the behavior of electric charges seemed to make it self-evident that charges were particulate. However, as more experimentalists began to turn their attention to explaining the interaction between light and matter, the inadequacies of the great prejudices became more and more evident. To explain the energy distribution of light emitted by a black body, Planck (1900) was led to postulate the quantization of the energy of the oscillators producing the light. But so ingrained was prejudice in favor of the wave nature of light, Planck was very slow to accept the conclusion of Einstein (1905), based on his interpretation of the photoelectric effect, that light itself is quantized. Meanwhile, thirty years had passed between the first orderly classifications of the lines in the spectrum of the hydrogen atom by Balmer (1885) and Rydberg (1900) and the proposal of an empirical, electrostatic model for the structure of the atom by Bohr (1913). In his perceptive leap, Bohr succeeded in shrugging off the prejudice in favor of the wave nature of light, but he could not break through the limitation of perceiving charges as particles. His theory therefore remained empirical, with quantization an unexplained postulate. We will leave the final triumphant storming of the Bastille of classical physics to the next section and conclude our brief historical introduction with a glimpse of what was happening to chemistry during this period.

The end of the nineteenth century saw chemistry equipped with a number of empirical but very powerful theoretical tools. The periodic table of Mendeleev (1869) (Figure 2.1) provided not only a framework for rationalizing and predicting the chemistry of the known and unknown elements but also the key to its own completion. While physicists experienced frustration at the heart of their discipline, chemists enjoyed the hunt for new elements, which were indicated by their periodic table. Meanwhile, the new ideas on bonding in carbon compounds developed by Van't Hoff (1874), LeBel (1874), and Kekulé (1865) and the coordination theory of Werner (1893) revealed a landscape of new synthetic and structural chemistry of almost unimaginable variety. The freedom that these new concepts provided—to explore and advance the ability to transform or to create new chemical compounds—relieved chemists of any pressing need to profoundly understand the theoretical underpinning of their science.

By the 1930s, the great revolution in atomic physics had largely worked itself out, and quantum theory and wave mechanics were the established bulwarks of the new order. Ironically, Einstein, who contributed so much to this new order, was unable to accept one of the fundamental new principles: the uncertainty principle of Heisenberg (1927). Two brilliant young American chemists, Pauling and Mulliken, had no such qualms. They both grasped the crucial importance of the great advances in atomic physics (of which they became aware during their studies in Europe) to an understanding of the structure and properties of chem-

Row	Group I — R_2O	Group II — RO	Group III — R_2O_3	Group IV RH_4 RO_2	Group V RH_3 R_2O_2	Group VI RH_2 RO_3	Group VII RH R_2O_3	Group VIII — RO_4
1	H = 1							
2	Li = 7	Be = 9.4	B = 11	C = 12	N = 14	O = 16	F = 19	
3	Na = 23	Mg = 24	Al = 27.3	Si = 28	P = 31	S = 32	Cl = 35.5	
4	K = 39	Ca = 40	— = 44	Ti = 48	V = 51	Cr = 52	Mn = 55	Fe = 56, Co = 59, Ni = 59, Cu = 63
5	(Cu = 63)	Zn = 65	— = 68	— = 72	As = 75	Se = 78	Br = 80	
6	Rb = 85	Sr = 87	?Yt = 88	Zr = 90	Nb = 94	Mo = 96	— = 100	Ru = 104, Rh = 104, Pd = 106, Ag = 108
7	(Ag = 108)	Cd = 112	In = 113	Sn = 118	Sb = 122	Te = 125	I = 127	
8	Cs = 133	Ba = 137	?Di = 138	?Ce = 140				
9								
10			?Er = 178	?La = 180	Ta = 182	W = 184		Os = 195, Ir = 197, Pt = 198, Au = 199
11	(Au = 199)	Hg = 200	Tl = 204	Pb = 207	Bi = 208			
12				Th = 231		U = 240		

Figure 2.1 Mendeleev's periodic table in its original English version (1871).

ical compounds. Following slightly different paths, they laid the basis for qualitative theories of chemical bonding that soon provided a rational interpretation of much of the enormous body of empirical knowledge gathered over the preceding century. Meanwhile, these theoretical advances pointed the way to new frontiers in experimental chemistry and the development of more quantitative theoretical treatments of bonding.

Until the early 1970s, the new theories of chemical bonding were severely limited in quantitative applications because of the very laborious computations required for even very simple problems. The advent and rapid development of electronic computers, together with the adoption of more sophisticated mathematical methods, have changed this situation; applied quantum mechanics is now a major branch of chemistry. Yet chemistry remains very much a practical science, closely linked to old and new technologies. Our intent in this book is to try to draw together the threads that connect the most fundamental principles of inorganic chemistry with the technological applications of the discipline. These two domains, principles and applications, are like the roots and leaves of a tree. The roots grow, but they change little from year to year; old leaves are shed and replaced by new, but each generation of leaves is indispensable to the sustenance and survival of the tree as a whole. Let us begin with the roots.

2.2 The Quantum Mechanical View of the Atom

The Bohr–Sommerfeld Model

Bohr (1913) devised a model for the hydrogen atom, based on a particulate electron in orbit about a central proton and bound by coulombic attraction. The goal of this model was to satisfactorily explain the line spectrum of the hydrogen atom. Bohr found that this could indeed be done if the following two conditions were postulated:

1. In the absence of any absorption or emission of radiation, the energy of the planetary electron remains constant (stationary state).

2. Absorption or emission of radiation can only occur in discrete amounts, corresponding to the passage of the electron from one stationary state, E_n, to another, E_m, so that

$$|E_n - E_m| = h\nu \tag{2.1}$$

where h = Planck's constant and ν = frequency of the radiation.

These two postulates incorporated Planck's ideas of the quantization of the energy of the oscillators emitting and absorbing electromagnetic radiation and the proportional relationship between the change in the oscillator's energy and the frequency of the emitted radiation into the model for the hydrogen atom. Combining these postulates with the classical electrostatic picture of an electron orbiting a proton yielded the Bohr equation for the difference in energy between two stationary states of the hydrogen atom:

$$|E_n - E_m| = h\nu = \frac{Z^2 m e^4}{8\varepsilon_0^2 h^2}\left(\frac{1}{n_n^2} - \frac{1}{n_m^2}\right) \tag{2.2}$$

where m = rest mass of the electron, e = electron charge, Z = proton charge, and ε_0 = permitivity of a vacuum. The Bohr equation may be compared to the empirical formula of Rydberg for the separation of any two lines of the hydrogen spectrum:

$$\frac{\nu}{c} = R\left(\frac{1}{n_m^2} - \frac{1}{n_n^2}\right) \tag{2.3}$$

where c = velocity of light in a vacuum and R = Rydberg constant. Combining Equations 2.2 and 2.3 gives

$$R = \frac{Z^2 m e^4}{8\varepsilon_0^2 h^3 c} \tag{2.4}$$

The remarkably close correspondence between the best experimental value of R (1.09678×10^7 m^{-1}) and that calculated from the simple Bohr equation must have aroused enormous excitement in its discoverer. (Try substituting the values for the universal constants in Equation 2.4 to see how close you come to the experimental value.) By substituting $n_n = 1$ and $n_m = \infty$ into Equation 2.2, the energy required to completely remove an electron from a ground-state hydrogen atom (1.3 MJ) was also easily calculated; this also agreed very well with the best experimental value. Finally, it should be mentioned that the Bohr equation can be rewritten in a form that contains the radius of the electron's orbit and that values for radii can be calculated from the resulting expression. In fact, the quantity r_n, the hypothetical radius of the orbit of an electron in quantum state n, could not be experimentally measured:

$$r_n = \frac{\varepsilon_0 h^2}{Z\pi m e^2} n^2 \tag{2.5}$$

However, attention is drawn to this "Bohr radius" because it will be referred to again when we discuss the wave mechanical view of the atom.

Following its early success, the Bohr theory was quickly tested by exciting new discoveries in spectroscopy. In Holland, Zeeman (1896) had shown that application of a magnetic field to hydrogen (and other elements) led to splitting of the field-free lines into multiplets. Similar effects due to electric fields were demonstrated by Stark (1913) in Germany.

The simple theory of Bohr was modified by Sommerfeld (1915) to account for the fact that externally applied fields would naturally perturb the otherwise circular orbits of planetary electrons. The perturbations would introduce elipticity into the orbits and modify the angular momentum properties of the electron. The splittings of the lines could be explained by using classical ideas that had long been applied in celestial mechanics but with coulombic rather than gravitational potentials. The splitting of all lines into doublets by an appropriate magnetic field was explained by the notion that the electron was spinning about its own axis and its interaction energy with the magnetic field was different for the two possible directions of spin. Later investigations postulated a coupling between the angular momenta of the electron's spin and of its orbital motion. Investigations of atoms with more than one electron further postulated the coupling of the spin and orbital momenta of the different electrons with each other. Again, these effects were all well established in celestial mechanics. There was, however, one persistent difference from classical, macroscopic mechanics that cropped up every time a new interaction was discovered in the atom: The energies of interaction in atomic systems are always quantized, whereas they are not (at least not measurably so) in macroscopic systems.

The failure of the Bohr theory to have a built-in explanation for quantization was a constant goad to theoretical physicists to continue their search for a more fundamental explanation for the behavior of atoms. The search ended through the brilliant insights of de Broglie (1924) and Schrödinger (1926) and with the birth of wave mechanics.

Wave Mechanics

By the beginning of the 1920s, the wave-particle duality of light had received much attention. The physics community in general was beginning to accept the validity of Einstein's photon and was becoming reconciled to the fact that the exhibition of wave and particle properties was not mutually exclusive. The wave-particle duality of the photon is most succinctly expressed by combining Planck's equation, which expresses energy in terms of wave properties

$$E = h\nu, \quad \text{or} \quad \frac{hc}{\lambda} \tag{2.6}$$

where λ = wavelength, and Einstein's equation for the energy equivalence of mass

$$E = m'c^2 \tag{2.7}$$

where m' = apparent mass of photon, to give Equation 2.8:

$$\nu = \frac{m'c^2}{h} \qquad \text{or} \qquad \lambda = \frac{h}{m'c} = \frac{h}{p'} \tag{2.8}$$

where p' = momentum of photon. All expressions relate a wave property (ν or λ) directly to a particle property (m' or p').

The next giant step was taken by de Broglie, who realized that this wave-particle duality is not necessarily unique to the photon but may also be exhibited by other particles, in particular the electron. He therefore suggested that expressions equivalent to Equation 2.8 could also apply to electrons. In this case, the wavelength would be given by the quotient of Planck's constant and the momentum of the electron (p):

$$\lambda = \frac{h}{mv} = \frac{h}{p} \tag{2.9}$$

A simple calculation shows that electrons accelerated through a potential of about 100 V have a wavelength of about 10 nm. The experimental demonstration of the wave property of electrons was shown by Davisson and Germer (1927), who recognized that this wavelength was in the order of lattice spacings in crystals, and the experiments showed that crystals diffract electrons in much the same way as they do X rays. Today, the diffraction of electrons by both fluids and solids has become an important tool in the determination of molecular structure.

De Broglie's expression (Equation 2.9) applies to electrons moving in field-free space, the kind of electrons that are easily produced by thermionic emission or photoemission. The condition of the electron in the atom is quite different because it is subject to intense electric fields from the nucleus or from other electrons. The electron in an atom is imprisoned by a three-dimensional force field that can only be broken by outside intervention. How can the wave property of such electrons be treated? The answer lies in recognizing that the particle waves in the atom are standing waves and have properties analogous to classical standing waves. Figure 2.2 illustrates an important property of a standing wave (such as a vibrating banjo string): Only those waves whose amplitude is zero at the boundaries can persist. All hypothetical waves with nonzero amplitude at the boundaries are dissipated by self-interference on reflection from the boundaries. The extension of the one-dimensional banjo string to two dimensions is also shown in Figure 2.2. In this case, only those waves that close exactly

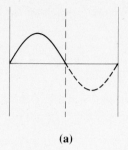

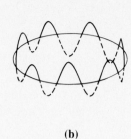

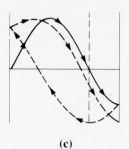

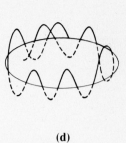

(a) (b) (c) (d)

Figure 2.2
Standing waves in one and two dimensions: (**a**) and (**b**) are waves that are closed and oscillate back and forth; (**c**) and (**d**) are nonclosing waves that continually cross each other out of phase and dissipate by interference.

on themselves survive as standing waves; those that do not close on themselves dissipate through self-interference.

The classical equation for a standing wave such as that shown in Figure 2.2 is

$$\Psi = A \sin \frac{2\pi x}{\lambda} \tag{2.10}$$

where Ψ = the displacement of the wave from the x axis and A = the maximum value of Ψ, or the amplitude. The first and second differentials of the equation are

$$\frac{d\Psi}{dx} = A \frac{2\pi}{\lambda} \cdot \cos \frac{2\pi x}{\lambda} \tag{2.11}$$

and

$$\frac{d^2\Psi}{dx^2} = -\frac{4\pi^2}{\lambda^2} \cdot \Psi \tag{2.12}$$

Using the same line of argument as de Broglie, we can now pursue the consequences of substituting a particle property for the wave property λ in Equation 2.12. A convenient quantity to use is the kinetic energy $T = \frac{1}{2} mv^2$. Using the de Broglie Equation 2.9 to replace v, we obtain

$$T = \frac{1}{2m} \left(\frac{h}{\lambda} \right)^2 \tag{2.13}$$

which allows us to substitute for λ^2 in Equation 2.12 to get

$$\frac{d^2\Psi}{dx^2} = -\frac{8\pi^2 m}{h^2} \cdot T \cdot \Psi \tag{2.14}$$

This is the wave equation for a particle with only kinetic energy, moving in field-free space. It could therefore be used to describe electrons moving in a vacuum, but not electrons moving in a strong electric field. If the electron is moving in a field, its total energy E is the sum of its kinetic energy and its potential energy V ($E = T + V$, or $T = E - V$). We can therefore replace T in Equation 2.14 by ($E - V$) to get

$$\frac{d^2\Psi}{dx^2} = -\frac{8\pi^2 m}{h^2}(E - V)\,\Psi \tag{2.15}$$

This equation has been derived for a one-dimensional wave like that shown in Figure 2.2a, but real systems are three-dimensional. The three-dimensional analog of Equation 2.15 is

$$\frac{\partial^2\Psi}{\partial x^2} + \frac{\partial^2\Psi}{\partial y^2} + \frac{\partial^2\Psi}{\partial z^2} = -\frac{8\pi^2 m}{h^2}(E - V)\,\Psi \tag{2.16}$$

This equation, developed by Schrödinger (1926), is the basis of the wave mechanical view of the atom. Because the constants and the potential function are known, the equation can in principle be solved for the unknowns Ψ (the wave function) and energy E. The beauty of the wave equation is that it provides a natural explanation for the quantization of energy in atoms. The energy solutions are not continuous but assume only certain allowed values. This is analogous to the banjo string, which can only vibrate at those frequencies that correspond to nodes at the frets.

The wave function has been a source of difficulty for several generations of students, but they have been in good company because Einstein even had difficulty with it up to the time he died. The problem is that the wave function has no perceivable physical significance, and Einstein had a strong intuitive distrust of anything that could not be reduced to a perceivable physical reality. However, the consensus of the theoretical physics community today is that Einstein's intuition in this case was ill-founded.

With the pragmatic view characteristic of chemists, we will not dwell on the nature of Ψ, beyond saying that it is a mathematical function that varies continuously with the dimensional variables, is finite for all values of the dimensional variables, and goes to zero when a dimensional variable goes to infinity. Each stationary state is characterized by its own wave function.

A critical advance in the use of the wave function in chemistry was the recognition by Born (1926) that the product of Ψ^2 and a small volume element $dx \cdot dy \cdot dz$ is proportional to the probability that the particle will be found in that volume element. It should be noted that this probability function for a classical Bohr-type atom would be zero in all regions of space inside or outside of the Bohr orbit. As we will see next, this is contrary to the predictions of the wave equation.

These two features of the solutions of the wave equation—sharp, discrete energies and wave functions that yield information on position in terms of statistical probabilities—are a manifestation of a broader and more fundamental property of microscopic systems. All communication between atomic matter is mediated by photons or by more massive particles. As long as the momentum of the communicating particle is small compared to the inertia of the sender and receiver of the message, the process of communication does not appreciably alter their states. However, if the momentum of the messenger is comparable to or greater than the inertia of either the sender or the receiver, the process of communication can profoundly alter the state of either in an unpredictable way. This fact finds expression in the uncertainty principle of Heisenberg (1927), one mathematical formulation of which is the following:

$$\Delta (x, y, z) \cdot \Delta (mv) \approx \frac{h}{4\pi} \tag{2.17}$$

where $\Delta (x, y, z)$ = the uncertainty in the position of a particle; $\Delta(mv)$ = the uncertainty in its momentum. We can slightly modify this equation:

$$\Delta (x, y, z) \cdot \Delta (v) \approx \frac{h}{4\pi m} \tag{2.18}$$

From this relation it is obvious that for macroscopic masses (for example, 1.0 g) the fundamental limitations on the precisions of position and velocity are infinitesimal compared to experimental imprecisions because $h = 10^{-35}$ Js. However, for an electron of rest mass of about 10^{-27} g, if we know the velocity within 1 ms^{-1}, we cannot know its position to better than about 10^{-9} m, that is, about one atomic radius. Similar limitations on the simultaneous definition of energy and time result from simple substitutions of the classical expression for energy in Equation 2.18 to give Equation 2.19:

$$\Delta E \cdot \Delta t \approx \frac{h}{4\pi} \tag{2.19}$$

The energy–time uncertainty relationship gives rise to the fact that nuclear decay processes, whose energetics can be defined precisely, cannot be precisely defined in terms of the timing of the decay. On the other hand, the line widths of atomic or molecular spectra are determined by the lifetimes of the excited states. In the UV/vis (visible) region of the spectrum, the limit of experimental resolution of lines is about 10^{-25} J, and uncertainty effects will not be detected for excited states with lifetimes longer than about 10^{-10} s. Very short-lived excited states with lifetimes of about 10^{-15} s cannot be defined energetically better than about 10^{-20} J. This uncertainty is approximately three orders of magnitude greater than the resolving power of most commercial IR or UV/vis spectrometers.

Thus, we can see that if we use wave mechanics in such a way that we derive precise values for energy or momentum for microscopic systems, the parameters of time or position will inevitably be imprecise. This imprecision is most conveniently quantified in terms of statistical quantities such as the half-life of a decay process or the most probable energy of a transition. Hence, this is the importance and usefulness of Ψ^2, which quantifies the probability of finding an electron in a particular position.

Solutions of the Wave Equation for the Hydrogen Atom

The hydrogen atom is important in testing the validity of the wave equation because it is one of the few cases for which an exact solution is possible. This limitation is not something peculiar to wave mechanics but is a general limitation to the solution of many-body problems. For example, whereas it is possible to calculate exactly and for all time the trajectory of a moon in orbit around a planet, it would not be possible to calculate exactly the trajectory of a third body such as a rocket ship voyaging between that planet and its moon. The trajectories of our space travelers are determined by using approximate solutions to the many-body equations of motion and by constantly improving the trajectory by iterative in-flight corrections. Part of the reason why the Bohr theory became intractable for atoms other than hydrogen was the inherent difficulty of solving the mechanics of many-body problems, even using a classical model.

For the case of the hydrogen atom, V in Equation 2.16 is simply the Coulomb potential $-Ze^2/4\pi\varepsilon_0 r$ (where Ze = the nuclear charge). The solution of this equation goes beyond the mathematical scope of this text but can be found in any good basic physical chemistry or quantum chemistry text. For mathematical simplicity, the equation is usually solved using polar rather than Cartesian coordinates; we will initially restrict our discussion of wave functions in terms of polar coordinates.

The first mathematical simplification in solving the full wave equation is to take advantage of the fact that the total wave function $\Psi\,(r, \theta, \phi)$ can be cast in the form of a product of three simpler functions $R(r)$ (the radial function), and $\Theta(\theta)$ and $\Phi(\phi)$ (the angular functions). Substitution of these functions in the wave equation allows separation into three new equations:

$$\frac{1}{R}\frac{d}{dr}\left(r^2\frac{dR}{dr}\right) + \frac{8\pi^2 m}{h^2}\left(E + \frac{Ze^2}{4\pi\varepsilon_0 r}\right)r^2 = u \tag{2.20}$$

$$\frac{1}{\sin\theta}\frac{d}{d\theta}\left(\sin\theta\cdot\frac{d\Theta}{d\theta}\right) - \frac{v^2\Theta}{\sin^2\theta} + u\Theta = 0 \tag{2.21}$$

and

$$\frac{1}{\Phi}\cdot\frac{d^2\Phi}{d\phi^2} = -v^2 \tag{2.22}$$

Attention is particularly drawn to the new constants u and v, which appear in these equations because these constants contain two of the quantum numbers that were identified empirically in the Bohr–Sommerfeld theory. In fact, $u = l(l + 1)$, where l = the *azimuthal quantum number*, and $v = m_l$, the *magnetic quantum number* of the Bohr-Sommerfeld theory. The third quantum number of this theory, the *principle quantum number*, n, appears in the full solution of Equation 2.20. The mathematics of Equations 2.20–2.22 require that n can only have integral values from unity to infinity, that l can only have positive values $(n - 1)$, $(n - 2)$, ..., 0 and that m_l can only have values l, $(l - 1)$, ..., 0, ..., $-(l - 1)$, ..., $-l$ exactly as concluded from experimental observation. There is an $R(r)$ solution of Equation 2.20 for each pair of values for n and l, a $\Theta(\theta)$ solution for each pair of l and $|m_l|$, and a $\Phi(\phi)$ solution for each value of m_l. These functions are readily available in tabular form in texts on physical chemistry. The full wave function for any stationary state of the hydrogen atom can be reconstructed by taking the appropriate product of $R(r) \cdot \Theta(\theta) \cdot \Phi(\phi)$. Some selected values for $\Psi_{n,l,m}$ for the hydrogen atom are listed in Table 2.1.

For each $\Psi_{n,l,m}$, there is a solution to Equation 2.16 for E. In fact, for a hydrogen atom unperturbed by external fields, the E values depend only on n, so that the values for E for Ψ_{200}, Ψ_{210}, Ψ_{211}, and Ψ_{21-1} are all identical, or *degenerate*. The first few energy levels of the hydrogen atom are illustrated in Figure 2.3.

Table 2.1. *Some Complete $\Psi_{n,l,m}$ for the Hydrogen Atom*

Quantum Numbers			Polar Function	Cartesian Function
n	l	m		
1	0	0	$\Psi_{100} = \dfrac{1}{\sqrt{\pi}}\left(\dfrac{1}{a_0}\right)^{3/2} \cdot e^{-r/a_0}$	Same
2	1	0	$\Psi_{210} = \dfrac{1}{4\sqrt{2\pi}}\left(\dfrac{1}{a_0}\right)^{3/2} \cdot \dfrac{1}{a_0} \cdot e^{-r/2a_0} \cdot r \cdot \cos\theta$	$r \cdot \cos\theta = z$
2	1	1	$\Psi_{211} = \dfrac{1}{4\sqrt{2\pi}}\left(\dfrac{1}{a_0}\right)^{3/2} \cdot \dfrac{1}{a_0} \cdot e^{-r/200}r \cdot \sin\theta \cdot \cos\phi$	$r \cdot \sin\theta \cdot \cos\phi = x$
3	2	-1	$\Psi_{3-10} = \dfrac{1}{81}\sqrt{\dfrac{2}{\pi}}\left(\dfrac{1}{a_0}\right)^{3/2} \cdot \left(\dfrac{1}{a_0}\right)^2 \cdot e^{-r/3a_0}$ $\cdot r^2 \cdot \sin\theta \cdot \cos\theta \cdot \sin\phi$	$r^2 \cdot \sin\theta \cdot \cos\theta \cdot \sin\phi = yz$

Figure 2.3
The first few energy levels of the hydrogen atom.

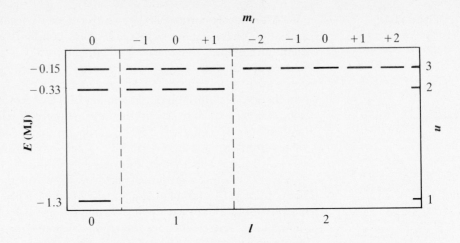

We have now at our disposal the solutions of the wave equation for both the wave functions and the energies of the infinite manifold of stationary states of the hydrogen atom. However, we alluded earlier to the difficulty of profiting from a knowledge of the wave function per se because it has no perceivable physical significance. This being the case, we must examine the properties of Ψ^2, or more exactly $\Psi^2 \cdot dx \cdot dy \cdot dz$, to find out what this information tells us about the location of the electron in the atom.

Figure 2.4 shows plots of the radial function $R(r)_{100}$, $R(r)^2_{100}$, and the function $4\pi r^2 \cdot R(r)^2_{100}$ (the integrated surface probability) all against r. It can be seen that both the $R(r)_{100}$ and the $R(r)^2_{100}$ plots go to a maximum at $r = 0$. This does not mean the electron is most likely to be found at $r = 0$, in the middle of the nucleus! The value of $R(r)_{100}$ refers to a point r, θ, ϕ (or x, y, z in Cartesian coordinates); whereas there is only one point at $r = 0$, there is a very large number of points on a sphere of radius r, and the number increases in proportion to r^2. This fact leads to a probability of finding the electron at $r \approx 0$ of essentially zero. The integrated probability rises with increasing r to a maximum at $r = a_0$. The constant a_0, which we encountered in the solutions $\Psi_{n,l,m}$, therefore corresponds to the radius of the sphere on which the electron is most likely to be located. It so happens that this radius corresponds exactly to the radius predicted by the Bohr theory for the orbit of the electron in the ground state of the hydrogen atom. But unlike the classical model, wave mechanics show that an electron whose energy is precisely defined cannot be pinpointed to be at a precise distance from the nucleus but has a significant probability of being either closer to or farther from the nucleus than the classical approach would predict.

The radial distribution curve of Figure 2.4 gives us a picture of the

Figure 2.4
Plots of radial distribution functions for the hydrogen atom. The quantity a_0 corresponds to the Bohr radius.

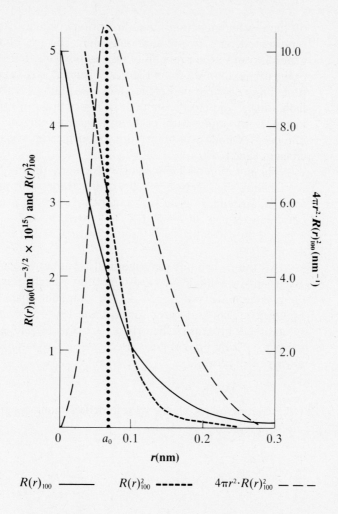

$R(r)_{100}$ ——— $R(r)^2_{100}$ ------ $4\pi r^2 \cdot R(r)^2_{100}$ — — —

electron in the $n, l, m = 1, 0, 0$ state of the hydrogen atom as a fuzzy sphere of negative charge that has its highest density at a distance a_0 from the nucleus and whose density falls off more or less rapidly for $a_0 < r < a_0$.

Because fuzzy spheres (or circles) are difficult to draw, chemists have agreed to used a stylized representation in place of the actual electron distribution. The actual form of the electron distribution is called an *orbital*. The stylized representation of an orbital, which has $l = 0$ and always has a spherical electron distribution, is a circle. This circle is taken to represent a sphere within which some large fraction of the electron probability is contained.

The radial distribution functions of some other orbitals are shown in Figure 2.5. It should be noted that for those orbitals with $l > 0$, only the radial part of the wave function is plotted. This is because the angular parts of these wave functions are direction-dependent and cannot be fully represented on a two-dimensional plot. The use of $R^2(r)$ does give the general form of the probability distribution function. The salient features of the distribution shown in Figures 2.4 and 2.5 are (a) the maximum of the probability distribution function moves farther away from the nucleus with increasing n and (b) the number of *nodes* in an orbital of a given l value increases by unity with each unit increase in the principal quantum number. (A node is a point, line, or surface at which the wave function changes sign.)

The notation for labeling orbitals comes directly from that developed for labeling the electronic states of the Bohr atom and is of the general form nl. Instead of using the numerical value, we use the classical alphabetic equivalence of $l = 0, 1, 2, 3, \ldots$ to $s, p, d, f, \ldots$. The orbital with $n = 1$, $l = 0$ is called a $1s$ orbital; with $n = 3$, $l = 2$ a $3d$ orbital; and so on.

All the ns orbitals are spherically symmetrical and, as stated above, are usually schematically represented by a circle. The orbitals with $l = 1$ (p orbitals) are not spherical but dumbbell-shaped, as illustrated in Figure 2.6. There are three such orbitals ($m_l = 1, 0, -1$) for each value of n, and the long axes of the orbitals are directed along the three Cartesian axes (p_x, p_y, and p_z orbitals). Because the wave function of a p orbital is always of the form

$$\Psi_{p\omega} = \text{const} \cdot \omega \cdot e^{\text{const}} \tag{2.23}$$

(where $\omega = x, y,$ or z), the wave function changes sign on passing through

Figure 2.5
Radial probability density-distribution functions:
(a) $2s$ orbital,
(b) $2p$ orbital,
(c) $3d$ orbital. Note the different scale for (c).

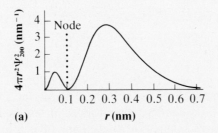

(a)

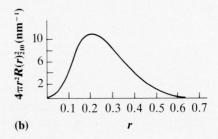

(b)

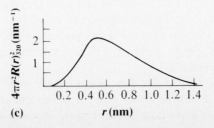

(c)

Figure 2.6
Hydrogenlike
orbitals.

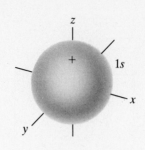

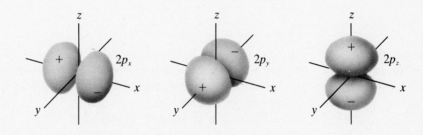

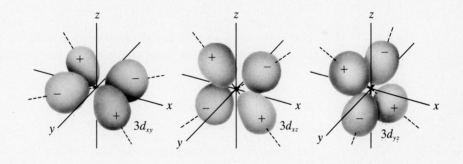

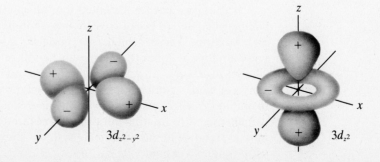

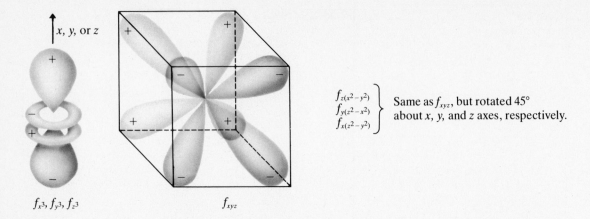

$x, y, \text{ or } z$

$\left. \begin{array}{l} f_{z(x^2-y^2)} \\ f_{y(z^2-x^2)} \\ f_{x(z^2-y^2)} \end{array} \right\}$ Same as f_{xyz}, but rotated 45° about x, y, and z axes, respectively.

$f_{x^3}, f_{y^3}, f_{z^3}$ f_{xyz}

Figure 2.7
Shapes of f orbitals.

the origin. This is an important property of p orbitals and one in which they differ from s orbitals. It is a common practice to note the sign of the wave function on the orbital, especially when combining the orbitals of different atoms to form molecules.

Figure 2.6 illustrates the set of d orbitals conventionally used by chemists. Three of these orbitals, designated d_{xy}, d_{xz}, and d_{yz}, have four lobes at right angles to each other, and these lobes are directed into the four quadrants of a Cartesian plane (xy, xz, and yz planes). A fourth orbital $d_{x^2-y^2}$ has the same shape as d_{xy}, d_{xz}, and d_{yz}, but its lobes are directed along the x and y axes ($\pm x$, $\pm y$). The fifth orbital d_{z^2} has a unique shape consisting of a dumbbell concentrated along the z axis with greater total integrated density than the lobes of the other d orbitals and an equatorial doughnut-shaped belt of electron density in the xy plane.

The d orbitals have wave functions whose directional parts contain the product of two coordinates:

$$\Psi_{d_{xy}} = \text{const} \cdot x \cdot y \cdot e^{\text{const}} \tag{2.24}$$

(or the same coordinate squared). This ensures that the sign of the wave function remains the same on passing through the origin. The d orbitals are therefore centrosymmetric like the s orbitals and unlike the p orbitals.

Because we have not treated the full solution of the wave equation, you may be puzzled because we do not mention the orbitals d_{x^2}, d_{y^2}, $d_{x^2-z^2}$, and so forth. These orbitals are perfectly good solutions to the wave equation for $l = 2$, but there can only be sets of five *independent* solutions. The set of d_{z^2}, $d_{x^2-y^2}$, d_{xy}, d_{xz}, and d_{yz} are independent solutions, but we can take linear combinations of these solutions to construct an infinite manifold of other sets of five independent solutions. That futile exercise would amount to rotating our original set through all possible orientations in space relative to our coordinate system. Thus, a rotation of our original set through 90° would give d_{x^2}, $d_{y^2-z^2}$, d_{xy}, d_{xz}, and d_{yz}.

Table 2.2 *Properties of Orbitals*			
Orbital Type	Nodal Planes	Coordinates in	Symmetry[1]
s	0	r	g
p_x	1	x	u
d_{xy}	2	xy	g
f_{xyz}	3	xyz	u

[1] No change in sign on reflection through the origin of the coordinate system is designated g. Inversion of sign on reflection through the origin is u. See Chapter 5.

The detailed shapes of the seven f orbitals are shown in Figure 2.7. It suffices to say that there is a rational continuation of the various properties outlined previously for the s, p, and d orbitals. These properties are summarized in Table 2.2.

2.3   Many-Electron Atoms and the Periodic Table

The Pauli Principle and the Aufbau Principle

In the preceding section, we described the properties of the exact solutions of the wave equation for Ψ and E for the case of a single electron under the influence of a central positive charge. These solutions correspond very precisely to the physical reality of the hydrogen atom and to the large amount of experimental data on hydrogen that had been gathered in the fifty or so years before the derivation of the wave equation.

Despite the evident success of this triumph in theoretical physics, by itself it did not greatly enhance the art of chemistry. Hydrogen is only one of the 100 or so elements known to chemists, and isolated atoms are rarely encountered in the laboratory. Chemists are more interested in the solutions of the wave equation for the other atoms with $Z > 1$ and with more than one electron. However, as discussed earlier, the equations of motion for interacting bodies in numbers greater than two defy exact solution. This problem is one of esthetics rather than practicality because much useful information can be derived from approximate solutions. Nonetheless, the approximate methods for solutions of the wave equation for many-electron atoms are highly mathematical and computationally laborious. Only in relatively recent years, with the development of powerful, more readily available computers, have such calculations become more or less routine. We will not dwell on how the solutions are obtained but only on the qualitative results indicated by those solutions.

In general, many-electron atoms do not differ greatly from hydrogen. Their electronic structures can be described using the same kinds of energy levels and orbitals as were derived for the hydrogen atom, except that the different l states within a given n state are no longer degenerate and all orbitals decrease in energy and size as the nuclear charge increases. The lifting of the degeneracy of the l states occurs so that for a given n, $E_s < E_p < E_d$ and so forth. This spreading of the l manifold of levels can lead to anomalies in the ordering of energy levels as the separation between n states gets to be comparable to the separations between l states within a given n, especially because these separations are very sensitive to effective nuclear charge. The manner in which this ordering can vary is shown in Figure 2.8.

Before we can describe the electronic structure of atoms with more than one electron, we must return to the phenomenon of electron spin to which we alluded briefly in our discussion of the Bohr–Sommerfeld model for many-electron atoms. The notion of electron spin was introduced by Pauli (1926) and by Uhlenbeck and Goudsmit (1926) to explain an experiment performed by Stern and Gerlach in 1921. This experiment showed

Figure 2.8
Variation in energy of atomic orbitals for many-electron atoms.

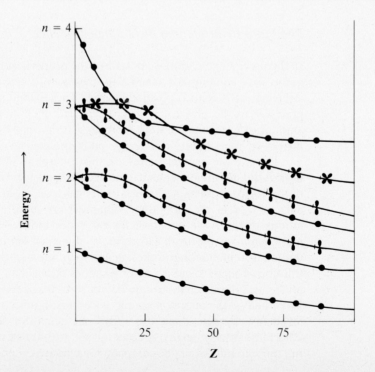

that a beam of silver atoms (the silver atom has a single unpaired electron) was split into two separate beams by passage through a strong, inhomogeneous magnetic field. A short time later, Dirac (1928) showed that inclusion of relativistic effects in the wave mechanical treatment of electrons gave an *ab initio* (from first principles) explanation of the spin phenomenon.

Chemists have frequently chosen, however, to retain the simple mechanical concept of a spinning electron with total spin angular momentum of $h/2\pi$ units of momentum. In a magnetic field (which has helical symmetry and therefore interacts differently with charges in clockwise or counterclockwise motion), the total spin momentum can have components in the field direction (usually defined as the z direction) of $m_s(h/2\pi)$, where m_s (the spin quantum number) can only have the values $\pm\frac{1}{2}$. The absolute necessity of taking account of the spin property in discussing atomic and molecular structures is embodied in the Pauli exclusion principle, one statement of which is: In an atom or molecule no two electrons may have the same set of four quantum numbers, n, l, m_l and m_s. An inevitable consequence of this principle is that a single orbital (n, l, and m_l defined) cannot accommodate more than two electrons (with $m_s = +\frac{1}{2}$ and $-\frac{1}{2}$).

The foregoing concepts—the ordering of orbital energies and the restriction of orbital occupancy to a maximum of two electrons—are combined to generate a picture of the electronic structure of the elements through what is known as the Aufbau principle. This principle assumes that the ordering of the orbital energies does not change with increasing nuclear charge, and so we construct the elements beyond hydrogen by the stepwise addition of protons (and neutrons) to the nucleus and of the accompanying electrons to the lowest-energy, vacant, hydrogenlike orbitals.

Hund's Rules and the Ground-State Electronic Configurations of the Elements

Application of the above principles to the first five elements leads to the following electronic structures: $H(1s^1)$, $He(1s^2)$, $Li(1s^22s^1)$, $Be(1s^22s^2)$, and $B(1s^22s^22p^1)$, where the number of electrons in an orbital is indicated by a superscript. Because there is no internal indicator of direction, all three p orbitals of B are equivalent, but in progressing from boron to carbon, we are faced with two choices as to where the next electron is located. It may go into the p orbital already occupied, or it may go into a different p orbital. Both experiment and intuition tell us the electron should go into a different orbital. Experiment shows us that the ground state of carbon has two unpaired electrons, and intuition tells us the electrons should go into different orbitals in order to be as far away from each other as possible to minimize coulombic repulsion. Besides coulombic repulsion, there is also a force of quantum mechanical origin known as *spin-exchange interaction*, which also tends to keep electrons as far away from each other as possible. As in so many other instances, this fact was recognized empirically before a theoretical explanation was adduced, and it finds its expression in a series of rules formulated by Hund to explain the multiplicity of lines observed in atomic spectra. The relevant rules can be summarized thus: For an atom with partially occupied, degenerate orbitals, the ground state is that with the maximum total spin momentum and the maximum orbital momentum. The maximum spin is achieved by aligning the spins in the same direction, and this excludes by the Pauli principle any double occupancy of orbitals until the subshell is half-filled. Thus, the remaining electron configurations of the elements up to a closed $2p$ subshell are

$$C(1s^2 2s^2 2p_x^1 2p_y^1)$$

$$N(1s^2 2s^2 2p_x^1 2p_y^1 2p_z^1)$$

$$O(1s^2 2s^2 2p_x^2 2p_y^1 2p_z^1)$$

$$F(1s^2 2s^2 2p_x^2 2p_y^2 2p_z^1)$$

$$Ne(1s^2 2s^2 2p^6)$$

In these notations, the designation of the coordinates of the p orbitals is completely arbitrary. The shorthand for writing electronic structure may be abbreviated further by designating closed inner shells by the appropriate noble gas symbol—for example, for Li we can write $[He]2s^1$.

The elements Na to Ar involve the filling of $3s$ and $3p$ orbitals in a manner exactly analogous to the elements Li to Ne, except that the outer shell is now $n = 3$. For example, the electronic structure of silicon is $[Ne]3s^2 3p_x^1 3p_y^1$. Beyond argon, it is not evident a priori whether the next element potassium would involve occupancy of the $3d$ orbital or of the $4s$ orbital. In fact, as can be seen from Figure 2.8, the energies of these two orbitals are very similar in the vicinity of potassium ($Z = 19$), but both calculations and experiment show that K and Ca have ground states of $[Ar]4s^1$ and $[Ar]4s^2$, respectively. Only after $4s$ is filled does $3d$ filling begin with scandium $[Ar]4s^2 3d^1$. The filling of the $3d$ orbitals then continues as follows:

Ti:	$[Ar]4s^2 3d^2$
V:	$[Ar]4s^2 3d^3$
Cr:	$[Ar]4s^1 3d^5$
Mn:	$[Ar]4s^2 3d^5$
Fe:	$[Ar]4s^2 3d^6$

Co: $[Ar]4s^23d^7$

Ni: $[Ar]4s^23d^8$

Cu: $[Ar]4s^13d^{10}$

Zn: $[Ar]4s^23d^{10}$

It should be noted that the ground states for Cr and Cu correspond to situations where the orbital vacancies are partitioned between the s and d orbitals in such a way that the subshells are half-empty or filled. In these cases, we see a situation where simple hydrogenlike orbitals are inadequate for the job of describing the energies of the electrons because the hydrogenlike orbitals are essentially one-electron orbitals and do not properly take account of electron–electron interactions. With increasing n, as the energies of the subshells bunch together, the interelectronic repulsion energies become comparable to or even greater than the subshell separation energies and so begin to play a dominant role in deciding the preferred electronic configuration. In the heavier elements with partially filled d or f shells, a number of anomalies similar to those occurring in Cr or Cu are observed. When they do occur, they are generally such as to provide the maximum amount of half-filling or filling of subshells. There are no simple rules to predict such anomalies, and elements in the same group may behave differently—for example, unlike Ni, Pd($[Kr]4d^{10}$) and Pt($[Xe]4f^{14}5d^96s^1$) exhibit anomalies but of two different kinds; Mo follows Cr but W-($[Xe]4f^{14}5d^46s^2$) does not. The ground-state electron configurations of all the known elements, determined from gas-phase atomic spectroscopy, are summarized in Table 2.3.

Beyond the filling of the $3d$ and $4s$ subshells, the $4p$ subshell fills up to krypton, and then the same sequence as that occurring from K to Kr is repeated from Rb to Xe, except that the n values of the orbitals are increased by one. From Cs to Ra, the sequence of orbital filling is approximately $6s$, $4f$, $5d$, $6p$ and from Fr to Lr, $7s$, $5f$, $6d$. However, by the time we reach the last of the elements, the exceptions are almost as common as the cases that follow the rule. A mnemonic diagram to help remember the order of orbital filling is shown in Figure 2.9.

The Periodic Table

The foregoing discussion of electronic configurations of elements in the form of atoms is somewhat academic because elements, other than noble gases, are rarely encountered as atoms. The same principles outlined for atoms are also useful for ionic compounds but have little relevance to the electron configurations of atoms in the covalently or metallically bound state. In the case of ionic compounds of the d-block and f-block elements,

Table 2.3 *Electron Configurations of the Elements*

Z	Element	Electron Configuration	Z	Element	Electron Configuration
1	H	$1s^1$	27	Co	$[Ar]3d^74s^2$
2	He	$1s^2$	28	Ni	$[Ar]3d^84s^2$
3	Li	$[He]2s^1$	29	Cu	$[Ar]3d^{10}4s^1$
4	Be	$[He]2s^2$	30	Zn	$[Ar]3d^{10}4s^2$
5	B	$[He]2s^22p^1$	31	Ga	$[Ar]3d^{10}4s^24p^1$
6	C	$[He]2s^22p^2$	32	Ge	$[Ar]3d^{10}4s^24p^2$
7	N	$[He]2s^22p^3$	33	As	$[Ar]3d^{10}4s^24p^3$
8	O	$[He]2s^22p^4$	34	Se	$[Ar]3d^{10}4s^24p^4$
9	F	$[He]2s^22p^5$	35	Br	$[Ar]3d^{10}4s^24p^5$
10	Ne	$[He]2s^22p^6$	36	Kr	$[Ar]3d^{10}4s^24p^6$
11	Na	$[Ne]3s^1$	37	Rb	$[Kr]5s^1$
12	Mg	$[Ne]3s^2$	38	Sr	$[Kr]5s^2$
13	Al	$[Ne]3s^23p^1$	39	Y	$[Kr]4d^15s^2$
14	Si	$[Ne]3s^23p^2$	40	Zr	$[Kr]4d^25s^2$
15	P	$[Ne]3s^23p^3$	41	Nb	$[Kr]4d^45s^1$
16	S	$[Ne]3s^23p^4$	42	Mo	$[Kr]4d^55s^1$
17	Cl	$[Ne]3s^23p^5$	43	Tc	$[Kr]4d^55s^2$
18	Ar	$[Ne]3s^23p^6$	44	Ru	$[Kr]4d^75s^1$
19	K	$[Ar]4s^1$	45	Rh	$[Kr]4d^85s^1$
20	Ca	$[Ar]4s^2$	46	Pd	$[Kr]4d^{10}$
21	Sc	$[Ar]3d^14s^2$	47	Ag	$[Kr]4d^{10}5s^1$
22	Ti	$[Ar]3d^24s^2$	48	Cd	$[Kr]4d^{10}5s^2$
23	V	$[Ar]3d^34s^2$	49	In	$[Kr]4d^{10}5s^25p^1$
24	Cr	$[Ar]3d^54s^1$	50	Sn	$[Kr]4d^{10}5s^25p^2$
25	Mn	$[Ar]3d^54s^2$	51	Sb	$[Kr]4d^{10}5s^25p^3$
26	Fe	$[Ar]3d^64s^2$	52	Te	$[Kr]4d^{10}5s^25p^4$

the nd or nf subshells are sufficiently lower in energy than are the $(n + 1) s$ shell so that the latter always remains unoccupied.

The great importance of the success of theory to predict electronic structures of atoms is the theoretical substantiation it provides for the periodic table of the elements. The development of the periodic table, shown in a widely used modern form in Figure 2.10, was based largely on chemical observations. For skilled chemists, this table contains a vast amount of information on how the elements do or should combine to form compounds, but it also summarizes all the qualitative theoretical principles on atomic structure developed in the preceding sections. We now know that the periodic relationship between atomic weight and chemical prop-

Table 2.3 (*Continued*)

Z	Element	Electron Configuration	Z	Element	Electron Configuration
53	I	$[Kr]4d^{10}5s^25p^5$	79	Au	$[Xe]4f^{14}5d^{10}6s^1$
54	Xe	$[Kr]4d^{10}5s^25p^6$	80	Hg	$[Xe]4f^{14}5d^{10}6s^2$
55	Cs	$[Xe]6s^1$	81	Tl	$[Xe]4f^{14}5d^{10}6s^26p$
56	Ba	$[Xe]6s^2$	82	Pb	$[Xe]4f^{14}5d^{10}6s^26p^2$
57	La	$[Xe]5d^16s^2$	83	Bi	$[Xe]4f^{14}5d^{10}6s^26p^3$
58	Ce	$[Xe]4f^15d^16s^2$	84	Po	$[Xe]4f^{14}5d^{10}6s^26p^4$
59	Pr	$[Xe]4f^36s^2$	85	At	$[Xe]4f^{14}5d^{10}6s^26p^5$
60	Nd	$[Xe]4f^46s^2$	86	Rn	$[Xe]4f^{14}5d^{10}6s^26p^6$
61	Pm	$[Xe]4f^56s^2$	87	Fr	$[Rn]7s^1$
62	Sm	$[Xe]4f^66s^2$	88	Ra	$[Rn]7s^2$
63	Eu	$[Xe]4f^76s^2$	89	Ac	$[Rn]6d7s^2$
64	Gd	$[Xe]4f^75d6s^2$	90	Th	$[Rn]6d^27s^2$
65	Tb	$[Xe]4f^96s^2$	91	Pa	$[Rn]5f^26d^17s^2$
66	Dy	$[Xe]4f^{10}6s^2$	92	U	$[Rn]5f^36d^17s^2$
67	Ho	$[Xe]4f^{11}6s^2$	93	Np	$[Rn]5f^46d^17s^2$
68	Er	$[Xe]4f^{12}6s^2$	94	Pu	$[Rn]5f^67s^2$
69	Tm	$[Xe]4f^{13}6s^2$	95	Am	$[Rn]5f^77s^2$
70	Yb	$[Xe]4f^{14}6s^2$	96	Cm	$[Rn]5f^76d^17s^2$
71	Lu	$[Xe]4f^{14}5d^16s^2$	97	Bk	$[Rn]5f^97s^2$
72	Hf	$[Xe]4f^{14}5d^26s2$	98	Cf	$[Rn]5f^{10}7s^2$
73	Ta	$[Xe]4f^{14}5d^36s^2$	99	Es	$[Rn]5f^{11}7s^2$
74	W	$[Xe]4f^{14}5d^46s^2$	100	Fm	$[Rn]5f^{12}7s^2$
75	Re	$[Xe]4f^{14}5d^56s^2$	101	Md	$[Rn]5f^{13}7s^2$
76	Os	$[Xe]4f^{14}5d^66s^2$	102	No	$[Rn]5f^{14}7s^2$
77	Ir	$[Xe]4f^{14}5d^76s^2$	103	Lr	$[Rn]5f^{14}6d^17s^2$
78	Pt	$[Xe]4f^{14}5d^96s^1$			

erties, first fully enunciated by Mendeleev in 1869, is a reflection of the shell structure of the electrons in atoms. The numbers of the horizontal rows (periods) in the table correspond to the principal quantum number of the outermost occupied shell. The vertical groups correspond to those elements with a common outer-shell electron configuration, differing only in the values of n. These groups fall naturally into blocks that correspond to the filling of a common subshell of orbitals with the same value.

Up until now, chemists have not cast a unique form of the periodic table in "tablets of stone." Most recently, the International Union of Pure and Applied Chemistry (IUPAC) has adopted the group-numbering system shown in Figure 2.10. It is hoped that this convention will be uni-

Figure 2.9
Mnemonic diagram
for remembering
orbital filling.

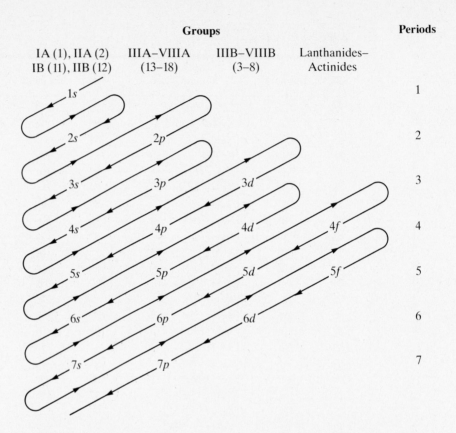

Groups				Periods
IA (1), IIA (2)	IIIA–VIIIA	IIIB–VIIIB	Lanthanides–	
IB (11), IIB (12)	(13–18)	(3–8)	Actinides	

versally accepted and that the earlier dichotomy of a European and an American convention, which differed in their preferences for A and B designators for the *s*-, *p*-, and *d*-block elements, will henceforth be avoided. We mention this because most textbooks and literature up to the present have used these older conventions. There are also several schools of thought about where the transition periods begin and end. These minor disputes are of little importance. The periodic table has its own internal, self-evident consistency, and its beauty transcends all efforts to smother it with nomenclature.

Energy States of Atoms with Partially Filled Shells

In the discussion of the electron configurations of many-electron atoms, we alluded to the difficulties arising from the use of one-electron, hydrogen-like orbitals to fully define the electronic structure of certain atoms where interelectronic repulsion has a determining effect on electron configuration (for example, Cr and Cu). This shortcoming is even more serious than might at first appear and is further exemplified by the difficulty of precisely

Group Number

s-Block

Transition Elements (*d*-Block)

p-Block

1	2	3	4	5	6	7	8	9	10	11	12	13	14	15	16	17	18
IA	IIA	IIIB	IVB	VB	VIB	VIIB	VIIIB	VIIIB	VIIIB	IB	IIB	IIIA	IVA	VA	VIA	VIIA	VIIIA
1 H																	2 He
3 Li	4 Be											5 B	6 C	7 N	8 O	9 F	10 Ne
11 Na	12 Mg											13 Al	14 Si	15 P	16 S	17 Cl	18 Ar
19 K	20 Ca	21 Sc	22 Ti	23 V	24 Cr	25 Mn	26 Fe	27 Co	28 Ni	29 Cu	30 Zn	31 Ga	32 Ge	33 As	34 Se	35 Br	36 Kr
37 Rb	38 Sr	39 Y	40 Zr	41 Nb	42 Mo	43 Tc	44 Ru	45 Rh	46 Pd	47 Ag	48 Cd	49 In	50 Sn	51 Sb	52 Te	53 I	54 Xe
55 Cs	56 Ba	57 La	72 Hf	73 Ta	74 W	75 Re	76 Os	77 Ir	78 Pt	79 Au	80 Hg	81 Tl	82 Pb	83 Bi	84 Po	85 At	86 Rn
87 Fr	88 Ra	89 Ac	104 Rf	105 Ha	106												

Lanthanides	58 Ce	59 Pr	60 Nd	61 Pm	62 Sm	63 Eu	64 Gd	65 Tb	66 Dy	67 Ho	68 Er	69 Tm	70 Yb	71 Lu
Actinides	90 Th	91 Pa	92 U	93 Np	94 Pu	95 Am	96 Cm	97 Bk	98 Cf	99 Es	100 Fm	101 Md	102 No	103 Lr

Figure 2.10 Modern version of the periodic table.

NOTE: The group numbers in Arabic numerals are those presently recommended by IUPAC. The Roman numerals are those generally used in North America until 1984. *Some pseudonyms for certain groups of elements:* Group 1: Alkali metals. Group 2: Alkaline earth metals. Group 15: Pnictogens. Group 16: Chalcogens. Group 17: Halogens. Group 18: Noble gases, inert gases, rare gases. Group 11: Coinage metals. Elements 44–46 and 76–78: Platinum metals. Groups 4, 5, and 6 are sometimes referred to as the refractory metals. The term *noble metal* is applied to those elements that are less electropositive than hydrogen—that is, the platinum metals, the coinage metals, and mercury.

identifying the ground state of the carbon atom on the basis of hydrogen-like orbitals alone. There are many distinct ways that two electrons can be assigned to three p orbitals, and they are illustrated in the so-called box diagrams (Figure 2.11).

In our introduction to Hund's rules, we indicated that the configuration labeled as ^3P was the ground state of carbon. Because in the absence of a magnetic field the three p orbitals are indistinguishable, the three different arrangements labeled ^3P will be degenerate (that is, have the same energy). The total spin of this ^3P state is $2 \times \frac{1}{2} = 1$ (by analogy with the one-electron symbolism, the total spin of many-electron atoms is given the quantum number symbol S). The total spin momentum is quantized with respect to direction such that components of total spin along the defined direction are M_s, where M_s has values $S, (S - 1), (S - 2), \ldots, 0, \ldots, -S$. Our ^3P state of C therefore has the M_s values $1, 0, -1$. We refer to this as a "spin triplet," and the degeneracy of the triplet will be lifted by an appropriate magnetic field.

The total orbital angular momentum (L) will also have components in a defined direction, $M_L = \Sigma m_l$. These components have the quantum numbers $L, (L - 1), \ldots, 0, \ldots, - L$, and for our ^3P state of carbon the M_L values are $1, 0, -1$. Again, in the absence of a magnetic field these states are degenerate, but their degeneracy is lifted in a magnetic field.

The above principles can be generalized in the following way: Any state with a total spin S will contain $(2S + 1)$ degenerate substates; $(2S + 1)$ is the *spin multiplicity* of the state. Any state with a total orbital angular momentum L will contain $(2L + 1)$ degenerate substates; $(2L + 1)$ is the *orbital multiplicity*. These degeneracies are the result of the

Figure 2.11
Box diagrams for the terms of the $2s^2 2p^2$ configuration of carbon.

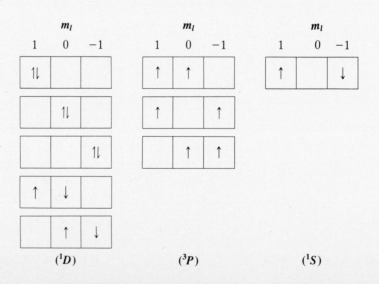

(^1D) (^3P) (^1S)

Figure 2.12
Vector
representations of the
quantization of
angular momenta
with respect to a
defined direction.

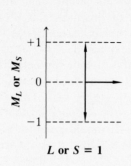

L or $S = 1$

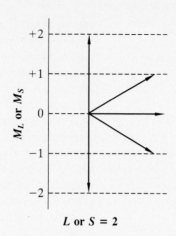

L or $S = 2$

quantization of the total momenta S or L, with respect to the defined direction. The nature of this quantization is illustrated in Figure 2.12.

We symbolize a state with spin multiplicity $(2S + 1)$ and total orbital momentum L by the symbol $^{(2S + 1)}L$. Instead of using the numeric value of L, we use the same alphabetic designation as we use for l, except capitalized. Thus, a state with $S = 1$ and $L = 1$ is given the symbol ^3P ("triplet pee"), which is why we started out applying this symbol to some of the states of the $2p^2$ configuration. The block of five substates on the left of Figure 2.11 has $S = 0$ and a maximum M_L, which corresponds to $L = 2$. This state is designated ^1D. Finally, there is the unique state shown in the right-hand box diagram which has $S = 0$ and $L = 0$, that is, a ^1S state. Hund's rules tell us that the state of highest spin multiplicity is the ground state. If there are two states with highest spin multiplicity, the one of highest orbital multiplicity will be of lower energy. There is no simple way of determining the ordering of higher-energy levels. In the case of carbon, the ordering is $^3P < {}^1D < {}^1S$. The energy states into which a particular electronic configuration separates are called *terms*, and the nomenclature we have just developed is called *term symbols*. In principle, the term symbols for any configuration can be deduced using the same line of reasoning outlined above. In the case of some d^n and f^n (n = number of electrons), the procedure can be laborious because of the very large number of substates and the terms for these configurations are usually looked up in a table. Some obvious terms are those of closed shells, always ^1S, and those of singly occupied subshells, where the L value is the same as l and the state is a doublet. It is also useful to note that the L value for a half-filled shell is 0.

In the above arguments, we made some implicit simplifying assumptions. For example, we assumed that the only important couplings of momenta were interelectronic and restricted to spin–spin and orbital–

orbital interactions. For very light atoms, such an approximation corresponds very closely to the experimental facts. For the elements in the middle of the periodic table, the approximation is rather poor and leaves many experimental results unexplained. A better approximation for these elements is that there is significant coupling between the total spin S and the total orbital momentum L to produce a new set of states describable by a new quantum number $J = |M_L + M_S|$. Thus, the ^3P state of carbon would be split by such a spin–orbit coupling into three new nondegenerate states with J values of $2, 1, 0$. [It should be noted that J can never be less than 0. So for cases where S is less than L, there are $(2S + 1)$ values of J; when L is less than S, there are $(2L + 1)$ values of J.] The new states are given the term symbols $^3P_2$ ("triplet pee two"), $^3P_1$, and $^3P_0$. Figure 2.13 shows a vector representation of how the L and S momenta add to give the J states.

Hund's third rule tells us that the ground state is that with the lowest J value for a less than half-filled shell and that with the highest J value for a more than half-filled shell. Thus, the ground term of carbon is $^3P_0$, while that for oxygen, which also has two unpaired p electrons, is $^3P_2$. The various interactions within the $2p^2$ configuration are summarized in Figure 2.14. The coupling scheme shown in Figure 2.14 is known as the Russell–Saunders, or L–S coupling scheme.

Some idea of the importance of spin–orbit coupling is given by the separation in J states. For carbon the order of magnitude is about 1 cm^{-1}, for titanium about 10 cm^{-1}, and for a $4f$ element about 100 cm^{-1}. In the case of hydrogen, there is a coupling between the spin of the electron and its own orbital angular momentum, which gives rise to a splitting in energy levels of about 0.01 cm^{-1}. The detection of such a small splitting was an experimental tour de force.

The coupling of an electron's spin with its own orbital motion remains negligible compared to L–S coupling for all but the heaviest elements. For the last-period elements, the L–S scheme does not work well and must be replaced by a scheme where the spin and orbital momenta of each electron

Figure 2.13
Vector addition of L and S momenta to give spin-orbit coupled J states.

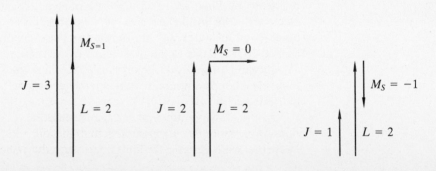

Figure 2.14
The splitting of p^2 configuration by interelectronic repulsion and by spin–orbit coupling. (The splitting by spin-orbit coupling is greatly exaggerated.)

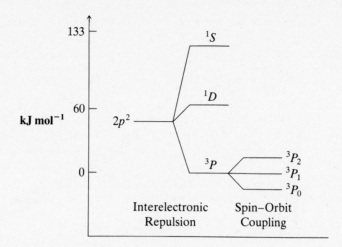

are first coupled. This gives for each electron a new total momentum described by a quantum number $j = l + s$. The total atomic momentum is then the sum of the individual electron j components. This scheme is known as the j–j coupling scheme and is not used in this text.

2.4   Some Important Trends in Atomic Properties

Trends in Atomic and Ionic Radii

The fact that the positions of electrons in atoms or ions cannot be precisely determined renders the definition of the radii of such species imprecise. Even if we had a precise definition, the measurement of the defined radius would give different answers for different measuring techniques because each technique may perturb the atom differently. Despite these difficulties, it is important to have knowledge of the trends in atomic sizes because the chemistry of atoms is intimately linked to size.

One possible criterion for size is the positions of the maxima for the radial distribution functions for the outermost electrons. A plot of such maxima, as a function of atomic number, is shown in Figure 2.15. From this figure, it is evident that the size of an atom is determined primarily by n, there being a substantial increase in the maxima of the radial distribution functions on passing from one shell to the next. Second, there is a fairly dramatic decrease in size as a subshell fills, the decrease being less steep per unit of atomic number as n increases.

The same trends are seen in experimentally measured atomic or ionic radii. It is essentially impossible to measure the radius of a gaseous atom

Figure 2.15
Relative sizes of atoms, expressed in terms of the maxima of the radial distribution functions.

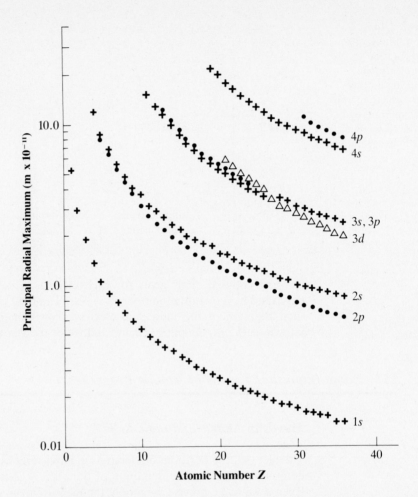

without profoundly disturbing it by collision. Yet, in the condensed state, few elements other than the noble gases exist in the form of free atoms. The value for the radius of xenon, obtained by crystallography on solid xenon, is quite a bit larger than the value obtained by gas-phase atomic-scattering experiments. Nevertheless, there is some validity in comparing values for radii obtained under similar or identical experimental conditions. Because, to date, crystallography provides the most direct means of determining atomic size, this method has furnished the most experimental measurements. The result is that most of the information we have on size is in the form of radii of atoms in the chemically combined state and in the condensed phase.

Of the various kinds of directly measured atomic size, the *covalent radius* is the most fully documented for comparative purposes. The covalent radius of an atom is defined as one-half the distance between the nuclei when two of the atoms are bonded covalently. If the covalent radius is not directly measurable because of no appropriate model compound, it can usually be estimated indirectly by measuring a bond length to an atom of known covalent radius and subtracting this known value from the total internuclear distance. A plot of some covalent radii versus atomic number is shown in Figure 2.16, and it can be seen that the trends are as expected from Figure 2.15. A more complete tabulation of covalent radii is given in Table 2.4.

Ionic radii would also be useful for showing size trend, but it is usually not possible to get complete series for ions of the same charge, which is important because charge has a large effect on size. Such data are available for the first-row d-block elements in the 2+ state and for the lanthanides, all of which exist in the 3+ state. Interpretation of d-block ionic radius trends is complicated because ions with partially filled d shells are usually nonspherical and the definition of radius is problematic. In the lanthanide ions, the $4f$ subshell is more or less buried in the outer $5s$ and $5p$ shells,

Figure 2.16
Some trends in covalent radii.

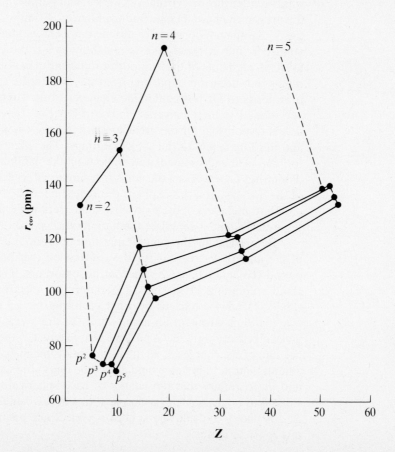

Table 2.4   *Covalent Radii of the Elements (pm)*

H																	He
37																	
Li	Be											B	C	N	O	F	Ne
134	90											82	77	75	73	72	
Na	Mg											Al	Si	P	S	Cl	Ar
154	130											118	117	106	102	99	
K	Ca	Sc	Ti	V	Cr	Mn	Fe	Co	Ni	Cu	Zn	Ga	Ge	As	Se	Br	Kr
196	174	144	132	125	127	146	120	126	120	138	131	126	122	120	116	114	115
Rb	Sr	Y	Zr	Nb	Mo	Tc	Ru	Rh	Pd	Ag	Cd	In	Sn	Sb	Te	I	Xe
211	192	162	148	137	145	156	126	135	131	153	148	144	141	140	136	133	126
Cs	Ba	La	Hf	Ta	W	Re	Os	Ir	Pt	Au	Hg	Tl	Pb	Bi			
225	198	169	149	138	146	159	128	137	128	143	151	152	147	146			

which render the ions virtually spherical and permit both the definition and the measurement of an ionic radius. Figure 2.17 shows a plot of the radii of the 3+ lanthanide ions versus atomic number. Although the radius decrement per unit increase in Z is relatively small, the contraction over the fourteen elements of the 4f group is sufficient to cancel out the increase in radius that occurred in passing from the fifth to the sixth period. This, in turn, leads to the postlanthanide elements of the sixth period having radii very similar to their congeners in the fifth period. This *lanthanide contraction* is responsible for the great similarity of the chemistries of the second and third members of the d-block groups. The effect is most pronounced for the elements immediately following the lanthanides and gradually diminishes on traversing the period. A similar effect accompanies the filling of the d-shells, and this explains the very similar chemistries of the second and third members of the p-groups. Some trends in ionic radii are shown in Figure 2.18, and a complete tabulation of ionic radii is listed in Table 2.5.

Before leaving the general question of atomic size, we examine some of the underlying reasons for the contraction that occurs on traversing a period. To understand why the contraction occurs, it is necessary to understand something about how the outermost electrons are screened or shielded from the nucleus by other electrons.

If the electrons of an atom were like the layers of an onion, then a given electron would feel a positive charge equal to the nuclear charge plus the sum of the charges of all the electrons of the inner layers. In the case of a neutral atom, this sum would always be unity. In fact, the onion model—and the corollary that the charge of each spherical electron can be considered to be the same as a charge sitting at the center of the sphere as far as any charge outside the sphere is concerned (Gauss' theorem)—does not apply to atoms.

Figure 2.17
Lanthanide 3+ ion radii.

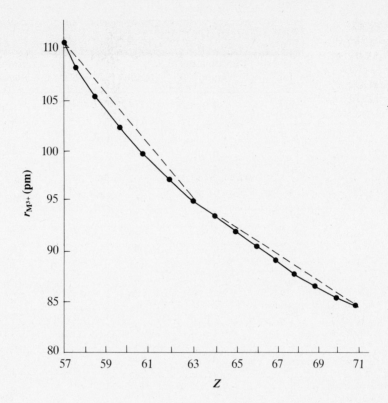

Figure 2.18
Some group trends in ionic radii, illustrating the effects of the lanthanide contraction. The scale on the right applies to the Group IA (1) ions, that on the left to all others.

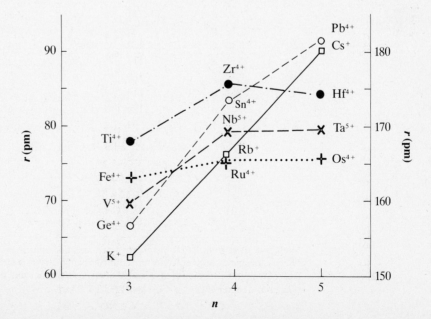

Table 2.5   *Ionic Radii[1] (pm)*

Ion	Coordination Number[2]	Radius	Ion	Coordination Number	Radius
Ag^+	4	114	Ce^{4+}	6	101
	4[a]	116		8	111
	5	123	Cl^-	6	167
	6	129	Co^{2+}	4	72
Al^{3+}	4	53		5	81
	5	62		6[d]	79
	6	68		6[c]	89
As^{3+}	6	72		8	104
As^{5+}	4	48	Co^{3+}	6[b]	69
	6	60		6	75
Au^+	6	151	Cr^{2+}	6[d]	87
Au^{3+}	4[a]	82		6[c]	94
	6	99	Cr^{3+}	6	76
B^{3+}	3	15	Cr^{4+}	4	55
	4	25		6	69
	6	41	Cr^{6+}	4	40
Ba^{2+}	6	149		6	58
	7	152	Cs^+	6	181
	8	156		8	188
Be^{2+}	3	30	Cu^+	2	60
	4	41		4	74
	6	59		6	91
Bi^{3+}	6	117	Cu^{2+}	4	71
	8	131		4[a]	71
Br^-	6	182		5	79
Ca^{2+}	6	114		6	87
	8	126	Cu^{3+}	6[d]	68
Cd^{2+}	4	92	F^-	2	115
	5	101		3	116
	6	109		4	117
	7	117		6	119
	8	124			
Ce^{3+}	6	115			
	8	128			

[1] Values derived from R. D. Shannon, *Acta Crystallographica*, A32, 751 (1976).
[2] Four-coordinate complexes are for tetrahedral geometry except for (a) square-planar and (b) pyramidal. Those values marked (c) are for high-spin and (d) low-spin configurations, see Chapter 13.

Table 2.5 (*Continued*)

Ion	Coordination Number	Radius	Ion	Coordination Number	Radius
Fe^{2+}	4^c	77	Li^+	4	73
	$4^{a,c}$	78		6	90
	6^d	75		8	106
	6^c	92	Lu^{3+}	6	100
	8^c	106		8	112
Fe^{3+}	4^c	63	Mg^{2+}	4	71
	6^d	69		5	80
Fe^{4+}	6	73		6	86
Fe^{6+}	4	39		8	103
Fr^+	6	194	Mn^{2+}	4^c	80
Ga^{3+}	4	61		5^c	89
	6	76		6^d	81
Ge^{2+}	6	87		6^c	97
Ge^{4+}	4	53		7^c	104
	6	67		8^c	110
H^-	4	122	Mn^{3+}	5	72
	6	140		6^d	72
Hf^{4+}	4	72		6^c	79
	6	85	Mn^{4+}	4	53
Hg^+	3	111		6	67
	6	133	Mo^{3+}	6	83
Hg^{2+}	2	83	Mo^{4+}	6	80
	4	110	Mo^{5+}	4	60
	6	116		6	75
	8	128	Mo^{6+}	4	55
I^-	6	206		6	73
In^{3+}	4	76	N^{3-}	4	132
	6	94	Na^+	4	113
Ir^{3+}	6	82		5	114
Ir^{4+}	6	77		6	116
Ir^{5+}	6	71		7	126
K^+	4	151		8	132
	6	152	Nb^{3+}	6	86
	7	160	Nb^{4+}	6	82
	8	165	Nb^{5+}	4	62
La^{3+}	6	117		6	78
	8	130		8	88

(*Continued*)

Table 2.5 (*Continued*)

Ion	Coordination Number	Radius	Ion	Coordination Number	Radius
Ni^{2+}	4	69	Sc^{3+}	6	89
	4^a	63		8	101
	5	77	Se^{2-}	6	184
	6	83	Si^{4+}	4	40
Ni^{3+}	6^d	70		6	54
	6^c	74	Sn^{2+}	8	136
Ni^{4+}	6^d	62	Sn^{4+}	4	69
O^{2-}	2	121		6	83
	3	122	Sr^{2+}	6	132
	4	124		7	135
	6	126		8	140
	8	128	Ta^{3+}	6	86
Os^{4+}	6	77	Ta^{4+}	6	82
Pb^{2+}	6	133	Ta^{5+}	6	78
Pb^{4+}	6	92		8	88
Pd^{2+}	4^a	78	Tc^{4+}	6	79
	6	100	Te^{2-}	6	207
Pd^{3+}	6	90	Th^{4+}	6	108
Pd^{4+}	6	76		8	119
Pt^{2+}	4^a	74		9	123
	6	94	Ti^{2+}	6	100
Pt^{4+}	6	77	Ti^{3+}	6	81
Ra^{2+}	8	162	Ti^{4+}	4	56
Rb^{+}	6	166		6	75
	7	170	Tl^{+}	6	164
	8	175	Tl^{3+}	4	89
Re^{4+}	6	77		6	103
Rh^{3+}	6	81	U^{3+}	6	117
Rh^{4+}	6	74	U^{4+}	6	103
Rh^{5+}	6	69		8	114
Ru^{3+}	6	82		9	105
Ru^{4+}	6	76	V^{2+}	6	93
Ru^{5+}	6	71	V^{3+}	6	78
S^{2-}	6	170	V^{4+}	6	72
Sb^{3+}	4^b	90	V^{5+}	4	50
	5	94		6	68
	6	90			

Table 2.5 (*Continued*)

Ion	Coordination Number	Radius	Ion	Coordination Number	Radius
W^{4+}	6	80	Zn^{2+}	4	74
W^{5+}	6	76		5	82
W^{6+}	4	56		6	88
	5	65	Zr^{4+}	4	73
	6	74		6	86
Xe^{8+}	4	54		7	92
	6	62		8	98
Y^{3+}	6	104			
	8	116			

The simplest case of filling a subshell is provided by hydrogen and helium. When a second electron is added to the $1s$ orbital, two new coulombic forces come into operation: a repulsion between the electrons, tending to expand the electron cloud, and an increased attraction to the nucleus, tending to shrink the electron cloud. Because so-called correlation effects (spin and charge) tend to keep the electrons as far apart as possible—that is, on opposite sides of the nucleus—the increased attraction to the nucleus dominates. An alternative statement of this effect is that the presence of the second electron does not effectively screen the first from the increased nuclear charge.

In the case of p, d, and f orbitals, each added electron goes into a new region of space, unoccupied by preceding electrons in the subshell, until the subshell is half-filled. Beyond half-filling, double occupancy of orbitals must occur; however, screening is minimized by the effects of correlation, keeping the paired electrons as far apart as possible.

The effects of screening are generally expressed in terms of an effective nuclear charge, Z^*, which can be viewed as some kind of average value of the nuclear charge that an electron in an orbital experiences after the screening effects of other electrons have been taken into account. A set of approximate, empirical rules were devised by Slater (1932) to calculate Z^*. To apply these rules, it is first necessary to write the electron configuration of the atom in the following form:

$$[1s][2s2p][3s3p][3d][4s4p][4d][4f][5s5p][5d][5f]$$

We will refer to orbitals in the same bracket as a group. Slater's rules may then be summarized as follows:

1. For an electron in an $[ns,np]$ group, electrons in groups to the right contribute nothing to shielding.

2. For an electron in an $[ns,np]$ group, other electrons in the same group contribute 0.35 charge units to the shielding, except for $1s$ electrons, which contribute 0.3.

3. Each electron in the $n - 1$ group contributes 0.85 to the shielding.

4. Each electron in the $n - 2$ or lower group contributes 1.0 to the shielding.

5. For an electron in an nd or nf group, rules 1 and 2 remain the same, and all electrons in groups to the left contribute 1.0 to the shielding.

As an example of how shielding constants are evaluated, it is interesting to consider the case of potassium. In this case, $Z = 19$, and it is necessary to consider the following orbital groups: $[1s][2s2p][3s3p][3d][4s4p]$. If we assume that the atom has the configuration $[Ar]3s^2 3p^6 3d^1$, the screening constant S is calculated as (8×1.0) for the $n = 3$ shell plus (10×1.0) for the remaining inner electrons. This gives a value of $S = 18.0$ charge units of screening. Because the nuclear charge is 19 charge units, the $3d$ electron will experience an effective nuclear charge $Z^* = 19 - 18 = 1.0$ charge units. Consider the case where the atom has the configu-

Table 2.6   *Some Selected Values for Effective Nuclear Charge*

	Z^*				
Z	$1s$	$2s$	$2p$	$4s$	$3d$
1	1.0				
2	1.7				
3	2.7	1.3			
4	3.7	1.9			
5	4.7	2.6	2.4		
6	5.7	3.2	3.1		
7	6.7	3.8	3.8		
8	7.7	4.5	4.5		
9	8.7	5.1	5.1		
10	9.6	5.8	5.8		
21	Argon core: $1s^2 2s^2 2p^6 3s^2 3p^6$			4.6	7.1
22				4.8	8.1
23				5.0	9.0

ration $[\text{Ar}]3s^23p^64s^1$. In this case, the screening constant is (8×0.85) for the $n = 3$ shell plus (10×1) for the inner electrons to give a total $S = 16.8$ charge units and a $Z^* = 19 - 16.8 = 2.2$. It is thus evident that the outermost electron of potassium experiences a much greater effective nuclear charge and is therefore more stable if it is in a $4s$ orbital rather than a $3d$ orbital.

Although Slater's rules are useful in giving a quantitative aspect to the concept of shielding, they are based on simplifying assumptions that lead to poor agreement between the calculated and the true values. More sophisticated models have been developed, which give better answers but which lose much of the conceptual simplicity of the Slater approach.

Some examples of effective nuclear charges are illustrated in Table 2.6. It should be stressed that these numbers have little absolute significance, but they give a clear idea of how dramatic the change in effective nuclear charge can be on filling a subshell. The tendency for the electron cloud to shrink is therefore not surprising.

Ionization Energy

The formation of compounds from atoms is a transaction in electrons. Electrons are either given or taken, in whole or in part, by the various atoms forming the compound. The ease with which atoms either acquire or release electrons is therefore intimately linked to their chemical reactivity.

The ease of removing an electron from an atom is expressed quantitatively in terms of the ionization energy of the atom. This quantity may be defined as the energy required to remove an electron from the highest occupied orbital of the gaseous atom to infinity, the electron and the positive ion of the final state being at rest relative to each other. The ionization process is always endothermic as illustrated in Equation 2.25:

$$\text{He} \longrightarrow \text{He}^+ + e^- - 2.39 \text{ MJ mol}^{-1} \qquad \textbf{(2.25)}$$

The ionization energies of atoms have been easily measurable for many years, using various kinds of mass spectrometry and the convergence limits of line series in electronic spectra. The energies are all positive quantities, following the normal chemical convention that energy put into the system is positive. All the ionization energies cited in this text are experimentally determined, but it is worth noting that this is one of the quantities that can be calculated with some precision and confidence using quantum mechanics.

The first ionization energies of all elements are charted in Figure 2.19. A number of things are very evident from this figure. First, there is a general decline in ionization energy within groups as the principal quantum

Figure 2.19
First ionization
energies of the
elements.

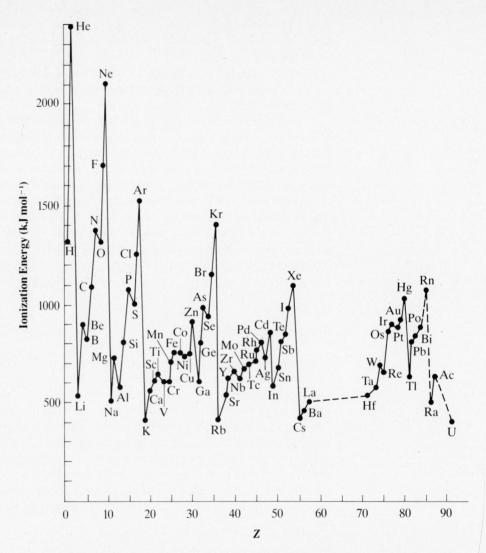

number increases. This is not too surprising because the coulombic attraction of the nucleus for the electron falls off inversely with distance. Superimposed on this general trend are a number of features that are associated with the particular stability of closed shells and half-filled subshells, for which the spin-exchange stabilization is particularly high. These effects may be listed as follows:

1. The highest ionization energies are for removal of an electron from a closed shell (noble gases).

2. The lowest ionization energies are for atoms that achieve a closed shell by loss of a single electron (alkali metals).

3. A discontinuity occurs between Group IIA (2) (breaking a closed *s* subshell) and Group IIIA (13) (creating a closed *s* subshell ion).

4. A discontinuity appears between Group VA (15) (breaking a half-filled *p* subshell) and Group VIA (16) (creating a half-filled *p* subshell).

The effects listed under numbers 3 and 4 are only evident for the earlier periods and are washed out by the intervention of *d*- and *f*-shell effects in later periods. Trends for the heavier elements, particularly *d*- and *f*-block elements, are complicated by spin–orbit coupling effects and electron spin–exchange energy effects imposed on relatively small changes in ionization energy.

It is also possible for atoms to lose several electrons under the influence of various kinds of excitation. Ionization energies rise quite steeply with increasing positive charge, but the abrupt rise on passing a closed-shell configuration is particularly dramatic. You can verify this by studying the multiple ionization energies listed in Table 2.7.

Table 2.7 *Some Multiple Ionization Energies of Selected Elements ($MJ\ mol^{-1}$)*

Element	1st	2nd	3rd	4th	5th	6th	7th
He	2.37	5.25					
Li	0.52	7.30	11.81				
Be	0.90	1.76	14.8	21.0			
C	1.09	2.35	4.62	6.22	37.83		
N	1.40	2.86	4.58	7.48	9.44	53.27	
O	1.31	3.39	5.30	7.47	10.99	13.32	71.33
F	1.68	3.37	6.05	8.41			
Ne	2.08	3.95					
Na	0.49	4.56	6.9				
Ca	0.59	1.15	4.91				
Se	0.63	1.24	2.39	7.08			
Ti	0.66	1.31	2.65	4.17	9.57		
Cr	0.65	1.50	2.99	4.74	6.69	8.74	15.54
Mo	0.69	1.56	2.62	4.48	5.91	6.6	12.23
Ni	0.74	1.75	3.39	5.30			
Cu	0.75	1.96	3.55	5.33			
Ag	0.73	2.07	3.36				
Au	0.89	1.98					
P	1.01	1.90	2.91	4.96	6.27	21.269	
Sn	0.71	1.41	2.94	3.93	6.97		

In conclusion, it is worth mentioning that the ionization energies up to the closed-shell configuration of most elements are of a magnitude that can be furnished through the formation of chemical bonds. The abrupt rise that occurs at the closed shell of positive ions is one of the most significant boundaries beyond which chemistry is unlikely to go. Only about twenty years ago, it was widely believed that the closed shell of neutral atoms, namely the noble gases, was chemically unbreachable. That belief was laid to rest with the synthesis of a number of noble gas compounds, and caution should perhaps be exercised in predicting that Cs^{2+} or Fr^{2+} compounds will never be made. Nevertheless, given that the second ionization energy of Cs is roughly twice that of Xe and considering the marginal stability of known xenon compounds, it will be very surprising indeed if Cs^{2+} compounds are ever made.

Electron Affinity

The complementary process to ionization, bringing an electron at rest from infinity to occupy the lowest-energy unoccupied orbital of an atom, is expressed quantitatively as an *electron affinity*. The values for electron affinity have been subject to an irritatingly eccentric treatment in that the sign convention most frequently used is the reverse of that normally used in thermodynamics; the sign for the exothermic process is positive. We use the normal thermodynamic convention, negative sign for exothermic process, as in Equation 2.26:

$$X + e^- \; \Delta H \to X^- \tag{2.26}$$

This means that we are strictly using the enthalpy of the process, but you are warned to expect a reversal of sign when consulting other literature for electron-affinity values.

Unlike ionization energies, which are always endothermic, electron affinities may be either endothermic or exothermic. They are also usually about an order of magnitude smaller than ionization energies for the one-electron processes. Because electron affinities are much more difficult to measure than are ionization energies, both the quantity and quality of data are lower for the former. Nevertheless, there are sufficient data to indicate that electron-affinity trends parallel those of ionization energies, as expected, especially regarding the resistance to breaking closed-shell or half-filled subshell configurations. In general, electron affinities become more positive on descending a group, but there seems to be a consistent anomaly between the second and third period. This has been explained in terms of the anomalously high interelectronic repulsion in the small atoms of the second period.

The attachment of more than one electron to an atom is always endothermic, and the values for second and third electron affinities are comparable to first ionization energies. Table 2.8 lists some selected electron affinities.

The Concept of Electronegativity

Chemists have long sought simple, quantitative measures of the tendency of elements to undergo reactions. Since the recognition of the importance of electrons in atomic structure and chemical bonding, one of the chief foci of this concern has been the quantification of the tendency of atoms in compounds to acquire electrons. This tendency is referred to as the *electronegativity* of the element.

When two atoms are close to one another, the electrons of each atom come under the influence of the other. We might conclude that the displacement of electrons would be largely the result of the balance of the four energy parameters, which define the tendency of the two atoms to lose an electron (ionization energies) and to gain an electron (electron affinities). In fact, one of the simpler proposals for electronegativity quantification, posited by Mulliken (1934), uses an average of the first ionization energy of an element and its electron affinity as a definition of electronegativity. Multiplied by an appropriate scaling factor, these electronegativities correspond remarkably well with those arrived at by more indirect means. The most serious fault with the Mulliken definition has been the paucity and poor quality of electron-affinity data, already referred to above.

The definition and the early advances in quantifying electronegativity were largely because of Pauling (1932). By intuition and empirical reasoning, he arrived at the conclusion that differences in electronegativity could

Table 2.8 *Some Selected First Electron-Affinity Values (kJ mol^{-1})*

				Group Number			
IA (1)	IIA (2)	VIB (6)	IIIA (13)	IVA (14)	VA (15)	VIA (16)	VIIA (17)
H: −73							
Li: −60	Be: +ve		B: −23	C: −122	N: 0	O: −141	F: −322
Na: −53	Mg: 0		Al: −44	Si: −120	P: −74	S: −200	Cl: −350
K: −48		Cr: −63	Ga: −36	Ge: −116	As: −77	Se: −195	Br: −325
		Mo: −96		Sn: −121		Te: −190	I: −295
		W: −50		Pb: −100			

be expressed in terms of an appropriate combination of homoatomic and heteroatomic bond energies.

Because bond dissociation energies were widely available, this scale of electronegativity was much more applicable than was the Mulliken scale. To use this scale, one electronegativity must be arbitrarily defined (in practice, $\chi_F = 4.0$), and so the Pauling scale is implicitly a relative scale.

Another commonly used scale, proposed by Allred and Rochow (1958), is based on the definition of electronegativity in terms of the electrostatic force operating on the valence electrons due to the effective nuclear charge. This scale is based partly on computed parameters and partly on experimental quantities, unlike the Pauling and Mulliken scales that are based entirely on experimental quantities. Table 2.9 lists some Allred–Rochow electronegativities of the elements.

In the foregoing discussion, we have been implicitly assuming that a unique electronegativity could be assigned to each element. This was recognized to be a gross oversimplification by Mulliken (1934) and by Jaffé and co-workers (1962) who developed the notion of "orbital electronegativity." In essence, this notion recognizes that different kinds of binding can result in the appearance of different electronegativities for the same atom. Most of the classical work dealt with the different valence states of the *p*-block elements. For example, carbon can be tetravalent, trivalent, or bivalent, as in ethane (C_2H_6), ethene (C_2H_4), and ethyne (C_2H_2), respectively. In these three different states, the electronegativity of carbon is computed to be 2.48, 2.75, and 3.29 (on the Pauling scale). A range of electronegativities will also be manifest by compounds in different oxidation states, although this area has been relatively little explored.

We will go no further into the development of electronegativity concepts now but will develop details in the text as the need arises.

BIBLIOGRAPHY

General References

Cotton, F. A., and G. Wilkinson. *Advanced Inorganic Chemistry*, 5th ed. New York: Wiley, 1988.

DeKock, R. L., and H. B. Gray. *Chemical Structure and Bonding*. Menlo Park, Calif.: Benjamin/Cummings, 1980.

Demitras, G. C., C. R. Russ, J. F. Salmon, J. H. Weber, and G. R. Weiss. *Inorganic Chemistry*. Englewood Cliffs, N.J.: Prentice-Hall, 1972.

Table 2.9 Allred–Rochow Electronegativities of the Elements

n	1 IA	2 IIA	3 IIIB	4 IVB	5 VB	6 VIB	7 VIIB	8	9 VIIIB	10	11 IB	12 IIB	13 IIIA	14 IVA	15 VA	16 VIA	17 VIIA	18 VIIIA
1	2.2 H																	He
2	0.97 Li	1.47 Be											2.01 B	2.50 C	3.07 N	3.5 O	4.10 F	Ne
3	1.01 Na	1.23 Mg											1.47 Al	1.74 Si	2.06 P	2.44 S	2.83 Cl	Ar
4	0.91 K	1.04 Ca	1.2 Sc	1.32 Ti(II)	1.45 V(II)	1.56 Cr(II)	1.60 Mn(II)	1.64 Fe(II)	1.70 Co(II)	1.75 Ni(II)	1.75 Cu(II)	1.66 Zn	1.82 Ga	2.02 Ge	2.20 As	2.48 Se	2.74 Br	Kr
5	0.89 Rb	0.99 Sr	1.11 Y	1.22 Zr	1.23 Nb	1.30 Mo	1.36 Tc	1.42 Ru	1.45 Rh	1.35 Pd	1.42 Ag	1.46 Cd	1.49 In	1.72 Sn	1.82 Sb	2.01 Te	2.21 I	Xe
6	0.86 Cs	0.97 Ba	1.08 La	1.23 Hf	1.33 Ta	1.40 W	1.46 Re	1.52 Os	1.55 Ir	1.44 Pt	1.42 Au	1.44 Hg	1.44 Tl	1.55 Pb	1.67 Bi	1.76 Po	1.90 At	Rn
7	0.86 Fr	0.97 Ra	1.00 Ac	Rf	Ha													

Lanthanide series	1.08 Ce	1.07 Pr	1.07 Nd	1.07 Pm	1.07 Sm	1.01 Eu	1.11 Gd	1.10 Tb	1.10 Dy	1.10 Ho	1.11 Er	1.11 Tm	1.06 Yb	1.14 Lu
Actinide series	1.11 Th	1.14 Pa	1.22 U	1.22 Np	1.22 Pu	Am	Cm	Bk	Cf	Es	Fm	Md	No	Lr

Douglas, B., D. H. McDaniel, J. J. Alexander. *Concepts and Models of Inorganic Chemistry*, 2d ed. Reading, Mass.: Addison-Wesley, 1983.

Gray, H. B. *Electrons and Chemical Bonding*. Menlo Park, Calif.: Benjamin/Cummings, 1965.

Huheey, J. E. *Inorganic Chemistry: Principles of Structure and Reactivity*. New York: Harper & Row, 1983.

Jolly, W. L. *Modern Inorganic Chemistry*. New York: McGraw-Hill, 1984.

Moeller, T. *Inorganic Chemistry: A Modern Introduction*. New York: Wiley, 1982.

Porterfield, W. W. *Inorganic Chemistry: A Unified Approach*. Reading, Mass.: Addison-Wesley, 1984.

Purcell, K. F., and J. Kotz. *Inorganic Chemistry*. Philadelphia: Saunders, 1980.

Sanderson, R. T. *Inorganic Chemistry*. New York: Van Nostrand-Reinhold, 1967.

Sebera, D. K. *Electronic Structure and Chemical Bonding*. Waltham, Mass.: Blaisdell, 1964.

Suggested Reading

Gray, H. B. *Chemical Bonds*. Menlo Park, Calif.: Benjamin/Cummings, 1973.

Heitler, W. *Elementary Quantum Mechanics*. Oxford, England: Clarendon Press, 1956.

Herzberg, G. *Atomic Spectra and Atomic Structure*. New York: Dover, 1944.

Linnett, J. W. *Wave Mechanics and Valency*. London: England: Methuen, 1960.

Weeks, M. E., and H. M. Leicester. *Discovery of the Elements*, 7th ed. Easton, Penn.: Chemical Education Publishing, 1968.

Landmark Publications

Bohr, N. *Philos. Mag.* 6, no. 26 (1913): 1, 476, 857.

de Broglie, L. V. "Researches on Quantum Theory." Thesis, University of Paris, France, 1924; *Philos. Mag.* 47 (1924): 446.

Heisenberg, W. Z. *Physik*. 33 (1925): 879.

Mendeleev, D. I. *J. Russ. Phys. Chem. Soc.* 1 (1869): 60.

Pauling, L. *The Nature of the Chemical Bond*, 3d ed. Ithaca, N.Y.: Cornell University Press, 1960.

Planck, M. *Ann. Physik*. (Leipzig) 1 (1900): 69; 4 (1901): 553.

Schrödinger, E. *Ann. Physik*. 81 (1926): 109; *Phys. Rev.* 28 (1926): 1049.

Slater, J. C. *Phys. Rev.* 42 (1932): 33.

PROBLEMS

Note: Tables of fundamental constants and conversion factors are given in Appendix II.

2.1 Calculate the wavelengths of the following sources of electromagnetic radiation, given their characteristic frequencies:
   a. Helium–neon laser, 4.738×10^{14} Hz
   b. Gamma-ray source, 3.0×10^{26} Hz
   c. Sunlamp, 1.2×10^{15} Hz
   d. Radiotelescope, 3.0×10^{11} Hz

2.2 What are the energies associated with the blue (488.0 nm) and green (514.5 nm) lines emitted by an argon-ion laser?

2.3 A typical star is believed to radiate energy equivalent to about 10^{40} kWh. What is this energy in kJ?

2.4 The entrance doors at airports are often controlled by photoelectric cells. What is the maximum wavelength of light that could be used for such systems with cesium cathodes if electrons are ejected from cesium with a kinetic energy of 9.6×10^{-20} J upon irradiation with 500 nm light?

2.5 The mid-IR region is approximately 4000–400 cm^{-1}. Would you expect to see the spectral line resulting from the electronic transition from the fifth to the eleventh electronic level in the hydrogen atom in this region of the electromagnetic spectrum?

2.6 How many spectral lines would you expect to detect in a spectroscope for atomic hydrogen if the initial electronic level is $n = 2$ and the final one is $n = 7$?

2.7 Given that the first ionization energy of H is 13.6 eV, to what electronic transition would you assign the bright red line at 1.14 eV in the emission spectrum of hydrogen?

2.8 Calculate the de Broglie wavelength of a steam locomotive weighing 5.0×10^4 kg and traveling at a velocity of 100 km h^{-1}. Then, do a similar calculation for a 1.0-g earthworm crawling through the grass at 0.01 m s^{-1}.

2.9 What is the uncertainty in the position of a proton traveling at 1.0×10^6 m s^{-1}?

2.10   What are the names, symbols, and roles of the four quantum numbers associated with the hydrogen atom?

2.11   What are the appropriate values of the quantum numbers for the $4s$, $6p$, $5f$, and $3d$ orbitals of the hydrogen atom?

2.12   Sketch the orbitals of atomic hydrogen for the first three quantum shells. In which of these orbitals are the signs associated with the wave functions arranged symmetrically?

2.13   What orbitals do the following quantum number combinations describe? Are they all permissible solutions of the Schrödinger equation?

	n	l	m_l
a.	5	3	-3
b.	3	2	0
c.	4	4	2
d.	2	1	-1
e.	1	0	$-\frac{1}{2}$

2.14   Which of the following orbitals are allowed for the hydrogen atom on the basis of wave mechanics?

   a. $7s$ **d.** $5p$
   b. $6f$ **e.** $5h$
   c. $2d$ **f.** $6d$

2.15   What is meant by the following?

   a. Bohr radius   **d.** Wave function
   b. Node   **e.** Bohr–Sommerfeld model
   c. Rydberg constant

2.16   What is meant by the Aufbau principle for neutral atoms? Are there any exceptions to this principle? Comment on the Aufbau principle in the case of transition metal and lanthanide metal cations.

2.17   What is the maximum electron capacity of the fourth quantum shell ($n = 4$)?

2.18   Consider the following radial probability density-distribution plot and respond to the associated questions.

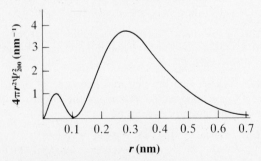

   a. How many nodes are there?
   b. What type of orbital is involved?
   c. Which orbital would it be if there were one more node?

2.19 What are the ground-state electronic configurations of the following atoms on the basis of the Aufbau principle?

a. Antimony (Sb)	**f.** Iridium (Ir)
b. Bromine (Br)	**g.** Chromium (Cr)
c. Cadmium (Cd)	**h.** Copper (Cu)
d. Europium (Eu)	**i.** Osmium (Os)
e. Hafnium (Hf)	**j.** Plutonium (Pu)

2.20 What are the symbols and names of the elements with the following ground-state electronic configurations?

a. $1s^2 2s^2 2p^6 3s^2 3p^4$
b. $1s^2 2s^2 2p^6 3s^2 3p^6 4s^2 3d^8$
c. $[Ar]4s^2 3d^{10} 4p^2$
d. $[Rn]7s^2 5f^4$
e. $[Ne]3s^2 3p^3$

2.21 Predict the ground-state electronic configurations of the following ions:

a. Ge^{4+}	**e.** Ir^+
b. S^{2-}	**f.** Pd^{2+}
c. Ag^+	**g.** Mn^{3-}
d. Cu^{2+}	

2.22 Predict the number of unpaired electrons in the following species:

a. Al^{3+}
b. Mn^{5+}
c. Cu^+
d. Zr^{3+}
e. Na^-

2.23 B^{4+} and C^{5+} are isoelectronic with He^+ and the hydrogen atom. How would you expect the ionization energies of B^{4+} and C^{5+} to compare with those of He^+ and H?

2.24 Which of the following cations do not have a ground-state electronic configuration of $[Kr]4d^6$?

a. Tc^+	**d.** Pd^{4+}
b. Ru^{2+}	**e.** Cd^{2+}
c. Rh^{3+}	

2.25 Classify the following elements according to their family names:

a. Cesium	**f.** Iridium
b. Calcium	**g.** Terbium
c. Radon	**h.** Einsteinium
d. Oxygen	**i.** Gold
e. Phosphorus	

2.26 Do you agree with the following statements concerning the periodic table?

a. Hydrogen is a member of the alkali metal family.
b. There are more metallic elements than nonmetallic ones.
c. There is an increase in metallic behavior in going across a period.
d. Some elements are difficult to classify as metals or nonmetals.

2.27   What are the general, ground-state, valence-shell electronic configurations for the following periodic groups?
    a. Coinage metals   **e.** Pnictogens
    b. Alkaline earths   **f.** Halogens
    c. Manganese family   **g.** Group IIB (12).
    d. Noble gases

2.28   What would be the ground-state terms for d^1 and f^1 configurations? For p^3 and d^5?

2.29   Derive the spectral term symbols for p^4, d^4, d^8, and f^3 configurations. What are the ground-state terms for these three systems? (For the f^3 case, you might put aside a weekend!)

2.30   Arrange the following elements in order of *increasing* first ionization energy:
    a. Sodium   **d.** Cesium
    b. Fluorine   **e.** Argon
    c. Iodine

2.31   Arrange the following elements in order of *decreasing* electron affinity:
    a. Chlorine   **d.** Sodium
    b. Phosphorus   **e.** Cesium
    c. Sulfur

2.32   How would you expect the first ionization energies of H, He^+, and Li^{2+} to compare?

2.33   Given that the electron affinity for a gaseous oxygen atom is 141 kJ mol^{-1} and the heat of reaction for the process $O(g) + 2e^- \rightarrow O^{2-}(g)$ is 639 kJ mol^{-1}, what is the electron affinity for gaseous O^-?

2.34   Explain with the aid of a suitable example in each case what is meant by the following:
    a. First ionization energy   **e.** Covalent radius
    b. Electron affinity   **f.** Ionic radius
    c. Electronegativity   **g.** Lanthanide contraction
    d. Effective nuclear charge   **h.** Spin multiplicity

2.35   In one sentence, describe the contributions that each of the following persons made to the development of the modern theory of atomic structure:
    a. Mendeleev   **f.** Planck
    b. Bohr   **g.** Pauli
    c. de Broglie   **h.** Hund
    d. Heisenberg   **i.** Russell–Saunders
    e. Schrödinger   **j.** Slater

2.36   Arrange the following isoelectronic ions in order of decreasing ionic radius:
    a. Ti^{4+}
    b. P^{3-}
    c. Sc^{3+}
    d. S^{2-}
    e. Mn^{7+}

2.37 Consider each of the following statements carefully and comment on their validity:
  a. Fluorine is the least electropositive element.
  b. Cesium is the most electropositive of the stable elements.
  c. Elements with a large difference in electronegativity will tend to form ionic bonds with one another.
  d. Covalent bonding implies similar electronegativities for the elements concerned.

2.38 Rationalize the seven ionization energies of chromium found in Table 2.7 on the basis of the ground-state electronic configuration of chromium.

2.39 Bearing in mind the effects of charge and atomic size on the spacing of shells, which, if any, of the following might have an anomalous ground-state electronic configuration?
  a. Sn^{2+} **c.** Ce^{4+}
  b. Fe^{3+} **d.** Pb^-

2.40 The C—C bond length in ethane is 154 pm and that of the Cl—Cl bond is 198.8 pm. Estimate the length of the C—Cl bond in chloroethane and compare it with the experimentally determined value of 177 pm.

2.41 By consulting Table 2.7, see how the following chemical facts conform to the patterns of ionization energy:
  a. Noble gases form few compounds.
  b. Transition elements tend to give a wider range of oxidation states than do main-group elements with the same number of valence electrons.
  c. Oxidation states IV, V, and VI are more stable for Mo than for Cr.
  d. The most stable oxidation states of Cu, Ag, and Au are II, I, and III, respectively.

2.42 Unlike the electronegativities of the other main groups, those of Groups IIIA (13) and IVA (14) pass through a minimum on descending the group. To what do you attribute this trend?

3

Theories of the Chemical Bond

3.1 *Empirical Theories*

Historical Background

The modern era in the understanding of chemical bonding began with recognizing the key role played by electron sharing between two atoms. Following this event, theory, driven by a large accumulation of empirical evidence, evolved rapidly toward a belief that electron pairs were crucial in chemical bonding. This evolution was climaxed with the formulation, particularly by Lewis (1916) and Langmuir (1920), of the electron-pair bond and the octet rule. We now know from quantum mechanics and from masses of new experimental evidence that the attribution of special qualities to the electron pair, both bonding and nonbonding, can lead to a severe distortion of the realities of electron distributions and of sources of binding energy in molecules. Nevertheless, the electron-pair concept has a beguiling simplicity that few chemists can resist. It works so wonderfully well in so many cases, particularly in main-group chemistry, that even if it were shown to be totally without theoretical justification, chemists would probably continue to use it as a kind of symbolic gesture.

The early notions of Lewis and Langmuir were founded largely on chemical facts derived from compounds of the main groups, particularly those containing light elements. In such compounds, the separations in energy levels are usually large compared to interelectronic repulsion energies, and excitation or ionization energies are high. These factors, coupled with the tendency of atoms to use as many of their valence-shell orbitals as possible in bonding, leads to stable molecules where all electrons tend to be paired and the number of pairs in the valence shell equals the number of valence orbitals—hence, the close association between molecular stability and the closed shell (or noble gas configuration). However, it is quite evident that cases where the number of electrons is other than twice the number of available valence orbitals present serious problems to theories that attribute a special role to electron pairs in bonding. Some examples of molecules from the main group that were known to disobey the electron-pair and octet rules at the time they were formulated are shown below:

	O_2	NO	NO_2	$(C_6H_5)_3C^{\cdot}$	KO_2	ClO_2	BF_3
Unpaired electrons:	2	1	1	1	1	1	0

It was also clear that the compounds of the *d*- and *f*-block elements presented more exceptions than conformities to the Lewis–Langmuir ideas.

Because the electron-pair view of bonding works so well for much of main-group chemistry, it has been retained and developed as a useful empirical formalism by chemists whose interests lie in that area. In the early days when very few detailed structures of molecules were known, the chief preoccupation was the reconciliation of the octet rule, which required that each atom have a shell of eight-valence electrons with known linkage patterns. For example, what are the allowed linkages in a molecule with the empirical formula $POCl_3$? Using the electron-pair and octet rules, structures (a) and (b) can be excluded, but (c) and (d) are possible (these four structures only represent a few of the possibilities):

      (a)   (b)   (c)   (d)

Another extremely important aspect of this simple view of bonding was the systematization it brought to one of the most fundamental of chemical reactions: the acid–base reaction. Lewis generalized the idea of acids and bases as classes of molecules or ions, which formed bonds by either accepting (an acid) or donating (a base) a pair of electrons. Initially, the tendency was to view this kind of sharing as qualitatively different from the situation where each atom contributed one electron to the bonding pair. However, we now know that this conception is false and that the electrons in a molecule carry no record of their atomic origins.

Some serious inadequacies of the electron-pair octet (EPO) model were removed or reduced by a series of rapid improvements on the original postulates. Langmuir (1920) quickly recognized that quasi–inert gas configurations other than neon (eight-electron shell) must be considered to explain connectivities (in modern parlance: coordination numbers) greater than four, for example, PCl_5 or SF_6. This concept was pursued by Sidgwick (1926), rather unsuccessfully, to account for the electron counts and structures of coordination compounds. Although many particularly stable complexes (for example, $[Fe(CN)_6]^{4-}$ or $[Co(NH_3)_6]^{3+}$) have valence shells with the same number of electrons as the next noble gas, many others (for example, $[Fe(CN)_6]^{3-}$ or $[Cr(NH_3)_6]^{3+}$) do not. Such inconsistencies have been instrumental in the almost complete abandonment of the electron pair–bond concept in transition metal chemistry.

Another inevitable consequence of the inadequacy of the localized EPO model was the necessity for evolving the clumsy and confusing notion of resonance. For example, in the case of nitric acid, physical and chemical evidence clearly shows that one of the oxygens is different from the other two, but EPO theory requires resonance between two structures to explain this result:

$$
\begin{array}{ccc}
\ddot{\text{O}}: & & \\
\text{H}:\ddot{\text{O}}\!\times\!\!\times\!\text{N} & \text{or} & \text{HO}-\text{N} \longleftrightarrow \text{HO}-\text{N} \\
\ddot{\text{O}}: & &
\end{array}
$$

On the other hand, the theory suggests that the nitrate ion, for which all theoretical and chemical evidence indicate the three oxygens to be indistinguishable, has three different kinds of oxygen. The EPO theory is rescued by the assertion that the true structure is in fact an average of the six possible equivalent EPO structures:

$$
^{-}\text{O}-\text{N} \longleftrightarrow {}^{-}\text{O}-\text{N} \longleftrightarrow \text{O}\!\leftarrow\!\text{N} \longleftrightarrow \text{O}\!=\!\text{N}
$$

and so on. In the above structures the sign ↔ symbolizes the simultaneous contribution of the two forms and not a dynamic change from one to another. Although this idea can be quite useful and has advanced theoretical understanding of chemical bonding, it also leads to much philosophical perplexity in the mind of the apprentice chemist, akin to that experienced by novices struggling with the oneness of the Holy Trinity.

Molecular Shape:
The Valence-Shell Electron-Pair Repulsion Theory

By the 1950s, a significant number of molecular structures had been determined in detail, and since then the trickle of structure determinations has become a deluge. This knowledge of molecular structure brought more refined pressure to bear on theory to satisfactorily explain the fine details of structure. An early advance was made by Sidgwick and Powell (1940) who concluded from analysis of the molecular structures known at the time that molecules or ions with only single bonds adopt those structures that allow the valence electron pairs to be as far away from each other as possible. In the cases of molecules AB_n with no lone pairs, the resulting geometries would be those corresponding to high-symmetry bodies of n apices, for example, $n = 4$ tetrahedron, $n = 5$ trigonal bipyramid, $n = 6$ octahedron. In the case of molecules with lone pairs, it was concluded that the lone pairs would be stereochemically active (that is, they would behave as though they were a substituent attached to the central atom A). However, in cases where geometrical isomers were possible, it was not clear which one, if any, would be favored energetically. For example, there was no problem in assigning an approximate location for the lone pairs in pseudotetrahedral NCl_3 or Cl_2O, or in pseudooctahedral BrF_5:

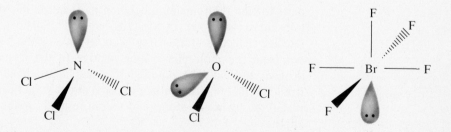

But why should the lone pair occupy an equatorial rather than an axial position in pseudotrigonal bipyramidal SF_4, and why should the lone pairs be trans, rather than cis, in pseudooctahedral IF_4?

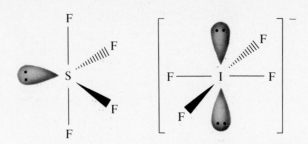

Notice that we use the prefix *pseudo* for each of the shapes adopted by these molecules carrying lone pairs. This is because few of them exhibit the exact angles required for the ideal body. These deviations from ideal angles also afforded an impetus to provide a more comprehensive model for rationalizing the fine details of molecular structure.

A set of simple postulates, which together constitute the valence-shell electron-pair repulsion (VSEPR) theory, were proposed by Nyholm and Gillespie (1957) to resolve many of these outstanding stereochemical questions. The postulates may be summarized as follows:

1. Each electron pair in the valence shell is stereochemically active.

2. The interpair repulsions are the sequence lone pair/lone pair > lone pair/bond pair > bond pair/bond pair.

3. In the case of molecules with multiple bonds, the interbond repulsions are the sequence triple/single > double/single > single/single.

4. The interpair repulsion between bonding pairs of a molecule AB_n decreases with increasing electronegativity of B.

Applying these simple postulates provides a rationale for an extraordinary number of structures. Thus, in the isoelectronic series CH_4 (109.4°), NH_3 (107.3°), and H_2O (104.5°), we see the increasing repulsion of lone pairs forcing the remaining bonding pairs into a decreasing angular relationship to each other. In NF_3 (102°) and NH_3 (104.5°), we see the effect of fluorine reducing interpair repulsion by moving the bonding pairs farther out from the molecular center relative to the less electronegative hydrogen. It should be noted that a first approximation steric argument would predict that the bulkier fluorine atoms would need more space than would the hydrogen atoms and therefore might prefer a larger F—N—F angle.

An outstanding success of VSEPR theory was the correct prediction of the structures of the noble gas compounds XeF_2 (linear) and XeF_4 (square planar), which were not synthesized until after the theory was formulated. The structure of XeF_6, although not yet unequivocally established in the simple molecular state, is almost certainly a distorted octahedron, in conformity with the expectations of VSEPR theory. Although some cases of crystalline solids have been reported where no distortion from ideal geometry due to lone pairs occurs (for example, $[SbX_6]^{3-}$ and $[TeX_6]^{2-}$; X = Cl, Br, or I), there are remarkably few cases where the VSEPR theory does not work for isolated molecules.

The one glaring example where VSEPR and any other explanation based on the localized electron-pair bond breaks down is Group IIA(2) halides. These compounds exist as ionic lattices in the solid state but can exist as triatomic molecules, MX_2, in the vapor phase. Some of these molecules are linear, as both VSEPR theory and simple electrostatics would predict, but at least as many are bent (for example, SrF_2 and BaF_2). As is so often the case in science, this apparently small and insignificant exception is the tip of a rather large iceberg of fundamental misrepresentation. This does not mean we must totally reject our simple theories any more than the inability of Newtonian mechanics to explain the precession of the planet Mercury required total rejection in favor of relativistic mechanics.

As noted in the introduction to Chapter 2, science satisfies two basic human tendencies: the desire to understand the fundamental workings of the universe and the desire to put that understanding to practical use. Often, fundamentally inadequate but simple theories are of much greater practical use than theories that are more fundamentally correct but so complex that they are inaccessible to most mortals.

Similarly, an insignificant exception at the practical level may prove more important to theoretical progress than all the generally obeyed rules. So it is with the localized electron-pair bond, so beloved by all practical chemists. It allows us to systematize and manipulate a vast amount of chemical information and is therefore justifiable in terms of its usefulness. The very thoroughness with which the ideas implicit in VSEPR theory have been applied has uncovered its few weaknesses at the systematic qualitative level. However, the theory does not lend itself at all to quantitative explanation or prediction, which is an even more serious objection at the fundamental level. Even worse is that recent theoretical calculations indicate that the relative importance of bonding and nonbonding electron-pair repulsions may in fact be the reverse of that assumed in the VSEPR theory!

3.2 Wave Mechanical Theories

The Many-Body Problem, Again

The Schrödinger equation is applicable to any system of interacting particles for which the interaction potential can be defined. In applying the equation to molecules, the many-body problem severely restricts the possibility for exact solution, analogous to the situation with atoms or ions with more than one electron. The simplest conceivable molecules must have at least three particles, two nuclei and one electron, for example, H_2^+ and HHe^{2+}. If the three particles are simultaneously in motion, there is no known way of exactly solving the equations of motion. Fortunately, a simplifying assumption—known as the Born–Oppenheimer approximation, which states that nuclear motion is negligibly slow compared to electron motion—allows an exact solution for the one-electron molecule. (We discuss how this approximation can be used later.)

For all real and chemically interesting molecules, however, solutions of the equation must be sought by approximate methods; the most common approach is to find ways of accurately guessing the wave functions for the various states of the molecule. Some very useful tests can guide the theoretician in the search for good wave functions (for example, an improvement in the quality of the wave function will always lead to a lowering of the calculated energy), but the *ab initio* calculation of the properties of many-electron molecules remains a precarious business. For example, until cyclobutadiene was prepared (1968) and its structure determined by spectroscopic techniques, a debate raged for many years among theoreticians as to whether the ground-state molecule would be square or rectangular. There is still debate between theoreticians and between theoreticians and experimentalists about the existence of hexazine as a stable molecule. This hypothetical molecule is a six-membered ring of nitrogen atoms, isostructural and isoelectronic with benzene:

Despite the great difficulties and uncertainties involved in quantitative solution of many-electron molecule problems, quantum mechanics has expanded greatly our understanding and systematization of molecular properties. For most experimental chemists, the qualitative results of quantum mechanical theories provide guidance in the laboratory. Our present treatment of the various theories will remain at a very empirical level.

The two wave mechanics–based views of bonding, which are widely used in chemistry today, have their roots in the question of how to choose a good molecular wave function. The mathematical nature of wave functions can be continually improved by addition to or multiplication by other wave functions. This fact leads to an inevitable convergence of theories that at the first approximation may start from very different viewpoints. This can be like the problem of describing an object with a limited number of words. Given a single word, one school of thought may choose *leaf* as the best way to describe a tree, whereas others may prefer *root*. As the number of words allowed for the description is increased, the tree descriptions of the different schools of thought may converge. So it is with *valence-bond (VB) theory* and *molecular orbital (MO) theory*. However, at the primitive level—the level most widely used in descriptive chemistry—the approaches are quite different and subject to their own serious flaws. Pauling's enormous influence has entrenched VB theory to a degree that it still receives consideration and deference, which some believe excessive. In the final analysis, only the MO theory (at the first approximation level) provides a unified, self-consistent view of bonding that is equally applicable across the periodic table.

Valence-Bond Theory: The Hydrogen Molecule

In 1927 Heitler and London applied quantum mechanics to the electron-pair bond. They considered the case of forming a covalent bond between two electrons of the simplest atom, hydrogen. In their approach to devising a "good" wave function for the H_2 molecule, they considered uniting two hydrogen atoms, A and B. In isolated atoms, we can assign a wave function to electron (1) on hydrogen atom A, $\psi_{A(1)}$, and a separate wave function to electron (2) on hydrogen atom B, $\psi_{B(2)}$. The central question is then: What happens to the wave functions when the two atoms are at a distance where they influence each other?

   The critical insight of Heitler and London was the recognition that once electrons come under the influence of both nuclei they can no longer be identified specifically with a single nucleus. Thus, whereas the total wave function for the isolated atoms is given by

$$\psi = \psi_{A(1)} \cdot \psi_{A(2)} \tag{3.1}$$

a better wave function for the interacting atoms would be one that recognizes that the two electrons can be interchanged without altering the system. Such a wave function is Equation 3.2:

$$\psi = \psi_{A(1)} \cdot \psi_{B(2)} + \psi_{A(2)} \cdot \psi_{B(1)} \tag{3.2}$$

   Using this function to solve the wave equation for the electronic energy of H_2 gives a value for the energy 280 kJ mol^{-1} lower than the function in Equation 3.1, but this value is still about 150 kJ mol^{-1} above the experimentally determined value of about -460 kJ mol^{-1}. A further improvement in ψ may be achieved by inclusion of wave functions for states where both electrons are associated with the same nucleus. The complete function, including ionic terms for $H_A^+H_B^-$ and $H_A^-H_B^+$, then becomes

$$\psi = \psi_{A(1)} \cdot \psi_{B(2)} + \psi_{A(2)} \cdot \psi_{B(1)} + \lambda\psi_{A(1)} \cdot \psi_{A(2)} + \lambda\psi_{B(1)} \cdot \psi_{B(2)} \tag{3.3}$$

The weighting coefficent $\lambda \ll 1$ is included because the ionic contributions are much less important than the covalent contributions in a homopolar molecule.

The process of "improving" the molecular wave function for H_2 has gone as far as to include 100 terms. Such a function can reproduce the experimental bond energy to within 0.05% (± 0.2 kJ mol^{-1}). Another mathematical guideline in quantum mechanics, the *variation principle*, states that no wave function can yield an energy lower than the experimental one, so only those corrections that *lower* the calculated energy remain in the set of wave functions and the experimental value cannot be surpassed. Although having enormous computational value, these properties of wave functions diminish the credibility of any physical implications one may wish to impute to components of the total wave function. For example, because there are many functions that when added to ψ lead to a diminution of E, the fact that $\psi_{A(1)} \cdot \psi_{A(2)}$ does so could be the result of ionic contributions having real importance, or it could be a mathematical coincidence. In the same vein, a chimpanzee equipped with a computer capable of generating linear combinations of all manner of algebraic functions could type out a function of no physical significance whatsoever, but when added to Equation 3.1 would produce a lowering of the calculated energy for the hydrogen molecule!

In our discussion of the bond in the hydrogen molecule, we have made no explicit mention of spin. In fact, all the foregoing discussion contained the implicit assumption that the two electrons have opposite spins. Only under these conditions does the solution of the wave equation result in an energy that is lower than that of the isolated atoms, that is, there is an attractive potential. The complete wave function must therefore also contain a spin function that describes the spin state of the two electrons (α or β):

$$\psi = [\psi_{A(1)} \cdot \psi_{B(2)} + \psi_{A(2)} \cdot \psi_{B(1)}, \ldots][\alpha_1\beta_2 - \alpha_2\beta_1] \tag{3.4}$$

Equation 3.4 embodies all the properties of the Lewis electron-pair bond.

There are also wave functions corresponding to the case where the electrons have the same spin. Three such functions provide three degenerate-energy solutions, all *higher* in energy than the isolated atoms. This repulsive state is the excited triplet state

$$\psi = [\psi_{A(1)} \cdot \psi_{B(2)} - \psi_{A(2)} \cdot \psi_{B(1)}] \begin{cases} [\alpha_1\alpha_2] \\ [\beta_1\beta_2] \\ [\alpha_1\beta_2 + \alpha_2\beta_1] \end{cases} \tag{3.5}$$

The difference in signs in the spin and orbital wave-function combinations in Equations 3.4 and 3.5 arises because the total wave function must change sign when the electrons are interchanged (antisymmetry principle). In hydrogen the forcing of electrons to a parallel-spin arrangement by the injection of an appropriate amount of energy would cause the molecule to fly apart because it is held together by only a single bond. In more complicated molecules, the excitation of one electron pair to a triplet state does not necessarily cause the molecule to dissociate (for example, see the O_2 molecule, discussed later in this chapter).

Another assumption implicit in the derivation of Equations 3.1–3.5 was that the functions ψ_A and ψ_B are 1s functions. Only when this is the case are the ground-state properties of the hydrogen molecule reproduced. However, excited states of the H_2 molecule may be generated by using combinations of all other atomic wave functions; furthermore, the ground-state wave function can be improved by mixing in small contributions from those excited states.

The principles behind the Heitler–London approach to the hydrogen molecule are summarized as follows:

1. A ground-state wave function is constructed by combination of an orbital part, which takes into account the interchangeability of electrons, and a spin part, which takes into account that a bond is formed by an electron pair with antiparallel spins.

2. For each antiparallel-spin state (singlet), there will be a higher-energy, parallel-spin (triplet) state.

3. A separate wave function can be constructed for each of an infinite manifold of excited singlets and triplets by using appropriate combinations of atomic wave functions with $n > 1$. In fact, such excited states for the H_2 molecule are purely hypothetical, but they are of great importance in more complicated molecules.

Valence-Bond Theory: Polyatomic Molecules

The widespread acceptance of VB theory has been in some measure the result of its intrinsic merits but is also the result of its brilliant development and advocacy by Pauling. Besides his fundamental contributions to research literature, Pauling's text *The Nature of the Chemical Bond* (1967) is and doubtlessly will remain one of the great classics of chemistry. By combining the approach developed by Heitler and London for H_2 with the concept of atomic orbital (AO) hybridization, Pauling (and the many theoreticians who followed in his footsteps) constructed a quantum mechanical justification of the Lewis–Langmuir theory of the two-electron bond and provided an elegantly pictorial, empirical description of the electronic structure of molecules. We cannot do justice to VB theory here and recommend that you read Pauling's book for a more detailed account. However, an understanding of modern chemical literature requires that you be familiar with the subject.

The more important basic features of polyatomic molecules explained by theory are stereochemistry (shape) and energetics (bond energies, electronic excitation energies). When quantum mechanics was first applied to molecules, stereochemistry was not well developed outside of the chemistry of the lighter main-group compounds and organic chemistry in particular. The greatest challenges in the early days lay therefore in those realms.

Dipole moment measurements had clearly established that molecules such as H_2O and NH_3 were not linear, or flat, as might be expected if their structures were determined by purely ionic forces. In the case of CH_4, although an ionic structure would explain the geometry, it could not satisfactorily account for the bond energy. Finally, the angular relationships and equivalences of the bonds in these molecules did not seem to bear any obvious relationship to the numbers and angular relationships of the AOs of the constituent atoms. This being the case, it was not immediately evident which AO from the heavier atom should be used to construct a wave function for a two-electron C—H or N—H bond. The concept of *orbital hybridization* provided the way out of this dilemma.

We start our discussion by considering the methane molecule CH_4. If we were to begin construction of CH_4 by overlapping hydrogen atom $1s$ orbitals with half-occupied orbitals on the carbon atom, the normal ground state of carbon $1s^2 2s^2 2p_x^1 2p_y^1$ presents a problem because it only has two unpaired electrons. It would not be expected that the stable hydride of monocarbon would be H_2C: because this species, carbene, does not have a completed octet. The deficiency in unpaired electrons could be circumvented by using the electronically excited state $1s^2 2s^1 2p_x^1 2p_y^1 2p_z^1$ and recouping the excitation energy from the formation of strong bonds to hydrogen. If constrained to think of bonds being formed by localized electron pairs in maximally overlapping orbital pairs, this excited state does not seem to satisfy the stereochemical evidence of the tetrahedral CH_4 molecule; the three p orbitals would require a mutually 90° set of bonds for maximum overlap, whereas the s orbital would give a maximum overlap independent of direction.

Pauling recognized that the mathematics of the wave function provided a solution to this problem through the process known as hybridization. Besides the infinite number of wave-equation solutions that correspond to the allowed physical states of an atom, there is also another infinite set that corresponds only to hypothetical states. We know of the existence of these solutions from the pure mathematics of wave mechanics, which tells us that any linear combination of good wave functions is also a good solution to the wave equation (subject to a well-defined set of mathematical rules). Thus, by taking appropriate linear combinations of ψ_{2s} and ψ_{2p}, Pauling constructed a hypothetical state (the valence state) of carbon that is characterized by a set of four degenerate (and therefore singly occupied) orbitals, each directed toward the corner of a tetrahedron.

Figure 3.1
sp^n hybrid orbitals.
(a) sp^3. One of four identical orbitals directed to the corners of a tetrahedron.
(b) sp^2. One of three directed to the corners of an equilateral triangle.
(c) sp. One of two directed colinearly and in opposite directions.

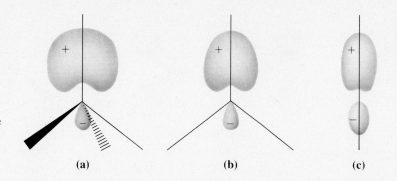

(a)   (b)   (c)

The wave functions for the four hybrid orbitals are as follows:

$$\psi_{sp^3} = \begin{cases} \frac{1}{2}(\psi_{2s} + \psi_{2p_x} + \psi_{2p_y} + \psi_{2p_z}) & \text{(3.6)} \\ \frac{1}{2}(\psi_{2s} + \psi_{2p_x} - \psi_{2p_y} - \psi_{2p_z}) & \text{(3.7)} \\ \frac{1}{2}(\psi_{2s} - \psi_{2p_x} + \psi_{2p_y} - \psi_{2p_z}) & \text{(3.8)} \\ \frac{1}{2}(\psi_{2s} - \psi_{2p_x} - \psi_{2p_y} + \psi_{2p_z}) & \text{(3.9)} \end{cases}$$

Because these orbitals are constructed from linear combinations of s and p orbitals in a ratio of $1:3$, we use the symbol sp^3 to designate this kind of hybrid. Figure 3.1 shows a picture of such a hybrid orbital. Despite the fact that wave functions of Equations 3.6–3.9 satisfy all the mathematical criteria for good solutions of the wave equation and despite the fact that one can calculate many physical properties for the sp^3 state, including its energy above the ground state and the electron-density distribution, the valence state has no real physical existence *as far as the free atom is concerned*. However, this valence state can be used to form the methane molecule by overlapping the sp^3 hybrids, each containing a single electron, with the four $1s^1$ hydrogen atoms. Maximum overlap occurs when the hydrogen atoms are at the corners of a regular tetrahedron and calculations of the ground-state energy of the molecule are similar in their correspondence to experimental values to those obtained for H_2. Again, as in the case of hydrogen, addition of many more corrective terms to the total wave function gives increasingly improved calculated properties.

The mathematics of the wave equation do not require using all three p orbitals for hybridization—it is equally possible to mix the s function with two p functions or with one p function. In these cases, there will be three (sp^2) or two (sp) hybrids and one or two unhybridized p orbitals, respectively. Figure 3.1 shows the shapes and spatial orientations of sp^2 and sp orbitals. That carbon exhibits trigonal and linear bonding in many of its unsaturated compounds conforms spectacularly well with the assumption that all of these modes of hybridization are used in the various compounds of carbon. Thus, in ethylene, all atoms are coplanar and all bond angles are *close to* $120°$ ($H\hat{C}H = 119°\ 55'$; $H\hat{C}C = 120°\ 22'$). The VB theory

Figure 3.2
Bonding in ethylene.
(a) Overlapping of
sp^2 hybrids.
(b) Overlapping of
unhybridized p
orbitals.

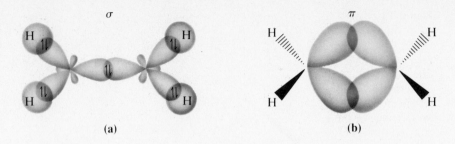

(a)   (b)

rationalizes such a structure by assuming the overlap of three sp^2 hybrids
on each carbon with two $1s$ hydrogen orbitals and with an sp^2 orbital of the
other carbon atom, as shown in Figure 3.2a. The unhybridized p orbitals
on each carbon atom overlap as shown in Figure 3.2b to form a second
bond between the two carbon atoms.

The overlapping orbitals in Figure 3.2a have the property of being
circular in cross section and without nodes. By analogy with s orbitals,
which have the same property, we call this type of overlap (or bond) *sigma*
(σ). A cross section through the overlap region of the overlapping
p orbitals (Figure 3.2b) has a node along the bond axis and resembles an
atomic p orbital in shape. Such an overlap (or bond) is labeled *pi* (π).

The benzene molecule provides an interesting example of the manner
in which chemists choose pictorial conventions that most suit their practical
ends. The strict application of localized valence-bond principles to benzene
would lead to an electron-pair picture analogous to that derived for
ethylene. However, the known equivalence of the six C—C bonds, whose
length (140 pm) is intermediate between that of ethylene (133 pm) and
ethane (154 pm), must be accommodated by invoking resonance between a
minimum of two structures:

To construct a good wave function for the benzene ring, we must use a
linear combination that includes equal contributions from these two structures. In fact, a really good wave function would include many other terms
as well, but for our present purposes we will deal with the minimum case.

To avoid having to deal with the annoying question of resonance, even
the most ardent VB adherents resort to a delocalized description of π
bonds in benzene (see MO theory below) while retaining the localized view
of σ bonds.

Although we have used the carbon atom to develop the principles of
hybridization, the concept is generally applicable. Thus, the $BeCl_2$ molecule is linear in the gas phase, a fact that accords with sp hybridization of
the beryllium atom, the bonds being formed by overlap of the two sp

Table 3.1 *Some Examples of Orbital Hybrids for Coordination Numbers Greater Than Four*

Coordination Number	Hybrid Orbitals	Geometry	Example
5	dsp^3 or d^3sp	Trigonal bipyramid	PF_5
5	d^2sp^2, d^4s, d^4p	Tetragonal pyramid*	IF_5
6	d^2sp^3	Octahedral	SF_6
7	d^5sp, d^3sp^3	Pentagonal bipyramid	$[ZrF_7]^{3-}$; IF_7
8	d^4sp^3	Dodecahedral	$[Mo(CN)_8]^{4-}$
9	d^5sp^3	Capped prism	$[ReH_9]^{2-}$

* Also referred to as square pyramid.

hybrids with the half-filled p orbitals of the chlorine atoms. It should be emphasized, however, that this molecule has only four-valence electrons and virtually any bonding theory would predict it to be linear.

The nonlinearity of H_2O and the nonplanarity of NH_3, besides the close correspondence of the bond angles in these compounds to the tetrahedral angle, are attributed in VB theory to the sp^3 hybridization of the central atom. In these cases, the sp^3 hybrids not involved in bonding are occupied by lone pairs. The slight reductions in bond angle would require an adjustment in hybridization, the bonding-pair hybrids having slightly more p character and the lone-pair hybrids slightly less p character than pure sp^3 hybrids. The congeners of H_2O and NH_3, namely H_2S and PH_3, have bond angles very close to 90°. In these molecules, it must be assumed that the orbitals used in bonding must be nearly pure p orbitals and the lone pair of PH_3 must occupy an orbital with very high s character.

The hybridization concept requires a further development to deal with molecules where the coordination number (number of surrounding atoms to which the central atom is attached) is greater than four. The solution to this problem was partially realized through the inclusion of d orbitals in the set from which hybrids are formed. Some hybrids based on s, p, and d orbitals and the associated geometries are listed in Table 3.1. Using a full set of s, p, and d orbitals allows rationalization of coordination numbers up to nine.

Molecular Orbital Theory: The Hydrogen Molecule and Other Simple Diatomic Molecules

When Heitler and London and Pauling were developing the VB theory of the covalent bond, a different wave mechanical approach was being taken by other physicists and physical chemists, notably Hund and Mulliken. This approach is now generally known as the molecular orbital (MO) theory, and it differs from the VB theory in two fundamental respects. First, it puts no special emphasis on electron pairs; second, it focuses more attention on the electron energies and orbital symmetries than does VB theory.

In the early days of the development of MO theory, the initial problem was, like all wave mechanical approaches to bonding, guessing a suitable wave function to facilitate a solution of the wave equation. Mulliken assumed that a simple linear combination of the isolated atomic functions might be a good choice. He applied this function not directly to the neutral hydrogen molecule but to the simplest possible molecule H_2^+. This one-electron molecule resembles in many respects the hydrogen and hydrogenlike atoms in that a minimum of simplifying assumptions is needed to render the problem tractable. The application of the Born–Oppenheimer approximation, which assumes nuclear motions to be so relatively slow compared to electron motions as to be negligible, reduces the H_2^+ case to a pseudo two-body problem and therefore amenable to solution.

Like the hydrogen atom, the solutions for H_2^+ correspond to a manifold of stationary states of ever-increasing energy. The most important difference from the hydrogen atom is a doubling of all energy states, as illustrated in Figure 3.3. This doubling arises naturally from the so-called linear combination of AOs. It is the mathematical nature of a linear combination such as a simple sum that both addition and subtraction are

Figure 3.3
Energy states of H and H_2^+.

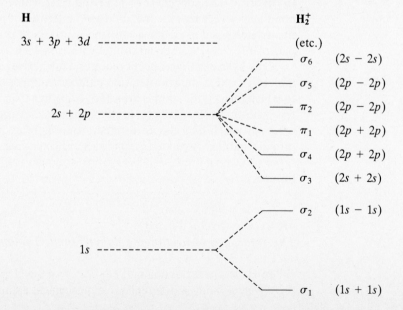

H	H_2^+	
$3s + 3p + 3d$ -----------------	(etc.)	
	σ_6	$(2s - 2s)$
	σ_5	$(2p - 2p)$
	π_2	$(2p - 2p)$
$2s + 2p$ -----------------	π_1	$(2p + 2p)$
	σ_4	$(2p + 2p)$
	σ_3	$(2s + 2s)$
	σ_2	$(1s - 1s)$
$1s$ -----------------		
	σ_1	$(1s + 1s)$

equally valid. In the physical world of waves, this property translates into the constructive and destructive interference modes of two interacting waves; it is this phenomenon that is strongly brought forward by the choice of a simple linear combination of AOs. If we consider the $1s$ wave functions of the two nuclei of H_2^+ [labeled (1) and (2) for convenient identification], we obtain the two linear combinations

$$\Phi^b = \Psi_{1s(1)} + \Psi_{1s(2)} \tag{3.10}$$

$$\Phi^a = \Psi_{1s(1)} - \Psi_{1s(2)} \tag{3.11}$$

The energy solutions for equilibrium internuclear separation corresponding to these two functions are those labeled σ_1 and σ_2 in Figure 3.3. One of these solutions is lower than the $1s$ hydrogen level, and the other is higher. Similar linear combinations of $2s$ and $2p$ AOs give rise to the higher-energy levels depicted for H_2^+ in Figure 3.3. The full degeneracy of the ns and np AOs does not persist in the molecule because the s–s and p–p orbital overlaps are not the same.

The interaction of AOs, represented by mathematical expressions like those in Equations 3.10 and 3.11, can also be shown pictorially in a simple way that conveys much information about electron distributions. Figure 3.4 shows such a pictorial representation of interacting s and p orbitals. In these pictures, two important pieces of information are conveyed. The outline of the orbital, which is related to electron distribution and to Φ^2, and the sign of the wave function, which is only a relative property but is of paramount importance to the way in which the electron waves interact. Figure 3.3 and 3.4 use the symbols σ and π to denote two different kinds of orbital. In the case of σ orbitals, there are no nodal planes parallel to the internuclear axis, whereas π orbitals have a single nodal plane parallel to the internuclear axis. For overlapping d orbitals, it is possible to produce MOs with two mutually perpendicular nodal planes parallel to the bond axis; such orbitals are designated δ orbitals. Such orbitals will be important later in discussing transition metal–metal bonding.

Equations 3.10 and 3.11 use the superscripts b and a in conjunction with the molecular wave functions. These superscripts are used to designate orbitals, or energy levels, that are *bonding* (lower energy than the isolated AOs) or *antibonding* (higher energy than the isolated AOs). From the schematic diagrams of Figure 3.4, it is clear that bonding orbitals are characterized by an augmentation of electron density between the bound nuclei, whereas antibonding orbitals show a depletion of electron density between the nuclei and a large internuclear repulsion. The reverse is true of electron densities in the regions outside the bonding regions. We will see later that there is a third kind of orbital that remains unaffected by incorporation into a molecule. Such an orbital is designated *nonbonding*.

Our discussion so far has revolved around the solutions of the wave

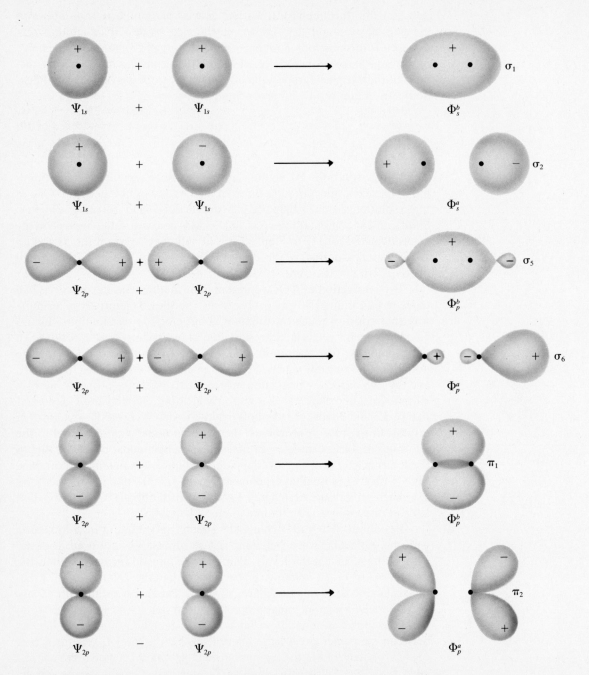

Figure 3.4   Pictorial representations of the linear combinations of *s* and *p* orbitals.

equation for the one-electron case. These solutions give us a series of energies corresponding to the one-electron stationary states and a set of electron-distribution functions, or orbitals. These orbitals can exist as a mathematical abstraction whether or not they are occupied by an electron, but physically real systems require a consideration of the electron-occupied orbitals.

The ground state for H_2^+ would involve accommodation of a single electron in the σ_1 orbital. Because the energy of the electron is lower in this situation than it would be with the nuclei separated, there is an attractive force holding the molecule together. If the electron were excited to the σ_2 level, its energy would be higher than with the nuclei separated, and a repulsive force results. Although the H_2^+ ion cannot be obtained as a stable species, it is readily observed in the gas phase using spectroscopic techniques, and its bond energy can be measured. The simple wave function, Equation 3.10, gives a bond-dissociation energy of about 170 kJ mol^{-1}, compared to an experimental value of about 270 kJ mol^{-1}. More complicated functions that take account of mixing in higher orbitals into the ground state give excellent agreement with the experimental value. This notion of mixing in other orbitals will be developed further in our consideration of many-electron diatomic molecules.

In the same way that electron configurations of the many-electron atoms can be constructed by applying the Aufbau principle to the one-electron hydrogen AOs, by using H_2^+ molecular orbitals we can construct electron configurations for many-electron diatomic molecules. Thus, for the H_2 molecule we would place the two electrons in the σ_1 orbital, and we would conclude that the H_2 molecule is stable. Because the H_2^+ orbital scheme takes no account of interelectronic coulombic repulsion and spin correlation, it would not give a good quantitative account of the properties of H_2. The proper accounting for such correlation is one of the more difficult problems in applying MO theory at the quantitative level. For our purposes, it is sufficient to point out that correlation effects tend to raise the energy of both bonding and antibonding orbitals, as illustrated in Figure 3.5. One consequence of this effect is that promotion of an electron from σ_1 to σ_2 in H_2 results in a definitively unstable species because the

Figure 3.5
Effect of correlation on orbital splitting.

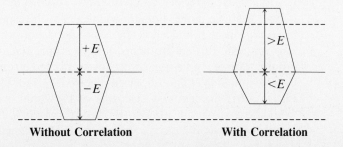

Without Correlation   **With Correlation**

antibonding orbital is destabilized more than the bonding orbital is stabilized.

Although some three-electron molecules can exist in the gas phase, like He_2^+, we will not deal with them here. No four-electron molecules (most notably He_2) have ever been observed. This is expected because four electrons would fill σ_1 and σ_2, resulting in a higher energy for He_2 than for two isolated He atoms.

The Li_2 molecule with six electrons would have σ_1, σ_2, and σ_3 filled. With a surplus of two electrons in a bonding orbital, this molecule should be stable and is indeed easily produced by vaporization of lithium metal. The bond energy of Li_2 (105 kJ mol^{-1}) is only about one-quarter that of H_2, in part due to the coulombic repulsion of the σ_1 and σ_2 electrons and in part due to the reduced concentration of σ_3 into the internuclear region as compared to σ_1. These trends persist with increasing n values for the constituent AOs and are responsible for the characteristic decline in bond strength that occurs on descending a group.

Before continuing with the filling of our H_2^+-like orbitals, we must make a correction to avoid a serious error that would be committed if we applied Figure 3.3 to Be_2. In our consideration of atomic structure, it was pointed out that, whereas the subshells of a particular n shell are degenerate in the hydrogen atom, shielding effects lift this degeneracy in atoms with more than one electron. Such shielding effects are also in operation in our many-electron diatomic molecules, and unless we correct our orbital scheme we will be seriously misled. The appropriate modification is to use orbitals that have been corrected for shielding, that is, the degeneracy of s and p has been lifted. The energy separation of $2s$ and $2p$ at Be_2 is such that the next orbital after σ_3 now has the character of an antibonding orbital arising from combination of $2s$ orbitals. Figure 3.6a shows the modified orbital scheme. Using this orbital scheme, the Be_2 molecule is predicted to have equal occupancy of bonding and antibonding orbitals and therefore unstable. In fact, the Be_2 molecule has never been detected, and the best calculations also predict it will not exist. It should be noted, however, that on the basis of qualitative assessments of the kind we are using here we would also not expect metallic beryllium to exist, a point we will return to in our discussion of metallic bonding.

When we consider what is known of the molecules B_2 and C_2— neither is stable under ambient conditions with respect to polymerization, but both can be produced in the gas phase—our orbital scheme of Figure 3.6a runs into further problems. Spectroscopic evidence indicates that the ground state of B_2 is a spin triplet (two unpaired electrons), whereas that of C_2 is a spin singlet (no unpaired electrons). According to the scheme of Figure 3.6a, the last two electrons of B_2 would enter σ_5 and therefore would be spin-paired, whereas those of C_2 would half-fill π_1 and following Hund's rule would have parallel spins. This evidence clearly points to the

Figure 3.6
Corrected MO
schemes for
homopolar
diatomics.
(a) without CI and
(b) with CI.

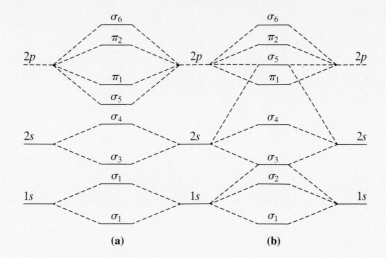

(a)   (b)

possibility that we have somehow reversed σ_5 and π_1 with respect to their relative energies.

The important question is, Why does the above reversal take place? The answer is that we were overly simplistic in our recognition of orbital overlaps in Figure 3.4. In this figure and the associated energy-level diagrams in Figure 3.3 and 3.6, we only admitted to the possibility that orbitals of the same quantum state on different atoms could combine. No such constraint exists in reality; Figure 3.7 illustrates a number of other possible linear combinations. It is important to recognize that because all orbitals die off asymptotically at an infinite distance from the nucleus all these overlaps (and others not mentioned) are possible. However, the degree to which orbitals overlap is strongly dependent on their difference in energy and falls off rapidly as this energy difference increases. Consequently, the interactions of orbitals of a different kind can generally be considered to be a perturbation on the energy changes produced by interactions of orbitals of the same kind. This can be translated into an interaction between MOs such that the first-order MO diagram is produced by overlapping AOs of the same quantum state and a second-order MO diagram is derived by mixing primary MOs whose constituent AOs are different but that can be combined with net overlap, as shown in Figure 3.7. Thus, the MO σ_3 ($2s + 2s$) can be further mixed with σ_5 ($2p + 2p$) because there is an overlap for the linear combination ($2s + 2p$). This kind of mixing, or interaction, is called *configuration interaction* (CI); Figure 3.6b shows a CI-corrected MO energy-level diagram for diatomic molecules. The most important interaction in this case is the mixing of σ_4 and σ_5.

In general, mixing two MOs leads to lower energy of one and higher energy of the other, in exactly the same way as mixing of two AOs.

Figure 3.7
Orbital overlaps that
lead to CI.

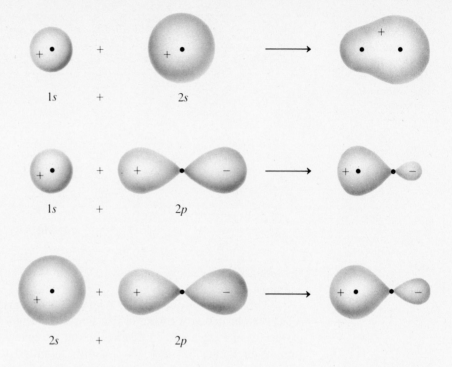

Because the primary overlaps of AOs are such as to make σ_4 and σ_5 closer together in energy, the primary overlap contributes to a strong interaction between σ_4 and σ_5. Because the overlap between $2s$ orbitals is greater than that between $2p$ orbitals, σ_4 resists being lowered in energy, and the dominant CI consequence in this case is to raise the energy of σ_5. In the cases of B_2 and C_2, σ_5 is raised above π_1, which is not subject to CI effects in these molecules, and the spin multiplicities of the ground states of both molecules are rationalized.

In N_2, photoelectron spectroscopy (PES) reveals the same ordering of energy levels as that shown in Figure 3.6b, and all levels up to and including σ_5 are filled.

The paramagnetism of O_2 is very easily explained on the basis of our simple MO diagram because either Figure 3.6a or 3.6b would predict that π_2 will be half-filled and the two electrons will adopt parallel spins. In the oxygen and fluorine atoms, the $2s$ and $2p$ levels are separated by a considerable difference in energy, thus reducing the CI between σ_4 and σ_5. PES confirms that O_2 conforms Figure 3.6a rather than 3.6b.

The multiplicity of chemical bonds, simply defined as the number of electron pairs located between bonded nuclei in the Lewis or Pauling descriptions, requires a slightly different interpretation in the MO theory. In fact, it is more normal to refer to the *bond order*, which is defined

Table 3.2	*Bond Orders and Bond Energies of Some Homonuclear Diatomic Molecules*		
Molecule	Bond Order	Bond Length (pm)	Bond Energy (kJ mol^{-1})
H_2^+	$\frac{1}{2}$	—	269
H_2	1	74.2	458
He_2	0	—	—
Li_2	1	267.2	105
Be_2	0	—	—
B_2	1	—	289
C_2	2	134	630
N_2	3	109.8	941
O_2	2	120.7	494
O_2^+	$2\frac{1}{2}$	—	—
F_2	1	141.8	153

as one-half the excess of electrons located in bonding orbitals, or bond order = (number of bonding electrons − number of antibonding electrons)/2.

Table 3.2 lists the bond orders and bond energies of some homonuclear diatomic molecules. It should be noted that in the cases of molecules like B_2 and O_2 it is irrelevant to the estimation of bond order that electrons are unpaired. Another set of predictions that derive very simply from the MO treatment of diatomic molecules is the strengthening of certain bonds by loss of an electron, for example, $O_2 \rightarrow O_2^+$. This is predicted whenever ionization occurs by loss of antibonding electrons. Loss of bonding electrons will, of course, have the reverse effect.

In general, the bond order corresponds to the bond multiplicity for simpler molecules. For more complicated molecules where it is difficult to decide whether orbitals are bonding, nonbonding, or antibonding, the clear definition of bond order is often impossible.

Some Heterodiatomic Molecules and the Molecular Orbital View of Polar Bonds

The MO treatment of heteropolar molecules follows precisely the same principles as that for homopolar molecules, but there is one important result that must be factored in to any discussion of heteropolar bonds. This result is that the bonding MO has more of the character of the AO of the more electronegative atom, whereas the antibonding MO has more of the character of the AO of the less electronegative atom. In terms of electron-density distribution, this means that bonding electrons tend to be associated more with the more electronegative atom and antibonding electrons are associated more with the less electronegative atom. In constructing a qualitative MO energy-level diagram for heteropolar diatomic molecules, it should be remembered that the valence AOs tend to be lower in energy, the higher the electronegativity of the atom. With these considerations in mind, the energy-level diagram shown in Figure 3.8 can be written.

There are a number of simple diatomic molecules whose electronic structures can be easily rationalized by this energy-level diagram. In particular, the three isoelectronic species CN^-, CO, and NO^+, each with a total of fourteen electrons will have all levels up to and including π_1 filled. These species are also isoelectronic with dinitrogen and similarly have a bond order of three. However, whereas each of the MOs in dinitrogen is equally distributed over the two nitrogen atoms, in the case of heteroatomic species there is an alternation of the electron distribution between the two atoms as one passes from one energy level to the next. The bonding orbitals, which are concentrated in the internuclear region, are biased toward the more electronegative atom, and the antibonding orbitals, which are concentrated in the extranuclear region, are biased toward the less electronegative atom. This tendency accounts for the near-zero

Figure 3.8
MO diagram for heterodiatomic molecules (excluding CI).

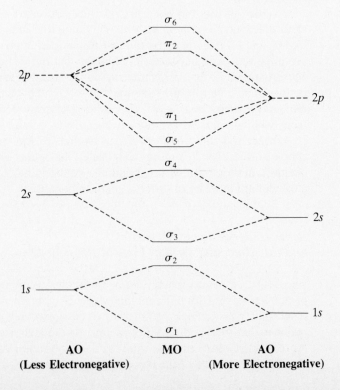

dipole moment of CO because, although the oxygen is more effective overall at attracting electron density than is the carbon, electron density that projects beyond the end of the molecule has a proportionately larger effect on the molecular dipole. This extranuclear electron density is mainly associated with σ_4, and it is the pair of electrons in this orbital that confers σ-donor properties on CN^-, CO, and NO^+ when they form metal complexes. Other fourteen-electron species such as CF^+, BF, and BO^- might also form stable metal complexes. To our knowledge, no stable compounds have been prepared, but the mass spectrum of $CF_3Mn(CO)_5$ has important peaks that are assigned to $[CF—Mn(CO)_n]^+$ ($n = 4, 3$).

The NO molecule has a single electron in the doubly degenerate π_2 orbital. This electron is antibonding, which accounts for the relative ease with which it is lost to form NO^+. The NO molecule also gains an electron fairly easily, but the NO^- ion is unstable with respect to dimerization. The loss of electrons by O_2 is more difficult than might be anticipated for a molecule with populated antibonding levels because of the large spin-exchange stability associated with the half-filled subshell configuration.

Molecular Orbitals in Polyatomic Molecules

The linear combination of AOs is not restricted to two-center interactions but can involve any number of orbitals, provided that there is indeed a net overlap. The only condition where there is no overlap is when the interacting orbitals satisfy the condition of orthogonality. This condition is satisfied when there is a change of sign in the overlapping region of one of the orbitals such that the regions of opposite sign are equivalent to each other. In this situation, constructive interference in one-half of the overlap is exactly canceled by destructive interference in the other. All overlaps illustrated in Figures 3.4 and 3.7 are nonorthogonal. Some orthogonal and some slightly nonorthogonal overlaps are shown in Figure 3.9. Note that moving three nuclei from a colinear to an angular relationship can transform noninteracting orthogonal orbital sets into interacting ones. This makes at least one MO more stable and one less stable and, depending on the degree to which the orbitals are occupied, may result in an increase or a decrease in molecular energy.

The overlapping of orbitals may occur directly or through the intervention of other orbitals. Four hypothetical polyhydrogen molecules are illustrated in Figure 3.10, together with their corresponding linear combinations (LCAOs) of Φ_{1s} and energy-level schemes. Although some of these H_n molecules show absolute stability according to calculations, none of them is stable relative to H_2. They do, however, illustrate the way in which shapes and energy relationships of MOs are strongly dependent on the spacial arrangement (or more particularly, the symmetry) of atoms in

Figure 3.9
Some orthogonal
(a) and some slightly
nonorthogonal
(b) overlaps.

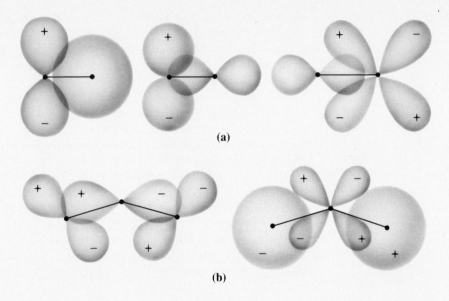

(a)

(b)

molecules. The following features of Figure 3.10 should be particularly noted:

1. The number of mutually independent linear combinations (MOs) is equal to the number of AOs being combined.

2. The energy of the molecular states increases with the number of nodes in the MO.

3. In Φ_2 of linear H_3, the central orbital is omitted because it is orthogonal to the terminal orbitals.

4. In triangular H_3, the proportions of the different AOs in the linear combinations are not equal in Φ_2 and Φ_3.

Characteristics 3 and 4 are necessary in chains or rings of odd numbers of orbitals, but not for even numbers.

Bonding in Triatomic Molecules

If we feed three electrons into the energy levels of H_3 molecules, two of them will pair in the bonding orbital and the third will occupy the nonbonding orbital. Thus, for linear molecules, there is one bonding pair for two bonds (bond order, $\frac{1}{2}$ per bond); for triangular molecules, there is one bonding pair for three bonds (bond order, $\frac{1}{3}$ per bond). It is therefore not surprising that these molecules do not exist under ordinary laboratory

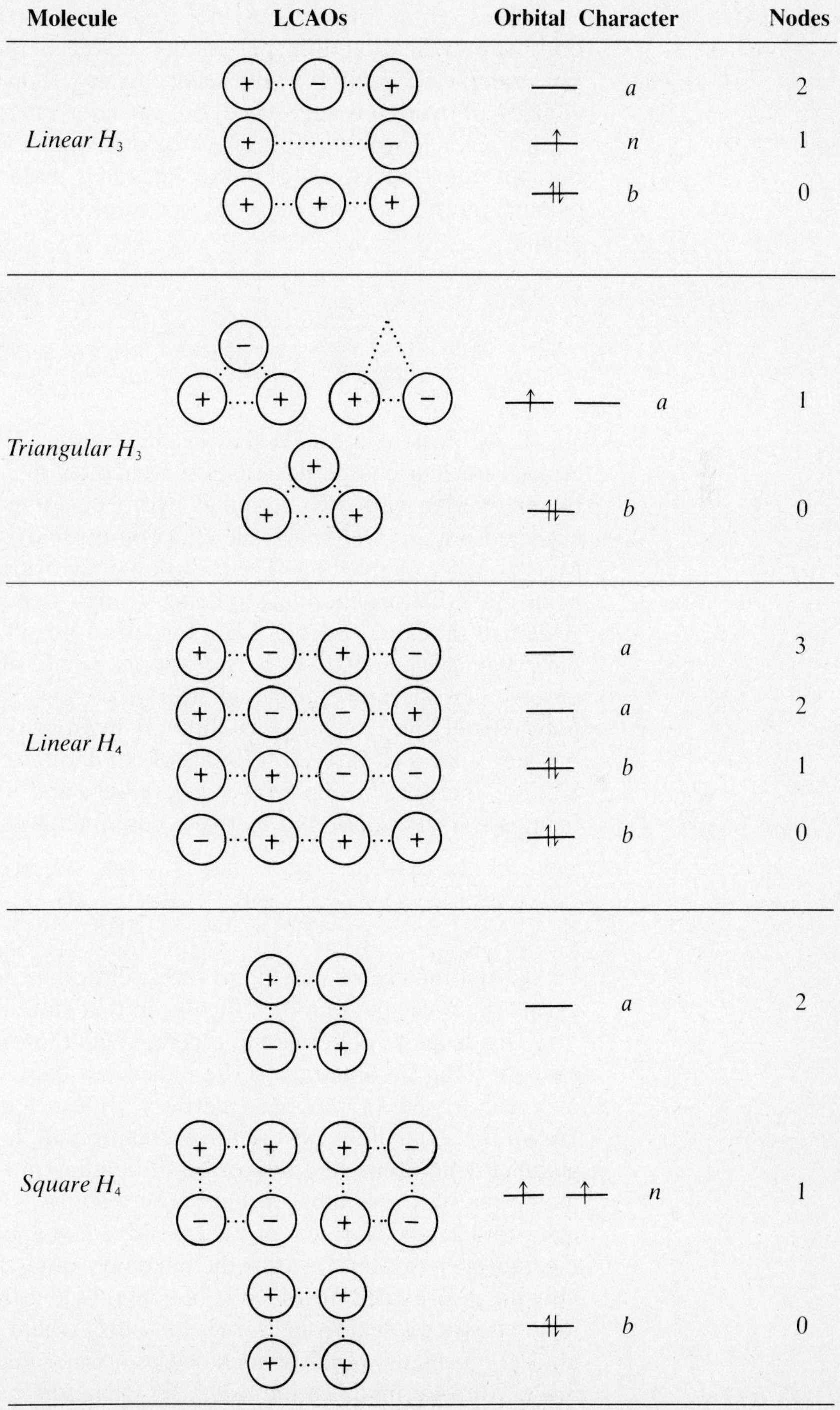

Figure 3.10
Some hypothetical
H_n molecules.

Molecule	LCAOs	Orbital Character		Nodes
Linear H_3		— a		2
		n		1
		b		0
Triangular H_3		— a		1
		b		0
Linear H_4		— a		3
		— a		2
		b		1
		b		0
Square H_4		— a		2
		n		1
		b		0

conditions. However, we must not forget that we have not included the possible participation of higher-energy orbitals in the linear combinations. For example, $2s$ orbitals would serve just as well as $1s$ as far as nonorthogonality of overlap is concerned, but the large energy separation between $1s$ and $2s$ for H would result in only a very small effect. More interesting, but still only a small effect in our linear H_3 molecule, is that the linear combination Φ_2 is nonorthogonal to one of the $2p$ orbitals on the central atom:

If we pass to the next hydrogenlike atom, lithium $(1s^2 2s^1)$, the increased nuclear charge decreases the energies of $2s$ and $2p$, and a much better overlap with two terminal hydrogen atom $1s$ orbitals would be expected. Figure 3.11 shows the effect of introducing overlap of $2p$ on the MO energy-level diagram. The inclusion of $2p$ overlap transforms Φ_2 from being essentially nonbonding to being strongly bonding in LiH_2 and BeH_2. Although the bond order of LiH_2 suggests a possible stable molecule, the high chemical reactivity of such a species would make it very difficult to observe. Furthermore, although BeH_2 is known as a stable solid, it is polymerized into long chains through hydrogen bridges. This type of bridging allows all three $2p$ orbitals of beryllium to participate in bonding because the relationship between beryllium and hydrogen is tetrahedral (compare with the MOs of methane described below):

$$\begin{array}{c} \text{H}\diagdown\quad\diagup\text{H}\diagdown\quad\diagup\text{H}\diagdown\quad\diagup\text{H} \\ \text{Be}\quad\text{Be}\quad\text{Be} \\ \diagdown\text{H}\diagup\quad\diagdown\text{H}\diagup\quad\diagdown\text{H}\diagup\quad\diagdown\text{H} \end{array}$$

Unlike the halides of beryllium, the hydride has not been observed as a gas-phase molecule, and its structure in that state is not known.

The dihalides of beryllium, together with those of other Group IIA (2) elements, can be obtained in the molecular state either in the gas phase or cocondensed in rare gas matrices at low temperature (15 K). All Group IIA (2) halides exist as three-dimensional, highly ionic solids under ambient conditions. In a discussion of bonding in such compounds in the molecular state, we must include both s and p AOs of both the central metal and halogens. If we consider that to a first approximation only one of the fluorine p orbitals (assume the bond lays along the z axis and therefore only the p_z orbital) is involved in bonding. Under this assumption, an MO diagram strictly analogous to that for BeH_2 would result, except that the unused p orbitals would be included as nonbonding orbitals, as in Figure 3.12. A further modification to the MO diagram in Figure 3.12a would be

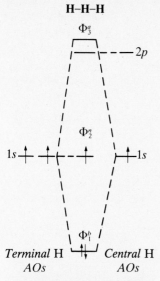

H–H–H

(a) H$_3$ molecule: Negligible mixing of Φ_2 and $2p$ due to large energy separation. Unpaired electron confers reactive free-radical character. NOTE: Φ_2^q means non-bonding orbital.

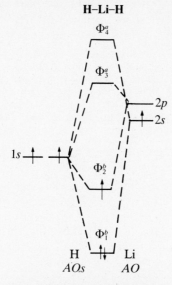

H–Li–H

(b) LiH$_2$ molecule: Mixing of Φ_2 and $2p$, but molecule highly reactive and unstable with respect to LiH and H$_2$.

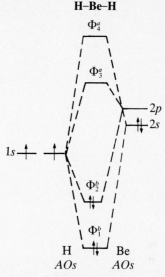

H–Be–H

(c) BeH$_2$ molecule: Mixing of Φ_2 and $2p$ confers bonding character on Φ_2. Filling of Φ_2^b confers stability. Bending and polymerization through H bridges confers even more stability through involvement of other $2p$ orbitals.

Figure 3.11
Mixing of $2p$ atomic orbital into Φ_2 of H$_3$, LiH$_2$, and BeH$_2$.

to include π overlaps of the $2p_x$ and $2p_y$ orbitals on fluorine and beryllium. The rules applying to linear combination of orbitals are the same, without respect to whether the overlaps are σ or π. Thus, linear combination of three parallel p_x orbitals in the π sense gives rise to a bonding, a nonbonding, and an antibonding MO in the same way that linear combination in a σ sense does. As a general rule, π overlaps are less effective than σ overlaps. Figure 3.12 shows such a π-orbital combination for BeF$_2$. Because each fluorine contributes two electrons and the beryllium none, the four π-bonding electrons are accommodated in a bonding and a nonbonding orbital, resulting in a net contribution to bonding. Because both p_x and p_y orbitals can give π overlaps, the scheme in Figure 3.13 also applies to p_y. The net result of π bonding is that two of the filled nonbonding p_x and p_y orbitals of Figure 3.12a become bonding while two remain nonbonding and that the two empty nonbonding p_x and p_y orbitals become antibonding.

Essentially, the same approach to bonding may be applied to all other Group IIA (2) halides by simply adjusting the principal quantum numbers of the outermost orbitals and, at least qualitatively, making adjustments for changes in electronegativity or relative energies of orbitals. Never overlook, however, that the real situation with heavier atoms becomes

Figure 3.12
(**a**) MO diagram for BeF$_2$.
(**b**) Overlaps producing
σ bonding.

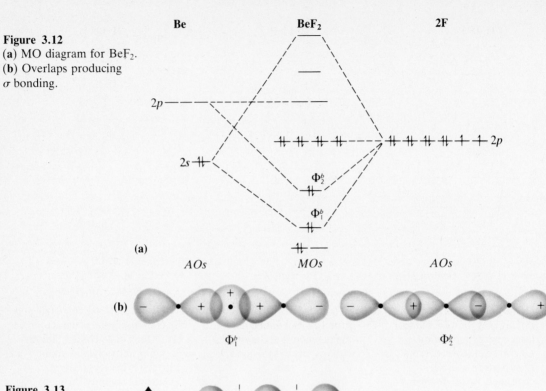

(**a**)

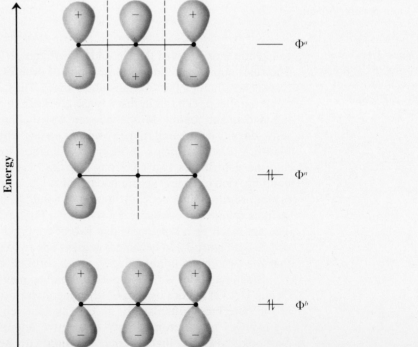

(**b**)

Figure 3.13
Linear combination
of three parallel p_x
orbitals.

complicated by the bunching of electron-shell energies and by the presence of filled and empty *d* subshells. On top of those uncertainties, we must add that hydrogenlike orbitals may just not be a very good approximation in the case of heavier atoms.

At this point in our development of qualitative MO theory, we must ask a very important question about Group IIA (2) halide molecules. Thus far, we have restricted our discussion to the possibility that such molecules would be linear. Is this necessarily so? Or, more importantly, does our theory make a prediction regarding the shape of such molecules? Before answering this question, let us look at what happens to the MO energies if we bend a BeF_2 molecule. Figure 3.14 shows what happens to the *psp* and the *ppp* overlaps as the bond angle is reduced from 180–90°. In this diagram, we have arbitrarily chosen to maintain the local coordinates constant. This is not a necessary condition because rotation of the local-coordinate system of any particular atom does not alter the total overlap of a complete orbital set, although it may alter the overlap of a particular orbital. If we had maintained the local coordinates of other atoms in a constant angular relationship to the internuclear lines, we would still obtain the same result for the complete orbital set. This result is in some ways disappointing because we find that at the qualitative level, although the detailed electron distribution is greatly altered by bending, there is no obvious loss or gain in bonding. The main changes are that the σ- and π-bonding overlaps (Figure 3.14) decrease while the nonbonding interaction becomes bonding as the angle decreases. Although there is no doubt that these two angular extremes are of different energy, we are in no position to make a judgment as to what their relative energies are nor to guess whether the lowest-energy configuration is at some intermediate angle.

Figure 3.14
Influence of bond angle on orbital overlaps in MX_2 molecules.

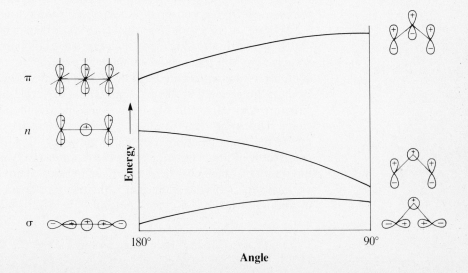

Despite the disappointing vagueness of the above conclusion, it conforms rather well to experimental observation. It has been observed that, of the sixteen easily accessible halides of Group IIA (2) at least one-half are nonlinear on the basis of two independent physicochemical methods (vibrational spectroscopy of molecules frozen in noble gas matrices and electric field deflection of molecular beams). The fluorides span the range of angles: MgF_2 (160°), CaF_2 (140°), SrF_2 (108°), BaF_2 (100°). It has also been found by vibrational spectroscopy that resistance to bending of the molecules is quite low. All this leads to the conclusion that although the crude first-order overlap picture we have developed leads to the correct overall conclusion that the MX_2 molecules do not show a clear preference for either linear or angular shape, the actual bond angles are determined by subtle second-order effects beyond the scope of easy analysis.

The water molecule has the same basic orbital set as BeH_2 but has four more electrons, and the relative electronegativities of the two constituent atoms are reversed. Figure 3.15 shows the MO energy-level diagrams for linear and bent H_2O. Once again, we are faced with difficulty in making a qualitative prediction in favor of either a linear or bent structure. In the process of bending, some overlap is gained, with one oxygen p orbital that was orthogonal to the hydrogen orbitals in the linear molecule, but some overlap is lost with the p orbital that lay along the bond axes in the linear molecule. From experiment, we know that the increased bonding energy, gained from making one of the occupied nonbonding orbitals more bonding, more than compensates for the reduction in other overlaps and that the molecule strongly prefers to be bent. Detailed MO calculations order the energies of the MOs of water (relative to the ionization continuum) as being: $1s^2$ (-55.7), $2s^2$ (-31.5), $1b_2^2$ (-1.78), $3a_1^2$ (-1.27), $2b_1^2$ (-1.13 MJ mol^{-1}).[1] The essential correctness of these calculations is borne out by the photoelectron spectrum of water, which shows three ionizations above -2 MJ mol^{-1}, corresponding roughly to the calculated energies.

It is worth stressing here that the electronic structure of water as revealed by MO theory and by modern experimental methods does not conform to that represented by Lewis octet theory or hybridization theory. Both of these views, at the first approximation level, present an electronic structure for water with two equivalent lone (nonbonding) pairs of electrons. This view lends itself to very easy pictorial representation and intellectual manipulation. For most purposes, that it is fundamentally incorrect is of little consequence because water rarely uses more than one lone pair at a time in its chemistry. Even in cases where two lone pairs may be involved, it is the molecular structure of the product that governs electronic structure, not the molecular or electronic structure of the

---

[1] The significance of the labels b_2 and a_1 used here and in Figure 3.15 will become evident in Chapter 5.

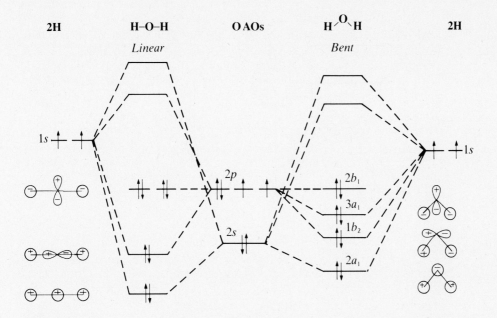

Figure 3.15
MO diagram for bent and linear H_2O.

antecedent water molecule. Although we can get away with using the classical structure of water or other similar molecules without negative consequences in most instances, we must always be aware of the approximations we are making. It is an interesting reflection of chemical psychology that very few chemists would now insist that π-electron pairs of benzene all have the same energy, but many still believe that the lone pairs of water do.

Molecular Orbitals of Methane

We continue our discussion of the application of MO theory to polyatomic molecules with a description of the bonding scheme for methane. The methane molecule is important because it is a prototype for discussion of bonding in organic chemistry. It also provides a useful starting point for the discussion of bonding in tetrahedral compounds of the heavier elements, particularly those of the transition elements. The problem reduces to one of finding a suitable set of linear combinations of four $1s$ hydrogen orbitals at the apices of a tetrahedron to overlap nonorthogonally with the valence orbitals of a central carbon atom. Figure 3.16 shows four such linear combinations where each of the overlaps represents a bonding overlap. The corresponding antibonding overlaps are produced by reversing the signs of either the central carbon or the hydrogen orbitals. In Figure 3.17, these linear combinations are translated into an energy-level diagram,

Figure 3.16
Linear combinations
for CH_4 molecule.

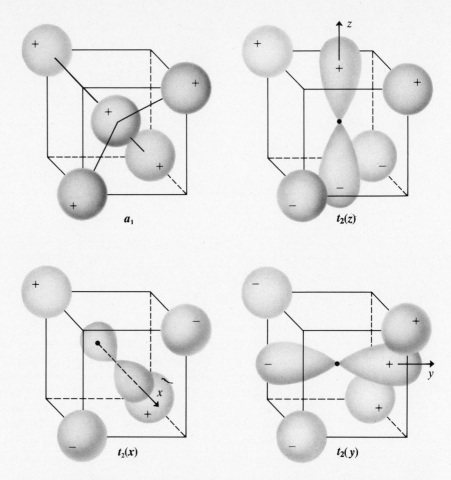

which shows how the eight-valence electrons are accommodated in four strongly bonding MOs, one of which has an energy different from the other three. Each of the bonding MOs spreads over the whole molecule, with the regions of greatest electron density being located in the four C—H bond axes. Thus, one-quarter of each of the MOs contributes to each C—H bond, and the bond order for each bond is unity.

Again, this view conflicts with the first-order VB approach, which strives to produce four equivalent bonds by producing four equivalent electron pairs. The process of mixing the *s* and *p* orbitals to produce sp^3 hybrids, although mathematically valid, does not conform to physical reality. It is now generally recognized that only orbitals that have the same symmetry relationship to the overall symmetry of the molecule can mix. This is simply a reexpression of the orthogonality rule because orbitals that do not have the same symmetry properties are always orthogonal. We have already encountered this phenomenon, without identifying it as such, when

Figure 3.17
Energy-level
diagram for CH_4.

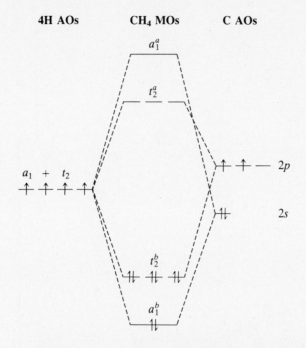

we discussed CI in diatomic molecules (page 87). While configuration interaction operated on all of the σ orbitals of the diatomic molecules, the π orbitals were left unaffected. This is because *in a diatomic molecule* all σ orbitals have the same symmetry properties, but π orbitals have symmetry different from σ orbitals. For more complicated molecules, different σ orbitals may belong to different symmetry classes, for example, the a_1 and t_2 orbitals of CH_4. Note that there are no nodal planes in the bonding regions for any of the bonding MOs of CH_4, and they are therefore all σ orbitals.

Molecular Orbitals of Octahedral Molecules

With the tetrahedral case, we reached the limit of coordination number possible with four-valence orbitals and with the preservation of classical two-electron bonds. Molecules with more bonds than bonding-electron pairs, known as *hypervalent molecules*, are possible, and a number of cases are known. However, it is more conventional to invoke the participation of d orbitals in bonding to rationalize coordination numbers greater than four. Although this is widely done, it should always be remembered that there is very little unequivocal evidence, either experimental or theoretical, in favor of significant participation of $3d$ orbitals in the bonding of third-period main-group elements. The participation of d orbitals in

Figure 3.18
Linear combinations
of an octahedral set
of σ orbitals with
the nine-valence
orbital set $(n-1)d$,
ns, np. The shaded
ligand orbitals are
nonorthogonal to
the central orbital.

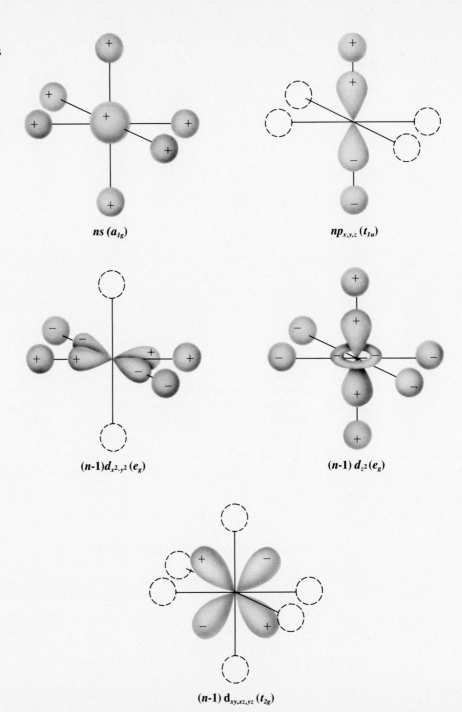

ns (a_{1g})

$np_{x,y,z}$ (t_{1u})

$(n-1)d_{x^2-y^2}$ (e_g)

$(n-1)\, d_{z^2}$ (e_g)

$(n-1)\, d_{xy,xz,yz}$ (t_{2g})

bonding in the compounds of the transition elements is, however, well established.

The main difference between the cases of main-group and transition compounds is in the energy of the d orbitals relative to that of the s- and p-valence orbitals. In main-group compounds, the relevant nd orbitals are in the same quantum shell as the other valence orbitals (ns and np) and quite a bit higher in energy. In transition compounds, the relevant $(n - 1)d$ orbitals are in an inner quantum shell relative to the other valence orbitals (ns and np) and lower in energy.

The linear combinations of an octahedral set of σ orbitals that are nonorthogonal to the $(n - 1)d$, ns, np set of the central atom are shown in Figure 3.18. It should be noted that there is no nonorthogonal combination of σ orbitals with the same symmetry as the d_{xy}, d_{xz}, and d_{yz} set, and these remain nonbonding (they are labeled t_{2g}^n in Figure 3.18). This still leaves six bonding overlaps, sufficient for the six two-electron bonds of an octahedral compound. Figure 3.19 shows the qualitative energy-level diagram for these overlaps.

A similar treatment of the main-group case would require placing the d-orbital set above the ns and np orbitals, but the final outcome would be qualitatively the same. For most cases, it appears that the contribution of the d orbitals to bonding is small, that is, the lifting of degeneracy of the d-orbital set into the orbitals labeled t_{2g} and e_g^a in Figure 3.19 is very small.

Although the t_{2g} orbitals are nonbonding in terms of σ bonding, they can participate in π bonding. This aspect is treated in greater detail in Chapter 10.

Figure 3.19
Qualitative energy-level diagram for the MOs of an octahedral compound.

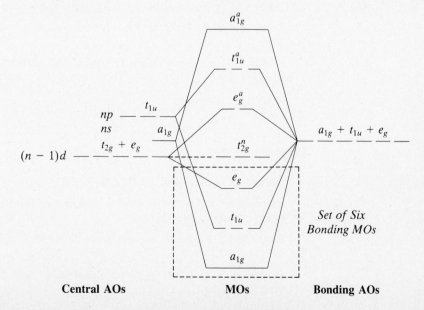

Central AOs   MOs   Bonding AOs

BIBLIOGRAPHY

General References

Cotton, F. A., and G. Wilkinson. *Advanced Inorganic Chemistry*, 5th ed. New York: Wiley, 1988.

DeKock, R. L., and H. B. Gray. *Chemical Structure and Bonding*. Menlo Park, Calif.: Benjamin/Cummings, 1980.

Demitras, G. C., C. R. Russ, J. F. Salmon, J. H. Weber, and G. R. Weiss. *Inorganic Chemistry*. Englewood Cliffs, N. J.: Prentice-Hall, 1972.

Douglas, B., D. H. McDaniel, and J. J. Alexander. *Concepts and Models of Inorganic Chemistry*, 2nd ed. New York: Wiley, 1982.

Gray, H. B. *Electrons and Chemical Bonding*. Menlo Park, Calif.: Benjamin/Cummings, 1965.

Huheey, J. E. *Inorganic Chemistry: Principles of Structure and Reactivity*. New York: Harper & Row, 1987.

Jolly, W. L. *Modern Inorganic Chemistry*. New York: McGraw-Hill, 1984.

Moeller, T. *Inorganic Chemistry: A Modern Introduction*. New York: Wiley, 1982.

Porterfield, W. W. *Inorganic Chemistry: A Unified Approach*. Reading, Mass.: Addison-Wesley, 1984.

Purcell, K. F., and J. Kotz. *Inorganic Chemistry*. Philadelphia: Saunders, 1980.

Sanderson, R. T. *Inorganic Chemistry*. New York: Van Nostrand Reinhold, 1967.

Sebera, D. K. *Electronic Structure and Chemical Bonding*. Waltham, Mass.: Blaisdell, 1964.

Suggested Reading

Benson, S. W. *J. Chem. Ed.* 42 (1965): 502.

Cartmell, E., and G. W. A. Fowles. *Valency and Molecular Structure*, 4th ed. Boston: Butterworths, 1977.

Gray, H. B. *Chemical Bonds*. Menlo Park, Calif.: Benjamin/Cummings, 1973.

Sanderson, R. J. *Chemical Bonds and Bond Energy*. New York: Academic Press, 1976.

Landmark Publications

Born, M. Z. *Phys.* 7 (1921): 124.

Drude, P. *Ann. Phys.* (Leipzig) 3 (1900): 369.

Heitler, W., and F. London. *Z. Phys.* 44 (1927): 455.

Lewis, G. N. *J. Am. Chem. Soc.* 38 (1916): 762.

Madelung, E. *Phys. Z.* 19 (1918): 524.

Mulliken, R. S. *Phys. Rev.* 40 (1932): 55.

Pauling, L. *The Nature of the Chemical Bond.* Ithaca, N.Y.: Cornell University Press, 1967.

PROBLEMS

3.1 With the aid of a suitable example in each case, explain what is meant by the following:

    a. Resonance   **d.** Octet rule
    b. Bonding MO   **e.** Configuration interaction
    c. Antibonding MO   **f.** Nonorthogonal overlap

3.2 Calculate the net bond order for each of the following diatomic species:

    a. Na_2 **e.** Br_2
    b. Mg_2 **f.** Ar_2
    c. C_2 **g.** IBr
    d. O_2^- **h.** NO^+

    Which of these species are diamagnetic and which are paramagnetic?

3.3 Write down the ground-state electronic configuration of O_2^{4-}, SN, F_2, CS, and PO^+ in terms of MOs. Give the bond order and the number of unpaired electrons for each species.

3.4 Describe the bonding in MO terms of LiH, HF, and CH.

3.5 Would you expect the ionic character of LiH to be more or less than that of HCl (17%)? Given that the separation between the atoms in LiH is 160 pm and the observed dipole moment is 1.96×10^{-29} C m, see if your prediction is correct.

3.6 Use the VSEPR theory to predict the structures of the following species. (The central atom in each case is the first one.)

    a. IF_7 **h.** XeF_2
    b. SiF_2 **i.** BrF_5
    c. SeF_6 **j.** XeF_4
    d. $TeCl_4$ **k.** XeF_6
    e. $[SiF_6]^{2-}$ **l.** $[SbBr_6]^{3-}$
    f. ClF_3 **m.** $POCl_3$
    g. $[SiF_5]^-$ **n.** $XeOF_4$

3.7 Sketch Lewis–Langmuir diagrams for the following species:
   a. CTe_2
   b. H_2NNH_2
   c. $PbCl_4$
   d. $[PCl_4]^+$
   e. $(CH_3)_2CO$

3.8 Can you draw Lewis–Langmuir diagrams for XeF_6, $[SnCl_5]^-$, SeO_2Cl_2?

3.9 Compare the bonding in benzene (C_6H_6) and borazine ($B_3N_3H_6$). Would you expect these molecules to have similar chemical properties?

3.10 Predict the order of *decreasing* bond length and *increasing* bond strength for the series S_2^+, S_2, S_2^-. How would you expect these properties to compare with those for the series O_2^+, O_2, O_2^-? Will any of these six species be paramagnetic?

3.11 How many Lewis–Langmuir structures can you write for OCN^-, CSe_2, NO_3^-, and SO_2?

3.12 Criticize the structural information given for the six molecules below:

Molecule	Structure
a. CSe_2	Bent molecule with one unpaired electron
b. NO	Diamagnetic molecule
c. HCN	Angular molecule with one triple bond
d. N_3^-	Linear molecule with two double bonds
e. $[PCl_4]^+$	Square planar
f. $XeOF_4$	Tetrahedral

3.13 Predict the bond angles for NO_2^+ and NO_2^-. How will they compare with that for NO_2?

3.14 Predict the structures using VSEPR theory of the following, as yet, undiscovered species: $[KrF_5]^+$, KrO_3, KrF_6.

3.15 List, with suitable examples, the main geometries found for three, four, five, and six coordination.

3.16 What hybridizations would you associate with the valence orbitals of the central atoms in the following species:
   a. SeO_3
   b. $[SnCl_5]^-$
   c. IF_7
   d. $[GaF_6]^{3-}$
   e. $TeCl_4$
   f. XeO_3

3.17 Describe the bonding in silane (SiH_4) in MO terms.

3.18 By what experimental technique could the H_2^+ ion be observed? Could this technique also provide information on the bond energy?

3.19 Would you expect He_2^{2+} to have a higher or a lower bond energy than H_2? (HINT: Consider the difference in the potential energy functions for the two molecules.)

3.20 Every transition metal ion forms cyanide (CN^-) compounds over a wide range of oxidation states. What factors do you think are important in the failure to observe a single BO^- compound?

3.21 The ions H_2F^+ and HF_2^- are both stable and have been identified in a number of compounds. Compare the MO diagrams for these two ions.

3.22 Can you draw some structures other than the Kekulé ones for the benzene molecule that may contribute significantly to the ground state?

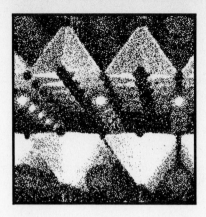

4

The Ionic Bond

4.1 Formation of Ions and Lattices

Recall from the discussion of atomic structure, that given the appropriate amount of energy atoms can either lose or gain electrons to form ions. Electron loss from a neutral atom to produce a positively charged ion always requires an input of energy (endoergic), but electron capture may be either endoergic or exoergic. Ions in the free gaseous state are commonly produced by physical processes such as irradiation with photons, very high electric field gradients, or bombardment with electron or ion beams. It is also possible to generate ions by chemical interaction of one species with a low first ionization energy with another species of high electron affinity. This may not be immediately obvious if we consider the first ionization energies of Group IA(1) (the most easily ionizable of all the elements), which span a range of about 400 to 500 kJ mol^{-1}, and the halogens (the elements with the greatest affinities for electrons), whose electron affinities range from -300 to -350 kJ mol^{-1}. Both ionization energy and electron affinity are cited here as ΔH values for the following processes (Equation 4.1):

$$M(g) \rightarrow M^+(g) + e^- \qquad \text{and} \qquad X(g) + e^- \rightarrow X^-(g) \qquad \textbf{(4.1)}$$

It seems therefore that there is an energy deficit of at least 50 kJ mol^{-1}, even for the most favorable case, for the process

$$M(g) + X(g) \rightarrow M^+(g) + X^-(g) \qquad \textbf{(4.2)}$$

However, this deficit is illusory because we have ignored an important source of energy income in our balance sheet, namely the coulomb potential energy available from the attraction between the two product ions. Our deficit of 50 kJ mol^{-1} would only hold if we insisted that the atoms and ions in Equation 4.2 remain at infinite separation. If we allow the ions to approach each other, they will yield up their potential energy in the amount $q^+q^-/4\pi\varepsilon_0 r$ (where q^+ and q^- = the charges on the cation and anion, respectively; ε_0 = the permittivity of a vacuum; and r = the interion distance). And we have only to ask, at what value of r would this energy equal 50 kJ mol^{-1}? The answer (which you should verify) is approximately 3000 pm, a distance well beyond that expected for a chemical bond—for example, the equilibrium Cs$^+$F$^-$ separation in the solid salt is about 300 pm. In other words, as long as our Cs and F atoms were less than 3000 pm apart, the electron transfer depicted in Equation 4.2 would be spontaneous. Even in the case of LiF, where the energy deficit is roughly four times as large as for CsF (because of the higher ionization energy of Li), the break-even distance is still well-beyond the solid-state value of interionic separation.

Of course, because the ion pairs described can yield up further energy by getting as close together as possible, they would in the gas phase move spontaneously to form tightly bound ion pairs. The limit of approach is determined by the balance point at which the repulsion between the electron clouds (or, in the case of the proton, repulsion between nuclei) of the two ions just equals the attraction between the charges of the ions. Because this balance point corresponds to approximately 300 pm for Cs$^+$F$^-$, the enthalpy excess for formation of this ion pair—the difference between the energy deficit for ion formation and the potential energy liberated on bringing the ions from a separation of infinity to 300 pm—is readily calculated to be about 450 kJ mol^{-1}.

Although the computational demonstration is more difficult (we will not describe intermediate cases here), a gas-phase assembly of ion pairs is unstable with respect to clustering, and the energy of the systems falls progressively with aggregation. An asymptotic limit is reached when ion pairs form an infinite assembly; the energy of the infinite assembly is minimized by the adoption of a highly ordered structure, or *lattice*, in

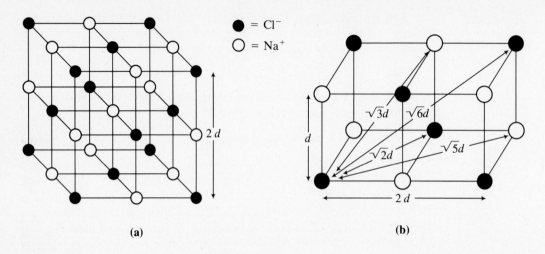

= Cl⁻

= Na⁺

(a)   (b)

Figure 4.1
The sodium chloride lattice.
(a) A general view of the lattice.
(b) Distances between a cation and some of its nearer neighbors.

which any given ion is in contact with the maximum possible number of ions of opposite charge. CsF forms such a lattice in which each ion is surrounded by six nearest neighbors of opposite charge, disposed at the corners of a regular octahedron. The excess enthalpy for the formation of such a CsF lattice is about 700 kJ mol^{-1}, compared to the 450 kJ mol^{-1} for gas-phase ion pairs.

The lattice described for CsF is, in fact, the one adopted by most Group IA(1) halides and is usually referred to as the sodium chloride lattice or structure. In such a structure, each ion forms a sublattice, which is face-centered cubic (or cubic close-packed). We will develop these concepts in more detail in Section 4.2. Figure 4.1 shows a diagram of the sodium chloride structure. In the actual crystal, this structure extends—to all intents and purposes—to infinity in all directions, and the constituent ions would be more or less touching all of their neighbors of opposite charge.

From the lattice shown in Figure 4.1 it is relatively easy to compute the coulomb interaction for any given ion in the crystal. If we call the length of the cube edge $2d$, it is clear that there are six anions at a distance d from the central cation, twelve cations (at the edge centers) at a distance $\sqrt{2}d$, eight anions (at the corners) at a distance $\sqrt{3}d$, six cations (at the centers of adjacent cubes) at a distance $2d$, and so on. The total electrostatic potential can be written as a sum:

$$V_{\text{lattice}} = 6\frac{q^+ \cdot q^-}{4\pi\varepsilon_0 d} + 12\frac{(q^+)^2}{4\pi\varepsilon_0 \cdot \sqrt{2}d} + 8\frac{q^+ \cdot q^-}{4\pi\varepsilon_0 \cdot \sqrt{3}d}$$

$$+ 6\frac{(q^+)^2}{4\pi\varepsilon_0 \cdot 2d} + \text{etc.} \tag{4.3}$$

and Equation 4.3 can be rewritten as,

$$V_{\text{lattice}} = \frac{q^+ \cdot q^-}{4\pi\varepsilon_0 d}\left[6 + \frac{12}{\sqrt{2}} \cdot \frac{q^+}{q^-} + \frac{8}{\sqrt{3}} + 3\frac{q^+}{q^-} \text{ etc.}\right] \qquad (4.4)$$

Since q^+/q^- is a constant for any particular structure, the bracketed series in Equation 3.14 is dependent only on geometry. For other lattices of different geometry, it is only this part of the potential equation that changes and the sum of this series is called the "Madelung constant" of the lattice, M.

In our computation, we only considered the potential energy due to the interaction of a single ion with its neighbors. This energy is converted to a molar quantity in Equation 4.5:

$$V_{\text{lattice}} = \frac{q^+ \cdot q^-}{d}M \cdot N \qquad (4.5)$$

where N = Avogadro's number.

If Equation 4.5 were a complete description of the lattice energetics, the crystal would collapse to a point because V_{lattice} is always negative and its magnitude increases with decreasing d. The factor that prevents this collapse is the large repulsion that develops as electron clouds encroach on each other. Unfortunately, this repulsion is much more difficult to deal with than the attractive potential, and we will not go into detail of how the problem has been treated. A common and very simple modification of Equation 4.5, which includes a repulsive term, is shown in Equation 4.6, the Born–Mayer equation:

$$U = \frac{q^+ \cdot q^-}{4\pi\varepsilon_0 d_e}\left(1 - \frac{\rho}{d_e}\right)M \cdot N \qquad (4.6)$$

where U = the lattice energy, d_e = equilibrium interionic separation, and ρ = a parameter that does not vary greatly for different lattices and is usually about 30–35 pm. Table 4.1 lists some values of M for some of the more common lattices.

The direct experimental determination of U is extremely difficult and has only been attempted in a few cases. More commonly, crystal-lattice energies are determined by using a thermochemical cycle, known as the Born–Haber cycle. The cycle for NaCl is the following:

$$Na(s) \xrightarrow{\Delta H_{\text{sub}}} Na(g) \xrightarrow{I} Na^+(g) + e^-$$

$$U_{\text{NaCl}} \longrightarrow NaCl(s)$$

$$\tfrac{1}{2}Cl_2(g) \xrightarrow{1/2 D_{Cl-Cl}} Cl(g) \xrightarrow[+e]{A} Cl^-(g)$$

Table 4.1	*Some Values for the Madelung Constant (M)*
Structure	M
NaCl	1.748
CsCl	1.763
CaF$_2$	2.519
ZnS (blende)	1.638
ZnS (wurtzite)	1.641

where ΔH_{sub} = heat of sublimation of Na metal, I = ionization energy of Na, D_{Cl-Cl} = bond-dissociation energy of Cl$_2$, A = electron affinity of Cl, and U_{NaCl} = lattice energy of NaCl. The arithmetic sum of the energies of the individual steps in the cycle is equal to the energy (ΔH_R) for the overall reaction

$$Na(s) + \tfrac{1}{2}Cl_2(g) \rightarrow NaCl(s) + \Delta H_R$$

Because all quantities except U_{NaCl} are experimentally available, U_{NaCl} can be derived. The relevant numbers are (kJ mol^{-1}): $\Delta H_R = -411$; $D_{Cl-Cl} = 242.6$; $A = -349$; $I = 495.4$; $\Delta H_{sub} = 107.8$. The heat of reaction is then calculated to be

$$\Delta H_R = -411 = U + 242.6/2 + 495.4 + 107.8 - 349$$

which gives a value for U of 786.5 kJ mol^{-1}.

4.2 Relationship Between Ionic and Covalent Bonding

The homopolar covalent bond, such as that of the hydrogen molecule, and the ionic bond, such as described for the alkali metal halides, represent two extreme perspectives of what is basically the same phenomenon. We could try to treat the interaction of two hydrogen atoms in the same way that we treated the Cs and F atoms, and we would conclude that the energy deficit due to electron transfer could be balanced by placing the resulting H$^+$ and H$^-$ at a separation of about 100 pm. Because the experimentally observed internuclear separation in H$_2$ is about 75 pm, this may not seem too absurd. However, when we consider that the 1s orbital of a hydrogen atom has appreciable density out to 200 pm (the hydride ion will be even more expanded), the whole notion of considering H$^-$ and H$^+$, at a separation of 100 pm, as separate charged particles falls apart. In the case of F$_2$ or the other halogen molecules a similar interpenetration of the two ions would arise, and again the perceptual basis of the ionic model is lost.

At the other extreme of the almost perfectly ionic species such as the alkali halides, one could in fact treat the bonding using a full-blown quantum mechanical treatment. Such an approach is not generally worth the effort because the small effects due to covalency are negligible and the results of a quantum mechanical calculation involving hours of computer time would be no more informative than a back-of-an-envelope calculation using the electrostatic–ionic model. In its simplest expression, the ionic model assumes that the formation of ions does not seriously perturb the AOs beyond adjustment of the electron cloud to a change in the effective centrosymmetrical nuclear charge. It assumes that the closed-shell ions are spherical and remain so even when in contact with other ions in the crystal. Figure 4.2 shows an energy-level diagram illustrating the energy changes of the AOs of Li and F upon formation of an LiF–ion pair.

These readjustments of electron-energy levels are the source of the net energy–change attributable to the electron-transfer process, but this diagram tells us nothing of the coulombic interaction potential. The assumption that the electrons on Li^+ are uninfluenced by the electrons and nucleus of F^- and vice versa is clearly one of limited validity. It happens to be a very good one for most of the alkali metal and alkaline earth metal halides.

As we will see next, many qualitative aspects of the ionic model, particularly descriptions of structure, can be carried over to systems where there is considerable covalency, but the discrepancy between computed and observed physical properties gets larger as covalency increases. Like all theories, the ionic theory can be extended computationally to fit more cases by the introduction of more parameters. Although it was fashionable in the early days of ionic-bonding theory to try to extend the theory as far

Figure 4.2
Energy of orbitals in Li^+F^-. In the absence of covalency, the changes in orbital energy are entirely due to changes in effective nuclear charge.

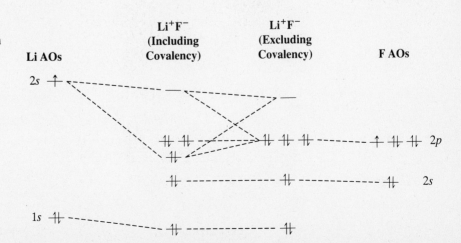

as possible by further parameterization and many rather complicated modifications to the basic model were attempted, the growth in the understanding of covalency largely superceded such efforts. However, one of the extensions of the simple ionic model, the introduction of the concept of *polarization*, has persisted, largely because of its conceptual simplicity and usefulness.

Polarization recognizes that ions may not be perfect undeformable spheres but, on the contrary, may have their shapes distorted by the fields of neighboring ions. Figure 4.3 shows an example in two dimensions of how a set of positive charges might distort the electron cloud of an anion by attracting the electrons. Such polarization introduces much flexibility because it can alter both effective charge and effective internuclear distance. The problem is that *ab initio* predictions of polarization effects are no easier to pin down than covalency, a fact that is not at all surprising when it is realized that polarization is an explicit recognition of covalency—that is, that electrons in the system are under the influence of more than one nucleus. Thus, the polarization concept only preserves the computational simplicity of the ionic model if polarization is incorporated as an adjustable parameter to help make the calculation fit the data. Despite its inadequacy as a fundamental concept, the idea of polarization is useful because it parallels the phenomenon of covalency and many of the guidelines developed in relation to polarization can be used to estimate the

Figure 4.3
Schematic representation of the influence of neighboring ions on ion shape.

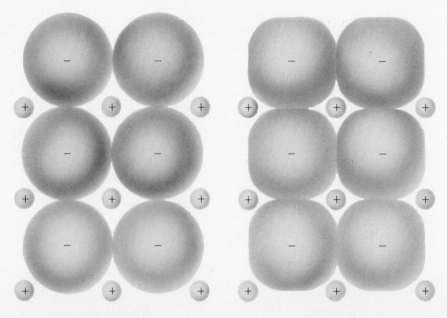

No Polarization   **Polarization**

relative importance of covalency in highly polar systems. These guidelines are summarized in a set of rules, first enunciated by Fajans. These rules are as follows:

1. Polarization is favored for small, highly charged cations (small, highly charged cations are *polarizing*).

2. Polarization is favored for large, highly charged anions (large, highly charged anions are *polarizable*).

3. Cations with filled outer *d* shells are more polarizing than are similar ions with noble gas configurations.

Insofar as polarization is equivalent to covalency, these rules also provide guidelines as to when the latter is likely to be important in highly polar compounds. In terms of periodic properties, the rules may be reexpressed in the form that covalency tends to increase

1. On ascending a group for cations or descending a group for anions (decreasing size for cations, increasing size for anions).

2. On going from left to right across a period for cations, and right-to-left for anions (for example, Na^+, Mg^{2+}, and Al^{3+}; F^-, O^{2-}, and N^{3-}).

3. In the posttransition triads relative to pretransition triads of the same group number (for example, Cu^+, Ag^+, and Au^+, relative to K^+, Rb^+, and Cs^+).

4. With increasing oxidation state of cations and decreasing oxidation state of anions.

4.3 Close Packing of Spheres

For purely ionic solids, it is self-evident that the structure of lowest energy will be that which allows ions of opposite charge to be as close to each other as possible, while keeping ions of like charge as far apart as possible. This optimum state is usually achieved in highly symmetrical arrays based on the closest packing of spheres.

To begin, consider what happens when we shake an assembly of spheres until they assume their most densely packed array in a two-dimensional plane—for example, as we would do in setting up balls for a game of pool (Figure 4.4). The shaded ball in this figure is representative of the situation of every ball if the array were continued indefinitely in an infinite plane. In such a plane, each ball has six nearest neighbors (coordination number six) arranged in a regular hexagon.

Figure 4.4
Close packing of
spheres of equal
size. *A* indicates a
tetrahedral hole, *B*
an octahedral hole.

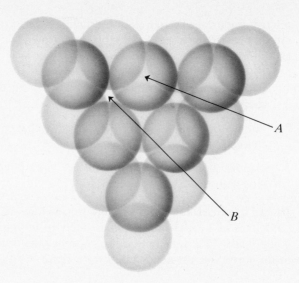

If we now place a second close-packed layer on top of the first so that each sphere of the second layer sits comfortably over a hole in the first (see dark circles in Figure 4.4), a number of features are immediately evident:

1. We can only cover half the holes in the first layer (hole A is covered; hole B is not).

2. The uncovered holes of the first layer coincide with half of the holes in the second layer.

3. When we place a sphere over a hole in the first layer, we create a cavity surrounded by a tetrahedron of spheres (Figure 4.4, hole A). We refer to the cavity within this tetrahedron of spheres as a *tetrahedral hole*.

4. When holes in the two layers coincide, they form a cluster of six spheres, in the form of a regular octahedron. The cavity within this octahedron is called an *octahedral hole* (Figure 4.4, hole B). There are in fact two tetrahedral holes for each octahedral hole.

When a third layer is placed over the first two, there are two nonequivalent choices for the location of the third layer. The first, over the *holes* of the second layer that are coincident with the *spheres* of the first layer, gives the structure known as the hexagonal close-packed (HCP) arrangement (Figure 4.5a). The second possibility is to place the third layer over the *holes* of the second layer that are coincident with the *holes* of the first layer to

Figure 4.5
(**a**) Hexagonal and
(**b**) cubic close
packing of spheres.
(Adapted with
permission from L.
Pauling. *The Nature
of the Chemical
Bond*, 3d ed. Ithica,
N.Y.: Cornell
University Press,
1960.)

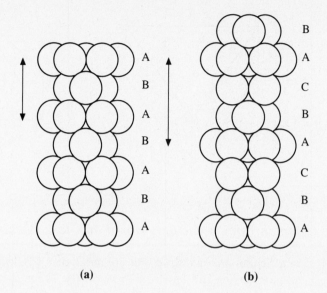

(**a**)   (**b**)

give the cubic close-packed (CCP) structure (Figure 4.5b). Thus, in the
HCP structure, the position of the first layer is repeated in the third layer,
whereas in the CCP structure repetition does not occur until the fourth
layer. In both structures, for a large three-dimensional array, the number
of octahedral holes is equal to the number of spheres, and the number of
tetrahedral holes is twice that number. The coordination number in both
cases is twelve, and the filling of space by spheres is about 52% (you should
verify this).

The CCP structure has the special property that the lattice points (the
centers of the spheres) have, precisely the same special distribution as the
octahedral interstitial points (the centers of the octahedral holes). Thus, a
CCP lattice consists of a CCP array of spheres *and* an interpenetrating CCP
array of octahedral holes. This does not hold true for the HCP lattice. In
a CCP lattice, each sphere is surrounded by six octahedral holes, the
centers of which also form a regular octahedron, whereas the sphere of a
HCP lattice is surrounded by six octahedral holes that form a regular
trigonal prism. This fact is quite evident from Figure 4.5, especially when it
is realized that a trigonal prism is converted into an octahedron by a
rotation of two opposed triangular faces through 60° relative to each other,
as illustrated in Figure 4.6. This difference can have important consequences if there is a force acting between lattice components because the
distances separating the corners of a regular octahedron or a regular prism
(for a fixed bond length) are different.

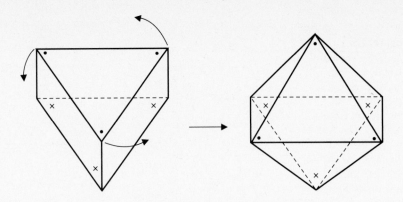

Figure 4.6
Conversion of a
trigonal prism to an
octahedron by
rotation of opposed
trigonal faces
through 60°.

4.4   *Some Common Ionic Structures Based on Close Packing*

Many common ionic structures are based on CCP or HCP, with one of the ions occupying the lattice points and the other the holes or interstitial points. The NaCl lattice is such a case, where the ions of one kind occupy a CCP array, while those of the other occupy all octahedral holes. It is not necessary to specify which ion occupies which site because the sites are equivalent except for a translation in space. It is reasonable to assign the anions to the sphere positions and the cations to the interstitial positions because the former are usually the larger and assumed to be in contact. Besides the alkali halides, the NaCl structure is also commonly adopted by the monoxides of Group IIA(2) and the first-row transition elements. It is common to generalize by referring to the cation as A and the anion as B. The NaCl structure is therefore one of the common structures adopted by AB compounds.

The AB case of a HCP lattice with the octahedral holes occupied is referred to as the nickel arsenide (NiAs) structure. In this situation, the Ni atoms occupy all octahedral holes in the HCP lattice of As atoms. Thus, the Ni atoms are six-coordinated octahedral, but the As atoms are six-coordinate to a trigonal prism of nickel atoms. We refer here to Ni and As atoms, rather than ions, because the bonding is far from purely ionic. The adoption of the NiAs structure may in fact be taken as symptomatic of bonding forces other than the simple attraction and repulsion of charged spheres, which would lead to the NaCl structure. The trigonal prism is only more stable than is the octahedron if there is an attractive force between the like species at the corners as well as for the unlike species at the center.

Two types of close-packed AB structure, in which one-half of the tetrahedral holes are occupied, are exemplified by two mineral forms of zinc sulfide: zinc blende (CCP) and wurzite (HCP) (Figure 4.7). Unlike the NiAs structure, the anion–cation sites are equivalent in the wurzite struc-

Figure 4.7
(**a**) Zinc blende
(CCP) and (**b**)
wurtzite structures
(HCP).

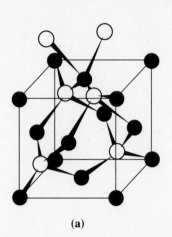

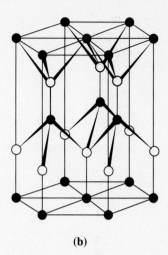

(a)   (b)

ture as they are in the zinc blende structure. The energy difference between these structures is very small, and this explains why both structures are readily adopted by the same compound. The zinc blende and wurtzite structures are common to the chalcogenides of the zinc group (2/6 compounds) and to the pnictides of the boron group (3/5 compounds). The electronic structure of these important classes of compounds are discussed further in Chapter 19.

The simple close-packing-of-spheres description accounts for the overwhelming majority of compounds of 1 : 1 stoichiometry composed of spherical (or effectively spherical) ions. Such close-packing descriptions also apply to very large numbers of compounds of other stoichiometries.

Most dihalides and dioxides belong to one of four structural types (either ideal or distorted), three of them simple close-packed. Nearly all dichlorides, dibromides, and di-iodides of the lighter transition elements and of the p-group elements adopt structures with half the octahedral holes of a close-packed anion lattice occupied by the metal cation. The CCP type is exemplified by the $CdCl_2$ structure and the HCP type by CdI_2. The $CdCl_2$ structure resembles the NaCl structure but with half the cations removed in the form of complete layers. In Figure 4.8a, the NaCl structure is redrawn in a schematic way to emphasize that it is made up of alternating sheets of cations and anions. The $CdCl_2$ structure is achieved by simply removing every other sheet of cations (Figure 4.8b). The crystal cohesion along the cation vacant planes (see arrow in Figure 4.8b) is very low, and crystals of this type exhibit extremely low sheer strength. They also tend to grow as thin wafers. Because the cation vacancies in this structural type emphasize that ions are present in layers, they are often referred to as

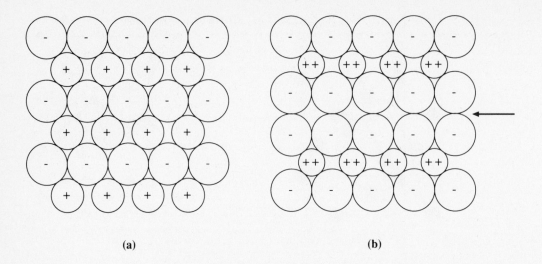

(a)   (b)

Figure 4.8
A comparison of the
(a) NaCl and (b)
CdCl$_2$ structures.

layer-lattice compounds. Another way of viewing the CdCl$_2$ structure is as a stack of CdCl$_2$ sandwiches, with the anions serving as the bread and the cations as the filling. The only way in which the CdCl$_2$ and the CdI$_2$ structure differ is in the superposition of the anion layers. In the CdCl$_2$ structure (CCP), the anions of the fourth layer superpose on those of the first layer; whereas in the CdI$_2$ (HCP) structure, superposition is between the third and first layers of anions. Thus, the only difference is a small translation of one sandwich relative to the other.

The third AB$_2$ structure based on close packing is exemplified by fluorite (CaF$_2$) and is less common than the CdX$_2$ types. The CaF$_2$ structure may be viewed as a CCP array of cations, with tetrahedral holes occupied by anions. In such a structure, the coordination number of the cation is eight and that of the anion is four. Charge balance always requires the coordination numbers to be in proportion to the ion charges (in the CdX$_2$ structures, because the cations are six-coordinate, the anions are necessarily three-coordinate). The CaF$_2$ structure is to be expected for AB$_2$ compounds where the cation is exceptionally large and/or the anion is extremely small. Some compounds of the type A$_2$B, with a very small cation-to-anion size ratio, adopt the *antifluorite* structure (Li$_2$O, Cu$_2$O) where all tetrahedral holes of a close-packed anion lattice are occupied by cations. In this case, the cation is four-coordinate, and the anion is eight-coordinate.

A number of trihalides of the lighter transition elements (for example, TiCl$_3$ and VCl$_3$) adopt a layer-lattice similar to the CdX$_2$ structures, but those cation layers that are completed in the CdX$_2$ structures are further depleted by one-third in the MX$_3$ layer lattices.

4.5 Some Simple Structures Not Based on Close Packing

Close-packed structures are very common and because of their high symmetry relatively easy to visualize. There are, however, many other structures adopted, even by relatively simple, spherical ionic systems. The close-packed arrangement is limited to some degree by the considerable difference in size between the spheres and the interstitial holes of a close-packed structure. Because cations are normally smaller than anions, the optimum lattice energy will be realized with the anions forming the close-packed lattice and the cations occupying the interstitial holes. Furthermore, the ideal structure is that in which the cations just touch the anions surrounding them and the anions are also mutually in contact. This ideal state is, of course, never really achieved, particularly because real ions are not hard spheres with well-defined boundaries and the possibility exists for significant deviations in either direction from the ideal state.

Deviations in the direction of the cation becoming much too small to fill the interstitial hole are complicated by the intrusion of covalency. For instance, the halides of Li and Be have anomalously low melting points (Table 4.2). Part of this anomaly may be because small cations do not fill the holes and on passing from the first to the second member of the group there is no expansion of the close-packed anion lattice. This being the case, the electrostatic energy does not change because d in Equation 4.6 does not change, but the reduced first ionization energy on descending the group increases the overall stability of the lattice for the second members.

Table 4.2 *Melting Points of Some Group IA (1) and IIA (2) Halides (°C)*

	F	Cl	Br	I
Li	870	613	547	446
Na	997	801	755	651
K	880	776	730	723
Rb	760	715	682	642
Cs	684	646	636	621
Be	800	450	490	510
Mg	1396	708	700	>700
Ca	1360	772	765	575
Sr	1190	873	643	402
Ba	1280	>925	847	740

On further descending the group, cations become larger than the close-packed interstitial sites and force the anion lattice to expand. The resulting increase in d overrides the slight decrease in ionization energy, and there is a net drop in lattice energy, contributing to the decline in melting point. An improvement in this grossly oversimplified view would be the inclusion of the properties of the melt because a phase transition such as melting is dependent on the energetics of both the initial and final state. Because melting removes the rigid constraints of the solid state, it is to be expected that the cations that are too small to fill close-packed cavities would succeed in reducing the interionic separation in the simpler species of the melt (ion pairs and clusters) and thereby favor the melting process. Such favoring of the molten state is absent when anion–cation contact is already present in the solid.

Although the close-packed structure will tolerate a certain amount of expansion, beyond a certain point other structures become energetically preferred. In the cases of most of the alkali halides, the NaCl structure is the most stable, but cesium chloride (CsCl) adopts a structure based on the simple cubic lattice. The simple cubic lattice is an infinite array of cubes sharing common faces (Figure 4.9). In the CsCl structure, one ion occupies the corners of the cubes, and the counter ion occupies the center of the cube. Because eight cubes share each corner, the ions at the corners are also eight-coordinate, and the two lattices are transformed into each other by a simple translation, as in the case of NaCl. The simple cubic CsCl lattice is transformed into the close-packed CaF_2 lattice by removal of every other cation. It thus becomes evident that the tetrahedral holes in a CCP lattice form a simple cubic sublattice.

A common structure for oxides and fluorides of empirical formulas MO_2 and MF_2 is the *rutile structure*, named after the exemplary TiO_2 in one of its mineral forms. The rutile structure, which allows six coordination of the cation and three coordination of the anion, cannot, except by assuming unrealistic distortions, be considered in terms of close packing of spheres. Figure 4.10 shows the rutile structure.

Figure 4.9
The CsCl structure.

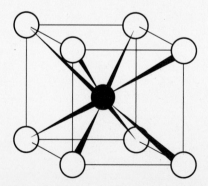

Figure 4.10
The rutile structure.

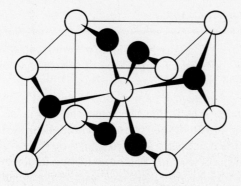

The transition metals and the Group IIIA (13) elements commonly give oxides, M_2O_3, which have the *corundum structure*, named after a form of Al_2O_3. In this structure, oxide ions form a highly distorted HCP lattice, and two-thirds of the octahedral holes are symmetrically occupied by the cations. Although this structure is difficult to visualize, it is worth pointing out that the vacancy pattern in the corundum structure does not lead to a layer lattice and that corundum is an extremely hard substance, which is widely used as an abrasive.

A final binary structure worthy of note is the rhenium trioxide (ReO_3) structure. This structure is common for the heavy transition metal trioxides and is closely related to the important ternary oxide structure known as the perovskite structure, discussed in Section 4.6. In the ReO_3 structure, the cations occupy a simple cubic lattice, and the anions are at the centers of the edges of the cubes. Thus, the cations are six-coordinate and the anions two-coordinate (Figure 4.11). The ReO_3 structure is a particularly open one because of the relatively large size of the unoccupied cubic cavity.

Figure 4.11
The ReO_3 structure.

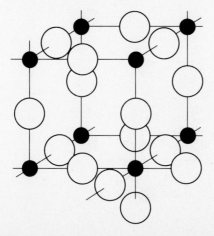

4.6   Some Ternary Oxide Structures

The structures of a number of very important ternary oxides can be described in terms of close packing of spheres and are closely related to some of the binary structures described.

The *spinel structure*, named after a mineral form of $MgAl_2O_4$ and of generic formula AB_2O_4 may be approximated as a CCP lattice of oxide ions with one-eighth of the tetrahedral holes occupied by the A^{2+} ions and one-half of the octahedral holes occupied by the B^{3+} ions. Closely related and of great interest as far as a fundamental understanding of structure is concerned is the *inverse spinel structure*. In the idealized form, this structure involves a site exchange between the A^{2+} ions and half of the B^{3+} ions of the spinel structure. Given the fact that this occurs, it is evident that the energy factors directing the two different ions to the two different sites are not overwhelmingly large, and it is not surprising that such structures are highly susceptible to defects in actual crystals. One important factor that can influence this site selectivity is the crystal field stabilization energy of transition metal ions. This phenomenon is discussed in greater detail in Chapter 13.

The *perovskite structure*, named after $CaTiO_3$, is very common for mixed oxides of general formula $A^{n+}B^{m+}O_3$ ($n = 1$, $m = 5$; $n = 2$, $m = 4$; $n = 3$, $m = 3$). This structure can be derived from the ReO_3 structure, previously described, by placing the larger cation (for example, Ca^{2+}) in the cubic cavity and the smaller cation at the cube corners (Figure 4.12). In this arrangement, the oxide ions and the large cations together form a CCP array, and the small cations occupy octahedral holes bounded exclusively by oxide ions. The large cation achieves the very high anion-coordination number of twelve in this structure. The recent discov-

Figure 4.12
The perovskite structure.

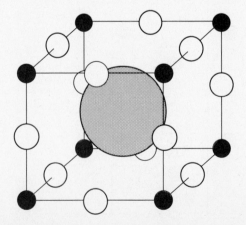

ery of perovskite-type mixed oxides with extraordinary superconducting properties is discussed further in Chapter 19.

The *ilmenites* (named after $FeTiO_3$) also have the general formula $A^{n+}B^{m+}O_3$ [$n = 2, m = 4$ (most common) and $n = 1, n = 5$]. This type of structure is preferred when the two cations are more similar in size and may be approximated to a HCP lattice of oxide ions with the cations distributed symmetrically among two-thirds of the octahedral holes. You should recognize the relationship that exists between the ilmenite structure and the corundum structure. Like other mixed oxides, ilmenites are highly susceptible to disordering by interchange of the two cations or by occupation of holes that would be vacant in the ideal structure.

4.7 Factors Influencing Structure

For purely ionic compounds, a structure is adopted that minimizes the distances between unlike ions and maximizes the distances between like ions. At the same time, the ions achieve a maximum coordination between unlike ions, within the constraints of the distance requirement. For spherical ions, these effects are manifest in a number of simple relationships between the radius ratio r^+/r^- and the coordination numbers of the ions.

In general, once the radius ratio falls below the value at which all ions are in contact and the cation no longer fills the cavity available for a particular structure, the structure changes to one that affords a smaller cavity.

The most interesting structures are those derived from the simple cubic and close-packed arrangements of spheres. In these structures, we are limited to a consideration of coordination numbers four, six, and eight and tetrahedral, octahedral, and cubic geometries. It is easy using elementary geometry to calculate r^+/r^- for all these geometries. Table 4.3 lists the values.

Table 4.3 *Radius Ratios for Some Common Structures*

Coordination Number	Geometry	r^+/r^-
4	Tetrahedral	0.23
6	Octahedral	0.41
8	Cubic	0.73

To calculate the radius ratio for tetrahedral coordination, remember that the vertices of a tetrahedron occupy the alternate corners of a cube. If a tetrahedron of spheres of equal size are placed at the alternate corners of a cube so that they are in contact across the cube-face diagonal, the cube dimensions are face diagonal = $2r$; cube edge = $\sqrt{2}r$; cube-body diagonal = $2r + 2r'$, where r' is the radius of the cavity at the center. Figure 4.13 shows these relationships. From Pythagoras' theorem,

$$(\text{Body Diagonal})^2 = 4r^2 + 2r^2 = 6r^2$$

therefore,

$$2r + 2r' = \sqrt{6}r$$

and

$$r'/r = (\sqrt{6} - 2)/2 = 0.23$$

The critical radius ratios shown in Table 4.3 represent the lower limit of stability for a particular structure. Coordination numbers higher than eight are geometrically impossible for simple binary compounds.

The best estimate radius ratio for LiI (about the smallest practically accessible value for a unipositive ion) is 0.436. Thus, the six-coordinate octahedral case represents a practical lower limit of coordination for alkali halides. For more highly charged ions, values below 0.436 are possible, but as the values of radius ratios enter the domain of favoring tetrahedral coordination, there is an inevitable increase in polarization and covalency. By the time the ratio reaches the lower limit of tetrahedral stability, application of the ionic model is no longer valid.

Figure 4.13
Cube dimensions in relation to the radii of a tetrahedron of spheres and of the cavity in the center.

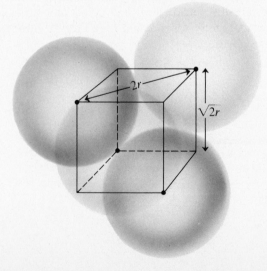

For higher oxidation states, it is common for the coordination number dictated by the radius ratio to be equal to or less than that dictated by the stoichiometry. In such situations, compounds tend to form simple molecules rather than extended two- or three-dimensional structures. Table 4.4 shows some examples of this effect. Although relative sizes of atoms and ions are clearly important in determining the tendency of molecules to expand their coordination spheres and develop extended structures, it must be stressed that the radius ratio is only one influence among many.

It is also important to remember that ions and atoms are not rigid bodies of clearly defined dimensions. Because electron clouds are diffuse, the dimensions of atoms and ions are not unequivocally definable or measurable (this is perhaps more important scientifically). The apparent dimensions may depend on the particular local environment, the physical conditions, or the method of measurement. All radius ratios cited in this section are derived from what are currently taken to be the best values available. Because the boundaries of ions cannot be seen even by the most sophisticated methods of observation, the partitioning of the internuclear separation (which can be precisely measured) between the two ions involves a number of hypotheses and assumptions. Once certain basic radii have been assigned (for example, O^{2-} and F^-), it is then a relatively easy task to determine the radii of all those cations that participate in binary compound formation with O^{2-} and F^- and then to use the cation values to determine other anion values. Finally, an empirical analysis of a large number of determinations of the same ion in different compounds can yield a "best value" for that particular ion.

Table 4.4 *Relationship Between Radius Ratio and Intermolecular Bonding*

Compound	r^+/r^-	Expected Coordination Number	Molecular Coordination Number	State Under Ambient Conditions
BF_3	0.134	2	3	Gas
AlF_3	0.600	6	3	Solid
SiF_4	0.36	4	4	Gas
GeF_4	0.47	6	4	Liquid
SnF_4	0.616	6	4	Solid
TiF_4	0.50	6	4	Solid
$TiCl_4$	0.35	4	4	Liquid
$ZrCl_4$	0.46	6	4	Solid
VF_5	0.54	6	5	Solid dimer
CrF_5	0.49	6	5	mp 30° C
WF_6	0.67	6	6	Volatile liquid

Earlier estimates of radii tended to underestimate cation and over-estimate anion radii. One result of this was the apparent occurrence of many exceptions to radius ratio rules (for example, all lithium halides were predicted to be tetrahedrally coordinated). With improved estimates of ionic radii, the number of apparent exceptions in simple, highly ionic cases has greatly declined. However, given the approximate nature of the hard-sphere model, do not expect that radius ratio rules will be an infallible guide to structure, especially for radius ratios that are close to the boundary values for transition from one structure to another.

4.8   Friction, Lubrication, and Wear

The microscopic structure of crystals has attracted human attention for millenia because of the intrinsic esthetic appeal of all the possible shapes and forms of crystals. In the nineteenth century, mathematicians used the external forms of crystals to deduce most of the mathematics necessary to deal with understanding the internal structure, long before experimental methods for probing the internal structure became available. Beyond this enormous appeal that shape and form exert on our esthetic sense and our scientific curiosity, there are also some very important practical considerations. The physical properties of macroscopic solids are intimately related to their microscopic structure. As described later in Chapter 18, much of the modern technological revolution has been brought about through a detailed understanding of the microscopic structure of solids, some of them very simple and some very complicated. Here, we draw attention to a few ways in which the structures of some ionic or virtually ionic solids manifest themselves in everyday life.

The pencil that is used to write the manuscript for this book glides smoothly across the page, leaving its graphite imprint on the paper wherever it touches. Both the ease of sliding and the transfer of material from the pencil point to the paper are due to the layer-lattice structure of graphite. In this case, we are dealing with a covalent rather than ionic crystal, but the layer-lattice structures of more ionic materials such as CdI_2 confer more or less the same properties. It is the easy mechanical disruption of the weak bonding between layers that leads to both the lubricity of layer-lattice compounds and to the ease with which they can be worn down (excuse me, while I stop to sharpen my pencil).

All but a few metals are thermodynamically unstable in oxygen. Fortunately, there are many kinetic barriers that greatly slow the reactions of metals with oxygen when they are exposed to air. Nevertheless, even though we get useful lifetimes out of many reactive metals, they will all, in the end, be converted to the oxide. Unfortunately, because of their ionic character and structure, oxides possess none of the mechanical properties (malleability, ductility, impact resistance, plastic flow) that recommend metals for structural applications, and the conversion to the oxide represents a deterioration of the metal as a structural entity.

In daily applications, there is no such thing as a clean metal surface. Most commonly, metals will be covered with a thin layer of oxide, and it is this layer of oxide that protects the metal from further rapid attack by oxygen. The character of these oxide films varies enormously from metal to metal, and the degree to which they protect the metal depends on their mechanical and chemical stability and on their adhesion to the metal. They all share the common property of having nonlayer lattices, which may be advantageous because this makes them difficult to rub off, but it can be a cause of great problems when metals are in sliding contact.

When two macroscopic surfaces are brought into contact, the roughness of the surface at the atomic level can produce enormous mechanical stresses at the points of contact. These stresses are sufficient to cause disruption of protective oxide layers and welding of exposed metal surfaces. When the two surfaces are in relative motion, these effects are greatly amplified. The dislodged oxide particles with their unyielding crystal edges exert an abrasive action that gouges out more oxide layer and pieces of finely divided metal that because of their small size oxidize much more rapidly than the bulk metal. These processes, which appear as frictional resistance to sliding, may quickly deteriorate into a complete welding of the two surfaces and lead to seizure, or galling. The seriousness of this problem depends on many factors including the mechanical strengths of both the oxide and the metal.

Fortunately, the most widely used metal, iron, has a combination of properties that allows the problem to be solved relatively easily by interposing a thin layer of a liquid to keep the two surfaces slightly separated (boundary lubrication). The internal cohesion of many softer steels is low enough that under conditions of boundary lubrication with a suitable fluid the welds between the two surfaces are small enough in number and strength that galling, or seizure, do not occur. With high-performance stainless steels, pure nickel, chromium, vanadium, and titanium, and most notably with aluminum, this is not the case. All these metals are notoriously difficult to lubricate, and the patent literature abounds in oil additives to help facilitate their machining (many of the suggested remedies have little effect).

An interesting and relatively general solution to the problem with transition metals is the inclusion of molecular iodine in the lubricant formulation. In the presence of such lubricants, the freshly exposed metal surfaces produced by sliding friction react to form the metal iodides, all of which have either the CdI_2 or some closely related structure. The iodides then act as graphitelike lubricants, absorbing mechanical energy, which would otherwise be devoted to gouging metal through shearing of the layer lattice. Even if the iodides did not have a layer lattice, they would still be less abrasive than would the oxides because of their generally lower lattice energies (greater interionic separation). A dramatic but not very practical example of the effectiveness of such iodine-modified lubricants was demonstrated by the inventors. They showed that a small internal-combustion engine made entirely of nickel would not run for more than a few minutes on conventional motor oil; however, the engine ran for 1000 hours after the addition of a few percent of iodine to the lubricant. This demonstration was not of practical value because the hydriodic acid produced in the combustion of the oil would not be acceptable as an automobile effluent. Even worse, the very process that leads to lubrication also leads to serious corrosion in the long term. The real practical advantages of such lubricants are realized in machining and cutting operations where contact between the metal and the corrosive lubricant is relatively short-lived.

Unlike oxides, many sulfides, particularly the disulfides, have layer lattices. Although some disulfides consisting of M^{4+} and S^{2-} ions have the $CdCl_2$ or CdI_2 structure, the ones that have found application as high-performance lubricants in combustion-engine application, MoS_2 and WS_2, have a more unusual structure. To begin with, these compounds are quite covalent, and the bonding interactions are complicated. They are best approximated as compounds of M^{2+} and S_2^{2-} ions. The disulfide ion consists of two sulfur atoms bonded together by a single covalent bond and is therefore not a spherical ion. Within this approximation, MoS_2 and WS_2 may be viewed as having the NiAs structure; the metal ions form a HCP lattice, and the octahedral holes are occupied by S_2^{2-} ions. Each metal ion is therefore surrounded by a trigonal prism of disulfide ions. Figure 4.14 shows a section of the MoS_2 structure in highly idealized form.

Figure 4.14
Structure of MoS_2.
The S_2^{2-} ions are
tilted at 45° to the
plane of the page.

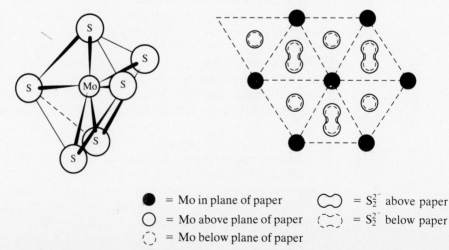

● = Mo in plane of paper	⬭ = S_2^{2-} above paper
○ = Mo above plane of paper	⬭ = S_2^{2-} below paper
○ = Mo below plane of paper	

From these illustrations, it can be seen that because of their nonspherical shape and their being tilted with respect to the close-packed planes of metal ions, three of the six disulfides interact much more intimately with the central cation than do the other three. The three closely coordinated disulfide ions enclose the central cation in a trigonal prism of six sulfur atoms. These trigonally prismatic units form continuous sheets throughout the crystal, and as in the case of the CdI_2 lattice, there are no cations between them. Thus, the layers are mainly held together by relatively weak S_2^{2-}/S_2^{2-} interactions, and the crystal shows very low sheer strength along these planes. Layer lattices of the sulfide family show many other useful properties to which we will return in Chapter 18.

Before closing, it is worth mentioning that halide- and sulfide-containing media are notoriously corrosive to transition metals. Among several contributing factors to this effect, the low mechanical integrity of the halide or sulfide layer at the surface of a metal immersed in such media is no doubt very important.

BIBLIOGRAPHY

The reading lists given in Chapters 2 and 3 also serve for this chapter.

PROBLEMS

4.1 Calculate the radius ratios for CsBr and RbCl using the data available in Table 2.5.

4.2 What are the coordination numbers (that is, the number of nearest neighbors) in face-centered cubic, body-centered cubic, and hexagonal close-packed structures?

4.3 What percentages of the space are occupied in face-centered and HCP arrangements of hard spheres? How do these percentages compare with that in a body-centered arrangement?

4.4 Using the ionic radii given in Table 2.5, predict the crystal structures of the following ionic compounds:
   a. CsF   **c.** $ZnBr_2$
   b. MgS   **d.** KI

4.5 Using the data given in Tables 2.5 and 4.1, calculate the lattice energies for zinc blende (ZnS) and fluorite (CaF_2). (Assume a value of 30 pm for ρ.)

4.6 Write a Born–Haber cycle for the formation of CaF_2 from the elements in their standard states; given the following information, calculate the lattice energy: $\Delta H_{sub}[Ca(s)] = 172$ kJ mol^{-1}; $I[Ca(g)] = 1.64$ MJ mol^{-1} (to Ca^{2+}); $D[F–F(g)] = 165$ kJ mol^{-1}; $A[F(g)] = 322$ kJ mol^{-1}; ΔH_f^0 [CaF$_2$(s)] = -1.2 MJ mol^{-1}. How does this compare with the value obtained from the previous question?

4.7   Given the following room-temperature data, calculate the lattice energy of RbCl: $\Delta H_f^0[RbCl(s)]$, -439 kJ mol^{-1}; $\Delta H_{sub}[Rb(s)]$, 86 kJ mol^{-1}; $D[Cl–Cl(g)]$, 241 kJ mol^{-1}; $I[Rb(g)]$, 401 kJ mol^{-1}; $A[Cl(g)]$, -400 kJ mol^{-1}.

4.8   Explain what is meant by Fajan's rules and their relationship to covalency.

4.9   Give examples for each of the following structural types and describe the main features of each type.

a. Sodium chloride	**g.** Rhenium trioxide
b. Cesium chloride	**h.** Spinel
c. Zinc blende	**i.** Inverse spinel
d. Wurtzite	**j.** Ilmenite
e. Rutile	**k.** Layer lattice
f. Corundum	

What effects would you expect on the application of high pressures to these systems?

4.10   Derive radius ratios for the following structures:
   a. Wurtzite
   b. Cesium chloride
   c. Rutile

4.11   Predict the empirical formula of a compound M$_a$X$_b$ that crystallizes in a face-centered cubic lattice such that the cations M occupy all the tetrahedral holes.

4.12   The vapors of CsF and CaS were mixed in a hot chamber. As the walls of the chamber slowly cooled, crystals began to form. What do you think the composition of these crystals would be?

4.13   See if you can verify the r^+/r^- values given in Table 4.3. In the tetrahedral case, the anion centers are at alternate corners of a cube, and the anions make contact halfway along a diagonal of the cube face. The cube-body diagonal is the sum of the diameters of the anion and cation.

4.14   What is the empirical formula of a compound M$_a$X$_b$ that has a face-centered cubic structure in which the M atoms occupy the corners and the X atoms are located at the face centers? What would the empirical formula be if the M atoms were situated at the corners of a body-centered cube and the X atoms were the center atoms?

4.15   If the radii of K$^+$ and Br$^-$ are 133 and 196 pm, respectively, what would you expect to be the composition of the vapor above boiling KBr?

4.16   What is the coordination number and coordination geometry of the *anion* in the following structures:

a. Wurtzite	**c.** Li$_2$O
b. CdI$_2$	**d.** CaF$_2$

4.17 The lattice energy of NaF was found to be 920 kJ mol^{-1}. If the fluoride ion radius is 136 pm, what is the radius of Na$^+$? ($M = 1.75$, $\rho = 31$ pm)

4.18 Compute the Madelung constant for a one-dimensional ionic crystal with alternating positive and negative ions separated by a distance d_e.

4.19 Discuss briefly the following facts: In salts with medium-sized cations, the F$^-$ ion shows consistently a radius of 133 pm. HF is a gas, and if it is assumed to consist of ion pairs with a proton sitting on the surface of a spherical F$^-$ ion, the calculated bond energy is 55 kJ mol^{-1}. The experimental bond energy for HF is 567 kJ mol^{-1}.

4.20 Using the data given in Problem 4.19, what would the F$^-$ ion radius have to be in order to fit the observed bond energy using an ionic model?

4.21 Suggest explanations for the following observations:
  a. Lavoisier was able to produce O$_2$ from the thermal dissociation of HgO but not ZnO.
  b. An all-nickel internal-combustion engine seized up after running for 5 min on standard automotive lubricating oil.
  c. A similar engine with 5% iodine added to the oil successfully passed a 1000-h running test.

4.22 Why do you think most of the calcium and carbon on earth is present in chalk and limestone deposits?

4.23 Account for the following:
  a. NiO is very abrasive, but NiI$_2$ is not.
  b. The melting point of BeF$_2$ is lower than that of MgF$_2$.

4.24 Comment on the validity of the following:
  a. CeCl$_3$ has a lower melting point than EuCl$_3$.
  b. MnO$_2$ is more covalent than MnO.

PART TWO

Molecular Structure and Analytical Methods

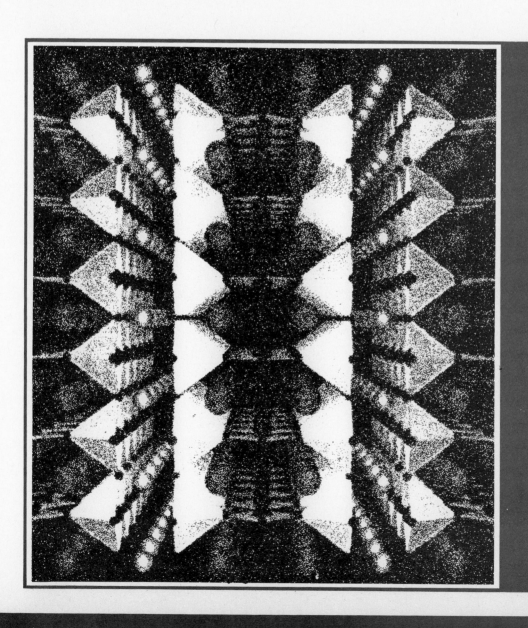

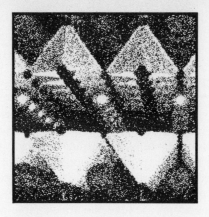

5

Molecular Symmetry in Inorganic Chemistry

Symmetry is a part of our everyday lives in countless different ways, and although we may not have appreciated it, we are accustomed to being surrounded by regular or symmetric objects and arrangements of objects. In the chemical sense, symmetry implies that two or more parts of an object are indistinguishable from one another in that they have the same size and shape. The human body is a good example—one-half is mimicked by the other, with the hands and feet being mirror images. The windows in our homes also have symmetry to them; they are usually regularly shaped, squares, rectangles, circles, diamonds. These windows are most often arranged symmetrically with respect to one another on the sides of the building.

In this chapter, we give you a brief introduction to molecular symmetry and the chemical applications of group theory and then show you how to identify the symmetries of atomic orbitals (AOs).

5.1 Molecular Symmetry and Group Theory

It is *molecular symmetry* that is most interesting to chemists. If a molecule can be rotated a certain number of degrees about an axis or if some of its constituent atoms can be reflected through a mirror plane from one side of the molecule to the other without changing the appearance and orientation of the molecule, then the molecule is said to possess symmetry. The axis about which the molecule is rotated and the mirror plane through which some of the atoms are reflected are examples of *symmetry elements*, whereas the actual transformations involved are referred to as *symmetry operations*. The rigorous definitions of these terms are as follows:

1. *Symmetry element*: A line, point, or plane, with respect to which one or more symmetry operations may be performed.

2. *Symmetry operation*: The movement of a molecule relative to some symmetry element such that every atom in the molecule before the operation is performed coincides with an equivalent atom (or itself) after the operation.

Some molecules have very few symmetry elements, whereas others have many; for example, two of the more symmetrical shapes occurring in nature are the *octahedron* and the *icosahedron*:

6 Vertices,
8 Equilateral Triangular Faces

12 Vertices,
20 Equilateral Triangular Faces

Figure 5.1 shows some specific examples of inorganic species with highly symmetrical shapes, including silane (SiH_4, tetrahedral), sulfur hexafluoride (SF_6, octahedral), and the dodecahydroborate anion ($[B_{12}H_{12}]^{2-}$, icosahedral).

The rigorous treatment of symmetry requires a knowledge of the mathematics of *group theory*, which we cannot adequately treat here due

Figure 5.1
Examples of some of the important geometrical shapes found in inorganic chemistry.

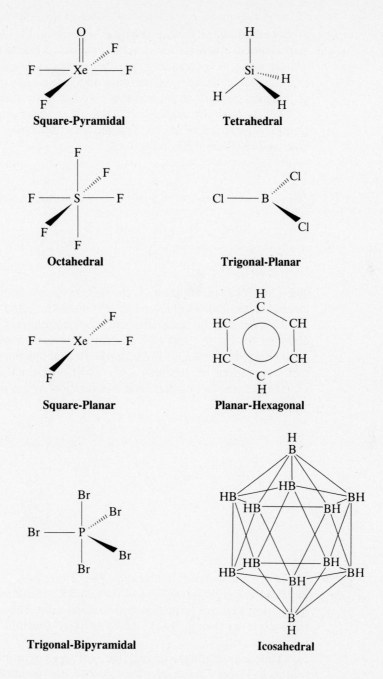

Square-Pyramidal

Tetrahedral

Octahedral

Trigonal-Planar

Square-Planar

Planar-Hexagonal

Trigonal-Bipyramidal

Icosahedral

to space limitations. Group theory is a powerful tool for solving complex mathematical problems and now has many uses. For more complete details on this topic, refer to the group theory books listed at the end of the chapter. It is, however, the chemical applications of group theory that are of principal concern to us, and these we can understand qualitatively.

Let us begin by taking a closer look at some symmetry operations. Consider the rotations about the z axis indicated below for the tetrahedral SiH_4 and octahedral SF_6 molecules:

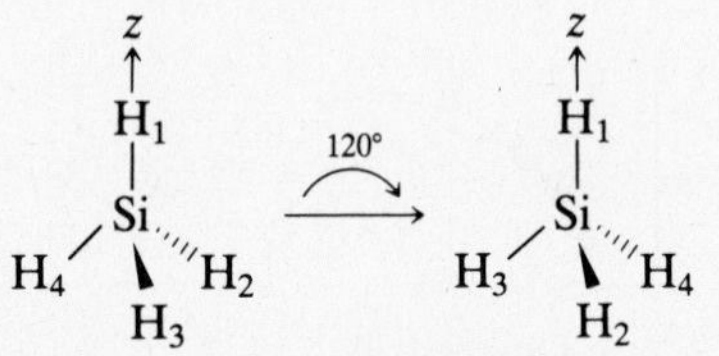

The 120° clockwise rotation for the SiH_4 molecule about the Si—H_1 bond located on the z axis will interchange H atoms 2, 3, and 4, but the molecule will look exactly the same after such a transformation. Similar results are obtained for the 90° and 180° clockwise rotations about the z axis in the octahedral SF_6 molecule. In group theoretical language, these three symmetry operations are labeled C_3, C_4, and C_2 rotations, respectively, where C_n (the general expression) indicates rotation by $2\pi/n$ radians:

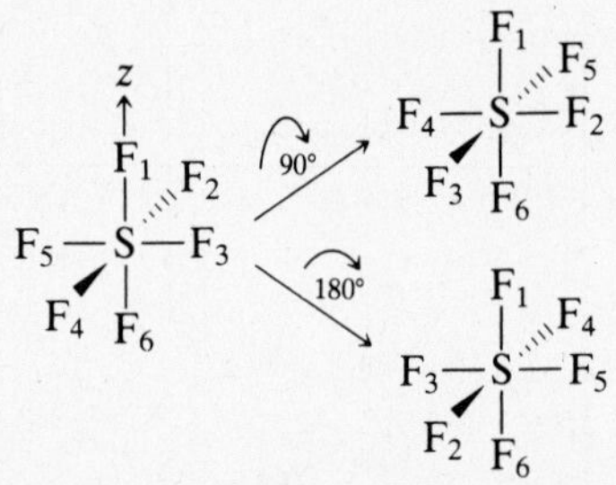

where n = the order of the rotation axis. Therefore, C_3 means rotation by $2\pi/3$ radians (120°), C_4 ($2\pi/4 = 90°$), and C_2 ($2\pi/2 = 180°$). These types of rotation axes are termed *proper rotation axes*. They are also sometimes referred to as twofold (C_2), threefold (C_3), fourfold (C_4), and so forth rotational axes.

Molecules that have several different types of proper rotation axes are said to have *high symmetry*; for example, the planar-hexagonal benzene (C_6H_6) molecule has C_2, C_3, and C_6 proper rotation axes colinear with the z axis and normal to the molecular plane and also several C_2 axes lying in the molecular plane (Figure 5.2). The C_n rotation axis with the highest

Figure 5.2
Examples of C_6, C_3, and C_2 proper rotation axes in benzene. Note that the three C_2 axes indicated differ in their location and are labeled C_2, C_2', and C_2''.

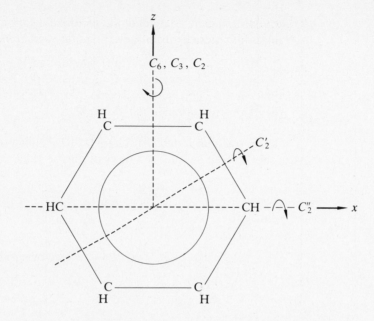

value of n is called the *principal rotation axis* and is usually located on the z axis. For benzene, C_6 is the principal rotation axis. If there is only one rotation axis present, then this axis is the principal axis.

In all, there are $6C_4$ symmetry operations possible in SF$_6$: $3C_4$ rotations (90° rotations) and $3C_4^3$ rotations (superscript 3 indicates three successive 90° rotations = 270°) about the three independent C_4 proper rotation axes. Note that the C_4 (90°) and C_4^3 (270°) rotations are equivalent to clockwise and counterclockwise rotation by 90°, respectively. The SF$_6$ molecule also has three possible C_4^2 symmetry operations about the C_4 axes. These C_4^2 symmetry operations are labeled C_2, however, because two successive 90° rotations are equivalent to one 180° rotation. The complete, general expression for a proper rotational symmetry operation is C_n^m, which means rotation by $2\pi m/n$ radians. Moreover, if m is an exact multiple of n, then the m and n values are reduced accordingly. For example, in benzene, the C_6^2, C_6^3, and C_6^4 symmetry operations normal to the molecular plane are labeled C_3, C_2, and C_3^2, respectively.

Another important symmetry element to be considered is a *mirror plane* (σ). The associated symmetry operation involves reflection through the mirror plane from one part of the molecule to another. There are three different types of mirror plane, which are labeled according to their location in the molecule:

1. σ_h: Mirror plane normal to the principal proper rotation axis, where the subscript h derives from the expression *horizontal mirror plane*.

2. σ_v: Mirror plane containing the principal rotation axis C_n, where the subscript v derives from the expression *vertical mirror plane*.

3. σ_d: Mirror plane that bisects the dihedral angle made by the principal rotation axis and two adjacent C_2 axes perpendicular to the principal rotation axis. This is a special case of σ_v. The subscript d derives from the expression *dihedral mirror plane*.

The tetrachloroiodate(III) anion $[ICl_4]^-$ and any other square-planar MX_4 species contain all three types of mirror plane: σ_h, $2\sigma_v$, and $2\sigma_d$:

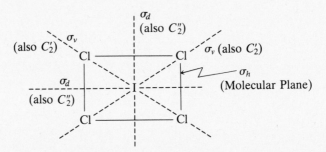

Both SiH_4 and SF_6 are molecules that possess other kinds of rotation axes that are termed *improper rotation*, or *rotation-reflection axes*, S_n, because they do not involve just rotation about an axis. The general expression for the associated symmetry operation is S_n^m. Tetrahedral SiH_4 has $3S_4$ and $3S_4^3$ symmetry operations, whereas octahedral SF_6 has $3S_4$, $3S_4^3$, $4S_6$, and $4S_6^5$ improper rotations. These types of symmetry operations are generated by combining a proper rotation (C_n^m) and a mirror plane.

The $2S_4$ (S_4 and S_4^3) operations in $[Re_2Cl_8]^{2-}$ are generated either by performing a C_4 (or C_4^3) rotation followed by reflection through a mirror plane normal to the C_4 proper rotation axis or *vice versa* (Figure 5.3a). A similar situation exists for the $6S_4$ ($3S_4$ and $3S_4^3$) operations in SiH_4 (Figure 5.3b). For $[Re_2Cl_8]^{2-}$, the σ_h mirror plane is a symmetry element associated with the molecular structure, whereas σ_h is not a symmetry element for SiH_4. The S_4 and S_4^3 symmetry operations are expressed mathematically by Equation 5.1, where the two symmetry operations are performed successively in the order indicated:

$$C_4\sigma_h = \sigma_h C_4 = S_4 \qquad \text{or} \qquad C_4^3\sigma_h = \sigma_h C_4^3 = S_4^3 \qquad \textbf{(5.1)}$$

It should now be clear that it is not essential for both C_n and σ_h to be present individually for an S_n^m symmetry element to exist. The Si_2H_6 molecule affords another example of this situation: There are $2S_6$ (S_6 and S_6^5) rotations possible, but there is no σ_h present. The S_6 operation moves H_1 into position H_5, H_2 into position H_6, and so on. On the other hand, the S_6^5 operation moves H_1 into position H_4, H_2 into position H_5, and so on.

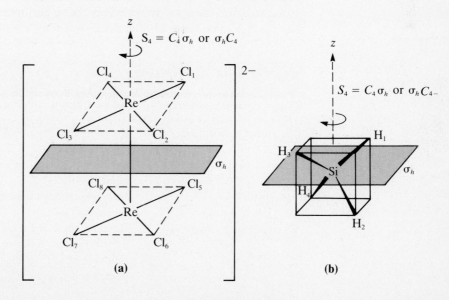

There is one other symmetry element that remains to be considered in this brief introduction to molecular symmetry: the *center of symmetry*, or *inversion center* (*i*), which involves inversion through the central point of the molecule such that each atom has a counterpart with identical Cartesian (x, y, z) coordinates, but of opposite sign. The sulfur atom in SF_6 is an example of an atom located at a center of symmetry. The Cartesian origin $(0,0,0)$ coincides with the S atom. Inversion through the S atom interchanges fluorines 1 and 6, 2 and 4, and 3 and 5.

Figure 5.3
Examples of S_4 improper rotations in (**a**) $[Re_2Cl_8]^{2-}$ and (**b**) SiH_4.
In (**a**), the clockwise S_4 operation takes Cl_1 into position Cl_6, whereas in (**b**), H_1 moves to the position previously occupied by H_2. Similarly, for (**a**) the clockwise S_4^3 operation would take Cl_1 into position Cl_7; for (**b**), an S_4^3 operation would take H_1 into position H_4.

$$
\begin{array}{c}
\underset{F_4}{\overset{\displaystyle F_1}{\underset{|}{F_5 - \overset{F_2}{\underset{|}{S}} - F_3}}} \quad \overset{i}{\longrightarrow} \quad \underset{F_2}{\overset{\displaystyle F_6}{\underset{|}{F_3 - \overset{F_4}{\underset{|}{S}} - F_5}}}
\end{array}
$$

The SF_6 molecule has three σ_h mirror planes through which the three pairs of F atoms seem to invert. It must be emphasized, however, that such mirror planes are not a prerequisite for a molecule to possess a center of symmetry. For instance, the Si_2H_6 molecule already mentioned, is centrosymmetric. Inversion through the midpoint of the Si—Si bond interchanges H_1 and H_6, H_2 and H_4, and H_3 and H_5.

The final thing needed for mathematical completion of a group is the *identity E*; the associated operation results in doing obsolutely nothing to the molecule. A useful analogy for the identity operation is multiplication of a number by one. There is no corresponding symmetry element because there is no geometric entity such as a line, point, or plane, with respect to which the identity operation can be performed. Table 5.1 defines fully the five main types of molecular symmetry operations.

Table 5.1 *Molecular Symmetry Operations*

Type of Symmetry Element	Actual Operation Performed
C_n^m, proper rotation axis	Rotation about the n-fold axis by $2\pi m/n$ radians
σ, mirror plane	Reflection through the symmetry plane
i, inversion	Inversion through the center of symmetry
S_n^m, improper rotation axis	Rotation about the n-fold axis by $2\pi m/n$ radians, followed by reflection in the mirror plane perpendicular to the n-fold axis or *vice versa*
E, identity	None

The shape of every molecule can be described in terms of the collection of symmetry elements that it possesses. It can be demonstrated that only a limited number (32) of such collections is chemically important because only shapes with certain symmetries can be fitted to form an infinitely repeating lattice, and each such combination is called a *molecular point group*.

In the case of SiH_4, the molecule belongs to the T_d point group and the complete set of 24 symmetry elements involved is E, $8C_3$, $3C_2$, $6S_4$, and $6\sigma_d$. By following a series of established mathematical rules for groups, it is a straightforward (but sometimes quite tedious) process to construct the *character tables* associated with the molecular point groups, ranging from that for molecules with no symmetry elements at all (C_1 symmetry, for example, CFClBrI) to those for highly symmetrical molecules [for example, icosahedral, I_h (120); octahedral, O_h (48); tetrahedral, T_d (24)]. The numbers in parentheses indicate the total number of different symmetry elements present in each case. Table 5.2 gives examples of some of the character tables that are important in inorganic chemistry. These character tables have been constructed assuming that the principal rotation axis is colinear with the z axis.

The symmetry elements involved in each molecular point group are given at the top of the character table concerned, and the total number of them is referred to as the *order of the (point) group*. Therefore, the orders of the T_d and O_h point groups are 24 and 48, respectively. Note that the order for the C_1 point group is 1 because there is only 1 symmetry element possible for such molecules, the identity E.

The alphanumeric labels (A_1, B_2, E, and so on) given on the left-hand side of the tables are known in group theoretical language as *irreducible representations*. This mathematical terminology has its origin in matrix algebra; you will discover shortly that it is usually possible to reduce mathematically *reducible representations* into their constituent *irreducible representations*. The numbers in the main bodies of the tables are the *characters* (χ) associated with the different types of symmetry elements indicated directly above them. The mathematical basis for these characters is not important here, but we will make use of the numbers themselves extensively in our applications of group theory to chemical situations.

The alphanumeric labels are also often referred to as *Mulliken labels* where

1. *A* and *B* are nondegenerate, irreducible representations such that $\chi(E) = 1$ and $\chi(C_n) = 1$ (symmetric) or -1 (antisymmetric) with respect to the principal proper rotation axis (C_n). There are also degenerate, irreducible representations possible: $E[\chi(E) = 2]$, doubly degenerate; T or $F[\chi(E) = 3]$, triply degenerate; $G[\chi(E) = 4]$, quadruply degenerate; $H[\chi(E) = 5]$, quintuply degenerate. [NOTE: G and H irreducible representations are extremely rare and only occur in very special cases such as the icosahedral (I_h) character table.]

Table 5.2 Selection of the Character Tables for the Most Commonly Encountered Point Groups in Inorganic Chemistry

C_s	E	σ_h		
A'	1	1	x, y, R_z	x^2, y^2, z^2, xy
A''	1	-1	z, R_z, R_y	yz, xz

D_{2d}	E	$2S_4(z)$	$C_2(z)$	$2C_2'$	$2\sigma_d$		
A_1	1	1	1	1	1		x^2+y^2, z^2
A_2	1	1	1	-1	-1	R_z	
B_1	1	-1	1	1	-1		x^2-y^2
B_2	1	-1	1	-1	1	z	xy
E	2	0	-2	0	0	$(x, y); (R_x, R_y)$	(xz, yz)

C_{3v}	E	$2C_3(z)$	$3\sigma_v$		
A_1	1	1	1	z	x^2+y^2, z^2
A_2	1	1	-1	R_z	
E	2	-1	0	(x, y) (R_z, R_y)	(x^2-y^2, xy) (xz, yz)

C_{2v}	E	$C_2(z)$	$\sigma_v(xz)$	$\sigma_v'(yz)$		
A_1	1	1	1	1	z	x^2, y^2, z^2
A_2	1	1	-1	-1	R_z	xy
B_1	1	-1	1	-1	x, R_y	xz
B_2	1	-1	-1	1	y, R_x	yz

C_{4v}	E	$2C_4(z)$	$C_2(z)$	$2\sigma_v$	$2\sigma_d$		
A_1	1	1	1	1	1	z	x^2+y^2, z^2
A_2	1	1	1	-1	-1	R_z	
B_1	1	-1	1	1	-1		x^2-y^2
B_2	1	-1	1	-1	1		xy
E	2	0	-2	0	0	(x, y) R_x, R_y	(xz, yz)

D_{3h}	E	$2C_3(z)$	$3C_2$	σ_h	$2S_3(z)$	$3\sigma_v$		
A_1'	1	1	1	1	1	1		x^2+y^2, z^2
A_2'	1	1	-1	1	1	-1	R_z	
E'	2	-1	0	2	-1	0	(x, y)	(x^2-y^2, xy)
A_1''	1	1	1	-1	-1	-1		
A_2''	1	1	-1	-1	-1	1	z	
E''	2	-1	0	-2	1	0	(R_x, R_y)	(xz, yz)

D_{4h}	E	$2C_4(z)$	$C_2(z)$	$2C_2'$	$2C_2''$	i	$2S_4(z)$	σ_h	$2\sigma_v$	$2\sigma_d$		
A_{1g}	1	1	1	1	1	1	1	1	1	1		$x^2+y^2,\ z^2$
A_{2g}	1	1	1	-1	-1	1	1	1	-1	-1	R_z	
B_{1g}	1	-1	1	1	-1	1	-1	1	1	-1		x^2-y^2
B_{2g}	1	-1	1	-1	1	1	-1	1	-1	1		xy
E_g	2	0	-2	0	0	2	0	-2	0	0	(R_x, R_y)	(xz, yz)
A_{1u}	1	1	1	1	1	-1	-1	-1	-1	-1		
A_{2u}	1	1	1	-1	-1	-1	-1	-1	1	1	z	
B_{1u}	1	-1	1	1	-1	-1	1	-1	-1	1		
B_{2u}	1	-1	1	-1	1	-1	1	-1	1	-1		
E_u	2	0	-2	0	0	-2	0	2	0	0	(x, y)	

T_d	E	$8C_3$	$3C_2$	$6S_4$	$6\sigma_d$		
A_1	1	1	1	1	1		$x^2+y^2+z^2$
A_2	1	1	1	-1	-1		
E	2	-1	2	0	0		$(2z^2-x^2-y^2,\ x^2-y^2)$
T_1	3	0	-1	1	-1	(R_x, R_y, R_z)	
T_2	3	0	-1	-1	1	(x, y, z)	(xy, xz, yz)

O_h	E	$8C_3$	$6C_2$	$6C_4$	$3C_2(=C_4^2)$	i	$6S_4$	$8S_6$	$3\sigma_h$	$6\sigma_d$		
A_{1g}	1	1	1	1	1	1	1	1	1	1		$x^2+y^2+z^2$
A_{2g}	1	1	-1	-1	1	1	-1	1	1	-1		
E_g	2	-1	0	0	2	2	0	-1	2	0		$(2z^2-x^2-y^2,\ x^2-y^2)$
T_{1g}	3	0	-1	1	-1	3	1	0	-1	-1	(R_x, R_y, R_z)	
T_{2g}	3	0	1	-1	-1	3	-1	0	-1	1		(xz, yz, xy)
A_{1u}	1	1	1	1	1	-1	-1	-1	-1	-1		
A_{2u}	1	1	-1	-1	1	-1	1	-1	-1	1		
E_u	2	-1	0	0	2	-2	0	1	-2	0		
T_{1u}	3	0	-1	1	-1	-3	-1	0	1	1	(x, y, z)	
T_{2u}	3	0	1	-1	-1	-3	1	0	1	-1		

2. Subscript 1 is symmetric with respect to a C_2 axis perpendicular to the principal rotation axis, that is, $\chi(C_2) = 1$. However, if no such C_2 axis is present, then it is symmetric with respect to a vertical mirror plane, that is, $\chi(\sigma_v) = 1$. For subscript 2: $\chi(C_2) = -1$ or, if no C_2 present, $\chi(\sigma_v) = -1$.

3. Subscript g (*Ger.* "gerade") is symmetric with respect to a center of symmetry, that is $\chi(i) = 1$. For subscript u (*Ger.* "ungerade"): $\chi(i) = -1$.

4. Superscript $'$ is symmetric with respect to a horizontal mirror plane, $\chi(\sigma_h) = 1$. For superscript $''$: $\chi(\sigma_h) = -1$.

The letters in the two columns on the right-hand side of the character tables will prove to be extremely important in our discussions of spectroscopic and other group theoretical applications. The letters x, y, and z refer to *translations* in the x, y, and z directions, respectively. In some books, the letters are replaced by T_x, T_y, and T_z, in which the capital Ts are meant to emphasize the translations more clearly. Let us consider a translation in the x direction under C_{2v} symmetry. How is the direction of this translation affected by the four different symmetry operations for C_{2v} symmetry? For the identity E and $\sigma_v(xz)$ operations, the direction is unchanged, whereas for $C_2(z)$ and $\sigma_v(yz)$, the direction is reversed (Figure 5.4).

The resulting characters for the four symmetry operations are $\chi(E) = 1$, $\chi[C_2(z)] = -1$, $\chi[\sigma_v(xz)] = 1$, and $\chi[\sigma_v'(yz)] = -1$. These characters are obtained from the *traces* (sums of the diagonal elements) of the following transformation matrices, which are generated by considering the effect of the different symmetry operations on the original Cartesian coordinates x, y, and z. The new coordinates, following the symmetry operations, are designated x', y', and z'.

E	x	y	z
x'	1	0	0
y'	0	0	0
z'	0	0	0

C_2	x	y	z
x'	-1	0	0
y'	0	0	0
z'	0	0	0

$\sigma_v(xz)$	x	y	z
x'	1	0	0
y'	0	0	0
z'	0	0	0

$\sigma_v'(yz)$	x	y	z
x'	-1	0	0
y'	0	0	0
z'	0	0	0

Figure 5.4
Effect of the four
symmetry operations
of the C_{2v} point
group on a
translation in the x
direction.

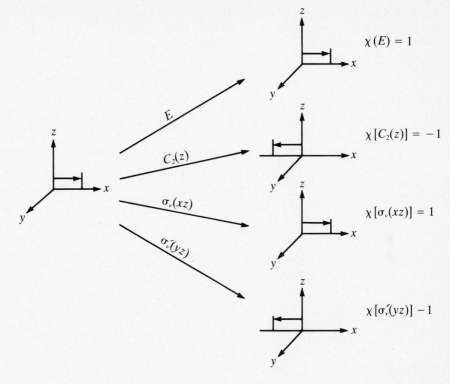

$\chi(E) = 1$

$\chi[C_2(z)] = -1$

$\chi[\sigma_v(xz)] = 1$

$\chi[\sigma_v'(yz)] - 1$

The four characters match those exactly for the B_1 irreducible representation under C_{2v} symmetry. In group theoretical language, translation in the x direction is said to *transform* as the B_1 irreducible representation. And this is why the letter x (or T_x in some character tables) is placed on the same line as the B_1 Mulliken label.

The labels R_x, R_y, and R_z mean *rotation* about the Cartesian axes indicated. Thus, rotation about the y axis (R_y) also transforms as B_1 because $C_2(z)$ and $\sigma_v'(yz)$ cause the direction of rotation to reverse leading to -1 characters, whereas E and $\sigma_v(xz)$ leave the rotational direction unchanged and characters of 1 (Figure 5.5).

For a full treatment of the translations and rotations associated with *degenerate* irreducible representations, a more detailed knowledge of matrix algebra is necessary. We will not discuss this mathematics, however, but the interested reader is referred to the books on group theory listed at the end of the chapter for more information.

In certain character tables, the translations and/or rotations appear in *parentheses* opposite the Mulliken labels for degenerate irreducible representations. For example, in the T_d character table, (R_x, R_y, R_z) and

Figure 5.5
Effect of the four symmetry operations of the C_{2v} point group on a clockwise rotation about the y axis.

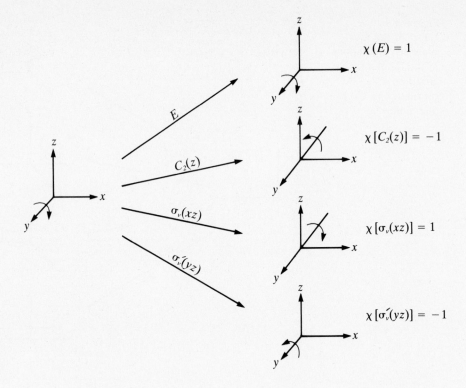

(x, y, z) are located opposite the T_1 and T_2 irreducible representations, respectively. These parenthetical labels mean that all three rotations and all three translations together transform as the T_1 and T_2 irreducible representations, respectively.

The remaining labels in the character tables, x^2, xy, and so forth are *binary products*, which are obtained by multiplying together the characters for the appropriate individual translations. For example, consider xy under C_{2v} symmetry. The characters corresponding to the translations in the x and y directions are multiplied together:

C_{2v}	$\chi(E)$	$\chi(C_2)$	$\chi[\sigma_v(xz)]$	$\chi[\sigma'_v(yz)]$
$B_1(x)$	1	-1	1	-1
$B_2(y)$	1	-1	-1	1
Multiply				
	1	1	-1	-1

The resulting new characters match those for A_2 exactly, and this is why the binary product xy is listed opposite A_2 in the C_{2v} character table. For degenerate irreducible representations, the situation is again more

Table 5.3 *Inorganic Examples of the Common Point Groups*

Point-Group Symmetry	Examples[1]
$C_{\infty v}$	HCl, [SCN]$^-$
$D_{\infty h}$	CO_2, CS_2
C_1	$(o\text{-}C_6H_4Me_2)Cr(CO)(CS)L$
C_s	$SOCl_2$, *cis*-$Fe(CO)_4IBr$
C_2	H_2O_2, S_2Cl_2
C_{2v}	SF_4, H_2O, SO_2, *cis*-$Fe(CO)_4Cl_2$
C_{3v}	NH_3, *fac*-$Mo(CO)_3L_3$
C_{4v}	$XeOF_4$, BrF_5, $Mn(CO)_5Br$
C_{3h}	$B(OH)_3$
D_3	$[Cr(en)_3]^{3+}$
D_{4d}	$Mn_2(CO)_{10}$
D_{5d}	$(C_5H_5)_2Fe$ (staggered)
D_{2h}	*trans*-$PtCl_2(NH_3)_2$
D_{3h}	BF_3, $Fe(CO)_5$
D_{4h}	$[PtCl_4]^{2-}$, $[Re_2Cl_8]^{2-}$, *trans*-SiF_4L_2
D_{5h}	$(C_5H_5)_2Fe$ (eclipsed)
D_{6h}	$(C_6H_6)_2Cr$ (eclipsed)
T_d	$GeCl_4$, $[SiO_4]^{4-}$, $Ni(CO)_4$
O_h	$Mo(CO)_6$, SF_6, $[IrCl_6]^{3-}$
I_h	$[B_{12}H_{12}]^{2-}$, $[B_{12}Cl_{12}]^{2-}$

[1] L = monodentate ligand such as PMe_3, $P(OMe)_3$, and so on.
en = H_2N—CH_2CH_2—NH_2.

complicated, necessitating the introduction of matrix algebra (we will not discuss this further).

Table 5.3 lists a few examples of molecules adopting the more common point-group symmetries. To conclude this introductory section on group theory, you must know how to determine quickly the point-group symmetry of any given molecule. There are several correlation schemes available for this purpose and the one illustrated in Figure 5.6 is a simplified version of the earlier approaches.

To use the flowchart, simply follow the appropriate arrows depending on your responses to the questions. The special point groups are $D_{\infty h}$ (linear molecules with a center of symmetry), $C_{\infty v}$ (linear molecules without a center of symmetry), T_d (molecules with the same symmetry as a tetrahedron), O_h (molecules with the same symmetry as an octahedron), and I_h (molecules with the same symmetry as an icosahedron). The infinity sign in $D_{\infty h}$ and $C_{\infty v}$ is because the molecular axis in species with these symmetries is an infinite-order rotation axis (C_∞).

Figure 5.6
Flowchart to
determine the
point-group
symmetry of a
molecule or ion.
Adapted from M.
Zeldin, *J. Chem.
Ed.* 43(1966): 17.

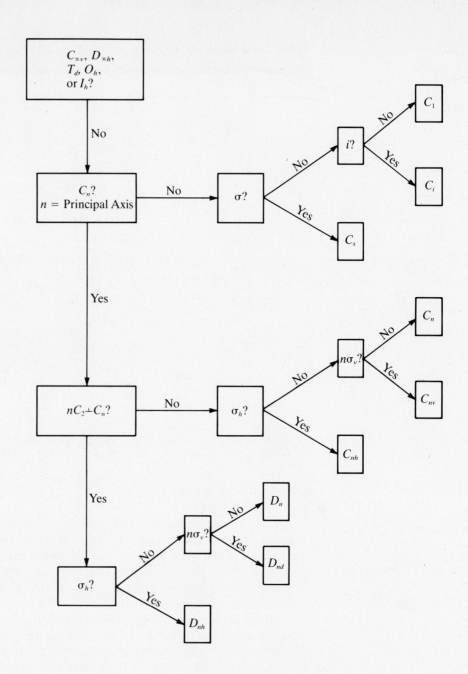

Let us first consider BF_3, a typical example of an inorganic molecule with trigonal-planar geometry:

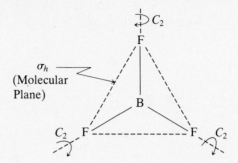

Clearly, BF_3 does not belong to one of the special point groups. It does, however, have a principal proper rotation axis (C_3, $n = 3$), and there are three C_2 proper rotation axes and a σ_h mirror plane perpendicular to the C_3 axis. By following the arrows on the flowchart (Figure 5.6), you can see that the BF_3 molecule has D_{3h} symmetry. What would its symmetry have been if it had been trigonal-pyramidal, like NH_3 and PF_3, rather than trigonal-planar? There would be no C_2 rotation axes nor a σ_h mirror plane, but there would now be three σ_v mirror planes. In this case, the molecular symmetry would be C_{3v}.

Sometimes it is hard to visualize all the axes in a given molecule, and so it is difficult to decide on the molecular point group. For example, from X-ray diffraction studies, the three dimeric metal carbonyl complexes, $M_2(CO)_{10}$ (M = Mn, Tc, Re), have the D_{4d} structure shown in Figure 5.7, in which the CO groups are arranged so as to minimize steric interactions between themselves. There are 16 symmetry elements in all: E, $2S_8$, $2C_4$, $2S_8^3$, C_2, $4C_2'$, and $4\sigma_d$, a few of which are indicated on the structure.

Figure 5.7
Molecular structure (D_{4d} symmetry) of the $M_2(CO)_{10}$ (M = Mn, Tc, Re) complexes.

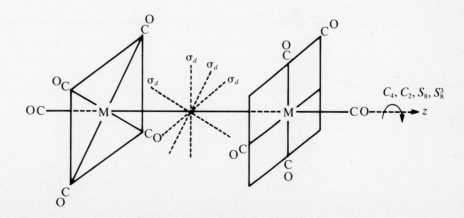

The $2C_4$ (C_4, C_4^3), C_2, $2S_8$ (S_8, S_8^7), and $2S_8^3(S_8^3, S_8^5)$ symmetry elements are all associated with the molecular axis and are quite easy to locate. But what about the $4C_2'$ elements? In such a case, it helps to sketch a projection diagram in which you look down the M—M bond axis of the molecule:

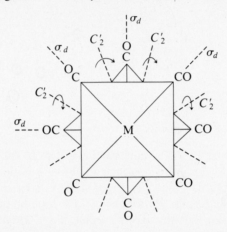

The $4C_2'$ proper rotation axes are indicated by the bisectors of the two squares in the projection diagram and pass perpendicularly through the center of gravity of the M—M bond linking the two $M(CO)_5$ fragments. Therefore, referring back to the molecular symmetry flowchart (Figure 5.6), it should now be apparent that the symmetry elements of the $M_2(CO)_{10}$ molecules do indeed satisfy the requirements of the D_{4d} point group: principal proper rotation axis, C_4 ($n = 4$); $4C_2'$ axes perpendicular to C_4; no σ_h mirror plane, but $4\sigma_d$ mirror planes. At this stage, it would be well worth your while to verify the point groups attributed to the species listed in Table 5.3.

5.2 Orbital Symmetry

Chemical bonding depends crucially on matching the symmetries of orbitals and maximizing their overlap with one another. Group theory and molecular symmetry allow us to determine the symmetries of atomic, hybrid, and molecular orbitals. You have already been introduced to some of the group theoretical labels used in MO diagrams. You will learn in Chapter 13 that the five d orbitals of a transition metal ion are not degenerate when the ion is situated in an octahedral environment. In fact, the d orbitals split into the two degenerate sets, t_{2g} (d_{xy}, d_{xz}, d_{yz}) and e_g (d_{z^2}, $d_{x^2-y^2}$), where t_{2g} and e_g are group theoretical labels from the octahedral (O_h) character table. Notice that the symmetry labels for orbitals are usually written in lowercase letters rather than uppercase ones, as they are in the character tables.

The s AOs are always spherical, and so they must be symmetric with respect to all the possible symmetry elements for any given point-group situation. In group theory terms, this means that all characters must be equal to 1. The first row in every character table satisfies this condition, for example, a' in C_s symmetry, a_1 in C_{2v} and T_d symmetries, and a_{1g} in O_h symmetry. Refer to the character tables given in Appendix 1 to verify this.

What about the p and d AOs? Do these have any symmetry? Let us consider initially a p_x orbital under C_{2v} symmetry. The two lobes of electron density have wave functions of opposite sign—this is the reason for the plus and minus signs on the lobes. The behaviour of the p_x orbital with respect to the four C_{2v} symmetry operations is illustrated in Figure 5.8. Reversal of the wave-function signs on the orbital lobes

Figure 5.8
Effect of the symmetry operations of the C_{2v} point group on a p_x orbital.

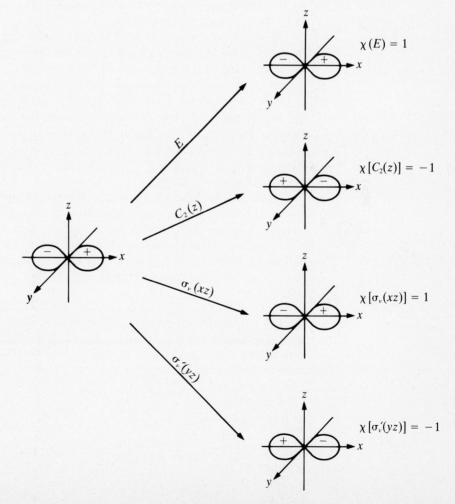

following application of a symmetry operation means that the character is -1, whereas no change in sign yields a character of 1.

The resulting characters for the four operations are $\chi(E) = 1$, $\chi(C_2) = 1$, $\chi[\sigma_v(xz)] = 1$, $\chi[\sigma_v'(yz)] = -1$, that is, the same characters as for translation in the x direction. Therefore, the p_x orbital is said to transform as the b_1 irreducible representation under C_{2v} symmetry. Similarly, it can be shown that the p_y and p_z AOs transform as the b_2 and a_1 irreducible representations, respectively. This information can be easily obtained by locating the x, y, and z translations in the C_{2v} character table. This observation is general, and so it is quite straightforward to decide how the p orbitals transform under a given symmetry. All that you must do is to identify the irreducible representations corresponding to the translations. In many cases, two or three of the p orbitals transform together as degenerate irreducible representations, for example e (x, y) under C_{3v} symmetry and t_2 (x, y, z) under T_d symmetry.

Similarly, the symmetries of the d orbitals can be located in the character tables by looking for the appropriate binary combinations on the right-hand side of the tables. It can be shown that $2z^2 - x^2 - y^2$ corresponds to z^2 (in orbital language). Table 5.4 summarizes how to find the symmetries of both the p and d orbitals. The mathematical treatment for the degenerate irreducible representations and the binary products is not directly important at this stage. However, it should now be clear, by referring to the O_h character table in Appendix 1, why there are two discrete sets of degenerate d orbitals corresponding to the e_g and t_{2g} irreducible representations.

Thus far, we have only considered AOs. A great deal of bonding theory in chemistry has been based on the notion of hybrid orbitals. For instance, you are familiar with the idea of sp^3 hybrid σ orbitals in tetrahedral carbon chemistry and $sp^3 d^2$ hybrid σ orbitals in octahedral compounds such as SF_6. It is quite simple to demonstrate that these hybrid orbitals are based on symmetry considerations. We can prove this from the

---

Table 5.4   *Orbital Symmetries*

Orbital type	Indicator in Character Table
s	Automatically corresponds to the total symmetric irreducible representation
p_x, p_y, p_z	Translations in the x, y, and z directions, respectively
d_{z^2}	Binary product z^2 (often written as $2z^2 - x^2 - y^2$)
$d_{x^2-y^2}$	Binary product $x^2 - y^2$
d_{xy}	Binary product xy
d_{xz}	Binary product xz
d_{yz}	Binary product yz

Figure 5.9
σ-bonding orbitals for a tetrahedral MX$_4$ species such as silane (SiH$_4$), with some of the T_d point-group symmetry elements indicated.

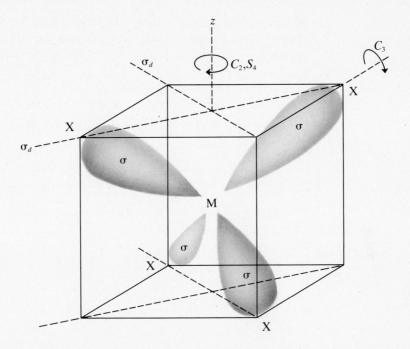

irreducible representations of the hybrid orbitals for both the T_d and O_h point groups.

For T_d symmetry, let us take silane (SiH$_4$) as our example. There are four hybrid σ orbitals indicated on the structure shown in Figure 5.9. How are these orbitals affected by the 24 symmetry operations of the T_d point group? Our aim is to find the so-called *reducible representation* (Γ_{red}) for the four hybrid σ orbitals. We must determine first the number of orbitals that are *unshifted* by each set of symmetry operations. The obtained numbers follow. The rationale for this apparently arbitrary approach is given in detail in any book dealing with group theory.

T_d	E	$8C_3$	$3C_2$	$6S_4$	$6\sigma_d$
Γ_{red}	4	1	0	0	2

Our next task is to determine how many of each type of irreducible representation are present in the reducible representation. To do this, we make use of a very important group theoretical formula (Equation 5.2):

$$a_i = \frac{1}{h} \sum_R \chi(R) \cdot \chi_i(R) \qquad \textbf{(5.2)}$$

where a_i = the number of times the ith irreducible representation appears in Γ_{red}, h = the order of the point group (that is, the total number of symmetry elements), $\chi(R)$ = the character of the Rth set of symmetry operations in Γ_{red}, and $\chi_i(R)$ = the character of the Rth set of symmetry operations in the ith irreducible representation. The T_d character table required for the mathematical analysis was given earlier in Table 5.2.

$$
\overbrace{E} \qquad \overbrace{8C_3} \qquad \overbrace{3C_2} \qquad \overbrace{6S_4} \qquad \overbrace{6\sigma_d}
$$

$$
a_{a_1} = \frac{1}{24}\left[(4)(1) + 8(1)(1) + 3(0)(1) + 6(0)(1) + 6(2)(1)\right] = 1
$$

$$
a_{a_2} = \frac{1}{24}\left[(4)(1) + 8(1)(1) + 3(0)(1) + 6(0)(-1) + 6(2)(-1)\right] = 0
$$

$$
a_e = \frac{1}{24}\left[(4)(2) + 8(1)(-1) + 3(0)(2) + 6(0)(0) + 6(2)(0)\right] = 0
$$

$$
a_{t_1} = \frac{1}{24}\left[(4)(3) + 8(1)(0) + 3(0)(-1) + 6(0)(1) + 6(2)(-1)\right] = 0
$$

$$
a_{t_2} = \frac{1}{24}\left[(4)(3) + 8(1)(0) + 3(0)(-1) + 6(0)(-1) + 6(2)(1)\right] = 1
$$

$$
\Gamma_{\text{red}} = a_1 + t_2
$$

This result indicates that the only orbitals allowed on symmetry grounds to participate in hybrid σ orbitals in tetrahedral MX_4 species must have the symmetries a_1 and t_2. Herein lies the real difficulty with hybridization theory because strictly only orbitals of the *same symmetry* type can mix. However, for our purposes, we can ignore this problem. Applying the orbital symmetry indicators given in Table 5.4 to the T_d character table indicates that the s orbital would transform as a_1, whereas the three p orbitals (p_x, p_y, p_z) or the three d orbitals (d_{xy}, d_{xz}, d_{yz}) would together transform as t_2. Thus, the only symmetry-allowed hybrid orbital combinations would be sp^3 or sd^3, where superscript 3 refers to the number of p or d orbitals, respectively.

For carbon, the lowest-energy d orbitals (the empty $3d$ levels) are not readily available for σ bonding, and so sp^3 is a reasonable description of the σ bonding in methane (CH_4). However, in the case of the transition metals, which do have available d orbitals, sd^3 hybridization is quite likely, and so the bonding in tetrahedral $TiCl_4$, for example, may be better described as sd^3 rather than sp^3 or perhaps some mixture of both hybridizations. A similar σ-bonding treatment is possible for all regular geometries. It should also be pointed out that the same group theoretical treatment also provides the symmetries of the σ-bonding MOs for the molecular symmetries concerned, in our case, for any tetrahedral species, $a_1 + t_2$.

The above approach can be extended to the symmetries of AOs involved in π bonding. The interested reader is referred especially to the book by Cotton to determine the procedure by which this is actually done.

BIBLIOGRAPHY

Suggested Reading

Bunker, P. R. *Molecular Symmetry and Spectroscopy*. New York: Academic Press, 1979.

Cotton, F. A. *Chemical Applications of Group Theory*, 2nd ed. New York: Wiley-Interscience, 1971.

Davidson, G. *Introductory Group Theory for Chemists*. London: Elsevier, 1971.

Douglas, B. E., and C. A. Hollingsworth. *Symmetry in Bonding and Spectra. An Introduction*. New York: Academic Press, 1985.

Fackler, J. P., Jr. *Symmetry in Coordination Chemistry*. New York: Academic Press, 1971.

Harris, D. C., and M. D. Bertolucci. *Symmetry and Spectroscopy*. New York: Oxford University Press, 1978.

Kettle, S. F. A. *Symmetry and Structure*. New York: Wiley, 1985.

Orchin, M. and H. H. Jaffe. *J. Chem. Ed.* 47 (1970): 246, 372, 510.

Salthouse, J. A., and M. J. Ware. *Point Group Character Tables and Related Data*. Cambridge, England: Cambridge University Press, 1972.

Schonland, D. S. *Molecular Structure: An Introduction to Group Theory and Its Uses in Chemistry*. New York: Van Nostrand, 1965.

Vincent, A. *Molecular Symmetry and Group Theory*. New York: Wiley, 1977.

Zeldin, M. *J. Chem. Ed.* 43 (1966): 17.

PROBLEMS

5.1 Locate the C_5 proper rotation axes in the icosahedral $[B_{12}H_{12}]^{2-}$ anion and the C_2 ones in SiH_4.

5.2 Locate the $6C_5$ (72°), $6C_5^2$ (144°), $6C_5^3$ (216°), and $6C_5^4$ (288°) symmetry elements in $[B_{12}H_{12}]^{2-}$.

5.3 Find the $6\sigma_d$ mirror planes in tetrahedral SiH_4 and the $6\sigma_d$ and $3\sigma_h$ mirror planes in octahedral SF_6.

5.4   Identify the $6S_4$ and $8S_6$ improper rotations in the SF_6 molecule. What types of S_n symmetry elements, if any, are there in $[B_{12}H_{12}]^{2-}$?

5.5   Does SiH_4 possess a center of symmetry?

5.6   Draw simple sketches to illustrate the following symmetry operations:
   a. C_7 **d.** i
   b. S_4 **e.** σ_v
   c. σ_h

5.7   List in their simplest forms the symmetry elements generated by an S_8 and an S_{10} axis.

5.8   Show that $S_2 = i$ and $iC_2 = \sigma_h$.

5.9   Draw a sketch indicating the location of the symmetry elements in a typical D_{3h} symmetry molecule.

5.10 Locate the symmetry elements for IF_7 given that it has a pentagonal-bipyramidal structure.

5.11 Justify the Mulliken labels given for the irreducible representations in the C_{2h} character table.

5.12 Assign the point-group symmetries of the following:
   a. S_8 **j.** $[NO_3]^-$
   b. H_2S_2 (Open-book structure)   **k.** SeO_2Cl_2
   c. C_3O_2 **l.** P_4
   d. $[Co(en)_3]^{3+}$ **m.** P_4O_6
   e. PH_3 **n.** $Ru(CO)_5$
   f. S_2Cl_2 **o.** $Ru_3(CO)_{12}$
   g. $[PF_6]^-$ **p.** $cis-[Mn(CO)_4IBr]^-$
   h. B_2H_6 **q.** $axial-Fe(CO)_4L$
   i. SiH_3D

5.13 Show that translations in the z and y directions transform as A_1 and B_2 respectively under C_{2v} point group symmetry.

5.14 What are the point-group symmetries of the following cage compounds? (NOTE: H atoms have been omitted for the sake of clarity.)

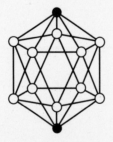

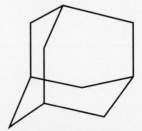

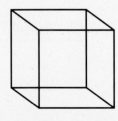

p - Carborane   Adamantane   Cubane

5.15 Give examples of molecules belonging to the following molecular point groups:

a. $D_{\infty h}$ **c.** D_{2d}

b. C_{2v} **d.** D_3

5.16 Under what symmetry species would the s, p, and d metal valence orbitals transform in GeI_4, $[AuCl_4]^-$, $[SnF_5]^-$, and $Mn(CO)_5Cl$?

5.17 If the osmium atom in trigonal-bipyramidal pentacarbonylosmium(0), $Os(CO)_5$, is dsp^3 hybridized, determine which combinations of irreducible representations would be permitted on symmetry grounds.

5.18 Use the VSEPR theory to predict the structures of the following species and then determine their point-group symmetries:

a. BrF_5 **g.** $[ICl_4]^-$

b. $[IBrCl]^-$ **h.** $TeCl_4$

c. ClF_3 **i.** $XeOF_4$

d. KrF_2 **j.** IF_7

e. $[H_3O]^+$ **k.** $PSCl_3$

f. $[SiF_6]^{2-}$ **l.** BeF_2

5.19 Write the point-group symmetry for each of the following species:

a. $[TeF_6]^-$, CTe_2, $SbCl_5$, and AtF_7, given that they have zero dipole moments.

b. $[SnClBr]^-$, SiF_2, and $Ru(CO)_3(PPh_3)_2$, given that they have measurable dipole moments.

5.20 Prove that sp^2 hybridization of the σ-bonding orbitals is symmetry-allowed for trigonal-planar molecules such as BF_3. Are there any other symmetry-allowed hybridizations possible for this geometry?

5.21 What are the symmetry-allowed hybridizations for square-pyramidal $Cr(CO)_5$?

5.22 Prove the following for octahedral SF_6:

a. The characters for the reducible representation (Γ_{red}) for the σ-bonding AOs are

O_h	E	$8C_3$	$6C_2^*$	$6C_4$	$3C_2'$	i	$6S_4$	$8S_6$	$3\sigma_h$	$6\sigma_d$
Γ_{red}	6	0	0	2	2	0	0	0	4	2

* These axes are located between the bond axes.

b. The only hybridization allowed on symmetry grounds is sp^3d^2.

c. The symmetry species associated with the σ-bonding MOs are $a_{1g} + e_g + t_{1u}$.

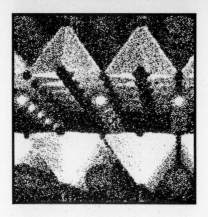

6

Infrared and Raman Spectroscopy

Infrared (IR) and Raman spectroscopy provide important information about the energies of molecular vibrations and hence the structures of molecules. Both techniques are extensively used by inorganic chemists because all kinds of samples (gases, pure liquids, solutions, and solids) can be easily examined over a wide range of temperatures and pressures. In this chapter, we present a brief introduction to the theoretical basis of vibrational spectroscopy and illustrate the role that molecular symmetry and group theory play in the IR and Raman spectra observed. Some applications of the two vibrational spectroscopic methods in inorganic chemistry are also described.

6.1 Theory of Infrared and Raman Spectroscopy

Diatomic molecules afford the best starting point for considering the vibrations of polyatomic molecules because these vibrations are often associated almost exclusively with the vibrations of pairs of bonded atoms, for example, the stretching motions of C—H, Pt—Cl, and C=O groups. An excellent mechanical model of a diatomic molecule consists of two balls of masses m_1 and m_2 joined by a weightless spring that has a resistance to stretching given by a force constant k.

Stretching the spring slightly and then releasing it causes the spring to vibrate with a frequency ν. The oscillations of the spring are simple harmonic, and the motion is described by the mathematical relationship given in Equation 6.1:

$$\nu = \frac{1}{2\pi}\sqrt{\frac{k}{\mu}} \tag{6.1}$$

where frequency ν is in cycles per second (s^{-1}), the force constant k is in Newtons per meter ($N\ m^{-1}$), and the reduced mass μ is in kilograms (kg). Note that μ is defined in Equations 6.2 and 6.3:

$$\frac{1}{\mu} = \frac{1}{m_1} + \frac{1}{m_2} \tag{6.2}$$

and so,

$$\mu = \frac{m_1 m_2}{m_1 + m_2} \tag{6.3}$$

Today, Equation 6.4 is used instead of Equation 6.1 because it takes into account that vibrational spectra are normally measured in wave numbers ($\bar{\nu}$, cm^{-1}), where $\bar{\nu} = 1/\lambda = \nu/c$. The velocity of light (c) is in m s^{-1}:

$$\bar{\nu} = \frac{1}{2\pi c}\sqrt{\frac{k}{\mu}} \tag{6.4}$$

Unfortunately, Equations 6.1 and 6.4 have led to considerable confusion throughout the scientific literature because the symbol ν is used interchangeably for both frequency and wave number in many books and journals. Although the symbol for wave number has been internationally accepted, it has yet to gain widespread approval.

The transition energy (E) between the vibrational ground state $(v = 0)$ of a molecule and its first excited vibrational level $(v = 1)$ is quantized $(E = hv = hc\bar{v})$, and the vibrational transition involved is referred to as a *fundamental*.

The utility of Equation 6.4 can be illustrated by determining the force constant for a typical diatomic species. For example, the wave number of the fundamental vibration of gaseous HF is 3962 cm^{-1}. Rearrangement of Equation 6.4 in this case leads to Equation 6.5:

$$k_{HF} = 4\pi^2 c^2 (\mu_{HF})(\bar{v}_{HF})^2 \tag{6.5}$$

where $\pi = 3.1416$, $c = 2.9979 \times 10^8$ m s^{-1}, $\bar{v} = 3962$ cm^{-1}, and

$$
\begin{aligned}
\mu_{HF} &= \frac{m_H m_F}{m_H + m_F} \\
&= \frac{(1.0079 \times 18.998)}{(1.0079 + 18.998)} \text{ g mol}^{-1} \times \frac{1 \times 10^{-3} \text{ kg g}^{-1}}{6.023 \times 10^{23} \text{ mol}^{-1}} \\
&= 1.5891 \times 10^{-27} \text{ kg}
\end{aligned}
$$

Incorporating these values into Equation 6.5, we obtain:

$$
\begin{aligned}
k_{HF} &= 4(3.1416)^2 (2.9979 \times 10^8 \text{ m s}^{-1})^2 (1.5891 \times 10^{-27} \text{ kg}) \\
&\quad \times (3962 \text{ cm}^{-1} \times 100 \text{ cm m}^{-1})^2 \\
&= 885 \text{ N m}^{-1} \quad [\text{where 1 Newton (N)} = 1 \text{ kg m s}^{-2}]
\end{aligned}
$$

Table 6.1 shows examples of some typical force constants. Notice that for the hydrogen halides (HX; X = F, Cl, Br, I) and the halogens (X$_2$), there is a steady decrease in both the position of the fundamental vibration and the calculated force constant. The reduced mass increases with increasing atomic weight of X, and this in part leads to a decrease in the vibrational wave number. However, the force constant is also crucial in determining the actual position of a band in the vibrational spectrum of a molecule. Notice also that when CO is coordinated to transition metals (in metal carbonyls), there is a definite lowering in both the wave number of the vibrational mode and the force constant by comparison with the data given for the free CO molecule. Similar effects have been found when NO, alkenes, and alkynes are coordinated to transition metals. These significant changes are now taken as direct evidence of π backbonding of $d\pi$ electrons from the central metals to the π^* orbitals on the ligands (see Chapter 22 for discussions of the bonding in various organometallic and related complexes).

The simple harmonic oscillator approach is, of course, only an approximation. Quantum mechanics yields the more rigorous relationship given in Equation 6.6:

$$E_v = \left(v + \frac{1}{2}\right) \frac{h}{2\pi} \sqrt{\frac{k}{\mu}} \qquad (v = 0, 1, 2 \ldots) \tag{6.6}$$

Table 6.1   *Wave Numbers and Associated Force Constants for the Fundamental Vibrations of Some Selected Diatomic Species*

Molecule	$\bar{\nu}\,(\mathrm{cm}^{-1})$	$k(\mathrm{N\ m}^{-1})$[a]
HF	3962	885
HCl	2886	480
HBr	2558	390
HI	2230	290
N_2	2331	2240
CO	2143	1860
C≡C in alkynes	~2150	1500
CO in metal carbonyl complexes	~2000	1700
C=C in alkenes	~1600	1000
NO	1876	1550
C—C in alkanes	~1350	800
F_2	892	450
Cl_2	557	320
Br_2	317	250
I_2	213	170

[a] Two older units for force constant that are still found in the literature are dyn cm^{-1} and mdyn A^{-1}; for example, for HF, the corresponding values are 8.85 dyn cm^{-1} or 8.85 mdyn A^{-1}. Data taken chiefly from J. B. Lambert et. al., *Organic Structural Analysis*. New York: Macmillan, 1976.

This equation is usually abbreviated as

$$E_v = \left(v + \frac{1}{2}\right)h\nu \qquad (v = 0, 1, 2 \ldots) \tag{6.7}$$

where

$$\frac{1}{2\pi}\sqrt{\frac{k}{\mu}} = \nu \tag{6.8}$$

For this quantum mechanical expression, the energy of the fundamental transition ($v = 0 \rightarrow v = 1$) still corresponds to $E = h\nu = hc\bar{\nu}$, so that

$$\bar{\nu} = \frac{1}{2\pi c}\sqrt{\frac{k}{\mu}} \tag{6.9}$$

in agreement with Equation 6.4 for the simple harmonic oscillator.

The chief difference between the quantum mechanical and simple harmonic oscillator models is that when $v = 0$ in Equation 6.8, the molecule still has some energy, the so-called *zero-point energy*:

$$E_0 = \frac{1}{2} h\nu \qquad \qquad (6.10)$$

The vibrational motions of all molecules, even those of diatomics, are not strictly harmonic. In the case of the quantum mechanical model, this situation results in the spacings between the levels ($v = 0 \rightarrow v = 1$, $v = 1 \rightarrow v = 2$, and so on) not being equal (as they are for the simple harmonic oscillator), and the upper levels become closer and closer with increasing values of the vibrational quantum number v.

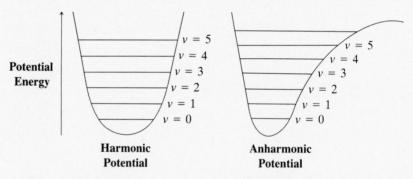

The anharmonicities of molecular vibrations range from about $+1 \ cm^{-1}$ for vibrations involving heavy atoms to over $+100 \ cm^{-1}$ for vibrations associated with H atoms (for example, H_2, $+234 \ cm^{-1}$; HCl, $+104 \ cm^{-1}$). Anharmonicities of up to $+30 \ cm^{-1}$ have been reported for the CO–stretching vibrations in metal carbonyl complexes; after adding the correction factors to the positions of the observed CO fundamental vibrations, the CO force constants are found to increase by about $+0.5 \ N \ m^{-1}$.

Vibrational Spectroscopy

The energies of vibrational motions in molecules are obtained experimentally from investigations of the IR and Raman spectra of the molecules. Although both spectroscopic methods provide similar (and sometimes complementary) information, they owe their origins to two distinctly different physical processes. IR spectroscopy is the result of *absorption* of IR radiation by vibrating molecules, whereas the Raman effect results from the *scattering* of electromagnetic radiation, principally in the visible region, by vibrating molecules. Usually, both techniques are concerned with the fundamental transitions, that is, those involving changes from the ground state ($v = 0$) to the $v = 1$ level. The rigorous *selection rule* for both IR and Raman activity is expressed mathematically as $\Delta v = \pm 1$, where $+1$ means absorption and -1 means emission. However, because many vibrations are quite anharmonic, this strict selection rule breaks down and other weaker features (namely, overtones and combinations) are also detected in the spectra.

Infrared Spectroscopy   Electromagnetic radiation in the IR region (4000–30 cm^{-1}) will be *absorbed* by a molecule when the incident IR energy matches that of one of the vibrational transitions of the molecule. IR radiation is absorbed from the oscillating electric field of the electromagnetic wave provided that the molecule has an oscillating dipole with a frequency that exactly matches that of the applied electric field. Because all molecules (except homonuclear diatomic species such as O_2 and Br_2) have some vibrations that result in changes in dipole moment, every molecule will have a characteristic IR spectrum.

The amount of IR energy absorbed during a vibration and the intensity of the band in the resulting IR spectrum depend on the probability of the vibrational transition, which in turn depends on the change in dipole moment associated with the transition. The significance of this dipole-moment dependence will become more apparent when we discuss the use of group theory in predicting IR activities of molecular vibrations later in this chapter.

Acetone $(CH_3)_2C{=}O$ is a solvent which is widely used in synthetic inorganic chemistry. Figure 6.1 shows the IR spectrum of a thin film of this liquid at room temperature. It is clear that acetone absorbs IR radiation at a variety of different energies because there are numerous peaks in the spectrum. Most importantly, these different IR–absorption peaks can be associated with specific vibrational motions of the acetone molecule, such as CH, CO, and CC stretching [$\nu(CH)$, $\nu(CO)$, $\nu(CC)$] and CCH bending [$\delta(CCH)$]. The motions involved in CO stretching and CCH bending are illustrated below; the H atoms have been omitted for the sake of clarity. The relative lengths of the arrows are intended to indicate approximately the relative displacements of the individual atoms during the vibration. Note that the center of gravity of the molecule must not shift during the vibrations because we are concerned with the vibrations and not the translations and rotations of the molecule. Drawings such as these illustrate what are known as the *normal modes of vibration*.

CO Stretching
$\nu(CO)$

CCH Bending
$\delta(CCH)$

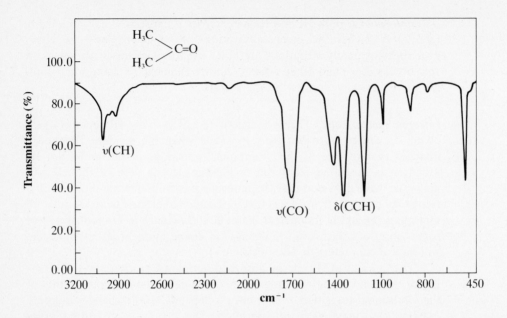

Figure 6.1
Room-temperature
IR absorption
spectrum of a thin
film of liquid
acetone
$(CH_3)_2C{=}O$. The
vibrations chiefly
responsible for the
main IR absorption
bands are indicated
on the spectrum
[ν = stretching;
δ = bending (or
deformation)].

The most important experimental observation about IR spectroscopy
is that many common chemical groups (particularly diatomic groupings,
such as C—H, C=O, C—N, and C—C) always absorb IR radiation at
approximately the same energies, irrespective of the molecules involved.
This observation has led to the concept of *group frequencies*, which is the
basis of the use of IR spectroscopy as a major analytical technique in the
rapid qualitative identification of chemical compounds. The origin of group
frequencies resides in the fact that Equation 6.4 is valid for the vibrations
of the diatomic fragments in polyatomic molecules because the force
constants (k) and the reduced masses (μ) remain virtually unchanged upon
going from molecule to molecule. In the inorganic field, considerable effort
has been expended in establishing the wave-number ranges expected for
IR absorptions of chemical groups often present in inorganic compounds.
Table 6.2 lists some selected examples of these IR absorption ranges.

There are two main types of IR spectrometers available commercially
at the present time: grating and Fourier transform (FT) IR spectrometers.
The former employ an optically ruled grating in order to select a monochromatic region of the IR spectrum for analysis. The heart of a modern
FT–IR instrument is a Michelson interferometer (Figure 6.2a). The optical principle involved is quite old—Michelson was awarded the 1907
Nobel Prize in Physics for his discovery. In a Michelson interferometer,
a beam of IR radiation is split into two components that after passage
through the optical system produce an interference pattern when they
are recombined. The splitting of the incident IR beam is achieved by
allowing it to impinge onto a beam splitter made of an optical substrate

Table 6.2 *Approximate IR Absorption Regions for Selected Chemical Groups Often Present in Inorganic Compounds*[a]

Chemical Group	IR Region (cm^{-1})
$\nu(CN)$ in terminal transition metal cyanides	2200–2000
$\nu(CO)$ in terminal transition metal carbonyls (M—CO)	2150–1850
$\nu(CO)$ in doubly bridged transition metal carbonyls	1850–1700
$\nu(CO)$ in triply bridged transition metal carbonyls	<1700
$\nu(M—H)$ in heavier main-group hydrides	2200–1800
$\nu(M—H)$ in terminal transition metal hydrides	1950–1750
$\nu(NO)$ in linear, terminal transition metal nitrosyls (M—NO)	1900–1800
$\nu(NO)$ in bent, terminal, transition metal nitrosyls	1700–1500
$\nu(NO)$ in doubly bridged transition metal nitrosyls	1550–1450
$\nu(NO)$ in triply bridged transition metal nitrosyls	<1400
$\nu(O—O)$ (superoxo)	1200–1100
$\nu(CS)$ in terminal transition metal thiocarbonyls (M—CS)	1400–1150
$\nu(CS)$ in doubly bridged transition metal thiocarbonyls	1150–1100
$\nu(O—O)$ (peroxo)	920–750
$\delta(MCO)$ in transition metal carbonyls	700–450
$\nu(M—C)$ in transition metal cyanides, carbonyls, thiocarbonyls, etc.	450–300
$\nu(M—X)$ in transition metal halides (X = Cl, Br, I)	400–150
$\nu(M—M)$ in metal–metal bonded compounds	250–100

[a] M = metal; ν = stretching, and δ = bending (or deformation) vibration.

Figure 6.2
(a) Optical layout of the Michelson interferometer in a FT–IR spectrometer. For the normal IR region (4000–300 cm^{-1}), the beam splitter is usually made of KBr. The longer the length of travel of the moving mirror M_2, the higher the resolution; typical resolutions commercially available are 0.005–8 cm^{-1}.
(b) A pair of light waves, in-phase and out-of-phase, leading to constructive and destructive interference, respectively.

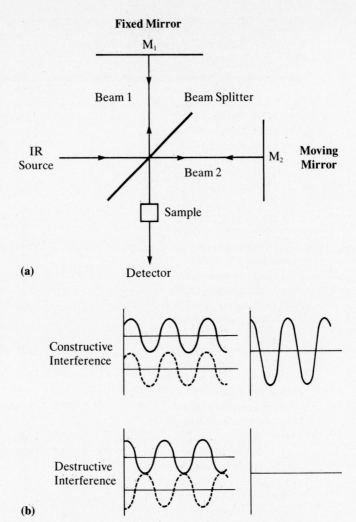

that transmits part of the beam and reflects the remainder, for example, a KBr crystal or thin Mylar polymer film. The beam splitter is positioned at 45° to the incident beam in order to maximize the intensities of the transmitted and relected parts of the incident beam. The transmitted part (beam two) strikes the moving mirror M_2, which is traveling at a constant velocity, and the beam is reflected back toward the beam splitter. The reflected part of the incident radiation (beam one) follows the other pathway and strikes the fixed mirror M_1 and is reflected back toward the beam splitter. When the two reflected beams are recombined at the beam splitter, they interfere constructively or destructively depending on whether they are in-phase or out-of-phase (Figure 6.2b), a situation that depends on the position of the moving mirror M_2 at the time that reflection occurs. The result is an interference pattern, or *interferogram*, which can subsequently be Fourier transformed mathematically on a microcomputer to produce an IR spectrum.

The complete FT–IR spectrum of a compound (from 4000–400 cm^{-1}) can be scanned in a fraction of a second. The FT–IR spectrum is actually obtained by first coadding the individual interferograms from several scans (typically thirty-two) of the background associated with the optical path in the spectrometer in the absence of a sample and then repeating the measurement with the sample positioned in the beam path, usually just before the detector. The resulting, coadded interferograms of the sample and the background are Fourier transformed and then ratioed against one another, thereby producing the desired FT–IR spectrum. For a full description of the theory, construction, and operation of a FT–IR spectrometer, the interested reader is encouraged to refer to the books listed at the end of the chapter.

Despite the rapid development of FT–IR spectroscopy over the past ten years, grating IR instruments still continue to be extremely popular, chiefly because of their significantly lower cost and easier operation and maintenance. Only a small spectral region can be examined at a time using a grating–IR spectrometer, however, and so the complete spectrum must be scanned wave-number increment by wave-number increment; this procedure can take from minutes to several hours depending on the spectral resolution employed. Grating instruments have also been interfaced to microcomputers, thereby making data collection and manipulation much easier than previously.

Raman Spectroscopy

Introduction   Unlike IR spectroscopy, the Raman effect is a *scattering*, rather than an absorption, process. When monochromatic, electromagnetic radiation (usually visible laser light) is allowed to impinge onto a vibrating molecule, most of the incident radiation is scattered *elastically* with the same frequency as the incident beam. This process is known as *Rayleigh scattering*. It is this type of scattering by water molecules in the upper atmosphere that is responsible for our sky appearing blue. A small fraction of the incident radiation, however, is scattered *inelastically*, and the frequency of the scattered radiation can be either more or less than the frequency of the incident radiation. The energy differences concerned are directly related to the energies of the vibrational transitions in the molecule. The Raman effect was first demonstrated experimentally in the late 1920s by Raman, who was awarded the 1930 Nobel Prize in Physics for his discovery.

Rayleigh and Raman scattering are illustrated schematically in Figure 6.3 for a simple homonuclear, diatomic molecule such as N_2. Scattering

Figure 6.3
Qualitative description of Rayleigh and Raman scattering by a homonuclear diatomic molecule that has a vibrational stretching frequency ν_1. The virtual states indicated by the dashed lines are not considered to be proper vibrational states because they have sufficient lifetimes for energy exchange only, if any, to occur.

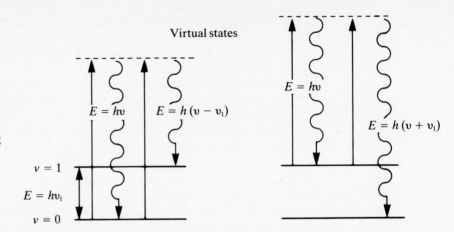

occurs in all directions of course, but for practical reasons, the scattered light is usually analyzed at 90° to the incident laser beam using a very sensitive monochromator that has an extremely low stray-light level.

The energy of the fundamental transition between the ground vibrational level ($\nu = 0$) and the first excited vibrational level ($\nu = 1$) of the diatomic molecule is $h\nu_1$. Rayleigh scattering can occur from both the $\nu = 0$ and $\nu = 1$ levels, and there is no energy exchange. However, the scattering process that originates from the $\nu = 0$ level and ultimately results in a net energy *loss* of $h\nu_1$ (compared to the energy of the Rayleigh line) leads to a Raman spectrum. Another Raman scattering process starts from the $\nu = 1$ level and results in a net energy *gain* of $h\nu_1$ (compared to the Rayleigh line). Originally, Raman spectra were recorded on photographic plates and appeared as a series of lines of varying intensity on the plates. The more intense lines occurred at energies lower than that of the Rayleigh line and became known as *Stokes lines*, whereas the similar but lower-intensity lines detected above the Rayleigh line were called *anti-Stokes lines*. Because most molecules are in the $\nu = 0$ state at ambient temperatures (Boltzmann distribution), Stokes transitions are more probable than are anti-Stokes ones. Consequently, Raman spectra are normally recorded in the direction of decreasing energy (decreasing wave number) with respect to the energy of the laser line used to excite the spectra. Figure 6.4 shows the Stokes and anti-Stokes Raman lines of CCl_4. Note that there are four main peaks observed (ignoring the splitting on the weakest peak centered at ~776 cm^{-1}) on either side of the Rayleigh line. We will see shortly that such a four-band Raman spectrum is typical of a MX_4 species with T_d symmetry.

The IR and Raman (Stokes) spectra of the organometallic complex, $CpMn(CO)_2(CS)$ ($Cp = \eta^5\text{-}C_5H_5$), are compared in Figure 6.5. Notice the much greater variation in peak intensities in the Raman (Figure 6.5b) than

Figure 6.4
Room-temperature
Raman spectrum of
liquid CCl$_4$ showing
both Stokes and
anti-Stokes lines.

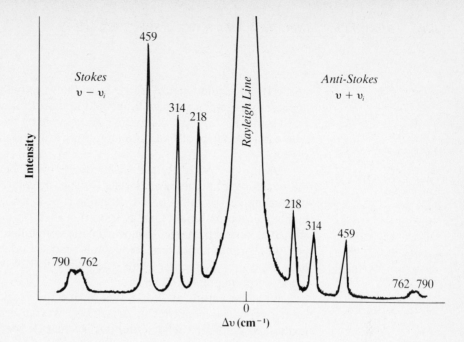

Figure 6.5
IR and Raman
spectra (cm^{-1}) of
CpMn(CO)$_2$(CS).
(**a**) IR spectrum of
vapor at ~100°C.
(**b**) Raman spectrum
of solid (Kr-laser
excitation,
641.7 nm, ~50 mW).
Reproduced with
permission from
G. G. Barna, I. S.
Butler, and K. R.
Plowman, *Can. J.
Chem.* 54 (1979): 10.

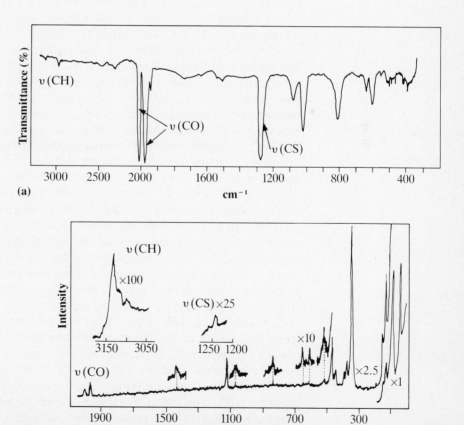

in the IR spectrum (Figure 6.5a). In addition, only the Raman spectrum apparently provides any significant information for low-energy vibrations (<300 cm^{-1}). This region is routinely accessible by the Raman technique, whereas a special far–IR spectrometer is normally required to examine this region.

During the 1930s, Raman spectroscopy was widely used to establish the structures of numerous colorless (or slightly colored) compounds, but the technique fell into disfavor because it was not especially useful for highly colored materials such as transition metal compounds. Even with the development of stable, monochromatic visible-light sources (for example, the Toronto arc), these compounds tended to decompose during the long exposures (sometimes days) necessary to detect the weak Raman peaks on photographic plates.

With the advent of World War II and the blockade on the natural rubber-producing countries in Asia, there was an urgent need for rapid analysis of synthetic rubbers, and this is when IR spectroscopy came into the forefront. IR spectroscopy continued to reign supreme in the structural analysis field until the early 1960s when the pendulum began to swing back toward Raman spectroscopy. The change was a result of the discovery of continuous-wave (CW) lasers. And there is now a wide variety of monochromatic, visible (and UV) CW laser sources commercially available that can be used to excite Raman spectra. The detection of Raman spectra is fairly straightforward now with the ready availability of cooled (-25 °C) photomultiplier tubes and other more rapidly photon-counting detection systems. Nevertheless, IR spectroscopy is expected to continue to dominate the vibrational field for some time to come. There are still two major problems associated with Raman spectroscopy that must be surmounted before the technique can truly become routine: The plasma tubes for the laser sources have limited lifetimes (1000–2000 h) and are prohibitively expensive to replace (for example, $15,000 for a 5-W argon-ion plasma tube); Raman spectroscopy is intrinsically difficult to do with colored samples, and most inorganic compounds are colored.

IR spectroscopy is now such an established analytical method that all university science students will encounter the technique during their courses. Raman spectroscopy, on the other hand, is still comparatively in its infancy, despite the fact that it is probably potentially more useful than IR spectroscopy, especially from the viewpoint of group frequencies and peak-intensity variations. In addition, water is only a weak Raman scatterer, thus a very useful solvent in which to study the Raman spectra of inorganic compounds. IR measurements using aqueous solutions are notoriously difficult because water absorbs so strongly in the IR region. The much longer data-acquisition times required for a normal Raman spectrum is certainly one of the reasons why the technique has not become widespread. However, the optics for both an IR and Raman spectrometer have recently been successfully mounted in the same instrument and both types of measurement can be controlled by the same microcomputer; Raman spectroscopy will undoubtedly become more popular in the not-too-distant future.

Theoretical Background   Raman scattering occurs because of the changes that take place in the *polarizability* (α) of a molecule during a vibration. The polarizability is a measure of the ease with which the electron clouds around the atoms in the molecule can be distorted. Heavier atoms have larger electron clouds and consequently larger individual polarizabilities. When a vibrating molecule interacts with the electric field of an incident beam of electromagnetic radiation (the laser beam), an oscillating dipole is induced (μ_{ind}) according to Equation 6.11:

$$\mu_{ind} = \alpha E \tag{6.11}$$

where E is the electric field of the incident beam. However, the electric field is oscillating, that is, it is time-dependent (Equation 6.12):

$$E = E_0 \cos 2\pi \nu t \tag{6.12}$$

Moreover, the polarizability of a molecule can be described by a Taylor expansion as shown in Equation 6.13:

$$\alpha = \alpha_0 + \left(\frac{d\alpha}{dQ}\right)_0 Q + \cdots \tag{6.13}$$

where Q refers to the *normal coordinate*, that is, the mathematical term for the complete set of bond stretches and angle bends that take place during a particular vibration and $(d\alpha/dQ)_0$ is the change in polarizability during the vibration described by the normal coordinate Q. The 0 subscripts indicate that the two quantities are evaluated at the equilibrium position of the vibration. Substituting the expressions in Equations 6.12 and 6.13 into Equation 6.11, we obtain

$$\mu_{ind} = \left[\alpha_0 + \left(\frac{d\alpha}{dQ}\right)_0 Q\right] E_0 \cos 2\pi \nu t \tag{6.14}$$

but the normal coordinate is also time-dependent (Equation 6.15):

$$Q = A \cos 2\pi \nu_i t \tag{6.15}$$

where A is a constant and ν_i is the frequency of the ith vibration. Therefore, Equation 6.14 now becomes

$$\mu_{ind} = \left[\alpha_0 + \left(\frac{d\alpha}{dQ}\right)_0 A \cos 2\pi \nu_i t\right] E_0 \cos 2\pi \nu t \tag{6.16}$$

From basic trigonometry, we know that

$$2\cos x \cos y = \cos(x + y) + \cos(x - y) \tag{6.17}$$

So, the final equation for Raman scattering is given in Equation 6.18:

$$\mu_{\text{ind}} = \alpha_0 E_0 \cos 2\pi\nu t + \frac{AE_0}{2}\left(\frac{d\alpha}{dQ}\right)_0$$

$$\times [\cos 2\pi(\nu + \nu_i)t + \cos 2\pi(\nu - \nu_i)t] \tag{6.18}$$

It should be pointed out, however, that the equilibrium polarizability (α_0) is actually a tensor because both the induced dipole and the electric field are vector quantities:

$$\begin{vmatrix} \mu_{\text{ind}}(x) \\ \mu_{\text{ind}}(y) \\ \mu_{\text{ind}}(z) \end{vmatrix} = \begin{vmatrix} \alpha_{xx} & \alpha_{xy} & \alpha_{xz} \\ \alpha_{yx} & \alpha_{yy} & \alpha_{yz} \\ \alpha_{zx} & \alpha_{zy} & \alpha_{zz} \end{vmatrix} \begin{vmatrix} E_x \\ E_y \\ E_z \end{vmatrix} \tag{6.19}$$

There are nine components to the polarizability tensor, but usually we only need be concerned with six of them because $\alpha_{xy} = \alpha_{yx}$, $\alpha_{xz} = \alpha_{zx}$, and $\alpha_{yz} = \alpha_{zy}$. The first term in Equation 6.18 describes the Rayleigh scattering, and because all molecules (including H_2) have a polarizability, every molecule exhibits Rayleigh scattering to some extent. The second and third terms in Equation 6.18 describe the conditions for Raman activity. For these derivative terms to be nonzero, at least one of the polarizability components α_{xx}, α_{xy}, and so on must change during the vibration described by the normal coordinate Q.

Raman-Band Intensities The intensity (I) of a Raman band depends on the square of the polarizability derivative because it depends on μ_{ind} according to Equation 6.20:

$$I = \left(\frac{16\pi^4\nu^4}{3c^3}\right)(\mu_{\text{ind}})^2 \tag{6.20}$$

Note that the fourth power of the incident frequency is important. This means that it should, in principle, be easier to detect the Raman spectrum of any compound using a high-frequency laser [for example, argon ion: 488 nm (blue) and 514.5 nm (green)] than with a low-frequency one [for example, krypton ion: 647.1 nm (red)]. However, other factors such as photomultiplier-tube response and thermal decomposition of the compound induced by absorption of energy from the laser beam come into play.

Depolarization Ratio Laser–Raman spectroscopy has an advantage over IR spectroscopy because the laser beam used to excite the spectra is plane-polarized in the vertical direction. This is an important feature

Figure 6.6
Schematic representation of the depolarization ratio measurements performed in a laser Raman experiment.

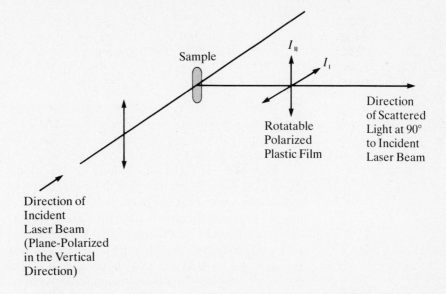

Sample

$I_{\parallel}$

$I_{\perp}$

Rotatable Polarized Plastic Film

Direction of Scattered Light at 90° to Incident Laser Beam

Direction of Incident Laser Beam (Plane-Polarized in the Vertical Direction)

because it means that for liquids, gases, and oriented single crystals (but not for polycrystalline or powdered materials), it is possible to analyze the scattered Raman light both parallel and perpendicular to this incident polarization by means of a polarized plastic sheet.. Such a situation is illustrated schematically in Figure 6.6. The ratio of the intensities of the scattered light in these two orientations is referred to as the *depolarization ratio* (ρ):

$$\rho = \frac{I_{\perp}}{I_{\parallel}} \qquad (6.21)$$

Polarized IR studies can be performed on thin crystals, but it is extremely difficult to cut crystals thin enough to allow sufficient IR radiation to be transmitted through them for such measurements to be made.

A complete treatment of the depolarization ratio is beyond the scope of this book, but the curious are referred to the text by Woodward listed at the end of the chapter for the full mathematics. The final relationship using a laser, plane-polarized in the vertical direction, is given in Equation 6.22:

$$\rho = \frac{I_{\perp}}{I_{\parallel}} = \frac{3\beta^2}{45\gamma^2 + 4\beta^2} \qquad (6.22)$$

where

$$\gamma = \frac{1}{3} \left(\alpha'_{xx} + \alpha'_{yy} + \alpha'_{zz} \right) \qquad (6.23)$$

and

$$\beta^2 = \frac{1}{2} [(\alpha'_{xx} - \alpha'_{yy})^2 + (\alpha'_{yy} - \alpha'_{zz})^2 +$$

$$(\alpha'_{zz} - \alpha'_{xx})^2 + 6(\alpha'_{xy} + \alpha'_{yz} + \alpha'_{zx})] \qquad \textbf{(6.24)}$$

The expressions for γ and β^2 represent the isotropic (diagonal components) and anisotropic (diagonal and off-diagonal components) parts of the polarizability, respectively. Note that α'_{xx} is just a convenient, short way of writing $(d\alpha_{xx}/dQ)_0$.

By looking at the possible limits for Equation 6.22, it is clear that if $\beta^2 = 0$, then $\rho = 0$. When such a situation occurs, the Raman band is said to be *completely* (or *fully*) *polarized*. In group theory (Section 6.2), only totally symmetric vibrations (those for which all characters of the irreducible representation in the appropriate character table are equal to +1) satisfy this condition. For all other (*depolarized*) vibrations, $\gamma^2 = 0$, and so Equation 6.22 will reduce to $\frac{3}{4}$. In general, it is relatively easy to identify the Raman bands due to totally symmetric vibrations by their polarization behavior because they exhibit values of between 0 and 0.75, depending on the magnitudes of β^2 and γ^2. This information provides another vital clue in assigning the vibrational spectra of molecules.

When the incident radiation is not plane-polarized, a similar analysis of the radiation scattered at 90° can be performed using a polarized plastic film and Equation 6.25, which is closely related to Equation 6.22, applies. The only change is that the $I_\perp/I_\parallel$ ratio for depolarized bands now becomes $\frac{6}{7}$ (0.857) rather than $\frac{3}{4}$ (0.75):

$$\rho = \frac{I_\perp}{I_\parallel} = \frac{6\beta^2}{45\gamma^2 + 7\beta^2} \qquad \textbf{(6.25)}$$

Liquid chloroform ($CHCl_3$) provides an excellent example of a recent Raman-polarization study using unpolarized incident radiation (Table 6.3). There are six Raman peaks observed, three of which are clearly polarized. The other three peaks give values very close to the expected 0.857. These observations are completely in accord with those expected for a five-atom, C_{3v}-symmetry molecule.

Normal-Coordinate Calculations Extension of the quantum mechanical treatment to polyatomic molecules leads eventually to Equation 6.26 for the energy of the vibrational modes:

$$E(v_1, v_2, \dots) = \sum_{i=1}^{3N-6} \left(v_i + \frac{1}{2}\right)h\nu_i \qquad (v_1, v_2, \dots = 0, 1, \dots) \quad \textbf{(6.26)}$$

The mathematical analysis of the vibrational motion of polyatomic molecules has been of interest to inorganic chemists for some time. And,

Table 6.3 *Raman Data for Liquid CHCl$_3$[a]*

$\bar{\nu}$ (cm^{-1})	ρ
3020	0.284 pol
1219	0.862 depol
761	0.841 depol
670	0.022 pol
368	0.160 pol
263	0.868 depol

[a] From D. Steele et al. *J. Raman Spectrosc.* 18 (1987): 373.

just as for diatomic molecules, it is possible to derive an expression relating the masses of the atoms in a polyatomic molecule and the various force constants involved (Equation 6.27):

$$| \, GF - E\lambda \, | = 0 \qquad\qquad (6.27)$$

where G is a matrix of the atomic masses, F is a force constant matrix, E is an identity matrix (necessary for the matrix algebra), and $\lambda = 4\pi^2 c^2 \bar{\nu}^2$. Solving Equation 6.27 exactly is difficult because there are usually insufficient vibrational wave numbers observed for the myriad of force constants necessary. A partial solution to this problem can be achieved by studying the vibrational spectra of isotopically labeled molecules. This procedure helps greatly in making detailed vibrational assignments for the normal modes.

In a now classic analysis by Jones and co-workers (1969), an especially thorough analysis of the IR and Raman spectra of $Cr(CO)_6$, $Cr(^{13}CO)_6$, and $Cr(C^{18}O)_6$ was made. The underlying assumption to these studies is that the force constants for the isotopic species are identical to those of the parent molecule. This is a good assumption, even in the case of H and D. The approach to these *normal-coordinate analyses* is to match (by means of an iterative least-squares computer program) the observed vibrational wave numbers to the values calculated while varying some of the force constants. A fortunate benefit of such normal-coordinate calculations is that it is possible to transfer certain force constants between structurally similar molecules and so reduce the enormity of the computer calculations. For example, the vibrational spectra of $Cr(CO)_5(CS)$, $Cr(^{13}CO)_5(CS)$, and $Cr(CO)_5(^{13}CS)$ have been assigned successfully by a normal-coordinate treatment using several force constants transferred from $Cr(CO)_6$. The final agreement between the observed and calculated band positions (after taking into account the anharmonicities in the CO–stretching modes) is excellent (± 1–2 cm^{-1}).

6.2   *Applying Group Theory to Molecular Vibrations*

Most of the vibrational work on inorganic compounds has been concerned with the determination of the molecular structures of the compounds by comparing their observed IR and Raman spectra with those predicted on the basis of group theoretical principles for different possible structures. We now show you how such spectral predictions can be easily made.

One of the earliest coordination compounds known is the four-coordinate, square-planar platinum(II) complex $Pt(NH_3)_2Cl_2$. This compound can exist in two *geometrical isomeric forms:* cis (two Cl groups at 90° to each other) and trans (two Cl groups 180° to each other).

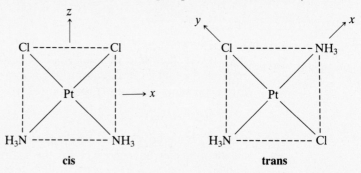

The cis isomer has been licensed as a chemotherapeutic drug (cisplatin) for use in the treatment of testicular and other cancers. The trans isomer, however, is therapeutically inactive. This is a particularly graphic example of how important it is to distinguish between the isomers, and one way this can be accomplished is by vibrational spectroscopy. Both isomers have two sets of vibrations that involve principally Pt—Cl stretching, namely, in-phase and out-of-phase Pt—Cl–stretching motions (Figure 6.7).

From Table 6.2, the Pt—Cl–stretching vibrations would be expected in the IR region at about 300 cm^{-1}. The point-group symmetries of the two isomers are different and, as it turns out, so are the vibrational selection rules.

Treating the NH_3 groups as point masses, the cis isomer has C_{2v} symmetry, whereas the trans isomer has D_{2h} symmetry. To determine the symmetries of the Pt—Cl–stretching modes in each case, we need to see how many Pt—Cl bonds are *unshifted* by the various symmetry operations for the two point groups. This generates a reducible representation that we can decompose into its component irreducible representations using the same group theoretical reduction formula that we gave earlier (Equation 5.2). The basis for this mechanical treatment is explained in all books on group theory but is beyond the scope of the present book. In reality, we are using what are known as *internal coordinates* to describe the vibrations of the Pt—Cl groups. Internal coordinates are defined as the changes in

Figure 6.7
In-phase and
out-of-phase
Pt—Cl–stretching
modes for cis- and
trans-Pt$(NH_3)_2Cl_2$.

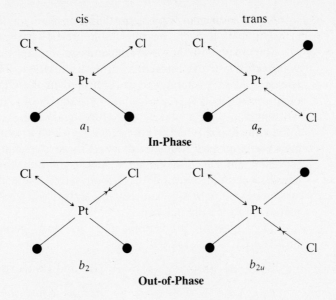

bond lengths (Δr) or bond angles ($\Delta \theta$) that occur during molecular vibrations. The symmetry properties of Δr and the associated bonds are identical, and it is usually easier to consider the effect of the various symmetry operations on the bonds themselves rather than on Δr.

C_{2v}	E	$C_2(z)$	$\sigma_v(xz)$	$\sigma'_v(yz)$
$\Gamma_{\text{Pt—Cl}}$	2	0	0	2

$$\Gamma_{\text{Pt—Cl}}(\text{cis}) = a_1 + b_2$$

D_{2h}	E	$C_2(z)$	$C_2(y)$	$C_2(x)$	i	$\sigma(xy)$	$\sigma(xz)$	$\sigma(yz)$
$\Gamma_{\text{Pt—Cl}}$	2	0	2	0	0	2	0	2

$$\Gamma_{\text{Pt—Cl}}(\text{trans}) = a_g + b_{2u}$$

(NOTE: Group theoretical symbols for the symmetries of vibrational modes are written as lowercase letters.)

This group theoretical analysis shows that there are indeed two Pt—Cl–stretching modes expected for each isomer, but they differ in spectral activity. Recall that there must be a change in the *dipole moment* of the molecule during a vibration for the vibration to be IR–active. Dipole moments are vector quantities, and a dipole moment vector in, say, the z direction exhibits the same symmetry-transformation properties as does a translation in the z direction (z or T_z in the character table). In other words, all molecular vibrations that transform in the same way as the translations in a given character table are IR–active.

Raman activity, however, depends on whether there is a change in the *polarizability* of the molecule during a vibration. The polarizability is a tensor, and its components are indicated in the character tables by the binary products (x^2, xy, and so on); in some books, these components are written as α_{xx}, α_{xy}, and so on. Once again, we have a straightforward (albeit mechanical) method of predicting which irreducible representations are associated with Raman activity. If there is a binary product located on the same line as the symmetry species of the vibrational mode being considered, then the mode is Raman active. Note that some vibrational modes can be *both* IR– and Raman-active, whereas others may be completely vibrationally inactive.

The spectral predictions for the two geometrical isomers of $Pt(NH_3)_2Cl_2$ are as follows:

cis	a_1 $(z; x^2, y^2, z^2)$	→ IR– and Raman-active
C_{2v}	a_2 $(-; xy)$	→ Raman-active *only*
	b_1 $(x; xz)$	→ IR– and Raman-active
	b_2 $(y; yz)$	→ IR– and Raman-active
trans	a_g $(-; x^2, y^2, z^2)$	→ Raman-active *only*
D_{2h}	b_{1g} $(-; xy)$	→ Raman-active *only*
	b_{2g} $(-; xz)$	→ Raman-active *only*
	b_{3g} $(-; yz)$	→ Raman-active *only*
	a_u $(-; -)$	→ inactive
	b_{1u} $(z; -)$	→ IR–active *only*
	b_{2u} $(y; -)$	→ IR–active *only*
	b_{3u} $(x; -)$	→ IR–active *only*

It is now clear why different spectroscopic activities are expected for the Pt—Cl–stretching modes of the two Pt(II) isomers. For the cis isomer, the ν(Pt—Cl) modes transform as $a_1 + b_2$, and both should be IR– and Raman-active, leading to two peaks in both spectra. On the other hand, for the trans isomer the ν(Pt—Cl) modes should transform as $a_g + b_{2u}$, indicating that there should be only one peak in the Raman and one peak in the IR. Table 6.4 shows vibrational data for the IR and Raman spectra of the two isomers, which are in accord with these predictions.

Notice that none of the irreducible representations for the D_{2h} point group leads to *both* IR and Raman activity, whereas the opposite is true for the C_{2v} point group. The D_{2h} point group and molecules with this sym-

Table 6.4 *Platinum–Chlorine–Stretching Modes (cm^{-1})*
for cis- *and* trans-[*$Pt(NH_3)_2Cl_2$*][a]

Compound	IR	Raman	Assignment
cis-[$Pt(NH_3)_2Cl_2$]	330	[b]	a_1
	323	[b]	b_2
trans-[$Pt(NH_3)_2Cl_2$]	365		b_{2u}
		318	a_g

[a] Data taken from K. Nakamoto. *Infrared Spectra of Inorganic and Coordination Compounds*, 4th ed. New York: Wiley, 1986.
[b] No measurements yet reported.

metry are said to be *centrosymmetric*, and the *rule of mutual exclusion* applies. This rules states that in centrosymmetric molecules, IR–active vibrational modes are Raman-inactive and vice versa. Moreover, the IR peaks of a centrosymmetric molecule should appear at different energies than those in the Raman spectrum. The $v(Pt—Cl)$ modes for the trans (D_{2h} symmetry) isomer in Table 6.5 are in agreement with the rule of mutual exclusion. In general, g modes are Raman-active, whereas u modes are IR–active. Occasionally, there are certain irreducible representations for which vibrational modes should not exhibit any vibrational spectroscopic activity, for example, any D_{2h} symmetry molecule that has a_u modes.

Molecular structures can often be predicted from observed vibrational spectra. An excellent example to illustrate the procedure involved is pentacarbonyliron(0), $Fe(CO)_5$. The structure of this yellow liquid was first established by an analysis of its IR and Raman spectra in the CO–stretching region. The two most probable structures for such a five-coordinate compound are square-pyramidal (C_{4v} symmetry) and trigonal-bipyramidal (D_{3h}) (Figure 6.8). The $v(CO)$ bands observed experimentally for the compound are given in Table 6.5.

The reducible representations shown below for the CO–stretching

Table 6.5 *Observed Vibrational Spectra (cm^{-1})*
in the $v(CO)$ Region
for Liquid $Fe(CO)_5$[a]

IR	Raman	Assignment
	2116 pol	a_1'
	2030 pol	a_1'
2002		a_2''
1979	1989 depol	e''

[a] From M. Bigorgne. *J. Organometal. Chem.* 24 (1970): 211.

Figure 6.8
Two possible structures for $Fe(CO)_5$. Note that even if the Fe atom in the square-pyramidal structure is not in the same plane as the four equatorial CO groups, the molecular symmetry is still C_{4v}.

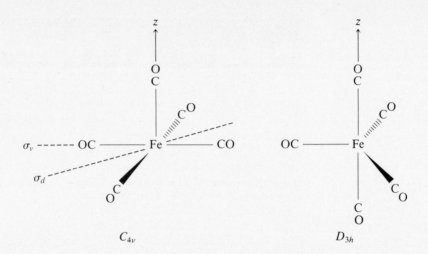

modes of the two structures were generated by determining how many CO groups [remember, strictly, changes in CO-bond lengths $\Delta(CO)$] are unchanged by applying the various symmetry operations. These reducible representations are then decomposed using the reduction formula (Equation 5.2).

C_{4v}	E	$2C_4(z)$	C_{2z}	$2\sigma_v$	$2\sigma_d$
Γ_{CO}	5	1	1	3	1

$$\Gamma_{CO} = 2a_1 \text{ (IR–Raman)} + b_1 \text{ (Raman)} + e \text{ (IR–Raman)}$$

and

D_{3h}	E	$2C_3(z)$	$3C_2$	σ_h	$2S_3$	$3\sigma_v$
Γ_{CO}	5	2	1	3	0	3

$$\Gamma_{CO} = 2a_1' \text{ (Raman)} + e' \text{ (IR–Raman)} + a_2'' \text{ (IR)}$$

The C_{4v} structure should exhibit three $\nu(CO)$ peaks in the IR and four in the Raman; the predictions for the D_{3h} structure are two $\nu(CO)$ peaks in the IR and three in the Raman. The observed data given in Table 6.5 are in excellent agreement with the trigonal-bipyramidal (D_{3h} symmetry) structure for liquid $Fe(CO)_5$. On the basis of an electron-diffraction study, this structure has been shown to persist in the vapor as well.

It is also possible to do a group theoretical calculation for the *complete* IR–Raman spectra of a polyatomic molecule containing N atoms and not just for selected pieces of it. It must be emphasized that the method we describe is applicable only to *nonlinear molecules*. A different procedure must be employed for linear molecules; see the article by Schäfer and

Cyvin (1971) listed at the end of the chapter for a discussion of the procedure in this case.

Three Cartesian coordinate vectors are assigned to each of the N atoms in the polyatomic molecule so that the resulting $3N-6$ Cartesian coordinate vectors form the basis of a reducible representation for the $3N-6$ degrees of vibrational freedom (*normal modes of vibration*). The six *nongenuine modes* are the three translations and three rotations. The full group theoretical analysis involves determining how these Cartesian coordinate vectors behave with respect to each of the symmetry operations for the point-group symmetry in question. This approach can be quite long; see any of the books on group theory. Fortunately, however, a shortcut has been established. We simply have to consider the number of atoms in the molecule that are *unshifted* by each symmetry operation of the point group. Then, the required reducible representation is generated by multiplying these numbers of unshifted atoms by the appropriate character contributions per unshifted atom for each of the various symmetry operations involved (Table 6.6). The reducible representation is broken down into irreducible representations using Equation 5.2 in the usual way. At this point, the complete set of irreducible representations that spans the vibrational, translational, and rotational degrees of freedom has been generated. To obtain the vibrational modes, the irreducible representations corresponding to the three translations and three rotations must be subtracted; these six nongenuine species can readily be identified from the right-hand side of the character tables.

Boron trifluoride (BF_3) is a trigonal-planar, D_{3h} symmetry molecule for which $3(4) - 6 = 6$ vibrational modes should be observed. The group theory calculation to determine the symmetries of these six modes and their expected spectral activity follows:

D_{3h}	E	$2C_3(z)$	$3C_2$	σ_h	$2S_3$	$3\sigma_v$
Number unshifted atoms	4	1	2	4	1	2
Character contribution per unshifted atom	3	0	−1	1	−2	1

Multiply

$\Gamma^{D_{3h}}_{\text{tot}}$	12	0	−2	4	−2	2

$$\Gamma^{D_{3h}}_{\text{tot}} = a'_1 + a'_2 + 3e' + 2a''_2 + e''$$

$$\Gamma^{D_{3h}}_{\text{trans}} + \Gamma^{D_{3h}}_{\text{rot}} = \qquad a'_2 + \ e' + \ a''_2 + e''$$

Subtract

$$\Gamma^{D_{3h}}_{\text{vib}} = a'_1 \qquad + 2e' + a''_2$$

Table 6.6 *Character Contributions per Unshifted Atom for Selected Symmetry Operations*

Symmetry Operation	Character Contribution per Unshifted Atom[a,b]
E	3
i	-3
σ	1
C_n	$2\cos(2\pi/n) + 1$
C_2	-1
C_3	0
C_4	1
C_6	2
S_n	$2\cos(2\pi/n) - 1$
S_3	-2
S_4	-1
S_6	0

[a] These numbers are derived from considerations of the effect of the various operations on the Cartesian coordinates *x*, *y*, *and z*; for example, in matrix notation, we have

$$E: \quad \begin{vmatrix} 1 & 0 & 0 \\ 0 & 1 & 0 \\ 0 & 0 & 1 \end{vmatrix} \begin{vmatrix} x \\ y \\ z \end{vmatrix} = \begin{vmatrix} x \\ y \\ z \end{vmatrix} \qquad \chi(E) = 3 \text{ (the trace of the matrix)}$$

$$\sigma(xy): \quad \begin{vmatrix} 1 & 0 & 0 \\ 0 & 1 & 0 \\ 0 & 0 & -1 \end{vmatrix} \begin{vmatrix} x \\ y \\ z \end{vmatrix} = \begin{vmatrix} x \\ y \\ z \end{vmatrix} \qquad \chi[\sigma(xy)] = 1$$

$$C_3(z): \quad \begin{vmatrix} \cos\dfrac{2\pi}{3} & \sin\dfrac{2\pi}{3} & 0 \\ -\sin\dfrac{2\pi}{3} & \cos\dfrac{2\pi}{3} & 0 \\ 0 & 0 & 1 \end{vmatrix} \begin{vmatrix} x_1 \\ y_1 \\ z_1 \end{vmatrix} = \begin{vmatrix} x_2 \\ y_2 \\ z_2 \end{vmatrix} \qquad \begin{aligned} &\chi[C_3(z)] = 0 \\ &\left(\cos\dfrac{2\pi}{3} = -\dfrac{1}{2}\right) \end{aligned}$$

[b] The general equations for rotation about the *z* axis by θ degrees are

$$x_2 = (\cos\theta)x_1 + (\sin\theta)y_1 + (0)z_1$$

$$y_2 = (-\sin\theta)x_1 + (\cos\theta)y_1 + (0)z_1$$

$$z_2 = (0)x_1 + (0)y_1 + (1)z_1$$

that is,

$$\begin{vmatrix} \cos\theta & \sin\theta & 0 \\ -\sin\theta & \cos\theta & 0 \\ 0 & 0 & 1 \end{vmatrix} \begin{vmatrix} x_1 \\ y_1 \\ z_1 \end{vmatrix} = \begin{vmatrix} x_2 \\ y_2 \\ z_2 \end{vmatrix} \qquad \therefore \ x\left[C\left(\frac{2\pi}{n}\right)\right] = 2\cos\frac{2\pi}{n} + 1$$

The selection rules for these six vibrational modes (the e' modes are doubly degenerate) are a_1' (Raman), e' (IR–Raman), and a_2'' (IR). This means that there should be three peaks in the IR and three in the Raman if the D_{3h} selection rules are obeyed. The observed data given in Table 6.7 for $^{10}BF_3$ and $^{11}BF_3$ are in excellent agreement with the predictions. Note that the positions of the e' modes are essentially coincident in the IR and Raman spectra, as expected for a noncentrosymmetric system. Figure 6.9 shows the six normal modes of vibration.

We are now ready to discuss the vibrational spectra mentioned earlier in this chapter for CCl_4 and $CHCl_3$. Carbon tetrachloride exhibits vibrational spectra characteristic of a T_d symmetry molecule. The group theoretical predictions are

$$\Gamma_{\text{vib}}^{T_d} = a_1(\text{Raman}) + e(\text{Raman}) + 2t_2(\text{IR–Raman})$$

The observed Raman data are $\sim$776(t_2), 459(a_1, pol), 314(t_2), and 218(e) cm^{-1} in agreement with the predicted spectrum. Moreover, the a_1 mode is polarized, as expected. On closer examination, the 776 cm^{-1} band is actually a weak doublet with component peaks at 790 and 762 cm^{-1} (see Figure 6.4) resulting from the close proximity of the t_2 fundamental and a combination mode of the same symmetry. The inherently weaker combination mode gains sufficient intensity from the t_2 fundamental to be observable; this phenomenon is known as *Fermi resonance*. The group theoretical predictions for the four fundamental vibrations of CCl_4 are not violated by the presence of this Fermi doublet.

Under high-resolution conditions (<1 cm^{-1}), the a_1 mode in the Raman spectrum of CCl_4 is split into several components (Figure 6.10). This situation arises because of the relative natural abundances of the ^{35}Cl

Table 6.7 *Vibrational Spectra (cm^{-1}) of $^{10}BF_3$ and $^{11}BF_3$*[a, b]

$^{10}BF_3$	$^{11}BF_3$	Assignment
1505	1454	ν_3, e' (IR–Raman; in-plane bend)
888	888	ν_1, a_1' (R; in-phase B—F–stretching)
791	691	ν_2, a_2'' (IR; out-of-plane bend)
482	480	ν_4, e' (IR–Raman; in-plane bend)

[a] Data from W. A. Guillory. *Introduction to Molecular Structure and Spectroscopy*. Boston: Allyn & Bacon, 1977, 219.
[b] The ν_i nomenclature given follows that now accepted by most vibrational spectroscopists: nondegenerate irreducible representations before degenerate ones and then in order of decreasing wave number within each class of irreducible representation.

Figure 6.9
Approximate description of the normal modes of vibration for BF_3. Notice that ν_1 should be independent of isotopic substitution at the boron, as is observed. ν_{3a} and ν_{3b}, and ν_{4a} and ν_{4b} are the pairs of components of the doubly-degenerate e' modes ν_3 and ν_4, respectively. Adapted from G. Herzberg, *Infrared and Raman Spectra of Polyatomic Molecules*. New York: Van Nostrand, 1945.

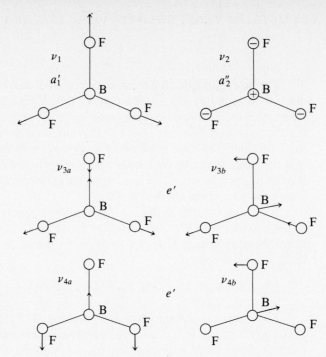

Figure 6.10
High-resolution Raman spectrum of liquid CCl_4 (488.0-nm argon-ion laser excitation), $1\ cm^{-1}$ resolution.

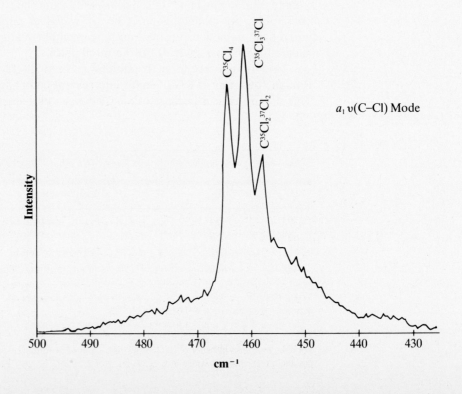

Table 6.8   *Infrared*[a] *and Raman*[b] *Data for Liquid CHCl₃*

IR (cm⁻¹)	Raman (cm⁻¹)	Vibrational Assignment
3033	3020 pol	ν_1 (a_1)
1205	1219 depol	ν_4 (e)
760	761 depol	ν_5 (e)
667	670 pol	ν_2 (a_1)
364	368 pol	ν_3 (a_1)
260	263 depol	ν_6 (e)

[a] From G. Herzberg. *Molecular Spectra and Molecular Structure*, Vol. II. New York: Van Nostrand, 1945, 316.
[b] From D. Steele et al. *J. Raman Spectrosc.* 18 (1987): 373.

and ^{37}Cl isotopes (3/1). There are five types of CCl_4 molecule present: $C^{35}Cl_4$ (31.6%), $C^{35}Cl_3{}^{37}Cl$ (42.2%), $C^{35}Cl_2{}^{37}Cl_2$ (21.1%), $C^{35}Cl^{37}Cl_3$ (4.7%), and $C^{37}Cl_4$ (0.4%). The relative percentage abundances of the molecules are shown in parentheses. The peaks shown in Figure 6.10 have been assigned accordingly.

In the case of $CHCl_3$ (C_{3v} symmetry), $\Gamma_{vib}^{C_{3v}}$ reduces as follows:

$$\Gamma_{vib}^{C_{3v}} = 3a_1(\text{IR–Raman}) + 3e(\text{IR–Raman})$$

The six peaks expected are observed in both the IR and Raman spectra (Table 6.8). Furthermore, three of the Raman peaks are fully polarized, and the other three are depolarized, as discussed earlier. Similar spectral results would be predicted for other closely related molecules such as $SiHCl_3$ and $GeHBr_3$.

6.3   Some Applications of Infrared and Raman Spectroscopy in Inorganic Chemistry

IR and Raman spectroscopy have been used to establish the type of the bonding for certain simple, tetrahedral inorganic anions, for example, $[SO_4]^{2-}$, $[ClO_4]^{-}$, and $[PO_4]^{3-}$. An excellent demonstration of this is for the $[SO_4]^{2-}$ ion in a series of cobalt(III) complexes. The IR spectra of three complexes in the S–O-stretching region (1200–900 cm⁻¹) are shown schematically in Figure 6.11. The $[SO_4]^{2-}$ ion can exist as a free ion (T_d symmetry), coordinated in a monodentate fashion through one oxygen atom (C_{3v} symmetry) or coordinated bidentately through two oxygen

Figure 6.11
Infrared spectra
(cm^{-1}) observed
in the $\nu(S-O)$
region for three
sulfato compounds:
(a)$[SO_4]^{2-}$,
(b)$[Co(NH_3)_5-SO_4]Br$, and
(c)$[NH_3)_4-Co(NH_2)(SO_4)-Co(NH_3)_4](NO_3)_3$.
Adapted with
permission from
K. Nakamoto.
*Infrared and Raman
Spectra of
Coordination
Compounds*, 4th ed.
New York: Wiley,
1986.

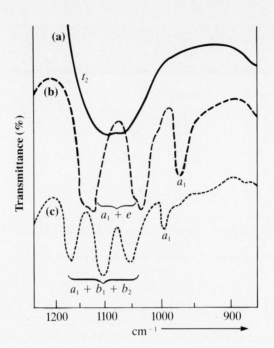

atoms—either to one metal or bridging between two metals (C_{2v} symmetry). Figure 6.12 shows the group theoretical predictions for $\nu(S-O)$ for these various possibilities.

Accordingly, only one IR–active $\nu(S-O)$ peak (t_2) is expected for the free $[SO_4]^{2-}$ ion. However, when one of the oxygen atoms is coordinated, as in the pentaamminesulfatocobalt(III) cation, a different situation exists. The molecular symmetry is reduced from T_d to C_{3v} such that the formally IR–inactive a_1 mode is now allowed and the t_2 mode splits into two IR–active components ($a_1 + e$). This means that there should be three $\nu(S-O)$ peaks observed for this type of coordination. If the $[SO_4]^{2-}$ ion is coordinated to two metal atoms, either in a chelated fashion or bridging two metal atoms, the symmetry is further reduced to C_{2v}. At this stage, the e mode should split into two IR–Raman–active components ($b_1 + b_2$), so that there should be four $\nu(S-O)$ peaks observed if either of these types of bidentate coordination exists.

Figure 6.12
Spectral activities
predicted for the
S—O-stretching
modes of $[SO_4]^{2-}$
under different
symmetry
conditions.

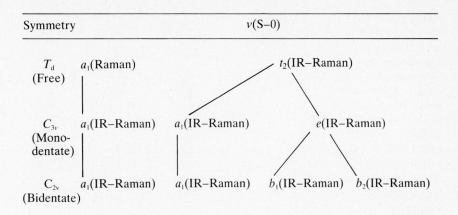

Symmetry			ν(S–0)	
T_d (Free)	a_1(Raman)			t_2(IR–Raman)
C_{3v} (Mono-dentate)	a_1(IR–Raman)	a_1(IR–Raman)		e(IR–Raman)
C_{2v} (Bidentate)	a_1(IR–Raman)	a_1(IR–Raman)	b_1(IR–Raman)	b_2(IR–Raman)

With these predictions in mind, it is quite straightforward to establish the coordination type for each of the three compounds indicated in Figure 6.11. Notice that the Raman spectra of these three complexes could be used equally well to distinguish between the three coordination possibilities. Finally, the same spectral arguments can be used for any other tetrahedral anions, for example, perchlorate ($[ClO_4]^-$) or phosphate ($[PO_4]^{3-}$).

Another major use of Raman spectroscopy over the past few years has been in the area of biologically important molecules such as hemoglobins and metalloporphyrins. The chief thrust of this work has been in identifying the active sites in these large biomolecules—that is, to which atoms are the metals coordinated? Raman spectroscopy has proven especially useful because the vibrations associated with the metals and the ligating atoms are often fairly easily identified. By deliberately irradiating the samples with laser energy that closely matches an electronic absorption band of the molecule, it is possible to excite specifically the metal–ligand vibrations through a *resonance Raman process*. There is a constant danger of thermally damaging the biological samples in this type of experiment because very dilute solutions are used and long irradiation times are necessary to collect sufficient Raman data. Ingenious methods have been devised to minimize the contact time of these dilute solutions with the incident laser beam. Some of these methods have involved sample spinning or rapidly flowing a solution past the point of laser focus.

A particularly simple example of a metalloprotein is rubredoxin, which has an active site consisting of an iron atom surrounded tetrahedrally by four sulfur atoms. Four Raman-active bands ($a_1 + e + 2t_2$) would be expected for such a tetrahedral arrangement (see CCl$_4$, discussed earlier). The resonance Raman spectrum shown in Figure 6.13 does display three distinct peaks and a shoulder in the region associated with ν(Fe—S) and δ(SFeS) vibrations, confirming the tetrahedral geometry of the FeS$_4$ site.

Figure 6.13
Resonance Raman
spectrum of oxidized
rubredoxin using
488.0-nm laser
excitation. The four
peaks indicated by
the arrows are due
to the T_d symmetry
of the FeS$_4$ moiety
of the rubredoxin
molecule. Adapted
with permission
from V. K.
Yachandra, J. Hare,
I. Moura, and T. G.
Spiro, *J. Am. Chem.
Soc.* 105 (1983):
6458.

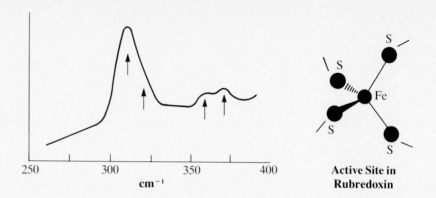

**Active Site in
Rubredoxin**

Considerable effort has been expended in studying the resonance
Raman spectra of carboxyhemoglobins and related compounds because
these measurements provide important information on the nature of the
active sites. A recent investigation was focused on a comparison of the
bonding in the isoelectronic Fe^{II}—CO and Mn^{II}—NO hemoglobins. Conventional Raman spectra of these materials are difficult to obtain. The
manganese–nitrosyl compound has an electronic absorption band centered
at 433 nm, and using the 413.1–nm line of a krypton-ion laser, it is possible
to excite a resonance Raman spectrum of the compound. The only peaks
observed are those associated with the MnNO bending and Mn—NO–and
NO–stretching vibrations at 573, 623, and 1721 cm^{-1}, respectively (Figure 6.14). The authenticity of the assignments given for these peaks was
established by their behavior on replacement of ^{14}NO by ^{15}NO. Introduction of the heavier isotope would be expected to result in the observed
small shifts to lower energies of the bands due to the increase in the
reduced mass of the NO group. The positions of the ν(NO) and
ν(Mn—NO) modes are higher and lower, respectively, than the corresponding ν(CO) and ν(Fe—CO) modes in carboxyhemoglobin, consistent
with the Mn—NO bonding being stronger than the Fe—CO bonding. This
conclusion is based on π-backbonding arguments, which are discussed fully
in Chapter 22.

Another interesting result of the use of resonance Raman spectroscopy is in the easier detection of overtones. Usually, such features are
significantly weaker than fundamentals and are difficult to observe experimentally. However, their intensities can be enhanced by resonance
Raman spectroscopy. For example, consider the three spectra shown in
Figure 6.15a and note especially the effect of changing the excitation
wavelength of the incident laser beam on the spectra. Using 514.5–nm
laser excitation, four overtones of the symmetric ν(Ti—I) mode (a_1) can be
easily detected. The differences between twice, three, and more times the
wave number of the fundamental vibration and the observed values $2\nu_1$,

Figure 6.14
Resonance Raman
spectra of
nitrosylmanganese
hemoglobins in
aqueous solution
using 413.1-nm
excitation (about
10 mW at the
samples, which were
spun in NMR
tubes):
(a) for the complex
containing natural
abundance NO;
(b) for the ^{15}NO
analog.
Adapted from
N. Parthasarathi and
T. G. Spiro. *Inorg.
Chem.* 26 (1987):
2280.

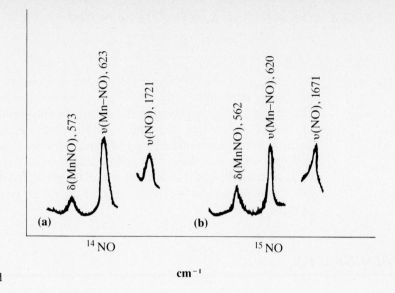

Figure 6.15
(a) Raman spectra
of TiI$_4$ in
cyclohexane solution
using three different
excitation
wavelengths. The
solvent peaks have
been omitted for the
sake of clarity. Note
the appearance of
four overtones ($2\nu_1$,
$3\nu_1$, $4\nu_1$, $5\nu_1$) in the
resonance Raman
spectrum obtained
using 514.5-nm laser
excitation.
(b) Electronic
absorption spectrum
of TiI$_4$. Adapted
with permission
from R. J. H. Clark
and R. D. Mitchell.
J. Am. Chem. Soc.
95 (1973): 8300.

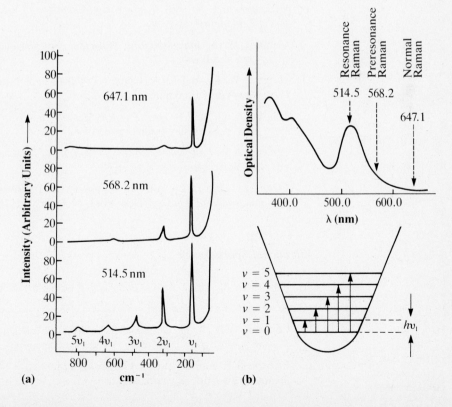

$3\nu_1$, and so on are directly related to the anharmonicity of the Ti—I–stretching vibration. The 514.5-nm excitation wavelength matches exactly that of the maximum in the electronic absorption spectrum of TiI_4 shown in Figure 6.15b. Excitation with the other two laser lines does not result in many overtones being observed. Strictly, the T_d symmetry TiI_4 molecule ought to exhibit four fundamentals in its Raman spectrum (see CCl_4, discussed earlier), but often not all the predicted peaks are observed in a resonance Raman spectrum, save for the totally symmetric vibrations, as is the case for TiI_4.

BIBLIOGRAPHY

Suggested Reading

Cotton, F. A., and G. Wilkinson. *Advanced Inorganic Chemistry. A Comprehensive Text*, 5th ed. New York: Wiley-Interscience, 1988.

Drago, R. S. *Physical Methods in Chemistry*. Philadelphia: Saunders, 1977.

Greenwood, N. N., and A. Earnshaw, *Chemistry of the Elements*. Elmsford, N.Y.: Pergamon Press, 1984.

Guillory, W. A. *Introduction to Molecular Structure and Spectroscopy*. Boston: Allyn and Bacon, 1977.

Hatfield, W. E., and R. A. Palmer. *Problems in Structural Inorganic Chemistry*. New York: Benjamin, 1971.

Hill, H. A. O., and P. Day. *Physical Methods in Advanced Inorganic Chemistry*. New York: Wiley-Interscience, 1968.

Huheey, J. E. *Inorganic Chemistry. Principles of Structure and Reactivity*, 3d ed. New York: Harper & Row, 1983.

Lambert, J. B., H. F. Shurvell, L. Verbit, K. J. Cooks, and G. H. Stout. *Organic Structural Analysis*. New York: Macmillan, 1976.

Vibrational Spectroscopy

Adams, D. M. *Metal–Ligand and Related Vibrations*. London: Arnold, 1967.

Braterman, P. S. *Metal Carbonyl Spectra*. New York: Academic Press, 1975.

Colthup, N. B., L. H. Daly, and S. E. Wiberly. *Introduction to Infrared and Raman Spectroscopy*. New York: Academic Press, 1964.

Ferraro, J. R. *Low-Frequency Vibrations of Inorganic and Coordination Compounds*. New York: Plenum, 1971.

Griffiths, P. R., and J. A. de Haseth. *Fourier Transform Infrared Spectrometry*. New York: Wiley-Interscience, 1986.

Herzberg, G. *Infrared and Raman Spectra of Polyatomic Molecules*. New York: Van Nostrand, 1945.

Jones, L. H. *Inorganic Vibrational Spectra*. New York: Dekker, 1971.

Nakamoto, K. *Infrared Spectra of Inorganic and Coordination Compounds*, 4th ed. New York: Wiley, 1986.

Tobin, M. C. *Laser Raman Spectroscopy*. New York: Wiley-Interscience, 1971.

Wilson, E. B., J. C. Decius, and P. C. Cross. *Molecular Vibrations*. New York: McGraw-Hill, 1955.

Woodward, L. A. *Introduction to the Theory of Molecular Vibrations and Vibrational Spectroscopy*. London: Oxford University Press, 1972.

Landmark Papers

Jones, L. H., R. H. McDowell, and M. Goldblatt. *Inorg. Chem.* 8 (1969): 2349.

Michaelson, A. A. *Phil. Mag.* 31 (1891): 256.

Raman, C. V., and K. S. Krishnan. *Nature* 122 (1928): 278, 882.

Schäfer, L., and S. J. Cyvin. "Complete Reduction of Representations of Infinite Point Groups." *J. Chem. Ed.* 48 (1971): 295.

PROBLEMS

6.1 Predict the position of the NO–stretching mode for $^{15}N^{16}O$ given that the corresponding value for $^{14}N^{16}O$ is 1876.11 cm^{-1}.

6.2 Prove that the CO force constant is essentially invariant to isotopic substitution given that the IR absorptions bands for $^{12}C^{16}O$ and $^{13}C^{16}O$ are at 2143.16 and 2096.07 cm^{-1}, respectively.

6.3 Show that the following vibrational data are consistent with a trigonal-bipyramidal geometry for PF_5 in the vapor state: 1026, 817, 640, 944, 575, 532, 300, and 514 cm^{-1}.

6.4 If removal of one electron from O_2 to give O_2^+ causes the k_{O-O} force constant to increase from 1176 to 1632 N m^{-1}, predict
    a. The position of the O—O–stretching mode in the Raman spectrum of O_2^+ given that ν(O—O) for O_2 is 1580 cm^{-1}.
    b. The value of the k_{O-O} force constant for O_2^- if the O—O–stretching mode of this species is at 1097 cm^{-1}.

6.5 Show that the following vibrational data are consistent with both $[InCl_2]^-$ and PbF_2 having bent XY_2 structures:

IR–Raman (cm^{-1})

$[InCl_2]^-$	PbF_2
328	531.2
177	165.0
291	507.2

6.6 On the basis of VSEPR theory, TiF_4 would be expected to have a tetrahedral structure. Do the following vibrational data for gaseous TiF_4 support this contention?

IR (cm^{-1})	Raman (cm^{-1})
793	793 depol
209	712 pol
	209 depol
	185 depol

6.7 Treatment of $Mn(CO)_5Br$ with Br^- yields the octahedral, anionic manganese carbonyl compound $[Mn(CO)_4Br_2]^-$, which exhibits four $\nu(CO)$ peaks in its IR spectrum in CH_2Cl_2 solution. What is the structure of this anionic complex?

6.8 Nitrate ion from various agriculture fertilizers containing KNO_3 has been identified as one of the major pollutants in water samples taken from rivers adjacent to farmland. IR spectroscopy has been proposed as a method of identifying not only the presence of $[NO_3]^-$ but also its mode of coordination to metal ions. Would you be in agreement with these suggestions? Explain the reasons for your decision.

6.9 Given the following vibrational data for KrF_2, predict whether the structure will be linear or bent: IR, 238, 588; Raman, 449 cm^{-1}.

6.10 Using the following vibrational data (cm^{-1}), prove that xenon tetrafluoride is square-planar not tetrahedral.

IR (Vapor)	Raman (Solid)
586	
	543
	502
	442
291	
	235
123	

6.11 Do the following vibrational data (cm^{-1}) for the dodecahydroborate anion $[B_{12}H_{12}]^{2-}$ satisfy the selection rules for I_h symmetry?

IR (Solid)	Raman (H$_2$O Solution)
	2518 pol
2480	
	2475
1070	
	949
	770
	743 pol
720	
	584

6.12 Explain why the vibrational wave numbers of the following species decrease in the order indicated:

$$\nu(C—H) > \nu(C\!\equiv\!O) > \nu(C\!\equiv\!Se) > \nu(Re\!\equiv\!Re) > \nu(Re—Re)$$

6.13 **a.** Predict the IR and Raman spectra of square-pyramidal Cr(CO$_4$)(CS) in the CO− and CS−stretching regions.

    b. Given that the force constants for C≡O and C≡Te are 1600 and 400 N m^{-1}, respectively, where approximately would $\nu(C\!\equiv\!Te)$ be observed in the IR region?

6.14 Predict the spectral activities of the three normal modes of vibration of a trigonal-bipyramidal MX$_5$ molecule that are shown below:

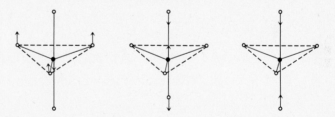

6.15 Outline briefly five advantages and five disadvantages of IR spectroscopy over Raman spectroscopy.

6.16 Predict the structure of SeF$_4$ in the vapor state given only that there are four polarized bands in its Raman spectrum.

6.17 Both [Co(CO)$_4$]$^-$ and [Fe(CO)$_4$]$^{2-}$ exhibit only one CO−stretching fundamental in their IR spectra. Would you expect these anionic complexes to be isostructural with tetrahedral Ni(CO)$_4$?

6.18 Under high-resolution conditions, the in-phase C—Cl−stretching mode of thiophosgene (CSCl$_2$) at about 500 cm^{-1} splits into three peaks due to chlorine-isotope effects. Given that the relative abundance of ^{35}Cl to ^{37}Cl is 3/1, predict the relative intensities of these three Raman peaks for CSCl$_2$.

6.19 On the basis of the information given in Problem 6.18, how many peaks could be observed theoretically in the C—Cl−stretching region of the Raman spectrum of liquid CHCl$_3$?

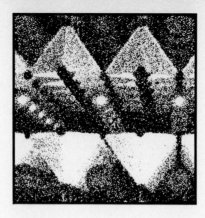

7

Resonance Spectroscopies

This chapter introduces you to the theoretical background of four spectroscopic techniques that can be considered to fall under the general title of resonance spectroscopies, namely, nuclear magnetic resonance, nuclear quadrupole resonance, Mössbauer, and electron-spin resonance spectroscopies. Several applications of these spectroscopic methods in inorganic chemistry are also described.

7.1 Nuclear Magnetic Resonance Spectroscopy

Nuclear magnetic resonance (NMR) spectroscopy provides information about the chemical environments of specific nuclei in compounds. In the case of organic and organometallic compounds, proton (^{1}H) and carbon-13 (^{13}C) NMR spectroscopy are especially informative. These nuclei were among the earliest investigated because of their relative ease of detection compared to most other nuclei and because hydrogen and carbon atoms are so widespread throughout organic chemistry. With the recent advent of Fourier-transform multinuclear NMR spectrometers equipped with super-conducting magnets, many other nuclei are now routinely accessible, for example, ^{17}O, ^{19}F, ^{23}Na, ^{29}Si, ^{31}P, ^{119}Sn, and ^{195}Pt. Phosphorus-31 and metal–NMR spectroscopy, in particular, are now extremely valuable tools for establishing the structures of transition metal complexes containing tertiary phosphine and phosphite ligands such as $P(CH_3)_3$ and $P(OC_6H_5)_3$. Another important application of NMR spectroscopy is in the investigation of the kinetics and mechanisms of chemical reactions. Reaction rates can be readily monitored by following the changes occurring in the NMR spectra of reaction mixtures.

NMR studies were initially applicable only to liquids where spinning the sample results in an averaged exposure to the applied magnetic field. Recently, however, NMR spectrometers equipped to study solids have become commercially available, and it is possible to obtain a high-quality spectrum of a solid spinning at about 2 kHz (2000 s^{-1}). For further information on this rapidly developing technique, see the book by Fyfe listed in the references at the end of the chapter.

The nuclei of interest to NMR spectroscopy have spin angular momenta ($I \neq 0$). Rotation of these positively charged particles creates small magnetic fields whose alignment either with or against an externally imposed magnetic field forms the basis of the NMR experiment. The spin angular momentum for a given nucleus is the resultant of the spins associated with the protons and neutrons in the nuclei. These particles have the spin quantum number $\frac{1}{2}$. In several nuclei, the nuclear spins of the protons and neutrons are all paired, for example, in ^{12}C and ^{16}O, and $I = 0$. In fact, whenever both A and Z in a nucleus $^A_Z M$ are even, $I = 0$. For certain nuclei such as ^1H, ^{13}C, ^{19}F, and ^{31}P, $I = \frac{1}{2}$, whereas for others (for example, ^2H and ^{14}N), $I = 1$. For most nuclei, however, $I > 1$, for example, for ^{10}B ($I = 3$), ^{35}Cl ($I = \frac{3}{2}$), and ^{127}I ($I = \frac{5}{2}$). When $I > \frac{1}{2}$, the nucleus has a nuclear quadrupole moment: NMR experiments with quadrupolar nuclei are possible but are more difficult than those for which $I = \frac{1}{2}$ because the resultant NMR signals are appreciably broadened by the random interaction of the nuclear quadrupoles with one another.

When there is no strong external magnetic field, all orientations of nuclei with $I = \frac{1}{2}$ are equivalent. However, when an external magnetic field (B_0) is applied, the spinning nuclei align themselves with their magnetic fields either parallel or opposed (a less stable arrangement) to the direction of the external field. This situation arises because the energy of the nuclear spin is quantized; when $I = \frac{1}{2}$, there are two allowed values for the nuclear spin angular momentum quantum number, $m_I = \frac{1}{2}$ or $-\frac{1}{2}$. In general, the allowed values of m_I are $I, I - 1, I - 2, \ldots, -I + 1, -I$. The

Figure 7.1
Effect of an applied
external magnetic
field on the nuclear
spin states for a
nucleus with $I = \frac{1}{2}$.

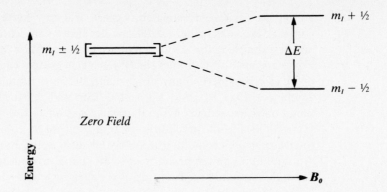

energy difference (ΔE) between these two nuclear states is very small
(Figure 7.1), and at normal temperatures, the lower-energy state
($m_I = -\frac{1}{2}$) is occupied only slightly more than the higher-energy one
($m_I = \frac{1}{2}$). The magnitude of ΔE increases linearly with the strength of the
applied external field (B_0), and the energy changes involved lie in the
radio-frequency region of the electromagnetic spectrum. Moreover, ΔE is
given by equation 7.1:

$$\Delta E = h\nu_0 = \frac{\gamma}{2\pi} B_0 \tag{7.1}$$

where γ is known as the *magnetogyric ratio* of the nucleus concerned.

The effect of magnetic field B_0 on the spinning nucleus is to cause it to
precess about the applied magnetic field axis at an angle θ to the direction
of the field (Figure 7.2). The Larmor (or precessional) frequency of this
motion, ω_0, is given by Equation 7.2:

$$\omega_0 = 2\pi\nu_0 = \gamma B_0 \tag{7.2}$$

Figure 7.2
Precession of a
spinning nucleus
about the direction
of an applied
magnetic field B_0.

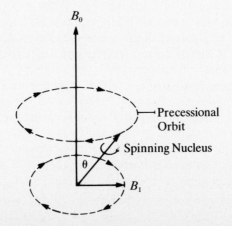

When the conditions for Equation 7.1 are met, nuclei in the lower-energy state absorb radio-frequency energy and are excited to the higher-energy state; in doing so, they flip from one orientation to another. This *resonance condition* is achieved experimentally by applying a second magnetic field B_1 at 90° to the main field B_0. This second field is rotating in a plane at right angles to the direction of B_0, in phase with the precessing nuclei. There are two ways of achieving resonance: either by varying the Larmor frequency at a fixed applied magnetic field B_0 or by holding the Larmor frequency fixed and sweeping the B_0 field. The latter procedure is the one usually employed in the classical continuous-wave (CW) NMR experiment described above.

CW NMR spectroscopy can be an extremely time-consuming process because each radio frequency in the resonance region being examined must be scanned individually. A much more efficient procedure is to monitor all frequencies simultaneously. This is an analogous situation to that described in Chapter 6 for grating and FT–IR spectroscopy. In a FT–NMR measurement, a short burst (pulse) of energy from a high-power radio-frequency field is applied to the sample at a frequency close to the resonances of interest. If the power is high enough, all nuclei absorb energy during the pulse and flip their orientations. Each set of chemically equivalent nuclei then precess at a characteristic Larmor frequency, which can be detected as a time-dependent voltage induced in a radio-frequency receiver coil. Coaddition of the data from several pulse cycles leads to a time-domain spectrum, which can be Fourier transformed on a microcomputer to give the conventional frequency-domain NMR spectrum.

If the external magnetic field (B_0) is constant, the resonance frequencies of different nuclei are expected to be directly proportional to their individual magnetogyric ratios. In practice, however, the applied field is modified somewhat by the local electronic environments of the nuclei. This leads into the concept of *chemical shielding* (σ in Equation 7.3) and *chemical shifts*:

$$\omega_0 = B_0(1 - \sigma) \tag{7.3}$$

From a chemist's viewpoint, chemical shifts are important quantities that can be obtained from an NMR experiment. Chemical shifts are quoted in parts per million (ppm) for the nuclei under investigation. These values are characteristic of the different shielding effects experienced by the nuclei due to the different chemical environments in the molecule being examined. The formal definition of chemical shift is given by Equation 7.4:

$$\delta(\text{in ppm}) = \frac{(\delta_S - \delta_R) \text{ in Hz}}{\nu_0 \text{ in MHz}} \tag{7.4}$$

where $(\delta_S - \delta_R)$ is the difference between the resonance frequencies of sample S and the reference compound R, and ν_0 is the constant radio-frequency of the probe (the name given to the attachment, into which the

Table 7.1 *Multinuclear Chemical Shifts (ppm)*
for Some Typical Inorganic Compounds and Ligands

Compound[a]	ppm	Compound	ppm
^{1}H			
TMS[b]	0.0	Acetylenic H	About 2.5
CH_3R	0.9	Alkenic H	About 5.3
CH_2R_2	1.4	Aromatic H	About 7.4
CHR_3	1.8	trans-$Pt(Et_3P)_2HCl$	16.9
^{11}B			
$NaBH_4$	−43	$(BH{-}NH)_3$	30
$(C_2H_5)O.BF_3$	0.0	BCl_3	47
$B(OH)_3$	19	$(CH_3)_3B$	86
^{13}C			
CH_3I	−21	RCN	113
TMS[b]	0.0	Aromatic C	128
Acetylenic C	75	CS_2	193
Alkenic C	110	$Cr(CO)_6$	212.1
^{15}N			
Anhydrous NH_3	0.0	$[S_4N_3]^+$	About 350
NH_4Cl	24	Concentrated HNO_3	367
KCN	279	$NaNO_2$	608
^{19}F			
MoF_6	−355	CF_3COOH	0.0
WF_6	−242	$[SiF_6]^{2-}$	49.8
NF_3	−219	SiF_4	83.3
SF_6	−127	GeF_4	99.0
^{31}P			
P_4	−488	Concentrated H_3PO_4	0.0
PH_3	−241	$P(OCH_3)_3$	141
$(CH_3)_3P$	−61	PBr_3	227

[a] R = alkyl.
[b] Tetramethylsilane $(CH_3)_4Si$. The chemical shifts for the other reference compounds in the table have also been arbitrarily set equal to 0.0 ppm. Data taken chiefly from J. B. Lambert et al., *Organic Structural Analysis*, New York: Macmillan, 1976.

sample is inserted, that is located at a precise position in the magnetic field). The quantity $(\delta_S - \delta_R)$ is obtained by varying the applied magnetic field strength. The choice of reference compound depends on the nucleus being examined, but tetramethylsilane $[(CH_3)_4Si, TMS]$ is particularly useful because it is the reference compound chosen for ^1H-, ^{13}C-, and ^{29}Si-NMR spectra. The chemical shifts of different nuclei vary widely, and representative examples of nuclei in some typical inorganic compounds and ligands are given in Table 7.1. By convention, δ is positive in the downfield direction (on going from right to left). Resonances that occur at higher fields than those for the reference compounds have negative δ values.

Shielding effects vary considerably depending on the type of bonding and the electronegativities of the bonded atoms, so that it is possible to distinguish equivalent sets of nuclei from one another. The nuclear spins associated with one set of nuclei can interact with those on an adjacent nucleus across a chemical bond (or bonds); two- and three-bond couplings are quite common—for example, $^3J_{^{31}P-^1H} \simeq 12$ Hz in quarternary phosphonium compounds such as $[P(C_6H_5)_4]^+I^-$.

Probably, the most famous example used to illustrate shielding effects is the ^1H–NMR spectrum of ethanol (CH_3—CH_2—OH). Three distinctly different signals are observed, with peak areas in the ratio of 3:2:1 corresponding, respectively, to the three hydrogens on the methyl group, the two hydrogens on the methylene group, and the single hydrogen on the hydroxyl group (Figure 7.3). The signals assigned to the methyl and methylene groups exhibit some fine structure due to nuclear spin–spin splitting. The same fine structure would be expected for any ethyl group, for example, in CH_3—CH_2—Br or $(CH_3$—$CH_2)_4Sn$. It is the chemical shielding and spin–spin splitting effects that make NMR spectroscopy such a useful structural tool for chemists.

The spin–spin splitting for the methylene group in any CH_2—CH_3 group arises because the protons on the neighboring methyl group cause

Figure 7.3
Proton–NMR
spectrum of slightly
acidified ethanol
CH_3—CH_2—OH;
$\delta(TMS) = 0.0$ ppm.

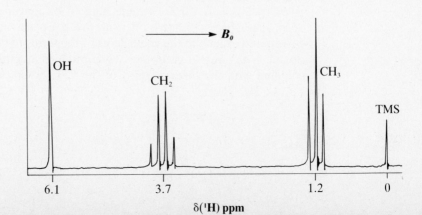

Table 7.2 *Analysis of the Spin–Spin Splitting in the ^1H–NMR Signal for the CH_2 Group of a CH_2CH_3 Group*

Spin Arrangement of the CH_3 protons	Spin Quantum Numbers	Number of Equivalent Spin Arrangements
↑ ↑ ↑	$\frac{1}{2}, \frac{1}{2}, \frac{1}{2}$	1
↑ ↑ ↓ ↑ ↓ ↑ ↓ ↑ ↑	$\frac{1}{2}, \frac{1}{2}, -\frac{1}{2}$ $\frac{1}{2}, -\frac{1}{2}, \frac{1}{2}$ $-\frac{1}{2}, \frac{1}{2}, \frac{1}{2}$	3
↑ ↓ ↓ ↓ ↑ ↓ ↓ ↓ ↑	$\frac{1}{2}, -\frac{1}{2}, -\frac{1}{2}$ $-\frac{1}{2}, \frac{1}{2}, -\frac{1}{2}$ $-\frac{1}{2}, -\frac{1}{2}, \frac{1}{2}$	3
↓ ↓ ↓	$-\frac{1}{2}, -\frac{1}{2}, -\frac{1}{2}$	1

the protons on the methylene group to resonate at slightly different field strengths. In general, such an effect is due to the neighboring nuclear magnetic moments causing a change in the effective magnetic field experienced by the nucleus (or nuclei) being examined. The magnitude of this magnetic field change depends on the orientation of the neighboring nuclei—in our case, the different possible orientations of the spins of the three methyl protons. If we assign the spin quantum numbers as $\frac{1}{2}$ (for up, ↑) and $-\frac{1}{2}$ (for down, ↓), the allowed arrangements of the spins of these three protons are shown in Table 7.2.

There are four different spin arrangements, and the probabilities of these occurring are in the ratio $1:3:3:1$, that is, the methylene signal should be split into a quartet with intensities $1:3:3:1$. Fortunately, we do not have to write all the possible spin permutations every time we want to predict spin–spin couplings. The number of peaks (or multiplicity) for a signal resulting from coupling with a nucleus (or set of equivalent nuclei) is given by $2nI + 1$, where n is the number of equivalent nuclei of spin I. Also, the expected relative intensities of the peaks in an NMR multiplet correspond exactly to the coefficients of the terms in the binominal expansion $(1 + x)^n$. In our case, the CH_2 protons in an ethyl group are split by coupling with the three protons of the CH_3 group, giving us $n = 3$ and $I = \frac{1}{2}$ (^1H). Thus,

$$2nI + 1 = (2 \times 3 \times \tfrac{1}{2}) + 1 = 4 \qquad (7.5)$$

and

$$(1 + x)^n = (1 + x)^3 = 1 + 3x + 3x^2 + x^3 \qquad (7.6)$$

Table 7.3 *Pascal's Triangle Indicating the Relative Intensities Expected for the Peaks in an NMR Multiplet Produced by Coupling* n *Equivalent Nuclei of Spin 1/2[a]*

n	Relative Intensities
0	1
1	1:1
2	1:2:1
3	1:3:3:1
4	1:4:6:4:1
5	1:5:10:10:5:1
6	1:6:15:20:15:6:1

[a] These are the coefficients for the various terms produced from the binomial expansion $(1 + x)^n$, as mentioned in the text.

From the coefficients in Equation 7.6, the intensities of the four peaks in the quartet for the CH_2 group should be in the ratio 1:3:3:1, as discussed earlier. Similarly, the signal for the three protons in the CH_3 group should be split by the two protons of the CH_2 group: $n = 2$, $2nI + 1 = (2 \times 2 \times \frac{1}{2}) + 1 = 3$, and $(1 + x)^n = (1 + x)^2 = 1 + 2x + x^2$. This means that there should be a 1:2:1 triplet observed for the CH_3 group. The actual splittings in hertz between the signals are referred to as *coupling constants*. For the ethyl group, the proton couplings are across three bonds, and this is expressed as $^3J_{^1H-^1H}$. It is quite common to detect such couplings, for example, $^2J_{^{29}Si-^1H}$ and $^3J_{^1H-^1H}$ have been observed for disilane $H_3Si-SiH_3$.

The relative intensities expected for NMR multiplets resulting from the coupling of n equivalent nuclei of spin $\frac{1}{2}$ are summarized in a *Pascal's triangle* in Table 7.3. The number of coefficients in each case follow the outline of an isosceles triangle.

Table 7.4 lists the spin–spin splittings usually observed for some typical inorganic and organometallic compounds. The detectability of these splittings critically depends on the sensitivity of the NMR nucleus being investigated. This sensitivity results from several factors including the relative natural abundance, the spin quantum number, whether the nucleus has a large quadrupole, and the concentration of nuclei in the sample (that is, pure liquids are better than solutions). The nuclei most often studied are those with $I = \frac{1}{2}$ that have natural abundances close to 100%, namely, ^1H (99.985%), ^{19}F (100%), and ^{31}P (100%). Although ^{13}C is an important NMR nucleus, its natural abundance is only 1.11%.

Table 7.4 *Spin-Spin Splitting Predicted (and in Most Cases Observed) for Selected Inorganic Compounds*[a]

Compound	Nucleus or Set of Equivalent Nuclei (A) Being Split	Nucleus or Set of Equivalent Nuclei (B) Causing Splitting	Number of Equivalent Nuclei B (n)	I_B	Multiplicity of Signal for Nuclei A ($2nI_B + 1$)
PF_3	^{31}P	$(^{19}F)_3$	3	$\frac{1}{2}$	4
	$(^{19}F)_3$	^{31}P	1	$\frac{1}{2}$	2
$(CH_3O)_3P$	^{31}P	$[(^1H)_3]_3$	9	$\frac{1}{2}$	10
	$[(^1H)_3]_3$	^{31}P	1	$\frac{1}{2}$	2
PF_5[b]	^{31}P	$(^{19}F)_2{}^{ax}$	2	$\frac{1}{2}$	3
	^{31}P	$(^{19}F)_3{}^{eq}$	3	$\frac{1}{2}$	4
	$(^{19}F)_5$	^{31}P	1	$\frac{1}{2}$	2
	$(^{19}F)_2{}^{ax}$	$(^{19}F)_3{}^{eq}$	3	$\frac{1}{2}$	4
	$(^{19}F)_3{}^{eq}$	$(^{19}F)_2{}^{ax}$	2	$\frac{1}{2}$	3
$Cr(CO)_5(CS)$[c]	$(^{13}CO)_4{}^{eq}$	$(^{13}CO)^{ax}$	1	$\frac{1}{2}$	2
	$(^{13}CO)^{ax}$	$(^{13}CO)_4{}^{eq}$	4	$\frac{1}{2}$	5
$[RhH(CN)_5]^{3-}$	^1H	^{103}Rh	1	$\frac{1}{2}$	2
$cis\text{-}TiF_4L_2$[d]	$(^{19}F)_2$	$(^{19}F)_2$	2	$\frac{1}{2}$	3
$[(CH_3)_4N]^+$	$(^1H)_3$	^{14}N	1	1	3
$PtHCl(PR_3)_2$	^1H	^{195}Pt	1	$\frac{1}{2}$	2
$B(CH_3)_3$	^{11}B	$(^1H_3)_3$	9	$\frac{1}{2}$	10
SiH_4	$(^1H)_4$	^{29}Si	1	$\frac{1}{2}$	2
	^{29}Si	$(^1H)_4$	4	$\frac{1}{2}$	5
H_2O	^{17}O	$(^1H)_2$	2	$\frac{1}{2}$	3

[a] Relative abundances of the nuclei involved: ^{19}F (100%), ^{17}O (0.037%), ^{31}P (100%), ^{14}N (99.6%), ^{13}C (1.11%), ^{11}B (81.2%), ^{29}Si (4.7%), ^{103}Rh (100%), ^{195}Pt (33.7%).
[b] ax = axial [trans (180°) to F atoms]; eq = equatorial (three F atoms in plane).
[c] eq = equatorial (trans to CO); ax = axial (trans to CS).
[d] L = monodentate ligand.

However, most ^{13}C resonances can now be located fairly easily because of the greater sensitivity associated with modern pulsed FT spectrometers, which are equipped with superconducting magnets that operate routinely at magnetic fields up to 500 MHz. Other nuclei that are frequently investigated by inorganic chemists are ^{10}B (19.6%, $I = 3$), ^{11}B (80.4%, $I = \frac{3}{2}$), ^{29}Si (4.7%, $I = \frac{1}{2}$), ^{33}S (0.76%, $I = \frac{3}{2}$), ^{81}Br (49.5%, $I = \frac{3}{2}$), ^{119}Sn (8.6%, $I = \frac{1}{2}$), ^{183}W (14.4%, $I = \frac{1}{2}$), and ^{195}Pt (33.8%, $I = \frac{1}{2}$).

A striking example of spin–spin splitting is shown in Figure 7.4 for the tetrafluoroborate(III) anion $[BF_4]^-$. There is a septet due to coupling of

Figure 7.4
Fluorine-19 NMR
spectrum of $[BF_4]^-$
in an aqueous
solution of NH_4BF_4.
The 0.047-ppm shift
between the two
multiplets results
from the different
isotopic masses of
^{10}B and ^{11}B.
Adapted with
permission from J.
W. Akitt. *NMR and
Chemistry*, 2nd ed.
New York:
Chapman & Hall,
1983.

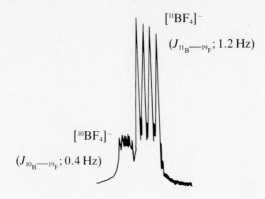

$[^{11}BF_4]^-$

$(J_{^{11}B}{-}_{^{19}F}; 1.2\ Hz)$

$[^{10}BF_4]^-$

$(J_{^{10}B}{-}_{^{19}F}; 0.4\ Hz)$

^{19}F ($I = \frac{1}{2}$) to ^{10}B ($I = 3$) and a more intense quartet due to coupling of ^{19}F ($I = \frac{1}{2}$) to ^{11}B ($I = \frac{3}{2}$).

Often, the predicted spin–spin splitting patterns of a multiplet are not fully resolved because of the low intensities of the outermost peaks. For example, the ^{31}P-NMR spectrum of liquid trimethylphosphite $P(OCH_3)_3$ should theoretically exhibit ten lines because of three-bond couplings of the ^{31}P nucleus ($I = \frac{1}{2}$) with the nine equivalent protons ($I = \frac{1}{2}$), but usually only the central eight peaks are intense enough to be detected.

There has been wide interest in the coordination chemistry of tertiary phosphite ligands such as $P(OCH_3)_3$, and complexes like the two chromium tricarbonyl isomers shown below can be isolated:

fac

CO
|
OC —— Cr —— P
|
P

CO

P

mer

CO
|
OC —— Cr —— CO
|
P

P

CO

$P = P(OCH_3)_3$

The cis (90°) and trans (180)° two-bond $^{31}P—^{31}P$ couplings, $^2J_{^{31}P-^{31}P}$, can be observed, but other couplings such as $^3J_{^{31}P-^1H}$ and $^2J_{^{13}C-^{13}C}$ are too weak to be detected. Figure 7.5 shows the room-temperature ^{31}P-NMR spectrum of an equilibrium mixture of fac- and mer-$Cr(CO)_3[P(OCH_3)_3]_3$ isomers. The three P atoms in the fac isomer are equivalent, and only a singlet is observed at about 186 ppm. The mer isomer displays the expected two-bond P—P couplings: a 1:1 doublet centered at 197 ppm for the two equivalent trans P atoms, which couple to the third nonequivalent P atom, and a 1:2:1 triplet centered at 188.7 ppm for the latter P atom coupled to the two equivalent trans P atoms.

Most of the recent inorganic NMR work has been directed toward obtaining metal–NMR spectra of transition metal coordination and organometallic complexes. (See also Chapter 22 for a discussion of ^{13}C-NMR

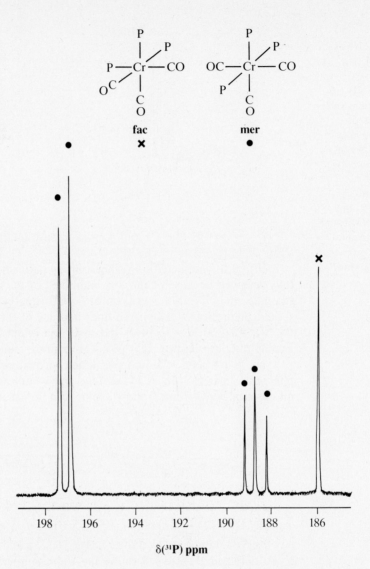

Figure 7.5
Room-temperature
^{31}P-NMR spectrum
(in deuterotoluene)
of an equilibrium
mixture of the fac
and mer isomers of
$Cr(CO)_3[P(OCH_3)_3]_3$.
Reproduced with
permission from
A. A. Ismail, Ph.D.
Thesis, McGill
University, 1985.

fac   mer

$\delta(^{31}P)$ ppm

spectra of organometallic complexes.) These studies have become possible because of the development of broadband radio-frequency oscillators that can be tuned over a wide range of radio frequencies so that the frequency associated with the resonance of a particular metal nucleus can be matched exactly.

Figure 7.6 shows an example of the application of metal–NMR spectroscopy for titanium in the $[TiF_6]^{2-}$ ion. Titanium has two NMR–active nuclei: ^{47}Ti (7.75%, $I = \frac{5}{2}$) and ^{49}Ti (5.5%, $I = \frac{7}{2}$). Coupling of these

Figure 7.6
Titanium-NMR resonances for a saturated solution of $(NH_4)_2[TiF_6]$ in H_2O/D_2O. The two sets of septets, at about -1160 ppm relative to $TiCl_4$, are separated by approximately 230 ppm and the $J_{49_{Ti}(47_{Ti})-19_F}$ coupling constant between the individual peaks in both septets is 34.0 Hz. Adapted with permission from N. Hao, B. G. Sayer, G. Denes. D. G. Bickley, C. Detellier, and M. J. McGlinchey. *J. Magn. Reson.* 50 (1982): 50.

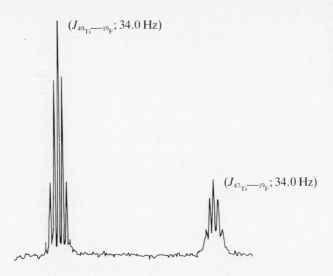

$(J_{49_{Ti}-19_F}; 34.0 \text{ Hz})$

$(J_{47_{Ti}-19_F}; 34.0 \text{ Hz})$

two types of titanium nucleus with the six equivalent F atoms ($I = \frac{1}{2}$) leads to the two sets of septets shown.

An excellent illustration of the power of spin–spin coupling is illustrated by the original determination of the ground-state molecular structure of BrF_5 from its ^{19}F-NMR spectrum. The two structures expected for such a five-coordinate compound are trigonal-bipyramidal or square-pyramidal:

For the trigonal-bipyramidal structure, there should be two sets of ^{19}F-NMR signals for the axial and equatorial F atoms in the intensity ratio of $2:3$. The signal from the two axial F atoms should be split into a $1:3:3:1$ quartet by coupling to the three equatorial F atoms, whereas that from the three equatorial F atoms themselves should appear as a $1:2:1$ triplet. A similar analysis of the square-pyramidal structure affords a $1:4:6:4:1$ quintet for the apical F atom and a $1:1$ doublet for the four basal F atoms. The ratio of the intensities of the two sets of signals should be $1:4$. The observed spectrum is completely in accord with the square-pyramidal geometry, as would be expected on the basis of VSEPR theory because there is one lone pair on the formally octahedrally coordinated, central Br atom.

Figure 7.7
^{17}O-NMR spectra of $[H_3{}^{17}O]^+$ ion in SO_2 at $-15°C$.
(a) Proton-coupled spectrum and
(b) proton-decoupled spectrum. Adapted with permission from G. D. Meteescu and G. M. Benedikt. *J. Am. Chem. Soc.* 101 (1979): 3959.

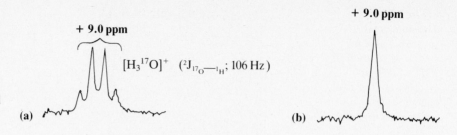

$[H_3{}^{17}O]^+$ $(^2J_{{}^{17}O-{}^1H};\ 106\,\text{Hz})$

The existence of the hydronium ion $[H_3O]^+$ in solution has been clearly demonstrated by the ^{17}O-NMR spectrum of a $H_2{}^{17}O$ solution in SO_2 at $-15\,°C$. A $1:3:3:1$ quartet ($J_{{}^{17}O-{}^1H} = 106$ Hz) is observed at $+9.0$ ppm with reference to bulk water (Figure 7.7) in accord with coupling to three equivalent protons. Enrichment is necessary because the natural abundance of ^{17}O is only 0.037%, and the signals are difficult to locate because of quadrupole broadening ($I = \frac{5}{2}$). The assignment was substantiated by employing a very powerful experimental technique known as *proton decoupling*. When the ^{17}O resonance is irradiated at 106 Hz, the quartet collapses to a singlet at $+9.0$ ppm because the ^{17}O—^{1}H coupling has been removed.

Coupling constants are also valuable in structure elucidation because they reflect the local chemical environments of nuclei. They are independent of the applied magnetic field (B_0). However, the spectra obtained at low fields are often difficult to interpret because of overlapping peaks. For this reason, multiplets are often examined at a higher field so that the peaks can be separated from one another. The geometrical isomers (cis and trans) of coordination compounds such as $PtCl_2(R_3P)_2$ provide a good example of the use of coupling constants in structure analysis. The ^{195}Pt—^{31}P coupling constants are significantly different for the isomers: cis, about 3.6 Hz; trans, about 2.5 Hz.

NMR spectroscopy has widespread application in the study of dynamic processes. Many compounds are stereochemically nonrigid on the NMR time scale, especially organometallic complexes (see also Chapter 22) and NMR techniques can be used to probe the rates and mechanisms of the isomerizations and exchange processes that occur. Variable-temperature NMR studies are relatively easy using modern spectrometers; thus, it is possible to extract information on the enthalpies and entropies of activation of the processes being investigated.

A great deal of this work has focused on analyses of the line shapes of NMR signals. Often, a signal that is broad at one temperature gradually splits into several components at lower temperatures. The variable-temperature ^{19}F-NMR spectra of sulfur tetrafluoride SF_4 illustrate well such spectral changes. This molecule has a distorted tetrahedral shape (C_{2v}

symmetry) in the gas phase:

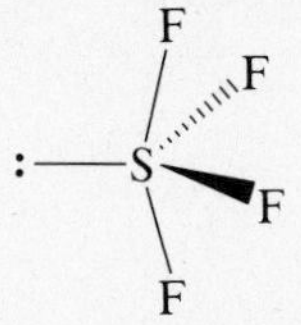

However, at room temperature, the four F atoms are equivalent, and only one sharp resonance is observed (Figure 7.8). Upon lowering the temperature, the singlet broadens, and two peaks begin to appear at about -50 °C, whereas at around -100 °C, two equally intense $1:2:1$ triplets are detected, as would be expected for two sets of pairs of nonequivalent F atoms. The alternative geometry with the lone pair in an axial position would lead to two sets of signals, a $1:1$ doublet and a $1:3:3:1$ quartet with a relative intensity ratio of $3:1$. The F-exchange rate is second-order in the SF_4 concentration, and it is probable that this exchange takes place through the F-bridged intermediate shown below:

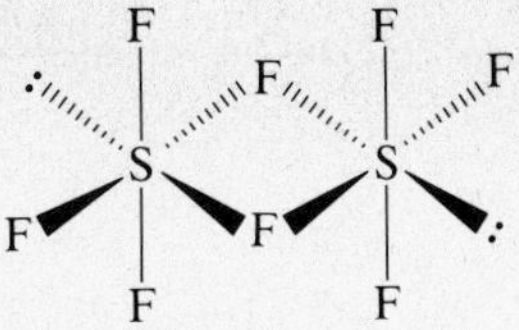

The mathematical derivation for the rate of chemical exchange between two equally populated sites is given in standard textbooks on NMR spectroscopy. Computer programs are available that simulate the observed spectra as a function of rate constant, and the calculated and observed spectra can be matched quite easily.

There are other, less accurate methods for determining exchange rates that are based on estimating the coalescence temperature — the temperature at which two (or more) signals collapse to a single peak. For example, the first-order rate constant at the coalescence temperature (k_c) of such a system with equally populated, uncoupled sites is given by Equation 7.7:

$$k_c = \frac{\pi \Delta \nu}{\sqrt{2}} \tag{7.7}$$

where $\Delta \nu$ is the separation between the two peaks at the slow-exchange limit (that is, when the peaks do not move any farther apart from each other). Other equations have been developed for coupled and nonequally populated sites.

An elegant example of site exchange in an inorganic system is the CO exchange in $Co_4(CO)_{12}$ (Figure 7.9). At about 85 °C, there is rapid exchange between the bridging and terminal CO groups, but on cooling to

Figure 7.8
Temperature
dependence of the
^{19}F-NMR spectrum
of SF$_4$. Reproduced
with permission
from E. Muetterties
and W. D. Phillips.
J. Am. Chem. Soc. 81
(1959): 1084.

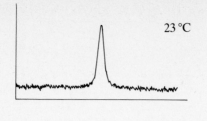

23 °C

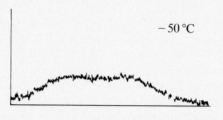

−50 °C

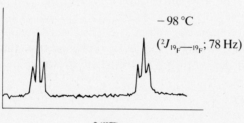

−98 °C

$(^2J_{^{19}\text{F}—^{19}\text{F}}; 78 \text{ Hz})$

δ(^{19}F)

Figure 7.9
Variable-
temperature
^{17}O-NMR spectrum
of ^{17}O-enriched
Co$_4$(CO)$_{12}$
(reference H$_2$O).
Adapted with
permission from
S. Aime, D. Osella,
L. Milone, G. E.
Hawkes, and E. W.
Randall. *J. Am.
Chem. Soc.* 103
(1981): 5920.

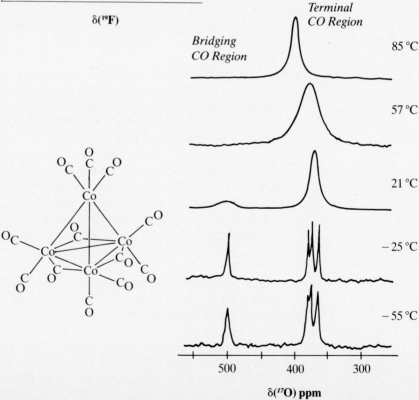

*Bridging
CO Region*

*Terminal
CO Region*

85 °C

57 °C

21 °C

−25 °C

−55 °C

δ(^{17}O) ppm

-55 °C the ^{17}O-NMR resonances due to the four different, equally populated CO environments present (apical, two types of basal terminal, and basal bridging) can readily be distinguished from one another.

Finally, in some five-coordinate, trigonal-bipyramidal systems—for example, PF_5—the axial and equatorial ligands exchange too quickly to be distinguished on the NMR time scale, even at -100 °C. For PF_5, all five fluorines appear to be equivalent, and a single ^{19}F resonance, split by coupling to ^{31}P, is observed.

7.2 *Nuclear Quadrupole Resonance Spectroscopy*

Nuclei for which I is greater than or equal to 1 have nuclear quadrupoles that often make their NMR signals too broad to detect. There is another type of molecular resonance spectroscopy, however, that takes advantage of nuclear quadrupoles: *nuclear quadrupole resonance (NQR) spectroscopy*. This technique is considered briefly in this section.

When a quadrupolar nucleus is placed in an inhomogeneous electric field, the nucleus interacts with the electric field in a variety of different ways, depending on the orientation of the nucleus with respect to the field. These different orientations are quantized, and there are $2I + 1$ allowed orientations that are distinguished by their *nuclear magnetic quantum numbers (m)*, where $m = I, I - 1, I - 2, \ldots 0, \ldots, -(I - 2), -(I - 1), -I$. For nuclei with $I = 0$ or $\frac{1}{2}$, the different orientations are degenerate; for all other nuclei, they are usually nondegenerate unless the nucleus has a symmetrical electronic environment such as in the halide ions (F^-, Cl^-, Br^-, I^-). The energy separations between the different quadrupolar states lie in the radio-frequency region of the electromagnetic spectrum. At normal temperatures, all these levels are populated to some extent. The quantum mechanical selection rule for transitions between quadrupolar energy levels is $\Delta|m| = \pm 1$.

In a NQR experiment, the sample is placed in an asymmetric electric field and then irradiated with radio-frequency energy. Absorption of energy of the requisite radio frequency leads to transitions between the different quadrupolar orientations. This spectroscopic technique is applicable only to solids because the nuclear orientations are averaged in gases, liquids, and solutions due to molecular rotation; thus, the quadrupolar energy levels are degenerate. Among the nuclei that have been investigated by NQR spectroscopy are ^{10}B, ^{11}B, ^{33}S, ^{35}Cl, ^{37}Cl, ^{79}Br, ^{81}Br, ^{123}Sn, ^{127}I, and ^{129}I.

An important quantity in NQR studies is the *quadrupole moment* (eQ), which is a measure of the asymmetry of the electric-charge distribution in the nucleus. There is an asymmetry induced in the nucleus because of the location of the negative charge due to the surrounding electrons

Table 7.5   *Number of Predicted NQR Transitions*
for Two Common Values of Nuclear Spin Quantum Number (*I*)

I	Examples of Nucleus	m Values	Allowed Transitions	Number of peaks	ΔE_m
$\frac{3}{2}$[a]	^{11}B, ^{33}S, ^{35}Cl, ^{81}Br	$\pm\frac{3}{2}$ $\pm\frac{1}{2}$	$E_{\pm 1/2} \rightarrow E_{\pm 3/2}$	1	$e^2Qq/2$
$\frac{5}{2}$	^{17}O, ^{55}Mn, ^{127}I	$\pm\frac{5}{2}$ $\pm\frac{3}{2}$ $\pm\frac{1}{2}$	$E_{\pm 1/2} \rightarrow E_{\pm 3/2}$ $E_{\pm 3/2} \rightarrow E_{\pm 5/2}$	1 1	$3e^2Qq/20$ $6e^2Qq/20$

[a] See also Figure 7.9 for the zero-field splitting diagram for this case.

influencing the location of the protonic charge in the nucleus. The lowest quadrupolar energy level is the one where the positive and negative charges are located the closest together. The distribution of the electrons around the nucleus determines the *field gradient* (*q*), which also has an effect on the splittings in the quadrupolar energy levels. The overall splitting is determined by the value of the *nuclear quadrupole coupling constant* e^2Qq). The field gradient is strictly defined in terms of Cartesian axes, q_{zz}, q_{yy}, and q_{xx}. If these three field gradients are equal, the resultant field gradient is spherical. By convention, the largest field gradient is usually located along the *z* axis. Many molecules such as Br_2 and SiH_3Cl exhibit axial symmetry, that is, they have $q_{yy} = q_{xx} < q_{zz}$. For such systems, the energies of the allowed quadrupolar states are given by the expression shown in Equation 7.8 (see also text by Drago):

$$E_m = \frac{e^2Qq[3m^2 - I(I + 1)]}{4I(2I - 1)} \tag{7.8}$$

The number of transitions can be predicted from Equation 7.8 by inserting the value of *I* for the nucleus being examined. Table 7.5 gives the number of predicted transitions for two typical *I* values. A singlet should be observed for nuclei with $I = \frac{3}{2}$, whereas doublets are expected for nuclei with $I = 2$ and $I = \frac{5}{2}$. Moreover, in the latter two cases, the higher-energy transition should appear at *twice* the energy of the lower-energy one. For nonsymmetric systems (that is, those for which $q_{zz} \neq q_{yy} \neq q_{xx}$), the spectral interpretation is too complicated to be considered here. Additional information can be obtained by removing the degeneracy of quadrupolar states by application of a weak magnetic field (Figure 7.10).

The e^2Qq values determined from NQR measurements can often provide some insight into the ionic character of bonds. For example,

Figure 7.10
Splittings predicted in the quadrupolar energy levels for a nucleus with $I = \frac{3}{2}$ (for example, ^{35}Cl) in an axially symmetric electric field and in the presence of a weak magnetic field.

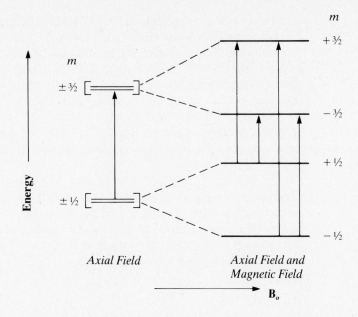

consider simple M—Cl bonds. The free Cl^- ion has a spherical electric field and is assigned arbitrarily an e^2Qq value of 0.00 MHz. When chlorine is bonded to a less electronegative atom such as lithium, there is relatively little asymmetry in the electric field, and the e^2Qq value for chlorine in LiCl is about -6 MHz. The asymmetry is much greater when chlorine is bonded to the more electronegative fluorine atom. The e^2Qq value for

Figure 7.11
First-derivative ^{35}Cl-NQR spectrum of crystalline $HgCl_2$ showing two resonances due to the different crystallographic locations of the two chemically equivalent Cl atoms. Adapted with permission from S. Melnick et al. *J. Mol. Struct.* 58 (1980): 337.

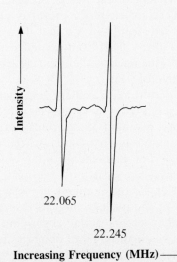

chlorine in ClF is approximately -145 MHz, affording evidence that the ClF bond is significantly more ionic than is the LiCl bond.

Chlorine-35 is probably the most widely studied NQR nucleus. For example, the ^{35}Cl-NQR spectrum of K_2PtCl_6 exhibits a single resonance in accord with the expected octahedral geometry of the hexachloroplatinate-(IV) anion. The analogous iodo-complex K_2PtI_6 is not isomorphous (that is, does not have the same crystal structure) with the chloro-derivative, however, because there are three ^{127}I-NQR resonances detected. This result suggests the possibility of three chemically nonequivalent iodine positions in the lattice.

At first sight, it would appear that NQR spectroscopy is a useful structural probe for solid materials. Although this is often the case, it must be used with caution because the crystal lattice can sometimes confuse the issue—chemically equivalent species are occasionally located at crystallographically nonequivalent positions in the lattice. Crystalline $HgCl_2$ is an example of such a situation (Figure 7.11).

7.3   *Mössbauer Spectroscopy*

Another type of nuclear spectroscopy that provides useful structural information for solids is *Mössbauer spectroscopy*. This spectroscopic technique owes its name to the discoverer R. L. Mössbauer (West Germany), who shared the 1961 Nobel Prize in Physics for his discovery. It is another resonance spectroscopy that this time results in transitions from low-energy nuclear states to higher-energy ones by the absorption of γ rays. The γ-ray source and sample nuclei are the same, although the γ-ray source is produced by the decay of a nucleus of a precursor element. The precursor in iron Mössbauer spectroscopy is ^{57}Co that decays initially by β decay (electron emission) to give a short-lived excited state of ^{57}Fe*, which has a half-life of 99.3 ns. This nucleus then undergoes γ decay (14.41 keV) to yield the ground state of ^{57}Fe.

When a shell is fired from an artillery gun, there is a visible recoil—you have certainly seen this effect in war movies. There is a similar recoil at the atomic level when a quantum of electromagnetic radiation is emitted from or absorbed by an atom. In most cases, the recoil energy is small enough to be ignored. For high-energy γ-ray emission or absorption, however, the quantum of energy is too large for the recoil energy to be neglected. The origin of the recoil energy in γ-ray emission must be the energy for the decay of the nucleus from its excited state to its ground state. A related situation occurs during absorption of a γ-ray quantum. Figure 7.12 illustrates schematically the conditions for γ-ray emission and absorption.

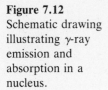

Figure 7.12
Schematic drawing illustrating γ-ray emission and absorption in a nucleus.

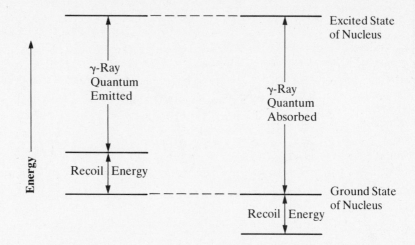

It appears from Figure 7.12 that a γ-ray quantum emitted by a nucleus can never be reabsorbed because of the different energy requirements. This is indeed true for gases and liquids. For a solid, however, the γ-ray emitter and absorber are in a rigid lattice, and the lattice vibrations are quantized. This means that there is a mechanism whereby the recoil energy can be dissipated to the whole lattice rather than to a single atom. The recoil energy can be considered negligible under these conditions.

During γ-ray absorption, the quantum of energy is usually sufficient to cause excitation of the vibrational lattice modes. Despite this, however, there is a finite probability of some of the nuclear absorption transitions taking place without any lattice-mode activation. The few such events that do occur are referred to as the *recoil-free fraction*. Mössbauer spectroscopy is the study of the recoil-free fractions of different nuclei.

A Mössbauer spectrometer consists of a γ-ray source, the sample (absorber), and two detectors. One detector measures the recoil energy, while the other records the decrease in energy when resonance occurs. The velocity of the source is varied until the energy absorbed by the sample is at a maximum. To effect sufficient absorption, it is almost always necessary to resort to measurements at liquid-nitrogen (77 K) or liquid-helium (4 K) temperatures. Under such conditions, the recoil energy is minimized, and the intensity of the Mössbauer signal should be at its greatest. The actual experiment consists of counting the number of γ rays passing through the sample for different source velocities (in mm s^{-1}). The source is moved both toward (positive velocities) and away (negative velocities) from the sample. The velocities corresponding to the peak maxima in a Mössbauer spectrum are called *isomer shifts*. These values reflect directly the differences in chemical environment between the nuclei in the source and in the sample.

Figure 7.13
^{57}Fe-Mössbauer spectrum of solid Fe(CO)$_5$ at 78 K. Adapted with permission from R. H. Herber, W. R. Kingston, and G. K. Wertheim. *Inorg. Chem.* 2 (1963): 153.

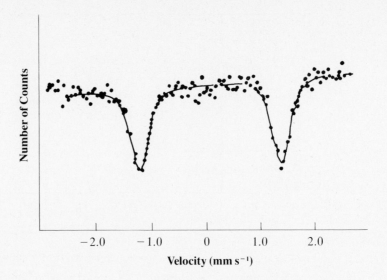

Although ^{57}Fe is the nucleus most often investigated in the Mössbauer effect, several other important nuclei have also been examined: ^{99}Ru, ^{119}Sn, ^{121}Sb, ^{125}Te, ^{129}I, ^{129}Xe, ^{182}W, and ^{197}Au. Some examples of the applications of Mössbauer spectroscopy are now given.

Mössbauer signals are often split because of asymmetry in the electric-field gradient around the nucleus due to nuclear quadrupole moment. The separation between the Mössbauer peaks is known as *quadrupole splitting*. Such a quadrupole splitting has been observed for Fe(CO)$_5$ at 78 K (Figure 7.13). This is because $I = \frac{3}{2}$ for the first excited nuclear state of ^{57}Fe and the degeneracy of this state is lifted in the presence of an asymmetric electric field, leading to two transitions from the ground state ($I = \frac{1}{2}$). In general, for structures that can be considered as having axially symmetric fields (for example, linear, trigonal-planar, square-planar, trigonal-bipyramidal, and octahedral geometries), the degeneracy of the excited state is given by $I + \frac{1}{2}$ for nuclei with spins $\frac{3}{2}$, $\frac{5}{2}$, and so on and by $I + 1$ for nuclei with spins 1, 2, and so on.

Certain ligands cause the $3d$ electrons in Fe(II) and Fe(III) compounds to be spin-paired (low-spin), while other ligands lead to the electrons being spin-free (high-spin) (see Chapter 13 for a detailed discussion of these effects). Table 7.6 lists the different possible electronic arrangements for Fe(II) and Fe(III). The most symmetric electronic arrangements are for low-spin Fe(II) and high-spin Fe(III). No quadrupole splittings would be expected for these symmetric arrangements. For the two other possibilities, high-spin Fe(II) and low-spin Fe(III), there should be some electronic asymmetry around the iron nuclei, and so quadrupole splittings

Table 7.6	*Possible Arrangements of 3d Electrons in Octahedral Iron(II) and Iron(III) Compounds*[a]	
Iron Species	Low-Spin	High-Spin
Fe(II)	$(t_{2g})^6$	$(t_{2g})^4(e_g)^2$
Fe(III)	$(t_{2g})^5$	$(t_{2g})^3(e_g)^2$

[a] The t_{2g} orbitals are the d_{xy}, d_{xz}, and d_{yz} orbitals, whereas the e_g orbitals are the $d_{x^2-y^2}$ and d_{z^2} orbitals. These group theoretical labels are always written in lower case when referring to orbitals (see Section 5.2). The superscripts refer to the number of electrons located in each set of orbitals.

are anticipated. These theoretical predictions are completely borne out by experiment. Iron-57 Mössbauer spectroscopy is therefore a convenient way of distinguishing between octahedral Fe(II) and Fe(III) complexes and those of lower symmetry. For this reason, there has been some Mössbauer work on iron-containing molecules of biological interest such as hemoglobins and peroxidases. The isomer shifts for ^{57}Fe are dependent on the oxidation state of the iron. The observed ranges are approximately as follows: Fe(II), +0.9 to +1.2; Fe(III), +0.15 to +0.7; Fe(IV), 0 to +0.3; Fe(VI), about −1.2 mm s^{-1}. One of the important species identified for horseradish peroxidase exhibits an isomer shift of +0.03 mm s^{-1}, suggesting the presence of Fe(IV).

Iron carbonyl chemistry has benefited greatly from Mössbauer studies. For example, a reasonable structure for $Fe_3(CO)_{12}$ would be the triangular shape shown below:

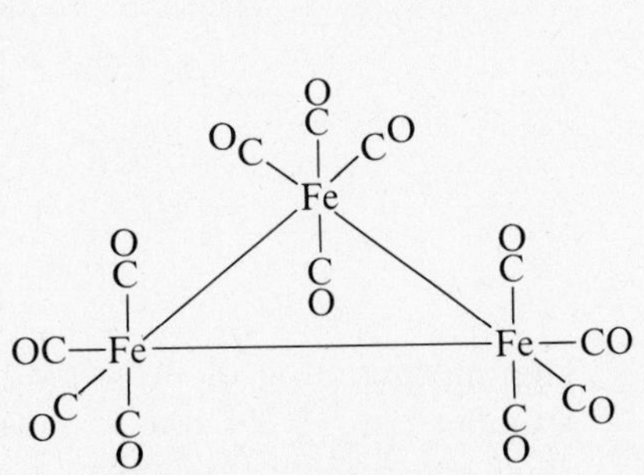

This symmetric geometry has been established for the related ruthenium and osmium dodecacarbonyls. The ^{57}Fe–Mössbauer spectrum of $Fe_3(CO)_{12}$ (Figure 7.14), however, provides evidence that the three Fe atoms cannot be equivalent. A single resonance would have been observed if this were

Figure 7.14
^{57}Fe–Mössbauer spectrum of solid $Fe_3(CO)_{12}$ at 78 K. Adapted with permission from R. H. Herber, W. R. Kingston, and G. K. Wertheim. *Inorg. Chem.* 2, (1963): 153.

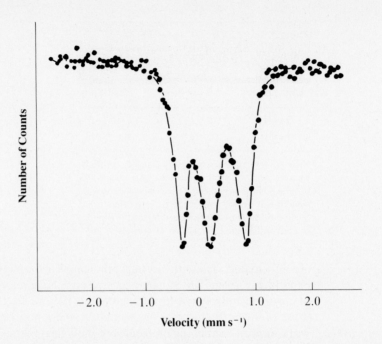

the case. Instead, there are three resonances of equal intensity. The structure of $Fe_3(CO)_{12}$ has now been firmly established from X-ray work (see below):

The interpretation of the Mössbauer spectrum is that the central, single resonance at $+0.208$ mm s^{-1} is due to the unique Fe atom, whereas the other two resonance maxima at $+0.847$ and -0.287 mm s^{-1} are the two halves of a quadrupole split line with an isomer shift of $+0.280$ mm s^{-1}. This doublet is due to the two equivalent Fe atoms.

As a final example of the applications of Mössbauer spectroscopy in inorganic chemistry, some gold-197 data have recently been reported for alkali metal–gold intermetallics such as CsAu. Figure 7.15 shows a typical spectrum of CsAu. The isomer shift (relative to gold metal) of the single

Figure 7.15
Gold-197 Mössbauer spectrum of CsAu at 4.2 K. Adapted with permission from R. J. Batchelor, T. Birchall, and R. C. Burns. *Inorg. Chem.* 25 (1986): 2009.

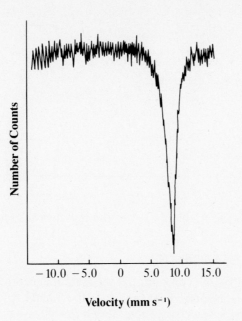

resonance is 8.2 mm s^{-1}, suggesting that the compound is best regarded as an ionic salt Cs^+Au^-—that is, gold is in the $-I$ oxidation state, not a very common occurrence. The usual oxidation states of gold are Au(I) and Au(III).

7.4 Electron-Spin Resonance Spectroscopy

The final resonance spectroscopy to be considered is *electron-spin resonance (ESR) spectroscopy*. [Sometimes this technique is referred to as electron paramagnetic resonance (EPR) spectroscopy.] In an ESR experiment, microwave energy is absorbed by paramagnetic molecular species, that is, those that contain unpaired electrons. The samples can be gases, liquids, or solids. It is difficult to obtain structural information on such paramagnetic materials by NMR spectroscopy because the signals are extremely broad. ESR spectroscopy is also useful for detecting short-lived, radical intermediates that are formed during many chemical reactions.

The spin angular momentum (m_s) associated with an electron of spin quantum number $\frac{1}{2}$ is quantized and has values of $\pm\frac{1}{2}$. In the presence of a magnetic field, there are two allowed spin states: a lower-energy one with $m_s = -\frac{1}{2}$ (spin magnetic moment aligned parallel to the direction of the applied field) and a higher-energy one with $m_s = +\frac{1}{2}$ (spin magnetic moment aligned directly opposed to the field) (Figure 7.16). The energy difference between the two spin states (ΔE) lies in the microwave region of the electromagnetic spectrum. A transition occurs between the two spin

Figure 7.16
Effect of applied
magnetic field on
the spin states of a
molecule containing
one unpaired
electron.

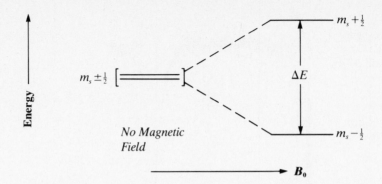

states when the frequency (ν) of the microwave radiation satisfies Equation 7.9:

$$\Delta E = h\nu = g\beta B_0 \tag{7.9}$$

where h is Planck's constant, β is the Bohr magneton, B_0 is the strength of the applied magnetic field, and g is the Landé splitting factor (usually simply referred to as the *g value*). For a free electron, the g value is 2.0023, but g can vary appreciably for different paramagnetic species. For gases and liquids, the magnitude of g is independent of the orientation of the sample with respect to the applied magnetic field, that is, it is *isotropic*. However, in the case of solids, it is very much dependent on this orientation, that is, it is *anisotropic*. Equations have been derived that relate the g values for paramagnetic species located at different sites in crystal-unit cells to the g values associated with the x, y, and z Cartesian axes of the paramagnetic species and their angular disposition with respect to the magnetic field. Because many molecules exhibit axial symmetry ($g_x = g_y \neq g_z$), the experimental g values most commonly found in the literature are *g parallel* ($g_{\parallel}$; that is, g_z parallel) and *g perpendicular* ($g_{\perp}$; that is, g_x and g_y perpendicular) to the magnetic field. A good organic example of a species with axial symmetry is the pyramidal methyl radical, $CH_3^{\bullet}$, which has C_{3v} point-group symmetry. The C_3 axis lies along the z axis.

Experimentally, an ESR spectrum is obtained by keeping the microwave frequency fixed and varying the applied magnetic field strength. The resulting spectra are usually displayed as first derivatives rather than absorption curves because ESR peaks are quite broad and contain many components. The number of derivative peaks (or lines as they are normally called) in an ESR spectrum depends on the *hyperfine splitting* that results from couplings between the spin of an unpaired electron and the spin of neighboring nuclei.

The interaction of an unpaired electron with a nucleus of spin I will result in $2I + 1$ lines of approximately equal intensity and spacing because the energy separations between the allowed levels are very small. A

Figure 7.17
ESR spectrum of
the methyl radical
CH$_3^{\cdot}$ in a methane
matrix at about
4 K. Adapted with
permission from
F. J. Adrian et al.
Adv. Chem. Ser. 36
(1962): 50.

4 Lines

B_0

detailed analysis of the selection rules for ESR spectra are not presented here, but some examples of spin–spin coupling are described.

Nitric oxide (NO) has one unpaired electron, and the most abundant nitrogen isotope is ^{14}N (99.6%, $I = 1$). This means that the ESR spectrum of ^{14}NO should have $2(1) + 1 = 3$ lines, provided the electron is localized on the nitrogen atom. This is exactly the type of spectrum obtained for the paramagnetic, low-spin Fe(II) $(3d^6)$, nitrosyl complex $[Fe(CN)_5(NO)]^{3-}$, indicating that the odd electron from the NO group resides on the nitrogen atom of the nitrosyl ligand.

In more complicated systems, it is possible to conceive of an electron being delocalized onto several different atoms. In such a case, if there are n-equivalent nuclei of spin I_i, the number of lines in the ESR spectrum is given by $2nI_i + 1$. When there are two types of coupling possible (with n- and m-equivalent nuclei of spins I_i and I_j, respectively), the general expression for the number of ESR lines becomes $(2nI_i + 1)(2mI_j + 1)$.

Both the methyl (CH$_3^{\cdot}$ and silyl (SiH$_3^{\cdot}$) radicals have three equivalent protons $(I = \frac{1}{2})$; thus, the ESR spectra would be expected to exhibit $[2(3)\frac{1}{2}] + 1 = 4$ lines in the absence of any other coupling. A four-line ESR spectrum is observed for the methyl radical (Figure 7.17) because there is no coupling with ^{12}C $(I = 0)$ and the natural abundance of ^{13}C $(I = \frac{1}{2})$ is only 1.11%. The ESR spectrum of the silyl radical should be more complex because ^{29}Si $(I = \frac{1}{2})$ has a reasonable natural abundance (4.7%), and so $4 \times \{[2(1)\frac{1}{2}] + 1\} = 8$ lines are expected.

The copper(II) ion has long been the mainstay of ESR studies because of its ubiquitous unpaired electron $(3d^9$ ground-state electronic configuration). A typical example of a Cu(II) complex is $[Cu(dipy)_3]^{2+}$ (dipy = 2,2'-dipyridyl) in which the metal is octahedrally coordinated to six nitrogen atoms. Copper has two naturally occurring nuclei suitable for ESR studies: ^{63}Cu (69.1%, $I = \frac{3}{2}$) and ^{65}Cu (30.9%, $I = \frac{3}{2}$). Let us consider coupling of the electron spin with the more abundant ^{63}Cu nucleus only—$[2(1)\frac{3}{2}] + 1 = 4$ lines would be expected. However, further coupling may occur with the six equivalent ^{14}N ($I = 1$) atoms, leading to each of the four major lines being split further into $[2(6)1 + 1] = 13$ additional lines. A total of $4 \times 13 = 52$ lines is therefore predicted for $[Cu(dipy)_3]^{2+}$. Whether all lines are actually observed depends on the relative magnitudes of the different coupling constants between the unpaired electron and the ^{63}Cu and ^{14}N nuclei.

There has been considerable interest in establishing whether there is electron delocalization from metal ions to ligands in paramagnetic complexes. Once again, Cu(II) chemistry has been the focus of attention. The elegant ESR work on bis(salicylaldiminato)(copper-63)(II) remains a classic in the field. The structure of the complex is shown below:

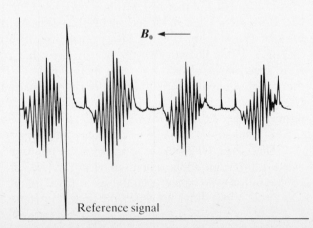

There are four main signals in the ESR spectrum, each split into 11 components (Figure 7.18). Substitution of the H_A atoms by deuterium (^2H, $I = 0$) produces no change in the spectrum, whereas replacement of the H_B atoms by CH_3 groups (^{12}C, $I = 0$) results in the splitting in each of the four major signals collapsing to five lines. These three results illustrate that the unpaired electron originally associated with the Cu(II) ion is indeed widely delocalized over the salicylaldiminato ligands. The splittings observed in the original spectrum arise from coupling of the electron spin to the ^{63}Cu

Figure 7.18
ESR spectrum of bis(salicylaldiminato)-copper(II) containing isotopically pure ^{63}Cu. Adapted with permission from A. H. Maki and B. R. McGarvey. *J. Chem. Phys.* 29 (1958): 35.

$B_0 \leftarrow$

Reference signal

$(I = \frac{3}{2})$, to two equivalent nitrogens (^{14}N, $I = 1$), and then to two equivalent protons ($I = \frac{1}{2}$), leading theoretically to $[2(2)1 + 1]\{[2(2)\frac{1}{2}] + 1\} = 5 \times 3 = 15$ lines. Some of these lines are presumably overlapped because only eleven lines were detected. When the proton coupling is removed by methyl group substitution, the predicted five-line spectrum is observed.

BIBLIOGRAPHY

Suggested Reading

Cotton, F. A., and G. Wilkinson. *Advanced Inorganic Chemistry. A Comprehensive Text*, 5th ed. New York: Wiley-Interscience, 1988.

Drago, R. S. *Physical Methods in Chemistry*. Philadelphia: Saunders, 1977.

Ebsworth, E. A. V., D. W. H. Rankin, and S. Cradock. *Structural Methods in Inorganic Chemistry*. Oxford, England: Blackwell, 1987.

Greenwood, N. N., and A. Earnshaw. *Chemistry of the Elements*. New York: Pergamon Press, 1984.

Guillory, W. A. *Introduction to Molecular Structure and Spectroscopy*. Boston: Allyn and Bacon, 1977.

Hatfield, W. E., and R. A. Palmer. *Problems in Structural Inorganic Chemistry*. New York: Benjamin, 1971.

Hill, H. A. O., and P. Day. *Physical Methods in Advanced Inorganic Chemistry*. New York: Wiley-Interscience, 1968.

Huheey, J. E. *Inorganic Chemistry. Principles of Structure and Reactivity*, 3rd ed. New York: Harper & Row, 1983.

Lambert, J. B., H. F. Shurvell, L. Verbit, K. J. Cooks, and G. H. Stout. *Organic Structural Analysis*. New York: Macmillan, 1976.

Nuclear Magnetic Resonance Spectroscopy

Akitt, J. W. *NMR and Chemistry. An Introduction to the Fourier Transform–Multinuclear Era*, 2nd ed. New York: Chapman & Hall, 1983.

Becker, E. D. *High Resolution NMR*. New York: Academic Press, 1969.

Carrington, A., and A. D. McLachlan. *Introduction to Magnetic Resonance*. New York: Harper & Row, 1971.

Fyfe, C. A. *Solid-State NMR for Chemists*. Guelph, Canada: CFC Press, 1983.

Gunther, H. *NMR Spectroscopy: An Introduction*. New York: Wiley, 1980.

Laszlo, P., ed. *NMR of Newly Accessible Nuclei*. New York: Academic Press, 1983.

Mason, J., ed. *Multinuclear NMR*. New York: Plenum, 1987.

Pople, J. A., W. G. Schneider, and H. J. Bernstein. *High-Resolution Nuclear Magnetic Resonance.* New York: McGraw-Hill, 1959.

Roberts, J. D. *Nuclear Magnetic Resonance*. New York: McGraw-Hill, 1959.

Nuclear Quadrupole Resonance Spectroscopy

Das, T. P., and E. L. Hahn. *Nuclear Quadrupole Resonance Spectroscopy*. New York: Academic Press, 1958.

Lucken, E. A. C. *Nuclear Quadrupole Coupling Constants*. New York: Academic Press, 1969.

Mössbauer Spectroscopy

Bancroft, G. M. *Mössbauer Spectroscopy*. London: McGraw-Hill, 1973.

Berry, F. J., and D. P. E. Dickson, eds. *Mössbauer Spectroscopy in Perspective*. Cambridge, Mass.: Cambridge University Press, 1985.

Gibb, T. C. *Principles of Mössbauer Spectroscopy*. London: Chapman & Hall, 1976.

Greenwood, N. N., and T. C. Gibb. *Mössbauer Spectroscopy*. London: Chapman & Hall, 1971.

Long, G. J., ed. *Mössbauer Spectroscopy Applied to Inorganic Chemistry, Vol. 1*. New York: Plenum, 1984.

Parish, R. V. "Mössbauer Spectroscopy and Chemistry." *Chem. Br.* 21 (1985): 546, 740.

Electron-Spin Resonance Spectroscopy

Poole, C. J. Jr. *Electron Spin Resonance*. New York: Wiley-Interscience, 1967.

Symons, M. C. R. *Chemical and Biochemical Aspects of Electron-Spin Resonance Spectroscopy*. New York: Van Nostrand Reinhold, 1978.

Wertz, J. E., and J. R. Bolton. *Electron Spin Resonance, Elementary Theory and Practical Applications*. New York: McGraw-Hill, 1972.

PROBLEMS

7.1   How many nuclear spin states would there be for nuclei with $I = 1$ and $I = \frac{3}{2}$, and what would the values of the nuclear spin angular momentum quantum numbers be in each case?

7.2 Compounds of the type XPF_5 (X = $MeNH_2$, Me_2O, Me_2SO, or C_5H_5N) exhibit ^{19}F-NMR spectra consisting of two sets of resonances, each having a quintet and a doublet with relative intensities 1:4. Predict the structure of these XPF_5 compounds. For further details, see E. L. Muetterties, T. A. Blither, M. W. Farlow, and D. D. Coffman. *J. Inorg. Nucl. Chem.* 16 (1960): 52.

7.3 Suggest a structure for the P_2NOCl_5 molecule that is produced upon treatment of PCl_3 with N_2O_4 at −10 °C, given that there are two resonances in its ^{31}P-NMR spectrum.

7.4 An anion of chemical stoichiometry $[W_2O_2F_9]^-$ displays a doublet and nonet in its ^{19}F-NMR spectrum with relative intensity 8:1. Suggest a structure consistent with these data. For further information, see W. McFarlane, A. M. Noble, and J. M. Winfield. *J. Chem. Soc. (A)* 948 (1971).

7.5 The ^{13}C-NMR spectrum of liquid $Fe(CO)_5$ exhibits a single peak that persists at lower temperatures. Would this observation be consistent with the known trigonal-bipyramidal structure for this compound? If not, explain why.

7.6 How would you account for the ^{35}Cl-NQR spectra observed for solid phosphorus pentachloride (two sets of signals consisting of 4 and 6 equal-intensity resonances, respectively) and $TeAlCl_7$ (two sets of signals consisting of 3 and 4 equal-intensity resonances, respectively)?

7.7 It has been claimed recently that the O_h symmetry of $[IF_6]^+$ suggested by its vibrational and ^{19}F-NMR spectra has been confirmed by the appearance of a single resonance in its ^{129}I-Mössbauer spectrum. Do you agree with this assertion?

7.8 Tellurium-125 ($I = \frac{1}{2}$; 7.0% natural abundance) is becoming a useful NMR and Mössbauer nucleus for probing the structure and bonding in tellurium compounds. Predict the number and relative intensities of the resonances expected in the ^{125}Te-NMR and −Mössbauer spectra of $[t-(CH_3)_3C]_3PTe$ and $\{[t-(CH_3)_3C]_2P\}_2Te$. For further details, see C. H. W. Jones and R. D. Sharma. *Organometallics* 6 (1986): 1419.

7.9 Explain why the ESR spectrum of NO_2 trapped in solid argon at very low temperatures exhibits three major lines.

7.10 Addition of sodium metal to naphthalene (see structure below) affords sodium naphthalide, $Na^+[C_{10}H_8^{\cdot}]^-$, in which the additional electron is delocalized over the entire naphthalene ring. Predict the number of lines expected in the ESR spectrum of sodium naphthalide.

7.11 Predict the ESR spectra of the borine radical anions $[^{10}BH_3^{\bullet}]^-$ and $[^{11}BH_3^{\bullet}]^-$.

7.12 How many lines would you expect to observe in the ESR spectrum of bis-(η^6-benzene)vanadium(0), $(\eta^6-C_6H_6)_2V$, if the natural abundance of ^{51}V ($I = \frac{7}{2}$) is 99.8%? For a detailed analysis of this spectrum, see M. P. Andrews, S. M. Mattar, and G. A. Ozin. *J. Phys. Chem.* 90 (1986): 1037.

7.13 The eight-coordinate octacyanomolybdate(IV) anion, $[Mo(CN)_8]^{4-}$, is believed to have a square-antiprismatic structure. Would the ESR spectrum of the anion and the totally ^{13}C-enriched species help in confirming this structure?

8

Photoelectron, X-Ray Absorption, and Mass Spectroscopies and X-Ray Crystallography

This is the final chapter in the brief introduction to some of the major modern analytical techniques used by inorganic chemists in molecular structure determination. With the exception of X-ray crystallography, all the methods discussed are based in some way on the ejection of electrons from atoms in molecules. Photoelectron spectroscopy provides quantitative data on the binding energies of electrons in AOs and MOs, whereas mass spectroscopy yields information about the masses of the cationic species produced upon the fragmentation of molecules upon impact by electrons and neutral atoms. Two other analytical methods to be described are the result of the absorption of X rays by atoms: extended X-ray absorption fine structure and X-ray absorption near-edge structure spectroscopy. These two techniques are becoming increasingly important because the former provides data on the distances between bonded atoms for liquids, solids, and sometimes even gases, whereas the latter yields diagnostic information on oxidation states. The last structural analysis method to be introduced in this chapter is X-ray crystallography, which has become firmly entrenched as the technique of choice when determining the molecular structures of crystalline solids.

8.1   Photoelectron Spectroscopy

Since its initial discovery in the earlier part of this century, *photoelectron spectroscopy* (*PES*) has received considerable Nobel Prize recognition in Physics. Modern PES was developed by K. M. Siegbahn (Uppsala, 1981) and is based on earlier work in two related fields: the photoelectric effect (A. Einstein, Berlin, 1921) and X-ray spectroscopy (M. Siegbahn, Uppsala, 1924).

In an effort to obtain a more quantitative understanding of chemical bonding, theoretical chemists continually perform more complicated MO calculations. To do this, however, the energies of the valence and inner-core orbitals must be known with some degree of accuracy. These energies can now be obtained for the valence orbitals by ultraviolet–photoelectron spectroscopy and for the inner-core orbitals by X-ray photoelectron spectroscopy. (The latter method is sometimes referred to by the acronym ESCA from electron spectroscopy for chemical analysis.) The binding energies of the inner-core electrons can often be taken as fingerprints for the chemical environment of an atom in a molecule. The only significant difference between an ultraviolet–photoelectron and an X-ray–photoelectron spectroscopy measurement is the ionization source: vacuum–UV radiation ($h\nu$ is less than 50 eV) for the ultraviolet-photoelectron experiments and more highly energetic X rays ($h\nu$ is more than 1000 eV) for X-ray photoelectron spectroscopy work.

Ultraviolet–Photoelectron Spectroscopy Studies

In a typical *ultraviolet–photoelectron spectroscopy* (*UPS*) experiment, a monochromatic beam of UV photons from a helium source emitting either the He(I) line ($h\nu = 21.22$ eV) or the He(II) line ($h\nu = 40.8$ eV) is directed onto a gaseous molecular sample (M), and electrons are ejected from the valence MOs with kinetic energies, $\frac{1}{2} m(v_{\text{eject}})^2$, leading to the formation of M^+ ions. (NOTE: M^{2+}, M^{3+}, and so on ions are not normally formed in a photoelectron experiment):

$$M + h\nu \rightarrow M^+ + e^- \tag{8.1}$$

The velocity, v_{eject}, of the ejected electrons is given by Einstein's photo-electric law (Equation 8.2):

$$\tfrac{1}{2} m(v_{eject})^2 = h\nu - IE \tag{8.2}$$

where IE = ionization energy.

The number of ejected electrons and their energies are measured by the PE spectrometer, and the resulting data are plotted in graphical form on either a strip-chart recorder or processed on a computer terminal. Some representative spectra for H, H_2, and He are illustrated schematically in Figure 8.1. The one-electron species H exhibits a single sharp peak at 13.6 eV due to ionization of the $1s^1$ electron. There is also one main band for H_2 centered at about 16.1 eV that results from the ionization of one electron from the σ_1–bonding MO. The difference between the PE band positions for these two species, $16.1 - 13.6 = 2.5$ eV, is a direct reflection of the greater stability of a H_2 molecule compared to a H atom.

Figure 8.1
UV–PE spectra of the one-electron species H and the two-electron species H_2 and He. Adapted with permission from R. L. Dekock, and H. B. Gray, *Chemical Structure and Bonding*, Menlo Park, California: Benjamin/Cummings, 1980.

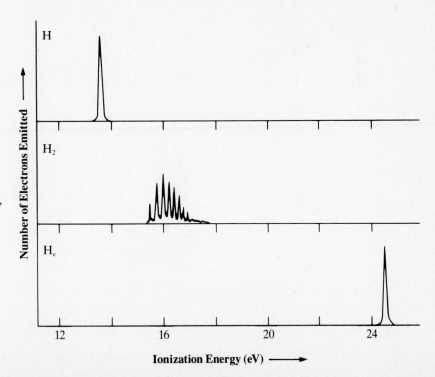

The additional peaks on the H_2 band are due to vibrational fine structure; the energy spacings are approximately 2200 cm^{-1}, corresponding to the vibrational energy of the H_2^+ molecule. Figure 8.2 illustrates the potential energy diagram, which shows the relationship between H_2 and H_2^+. There are a series of electronic transitions from the ground state of H_2 to the manifold of vibrational states in the electronic state of H_2^+. The most probable transition is that for which there is maximum overlap of the vibrational wave functions involved. This transition is referred to as the *vertical transition*, and on the diagram this is the transition between the

Figure 8.2
Qualitative potential-energy curves for H_2 and H_2^+ illustrating the origin of the vibrational fine structure on the band in the UV–PE spectrum of H_2.

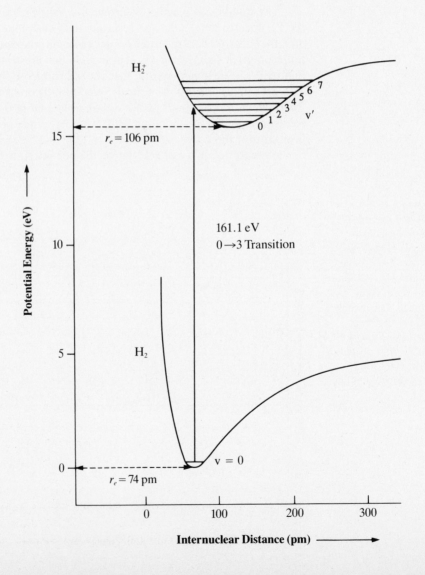

$v = 0$ state of H_2 to the fourth vibrational level ($v' = 3$) of H_2^+, the so-called $0 \rightarrow 3$ transition. This situation occurs because the bonding between H_2^+ is weaker than that in H_2, and consequently the internuclear separation is greater in H_2^+ ($r_e = 106$ pm) than in H_2 ($r_e = 74$ pm). Other transitions from the H_2 ground state to different vibrational levels of H_2^+ are also possible, for example, $0 \rightarrow 1$, $0 \rightarrow 2$, $0 \rightarrow 4$, and so on. The intensities of these transitions are lower than that of the main vertical one. All the various PE (or Franck–Condon) transitions are much more rapid than are molecular vibrations.

In the case of the He atom, there is a solitary sharp peak in Figure 8.1 at 24.6 eV. The ionization energy for the He atom is appreciably greater than that for the H atom (namely, $24.6 - 13.6 = 11.0$ eV) because the spins of the two electrons in the He $1s$ orbital are paired and because the effective nuclear charge on the He nucleus that is responsible for the electron attraction is much greater than that on the H nucleus.

The energy E_{1s} of the $1s$ orbital for all one-electron species is exactly equal to the negative of the measured ionization energy:

$$E_{1s} = -(IE)_{1s} \qquad \qquad (8.3)$$

Koopman was the first to suggest that Equation 8.3 ought also to apply to the energy of an orbital i in a many-electron atom or molecule (M)

$$E_i = -(IE)_i \qquad \qquad (8.4)$$

provided that the orbitals of the resulting M^+ ions can be assumed to be unchanged from those in the neutral M species—the so-called *frozen-orbital approximation*. This assumption is never going to be exactly true because the electron correlation in M and M^+ is not the same. Therefore, *Koopman's theorem* may not always aid in making assignments of the ionization energies observed in a UPS spectrum. The matching of these ionization energies $(IE)_i$ with the energies E_i derived theoretically for specific orbitals from MO calculations may not always be valid, especially for the closely spaced ionizations of a transition metal complex.

Because commercial PE spectrometers have become more widely available, many different types of molecules have been investigated. For example, consider the UPS results shown in Figure 8.3 for the dinitrogen (N_2) molecule. The PE spectrum clearly illustrates that the filled π_1-bonding orbitals are at higher energy than is the filled σ_5 orbital. This result is important when describing the chemical bonding in homonuclear diatomic molecules, as we discussed earlier.

Inorganic chemists are especially interested in the PE spectra of coordination compounds, and a good example is the spectrum of molybdenum hexacarbonyl $Mo(CO)_6$, shown in Figure 8.4. Although the bonding in metal carbonyls and other organometallic complexes is discussed in detail

Figure 8.3
He(II) (30.4 nm, 40.8 eV) PE spectrum of gaseous N_2. (See Figure 3.8.) Adapted with permission from J. L. Gardner and J. A. R. Samson. *J Chem. Phys.* 62 (1975): 1447.

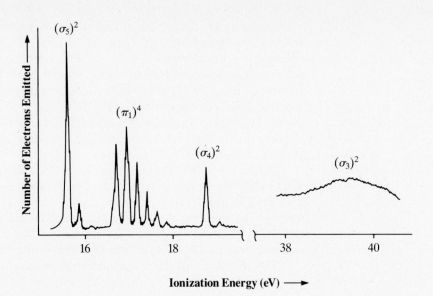

in Chapter 22, this PE spectrum is still informative at this stage: It is typical of the spectra found for all three group-VIB(6) metal hexacarbonyls, $M(CO)_6$ (M = Cr, Mo, W), and it illustrates the great utility of UPS in assigning the energies of MOs in many-atom molecules. There are two main features in the He(I) spectrum, a fairly sharp one at about 8.5 eV and a more complicated, much broader one in 12–17 eV region. The latter region is better resolved in the He(II) spectrum. Apart from this difference, the two spectra are quite similar.

The ground-state electronic arrangement for the valence electrons of Mo is $(t_{2g})^6$, in which the six electrons are paired in three degenerate orbitals (the reason for the spin pairing will become apparent in Chapter 13). Also, from our earlier discussion on orbital symmetry (Section 5.2), there are three sets of σ-bonding MOs for octahedral complexes such as $Mo(CO)_6$, with symmetries a_{1g}, e_g, and t_{1u}. The complete ground-state electronic configuration for $Mo(CO)_6$ with six electrons coming from the metal and twelve electrons from the six CO groups ($2e^-$ per CO) is $(e_g)^4$-$(a_{1g})^2 (t_{1u})^6 (t_{2g})^6$. The six σ-bonding MOs are Mo—CO in character, whereas the t_{2g} molecular orbitals are mainly $Mo_{d\pi}$. The t_{2g} orbitals are expected to be less stable than the σ-bonding orbitals; thus, the first peak in the PE spectrum is attributed to ionization of the t_{2g} electrons. The second peak is in the neighborhood of the ionization energy for the bonding orbital for CO itself (~14 eV), and consequently this peak is assigned to a mixture of ionizations from the σ-bonding Mo—CO orbitals and the CO ligands.

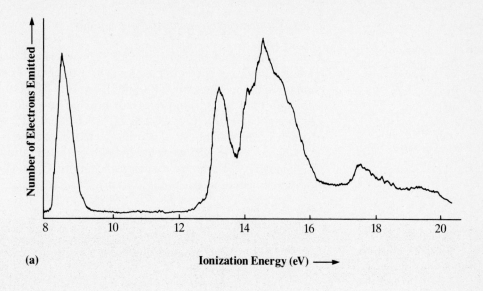

(a)

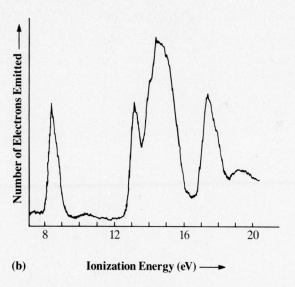

(b)

Figure 8.4 **(a)** He(I) (58.4 nm) and **(b)** He(II) (30.4 nm) UV–PE spectra of gaseous $Mo(CO)_6$. Similar spectra are obtained for the two other metal hexacarbonyls. Adapted with permission from B. R. Higginson, D. R. Lloyd, P. Burroughs, D. M. Gibson, and A. F. Orchard. *J. C. S., Faraday 2, 69* (1973): 1659.

X-Ray Photoelectron Spectroscopy Studies

X-ray photoelectron spectroscopy (*XPS*) provides a direct measure of the binding energies of the core electrons. The principle of an XPS measurement is similar to that for a UPS one. However, XPS is a lower-resolution technique than is UPS and does not provide any information about the vibrational band structure, as does UPS. The final XPS spectrum consists of a plot of the number of electrons ejected versus the binding energy.

A great deal of XPS work has focused on the variation of the $1s$-orbital binding energy for selected atoms (for example, C, O, and N) in a variety of different bonding situations. Often, there are detectable changes in

Figure 8.5
X-ray PE spectrum of *trans*-[Co(en)$_2$-(NO$_2$)$_2$]NO$_3$ in the N($1s$) region. Adapted with permission from J. M. Hollander and W. L. Jolly. *Acc. Chem. Res.* 3 (1970): 193.

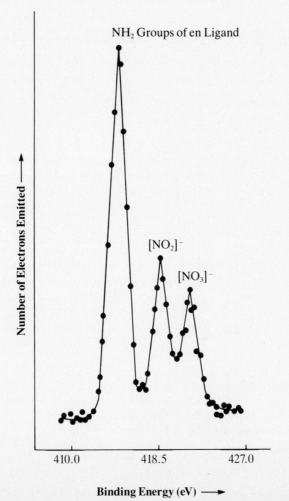

Figure 8.6
X-ray PE spectrum
(Al—Kα radiation,
1487 eV) of W(CO)$_6$
indicating the
energies associated
with several of the
W-core orbitals.
Adapted with
permission from
B. R. Higginson,
D. R. Lloyd, P.
Burroughs, D. M.
Gibson, and A. F.
Orchard. *J. C. S.,
Faraday 2*, 69 (1973):
1659.

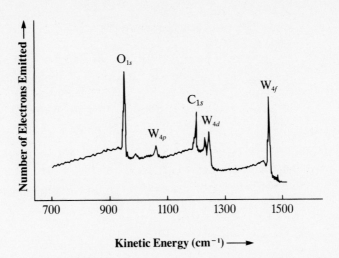

binding energy as the result of subtle changes in the chemical environment of the atom. For instance, in the classic example *trans*-[Co(en)$_2$(NO$_2$)$_2$]-NO$_3$ (en = ethylenediamine, H$_2$NCH$_2$CH$_2$NH$_2$), there are three diferent kinds of nitrogen atom present: N(III) in the four NH$_2$ groups of the en ligands, N(III) in [NO$_2$]$^-$, and N(V) in [NO$_3$]$^-$. Each of these different types of N atom exhibits a characteristic peak in the XPS spectrum of the complex (Figure 8.5). Similar spectra have been obtained for C and O 1s-binding energies in transition metal carbonyl complexes. As an example, the complete X-ray PE spectrum of W(CO)$_6$ is shown in Figure 8.6.

8.2 *Extended X-Ray Absorption Fine Structure and X-Ray Absorption Near-Edge Structure Spectroscopies*

In an XPS experiment, a core electron is ejected when an X-ray photon is absorbed by an atom. As a secondary effect of the electron ejection, an oscillation is introduced in the absorption coefficient on the high-energy side of the absorption edge. This oscillation is due to constructive and destructive interference of the scattered electron wave with other electron waves that result from the backscattering from neighboring atoms (Figure 8.7). The variations observed in the absorption coefficient are a direct reflection of the arrangement of the atoms in a molecule, and in many instances, it is possible to extract the distances of a given atom from its neighbors, irrespective of the nature of the sample.

The X-ray source in an extended X-ray absorption fine structure (EXAFS) experiment must be tunable because the absorption edges differ for each element. In most cases, the source used is high-intensity synchrotron radiation. The approach is to measure the total intensity associated

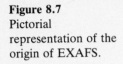

Figure 8.7
Pictorial
representation of the
origin of EXAFS.

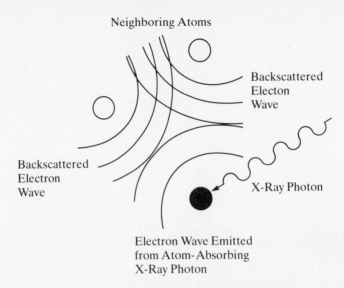

Neighboring Atoms

Backscattered
Electon
Wave

Backscattered
Electron
Wave

X-Ray Photon

Electron Wave Emitted
from Atom-Absorbing
X-Ray Photon

with the oscillating pattern at the absorption edge and then to refine a
proposed structural model to fit the observed intensity data. The mathematical function for the intensity is complex, but it critically depends on the
distances from the cental atom to its nearest neighbors. These distance
parameters are adjusted using a least-squares computer program to give the
best fit for the intensity data.

A relatively simple EXAFS spectrum, as shown in Figure 8.8, for the
uranium X-ray absorption edge in the solid uranium(VI) salt $Na[UO_2$-
$(CH_3COO)_3]$ serves to illustrate the type of spectrum typically obtained.
The first peak in the spectrum corresponds to the axial U—O distances
(175.5 pm) in the uranyl ion $[UO_2]^{2+}$, the second to the equatorial U—O_6
distances (245.9 pm) in which the three acetate ligands function as
bidentate ligands. There is no evidence of any equatorial U—C_3 (the carbon atoms of the carboxylate groups) interaction. In another example, the
EXAFS data for X-ray absorption by technetium in $[TcCl_6]^{2-}$ ion yield a
Tc—Cl distance of 237 pm compared to that of 235 pm from single-crystal
X-ray diffraction. Also, the number of equivalent Cl neighbors around the
central Tc atom is 6.6, in agreement with the anticipated octahedral
coordination.

X-ray absorption spectra also often exhibit some structure below the
absorption edge leading to a second type of X-ray absorption spectroscopy—*X-ray absorption near-edge structure (XANES)*. This edge structure
arises from excitation of some core electrons of an atom to vacant, higher-
energy orbitals. The observed spectral band shapes have been found to be
particularly useful in identifying the oxidation states of certain elements.
For example, the data given in Figure 8.9 reveal that Au(0), Au(I), and

Figure 8.8
EXAFS spectrum of
$Na[UO_2(CH_3COO)_3]$.
Adapted with
permission from
A. Dejean, P.
Charpin,
G. Folcher,
P. Rigny, A.
Navaza, and G.
Tsoucaris.
Polyhedron 6
(1987): 189.

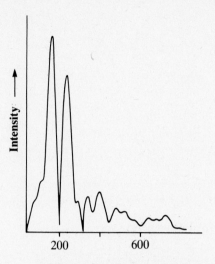

Distance from Central Atom (pm)

Figure 8.9
XANES spectra of
$[Au(NH_3)_4]^{3+}$,
$[Au(S_2O_3)_2]^{3-}$, and
gold-metal foil
illustrating the effect
of the respective
gold oxidation
states—Au(III),
Au(I), and
Au(0)—on the
observed spectra.
Adapted with
permission from
R. C. Elder and
M. K. Eldress,
Chem. Rev.
87 (1987): 1027.

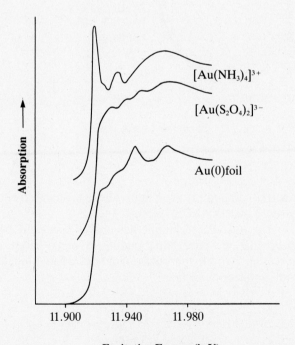

Excitation Energy (keV)

Figure 8.10
(a) XANES spectra of auranofin (———) and auranofin bound to bovine serum albumin (BSA) (– – –), indicating retention of the Au—P bond on binding to BSA. **(b)** Comparison of XANES spectra for a series of gold compounds containing different numbers of Au—P bonds. Note that the characteristic peak associated with Au—P bonding increases in intensity with increasing Au—P coordination number. Adapted with permission from R. C. Elder and M. K. Eidness. *Chem. Rev.* 87 (1987): 1027.

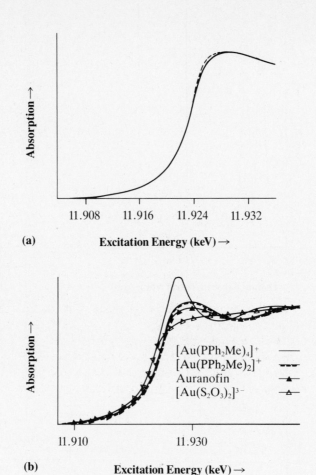

(a)

(b)

Au(III) compounds are readily distinguishable from one another on the basis of their XANES spectra.

The XANES technique has proved especially powerful in identifying the oxidation states and bonding in metal-based drugs. For example, *auranofin* [(triethylphosphine)gold tetraacetylthioglucose] is a licensed, oral antiarthritic drug.

$Ac = CH_3CO$

From a comparison of its XANES spectrum (Figure 8.10a) with Figure 8.9, it is clear that auranofin is indeed a Au(I) compound, as expected. Moreover, because auranofin contains both Au—S and Au—P bonds, it is important to know which of these bonds is retained when auranofin binds to protein under clinical conditions. Such information was unattainable before the development of XANES spectroscopy because crystals of the bound drug had never been isolated. A XANES spectrum of auranofin bound to bovine serum albumin (BSA) has now been reported (Figure 8.10b). This spectrum closely resembles those of the model coordination compounds containing Au—P bonds that are also shown in Figure 8.10a. The typical Au—P peak observed at around 11.928 keV is completely absent in the case of Au—S bonding.

8.3   Mass Spectroscopy

Mass spectroscopy is an extremely useful technique for obtaining accurate information on the molecular weights of compounds and on the structures of molecules. The most common procedure used is *electron-impact mass spectrometry* (*EIMS*), which involves bombardment of a molecule (M) with high-energy electrons (usually ~70 eV) and then a mass analysis of the fragment ions produced. The radical cation, $M^{\cdot+}$, which is produced by the following process, is known as the *parent molecular ion*:

$$M + e^- \rightarrow M^{\cdot+} + 2e^- \tag{8.5}$$

All types of samples—gases, liquids, or solids—can be analyzed by mass spectroscopy. The compound being examined is introduced into the sample chamber of the mass spectrometer, which is maintained at an extremely low pressure (less than 10^{-6} torr) and can be easily heated to 200–300 °C. Under these conditions, most compounds volatilize sufficiently for analysis. Electron bombardment results in the formation of a host of ions that are separated from one another according to their mass/charge (m/e) values by passing them through a strong external magnetic field. The m/e ratios for the ions can be measured extremely accurately, and the actual composition of the ions can often be established by comparison with the m/e values predicted for various molecular fragments.

Figure 8.11 shows some of the kinds of ionization processes that can occur in a mass spectrometer for a gaseous species such as tetracarbonylnickel(0) $Ni(CO)_4$. Note that *doubly charged ions* can also be produced and detected; these ions are usually of low abundance and appear at m/e values exactly half that of the singly charged ion:

$$M^+ + e^- \rightarrow M^{2+} + 2e^- \tag{8.6}$$

Figure 8.11
Part of the
fragmentation
scheme observed in
the EIM spectrum
of gaseous
tetracarbonylnickel(0)
$Ni(CO)_4$.

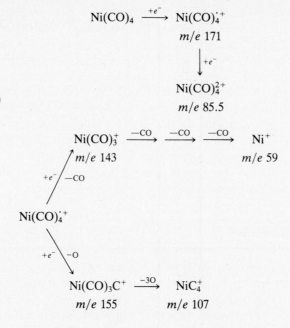

$$Ni(CO)_4 \xrightarrow{+e^-} Ni(CO)_4^{.+}$$
$$m/e\ 171$$

$$\downarrow {+e^-}$$

$$Ni(CO)_4^{2+}$$
$$m/e\ 85.5$$

$$Ni(CO)_3^+ \xrightarrow{-CO} \xrightarrow{-CO} \xrightarrow{-CO} Ni^+$$
$$m/e\ 143 \qquad\qquad\qquad m/e\ 59$$

$${+e^-}\nearrow {-CO}$$

$$Ni(CO)_4^{.+}$$

$${+e^-}\searrow {-O}$$

$$Ni(CO)_3C^+ \xrightarrow{-3O} NiC_4^+$$
$$m/e\ 155 \qquad\qquad m/e\ 107$$

Using a modern mass spectrometer, it is possible to determine the exact masses of parent molecular ions and molecular fragments extremely accurately. Moreover, if the parent molecular ion of a purely organic compound can be identified unambiguously from its exact mass, some scientific journals will accept this as proof of the proposed chemical composition of a new compound. This is not the case for inorganic compounds because of the wide variety of different atoms present; it is still necessary to rely on traditional elemental analysis for their proper identification.

New ionization techniques that permit mass spectra to be obtained for relatively involatile or thermally sensitive materials have been developed recently. One of these techniques is known as fast atom bombardment (FAB). In this process, molecules are bombarded with high-energy, neutral atoms such as argon, leading to the ejection of electrons and subsequent formation of a variety of ions inaccessible by electron impact, for example,

$$M + Ar \rightarrow M^+ + e^- + Ar \tag{8.7}$$

Compared to EIMS, the FAB technique is a relatively "soft" ionization process; unlike EIMS, the parent molecular ion is more often than not the base peak—that is, it is the major ion present. In this way, the molecular weights of relatively involatile species, including complex cations and anions, can now be obtained.

The EI mass spectrum of $Ni(CO)_4$ (mentioned above) is typical of most transition metal carbonyl complexes in that there is sequential loss of

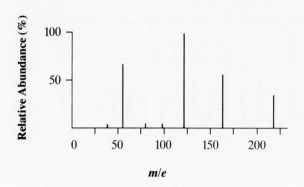

m/e	Ion	Relative Abundance (%)
220	CpMn(CO)$_2$(CS)$^+$	31
192	CpMn(CO)(CS)$^+$	1.4
164	CpMn(CS)$^+$	55
120	CpMn$^+$	100
99	Mn(CS)$^+$	6.5
93	MnC$_3$H$_2^+$	4.5
91	MnC$_3^+$	2.1
80	MnC$_2$H$^+$	6.2
66	C$_5$H$_6^+$	2.7
65	Cp$^+$	2.4
56	MnH$^+$	2.1
55	Mn$^+$	65
39	C$_3$H$_3^+$	6.5

Figure 8.12
EIM spectrum (70 eV) of CpMn(CO)$_2$(CS) at room temperature. Adapted with permission from A. E. Fenster. Ph.D. Thesis, McGill University, Montreal, 1972.

CO groups from the parent molecular ion to give a series of daughter ions and ultimately the bare metal ion. As a result of the large number of CO groups being fragmented off, one of the major peaks in the mass spectra is CO$^+$ (m/e 28).

In the organometallic chemistry field, there are literally hundreds of cyclopentadienyl complexes known, such as CpMn(CO)$_3$ (Cp = η^5-C$_5$H$_5$). The mass spectra of these complexes do exhibit parent molecular ions—for example, CpMn(CO)$_3^{\cdot+}$ (m/e 204)—but the base peak is usually CpM$^+$. Moreover, in cyclopentadienyl compounds that also contain metal carbonyl groups, it is unusual to detect M(CO)$^+$ in any appreciable abundance. However, in the mass spectrum of the thiocarbonyl derivative, CpMn(CO)$_2$(CS), a direct analog of the tricarbonyl complex already mentioned, one of the main fragment ions is Mn(CS)$^+$ (m/e 99), whereas the corresponding Mn(CO)$^+$ fragment ion (m/e 83) is almost nonexistent (Figure 8.12). This observation strongly suggests that the CS ligand is more strongly bonded to manganese than are the two CO groups, in agreement with other spectroscopic and structural evidence.

Figure 8.13 shows the mass spectrum of a cyclic cobalt carbonyl siloxane complex [CH$_3$(Co(CO)$_4$)SiO]$_5$; Table 8.1 gives the assignments proposed for a selection of the observed peaks. Note the stepwise loss of the last fifteen of the twenty CO groups and the presence of an appreciable number of doubly charged ions. Compounds such as this may eventually prove useful as high molecular weight mass markers.

Exact mass measurements can frequently be made to within a few millimass units—for example, the measured exact mass for a fragment ion of an organosilicon compound is 252.988, whereas that calculated for the most abundant isotopes of ^{12}C (12.00000), ^{1}H (1.00783), ^{28}Si (27.97693), and ^{16}O (15.99492) corresponding to the formula C$_4$H$_{13}$Si$_4$O$_5$ is 252.98411. The exact masses are determined relative to a calibration peak, often

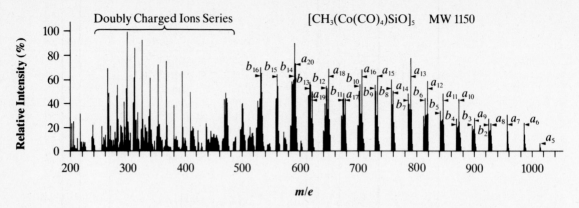

Figure 8.13
EIM spectrum (70 eV) of tetracarbonylcobalt-methylcyclopentasiloxane $[CH_3(Co(CO)_4)SiO]_5$ at 50 °C. Adapted with permission from E. Pelletier and J. F. Harrod. *Can. J. Chem.* 61 (1983): 762.

$^{13}CF_3^+$ (m/e 69.99855) in perfluorokerosine. The measured exact mass of a peak (most commonly, the parent molecular ion) is then matched by computer. The exact atomic weights and isotopes of the possible atoms must be programmed into the computer. Often, the computer generates several possible combinations of atoms, but the choices can be narrowed down by an examination of the fragmentation patterns and an analysis of the isotopic clusters. Chlorine- and bromine-containing compounds are quite readily identified by their relative abundances ($^{35}Cl:^{37}Cl$, 3:1; $^{79}Br:^{81}Br$, 1:1) and the absence of peaks for the intermediate mass numbers.

Metal-containing ions are especially easily identified by their distinctive isotope patterns. An excellent example of this is for transition metal cluster complexes—these complexes contain several metal atoms (sometimes of more than one element). Computer analysis of the metal isotope distribution enables the determination of the number of metal atoms. Such

Table 8.1 *Assignments Proposed for a Selection of Peaks Observed in the EI Mass Spectrum of $[CH_3(Co(CO)_4)SiO]_5$*

Peak Number	m/e	Assignment[a]
a_5	1010	M−5CO, $C_{20}H_{15}Si_5Co_5O_{20}^+$
a_{10}	870	M−10CO, $C_{15}H_{15}Si_5Co_5O_{15}^+$
a_{20}	590	M−20CO, $C_5H_{15}Si_5Co_5O_5^+$
b_2	923	M′−2CO, $C_{19}H_{15}Si_5Co_4O_{19}^+$
b_{16}	531	M′−16CO, $C_5H_{15}Si_5Co_4O_5^+$

[a] The parent molecular ion in the *a* series, $\{CH_3[Co(CO)_4]SiO\}_5^{\cdot+}$ ($M^{\cdot+}$), was too weak to be detected, as was the parent ion of the *b* series, $M'^{\cdot+} = [M-Co(CO)_4]^{\cdot+}$.

Table 8.2 *Comparison of Average Observed Multiplets[a]
with Theoretical Isotopic Multiplet Patterns
for $(\mu\text{-}H)_3Os_3Ni(\eta^5\text{-}C_5H_5)(CO)_9^+$*

Mass	Calculated Relative Intensity (%)	Observed Relative Intensity (%)	Difference
938	0.0067	0.0000	0.0067
939	0.0203	0.0275	−0.0071
940	0.0855	0.1242	−0.0388
941	0.2175	0.2346	−0.0171
942	0.6101	0.6486	−0.0385
943	1.1974	1.2834	−0.0860
944	2.5356	2.7737	−0.2382
945	3.8458	4.0704	−0.2245
946	6.4309	6.5824	−0.1516
947	7.8145	7.9209	−0.1063
948	11.5747	11.3976	0.1772
949	10.6817	10.4874	0.1943
950	14.0001	13.4674	0.5327
951	9.8484	9.6591	0.1893
952	13.1711	12.5573	0.6138
953	4.4981	4.6503	−0.1522
954	8.2784	8.3339	−0.0556
955	1.8052	1.8901	−0.0849
956	2.3781	2.5939	−0.2158
957	0.5148	0.5657	−0.0509
958	0.4216	0.5106	−0.0891
959	0.0652	0.1104	−0.0452
960	0.0737	0.1104	−0.0367
961	0.0113	0.0000	0.0113
Total	100.0868	100.0000	−0.0868

[a] Adapted with permission from G. Lavigne et al. *Inorg. Chem.*
22 (1983): 2485. The calculations were performed using the
program MASPAN.

information is difficult to obtain in the absence of a complete single-crystal
X-ray diffraction study. Table 8.2 illustrates the results of such a mass
spectrometric analysis for the parent molecular ion multiplet of the
trihydro mixed-metal tetranuclear cluster $(\mu\text{-}H)_3Os_3Ni(\eta^5\text{-}C_5H_5)(CO)_9$.
This chemical formulation has since been substantiated by single-crystal
X-ray diffraction.

8.4 X-Ray Crystallography

The importance of *X-ray crystallography* is clearly highlighted by the unusually large number of scientists who have received Nobel Prizes for their pioneering work in this field: von Laue (1914) and the Braggs (1915) for the discovery of X-ray diffraction by single crystals; Perutz, Crowfoot-Hodgkin, Kendrew, Sanger, and Pauling for later determining the structures of several proteins. Despite all this activity, it was rare, even twenty years ago, to find a synthetic inorganic chemist who contemplated determining the structure of a compound by single-crystal X-ray diffraction. Today, such structural analyses have become almost routine because of the ready commercial availability of automatic diffractometers and user-friendly, structure-determination software packages. The crystal structure of a molecule containing about 100 atoms can often be solved in a matter of days, and the complicated computer programs necessary for structural analysis can even be run on personal computers. The cost of an X-ray structure used to be prohibitive, and most inorganic chemists had to rely on the goodwill of their colleagues to have any structural work done at all. This is no longer true, and there are private companies that provide X-ray diffraction services for reasonable prices. Because X-ray crystallography has now become such a crucial technique in chemistry, students should have at least a rudimentary idea of the principles involved. In this section, we attempt to provide this, but the complex mathematics of X-ray diffraction are not discussed. The interested reader is referred to the specialized books listed at the end of the chapter.

Since the time of the Nobel Laureates M. von Laue (Frankfurt), W. H. Bragg (London) and W. L. Bragg (Manchester), it has been known that X rays are diffracted by the atoms in crystals. Today, we have an excellent idea of the scattering expected for the atoms of essentially every element. An X-ray crystallographic study is simply an attempt to match up the intensities of the diffracted X rays with those calculated by computer from the known scattering factors of each atom and an assumed spatial arrangement of the atoms in the crystal. The first part of such an X-ray investigation involves determining the space-group symmetry and the unit-cell dimensions of the crystal lattice.

A *unit cell* is defined as the simplest, three-dimensional arrangement of atoms that repeats itself throughout the whole crystal lattice—think of it as the repeating pattern of a wallpaper design in three, rather than two, dimensions. Six parameters are needed to define a unit cell mathematically: three cell edges (a, b, c) and three angles (α, β, γ). By convention, α is the angle between b and c, β between a and c, and γ between a and b. There are seven principal types of unit cell possible, each with its own interrelationship of cell edges and angles. These seven unit cells constitute the *seven crystal systems*, or *primitive lattices*, as they are more commonly

Table 8.3 *The Seven Crystal Systems and the Associated Thirty-Two Crystal Classes Allowed on the Basis of Symmetry Restrictions*

Crystal System	Axial/Angular Relationships	Crystal Classes[b]
Triclinic	$a \neq b \neq c$ $\alpha \neq \beta \neq \gamma \neq 90°$	C_1 C_i 1 $\bar{1}$
Monoclinic	$a \neq b \neq c$ $\alpha = \beta = 90° \neq \gamma$	C_2 C_s C_{2h} 2 m $2/m$
Orthorhombic	$a \neq b \neq c$ $\alpha = \beta = \gamma = 90°$	D_2 D_{2h} C_{2v} 222 mmm $2mm$
Tetragonal	$a = b \neq c$ $\alpha = \beta = \gamma = 90°$	S_4 C_4 C_{4v} C_{4h} D_4 D_{4h} D_{2d} $\bar{4}$ 4 $4mm$ $4/m$ 422 $4mmm$ $\bar{4}2m$
Rhombohedral[a]	$a = b = c$ $\alpha = \beta = \gamma < 120° \neq 90°$	C_3 C_{3v} C_{3h} D_3 D_{3h} D_{3d} 3 $3m$ $\bar{6}$ 32 $\bar{6}m2$ $\bar{3}m$
Hexagonal	$a = b \neq c$ $\alpha = \beta = 90°$ $\gamma = 120°$	C_6 C_{6v} C_{6h} D_6 D_{6h} C_{3i} 6 $6mm$ $6/m$ 622 $6/mmm$ $\bar{3}$
Cubic	$a = b = c$ $\alpha = \beta = \gamma = 90°$	T T_d T_h O O_h 23 $\bar{4}3m$ $m3$ 432 $m3m$

[a] Also known as trigonal.
[b] Schoenflies symbols above, Hermann-Maugin equivalences below.

termed (Table 8.3). Besides these, there are seven other important lattice structures for crystals, giving in all the so-called fourteen *Bravais lattices* (Figure 8.14). By considering all possible symmetry–element combinations within the Bravais lattices, it can be shown that there are in all thirty-two symmetry-allowed *crystal classes* (Table 8.3).

Atoms are located at different but often *symmetry-equivalent* positions in a unit cell. There are two special types of symmetry element that relate these equivalent positions to one another. Both of these symmetry elements involve translations within the unit cell; they are called *screw axes* and *glide planes*. A *n*-fold screw axis is designated n_m and involves a $2\pi/n$ rotation followed by a translation of m/n along the rotation axis. A glide plane σ^g combines a reflection and a translation. Figure 8.15 shows typical examples of a two-fold screw axis (2_1) and a glide plane.

When all possible molecular symmetry operations (proper rotations, mirror planes, centers of symmetry, and so on) and the screw axes and glide planes are incorporated into the thirty-two crystal classes, there are

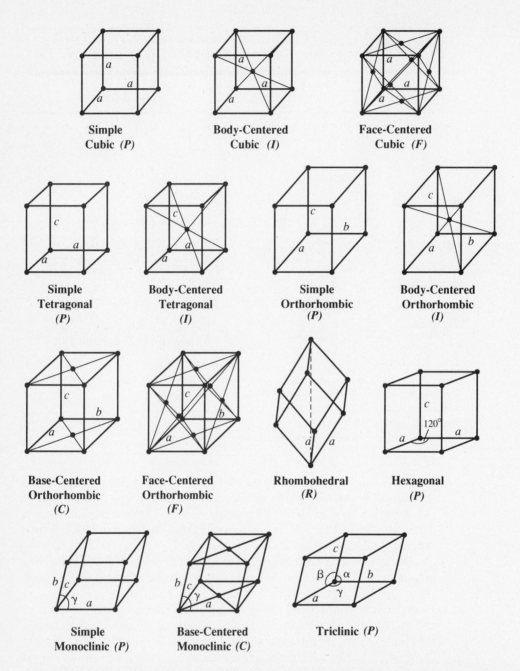

Figure 8.14 The fourteen Bravais lattices. Notice that each of these lattices has capital letter in parentheses associated with it. These letters are part of the Hermann–Maugin labeling system for crystallographic space groups. A simple (or primitive) lattice is labeled *P*; body-centered, *I* (meaning inner); side-centered, *A*, *B*, or *C*; face-centered, *F* (meaning all faces).

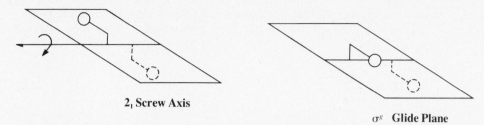

2₁ **Screw Axis**

σ^g **Glide Plane**

Figure 8.15
Examples of a 2₁ screw axis and a glide plane (σ^g). The screw axis involves 180° rotation about the axis labeled with the half-headed arrow, followed by a translation. The glide plane involves reflection through the plane indicated, followed by a translation.

only 230 discrete structural arrangements allowed. These are known as the 230 *space groups* and are formally distinguished from one another by their Hermann–Mauguin symmetry labels. These labels take into account the type of crystal lattice concerned, and some of the symbols have Schoenflies equivalents: $n = C_n$, $m = \sigma$, $\bar{1} = i$, $\bar{n} = S_n$. Many inorganic compounds crystallize in the monoclinic space group, $P2_1/c$. The Hermann–Mauguin labeling in this case indicates that the space group involves a primitive (P) lattice and that there is a two-fold screw axis perpendicular to a glide plane located on the c axis of the unit cell.

Once a single crystal suitable for X-ray investigation has been grown (not always a trivial task), it is glued onto the tip of a very thin glass rod and mounted on the head of a goniometer. This mechanical device permits rotation of the crystal in three orthogonol planes, allowing the crystal to be centered in the X-ray beam and the intensity of the scattered X rays to be maximized. All modern X-ray diffractometers are fully automated so that the computer searches for a specified number of strong reflections usually twenty-five to fifty) and adjusts the goniometer position so that the intensities of these reflections are maximized. The computer also uses these strong reflections to determine the unit-cell parameters and eventually the space group.

The next step is to collect a large enough data set to solve the structure. Often, several thousand independent reflections are collected, especially if there are many atoms to be located in the structure. This procedure may take several days, depending on the time interval allowed for measuring the intensity of the scattered X radiation for each reflection. The individual reflections are distinguished from one another by their *Miller indices* (*hkl*). These indices provide a useful way of labeling the planes that pass through the so-called reciprocal lattice (the details of this particular topic are dealt with thoroughly in any of the specialized references on X-ray crystallography listed at the end of the chapter). As an example, the cell faces of an orthorhombic lattice are designated by the Miller indices (100), (010), and (001).

The X-ray source used in the data-set collection is often a $CuK\alpha_1$ ($\lambda = 154.05$ pm) or $MoK\alpha_1$ ($\lambda = 70.926$ pm) X-ray tube, and the intensity of the scattered X radiation is measured by a scintillation counter. Once a suitable data set has been collected, the intensities are converted to *structure factors* ($|F_0|$) for use in the subsequent least-squares refinements. The structure factors are related to the observed intensities (I_0) by Equation 8.8:

$$|F_0| = \sqrt{\frac{KI_0}{Lp}} \qquad\qquad (8.8)$$

where K is a scaling factor and L and p are Lorentz and polarization factors, respectively.

The list of atoms (from the presumed elemental composition of the compound), the preliminary scattering factors for these atoms, and their x, y, z coordinates in three-dimensional space are input into the computer, together with information on the space group and the number of molecules per unit cell (Z). From this point on, the computational work involves recycling through a chain of computer programs until the discrepancy between the observed and calculated structure factor, $\Delta F = |F_0| - |F_c|$, for each reflection is minimized by means of an iterative, least-squares, refinement procedure. One of the outputs at the end of each refinement cycle is a discrepancy index, or *R factor*, such as that given by Equation 8.9:

$$R = \frac{\Sigma|\Delta F|}{\Sigma|F_0|} \qquad\qquad (8.9)$$

For an acceptable solution to a structural problem, the R factor is normally between 3–10%. If the R factor cannot be reduced to below about 12%, either there is statistical disorder in some atoms, there are some additional atoms present (possibly solvent molecules trapped in the lattice), or the compound is not what you think it is! The remainder of the crystal-structure analysis involves mapping the electron density in the unit cell and producing a perspective drawing of the molecule, which illustrates the bonding between the atoms and the relative thermal motions of the atoms. The thermal-motion drawings provide a convenient way of illustrating the uncertainties in the positions of the atoms and are most often depicted as either spheres or ellipsoids. Figure 8.16a shows such a drawing for the structure of $CpMn(CO)_3$. If the thermal motions are too large, then the errors in the calculated bond distances and bond angles are greatly increased. For this reason and to reduce the risk of decomposition in the X-ray beam, the structures of many compounds, especially organometallics, are now routinely determined at low temperatures (77 K). Under these conditions, hydrogen atoms can be located quite easily. In many cases, however, hydrogen atoms can even be found at room temperature, provided their thermal motions are not too large. In the case of $CpMn(CO)_3$, there are four molecules per crystallographic unit cell; Figure 8.16b shows a diagram of the packing.

Figure 8.16
(a) Perspective view of CpMn(CO)$_3$ down the molecular axis, illustrating the atomic numbering system. The thermal ellipsoids provide an indication of the relative motions of the atoms; the spheres shown for the hydrogen atoms are not realistic but simplify the drawing. **(b)** Packing arrangement of the four CpMn(CO)$_3$ molecules located in the orthorhombic crystallographic unit cell. Both drawings are reproduced with permission from P. J. Fitzpatrick, Y. Le Page, J. Sedman, and I. S. Butler. *Inorg. Chem.* 20 (1981): 2852.

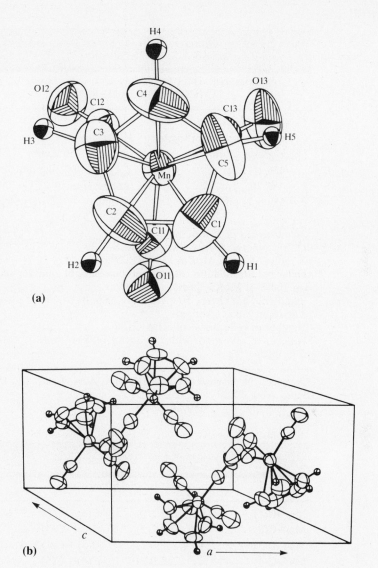

(a)

(b)

Complete lists of bond lengths and bond angles are also part of the output from the least-squares refinements. For a good structure involving not too many atoms, the accuracy of the bond lengths is about ±0.5 pm, while the angles can be measured to within 0.05°. Tables 8.4 and 8.5 show some typical data from an X-ray structural analysis. Table 8.4 is an example of how crystal-structure information is often published in scientific journals such as *Inorganic Chemistry* and *Organometallics*. Table 8.5 illustrates how bond lengths and bond angles are reported.

Table 8.4 *Typical Information Given for an X-Ray Structure Reported in a Scientific Journal*[a]

Crystallographic Data for X-Ray Diffraction Study of
$Cr(CO)_2(CS)[(CH_3O)_3P]_3$

Crystal Parameters

Crystal system: Orthorhombic	$Z = 8$
Space group: Pbca	$d_{calcd} = 1.466$ kg dm^{-3}
$a = 156.1(1)$ pm	$d_{obsd} = 1.40(2)$ kg dm^{-3}
$b = 183.2(2)$ pm	Temperature = 22°C
$c = 188.8(1)$ pm	Formula = $C_{12}H_{27}O_{11}SP_3Cr$
	M W = 542.0

Measurement of X-Ray Intensity Data
Diffractometer: Picker Nuclear FACS–1
X radiation used: $MoK\alpha_1$ ($\lambda = 71.069$ pm)
Detector: Scintillation counter and pulse-height analyzer set for 100% of $MoK\alpha_1$ peak
Standards (*hkl*) every 50 cycles: 430, 006, 043
Variation: ±3% (random)
Number of reflections collected: 2106
Number of reflections with $I > 3\sigma(I)$: 1453[b]
$R = 0.051$

[a] Adapted with permission from P. H. Bird, A. A. Ismail, and I. S. Butler. *Inorg. Chem.* 24 (1985): 2911.
[b] I = intensity; σ = standard deviation.

Table 8.5 *Bond Distances (pm) and Angles (°) for CpMn(CO)$_3$*[a]

Bond Distances

Mn—C(1)	213.1(2)	C(1)—C(2)	140.1(3)
Mn—C(2)	213.2(2)	C(2)—C(3)	136.5(3)
Mn—C(3)	212.8(2)	C(3)—C(4)	138.8(3)
Mn—C(4)	212.0(2)	C(4)—C(5)	138.2(3)
Mn—C(5)	212.1(2)	C(5)—C(1)	140.5(3)
Mn—C(11)	178.1(2)	C(1)—H(1)	80.3(16)
Mn—C(12)	179.2(2)	C(2)—H(2)	90.9(18)
Mn—C(13)	178.6(2)	C(3)—H(3)	100.5(17)
C(11)—O(11)	115.1(2)	C(4)—H(4)	96.4(17)
C(12)—O(12)	115.6(2)	C(5)—H(5)	89.7(18)
C(13)—O(13)	114.3(2)		

[a] The atom numbering refers to the labeling given in Figure 8.10. The numbers in parentheses are estimated standard deviations in the least significant digits. The data are adapted with permission from P. J. Fitzpatrick, Y. Le Page, J. Sedman, and I. S. Butler. *Inorg. Chem.* 20 (1981): 2852.

Table 8.5 (Continued)			
Bond Angles			
C(1)—Mn—C(2)	38.46(8)	C(11)—Mn—C(12)	91.59(9)
C(2)—Mn—C(3)	37.47(8)	C(11)—Mn—C(13)	92.52(9)
C(3)—Mn—C(4)	38.15(8)	C(12)—Ma—C(13)	91.96(8)
C(4)—Mn—C(5)	38.04(9)	C(5)—C(1)—H(1)	122.2(11)
C(5)—Mn—C(1)	38.59(9)	C(2)—C(1)—H(1)	131.7(11)
C(5)—C(1)—C(2)	106.05(17)	C(1)—C(2)—H(2)	124.6(11)
C(1)—C(2)—C(3)	109.02(18)	C(3)—C(2)—H(2)	126.3(11)
C(2)—C(3)—C(4)	108.61(18)	C(2)—C(3)—H(3)	126.2(10)
C(3)—C(4)—C(5)	107.59(18)	C(4)—C(3)—H(3)	124.8(10)
C(4)—C(5)—C(1)	108.72(17)	C(3)—C(4)—H(4)	125.7(10)
Mn—C(11)—O(11)	179.33(16)	C(5)—C(4)—H(4)	126.7(10)
Mn—C(12)—O(12)	178.83(15)	C(4)—C(5)—H(5)	132.1(11)
Mn—C(13)—O(13)	178.56(19)	C(1)—C(5)—H(5)	119.2(11)

8.5   Concluding Remarks

It has been possible in Chapters 5–8 to present only an overview of the main structural techniques used today by inorganic chemists. Whenever a new compound is synthesized, it is necessary to provide convincing evidence for its proposed structure before a claim may be made in the scientific literature. This evidence is almost certainly spectroscopic and includes data obtained by some of the techniques described in these four chapters. In addition, X-ray structures are often published as an integral part of papers dealing primarily with the synthesis of new inorganic compounds. This trend is likely to continue, especially now that X-ray structural analysis has become so widespread and relatively routine.

BIBLIOGRAPHY

Suggested Reading

Cotton, F. A., and G. Wilkinson. *Advanced Inorganic Chemistry. A Comprehensive Text*, 5th ed. New York: Wiley-Interscience, 1988.

Drago, R. S. *Physical Methods in Chemistry*. Philadelphia: Saunders, 1977.

Greenwood, N. N., and A. Earnshaw. *Chemistry of the Elements*. New York: Pergamon Press, 1984.

Guillory, W. A. *Introduction to Molecular Structure and Spectroscopy*. Boston: Allyn and Bacon, 1977.

Hatfield, W. E., and R. A. Palmer. *Problems in Structural Inorganic Chemistry*. New York: Benjamin, 1971.

Hill, H. A. O., and P. Day. *Physical Methods in Advanced Inorganic Chemistry*. New York: Wiley-Interscience, 1968.

Huheey, J. E. *Inorganic Chemistry. Principles of Structure and Reactivity*, 3rd ed. New York: Harper & Row, 1983.

Lambert, J. B., H. F. Shurvell, L. Verbit, K. J. Cooks, and G. H. Stout. *Organic Structural Analysis*. New York: Macmillan, 1976.

Photoelectron Spectroscopy

Brundle, C. R., and A. D. Baker, eds. *Electron Spectroscopy: Theory, Techniques and Applications*, Vols. I and II. London: Academic Press, 1977.

Ghosh, P. K. *Introduction to Photoelectron Spectroscopy*. New York: Wiley, 1983.

Lichtenberger, D. L., and G. E. Kellogg. "Experimental Quantum Chemistry: Photoelectron Spectroscopy of Organotransition-Metal Complexes." *Acc. Chem. Res.* 20 (1987): 379.

Shirely, D. A., ed. *Electron Spectroscopy*. Amsterdam: North-Holland Publishing Co., 1972.

Turner, D. W., C. Baker, A. D. Baker, and C. R. Brundle. *Molecular Photoelectron Spectroscopy*. London: Wiley, 1970.

EXAFS and XANES Spectroscopy

Bianconi, A., L. Incoccia, and S. Stiptich. *EXAFS and Near Edge Structure*. Berlin: Springer-Verlag, 1983.

Konigsberger, D. C. and R. Prins, eds. *X-Ray Absorbtion: Principles, Applications, and Techniques of EXAFS, SEXAFS, and XANES*, New York: Wiley, 1988.

Kramer, S. P., and K. O. Hodgson. *Prog. Inorg. Chem.* 25 (1979): 1.

Teo, B. K., and D. C. Joy, eds. *EXAFS Spectroscopy*. New York: Plenum, 1981.

Mass Spectroscopy

Litzow, M. R., and T. R. Spalding. *Mass Spectrometry of Inorganic and Organometallic Compounds*. Amsterdam: Elsevier, 1973.

McLafferty, F. W. *Interpretation of Mass Spectra*, 3rd ed. Mill Valley, Calif.: University Science Books, 1980.

X-Ray Crystallography

Bunn, C. W. *Chemical Crystallography: An Introduction to Optical and X-Ray Methods*, 2nd ed. Oxford, England: Clarendon Press, 1961.

Glasser, L. S. D. *Crystallography and Its Applications*. London: Van Nostrand Reinhold, 1977.

Glusker, J. P., and K. N. Trueblood. *Crystal Structure Analysis*. London: Oxford University Press, 1972.

Sands, D. E. *Introduction to Crystallography*. New York: Benjamin, 1969.

Stout, G. H., and L. H. Jensen. *X-Ray Structure Determination*. New York: Macmillan, 1968.

PROBLEMS

8.1   The pyridine ligand and its substituted derivatives shown below readily form complexes with various transition metals.

a. How would you expect the $N(1s)$-binding energies to be affected when the ligands coordinate to a metal?

b. Would you anticipate any binding-energy effects due to the differences in the ligands themselves?

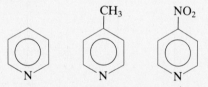

8.2   From a consideration of the XANES spectra shown below, what can you conclude about the nature of the binding of $(Et_3P)AuCl$ to bovine serum albumin (BSA)? For further details, see M. T. Coffer, C. F. Shaw III,

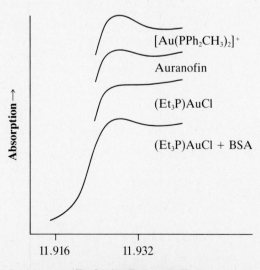

$[Au(PPh_2CH_3)_2]^+$

Auranofin

$(Et_3P)AuCl$

$(Et_3P)AuCl + BSA$

Absorption →

11.916   11.932

Excitation Energy (keV) →

M. K. Eidsness, J. W. Watkins II, and R. C. Elder. *Inorg. Chem.* 25 (1986): 333.

8.3   The He(I) valence-ionization spectra for $CpMn(CO)_3$ and $CpMn(CO)_2$ (CS) are compared in the figure shown below. Suggest possible qualitative assignments for the ionizations observed for the two compounds. For further details, see D. L. Lichtenberger and R. F. Fenske, *Inorg. Chem.* 15 (1976): 2015.

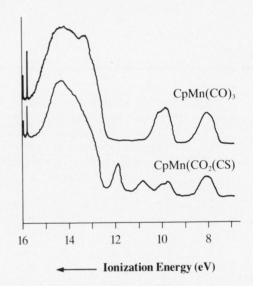

8.4   The prominent peaks in the mass spectrum of a compound of formula C_4NO_5Mn appear at m/e 197, 169, 141, 139, 113, 111, 85, 83, and 55 amu. What is the probable structure of the compound? To what molecular fragments do you attribute the observed peaks?

8.5   Given the following mass spectral data for an unknown compound X, make a sketch of the mass spectrum observed. If the peak at m/e 302 amu is the parent molecular ion and the only isotopes present are ^1H, ^{12}C, ^{16}O, ^{32}S, and ^{55}Mn, suggest a structure for compound X.

m/e	Relative Intensity (%)
302	11.7
274	0.8
230	7.1
192	0.8
164	100.0
120	73.0
66	5.7
65	3.3
55	48.9

8.6 One of the fragment ions observed in the mass spectrum of an unknown organosiloxane compound has an m/e value of 252.988 under high-resolution conditions. Verify that the molecular formula of this ion is $C_4H_{13}Si_4O_5^+$. If the ion is formed by the release of methylsilane (CH_3SiH_3) from the parent molecular ion, propose a structure for the organosiloxane compound.

8.7 The following unit-cell parameters were determined for two new inorganic compounds. To what crystal classes do these unit cells belong?
  a. $a = 1561$, $b = 1532$, $c = 1888$ pm;
  $\alpha = \beta = \gamma = 90.0°$
  b. $a = 2054.9$, $b = 1113.8$, $c = 965.8$ pm;
  $\alpha = \beta = 90.00$, $\gamma = 108.81°$

8.8 The general expression for the volume (V) of a unit cell with parameters a, b, c, α, β, γ is

$$V = abc(1 - \cos^2\alpha - \cos^2\beta - \cos^2\gamma + 2\cos\alpha \cos\beta \cos\gamma)^{1/2}$$

Given the following additional structural information for the two compounds in Problem 8.7, calculate the number of molecules per unit cell in each case.
  a. Molecular formula, $C_{12}H_{27}O_{11}SP_3Cr$
  Density (calculated), 1.466 kg dm^{-3}
  b. Molecular formula, $C_{10}H_{15}NH_3Cl$
  Density (calculated), 1.191 kg dm^{-3}

8.9 Calculate the molecular weight of a compound that crystallizes in a monoclinc unit ($a = 1207.7$, $b = 705.7$, $c = 1091.3$ pm; $\alpha = \beta = 90.00°$, $\gamma = 117.68°$) if its measured density is 1.64 kg dm^{-3} and there are four molecules per unit cell.

8.10 The locations of the planes in the reciprocal lattice may be represented either by fractional axis intercepts or their reciprocals (Miller indices). Given the following fractional axis intercepts, what are the corresponding Miller indices?
  a. $\frac{1}{4}$, 1, $\frac{1}{2}$ **c.** $\frac{1}{3}$, $\frac{3}{4}$, ∞
  b. $\frac{1}{3}$, $\frac{2}{3}$, $\frac{1}{2}$ **d.** $\frac{1}{2}$, ∞, 1

PART THREE

Periodic Trends for the Elements and Simple Compounds

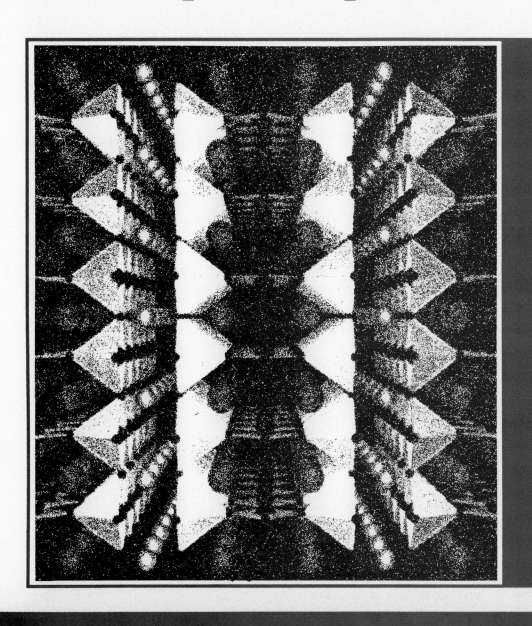

9

Hydrogen and Its Compounds

9.1 Hydrogen

Hydrogen is overwhelmingly the most abundant element in the universe. This is of little interest to practicing chemists, however, because most hydrogen in the universe is in the form of atoms, rather than molecules or compounds. Hydrogen still remains one of the more abundant elements found on earth, despite the great concentration of heavier elements; here it exists almost entirely in the form of compounds, notably water.

Three isotopes of hydrogen are known: ^1H, ^2H, and ^3H with atomic masses of 1.0078, 2.014, and 3.016 amu, respectively. Because of the proportionately large differences in their masses, these isotopes confer differences in physical and chemical properties on their compounds that are appreciably larger than similar effects with the isotopes of heavier elements. This great difference in behavior has encouraged the unique practice of conferring different names and symbols on each of the isotopes of hydrogen, namely: ^{1}H (hydrogen, H), ^{2}H (deuterium, D) and ^{3}H (tritium, T).

The most important effect of the different isotopic masses is on the zero-point vibrational energy of X—H (D or T) bonds. This energy is the residual energy of vibration, which persists in chemical bonds at absolute zero temperature, and it influences all vibrational modes and reactivities of X—H bonds. Because these effects are quite large for the isotopes of hydrogen, fractionation occurs fairly easily by either physical or chemical processes, and the abundance of the isotopes can vary widely, depending on the history of the compound. For example, the simple processes of evaporation of the oceans and recondensation and precipitation in the form of rain cause a measurable change in isotopic abundances.

The industrial concentration of D_2O is effected by a number of processes including distillation, electrolysis of H_2O to H_2 and O_2, and taking advantage of the isotope effect on an equilibrium such as

$$HSD(g) + H_2O(l) \rightleftharpoons H_2S(g) + HOD(l) \tag{9.1}$$

This type of equilibrium allows enrichment up to a few percent at low cost. Higher enrichments can then be achieved using the much more costly electrolysis of partially enriched water.

The large isotope effect on the X—H bond vibrational energies makes deuterium labeling of compounds an extremely useful tool in vibrational spectroscopy (see Chapter 6), both in assignment of vibrations and in the complete derivation of the dynamic mechanical properties of molecules containing hydrogen. The presence of a nuclear spin of $\frac{1}{2}$ on the proton, along with the ubiquity, the sensitivity, and high abundance of ^{1}H, makes it the most widely used nucleus in NMR spectroscopy. Although D and T also have nuclear spin (1 and $\frac{1}{2}$ respectively), their use in NMR spectroscopy is much more limited by virtue of their low natural abundance and the radioactivity of T, among other reasons.

Despite the low abundance of D relative to ^{1}H (0.015%), the large absolute abundance of ^{1}H in the universe and on earth make D abundant on an absolute scale. Deuterium is separated in large amounts, particularly for use in the form of D_2O, as a neutron moderator in nuclear reactors. Tritium is to all intents and purposes absent from natural hydrogen because of its short half-life (about twelve years). For the limited uses to which it is presently put, T is best prepared by nuclear chemistry through the slow neutron irradiation of ^{6}Li:

$$^6Li + {}^1n \rightarrow T + {}^4He \tag{9.2}$$

The chemical separation of T and ^{4}He is very easy because of the chemical inactivity of the latter.

Hydrogen is the energy-storage system par excellence in the universe. For a short time after the big bang, which is presently favored as the best model for the initiation of our current universe, conditions moved rapidly toward equilibrium in the intense energy flux then available. As the free energy either converted to mass or was dissipated in space, its availability decreased, and the move toward equilibrium stalled at the point where most of the available mass and energy was converted to hydrogen. The highly exothermic conversion of hydrogen to heavier elements could not take place spontaneously at the low temperature of the evolved universe and had to await the evolution of special ignition processes that could take place in the interiors of massive agglomerations of hydrogen gas. Within these massive agglomerations, thermonuclear fires were lit through the agency of gravitational energy. Relatively young stars, like our own sun, are still composed of mainly hydrogen and sustain themselves by such reactions as

$$^1H + {}^1H \rightarrow {}^2H + \beta^+ \tag{9.3}$$

$$^1H + {}^2H \rightarrow {}^3He \tag{9.4}$$

$$^3He + {}^3He \rightarrow {}^4He + 2{}^1H \tag{9.5}$$

Toward the ends of their lives, stars may derive energy from a variety of nuclear syntheses of heavier nuclei. But for our solar system, the great dynamo runs on hydrogen.

Because of its extremely low density and boiling point, hydrogen tends to be lost from the atmospheres of even the largest planets. Within our own solar system, we have a number of planets (either very large or not too close to the sun) that have retained enormous amounts of the simplest compounds of hydrogen, that is, CH_4 (Jupiter), NH_3 (Saturn), and H_2O (Earth).

In many respects, H_2O is close to being the graveyard for hydrogen. Because the O—H bond is one of the strongest single bonds, there is not a great deal of further free energy that can be extracted from H_2O. (Many hydrolyses can liberate much energy , but it is usually in the species being hydrolyzed that the free energy is stored.) In fact, much of the water on earth is probably the result of combustion of the prebiotic atmosphere (mainly, more reducing hydrogen compounds such as NH_3 and CH_4) by the oxygen resulting from primitive photosynthesis.

All planets derive energy from the sun in a passive way and store it as elevation of temperature, high-energy photochemistry, or the kinetic energy of motion of planetary fluids. The earth is unique in the solar system in possessing an *active* trap for the sun's energy, in the form of photosynthetic plants and bacteria. These photosynthetic life-forms use the sun's radiation directly to resurrect H_2O and CO_2 (the graveyard of carbon) by conversion to energy-rich carbohydrates and oxygen. For most of the time span since the evolution of photosynthetic life-forms, the amount of energy stored exceeded the amount used. The surplus accumulated just below the earth's surface and gradually transformed by geological effects into fossil fuels: coal, oil, and natural gas. It is from these fuels that we derive most of our hydrogen. From coal, hydrogen may be produced by the following reactions:

$$C \text{ (coal)} + H_2O \rightarrow CO + H_2 \tag{9.6}$$

$$CO + H_2O \rightarrow CO_2 + H_2 \tag{9.7}$$

We still have plenty of coal left to carry out this chemistry, but to carry it out *cleanly* is very costly: The degradation of coal liberates a wide range of toxic and radioactive inorganic materials besides some very disagreeable organic products. Those lucky enough to have lived in the second half of the twentieth century have had a very cheap ride indeed, using up most of the high-quality oil and gas that photosynthesis took eons to produce. The main source of cheap hydrogen during this period has come from the steam reformation of petroleum or natural gas:

$$C_nH_{2m} + nH_2O \xrightarrow{\text{Ni/Al}_2\text{O}_3} nCO + (n + m)H_2 \tag{9.8}$$

$$CH_4 + 2H_2O \xrightarrow{\text{Pt/Al}_2\text{O}_3/\text{SiO}_2} CO + 4H_2 \tag{9.9}$$

Coupled into this cheap source of hydrogen is the agricultural revolution, which has occurred through the massive application of synthetic ammonia as a fertilizer. The ammonia is produced through the Haber process (Equation 9.10), the largest industrial consumer of hydrogen:

$$N_2 + 3H_2 \xrightarrow{\text{Fe catalyst}} 2NH_3 \tag{9.10}$$

It is thus evident that we must find alternatives to our present cheap supplies of H_2 if we are to avoid a dramatic decline in world food production.

There is a further imperative to developing new, cheap ways of making hydrogen. Like the fossil fuels themselves, hydrogen is a form of stored portable energy by virtue of its highly exoergic reaction with oxygen. Hydrogen's major disadvantages are its low density and its low boiling and critical points. These disadvantages are offset to some degree by its nonpolluting chemistry and its ready application to use in fuel cells. A number of eminent scientists advocate the idea of a "hydrogen economy" to supplement and eventually to supercede the "fossil-fuel economy" of the industrial world. Although some of the necessary hydrogen could be derived by using "passive" solar energy as electricity (hydro, windmill, tidal) to electrolyze water, a full adoption of a hydrogen economy will require "active" collection of solar energy, either for direct photolysis of water (analogous to photosynthesis) to H_2 and O_2, by photoelectrolysis (driving electrolysis by photoactivation of an electrode), or by electrolysis (electricity from photoelectric collectors).

Hydrogen, because its electrons are at very low energy and its lowest unoccupied orbitals are at very high energy, shows little tendency to react with anything but the strongest acids and bases. For example, there is some evidence that H_2 can react directly with carbonium ions (very strong acids) and with tertbutoxide ion (very strong base). The reaction with base has some utility in catalyzing the H_2/H_2O equilibrium for deuterium enrichment. Most useful reactions of hydrogen require the participation of metal catalysts, either homogeneous or heterogeneous; this subject is treated in more detail in Chapter 23. A number of radical reactions involving formation of X—H bonds that are of comparable or greater strength than the H—H bond occur with ease. Notable among these are the reaction with O_2 (of legendary violence due to the branched-chain radical mechanism) and with the halogens. All these reactions involve atoms or radicals capable of effecting an abstraction of a hydrogen atom from H_2. The transition states for such abstractions resemble the three-atom/three-electron bonding arrangements, discussed earlier for H_3 (Chapter 3). The stronger the product X—H bond energy, the lower is the energy of this transition state. In the case of the H_2/I_2 reaction, the bond energy of HI (295 kJ mol^{-1}) is so much below that of H_2 (432 kJ mol^{-1}) that the activation energy is very high and the reaction proceeds by both a concerted and a radical mechanism under severe conditions.

There are also those who look forward to the day when we can have our own thermonuclear furnaces here on earth. Although it is unlikely that the reaction (Equation 9.3) will be achieved in the laboratory within a meaningful time frame, a number of reactions involving D and T look promising and, should the technology of power generation come to fruition, could provide electricity in sufficient abundance to underpin a hydrogen economy based on electrolysis of water.

9.2   Types of Binary Hydrogen Compounds

Binary hydrides are of great importance theoretically because they allow us to see periodic trends in compounds with the simplest kinds of bonds, in particular uncomplicated by multiple bonding. The electronegativity of hydrogen is greater than that of most of the elements except those of Groups VIA and VIIA (16 and 17) and some lighter members of Groups IVA and VA (14 and 15). This means that most binary hydrogen compounds should be polarized in the sense $M^{\delta+}$—$H^{\delta-}$ or that the hydrogen should possess hydridic character. In fact, only a few of the elements that are less electronegative than hydrogen actually form stable binary hydrides, for reasons discussed later. Figure 9.1 shows the periodic table, broken down approximately into those elements whose electronegativities are within 0.6 units of that of hydrogen and those that fall below and above

Figure 9.1
Electronegativity
trends in the
periodic table.

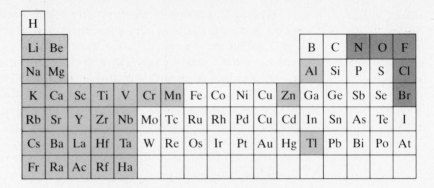

Electronegativity > 0.6 below H. (All lanthanides
and actinides also fall in this category.)

Electronegativity > 0.6 above H.

Electronegativity within 0.6 units of H.

that range. The value of 0.6 units was chosen because BeH_2 is the only hydride of Groups IA and IIA (1 and 2) that manifests considerable covalent character and the electronegativity of Be is about 0.6 units below that of hydrogen. By coincidence, those elements that are more than 0.6 units above hydrogen are those that characteristically lose H^+ fairly easily and form strong hydrogen bonds (see below). Those elements that are within 0.6 units of hydrogen give compounds that have considerable covalent character. Before proceeding further, we stress that this classification is very qualitative because of the imprecision of the concept of electronegativity and the lack of any real boundary between the different kinds of bonding.

Given the latter qualification, it is expected that the elements at the extreme left of the periodic table will give highly ionic compounds in which the hydrogen exists more or less as the H^- ion. In the middle of the table, we expect most hydrides to exhibit considerable covalency but still to behave as though the hydrogen carries a negative charge. Toward the right side of the table where the electronegativities of elements are very close to that of hydrogen (for example, C, Si, P), bonds will be almost homopolar covalent. Finally, one might expect for the elements at the extreme right, especially O and F, that ionic compounds of the kind H^+X^- might be formed. In fact, the essentially zero radius of H^+ makes it so polarizing that it never exists as the free ion in chemically significant situations. Its small radius also tends to make H^+ prefer a coordination number of two or one, a factor that mitigates against the formation of three-dimensional "ionic" solids containing H^+ (as with all assertions of this kind, there are some exceptions, which are discussed below). Although hydrogen compounds of the more electronegative elements do not furnish free H^+ ions, they do very readily transfer the proton to other suitable molecules, a factor of enormous importance in chemistry.

9.3   Hydrides of Groups IA and IIA (1 and 2)

With the exception of beryllium, all elements of Groups IA and IIA (1 and 2) react spontaneously when heated in hydrogen gas, to give the white solid hydrides MH [group IA (1)] or MH_2 [Group IIA (2)]. All MH hydrides have the NaCl structure, and their chemical and physical properties indicate that they are highly ionic. There is good agreement between calculated and experimental lattice energies, but these results must be viewed in the knowledge that the radius of the hydride ion is very difficult to measure and experimental values cover a rather large range. This is expected because the electrons of H^- are loosely held and the ion is very polarizable. Another experiment, which is often cited to support the presence of hydride ions, is the electrolysis of molten hydrides that leads to the evolution of hydrogen at the anode. Although such evidence strongly supports the presence of hydrogen in an anionic species, the species is not necessarily a free hydride ion but more likely a complex species, for example, $[LiH_2]^-$:

$$[LiH_2]^- \rightarrow LiH_2^{\cdot} + e^- \rightarrow LiH + H^{\cdot} \tag{9.11}$$

$$2H^{\cdot} \rightarrow H_2$$

(NOTE: See Chapter 3 for a discussion of bonding in $LiH_2^{\cdot}$.) However, when all evidence is considered, it would be foolish to deny the essential ionicity of these compounds.

The hydrides of Group IA (1) find a number of specialized uses as reducing agents, particularly in the form of complex hydrides (see Section 9.4). Lithium deuteride is a useful fuel for nuclear fusion reactions by virtue of the following reaction:

$$^6Li + {}^2H \rightarrow 2\,{}^4He + 22.4\ MeV \tag{9.12}$$

Unfortunately, this reaction presently is used in the manufacture of thermonuclear weapons, but there is hope that it may one day furnish a cheap and clean source of energy for nonmilitary use.

All Group IIA (2) hydrides, with the exception of BeH_2, adopt the fluorite structure in which the cation has a coordination number of eight and the hydride a coordination number of four, reflecting the rather small size of the hydride ion (comparable to F^-).

Beryllium hydride cannot be made by direct combination of the elements. It is best prepared by thermal decomposition of beryllium alkyls:

$$Be(\overset{|}{\underset{|}{C}}-\overset{|}{\underset{|}{C}}-H)_2 \longrightarrow BeH_2 + {>}C{=}C{<} \tag{9.13}$$

This type of reaction is quite general as a synthetic method for making hydrides and proceeds more easily in the order primary $<$ secondary $<$ tertiary alkyl. The hydride produced by this method is amorphous, as are beryllium hydrides produced by all known methods, and therefore its structure has not been determined by crystallographic methods. The structure is believed to consist of long chains of beryllium atoms doubly bridged by hydrogen (see Chapter 3).

The failure of BeH_2 to form a three-dimensional lattice is due to the small sizes of the beryllium and hydride ions (this terminology is used loosely because the charges on Be and H probably deviate substantially from the integral values for ions). The very strong tendency for Be^{2+} to prefer tetrahedral coordination obliges the hydride ion to be two-coordinate.

The chemistry of the saltlike hydrides is dominated by two kinds of reaction: (a) reactions with protonic reagents to liberate hydrogen and (b) reduction by either electron or hydrogen-atom transfer. Examples of these types of reaction are given in Equations 9.14 and 9.15:

$$2CaH_2 + 2NH_3 \rightarrow 2CaNH_2 + 3H_2 \tag{9.14}$$

$$NaH + 2FeCl_3 \rightarrow 2FeCl_2 + NaCl + HCl \tag{9.15}$$

Because of their general insolubility in nonprotic media and their reactivity with protic media (a number of these hydrides react violently with water), saltlike hydrides are rarely used directly, but their potential reactivity is realized through conversion to the more soluble complex hydrides to be discussed below. The reactivity of the hydrides increases on descending the groups, but even BeH_2, the least ionic of the Group IA and IIA (1 and 2) hydrides, reacts chemically as though it contains the H^- ion. The chemistry of BeH_2 has been little studied because the great toxicity of beryllium compounds mitigates against their use in practical applications.

9.4 Hydrides and Complex Hydrides of Group IIIA (13)

The hydrides of Group IIIA (13) are transitional in their properties, particularly those of the abundant elements B and Al. The simplest hydride of boron exists as a gaseous dimer B_2H_6, whereas that of aluminum is a solid polymer $(AlH_3)_n$. The original preparation of diborane (by Alfred

Stock, the pioneer of boron hydride chemistry) was effected through hydrolysis of magnesium boride. Today, it is routinely prepared by a variety of reactions of boron compounds with complex hydrides, the most convenient being the following:

$$3LiAlH_4 + 4BF_3 \cdot O(C_2H_5)_2 \xrightarrow{(C_2H_5)_2O}$$
$$2B_2H_6 + Li_3AlF_6 + 2AlF_3 + 4(C_2H_5)_2O \quad \textbf{(9.16)}$$

$$3NaBH_4 + 4BF_3 \cdot O(C_2H_5)_2 \xrightarrow{CH_3O(CH_2CH_2O)_2CH_3}$$
$$2B_2H_6 + 3NaBF_4 + 4(C_2H_5)_2O \quad \textbf{(9.17)}$$

Diborane, a colorless gas that boils at $-90\,°C$, is clearly a highly covalent compound. On the basis of electronegativity, the B—H bond would be expected to be relatively nonpolar. The structure of diborane, which was first determined by gas-phase electron-diffraction studies, is shown below:

The elucidation of the structures of B_2H_6 and of the higher-boron hydrides (discussed in Chapter 15) in the 1950s presented a staggering challenge to the theories of chemical bonding and was an important factor in the belated recognition of the advantages of MO theory. Today, with multicenter, electron-deficient bonds recognized as a commonplace phenomenon, it is difficult to imagine the ferment of quasi-theoretical ingenuity that was lavished on the structure of diborane. It seemed at one point that the "corner-grocery-store" school of chemical theory had won the day by using "bananas" and "sausages" to describe the bridge bonds of B_2H_6. Those heady days are gone, and we are left with the less appetizing but intellectually more nourishing linear combination of AOs. In 1976 Lipscomb was awarded the Nobel Prize for his development of a theoretical description of the bonding in the polyboranes.

The MOs of B_2H_6 can be vizualized by building the molecule from fragments. We can begin with two bent BH_2 fragments, which will have the same MO energy-level diagram as H_2O (see Figure 3.15). We can construct the terminal σ bonds of the BH_2 fragments as shown in Figure 9.2, and we assume for pictorial simplicity that the p_z orbitals (a_1 symmetry) do not mix with the a_1 σ orbitals. This approximation gives us two BH_2 fragments, each with two strongly bonding, two nonbonding, and two strongly antibonding orbitals. We can now overlap these fragments with two bridging hydrogen atoms, using the nonbonding $p_z(a_1)$ and $p_y(b_2)$ orbitals. These orbitals of the BH_2 fragments can participate in three-center overlaps with the symmetric and antisymmetric combinations of

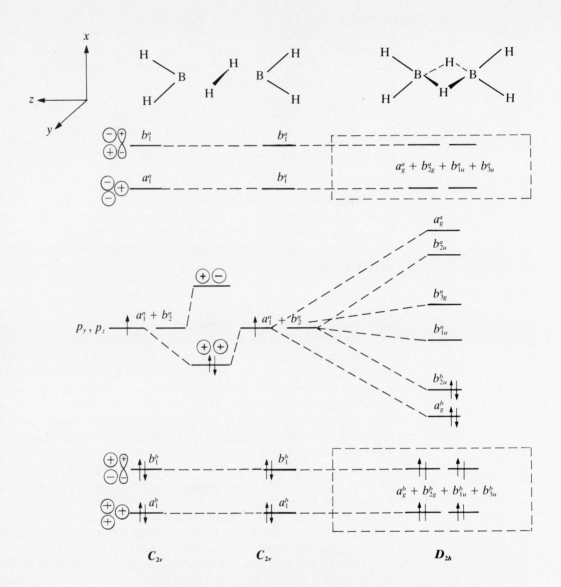

Figure 9.2
MO diagram for B_2H_6.

the bridging H(1s) orbitals, respectively, to produce a pair of bonding, a pair of nonbonding, and a pair of antibonding three-center orbitals (Figure 9.2). Thus, the B_2H_6 dimer is held together by two delocalized filled orbitals that extend over all four atoms participating in the bridge. The b_{2u} orbital is strictly orthogonal to the orbitals of the terminal B—H bonds, but the bridging a_g orbital is mixed with the a_g orbitals of the terminal B—H bonds.

By adopting the bridged structure, B_2H_6 achieves a condition where all the occupied orbitals are bonding in character and the lowest unoccupied MOs are nonbonding in character. If ethane were to adopt the same structure, the additional two electrons would occupy an essentially non-bonding orbital, a situation that might not seem energetically unfavorable. However, the translation of the D_{2h} orbital scheme of Figure 9.2 to the D_{3d} staggered (or D_{3h} eclipsed) symmetry of an ethanelike molecule converts the two nonbonding orbitals into a strongly bonding orbital and a strongly antibonding orbital. Because these orbitals are unoccupied in B_2H_6, there is no gain in energy, but in C_2H_6 a nonbonding electron pair becomes strongly bonding. Also, it should be noted that the lowering in energy of an empty orbital by the transformation from a bridged to a nonbridged B_2H_6 leads to an increase in the acidity, or electron-accepting power, of the molecule, a factor that is of some importance to the reactivity of B_2H_6 with bases. The chemistry of the higher boranes is treated in more detail in Chapter 15.

Aluminum hydride, or alane, does not form a simple dimer like B_2H_6. The greater size and diminished polarizing power of Al^{3+} allows a larger coordination number than four for the Al^{3+} ion, and it is probable that a coordination number of six is achieved in the various forms of aluminum hydride. One preparation of aluminum hydride, the reaction of $AlCl_3$ with three equivalents of LiH in tetrahydrofuran (dioxane), yields a colorless homogeneous solution in which the AlH_3 is probably coordinated to three ether ligands. In this form, alane is metastable with respect to either precipitation as an amorphous white polymer $(AlH_3)_n$ or decomposition into metallic aluminum and hydrogen in the presence of noble metal catalysts. Aluminum hydride can also be easily prepared by the thermal decomposition of trialkylalanes (see Equation 9.13). Because all known preparations of alane give amorphous materials, the structure in the solid is unknown.

The chemical reactivity of alane resembles that of BeH_2. Reaction with water is violent, to give hydrogen and aluminum oxide (or hydroxide, depending on the amount of water). Although alane must be covalent, its reactivity closely resembles that of the saltlike hydrides. One reaction that borane and alane share in common is the relative ease with which they react with unsaturated hydrocarbons. In the case of boron, reactions occur rapidly under ambient conditions and have found enormously varied and important uses in organic synthesis. The discoverer and the outstanding contributor to the development of the field of "hydroboration," Brown, received the Nobel Prize in 1979 for his work on this reaction. Borane is usually generated *in situ* by the reaction noted in Equation 9.17 and in ether solvents; the reactive species is probably a monomeric BH_3 loosely coordinated to a solvent molecule. The reaction with alkenes is shown in Equation 9.18:

$$BH_3 + 3C_6H_{13}CH{=}CH_2 \xrightarrow{\text{Ether solvent}} B(CH_2CH_2C_6H_{13})_3 \qquad (9.18)$$

Alane reacts less easily with unsaturated hydrocarbons than does borane. Because the resulting alkyl aluminum compounds are extremely reactive and sensitive to air, "hydroalation" has not found the same synthetic utility in fine organic synthesis as has hydroboration. Nevertheless, the reaction is of considerable industrial importance in the direct synthesis of aluminum alkyls. These important compounds, used for the synthesis of long-chain alkenes and alkanols and Ziegler–Natta catalysts (see Chapter 23), can be prepared directly from the metal, hydrogen, and alkene in the presence of an aluminum alkyl. The most important of these reactions is the synthesis of triethylalane, shown in Equation 9.19:

$$2Al + 3H_2 + 6C_2H_4 \xrightarrow{Al(C_2H_5)_3} 2Al(C_2H_5)_3 \qquad (9.19)$$

Although the hydride intermediates are not normally detected in the reaction shown in Equation 9.19, there is little doubt that they are implicated in the reaction. It is an interesting coincidence that Ziegler, the discoverer of the reaction shown in Equation 9.19, was also a Nobel Laureate (1963), although in his case the prize was accorded for the application of alkylalanes (and other organometallic compounds) to the polymerization of alkenes.

Although gallane and indane have also been prepared by reduction of the halides by active hydrides, relatively little is known of their chemistries. Gallane resembles alane quite closely in its properties, but indane differs in showing a facile loss of hydrogen to give a polymeric monohydride $(InH)_n$. The reaction, shown in Equation 9.20, is slow at room temperature but becomes rapid at 100 °C:

$$(InH_3)_n \xrightarrow{Ether} (InH)_n + nH_2 \qquad (9.20)$$

Thallane can also be produced as a polymeric trihydride in diethyl ether solution by reduction of the trihalides, but even at 0 °C the trihydride rapidly decomposes by loss of H_2 to give a brown polymeric monohydride $(TlH)_n$. The monohydrides also show an increasing ease of decomposition to the elements on descending the group ($[InH]_n$: decomposition at 400 °C; $[TlH]_n$: decomposition at 270 °C). This tendency toward a decreasing stability of a higher-oxidation state and a general weakening of bonds on descending a p-block group is of great importance to the chemistry of these elements. In the case of hydrogen compounds, the trend is hardly surprising in view of the small size of the hydrogen atom and the difficulty of concentrating electron density into the internuclear region (that is, orbital overlap) when it is bonded to large atoms. This question will be raised again in our discussion of the binary oxides and halides.

Despite the great synthetic utility of boranes and alanes, they are rarely seen or handled as binary compounds by most chemists. This is due to the availability of a broad family of complex hydrides of boron and aluminum that while retaining the reactivity of the binary compounds are much more convenient to use. The most widely used complex hydrides are lithium tetrahydridoaluminate(III) (commonly known as lithium aluminum hydride) and sodium tetrahydridoborate(III) (sodium borohydride). The former, prepared by reaction of $AlCl_3$ with LiH (Equation 9.21), typifies a highly reactive, "hydridic" complex hydride, which shows all the violent reactivity with water common to saltlike hydrides:

$$4LiH + AlCl_3 \xrightarrow{\text{Ether}} LiAlH_4 + 3LiCl \qquad \textbf{(9.21)}$$

Sodium borohydride, on the other hand, shows the inertness toward water that might be expected of a species isoelectronic with methane and with a nearly homopolar bond.[1] Despite this relative inertia, $NaBH_4$ undergoes a variety of useful organic reductions. It is also the preferred source of BH_3 for most hydroboration reactions. Sodium borohydride can be prepared using NaH and BF_3 (by analogy with Equation 9.21, BF_3 is preferred to BCl_3 because of its lower tendency to hydrolyze and ease of storage as the diethyl ether complex). An alternative synthesis is that shown in Equation 9.22:

$$4NaH + 4B(OCH_3)_3 \xrightarrow{CH_3O(CH_2CH_2O)_2CH_3 \text{ or THF}}$$
$$NaBH_4 + 3NaB(OCH_3)_4 \quad \textbf{(9.22)}$$

The formation of the complex metal hydrides is the prototype of the acid–base chemistry that constitutes the most important single feature of group IIIA (13) chemistry. The reaction may be generalized as in Equation 9.23:

$$MX_3 + B \rightarrow BMX_3 \qquad \textbf{(9.23)}$$

where M is a group IIIA (13) element, X is virtually any substituent, and B is a base. The hydride ion is a very strong base, and the equilibrium constants for production of $[MH_4]^-$ are very large for M = B and Al. On descending the group, the formation constants fall dramatically, and $[InH_4]^-$ and $[TlH_4]^-$ decompose to the polymeric hydrides in diethyl ether solution. This is a further manifestion of the bond weakening in heavier element compounds, referred to above.

9.5 Hydrides of Group IVA (14)

The characteristic hydrides of Group IVA (14) are those of general formula MH_4. These compounds are known for all members of the group and are thermodynamically stable for C, Si, and Ge. In the case of tin, the

---

[1] This statement is made with the full benefit of hindsight. The original synthesis of $NaBH_4$ by H. C. Brown was undertaken to find a simple storage compound for hydrogen. The discovery that $NaBH_4$ was unreactive to neutral water was initially a surprise and disappointment to Brown and his colleagues. They soon showed that a change in pH or the addition of catalysts would do the trick.

tetrahydride is very fragile, decomposing readily under ambient conditions to a transient SnH_2 species, which in turn decomposes to the elements. Plumbane PbH_4 is even more fragile and has only been prepared and observed as a transient species at low temperature. All five MH_4 molecules are colorless gases under ambient conditions and are highly covalent, tetrahedral species.

Methane, together with the other paraffins, is notable for its extreme chemical inertness. Apart from a few well-known reactions involving very reactive free radicals (for example, combustion with O_2 and halogenation), CH_4 undergoes very few reactions under mild conditions. This lack of reactivity is a result of the closed-shell electronic structure of CH_4: All bonding orbitals are at low energy and filled (little tendency to react with electrophiles), and the antibonding orbitals are at very high energy and empty (little tendency to react with nucleophiles). Besides these factors, the paraffinic C—H bond is strong (about 410 kJ mol^{-1}), and hydrogen atom abstraction can occur only with those relatively few atoms and radicals that give X—H bonds of comparable or greater strength. In particular, this applies to O_2 and the more reactive halogens (compare with the reactivity of H_2). In the case of CH_4, the reaction with O_2, which is highly exothermic, requires external intervention for initiation, but once initiated proceeds with greater or lesser violence depending on conditions. The other MH_4 species differ from CH_4 in that their reactions with O_2 occur spontaneously. This greater reactivity doubtless results from the greater ease of formation of peroxidic species, because of the easier interaction of electrophilic O_2 with the highest-occupied MO of MH_4, and the greater ease of abstraction of a H atom, because of the lower M—H bond energy (Si—H: about 320 kJ mol^{-1}).

In CH_4 the bond is slightly polarized in the sense $C^{\delta-}$—$H^{\delta+}$. For the remaining congeners of Group IVA (14), there is a slight polarization in the sense $M^{\delta+}$—$H^{\delta-}$. Unlike methane, the other MH_4 species are relatively easily attacked by moderate electrophiles and nucleophiles. Although they only react very slowly with neutral water, both SiH_4 and GeH_4 react readily with aqueous alkali and acid to give H_2 and either silicates (basic) or hydrated silica (acid). The reactions with halogens also occur much more readily than with CH_4. The greater reactivity of the lower members of the group can be attributed to a number of factors:

1. The availability of empty *d* orbitals at lower energy than the *s*- and *p*-derived antibonding orbitals facilitates electron acceptance.

2. The gradual rise in energy of the HOMO on descending the group facilitates electrophilic attack.

3. The greater size of the central atom facilitates an increase in the coordination number required for S_N2 reactions.

4. The decrease in bond energy decreases the energy demands for changing shape, changing coordination number, and departure of ligand.

Unlike borane and alane, Group IVA (14) hydrides show almost no tendency to increase their coordination number. Even F^-, which has a very high affinity for silicon, does not appear to form a stable complex with SiH_4. That the F^- ion in nonprotic media greatly increases the hydridic reducing power of substituted silanes [for example, $(C_2H_5)_3SiH$] suggests that short-lived reactive intermediates of higher coordination number may be present. Coordination of F^- to four-coordinate silicon raises the electron density on Si by a combination of σ and π donation and would be expected to increase the partial negative charge on the hydrogen atom of an Si—H bond.

The cheapness and easy synthesis of compounds containing the Si—H bond has led to a number of synthetic applications of substituted silanes in fine organic synthesis. For example, silyl enol ethers, which find wide use in the synthesis of pharmaceutically interesting compounds, may be synthesized according to Equation 9.24:

$$CO + R_3SiH + C_4H_9CH{=}CH_2 \xrightarrow{Co_2(CO)_8}$$
$$R_3SiOCH{=}CHCH_2C_4H_9 \quad (9.24)$$

The catalyzed addition of silanes to unsaturated compounds (the hydrosilation reaction, see Chapter 23), such as shown in Equation 9.25 is an important method for the functionalization of organosilicon compounds:

$$R_3SiH + C_4H_9CH{=}CH_2 \xrightarrow{PtCl_4} R_3SiCH_2CH_2C_4H_9 \qquad (9.25)$$

Although the germanes have a very similar chemistry, the low natural abundance and high cost of germanium has prohibited the wide application of germanium compounds. An enormous number of substituted silanes, germanes, and stannanes have been prepared and studied. The number and type of substituents possible for silanes and germanes seem to be almost without limit. The substituted stannanes with substituents X (of high electronegativity) tend to be very unstable with regard to loss of HX. Table 9.1 shows a selection of known reactions of substituted Group IVA (14) hydrides.

Silicon forms a large family of higher hydrides containing Si—Si bonds. Some of these are discussed in more detail in Chapter 15. The lower energy of Si—Si, relative to C—C bonds, renders these higher hydrides more reactive than the corresponding alkanes. The silicon analogs of simple alkenes and acetylenes are unstable with respect to polymerization to saturated products. The Si=Si bond is only stable when the silicon atoms

---

Table 9.1 *Some Reactions of Group IVA (14) Hydrides*

$BuSiH_3 + I_2 \rightarrow BuSiH_2I + HI$

$PhSiH_3 + CH_3OH \xrightarrow{Cu} PhSiH_2(OCH_3) + H_2$

$Et_3SiH + (CH_3)_2NH \rightarrow Et_3SiN(CH_3)_2 + H_2$

$Bu_2GeH_2 + 2Ph_3CCl \rightarrow Bu_2GeCl_2 + 2Ph_3CH$

$2Et_3GeH + H_2O \rightarrow (Et_3Ge)_2O + 2H_2$

$GeH_4 + NH_3 \rightarrow [NH_4]^+[GeH_3]^-$

$Cl_3GeH + HMn(CO)_5 \rightarrow Cl_3GeMn(CO)_5 + H_2$

$3Me_2SnH_2 \xrightarrow{120°C} 2Me_3SnH + Sn + 2H_2$

$EtSnH_3 \xrightarrow[HBr]{-78°C} EtSnH_2Br + (EtSnBr)_n + H_2$

$Bu_3SnH + BuI \rightarrow Bu_4Sn + HI$

---

are fully substituted with very sterically demanding groups:

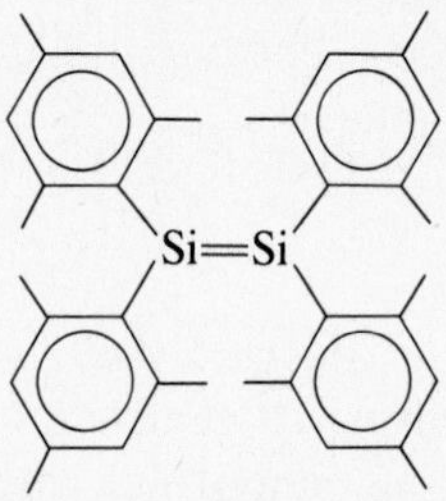

These trends become further exaggerated on descending the group, although the differences between Si and Ge are marginal.

In addition to the well-defined molecular hydrides of Si and Ge (and to a much lesser extent, Sn), there also exists a large number of solids of more or less indefinite composition $[SiH_n]$ ($n = 0–2$) or $[GeH_n]$. Such materials may be produced by pyrolysis of higher molecular hydrides or by decomposition of MH_4 in a microwave or electric discharge. Although known for a very long time, such materials have been largely ignored by chemists, probably because of their nonstoichiometry. However, recent developments in the preparation of amorphous silicon for electronic and photoelectric applications has aroused a great deal of interest in them.

For about the first thirty years (1940–1970) of the semiconductor revolution (see Chapter 19), it was firmly believed that the electronic behavior of semiconductor materials was intimately linked to high-level crystal perfection. Were this to be true, it would place a rather high lower limit on the cost of manufacture of such materials because the growth of single crystals is inherently expensive. This limitation is not very important in microelectronics where a few grams of single-crystal silicon can yield millions of circuits, but for large-scale use such as solar energy–collector panels, it is the primary limitation to economic feasibility. Fortunately, we now know that amorphous materials, which lend themselves to easier and cheaper large-scale manufacture, are not *necessarily* different in electronic structure from crystals.

One big difference between a crystal and a disordered amorphous material is that it is much easier to tie up all the "loose ends" in the former than in the latter. In a crystal of elemental silicon, almost all silicon atoms are coordinated to a regular tetrahedron of other silicon atoms, thus saturating the valence demands of each silicon atom. The constraints of geometry do not allow such saturation when the valence tetrahedra are randomly oriented, as in an amorphous material. In this case, many of the silicon atoms are left with only three bonds and an unpaired electron. Such radical species are of dramatically different energy from the saturated Si atoms and profoundly disturb the electronic structure of the material. Being chemically "hot," these radicals also destabilize the amorphous state and make the transformation to the thermodynamically more stable polycrystalline state easier. Both problems can be solved by tying up the dangling ends in the form of Si—H bonds. The Si—H bond is particularly attractive because the decomposition of silicon hydrides, a well-explored way of making thin-film amorphous silicon, provides them automatically (additional H_2 is needed to ensure complete saturation). In addition, the Si—H bond is electronically very similar to the Si—Si bond, and the small size of H makes its presence in the amorphous matrix easy to accommodate.

At the time of this writing, amorphous silicon solar-collector panels are still only in the development stage, but there are futurists who believe that the production of such panels could be one of the largest manufacturing industries by the year 2000.

9.6 Hydrides of Groups VA and VIA (15 and 16)

We treat the hydrides of Groups VA and VIA (15 and 16) together because they have many properties in common with each other or with the groups we have already discussed. The electronegativity of nitrogen is significantly higher than that of hydrogen, whereas the remaining elements of Group VA (15) have electronegativities very close to that of hydrogen. Thus, the N—H bond is polarized in the sense $N^{\delta-}$—$H^{\delta+}$, whereas the M—H bonds (M = P, As, Sb, and Bi) are almost nonpolar. In the case of Group VIA (16), all elements have electronegativities higher than hydrogen, and therefore their hydrides are all polarized in the sense $M^{\delta-}$—$H^{\delta+}$. With the exception of water, all binary hydrides of Groups VA and VIA (15 and 16) are colorless gases under ambient conditions.

And except for water and ammonia, they are all extremely poisonous. As with the preceding groups, the M—H bonds decline in strength on descending the group, and the thermal dissociation into the elements becomes easier. The deposition of an arsenic mirror on contacting gaseous arsine with the walls of a hot glass tube is the basis for a very important forensic test for arsenic (the Marsh test). Although arsenic has been a favorite weapon of poisoners throughout recorded history, the Marsh test was only developed in the nineteenth century. The reaction used in the Marsh test is shown in Equation 9.26:

$$Zn + H_2SO_4 + As \text{ compound} \longrightarrow AsH_3 \qquad (9.26)$$

$$AsH_3 \xrightarrow{\text{Heat}} As + 3/2 \; H_2$$

The compounds BiH_3 and PoH_3 have only been obtained in trace amounts and are very unstable.

The trend toward an increasingly stable lower oxidation state on descending the group, noted for Groups IIIA and IVA (13 and 14), becomes so dominant for the hydrides of subsequent groups that the mononuclear binary hydrides of all elements of Groups VA–VIIA (15–17) are known only in the lowest available oxidation states, MH_3, MH_2, and MH, respectively. The phenomenon is strongly related to the *relatively* low electronegativity of hydrogen and its limited ability to displace electron density away from the heteroatom, a difficulty that increases with increasing oxidation state of the heteroatom. This behavior contrasts with that of the halide and chalcogenide compounds of the elements, as we will see later.

Both ammonia and water may be synthesized directly from the elements. Under ambient conditions, the direct reaction of N_2 and H_2 is only slightly exothermic and very slow. In the commercial synthesis of NH_3 by the Haber process, the equilibrium (Equation 9.27) is pushed to the right by application of moderate pressure, and the rate of the reaction is increased by use of iron-based catalysts at moderate temperature:

$$3H_2 + N_2 \rightleftharpoons 2NH_3 \qquad (9.27)$$

The analogous reaction of hydrogen with oxygen is highly exothermic and, because of the unusual branched-chain radical mechanism by which the reaction occurs, may be dangerously explosive.

Because of the great stability of water, all hydrides, except those of the lighter halogens, undergo combustion in oxygen to give water and either the oxide or the free element of the heteroatom. Ammonia burns quietly in oxygen or air according to Equation 9.28:

$$4NH_3 + 3O_2 \rightarrow 2N_2 + 6H_2O \qquad (9.28)$$

All hydrides of Groups IVA and VA (14 and 15) combust more or less violently to give water and the very stable oxides of these elements.

The hydrides of the heavier members of Groups VA and VIA (15 and 16) are produced most easily by the acid hydrolysis of their binary compounds with electropositive metals:

$$Mg_3P_2 + 6H^+ \xrightarrow{\text{H}_2\text{O}} 3Mg^{2+} + 2PH_3 \tag{9.29}$$

$$FeS + 2H^+ \xrightarrow{\text{H}_2\text{O}} Fe^{2+} + H_2S \tag{9.30}$$

$$Zn_3Bi_2 + 6H^+ \xrightarrow{\text{H}_2\text{O}} 3Zn^{2+} + 2BiH_3 \tag{9.31}$$

Contamination of the product with H_2 can be a serious problem with the heavier congeners and the reaction in Equation 9.31 actually only leads to a trace of BiH_3 (ppm) in H_2. In the reaction in Equation 9.29, the product is also contaminated with P_2H_4, which renders the gas spontaneously inflammable. The same problem arises with the other classical route to PH_3, the reaction of yellow phosphorus with aqueous alkali. This reaction is represented in an idealized way by Equation 9.32:

$$P_4 + 6OH^- \longrightarrow PH_3 + 3H\!-\!P\!\!\begin{array}{c} O^- \\ \diagup \\ \diagdown \\ O^- \end{array} \tag{9.32}$$

Most of the heavier hydrides can also be prepared in a high degree of purity by reaction of a binary halide with an appropriate complex hydride.

Except for water, which is relatively inert to oxidation–reduction reactions, the hydrides of Groups VA and VIA (15 and 16) tend to be reducing agents. This reducing power can be the result of various types of reactions, including hydrogen-atom transfer and electron transfer. Many free-radical species abstract a hydrogen atom with great ease from the heavier hydrides to produce heteroatom radicals, which then dimerize. This behavior is extremely useful in moderating certain industrial oxidation processes. For example, the oxidative dehydrogenation of paraffins to the much more useful alkenes, as in Equation 9.33, is difficult to achieve as a direct reaction because the alkene is more susceptible to oxidation than is the alkane:

$$4CH_3CH_2CH_3 + O_2 \rightarrow 4CH_3CH{=}CH_2 + 2H_2O \tag{9.33}$$

However, the inclusion of small amounts of H_2S or HI in the feedstock suppresses the alkene oxidation and greatly increases the selectivity toward alkene production. Although all mechanistic details of this type of reaction are not understood, it is likely that the heavy hydride selectively terminates an important chain reaction in the autoxidation of the alkene by hydrogen atom transfer to a key radical:

$$R^{\displaystyle \cdot} + HI \rightarrow RH + I^{\displaystyle \cdot} \tag{9.34}$$

$$2I^{\displaystyle \cdot} \rightarrow I_2 \tag{9.35}$$

The HI would be regenerated by reaction of I_2 with a hydrocarbon species.

Another important mechanism for oxidation of the hydrides of the more electronegative elements is the transfer of an electron from the anion generated by proton loss; an example of such a reaction is the following:

$$R—SH \rightarrow R—S^- + H^+ \tag{9.36}$$

$$R—S^- + Cu^{2+} \rightarrow R—S^{\cdot} + Cu^+ \tag{9.37}$$

$$2R—S^{\cdot} \rightarrow R—S—S—R \tag{9.38}$$

Although the thiol group is a rather weak acid, this pathway is fast at neutral and higher pH because of the very high reactivity of the anion as opposed to the undissociated thiol. The reactions (Equations 9.36–9.38) provide a useful means for cross-linking protein molecules in biochemical systems. The permanent waving or straightening of hair involves the chemical rupture and reformation of the disulfide cross-links of the hair protein keratin. In this case, the RSH group is the amino acid residue cysteine.

The tendency of hydrogen compounds to dissociate by loss of a proton, a subject that is treated in greater depth in Chapter 21, follows approximately the trend of increasing electronegativity. Thus, for water such acid dissociation is extremely important; for ammonia, less so.[2] Although we often write reactions as though they involve free protons—for example, Equation 9.31—they are never actually produced in condensed phases and only under highly energetic conditions (relative to chemical energies) in the gas phase. In reactions such as that in Equation 9.36, the proton is always transferred to another molecule or ion. Thus, in water the proton is usually transferred to a solvent molecule to produce a solvated $[H_3O]^+$ ion.

Unlike boron and the Group IVA (14) elements, the hydrides of Groups VA and VIA (15 and 16) have little tendency to produce higher hydrides. The only higher binary hydride of nitrogen of any importance is hydrazine N_2H_4. A convenient synthesis of hydrazine involves the chlorination of liquid ammonia or the aqueous chlorination of ammonia with sodium hypochlorite (Raschig process). In both cases, the initial reaction is a rapid formation of chloramine (Equations 9.39 and 9.40), followed by nucleophilic displacement of chloride by ammonia (Equation 9.41):

$$2NH_3 + Cl_2 \rightarrow NH_2Cl + NH_4Cl \tag{9.39}$$

or

$$NH_3 + OCl^- \rightarrow NH_2Cl + OH^- \tag{9.40}$$

$$NH_2Cl + 2NH_3 \rightarrow N_2H_4 + NH_4Cl \tag{9.41}$$

---

[2] Important exceptions to this generalization are found in going from the first to second members of Groups VA–VIIA (15–17). In aqueous solution, the hydrides of the first members are weaker acids than are the second members. This arises because of the higher bond energies of the NH, OH, and HF bonds and the low electron affinities of N, O, and F, relative to P, S, and Cl.

In reality the reactions are more complicated than the above equations suggest. A side reaction of some importance is the reaction shown in Equation 9.42, which is strongly catalyzed by metal ions:

$$2NH_2Cl + N_2H_4 \rightarrow 2NH_4Cl + N_2 \tag{9.42}$$

This catalysis is usually suppressed in the commercial process by addition of gelatin to complex adventitious metal ions.

The high cost of oxidants such as Cl_2 and the great usefulness of hydrazine and its derivatives as fuels have encouraged a great deal of research into the catalytic synthesis of hydrazine from the elements, so far with little success. Unlike ammonia, hydrazine is a liquid under ambient conditions (melting point 2 °C; boiling point 114 °C). A 1:1 mixture of hydrazine and 1,1-dimethylhydrazine has found use as a fuel (Aerozine-50) for the attitude-control engines of space vehicles, and much work has been carried out on the development of a hydrazine–oxygen fuel cell.

The energy content of hydrazine is in part due to the anomalously low $=N-N=$ bond energy by comparison, for example, with the C—C bond. This anomaly becomes more exaggerated for the cases of —O—O— bonds and the F—F bond. This effect is usually attributed to the greater magnitude of lone pair-bonding pair repulsions, as compared to bonding pair-bonding pair repulsions (see Chapter 3) or, more generally, to the appearance of an important interelectronic repulsion term with increasing population of the bonding shells for small-radius, first-row species. This effect is evident in and contributes to the relatively great reactivity of all compounds containing bonds between the elements N, O, and F. Table 9.2 shows some bond energies illustrating these trends.

Table 9.2 *Some Approximate Bond Energies (kJ mol⁻¹)*

Bond	Bond Energy
C—C	350
Si—Si	220
N—N	160
P—P	240
O—O	140
S—S	250
O—F	190
S—F	280
F—F	155
Cl—Cl	240
Br—Br	190

Table 9.3 *Some Redox Reactions of N_2H_4 and H_2O_2*

$$N_2H_4 + 2Cl_2 \rightarrow 4HCl + N_2$$

$$H_2O_2 + Cl_2 \rightarrow 2HCl + O_2$$

$$N_2H_4 + O_2 \xrightarrow{\text{Combustion}} 2H_2O + N_2$$

$$N_2H_4 + 2O_2 \xrightarrow{\text{Fe(III)}} 2H_2O_2 + N_2$$

$$6H^+ + 5H_2O_2 + 2[MnO_4]^- \rightarrow 5O_2 + 2Mn^{2+} + 8H_2O$$

$$H_2O_2 + SO_2 \rightarrow H_2SO_4$$

The chemistry of hydrazine is rich in reactions that lead to stripping of H atoms and formation of the very stable N_2 molecule. Reactions that proceed by cleavage of the N—N bond are rarer. Hydrogen peroxide, on the other hand, exhibits ambivalent behavior—some reactions proceeding by H stripping to produce O_2 (H_2O_2 as a reducing agent), others proceeding by O—O bond cleavage to produce OH^- or H_2O (H_2O_2 as an oxidizing agent). A consequence of this ambivalent behavior is the ease with which H_2O_2 disproportionates under the influence of a wide range of catalysts:

$$2H_2O_2 \xrightarrow{\text{MnO}_2} 2H_2O + O_2 \tag{9.43}$$

Table 9.3 shows some typical redox reactions of N_2H_4 and H_2O_2.

The only other higher hydride from Group VA (15) of any significance is P_2H_4. This compound is usually produced as a side product in reactions that produce PH_3, and it may be condensed out from the latter as a pyrophoric yellow liquid.

Hydrogen peroxide is the only important higher hydride of oxygen. Although spectroscopic evidence for the existence of more highly catenated molecules (H_2O_3, H_2O_4) has been reported, they are so unstable that they are unlikely to achieve higher status than laboratory curiosities. Hydrogen peroxide may be synthesized by the partial reduction of O_2 with an electropositive metal. A classical synthesis involved the acid hydrolysis of barium peroxide with dilute sulfuric acid, Equation 9.44:

$$BaO_2 + H_2SO_4 \xrightarrow{\text{H}_2\text{O}} BaSO_4 + H_2O_2 \tag{9.44}$$

At one time, the main process for chlorine manufacture, the electrolysis of brine at a mercury cathode, made available dilute sodium amalgam (Na/Hg alloy) that was normally allowed to react with water to produce sodium hydroxide. However, air oxidation of such amalgam to sodium peroxide, followed by hydrolysis, provided a relatively cheap source of dilute aqueous H_2O_2. The replacement of mercury cathode electrolysis by alternative technologies following severe contamination of the environment by the mercury emissions has led to the demise of this source. A number of newer syntheses, which have the advantage of being catalytic, involve the autooxidation of a variety of organic molecules. A fairly successful process uses the hydroquinone–quinone redox couple. For example, with 2-alkylanthraquinones the reactions in Equation 9.45 and 9.46 provide the net catalytic cycle (Equation 9.47):

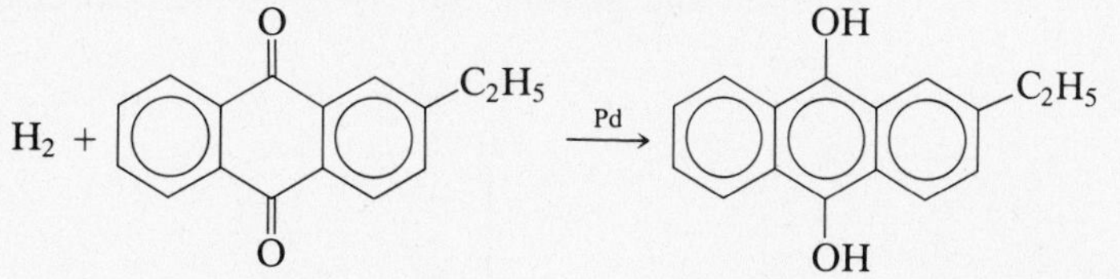

$$ (9.45) $$

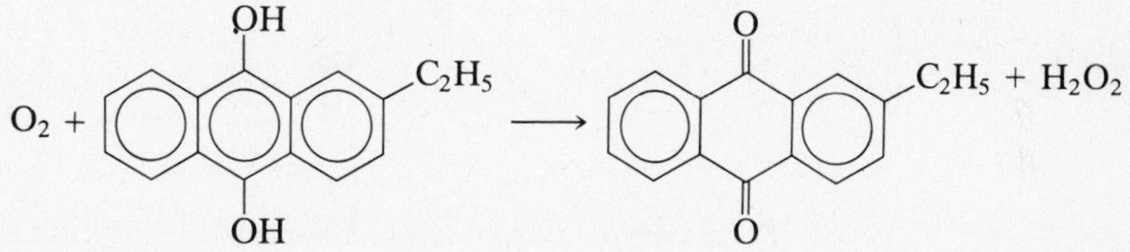

$$ (9.46) $$

Net Reaction

$$ H_2 + O_2 \rightarrow H_2O_2 \qquad (9.47) $$

The success of such a process hinges on reducing irreversible losses of the relatively expensive organic catalyst. A particularly impressive success for this technology is the anthroquinone-catalyzed bleaching of woodpulp, which is rapidly making inroads into the more traditional Cl_2– and ClO_2–based processes.

Most hydrogen peroxide is used in the form of sodium peroxoborate, the most important nonchlorine bleach. Although sodium peroxoborate can be made by the reaction of sodium peroxide with boric acid, it is usually made on an industrial scale by the electrolysis of an aqueous solution of sodium borate containing sodium carbonate. In this process, the peroxide is generated *in situ*. The product contains a mixture of peroxoborates, one of which contains the anion shown below:

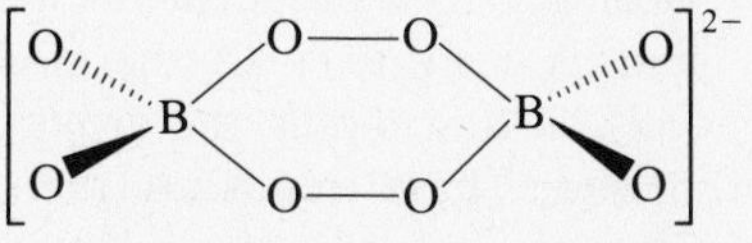

As will be seen in Chapter 24, reactions like that shown in Equations 9.45 and 9.46 occur in most living systems, and nature has developed sophisticated enzymes specifically designed to remove the toxic H_2O_2 before it can damage the cell. Pure hydrogen peroxide greatly resembles hydrazine in its liquid range (-0.4 °C to 150 °C) and its exothermic decomposition to the elements. On the other hand, hydrazine is predominantly basic in character, whereas H_2O_2 is a weak acid that exists mainly as the $[HO_2]^-$ ion in water above pH 12.

The structures of N_2H_4 and of H_2O_2 show slight differences in their gas-phase conformational preferences. Both N_2H_4 and P_2H_4 exhibit a conformation about the N—N or P—P axis, intermediate between eclipsed and gauche, but H_2O_2 is much closer to being eclipsed:

(The perspective of these structures is looking down the N—N and O—O bond axes, respectively.)

The conformations of N_2H_4 and H_2O_2 are frequently rationalized in terms of the lone pair-bonding pair repulsions, as suggested by the structures drawn above. It must be remembered however (see discussion of bonding in water, Chapter 3) that the lone pairs as drawn are a poor approximation to reality!

The higher hydrides of S and Se are more stable than those of O, and the linear polymers H_2S_n have been obtained as pure compounds up to $n = 8$. Aqueous alkali metal sulfides dissolve S_8 to give linear polymeric S_n^{2-} ions, which can be acidified to give a viscous yellow oil. This material may be cracked and fractionally distilled to give the lighter H_2S_n species. On standing, especially in the light, these compounds revert to the thermodynamically more stable mixture of H_2S and S_8.

Besides their tendency to transfer a proton and to constitute the archetypical group of Brønsted acids, the hydrides of Groups VA–VIIA (15–17) also possess nonbonding electron pairs that confer on them Brønsted-base properties as well. Of the neutral binary hydrides, ammonia and its derivatives are the most basic toward the proton, and the basicity falls on descending the groups and on passing from left to right along a period. These trends are the opposite of the trends in Brønsted acidity. The great family of simple inorganic bases also includes the conjugate anions, produced by deprotonation of the neutral hydrides, of which OH^- is the most obvious example. The nonbonding pairs on these hydrides and their derivatives can also be involved in bonding to metal ions in a more generalized acid–base interaction. Such interactions form the basis of much of coordination chemistry, to be presented in Chapter 12. The trends in the strengths of interactions of these bases with metals are much less systematic than those for proton interactions and are dealt with in more detail in Chapter 21.

Before leaving the discussion of the binary hydrides of Groups VA and VIA (15 and 16), mention should be made of the unique and unparalleled compound hydrazoic acid, or hydrogen azide, HN_3. The free acid itself may be prepared by distillation from a mixture of sodium azide and concentrated sulfuric acid, but few brave souls have risked this enterprise. The pure acid is extremely sensitive and dangerously explosive, as are many of the salts with heavy or transition metals. The sensitive salts, particularly lead(II) azide $Pb(N_3)_2$, are used for detonators and percussion caps. Hydrazoic acid is a weak acid, but considerably stronger than ammonia or hydrazine, and is significantly ionized in water. The driving force for this ionization is the formation of the conjugation-stabilized azide ion:

$$
\begin{array}{c}
\overset{\displaystyle 124\ \text{pm}\quad 113\ \text{pm}}{\text{N}\!-\!\!-\!\!-\!\text{N}\!-\!\!-\!\!-\!\text{N}} \\
\underset{\substack{\\ 101\ \text{pm}}}{\Big/}\quad 120^\circ \\
\text{H}
\end{array}
\qquad
\overset{\displaystyle 115\ \text{pm}\quad 115\ \text{pm}}{[\text{N}\!-\!\!-\!\!-\!\text{N}\!-\!\!-\!\!-\!\text{N}]^-} + \text{H}^+
$$

The azide ion is isoelectronic with the extremely stable CO_2. The highly ionic azides are quite stable (for example, the sodium salt is prepared by passing N_2O over molten $NaNH_2$ at 190 °C), but any derivatives where there is significant covalent bonding to the N_3 moiety are likely to be dangerous. This type of behavior is common to a number of very thermodynamically unstable species that are kinetically stabilized in ionic form by a highly symmetrical closed-shell structure. Some examples are H_2SO_4, $HClO_4$, and $HMnO_4$, which are progressively more dangerous as the undissociated acid but relatively innocuous in the form of highly ionic salts.

9.7 Hydrogen Halides

The hydrogen halides are all highly polar diatomic molecules whose electronic structure and molecular dynamics are sufficiently simple, thus playing an important role in the development and testing of bonding theories. They are all pungent colorless gases under ambient conditions with an extremely high solubility in water.

The hydrogen halides in aqueous solution are the simplest of Brønsted acids. Hydrogen fluoride is a rather weak acid, but all others are very strong. The weakness of HF as an acid in water is due to a combination of its high homolytic bond strength and to its anomalously low electron affinity. At the electronic level, the anomalously low electron affinity of fluorine is due to the high interelectronic repulsion arising from a nearly filled shell on a very small atom. The Brønsted acidity of a species X—H can be expressed in terms of a cycle:

$$
\begin{array}{ccc}
\text{X—H} & \xrightarrow{\quad D_{HX} \quad} & \text{X}^{\bullet} \quad \text{H}^{\bullet} \\[2pt]
\Big\downarrow {\scriptstyle \Delta H_{sol}} & & {\scriptstyle EA}\Big\downarrow \qquad \Big\downarrow {\scriptstyle I} \\[2pt]
\text{X}^{-}(aq) + \text{H}^{+}(aq) & \xleftarrow{\quad \Delta H_{hyd} \quad} & \text{X}^{-} + \text{H}^{+}
\end{array}
$$

In the case of fluorine, relative to the other members of Group VIIA (17), both the bond dissociation energy D_{HX} and the electron affinity EA disfavor the ionization of HF. The more favorable hydration energy of F^- is not sufficient to override the other factors.

Although all hydrogen halides can be made by direct combination of the elements, this does not constitute a particularly useful synthetic method for HF or HI. Both HF and HCl can be liberated from their salts by the action of concentrated sulfuric acid. A generally useful laboratory-scale synthesis for HI and HBr is the hydrolysis of PX_3 prepared *in situ* by reaction of the halogen with phosphorus. The chlorination or bromination of some simple organic molecules, for example, dioxane, is also a useful laboratory method for generating HCl or HBr:

$$
\begin{array}{ccc}
\begin{array}{c}
\text{H} \qquad \text{O} \qquad \text{H} \\
\text{H—C} \qquad \text{C—H} \\
\mid \qquad\qquad \mid \\
\text{H—C} \qquad \text{C—H} \\
\text{H} \qquad \text{O} \qquad \text{H}
\end{array}
& + 4Br_2 \longrightarrow &
\begin{array}{c}
\text{H} \qquad \text{O} \qquad \text{H} \\
\text{Br—C} \qquad \text{C—Br} \\
\mid \qquad\qquad \mid \\
\text{Br—C} \qquad \text{C—Br} \\
\text{H} \qquad \text{O} \qquad \text{H}
\end{array}
& + \, 4HBr \qquad \textbf{(9.48)}
\end{array}
$$

Besides their Brønsted acidity in water, the hydrogen halides, especially HF, can generate superstrong protonic acids through their reactions with strong Lewis acids. Perhaps the best known of these is $HSbF_6$, generated by dissolving SbF_5 in liquid HF:

$$
SbF_5 + 2HF \xrightarrow{\text{HF (liquid)}} [H_2F]^+[SbF_6]^- \qquad\qquad \textbf{(9.49)}
$$

The $[H_2F]^+$ ion is analogous to the $[H_3O]^+$ ion encountered in aqueous acid solutions (see Chapter 21). Aqueous solutions of HF attack glass and silicates because the extremely high Si—F bond energy renders the F^- ion an extremely strong base toward silicon; the same is true of the Si—O— bond and the OH^- ion, which is isoelectronic with F^-.

The ability of HF to etch SiO_2 is of key importance in the printing of microcircuits onto silicon chips. The pure silicon wafer is first heated in oxygen to produce a thin layer of insulating SiO_2 on the surface. An image of the circuit is then printed over the SiO_2, using the photopolymerization of an organic monomer to record the image. When the surface is washed with dilute aqueous HF, areas that were coated by organic polymer remain unaffected, but the areas where the SiO_2 is exposed are etched to reveal the pure silicon surface again, ready for doping with an appropriate impurity.

Hydrogen fluoride is quite inert and hydrogen chloride relatively inert to redox chemistry, whereas HBr and HI are fairly easily oxidized to the free element. Because oxygen is a strong oxidizing agent, stored solutions of HBr and HI are difficult to keep free of the halogen. This trend parallels the previously mentioned general trend for a weakening of the X—H bond energy on descending a group.

Hydrochloric acid is a major industrial inorganic chemical, which finds a wide range of uses. Besides its uses as a relatively cheap general acid, HCl is widely used in organic synthesis. The anhydrous hydrogen halides show a relatively high reactivity toward unsaturated organic molecules due to electrophilic attack of the highly polar hydrogen on the π electrons of the multiple bond. The hydrides of Groups IVA–VIA (14–16) are not generally acidic enough for such a mechanism to prevail and more commonly add to multiple bonds by free-radical mechanisms involving hydrogen atom transfer or by concerted addition.

9.8   *Hydrides of the* **d-** *and* **f-***Block Elements*

The evolution of our understanding of the *d-* and *f*-block hydrides has been slow, littered with confusion, and is still far from complete. A number of factors render the study of these materials much more difficult than the study of hydrides of the *s-* and *p*-block elements. Some of the more important difficulties are as follows:

1. Hydrides are usually produced by reaction of metal with hydrogen, and there are no selective solvents for extracting unreacted metal or impurities.

2. Many of the *d-* and *f*-block metals are tenaciously reactive with light elements and are extremely difficult to purify. These impurities have a large influence on the hydriding reactions.

3. Many of the hydride phases exhibit ranges of nonstoichiometry.

4. Besides the formation of true hydride phases, hydrogen also dissolves in the metal to give a solution phase.

5. Many of the hydride phases exhibit metallic properties such as high electronic conductivity and high optical reflectivity.

Together, these problems have presented a large barrier to chemists, steeped in their beliefs of stoichiometry and two-electron bonds, and these hydrides have generally been locked away in the family closet as being of little interest to the chemist. However, as methods of preparation and characterization improve and as chemists begin to approach the problem of synthesizing model molecular compounds that possess some of the same structural features as the lattice hydrides (see Chapter 18), their integration into the theoretical framework of chemistry becomes more possible.

It is now clear that the elements of the Ti group all form a more or less stoichiometric hydride of composition MH_2 when the metal is heated in hydrogen at 400 °C and 1 atm. At room temperature, the stable structure of TiH_2 is a *fcc* fluorite lattice. On the other hand, ZrH_2 and HfH_2 adopt a less symmetrical body-centered tetragonal structure. At low temperature, the latter structure is also adopted by TiH_2. When prepared carefully from highly purified metal, these hydrides have the same appearance as the metal. If the hydrides are heated in the absence of a hydrogen atmosphere, they decompose with evolution of hydrogen gas. Zirconium hydride has attracted a great deal of attention because its formation in zirconium alloys, used in nuclear reactors, can lead to a serious degradation of mechanical properties of the alloys.

Because of the high density of TiH_2 (about 4 kg dm^{-3}) and its relatively low molecular weight (about 50), 1 mole of H_2 can be stored in a volume of about 12 cm^3. This is the same volume as occupied by 1 mole of gaseous H_2 at about 2000 atm! This property makes TiH_2 of interest as a fuel-storage device for the hydrogen-powered automobile. The heat that is generated during the combustion of the hydrogen can be used to decompose the TiH_2 into hydrogen and Ti.

The elements of Group VB (5) behave very similarly to those of Group IVB (4), with the difference that the stoichiometries fall well short of MH_2 under 1 atm of hydrogen. Using higher hydrogen pressures or cathodic hydrogenation, higher H/M ratios can be achieved, but for none of these elements has the ideal MH_2 composition been achieved. The predominant structure of the MH_2 phases again appears to be a *fcc* fluorite structure. The compositions MH_n, n greater than 1, are unstable in the absence of hydrogen under pressure and spontaneously lose hydrogen on pumping at room temperature.

The trend toward lower stability continues into Group VI B (6), and only chromium has been shown to form stable hydride phases. There is definite evidence for CrH and more tentative evidence for CrH_2. Both of these phases are produced by electrolysis of chromic acid. Beyond chromium there occurs the so-called hydride gap, and none of the remaining transition elements appears to form a stable hydride phase under ambient conditions, with the exceptions of palladium, copper, and zinc (mercury and cadmium give stable hydrides at low temperatures).

Palladium hydride is a more or less typical transition hydride, showing variable nonstoichiometric composition and metallic behavior. The hydride consists of two phases: (a) a solution of hydrogen in the metal whose concentration may vary from 3–10% and (b) another whose ideal composition approaches PdH and that has a defective NaCl structure (some of the octahedral holes that are supposed to contain hydrogen are empty).

The hydrides of copper and zinc can be produced by chemical or electrolytic reduction of metal salts. Neither is thermodynamically stable, and they decompose at room temperature at a rate strongly dependent on the presence of catalytic or inhibitive impurities. These hydrides are akin to the slightly covalent ionic hydrides and have none of the characteristics of the transition hydrides described above.

All elements of Group IIIB (3) and the lanthanides combine spontaneously with hydrogen at modest temperatures to give phases of idealized composition MH_2 and (except for EuH_2 and YbH_2) possessing a fluorite structure. Most of these dihydrides take up further hydrogen at higher pressures to give phases of composition MH_3. This additional hydrogen is relatively easily lost when the pressure is removed. In both the MH_2 and MH_3 compositions, magnetic susceptibility measurements agree closely with the theoretical predictions for M^{3+} ions. The f electrons are not expected to participate significantly in bonding, and the apparent oxidation state of III for the MH_2 compounds is explained by the use of two electrons in bonding with hydrogen and the loss of a third to a delocalized conduction band (see metallic bonding in Chapter 19). This confers on the MH_2 compounds of Group IIIB(3) and the lanthanides the metallic properties typical of the transition hydrides. The stabilities of half-filled and filled subshells confer exceptional properties on EuH_2 (f^7 case) and YbH_2 (f^{14} case). For europium, it seems that the energy required to promote an electron from the Eu^{2+} ion is too prohibitive to permit population of a conduction band, and it behaves as a typical ionic hydride, for example, CaH_2. Although YbH_2 is also predominantly ionic, it does show a small paramagnetism and slight metallic properties.

Actinide hydrides have been less studied than those of lanthanides, but for the most part they resemble the latter in properties. With the exception of uranium, which gives UH_3, actinides all seem to give an MH_2 hydride spontaneously but less exothermically than the lanthanides. Trihydrides seem to be accessible at higher pressure. Because of their reactivity, actinide hydrides have proved useful in the synthesis of actinide compounds.

All hydrides discussed in this section are strong reducing agents, those of lanthanides and actinides being more reactive than those of d-block elements. They react more or less easily with oxygen. Uranium hydride is spontaneously inflammable in oxgyen, but TiH_2 is stable under ambient conditions, presumably protected by an impermeable oxide layer. A number of typical reactions of these hydrides are shown in the following equations:

$$TiH_2 + 3CuCl_2 \xrightarrow{100°} TiCl_4 + 2HCl + 3Cu \tag{9.50}$$

$$2UH_3 + 8HF \xrightarrow{100°} 2UF_4 + 7H_2 \tag{9.51}$$

$$UH_3 + 3HCl \longrightarrow UCl_3 + 3H_2 \tag{9.52}$$

$$2ScH_2 + 5CO_2 \longrightarrow Sc_2O_3 + 2H_2O + 5CO \tag{9.53}$$

Besides the binary hydrides described above, there is a very large number of ternary hydride systems that contain either two transition elements (including lanthanides and actinides) or a transition and a main-group metal. Most of these systems are very complicated, and we do not describe them here. We conclude our discussion of transition hydrides with a mention of the ternary systems Li_4RhH_4, Li_4Rh_5, and M_2RuH_6 (M = Ca, Sr, Eu, and Yb). All these compounds are synthesized by moderate heating of finely divided Rh or Ru with an excess of the binary hydride of the other metal. Although they only exist as solid-state materials, these compounds do not manifest the metallic properties of binary transition hydrides. Neutron diffraction suggests that the rhodium in Li_4RhH_4 has square-planar coordination, and in the M_2RuH_6 species the Ru is six-coordinate. It is most probable that these materials represent a transitional type between the binary metallic transition hydrides and the true ternary coordination complexes, of which only two examples are known at the time of this writing, K_2ReH_9 and Mg_2FeH_6. The nonahydridorhenate(VIII) ion was one of the earliest hydrido-complexes to be synthesized (although its structure was uncertain until the advent of NMR spectroscopy) whereas the hexahydroferrate ion is one of the most recent. The rather astonishing isolation of the latter from a fairly simple reaction raises the expectation that perhaps many other such species remain yet to be discovered. We will return to the discussion of coordination compounds containing hydrogen ligands in Chapter 22.

BIBLIOGRAPHY

Suggested Reading

Transition Metal Hydrides. Advances in Chemistry Series no. 167. Washington, D.C.: American Chemical Society, 1978.

Wiberg, E., and A. Amberger. *Hydrides of the Elements of the Main Groups I–IV.* New York: Elsevier, 1971.

PROBLEMS

9.1 Give three reactions for the industrial production of hydrogen.

9.2 Using ionization energies and electron affinities given in Chapter 2, which of the hydrides of Group IA (1) would be stable in the gas phase?

9.3 Give two pieces of experimental evidence to support the idea that CaH_2 contains the hydride ion.

9.4 Why is the structure of beryllium hydride not known?

9.5 The hydrides of Groups IA and IIA (1 and 2) are strong reducing agents. Why?

9.6 Diborane reacts explosively with oxygen. What is the driving force for this reaction?

9.7 In what important ways does ethane differ from borane?

9.8 Analyze the following reactions in terms of acid–base behavior:
a. $4LiH + AlCl_3 \rightarrow LiAlH_4 + 3LiCl$
b. $3NaBH_4 + 4BF_3 \rightarrow 2B_2H_6 + 3NaBF_4$
c. $4NaH + 4B(OCH_3)_3 \rightarrow NaBH_4 + 3NaB(OCH_3)_4$

9.9 The most widely used laboratory synthesis of covalent hydrides is the reduction of a covalent halide with a complex hydride. Give an example of such a reaction and explain why the reaction works.

9.10 Give two reasons each why the following molecules are more chemically reactive than is C_2H_6. Address your reasons to specific reactions.
a. B_2H_6
b. N_2H_4
c. Si_2H_6

9.11 Why do boron and carbon produce polymeric hydrides while oxygen and nitrogen do not?

9.12 How is hydrazine made industrially?

9.13 Give one way in which N_2H_4 and H_2O_2 resemble each other and one way in which they differ.

9.14 Devise a synthesis for hydrazoic acid and give balanced equations for the reactions involved.

9.15 Give one chemical and one physical reason why TiH_2 is a more practical storage compound for hydrogen than is CaH_2. (Densities of Ti, Ca, TiH_2, and CaH_2 are 4.5, 1.55, 3.9, and 1.7 kg dm^{-3}, respectively.)

9.16 Suggest a laboratory synthesis for high-purity HD.

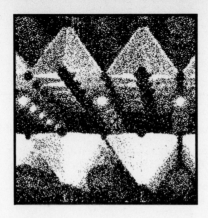

10

Halides and Oxides of the Main Groups

10.1 *Differences Between Hydrides, Oxides, and Halides*

The hydrogen atom is unique in that it possesses no core electrons. One consequence of this is the great difficulty we encounter in trying to describe the "size" of the hydrogen atom or in attributing a particular fraction of electron density to it. We did not elaborate on the theoretical description of the metallic transition hydrides because the issue is surrounded by uncertainty and because more or less equal success has been achieved using models in which the hydrogen is assumed to be H^+ and others in which the hydrogen is assumed to be H^-. The only consensus is that the truth probably lies somewhere between!

This kind of ambiguity is much less evident in the oxides and halides, where the bonding electrons are always to some degree shielded from exposure to a naked nucleus and interelectronic repulsions between electron cores help preserve the identities of the component atoms.

A second important difference between hydrogen and other elements is that bonding a hydrogen to another atom is essentially a one-way interaction between the $1s$ orbital of the hydrogen and an orbital of the heteroatom (for different compounds, however, the displacement of electron density may be in different directions). With the heavier atoms, the possibility exists for the return of electron density through π donation, for example, O^{2-} and F^-. And, for very heavy atoms, the relatively closely spaced, filled and empty orbitals of π symmetry give rise to the possibility of either π donation or π acceptance of electrons.

In terms of MO theory, we can extend our treatment of CH_4 (see Chapter 3) to CF_4 relatively easily if we assume that the $2s$ orbital of each fluorine is unaffected by bonding and that one of the $2p$ orbitals of each fluorine is involved in σ overlaps. The symmetries of the four linear combinations of these p orbitals are the same as the σ orbitals of methane, that is, $a_1 + t_2$. The remaining pairs of $2p$ orbitals on the fluorines can overlap, corresponding to a set of eight linear combinations of symmetries: $e + t_1 + t_2$. We will say no more of the $e + t_1$ combinations because they are unaffected by bonding. Figure 10.1 shows a combination of the t_2 type. Because this set of three orbitals has the same symmetry as the t_2 (σ) set, they can mix and cause a lowering of the energy of the system at the expense of making the antibonding t_2 orbitals higher in energy. Figure 10.2a summarizes these effects.

Figure 10.1
Linear combination of the t_2 (π) type.

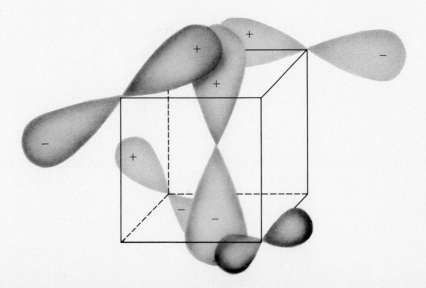

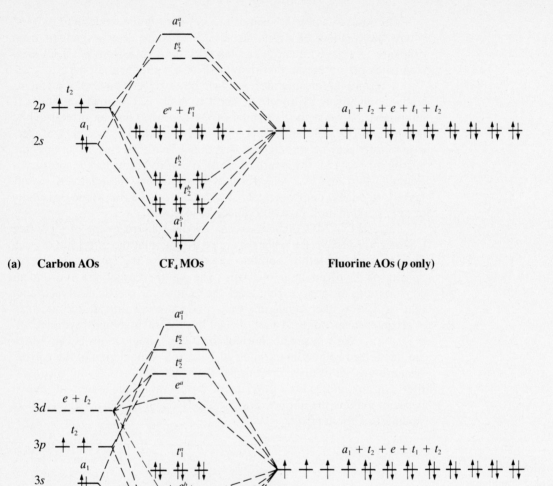

Figure 10.2   MO correlation diagrams for (**a**) CF_4 and (**b**) SiF_4.

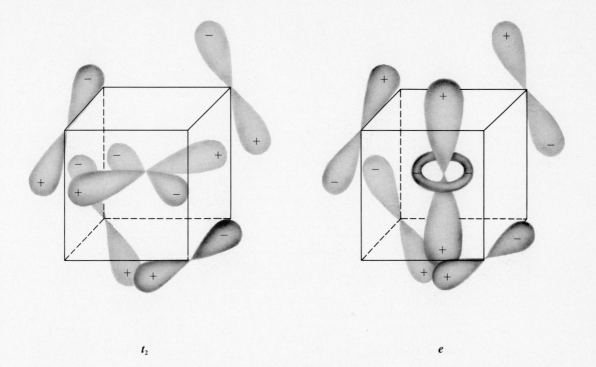

t_2 e

Figure 10.3
Overlaps of the *e*
and t_2 kinds with *d*
orbitals.

In physical terms, additional overlaps available to ligands with *p* orbitals may allow the spreading of charge in such a way as to reduce interelectronic repulsions or charge polarization. In the case of the CF_4 molecule, we can imagine the fluorine picking up charge from the less electronegative carbon through the simple σ-bonding framework while relieving interelectronic repulsion by spreading or expanding the other *p* orbitals into the interbond region.

These effects, which are rather limited in the case of CF_4 because of the competition of all fluorine *p* orbitals for the same set of orbitals on the carbon, become more important when there are empty, low-lying orbitals of appropriate symmetry on the central atom. This becomes the case for the heavier *p*-block elements, which possess a relatively low-lying set of empty *d* orbitals. In tetrahedral molecules, the five *d* orbitals have $t_2 + e$ symmetry, and so all five can participate in accepting electron density from fluorine *p* orbitals. Figure 10.3 shows an example of one each of the t_2 and *e* overlaps. These overlaps can only make a significant contribution to bonding if the energy separation between the fluorine *p* orbitals and the *d* orbitals is not too great. This situation is most likely to exist in cases where the central atom is in a high effective oxidation state because excess positive charge on the central atom contracts the orbitals and lowers their energies. Figure 10.2b shows the MO energy-level diagram for SiF_4.

Figure 10.4
MO energy-level diagram and orbital overlaps in an octahedral molecule. (The exact ordering of energy levels depends on the particular system.)

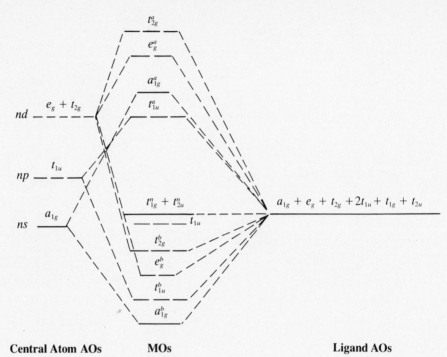

Central Atom AOs   MOs   Ligand AOs

Because F^- and O^{2-} are isoelectronic, there is a family of molecules with more or less the same MO diagram and that show the unusual stability common to molecules in which the maximum possible number of orbitals is used in bonding. Among these are XeO_4, $[ClO_4]^-$, $[SO_4]^{2-}$, and $[PO_4]^{3-}$. The increasing negative charge tends to diminish the stability of the simple ionic species with respect to polycondensation, but even in polyphosphate and polysilicate species, the tetrahedral coordination to oxygen is preserved, as is the potential for π bonding.

A similar state of affairs exists in molecules of other symmetries, but we will only consider octahedral compounds where only three of the d orbitals (the t_{2g} set) can participate in π bonding. Figure 10.4 shows an MO energy-level diagram and overlaps in an octahedral molecule. Such a diagram applies to SF_6 (an unusually stable molecule), $Te(OH)_6$, $[PF_6]^-$, and so on.

A third important difference between hydrogen and oxygen and the halogens is that the latter, because of their higher electronegativity, are much more effective at increasing the positive charge on the central metal and contracting the valence orbitals. This reduces but does not eliminate the tendency to form metallic bonds, increases the tendency to form ionic bonds, and gives rise to a wider range of oxidation states. Thus, relative to the hydrides, the chemistry of the transition metal oxides and halides is very much richer. To some extent, this is also true of the main-group elements, but the difference is not so striking.

10.2   Groups IA and IIA (1 and 2)

Oxides

Like most compounds of the alkali and alkaline earth elements, halides and oxides are predominantly ionic. The oxides of Group IA (1) can be made by direct reaction of the metals with oxygen. These reactions are complicated because oxygen can exist as a stable species in three anionic oxidation states, namely, O_2^-, O_2^{2-} and O^{2-}. (Although O^- can also exist in principle, it is of no chemical significance. Its protonated form, the hydroxyl radical $OH^{\bullet}$, does participate as a highly reactive species in many radical oxidation reactions.)

In the case of lithium, the main product is Li_2O when an abundant oxygen supply is available. Under the same conditions, sodium gives mainly peroxide, whereas the remaining elements of the group give superoxides. At high temperatures, both O_2^{2-} and O_2^- disporportionate to O^{2-} and O_2. Oxides of all Group IA (1) elements can therefore be produced by pyrolysis of the mixture produced by oxidation under milder conditions. The lattice energies for the three kinds of oxide increase in the order $MO_2 < M_2O_2 < MO$, the oxide being always the most stable. However, as cation size increases, the differences in lattice energies between the three types diminish, thus explaining the *relative* increase in stability of the intermediate oxidation states for the larger cations. Given the much greater lattice energies and smaller cation sizes, it is not surprising that peroxides and superoxides of lighter Group IIA (2) metals are not easily formed by direct combustion of the metal in oxygen. Because of the rather violent nature of the reaction, direct combustion is not usually used for these metals, except for beryllium. The decompositions of various oxyacid salts, particularly the carbonates, are preferred for other members of the group.

The difference in energy between the oxide and peroxide of a Group IIA (2) metal decreases with increasing temperature and increasing oxygen pressure. Thus, at high enough temperatures and oxygen pressures, the stability order may be reversed. Even at 1 atm of O_2, BaO is converted almost completely to BaO_2 at about 600 °C. The peroxide of strontium may be obtained similarly, but the conditions to achieve reaction with calcium are prohibitive. (Besides thermodynamic problems, the rates of reaction, which require diffusion in the solid state, decline with increasing lattice energy.)

The superoxide ion is a very powerful oxidizing agent, and dissolution of alkali superoxides in water or other protonic solvents leads to very rapid (if not explosive) decomposition. In water, the initial step is probably protonation to produce the unstable hydroperoxyl radical $HOO^{\cdot}$. This radical undergoes rapid disproportionation to O_2 and H_2O. Rather surprisingly, solutions of potassium superoxide complexed with multidentate, cyclic polyethers (known as crown ethers; see below for further discussion of alkali metal complexes) are relatively stable in certain organic solvents; the systematic chemistry of the superoxide ion with organic compounds has recently become accessible to study.

Because peroxide and hydroperoxide anions are relatively strong bases, both are protonated when the metal salts are dissolved in water, the former essentially completely. Hydrogen peroxide is slightly more acidic than is water, so solutions of alkali peroxides contain considerable amounts of HOO^-. Alkali peroxides are powerful oxidizing agents, and sodium peroxide, which is cheap and readily available, is widely used in both inorganic and organic chemistry as an oxidant or as a source of peroxo-compounds of other elements.

The free oxide ion is extremely basic; thus, the highly ionic oxides (basic oxides) tend to react readily with proton sources to give the less basic OH^- compounds or water plus the salt of the metal and the anion of the proton source. All Group IA (1) oxides dissolve in water to produce the corresponding hydroxides. Of Group IIA (2) oxides, beryllium oxide is completely insoluble in water, and finely divided MgO may be slowly converted to the hydroxide, but not coarsely crystalline material. The remaining Group IIA (2) oxides undergo easy protonation to form hydroxides. This difference in behavior is partly due to the large lattice energies of the oxides of beryllium and magnesium slowing the various surface-transport processes required for dissolution and partly due to the reduced basicity of the oxide under the influence of highly polarizing cations.

With beryllium oxide, we encounter the first example of amphoteric behavior—that is, BeO is soluble in both aqueous alkali and aqueous acid, even though it is not soluble in neutral water. In acid medium, the lattice is broken down by protonation of the oxide ions (remembering that in BeO, we are using the word *ion* in a loose sense). Once the oxide bridge has been doubly protonated, it loses its bridging properties and is easily replaced at one of the Be ions by a water molecule:

$$—Be—O—Be— + 2H^+ + H_2O \rightarrow —Be(H_2O)^+ + {}^+(H_2O)Be—$$

$$\textbf{(10.1)}$$

Repetition of this process eventually leads to the detachment of a soluble species in which the Be^{2+} ion is coordinated to four oxygen atoms, contained in either OH^- or H_2O ligands.

In strongly basic media, the surface Be^{2+} ions are hydroxylated by attack of OH^- ions. The species that detach from the solid consist of a variety of polymeric hydroxo-bridged, aquated cations, but in strong alkali the ion $[Be(OH)_4]^{2-}$ is the main product. The strong polarizing power of Be^{2+}, which reduces the basicity of O^{2-}, also enhances the acidity of coordinated water. Thus, solutions of the beryllium salts of strong acids in neutral water are quite acidic due to the process

$$[Be(H_2O)_4]^{2+} \rightarrow [Be(H_2O)_3(OH)]^+ + H^+ \qquad (10.2)$$

Although magnesium salts do undergo slight hydrolysis, the oxide is not sufficiently amphoteric to dissolve in aqueous base. Aqueous cations of other Group IIA (2) elements are not significantly hydrolyzed.

The very high melting points, resistance to water, lack of color, and high transmittancy of BeO and MgO give them highly desirable refractory properties (resistance to heat and corrosion). The former is not much used because of its toxicity, but magnesium oxide is widely used in a variety of high-temperature applications. Calcium oxide, which is cheaply available from roasting limestone, is used extensively in metallurgy as a slagging agent. The basic oxide combines with the more acidic oxides of such elements as silicon and boron and extracts them into a layer that is immiscible with and lighter than the molten metal, allowing easy removal. A great deal of calcium oxide is also used in the manufacture of sodium carbonate by the Solvay process. The cycle of reactions used is the following:

$$CaCO_3 \xrightarrow{\text{Heat}} CaO + CO_2 \qquad (10.3)$$

$$CaO + H_2O \longrightarrow Ca(OH)_2 \qquad (10.4)$$

$$Ca(OH)_2 + 2NH_4Cl \longrightarrow CaCl_2 + 2NH_3 + 2H_2O \qquad (10.5)$$

$$2NaCl + 2CO_2 + 2NH_3 + 2H_2O \longrightarrow 2NaHCO_3 + 2NH_4Cl \qquad (10.6)$$

$$2NaHCO_3 \xrightarrow{\text{Heat}} Na_2CO_3 + H_2O + CO_2 \qquad (10.7)$$

The overall reaction is

$$2NaCl + CaCO_3 \rightarrow Na_2CO_3 + CaCl_2 \qquad (10.8)$$

Because there is virtually no use for the vast amount of $CaCl_2$ produced in this process and because environmental laws are becoming very strict about industrial salt-waste disposal, there is a growing need for imaginative new developments in this old chemistry.

Halides

With the exception of beryllium and to a lesser extent lithium, halides of Groups IA and IIA (1 and 2) constitute the archetypal family of ionic compounds. All these compounds can be made by direct combination of the elements or by reaction of the metal with anhydrous hydrogen halide. In many cases, the violence of such reactions render them of questionable utility. Group IA (1) oxides dissolve in aqueous hydrohalic acids to give solutions from which (except for lithium) the anhydrous salts can be crystallized. Group IIA (2) oxides also dissolve in the appropriate acid, but crystallization gives hydrated salts. The hydrated salts of lithium, beryllium, and magnesium can be dehydrated by heating in a stream of dry hydrogen halide to suppress the hydrolysis that would otherwise occur. The remaining hydrated halides can be dehydrated by simple heating.

Most of Group IA (1) halides have the NaCl structure; a few have the CsCl structure under ambient conditions. Beryllium fluoride, which is produced as a glass by thermal decomposition of $[NH_4]_2[BeF_4]$, can be crystallized by slow cooling of the melt (melting point 803 °C). The resulting crystalline form has the crystobalite structure (SiO_2, see Section 10.3), which could indicate a high degree of covalency. On the other hand, a very strong tendency of the small Be^{2+} ion to be tetrahedrally coordinated could force such a structure, even in the absence of significant covalency. Beryllium chloride has an infinite doubly bridged chain structure in the solid state:

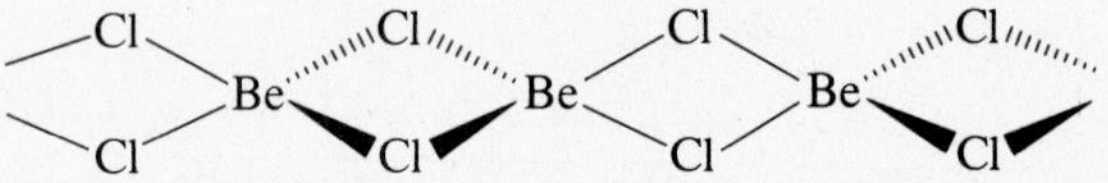

This structure is more favorable for the heavier halides because the large-size halide ion allows a reasonable separation of adjacent beryllium ions. The close approaches of ions of like charge that are more important for F^- bridges can be reduced by opening up the Be—F—Be bond angle, that is, going to a three-dimensional, singly bridged structure. We will see such effects again in the structures of transition metal halides. Magnesium fluoride prefers a structure in which the Mg^{2+} ion is six-coordinate and therefore adopts the rutile structure. The remaining fluorides adopt the fluorite structure with eight-coordinate cations. The heavier halides of Mg, Ca, Sr, and Ba adopt a variety of distorted layer structures, related to the $CdCl_2$ and CdI_2 structures.

All Group IA and IIA (1 and 2) halides are soluble in water, with solubilities ranging from moderate to extremely high. The energetics of dissolution are complicated, but clearly the criterion that the solvation energies of the ions must compensate for lost lattice energy must be an overwhelming factor. In the case of Li^+, Be^{2+}, and Mg^{2+}, the existence of a relatively long-lived coordination sphere of H_2O in aqueous solution has been demonstrated by ^{1}H-NMR techniques. In the case of the other ions, it is assumed that an inner-coordination sphere exists, but the exchange with free water is so fast that the distinction between coordinated and uncoordinated water is difficult to demonstrate. The strong binding of a sphere of hydration to Li^+ and Be^{2+} makes these ions behave in aqueous solution as though they were considerably larger than their crystallographic radii would suggest. This apparent increase in radius allows these ions to mimic their congeners Na^+ and Mg^{2+} to a degree that they can gravely disrupt the biochemistry of the latter. This leads to the unusual neuroactivity of Li^+ salts and to the extreme toxicity of Be^{2+}, a fact that might seem peculiar in such chemically different species.

It is worth noting the strong affinity of Be^{2+} for the fluoride ion, a property that is typical of small, highly charged cations. This feature is repeatedly encountered as we progress through the periodic table, but in Groups IA and IIA (1 and 2) only in the case of Be^{2+} can any other species compete successfully with H_2O for the coordination sphere of the ion in aqueous solution. In solutions fairly concentrated in fluoride, the species $[BeF_4]^{2-}$ is fully formed. It can be precipitated from water in the form of insoluble salts such as $[NH_4]_2[BeF_4]$.

10.3 Groups IIIA and IVA (13 and 14)

Oxides

The very small radius and high charge of B^{3+} results in very covalent compounds, even with the most electronegative elements. In its oxygen and halogen compounds, boron exhibits both three and four coordination, sometimes in isomers of the same compound. For example, the thermodynamically most stable structure of B_2O_3 consists of a three-dimensional array of interconnected planar BO_3 units, each oxygen being coordinated to two borons (Figure 10.5). However, at high temperature and pressure a

Figure 10.5
Structure of B_2O_3.

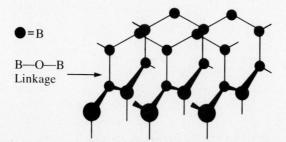

● = B

B—O—B
Linkage →

denser, metastable form of B_2O_3 can be produced in which the borons are tetrahedrally coordinated to oxygens. To satisfy the constraints of stoichiometry in this structure, there are two different kinds of oxygen: two-thirds being three-coordinate and one-third being two-coordinate.

The tendency to give polyoxyanions (or their conjugate acids), which has already been noted for beryllium, is very important for boron and for most members of Groups IVA and VA (14 and 15). Boric oxide is weakly acidic and exhibits no basic properties. It reacts sluggishly with aqueous bases, and the borate salts are best prepared by fusion of B_2O_3 with basic oxides. The relatively common mineral source of boron, borax, contains the anion $[B_4O_5(OH)_4]^{2-}$. Acidification of hot solutions of borax results in crystallization of orthoboric acid $B(OH)_3$ on cooling. The latter, which is planar and forms a two-dimensional sheet structure through hydrogen bonding in the solid, can be dehydrated by heating to give the planar-hexagonal metaboric acid $B_3O_3(OH)_3$. Figure 10.6 shows the structures of these species. Particular attention is drawn to the fact that $[B_4O_5(OH)_4]^{2-}$ contains both trigonally and tetrahedrally coordinated boron.

As might be expected from the geometric constraints of packing, the B—O bond length is shorter in planar-trigonal coordination (135 pm) than in tetrahedral coordination (150 pm). This difference in bond length is often attributed to the presence of π bonding in planar-trigonal B—X bonds and its absence in tetrahedral B—X bonds, but the presence of such π bonding is difficult to demonstrate.

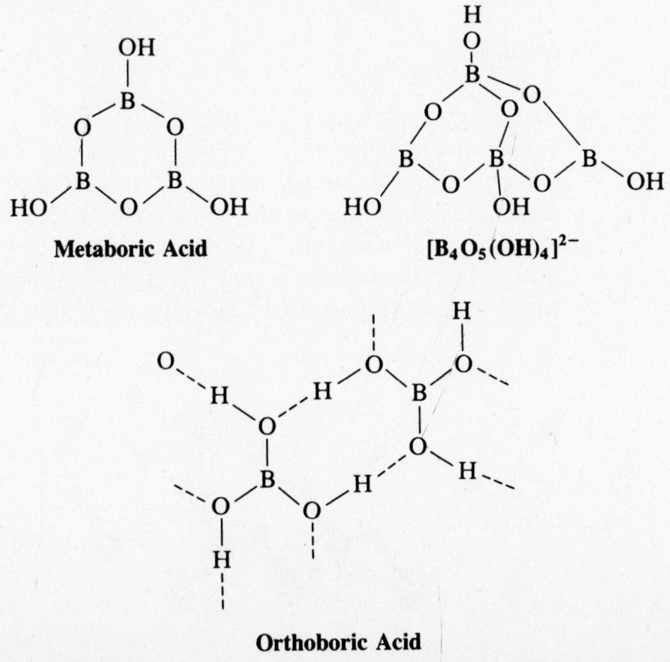

Figure 10.6
Structures of some oxoboron species.

Metaboric Acid

$[B_4O_5(OH)_4]^{2-}$

Orthoboric Acid

One might assume, if there were substantial backdonation from the OH group in $B(OH)_3$, that the acidity of the OH groups might be quite high, but the feeble acidity of orthoboric acid is due to coordination of water and not to ionization of B—OH:

$$B(OH)_3 + H_2O \rightarrow [B(OH)_4]^- + H^+ \tag{10.9}$$

Although B_2O_3 can be obtained with difficulty by oxidizing elemental boron or with great danger from oxidation of boranes, it is far more easily obtained by roasting $B(OH)_3$ at around 700–800 °C.

A suboxide of approximate formula BO can be obtained by heating B_2O_3 with B at 1000 °C (compare with SiO below). Although the structure of this material is unknown, its reaction with water yields the compound $B_2(OH)_4$, which contains a B—B bond (compare with B_2Cl_4 below).

Whereas B(III) shows a mixed preference for three and four coordination, Al(III) shows a mixed preference for four and six coordination. The most stable oxide, α-Al_2O_3 (corundum), has aluminum ions occupying two-thirds of the octahedral holes in a distorted HCP lattice of oxide ions. In coarse crystalline form, this oxide is very hard (it is used as an abrasive in sandpaper and grinding wheels) and chemically inert. Another form of Al_2O_3, the γ form, has the aluminum ions in both four and six coordinate sites. The γ-Al_2O_3 structure is much more open than is the α-Al_2O_3 structure and takes up substantial amounts of foreign ions or molecules. For this reason, it is kinetically much more reactive than α-alumina and is useful in chromatography and catalyst production. We will return briefly to this subject in the discussion of solid electrolytes in Chapter 19.

Aluminum oxide is amphoteric. In noncomplexing acids, it dissolves to give the $[Al(H_2O)_6]^{3+}$ ion at low pH. In basic solution, it dissolves to give "aluminates," about whose structures relatively little is known. Compounds formulated as $Al(OH)_3$ and $AlO(OH)$ can be precipitated from aqueous aluminum salt solutions by appropriate adjustment of pH. These species are very fragile with respect to condensation and loss of H_2O, and their pyrolysis provides the simplest syntheses of α and γ aluminas. The remaining Group IIIA (13) elements behave similarly to aluminum (particularly gallium), but the oxides become more basic on descending the group. In addition to Tl_2O_3, thallium also forms an oxide of unknown structure, Tl_2O.

The trend toward decreasing coordination number (C.N.) reaches its ultimate limit with carbon (excluding the rather uninteresting case of C.N. = 0). In the compounds of carbon, covalency effects are overwhelming, and the unusual stabilities of CO and CO_2 are entirely related to covalency and have little to do with size and polarity effects. Among oxides, CO and NO are a unique pair, with little relation to any other oxides. We have already discussed their unusual electronic structure (Chapter 3), and any further discussion will be confined to their chemistry.

Both CO and CO_2 can be made by reaction of carbon or organic compounds with oxygen. Oxygen deficiency favors CO, oxygen excess, CO_2. Carbon monoxide shows neither acid nor base properties in the Brønsted sense. Its feeble Lewis acidity is manifest through formation of H_3BCO with borane. Discussion of the remarkable affinity of CO for transition metals is deferred until Chapter 22. Although there are many reactions of CO, particularly those involving its oxidation, which have a favorable free energy, they usually proceed very slowly or not at all in the absence of catalysts. Reactions with halogens to produce COX_2 proceed fairly readily as do those with sulfur and selenium, but the exothermic reaction with O_2 at room temperature does not proceed spontaneously. Ignition initiates combustion, explosively if the gas composition is appropriate. The reaction of CO with water, the "water–gas shift reaction" (Equation 9.7), does not proceed at an appreciable rate in the absence of catalysts. Likewise, reduction of CO by hydrogen (which only becomes favored thermodynamically at moderate pressure) requires catalysis. The economic and strategic importance of these reactions has stimulated an enormous amount of research in the past century.

With carbon dioxide, we return to the mainstream of periodic properties in that we can make the connection to $B(OH)_3$ by considering the reaction product of the mildly acidic CO_2 with water (Equation 10.10):

$$CO_2 + H_2O \rightarrow [HCO_3]^- + H^+ \qquad \textbf{(10.10)}$$

In neutral water, the equilibrium hardly occurs at all, but aqueous bases pull the equilibrium fully to the right and farther to produce $[CO_3]^{2-}$. The relationship between these planar-trigonal anions and the planar oxo-boron compounds is obvious. In the case of carbon, the effects of p_π—p_π bonding are very striking in terms of the much greater acidity of C—OH relative to B—OH and the greatly reduced tendency of carbon to expand its coordination sphere to four oxygens. Although orthocarbonate esters are well known in organic chemistry, there are no known ionic salts of the hypothetical orthocarbonic acid $C(OH)_4$. The ability of carbon to saturate the oxygens bonded to it through π bonds also completely inhibits the formation of polyoxyacids and anions. The $[CO_3]^{2-}$ and $[HCO_3]^-$ ions are the only known stable anions in which carbon is attached only to oxygen.

Whereas the unreactivity of CO is largely a problem of kinetics, the low reactivity of CO_2 is a problem of thermodynamics. There are very few reactions of CO_2, other than its acid–base reactions (which move us even farther into the free-energy abyss), that proceed without great expense of energy. It can be reduced relatively easily by strongly electropositive metals, which themselves require much energy to produce, or by highly ionic organometallic reagents, but none of these reactions are interesting in themselves for the large-scale fixation of CO_2. Despite the difficulties involved, the reduction of CO_2 is the route by which carbon is introduced into the living world. Evolution was obliged to follow this route because almost all of the earth's available carbon was in the form of carbon dioxide or mineral carbonates. The same is true today, only a minor fraction of our carbon resources being in the form of fossil fuel. The carbon potential (not the energy potential) of our fossil fuels would be greatly increased if we were able to take advantage of abundant cheap energy to effect the reaction:

$$C + CO_2 \rightleftharpoons 2CO \qquad \begin{aligned} \Delta G^0_{273} &= 120 \text{ kJ mol}^{-1} \\ \Delta G^0_{700} &= 48 \text{ kJ mol}^{-1} \end{aligned} \qquad \textbf{(10.11)}$$

Carbon and silicon in their oxygen compounds (as in many other respects) bracket the behavior of boron. Although carbon prefers three coordination to oxygen and even goes to C.N. = 2 in CO_2, silicon is almost invariably four-coordinate to oxygen. The Si—O single bond is one of the stronger bonds encountered (466 kJ mol^{-1} compared to 359 kJ mol^{-1} for C—O, whereas the total bond energies for C=O and Si=O in gaseous MO_2 are 805 and 640 kJ mol^{-1}, respectively). It is this factor, combined with the relatively weak p_π—p_π bonds that silicon might form to the first members of the p-block groups, that is important in preventing silicon from forming stable compounds with p_π—p_π bonds.[1]

There is also an important kinetic factor involved. Many of the carbon compounds that contain p_π—p_π bonds are unstable relative to polymers that can be produced by opening up a double (σ, π) bond to generate two single (σ) bonds. However, the kinetic barriers to these processes are usually quite high. This does not apply to CO_2, the doubly bonded triatomic molecule being the most stable form. We are thus left with the extraordinary contrast between CO_2, a gas subliming at $-78\,°C$ at ambient pressure, and SiO_2, a solid melting at 1710 °C!

Silicon dioxide, or silica, may be produced by oxidation of silicon in oxygen at high temperature but is more appropriately and cheaply produced by thermal dehydration of hydrated silica (or silica gel), which can be precipitated from aqueous solutions of alkaline silicates by appropriate reduction of pH. Amorphous silicas of low-degree hydration can also be produced by combustion of organosilicates in oxygen or by combustion of $SiCl_4$ in a mixture of hydrogen and oxygen (Equation 10.12):

$$SiCl_4 + 2H_2 + O_2 \rightarrow SiO_2 + 4HCl \qquad \textbf{(10.12)}$$

Silicas produced in this way can be extremely finely divided and of exceptionally high-surface area. They are extensively used as fillers for silicone

---

[1] A few cases of Si=C and Si=Si have been reported and others will undoubtedly be discovered, but this does not seriously alter the essential validity of the above assertions.

rubbers. Silicas find a wide variety of industrial uses including catalyst supports, absorbents, and drying agents.

Amorphous silica melts to a highly viscous liquid when heated above 1500 °C. Because of the high viscosity of the melt, crystallization rate is extremely slow, and crystallization from the melt is very difficult. On the other hand, cooling the melt gives very easily a homogeneous glass, which, because of its low coefficient of thermal expansion and transparency to UV radiation, is useful in applications requiring stability to thermal shock and optical transparency.

Access to the crystalline forms of SiO_2 is most easily achieved by the hydrothermal conversion of amorphous SiO_2 to quartz. Under supercritical conditions, the solubility of amorphous SiO_2 in water is slighty greater than that of quartz; thus, crystal growth of a quartz seed occurs in the presence of a solution saturated over amorphous SiO_2. By applying heat and pressure to quartz, many other crystalline modifications of SiO_2 can be obtained. The two most important are tridymite and cristobalite. The transformations of quartz to tridymite ($\sim$900 °C) and of tridymite to cristobalite ($\sim$1500 °C) involve extensive breaking and remaking of Si—O bonds and only occur very slowly in the absence of catalysts. There are many other phase transitions, which are associated with conformational reorientations without bond cleavage, and these phase changes tend to occur rapidly.

All crystalline forms of SiO_2 contain silicon bonded tetrahedrally to oxygen. They differ in the way in which the apically interconnected tetrahedra are oriented with respect to each other. Quartz has a rather complicated structure that is difficult to visualize without the help of three-dimensional models. Figure 10.7 shows the structure. A special feature of the quartz structure is the helical interconnection of tetrahedra along one axis of the crystal.

Because the helix is chiral (a left-handed coil cannot be superimposed on a right-handed coil), quartz crystals are one of the rare examples of a macroscopic resolution of inorganic chirality in nature. Geothermal growth can lead to the formation of large optically active crystals. Although the total number of each enantiomeric form produced is the same, the chance of accidental resolution (that is, separation of an enantiomeric excess into two different physical locations) becomes greater with the physical size of the crystal. For this reason, there has been considerable speculation concerning the possible role of quartz in biogenesis and in the bioselection of one chirality over another in biomolecules. Imagine a rockfall or a volcanic expulsion in which there are two large quartz crystals of opposite chirality; one ends up in the shade, the other exposed directly to the sun. Photochemical synthesis occurring, perhaps catalytically, on the surface of the exposed quartz is subject to a chiral influence and produces small chiral organic molecules in which there is an enrichment of one enantiomer over

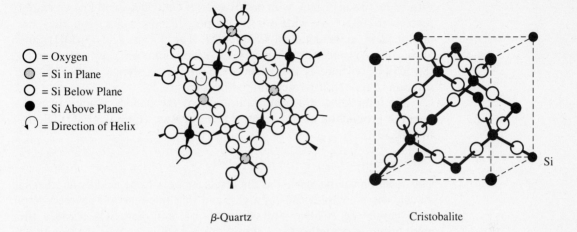

○ = Oxygen
◉ = Si in Plane
○ = Si Below Plane
● = Si Above Plane
↻ = Direction of Helix

β-Quartz   Cristobalite

Si

Figure 10.7
Structures of quartz
and cristobalite.

the other. Although highly speculative, this hypothesis provides a comfortingly simple explanation for monochiral genesis. It is important to recognize that chiral genesis like other forms of biogenesis does not need to be 100% selective. The process of natural selection operates on small biases and amplifies them.

The cristobalite structure is much simpler and more symmetrical than that of quartz. It is closely related to the zinc blende and the cubic-ice structures and to the diamond structure (see Chapter 19). Thus, silicon atoms occupy the lattice sites and one-half of the tetrahedral holes of a CCP lattice (that is, both the Zn and S sites of zinc blende), and each silicon has a tetrahedron of nearest neighbor silicon atoms. Oxygen atoms in cristobalite are situated half-way along and slightly displaced from the lines joining adjacent silicons. The interpolation of the oxygen atoms make cristobalite an open structure, relative to zinc blende or diamond, and in this respect it resembles cubic ice.

We will not discuss the structure of tridymite here but conclude with mention of a remarkable modification that is produced under conditions of high pressure and temperature and that has the rutile structure. Besides being synthesized in the laboratory, this form is also found occasionally in nature at sites of meteor impact. This forcing of silicon into an octahedral environment by high pressure is reminiscent of the high-pressure form of B_2O_3, where the boron is tetrahedrally coordinated as opposed to trigonally coordinated in the low-pressure forms.

The high surface–area forms of amorphous silica or freshly prepared silica gels readily manifest the weak acidity of silica by redissolving to form silicates. Crystalline and glassy silicas are much more resistant to chemical attack by aqueous bases, although they are slowly etched by very concentrated alkali. Silicas resist dissolution by aqueous acids, with the exception of HF, which forms the strong acid H_2SiF_6. Silicates are easily formed by fusion of SiO_2 with basic oxides (about 90% of all known oxides react with SiO_2). A wide variety of polysilicates exists, and they are described in more detail in Chapter 17. The parent acids $Si(OH)_4$ and $H_2Si(OH)_6$ (isoelectronic with H_2SiF_6) are unknown, as is the anion $[Si(OH)_6]^{2-}$. There is some doubt about the existence of $[SiO_4]^{4-}$ in solution, although it is encountered in some solids.

An interesting practical application of the general acidity of SiO_2 is in the production of sulfuric acid from gypsum ($CaSO_4 \cdot H_2O$) (Equation 10.13):

$$CaSO_4 + SiO_2 \rightarrow CaSiO_3 + SO_3 \qquad \textbf{(10.13)}$$

This reaction can be viewed as the displacement of one acid by another in an acid–base complex. SO_3 is a gas, and that this reaction proceeds says little about the relative acid strengths of SiO_2 and SO_3 because the equilibrium is drawn to the right by the departure of SO_3. At one time, when the world suffered sulfur shortages, this process looked promising as a replacement for the use of elemental sulfur for sulfuric acid production. However, the present oversupply of sulfur from the purification of natural gas and the recycling of effluent SO_2 has removed any advantage that the gypsum process may have offered. On the contrary, there is more calcium sulfate being produced by the scrubbing of SO_2–containing gas streams over CaO, to reduce SO_2 pollution.

Silica is generally resistant to reaction with strong oxidants, except F_2, with which it reacts to form SiF_4. Strongly reducing metals reduce SiO_2 to metal silicides, and impure silicon results from reduction of SiO_2 with carbon at very high temperature.

Germanium dioxide differs in subtle ways from silica. It exists in the quartz structure, but the rutile GeO_2 structure is much easier to obtain than is the SiO_2 analog. Although the only aqueous acid that SiO_2 reacts with is HF to give $[SiF_6]^{2-}$, GeO_2 also reacts with concentrated hydrochloric acid to give $[GeCl_6]^{2-}$, demonstrating a decline from the very strong preference for oxygen binding typical of silicon. These trends continue with SnO_2, existing preferentially in the rutile structure. Hydrous SnO_2 is quite basic and dissolves easily in acids to give complexes such as $[SnCl_6]^{2-}$. Concentrated alkalis also dissolve hydrous SnO_2 to give solutions containing $[Sn(OH)_6]^{2-}$ anions.

The increased stability of the lower oxidation state on passing down a group (previously alluded to in our discussion of the hydrides of the *p* block), becomes particularly evident in Group IVA (14). In the chemistry of stable compounds of carbon and silicon, oxidation state II is of little importance. The so-called silicon monoxide, produced by reaction of SiH_4 and O_2 in a microwave discharge and of some interest to the electronic industry as a masking material, probably contains Si—Si bonds (compare with BO, mentioned above). A monoxide of germanium, relatively well-defined but of unknown structure, is obtained by dehydration of a yellow precipitate that results from treatment of aqueous $GeCl_2$ with ammonia. Both SnO and PbO are well-defined compounds of known structure. Oxidation of tin in oxygen produces SnO_2, but oxidation of lead gives PbO. Production of PbO_2 requires strong chemical oxidants or electrochemical oxidation at high potential. The Pb(IV) "ion" is much too powerful an oxidant to coexist with many anions; therefore, attempts to dissolve PbO_2 in acids often lead to oxidation of the acid. As with most strongly oxidizing cations, fluoro-complexes and complexes of ligands binding through oxygen are most likely to survive. Thus, like SnO_2, PbO_2 also dissolves in concentrated alkali to give $[Pb(OH)_6]^{2-}$ species. All three monoxides are amphoteric.

A third oxide of lead Pb_3O_4 (known as red lead) may be obtained by roasting PbO in air or oxygen. This oxide has a complicated structure, but it is clear that there are two distinct types of lead; the formula is written $(Pb_2^{II}Pb^{IV})O_4$. The presence of two oxidation states of lead is manifest chemically by dissolution in glacial acetic acid. Both Pb(II) and Pb(IV) acetates are produced because the acetate ligand is highly resistant to oxidation.

At one time, oxides of lead were widely used as paint pigments, but concern about their toxicity has led to their replacement by nontoxic alternatives.

Halides

Among the oxides of Groups IIIA and IVA (13 and 14), only carbon can polarize oxygen to the point that it assumes a coordination number of one, thus producing a discrete molecular species. The heavier halogen atoms are more easily polarized, and many of the binary halides of the Group IIIA and IVA (13 and 14) elements are simple molecular species.

The boron halides are colorless, monomeric, trigonal-planar, molecular compounds. The trifluoride is a colorless gas (boiling point -99 °C), whereas BCl_3 and BBr_3 are liquids (boiling points 12 °C and 90 °C) and BI_3 is a solid (melting point 43 °C). It is commonly assumed that there is substantial p_π—p_π bonding in BF_3 resulting from a four-center overlap of p orbitals on all four atoms, as illustrated in Figure 10.8.

Boron trifluoride is best prepared by reaction of HF with a boron(III) compound. The most convenient boron compounds are the borates such as borax or boron oxide. HF is usually generated *in situ* by action of concentrated sulfuric acid on an ionic fluoride. The high volatility of BF_3 facilitates its removal from the reaction zone and avoids backreactions.

Figure 10.8
π-bonding orbitals
in BF_3.

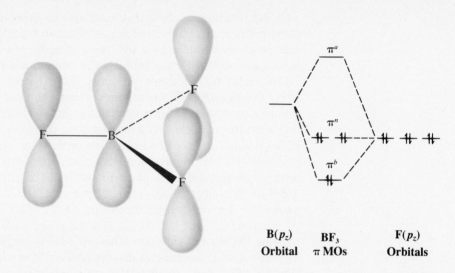

$B(p_z)$	BF_3	$F(p_z)$
Orbital	**π MOs**	**Orbitals**

The other trihalides can be synthesized by reaction of BF_3 [or $B(OCH_3)_3$] with the appropriate aluminum halide. In each case, the BX_3 product is the most volatile component in the system, and it is easily removed by distillation.

The boron halides are typical and strong Lewis acids, capable of forming complexes with most molecules that contain nonbonding electrons. Except for BF_3, they undergo solvolysis by most compounds containing acid hydrogen; all, again except for BF_3, undergo violent hydrolysis on contact with water. We see here again the extraordinary strength of a light element–fluorine bond when there is no lone-pair repulsion and the possibility for π bonding exists. With one equivalent of water, BF_3 forms the complex $F_3B \cdot OH_2$. This very strong Brønsted acid easily acquires a second water molecule to give $[H_3O]^+ [F_3B \cdot OH]^-$. In the presence of excess water, partial hydrolysis leads to the formation of $[BF_4]^-$ (Equation 10.14):

$$4BF_3 + 6H_2O \rightarrow 3[H_3O]^+ + 3[BF_4]^- + B(OH)_3 \qquad (10.14)$$

Molecular boron(III) compounds are isoelectronic with the corresponding carbonium ions, $[R_3C]^+$. When they combine with Lewis bases, they become isoelectronic and isostructural with the tetravalent carbon analogues. Thus, $[BF_4]^-$ is analogous to CF_4, whereas $[F_3BOH]^-$ is isostructural with the highly unstable CF_3OH. The latter loses HF easily, with formation of CO. The inability of B(III) to form B=O stabilizes the boron analogue, as does the extra negative charge.

The sequence of Lewis acid strength for the boron halides—BF_3 is less than BCl_3 and is less than BBr_3, which is opposite to what one would expect on the basis of induction effects—is usually attributed to the stabilization of the trigonal-planar structure in BF_3 by π bonding and a diminution in the importance of π bonding for the heavier halogens. It must be remembered however that trigonal-planar coordination allows closer approach between ligands and the central atom than does tetrahedral coordination. This can have a large effect on σ bonding and π bonding, and the relative sacrifice of σ-bond energy on going from a planar to a tetrahedral arrangement is probably greatest for BF_3.

Unlike $[BF_4]^-$ the other $[BX_4]^-$ species (X = Cl, Br, and I), although they can be made in nonaqueous media, react very rapidly with water. The driving force for these reactions is the formation of the much stronger B—O bond; the rate is high because the excess negative charge greatly facilitates the substitution mechanism, relative to the isoelectronic CX_4 species.

Boron trifluoride is widely used as an acid catalyst in organic chemistry and is conveniently handled as the diethyl ether complex $BF_3 \cdot O(C_2H_5)_2$. Such complexes are easily prepared by passing gaseous BF_3 as it is made into an ether. The diethyl ether complex is easily purified by fractional distillation (boiling point 126 °C). The etherate complex is also used to liberate BH_3 from $[BH_4]^-$ in hydroboration reactions (see Equations 9.16 and 9.17). Although the other halides are even stronger Lewis acids than is BF_3, their aggressive chemical reactivity renders them much less useful and used than BF_3.

There are a few higher halides of boron, containing B—B bonds. As with all polyboron compounds, they tend to be electron-deficient cage compounds, rather than the chains and rings typical of Group IVA (14) chemistry. These compounds are of little importance and are not discussed further.

The larger, less polarizing M^{3+} congeners of B^{3+} cannot sufficiently polarize the halogens to form monomeric molecular species. The fluorides are all colorless, high-melting solids with complicated but typically ionic structures. Aluminum is invariably six-coordinate to fluoride, both in AlF_3 and in ternary fluoride compounds such as cryolite $Na_3[AlF_6]$, an important aluminum-containing mineral. Your attention is drawn to the parallel between BF_3 (boiling point -99.9 °C) and AlF_3 (melting point 1290 °C) on one hand and CO_2 (sublimes -78 °C) and SiO_2 (melting point greater than 1600 °C) on the other. In these two pairs of compounds, we see the dramatic effect on physical properties that results from polymerization. Polymerization becomes possible when the radius ratio of the constituent elements is large enough to allow a coordination number greater than that indicated by the empirical formula of the compound and when the polarizing power of the central atom is not sufficiently high to prevent oxide or halide from bridging.

The remaining halides of Group IIIA (13) (except for thallium, see below) are relatively volatile solids that produce vapors or organic solutions of dimeric molecules, isostructural with B_2H_6.

This behavior reflects, on one hand, the combined effects of preference for a higher coordination number and the lower polarizing power of its congeners relative to boron, and on the other hand, the greater polarizability (ease of saturation) of the heavier halogens relative to fluorine.

Like the boron halides, all trihalides of Group IIIA (13) are more or less strong Lewis acids. The bridges of the dimers or of the lattice compounds are rather easily broken by electron-pair donors to give 1:1 complexes. Virtually all complexes of aluminum and gallium trihalides are tetrahedral or pseudotetrahedral, but indium and thallium can exhibit higher coordination numbers, for example, $[InCl_5]^{2-}$ and $[TlCl_6]^{3-}$. The abundance, cheapness, and general availability of the aluminum halides have encouraged their use in organic chemistry as acid catalysts; they are used in a number of large-scale industrial processes for alkylation and acylation of aromatic compounds, polymerization of alkenes, and so on. In general, carbonium ions are more acidic than their Group IIIA (13) analogs, and so the acid–base exchange equilibrium (Equation 10.15) normally lies to the left:

$$MX_3 + R_3CX \rightarrow [R_3C]^+ + [MX_4]^- \qquad \textbf{(10.15)}$$

In certain cases where the carbonium ion is unusually stable, the equilibrium can be shifted substantially to the right. For example, reaction of an acyl chloride with $AlCl_3$ can give the ion pair $[RCO]^+[AlCl_4]^-$.

All trihalides of aluminum, except AlF_3, are violently hydrolyzed by water, as are their complexes. An interesting property of the aluminum trihalides (particularly, $AlCl_3$) is their ability to form ternary compounds with other metal halides that possess similar properties of volatility and solubility in nonpolar media to the aluminum halides themselves. For example, most of the transition metal halides MX_2 and MX_3 can be dissolved in hexane in the presence of $AlCl_3$, and compounds of the type $M(AlCl_4)_2$ can be sublimed from mixtures of the chlorides. These compounds have the same type of halide bridged structure as AlX_3, for example:

In the case of palladium, however, the metal adopts a square-planar coordination rather than the tetrahedral coordination adopted by the aluminum.

Aluminum monohalides are known, but they only exist at high temperature (greater than 1000 °C). At lower temperatures, they disproportionate into metal and trihalides. The equilibrium is illustrated in Equation 10.16:

$$AlCl_3 + 2Al \underset{<1000\,°C}{\overset{>1000\,°C}{\rightleftharpoons}} 3AlCl \qquad (10.16)$$

A great deal of research and development has gone into using Equation 10.16 in a nonelectrolytic process for aluminum reduction. Such processes rely on production of an impure aluminum or an alloy from which AlCl can be vaporized at high temperature by reaction with $AlCl_3$. On cooling the AlCl vapor, pure aluminum condenses out, and $AlCl_3$ is recycled. Although simple in principle, this process presents formidable engineering problems due to the high reactivity of the reagents and products. Despite the large amount of development dedicated to the subhalide process, most aluminum is still manufactured by electrolysis of Al_2O_3/Na_3AlF_6.

The monochlorides of gallium and indium can be obtained by thermal decomposition of the trichlorides. Reducing trichlorides with metals yields the mixed-valence "dihalides," $M^+[MCl_4]^-$ (M = Ga and In). With thallium the stability order of the two oxidation states is reversed: the trichloride and the tribromide decomposing easily to TlCl and TlBr. The only iodides of thallium are TlI and $Tl^I[I_3]$. With halogen ligands, the Tl(III) oxidation state is stabilized by complexation, even to the point that the reaction in Equation 10.17 takes place spontaneously:

$$Tl^I[I_3] + NaI \rightarrow Na[Tl^{III}I_4] \qquad (10.17)$$

The decreased stability of higher oxidation states of elements in combination with heavier, relative to lighter, *p*-block elements is a general phenomenon. The trend is the result of several factors including general decreasing of bond energies with increasing *n*, lowering of lattice energies with increasing interionic separation, and declining electron affinity with increasing *n* for a given group.

Tetrahalides of carbon differ dramatically from the Group IIIA (13) halides in their resistance to hydrolysis by neutral water and in their failure to exhibit any Lewis acidity. The lack of Lewis acidity is because all four-valence orbitals are used fully in bonding and the lowest unoccupied orbitals (the antibonding MOs) are high in energy. Resistance to nucleophilic attack also can be attributed to the generally high negative charge on the surfaces of the molecules preventing access to the central carbon atom and the large deformation energy necessary to achieve a more favorable configuration for attack (for example, the overlaps in a square-planar CX_4 molecule give a much lower binding energy than can be achieved in a tetrahedral molecule). The availability of relatively low-lying empty *d*-orbitals on the other elements of Group IVA (14), especially in compounds with strongly electron-attracting substituents, brings their behavior much more into line with that of the Group IIIA (13) elements.

Despite the very favorable energetics of the reactions

$$CX_4 + H_2O \rightarrow COX_2 + 2HX \tag{10.18}$$

and

$$CX_4 + 2H_2O \rightarrow CO_2 + 4HX \tag{10.19}$$

they only occur under quite severe reaction conditions; to all intents and purposes, CF_4 does not react at all. This chemical inertness is a general property of CF_4; apart from reaction with strongly reducing metals at high temperatures, it has relatively little chemistry.

Other carbon tetrahalides do react relatively easily with strong aqueous alkali and with other strong bases. Reactions with organic bases under anhydrous conditions rarely proceed cleanly because of intervention of free-radical reactions. The rather low homolytic–bond energies of the carbon tetrahalides makes them very susceptible to the loss of a halogen atom. This makes them useful as mild halogenating agents for both organic and inorganic molecules. The addition of carbon tetrahalides to unsaturated organic molecules can be catalyzed by one-electron redox ions (for example, Cu^+/Cu^{2+} and Fe^{2+}/Fe^{3+}) in the following cycle:

$$CCl_4 + CuCl \longrightarrow CuCl_2 + Cl_3C^{\bullet} \tag{10.20}$$

$$Cl_3C^{\bullet} + \;\;{\searrow}C{=}C{\nearrow}\;\; \longrightarrow Cl_3C{-}\overset{|}{\underset{|}{C}}{-}\overset{|}{\underset{|}{C}}{}^{\bullet} \tag{10.21}$$

$$Cl_3C{-}\overset{|}{\underset{|}{C}}{-}\overset{|}{\underset{|}{C}}{}^{\bullet} + CuCl_2 \longrightarrow Cl_3C{-}\overset{|}{\underset{|}{C}}{-}\overset{|}{\underset{|}{C}}{-}Cl + CuCl \tag{10.22}$$

Carbon tetrahalides are also extremely useful reagents for the production of anhydrous metal halides through the metal oxide:

$$Cr_2O_3 + 3CCl_4 \rightarrow 2CrCl_3 + 3COCl_2 \tag{10.23}$$

Until recently carbon tetrachloride was widely used as a solvent. Evidence of carcinogenicity has greatly curbed its use, as is the case with many other organic halides. Small traces of CCl_4 get introduced from organic material during the chlorination of drinking water. There is therefore some impetus for replacing chlorine as a disinfectant for drinking water, and there has been some success using ozone. In view of the millions of lives that are saved every year by chlorination of public water, we must be careful not to overestimate the dangers.

Physically, the halides of silicon and germanium resemble closely the analogs of carbon and boron. The tetrafluorides are low boiling point gases, the chlorides and bromides are volatile liquids, and the iodides are low melting point solids. Silicon tetrafluoride is readily made by reaction of HF (generated *in situ* by reaction of H_2SO_4 with an ionic fluoride) with silica:

$$2CaF_2 + 2H_2SO_4 + SiO_2 \rightarrow SiF_4 + 2CaSO_4 \cdot H_2O \qquad \textbf{(10.24)}$$

Silicon tetrachloride is made by direct combination of the elements or by chlorination of ferrosilicon (an alloy of iron and silicon), followed by distillation of $SiCl_4$ from the less volatile $FeCl_3$. The bromides and iodides can be made by halide exchange between $SiCl_4$ and $AlBr_3$ or AlI_3. Similar reactions can be used for the germanium compounds, which strongly resemble their silicon analogs in both physical and chemical properties.

Tin tetrafluoride is an involatile solid (sublimes at about 700 °C), and in this respect differs dramatically from the tetrafluorides of the lighter Group IVA (14) elements. The structure consists of layers of tetragonally distorted SnF_6 octahedra sharing corners in the plane (Figure 10.9). Lead tetrafluoride is isostructural with SnF_4, but it is thermally unstable with respect to dissociation into PbF_2 and F_2. This dissociation bears witness to the extraordinary oxidizing character of Pb(IV). Other tetrahalides of tin are very similar to those of Ge and Si in properties: $SnCl_4$ and $SnBr_4$ being volatile liquids and SnI_4 being an easily sublimable solid. The only other tetrahalide of lead under ambient conditions is $PbCl_4$, a strongly oxidizing liquid that decomposes to $PbCl_2$ and Cl_2 on mild heating and is rapidly hydrolyzed by water.

Oxidation state II halides of Group IVA (14) have an extensive chemistry. In the case of carbon, CX_2 species (dihalocarbenes) are widely

Figure 10.9
Structure of SnF_4.

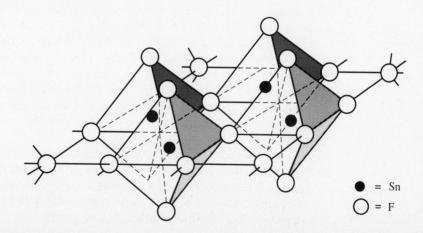

● = Sn
○ = F

encountered and used as unstable, reactive intermediates in organic chemistry. Dichlorocarbene is readily generated by thermal decomposition of trichloromethyl mercury compounds:

$$Hg(CCl_3)_2 \rightarrow HgCl_2 + 2:CCl_2 \qquad\qquad \textbf{(10.25)}$$

The species SiF_2 and $SiCl_2$ can also be obtained as unstable intermediates by reaction of tetrahalides with silicon at high temperature or by cocondensation of tetrahalides with a reactive metal (for example, Mg) in a low-temperature matrix. The silylenes polymerize to $(SiX_2)_n$ species at ambient temperatures.

Germanium difluoride and dichloride are also produced by the high-temperature reaction of GeX_4 with elemental germanium, but unlike the C and Si analogs, they are stable under ambient conditions. The dibromide and diiodide can be made by action of the acid HX on either GeO or $Ge(OH)_2$ prepared from $GeCl_2$. Compounds of Ge(II) are strongly reducing.

All tin dihalides are known and are stable quasi-ionic solids. Stannous fluoride has an interesting structure containing eight-membered $(Sn—F)_4$ rings in which each tin is also coordinated to a third terminal fluoride. The trigonal-pyramidal coordination of tin is usually taken to indicate the presence of a sterically active, nonbonding electron pair. Tin halides are soluble in water, from which they can be recrystallized as hydrated salts. Tin(II) salts are useful mild reducing agents. The lead dihalides are also all known. They are stable to hydrolysis but much less soluble in pure water than are the tin analogs.

All halides of Group IVA (14), except those of carbon, exhibit Lewis acidity and form complexes with electron donors. Silicon tetrafluoride forms both $[SiF_6]^{2-}$ and $[SiF_5]^-$ salts, the former even in aqueous solution, but other halosilicates have not yet been prepared. Both $[GeF_6]^{2-}$ and $[GeCl_6]^{2-}$ salts can be made in aqueous solution at low pH and in the presence of high halide–ion concentrations. All hexahalostannate(IV) ions are known. The yellow salt $[NH_4]_2[PbCl_6]$ is produced by oxidation of a concentrated aqueous solution of $PbCl_2$ and NH_4Cl with chlorine. It is considerably more stable than $PbCl_4$. This is a good example of how the oxidizing power of an ion can be reduced by coordination with an oxidation-resistant electron donor. The lowest-lying empty orbitals, which would accept electrons on reduction of the central ion, must become more antibonding with the formation of additional bonds to a two-electron donor.

The dihalides of germanium and tin show an affinity for coordination to a third halide ion to give the species $[MX_3]^-$, but the dihalides are in general much milder Lewis acids than are the tetrahalides.

Elements C and Si form many compounds with M—M bonds. These compounds are discussed further in Chapter 15.

10.4   Group VA (15)

Oxides

As we have moved across the periodic table, we have witnessed a gradual diminution in the initially large differences between the electronegativities of the elements of the earlier groups and oxygen. With Group VA and VIA (15 and 16), this trend has reached the point where even oxygen and the halogens are no longer capable of giving compounds of substantial ionic character; on the contrary, from now on all oxides and halides are essentially covalent.

Like carbon, nitrogen gives a family of oxides whose structures and chemistries are unique. The simplest of these, NO, has been encountered in the discussion of the electronic structures of diatomic molecules. Nitric oxide is a colorless gas and is easily produced in the laboratory by reaction of nitric acid with various reducing agents. Reaction of copper with aqueous nitric acid proceeds thus:

$$3Cu + 8HNO_3 \rightarrow 3Cu(NO_3)_2 + 2NO + 4H_2O \qquad (10.26)$$

The product is not very pure due to the secondary reaction, Equation 10.27, which becomes dominant in concentrated nitric acid:

$$Cu + 4HNO_3 \rightarrow Cu(NO_3)_2 + 2H_2O + 2NO_2 \qquad (10.27)$$

On an industrial scale, NO is produced by oxidation of ammonia over platinum. In the absence of catalyst, ammonia burns to give molecular nitrogen and water. Industrially produced NO is mostly used for production of nitric acid by the sequence of reactions (among others):

$$2NO + O_2 \rightarrow 2NO_2 \qquad (10.28)$$

$$2NO_2 \rightarrow N_2O_4 \qquad (10.29)$$

$$N_2O_4 + H_2O \rightarrow HNO_3 + HNO_2 \qquad (10.30)$$

$$2HNO_2 \rightarrow NO + NO_2 + H_2O \qquad (10.31)$$

The odd electron of NO resides in an antibonding π orbital of comparatively high energy. Thus, NO shows little tendency to dimerize but does tend to be oxidized fairly easily to either NO^+ or NOX.

Reaction with halogens, X_2, gives spin-saturated, bent nitrosyl compounds, for example:

$$2NO + Cl_2 \longrightarrow 2ClNO \qquad \qquad (10.32)$$

We may envisage the formation of the N—X bond as being due to the overlap of a low-lying, singly occupied $X^{\bullet}$ orbital (which is lowered in

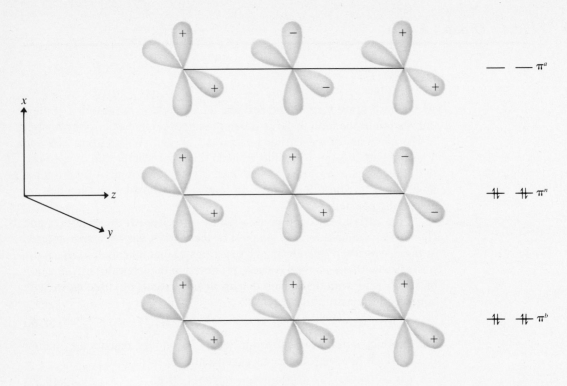

x

z

y

π^a

π^n

π^b

Figure 10.10
MO diagram for
CO_2 (π bonding).

energy), with the singly occupied π^a orbital of NO (which is raised in energy). If the orbital of the X$^{\cdot}$ is to be directed along the bond axis (that is, a σ-type orbital), the molecule must be bent to give a nonorthogonal overlap.

NO is rapidly oxidized to NO_2 in the presence of oxygen at room temperature. The reaction is substantially reversed at temperatures above 200 °C. The molecule NO_2, like NO, is also an odd-electron molecule. Although it shows a slightly greater tendency to dimerize than does NO, the dimer bond is still rather weak. We can better understand the electronic structure of NO_2 by comparing it to CO_2. The highest-energy occupied MOs of CO_2 result from three-center overlaps of p_π orbitals on each atom. These overlaps occur simultaneously in two perpendicular planes to produce a doubly degenerate set of π^b, π^n, and π^a orbitals. Figure 10.10 shows the formation and electron occupancy of these orbitals. To make a linear NO_2 molecule, we would have to add one additional electron to the energy-level scheme of Figure 10.10, and it would have to enter the doubly degenerate π^a orbital. The existence of multiple degeneracy in a molecular ground state for a polyatomic molecule is often a signal of metastability, and it is very frequently the case that a deformation of the molecule to lower symmetry will lift the degeneracy and lower the energy of the system. This phenomenon finds general expression in the Jahn–Teller theorem (discussed in Chapter 13).

Figure 10.11
Bonding and antibonding three-center overlaps in a bent triatomic molecule.
(a) Bonding overlap shows increased electron density in the concave region.
(b) Antibonding overlap shows decreased electron density in the concave region.

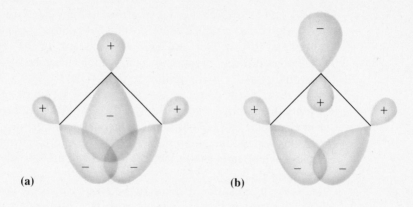

(a)   (b)

In the case of our linear XY_2 molecules, if we bend them in one π plane, we cause considerable changes in the orbitals occupying that plane, while leaving the overlaps in the perpendicular plane relatively unaffected. The most important effect of bending the molecule is that it allows a decrease in the energy of the antibonding electron by spreading it into the convex region of the molecule, while the bonding electrons occupy the concave region (Figure 10.11). Clearly, there are other adjustments as well, but the latter are the most important. The sum of these effects is to cause a bending of the molecule (O$\hat{\text{N}}$O = 134°) and to push the unpaired electron into the region of space farthest from the oxygen ligands. The relatively high spatial dispersity of the odd-electron orbital is largely responsible for the weakness of the N—N bond of the N_2O_4 dimer.

The solid (melting point −9 °C) is entirely dimerized, but dissociation begins on melting. At the liquid boiling point (21 °C), the vapor is golden brown and contains about 85–90% N_2O_4. Dissociation becomes progressively more extensive, and the brown color intensifies as it approaches completion at about 150 °C. The oxalate ion $[C_2O_4]^{2-}$ is isostructural and isoelectronic with N_2O_4, and the C—C bond shows considerably greater stability. The oxalate ion is mildly reducing, reflecting the tendency of CO_2^- to lose an electron and transform to the very stable CO_2. Nitrogen dioxide is more ambivalent, oxidizing strong reductants and reducing strong oxidants. This ambivalence is particularly evident in its disproportionation reactions (Equations 10.30 and 10.33). In Equation 10.33, all cationic products have particularly stable electronic configurations:

$$N_2O_4 + 3H_2SO_4 \rightarrow NO^+ + [NO_2]^+ + [H_3O]^+ + [3HSO_4]^- \quad \textbf{(10.33)}$$

The NO_2 molecule acquires an electron with ease because it has already paid the energy cost of breaking a conjugated π bond, and it easily loses an electron to give the conjugated linear system.

A very fragile codimer of NO and NO_2 exists at very low temperature as a blue liquid. This oxide, N_2O_3, is in principle the anhydride of nitrous acid HO—N=O. The anhydride of nitric acid N_2O_5 is also known, a colorless solid which is quite unstable at room temperature but relatively stable at 0 °C. It is a very aggressive oxidant. In the vapor and in certain condensed-phase preparations, it is molecular, but in the solid it has a strong tendency to autoionize to $[NO_2]^+[NO_3]^-$.

In all oxides of nitrogen discussed, we see an extreme chemical reactivity that contrasts dramatically with all oxides that have gone before, but in particular those of carbon and silicon. One of the most important causes of this general increase in reactivity is the interelectronic repulsion effects that become important as we combine small atoms with substantially filled valence shells.

An interesting exception to the above rule is the last oxide of nitrogen to be mentioned, N_2O. Nitrous oxide is a colorless gas consisting of linear molecules, and it is slightly soluble in water to give a neutral solution. It is linear and isoelectronic with CO_2, a factor that correlates with its exceptional lack of reactivity (for an oxide of nitrogen, that is). Although N_2O can support combustion of many reducing materials, ignition is necessary to initiate reaction, and in general N_2O undergoes remarkably few reactions under mild conditions. Its poorer performance as a Lewis acid, compared to CO_2, is again probably the result of the greater interelectronic repulsion generated by attaching O to N, as opposed to C.

Nitrous oxide is most commonly prepared by controlled thermal decomposition of ammonium nitrate:

$$NH_4NO_3 \rightarrow N_2O + 2H_2O \tag{10.34}$$

It finds extensive use as a general anaesthetic in hospitals and as an inert propellant in canned foams (shaving cream, crème Chantilly).

The other oxides of nitrogen, particularly in the form of inorganic and organic derivatives, are one of the largest volume chemical commodities. Ammonium nitrate is used in huge quantities as a fertilizer and as a slow explosive. Nitric acid is used massively for nitration of organic materials, to produce explosives (nitrocellulose, TNT), or to serve as intermediates to other nitrogen-functionalized organics for dyestuffs synthesis.

The remaining elements of Group VA (15) have a relatively orderly chemistry characterized by oxidation states III and V. All but bismuth give stable pentoxides, and these become relatively more oxidizing on descending the group. All pentoxides are strongly acidic, but the acidity also declines on descending the group.

Phosphorus pentoxide can be produced easily by burning phosphorus in an abundant oxygen supply. Oxygen deficiency leads to predominance of phosphorus(III) oxide (P_2O_3), and the oxides of arsenic(III) and antimony(III) are produced under all conditions of oxidation in oxygen. The latter two oxides can be oxidized to the (V) oxides with nitric acid, ozone, or hydrogen peroxide.

Figure 10.12
Structures of P_4O_6
and P_4O_{10}.

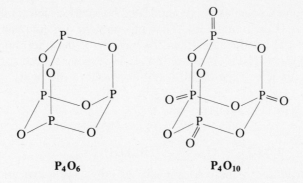

P_4O_6   **P_4O_{10}**

Phosphorus pentoxide and phosphorus trioxide exist as tetrameric molecules in the vapor and can be condensed to white solids based on the same molecules. Figure 10.12 shows the structures, and it is obvious that the only difference between the two structures is the presence of a P=O on each phosphorous atom of P_4O_{10}. This is a very common dichotomy in phosphorus chemistry; that is, virtually any P(III) compound can be transformed into a P^V=O analog, and that is due to the very high bond energy of the P^V=O bond. Although this bond is usually represented as a double bond, it should be evident that there are in fact more than two possible sets of overlaps in this bond. The basic structural unit of both P_4O_6 and P_4O_{10} is a tetrahedron of P atoms, with a bridging oxygen along and slightly offset from each of the six edges of the tetrahedrons.

The highly symmetrical P_4O_{10} structure is actually metastable under ambient temperature conditions. If it is annealed by heating for some time and slow cooling, other polymerized forms can be obtained. The P_4O_{10} structure can be viewed as a set of PO_4 pseudotetrahedra, each of which shares each corner of a trigonal face with another pseudotetrahedron. This molecule is favored in the vapor phase by entropy, which encourages the smallest possible molecules (maximum number), and bond strain, which makes molecules smaller than P_4O_{10} enthalpically unstable. By rapid condensation to room temperature these molecules can be "frozen" into the solid state. At higher temperatures, bond breaking and remaking allow reorganization to give a slightly more stable polymerized form. The simplest such structure consists of infinite sheets of PO_4 tetrahedra singly sharing the corners of one trigonal face and capped with P=O. The other M_4O_6 and M_4O_{10} (M = As and Sb) compounds have very much the same structural properties as the phosphorus compounds, but the metastable molecular forms become less stable on descending the group. Bismuth trioxide has a complicated structure, which is not described here.

Phosphorus trioxide dissolves in water with hydrolysis, and evaporation of the resulting solution yields crystalline H_3PO_3, known as phosphorous acid. This compound is a weak acid in aqueous solution, but its aqueous chemistry is complicated by a curious intramolecular proton transfer (Equation 10.35). That this process is a function of the very strong $P{=}O$ bond, rather than the basicity of the phosphorus atom, is evident because H_3PO_3 is not protonated by strong aqueous acids [that is, equilibrium (Equation 10.36) is not displaced to the right]:

$$P(OH)_3 \longrightarrow \begin{array}{c} H \diagdown \quad {}^{\prime\prime\prime}OH \\ P \\ O \diagup \quad \diagdown OH \end{array} \qquad (10.35)$$

$$P(OH)_3 + H^+ \rightleftharpoons [HP(OH)_3]^+ \qquad (10.36)$$

Addition of base to aqueous H_3PO_3 causes ionization, but the anionic species are even more prone to intramolecular protonation, and the salt isolated from neutralization of H_3PO_3 by alkali is $M_2P(H)O_3$, a phosphonate. The reaction (Equation 10.35) is closely related to the Arbusov rearrangement of trialkylphosphites $P(OR)_3$ to alkyldiacylphosphonates $PR(O)(OR)_2$.

Phosphorus pentoxide reacts violently with water to produce the strong acid H_3PO_4, orthophosphoric acid. The affinity of P_4O_{10} for water is sufficiently great that it is widely used as a drying agent for gases and liquids. At 25 °C, the three pK_a values for H_3PO_4 are 2.1, 7.1, and 12.4. The $[PO_4]^{3-}$ ion can therefore only be produced in strongly basic aqueous solution, and salts of this ion can be easily made, as can salts of $[HPO_4]^{2-}$ and $[H_2PO_4]^-$, at lower pH.

Like silicon, phosphorus(V) gives rise to a large family of condensed polymeric acids and salts. These compounds are discussed further in Chapter 17. For now, we will conclude with mention of the large-scale use of phosphates as fertilizers and the former widespread use of polyphosphates, which solubilize metal ions by chelate formation, as "builders" in detergents. The serious problems associated with the overfertilization of natural waters by detergent-derived phosphates has led to replacement of phosphate by less troublesome chelating agents.

Arsenic(III) and antimony(III) oxides dissolve in water, but are recovered as such on evaporation. The weakly acidic aqueous solutions react with alkali to give arsenites and antimonites, which mostly contain polymeric anions. These oxides are amphoteric and dissolve in strong mineral acids. Arsenic and antimony oxides find relatively few uses—their high toxicity, volatility, and solubility create environmental problems. Most heavy-metal sulfide ores are contaminated with these elements, which get in the environment by leaching of mine tailings and smelter slags and by fume emissions during oxidative roasting of ores.

Arsenic and antimony(V) oxides are much more strongly acidic and give the corresponding orthoarsenate and orthoantimonate salts with alkali. In the few cases where structures are known, the orthoantimonate occurs in the form of the hydrated $[Sb(OH)_6]^-$ ion, showing the characteristic tendency of heavier atoms to have a higher coordination number. Acid solutions of As(V) and Sb(V) are strong oxidants, but the oxidation potential falls with increasing pH, as is evident from Equation 10.37:

$$H_3AsO_4 + 2H^+ + 2e^- \rightarrow H_3AsO_3 + H_2O \qquad E^0 = +0.56 \text{ V}$$

$$(10.37)$$

In the presence of complexing anions, the system is, of course, more complicated.

Bismuth(III) oxide is a basic oxide, and its chemistry is of little importance beyond its use as a component in a number of mixed-oxide catalysts (for example, the so-called bismuth molybdate, a widely used hydrocarbon oxidation catalyst). Although Bi_2O_5 is unknown, some derivative "bismuthate" metal salts are known. Fusion of Bi_2O_3 with an alkali metal hydroxide in oxygen or with an alkali metal peroxide yields the alkali metal bismuthate of empirical formula $MBiO_3$. Sodium bismuthate is a useful reagent for preparing some extremely oxidizing aqueous species. For example, aqueous $NaBiO_3$ is one of the few reagents that will oxidize iron to iron(VI):

$$6BiO_3^- + 4Fe^{3+} + 14OH^- \rightarrow 4[FeO_4]^{2-} + 3Bi_2O_3 + 7H_2O$$

$$(10.38)$$

Halides

With the exception of ONF_3, nitrogen only gives halides in the oxidation state (III). The trifluoride is chemically inert, being neither easily hydrolyzed nor exhibiting any significant Lewis basicity. It is a colorless gas (boiling point $-130 \,°C$), and its structure is a trigonal-pyramid with an F—N—F angle of 102°. This angle is less than the H—N—H angle of ammonia and is usually attributed to a large lone-pair/bonding-pair repulsion. The effect is exaggerated by the inductive effect of the fluorine atoms causing the nonbonding pair of the nitrogen to be drawn in and to become more repulsive. This inductive effect also greatly reduces the basicity of the nonbonding electrons and reverses the direction of the molecular dipole relative to NH_3. The trifluoride is the only exothermic binary halide of nitrogen. The trichloride is strongly endothermic and can be explosively unstable. It is a powerful oxidant and is violently hydrolyzed by water to produce ammonia rather than nitrous acid, the product expected from nucleophilic attack at nitrogen (Equation 10.39):

$$NCl_3 + 3OH^- \rightarrow NH_3 + 3OCl^- \qquad (10.39)$$

The mechanism of this reaction is not a nucleophilic attack at the chlorine. That the reaction proceeds anomalously has been attributed to the inaccessibility of the central nitrogen atom to nucleophiles.

Both NCl_3 and NBr_3 are prepared by halogenation of concentrated aqueous ammonium chloride. The compound resulting from reaction of I_2 with concentrated ammonia and known as "nitrogen triiodide" is in fact a compound of formula: $NI_3 \cdot NH_3$. This shock-sensitive, explosive compound has a crystal structure consisting of chains of NI_4 tetrahedra sharing two corners. The iodine atoms at the other two corners are bonded to the nitrogen atoms of adjacent ammonia molecules. Similar structures are observed for other NI_3–base adducts.

Phosphorus trifluoride is a colorless, odorless, toxic gas and resembles NF_3 in its lack of reactivity and its very low Lewis basicity. It differs in the important respect that it, unlike NF_3, forms a wide range of coordination compounds with electron-rich transition metal atoms and ions. This difference extends to a wide range of substituted N(III) and P(III) compounds in their coordination chemistry and is attributed to the participation of the phosphorus $3d$ orbitals in bonding. The other phosphorus(III) halides, which are easily prepared by halogenation of phosphorus, react violently with water to produce $P(OH)_3$ and the hydrogen halide. They are oxidized by oxygen or oxygen-atom donors to give the phosphorus(V) oxyhalides, POX_3.

Although a great deal of chemical evidence exists for the participation of tetrasubstituted intermediates in the substitution reactions of the phosphorus trihalides, they form very few stable complexes. The trichloride forms a trimethylamine adduct in which the nitrogen seems to be bonded to phosphorus rather than to the halogen as in $NH_3 \cdot NI_3$. Only a few rare examples of $[PX_4]^-$ anions are known, for example, $[(C_3H_7)_4N][PBr_4]$.

Trihalides of arsenic closely resemble, physically and chemically, those of phosphorus. The trifluoride resembles SiF_4 and BF_3 in its reactions with water. Although hydrolyzed by neutral water, the hydrolysis is not complete in the presence of excess F^- or HF. Also, AsF_3 can be synthesized by the reaction of HF, generated *in situ*, with As_2O_3 (compare with SiF_4 and BF_3). Even $AsCl_3$ can be distilled from a mixture of As_2O_3 and concentrated aqueous hydrochloric acid. Antimony trihalides, which are all low melting point solids with polymeric molecular crystal structures, can also be prepared by reaction of Sb_2O_3 with concentrated aqueous mineral acids.

Both the arsenic and antimony trihalides exhibit a tendency to self-ionize. In certain condensed phases, they exhibit ionic conductivity, believed to be due to the reaction shown in Equation 10.40. Salts of the $[AsF_2]^+$ ion can be prepared by reaction of strong fluoride acceptors with AsF_3 (Equation 10.41), and many $[AsX_4]^-$ complexes are known:

$$2MX_3 \;\rightarrow\; [MX_2]^+ + [MX_4]^- \qquad\qquad \textbf{(10.40)}$$

$$AsF_3 + SbF_5 \;\rightarrow\; [AsF_2]^+\,[SbF_6]^- \qquad\qquad \textbf{(10.41)}$$

Arsenic(III) and antimony(III) halides show a much greater tendency to form coordination complexes than do the phosphorus(III) halides. Besides the $[MX_4]^-$ ions, a number of species of higher coordination number are known. The $[MX_4]^-$ ions always have the pseudotrigonal-bipyramidal structure predicted by VSEPR theory. The $[SbF_5]^{2-}$ ion, obtained by reaction of SbF_3 with alkali metal fluorides, has the expected square-pyramidal (pseudooctahedral) structure, but ions of the type $[SbCl_6]^{3-}$ show relatively undistorted octahedral structures. This discrepancy between experimental observation and the predictions of VSEPR theory may be papered over by a number of *ad hoc* assumptions such as assigning the lone pair to a spherically symmetrical *s* orbital or allowing that steric effects might override the VSEPR requirement of a pseudo-seven-coordinate structure. However, when all is said and done, it would be surprising if such a simple qualitative model as VSEPR would provide a universal explanation of chemical structure, particularly when dealing with the $n = 5$ valence shell. Figure 10.13 shows an MO diagram, which qualitatively satisfies the requirements of O_h symmetry and diamagnetism . This MO diagram is qualitative but provides a picture that is strictly analogous to the VSEPR explanation—namely, that the lone pair is in a spherical 5*s* orbital. In Figure 10.13, if the ligands are more electronegative than the central atom, a_{1g}^a approximates the central metal *s* orbital.

All pentafluorides of P, As, Sb, and Bi are known. They become increasingly strong fluorinating agents on descending the group and are all

Figure 10.13
Qualitative MO diagram for a fourteen-electron O_h molecule.

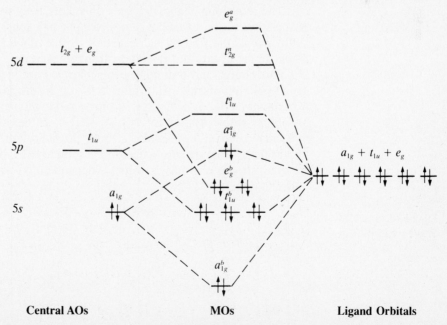

Figure 10.14
Fluxional
interchange of
ligands in PF$_5$
(a = axial,
e = equatorial).

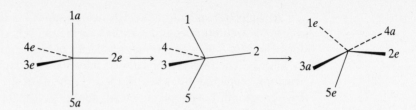

very strong Lewis acids. In combination with liquid HF, they form some of the strongest known Brønsted acids. The [MF$_6$]$^-$ ions (M = P, As, and Sb) are easily made by reactions of MF$_5$ with fluoride donors, and the [PF$_6$]$^-$ anion is a particularly useful noncomplexing anion in coordination chemistry.

Phosphorus pentafluoride is a trigonal-bipyramidal molecule, but it is relatively flexible. It can be shown by ^{19}F-NMR spectroscopy that there is a fairly rapid interchange of fluoride ligands between axial and equatorial positions. This fluxionality is attributed to the small energy difference between the trigonal-bipyramidal and square pyramidal geometries and to the low-energy barrier to isomerization between these two geometries. Figure 10.14 shows the manner in which axial-equatorial interchange can occur.

Antimony pentafluoride exists as tetramers in the solid state. The structure, which is very common for other pentafluorides, consists of four SbF$_6$ octahedra, each sharing two adjacent corners with its neighbors (Figure 10.15).

The remaining pentahalides are only marginally stable with respect to dissociation to the trihalides and molecular halogens. Those that can be made are synthesized by halogenation of the trihalide. They are strong halogenating agents and rapidly hydrolyzed by water.

Phosphorus and antimony pentachlorides are stable under ambient conditions, but arsenic pentachloride is not. In the vapor, PCl$_5$ is molecular, monomeric, and of trigonal-bipyramidal geometry. In the solid, it undergoes autoionization to [PCl$_4$]$^+$[PCl$_6$]$^-$. On the other hand, SbCl$_5$ is molecular trigonal-bipyramidal in both the solid and liquid state. Phosphorus pentabromide autoionizes in the solid to [PBr$_4$]$^+$Br$^-$. The mode of autoionization is obviously sensitively dependent on bond energies and on lattice energies.

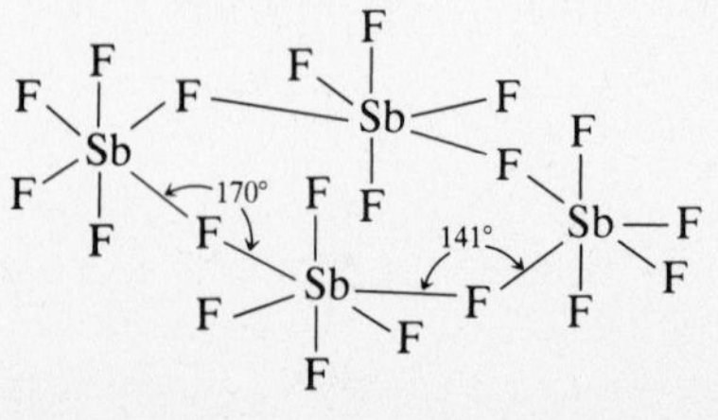

Figure 10.15
Structure of [SbF$_5$]$_4$.

10.5   Oxides and Halides of Group VIA (16)

Although we cannot reasonably discuss the "oxides of oxygen," it is worthwhile beginning this section with a mention of ozone O_3. This very endothermic, reactive molecule is isoelectronic with $[NO_2]^-$, whose electronic structure was discussed in some detail, and with the much more stable SO_2. Ozone is usually produced by subjecting O_2 to a silent electric discharge, but it also results in highly variable amounts on submission of O_2 to many high-energy treatments such as electric discharge, hard UV, or ionizing radiation. The photochemical production of ozone in the upper atmosphere is responsible to a large degree for blocking the UV component of the sun's radiation. The steady-state concentration of ozone depends on a delicate balance between a number of simple reactions leading to its production and destruction. Considerable publicity and controversy have surrounded the danger that may be associated with catalytic destruction of ozone under the influence of certain pollutants that find their way into the upper atmosphere. Two classes of pollutants are oxides of nitrogen (from highflying aircraft exhaust) and halogen atoms (from photolysis of chlorofluorocarbons used as refrigerants and propellants in spray-can products). Some of the important elementary gas-phase reactions resulting in the destruction of ozone are the following:

$$O_3 + NO \longrightarrow O_2 + NO_2 \tag{10.42}$$

$$O_3 + C \longrightarrow O_2 + CO \tag{10.43}$$

$$O_3 \xrightarrow[220-230 \text{ nm}]{h\nu} O_2 + O \tag{10.44}$$

$$NO_2 + O \longrightarrow NO + O_2 \tag{10.45}$$

$$CCl_2F_2 \longrightarrow Cl + CClF_2 \tag{10.46}$$

$$Cl + O_3 \longrightarrow ClO + O_2 \tag{10.47}$$

$$ClO + O \longrightarrow Cl + O_2 \tag{10.48}$$

In general, oxygen halides are both thermodynamically and kinetically unstable. Oxygen difluoride is exceptional. When pure it is relatively stable to heat but is a good oxidizing and fluorinating agent. It is relatively resistant to hydrolysis by neutral water but rapidly hydrolyzed by strong alkali. Despite the latter, the best method of synthesis for this pale yellow gas is fluorination of dilute aqueous sodium hydroxide. The hydrolysis products are HF and O_2, the same products that result from reaction of fluorine with water.

The chlorine analog, Cl_2O, a yellowish brown gas (boiling point about 4 °C), is unstable and presents an explosion hazard at room temperature. Chlorine dioxide ClO_2, a similarly unstable yellow gas, is of interest because of its unusual electronic structure and its fairly large-scale industrial use. The molecule has an odd number of electrons but shows no tendency to dimerize, an observation that suggests the unpaired electron is in an orbital that is antibonding in character. Because ClO_2 is a catenated triatomic molecule like CO_2 and NO_2, we can view the bonding in the same way that we did for NO_2 (Figure 6.7). In going from NO_2 to $[NO_2]^-$, it was possible by accommodating the additional electron in the slightly antibonding orbital in the molecular plane to leave the three-atom delocalized π system perpendicular to the molecular plane intact. Addition of another electron to the orbital scheme must be to the antibonding orbital of the latter π system. The molecule thus remains bent (ClO_2 has a bond angle of about 117°), and there is a progressive decrease in bond order in going through the series CO_2, NO_2, $[NO_2]^-$, ClO_2.

Despite its disagreeable properties, ClO_2 is relatively easy to generate in place and is extensively used for bleaching wood pulp and flour and for disinfecting public-water supplies. It may be made from aqueous sodium chlorate by the reaction shown in Equation 10.49, and sodium chlorate is easily generated on site by electrolysis of brine.

$$2NaClO_3 + SO_2 + H_2SO_4 \rightarrow 2NaHSO_4 + 2ClO_2 \qquad \textbf{(10.49)}$$

The explosive liquid Cl_2O_7 is the anhydride of perchloric acid, from which the courageous and the foolish may obtain it by dehydration with P_2O_5. It is isostructural with the similarly dangerous Mn_2O_7, the anhydride of permanganic acid, and the more benign dichromate ion $[Cr_2O_7]^{2-}$. All these species consist of two tetrahedra sharing a common corner. The only other halide of oxygen we will mention is iodine pentoxide I_2O_5. This remarkable compound, formally the anhydride of iodic acid, is a white solid, stable up to about 300 °C. Its structure is unknown but is probably either a molecular polymer (compare with P_2O_5) or perhaps even quasi-ionic $[IO_2]^+ [IO_3]^-$.

The important oxides of sulfur, selenium and tellurium, are the dioxides and the trioxides. Sulfur dioxide is an odorous, pungent gas (boiling point -10 °C) produced by oxidation or combustion of most compounds of sulfur and the element itself. It dissolves in water to give traces of sulfurous acid and in alkaline solutions to give sulfite and bisulfite salts. It is a moderately strong reducing agent, and in aqueous solutions is easily oxidized to sulfate. Reduction to elemental sulfur also occurs very easily, and its redox reaction with H_2S (Equation 10.50) is of great economic importance in removing H_2S from natural gas.

$$16H_2S + 8SO_2 \rightarrow 16H_2O + 3S_8 \qquad \textbf{(10.50)}$$

Oxidation of SO_2 to SO_3 by oxygen occurs only slowly in the absence of catalysts. In early versions of the "contact process," this oxidation was carried out over platinum catalysts, but the modern process uses a much cheaper vanadium pentoxide catalyst. Besides its massive use in sulfuric acid production, the oxidation of SO_2 by oxygen has received a fair amount of attention as a possible solar energy–storage process. This concept uses the endothermic dissociation of SO_3 into SO_2 and O_2 as a trap for the energy produced in a solar oven. The quenched mixture of SO_2 and O_2 does not recombine or give up stored energy until exposed to a catalyst. The energy can therefore in principle be stored, transported, and recovered fairly conveniently.

Sulfur trioxide is mainly produced as an intermediate in sulfuric acid manufacture. It is the anhydride of sulfuric acid, but although the reaction is very favorable thermodynamically, its reaction with water is inconveniently slow. The hydrolysis is therefore effected by absorption of the SO_3 into concentrated sulfuric acid.

The kinetic stability of SO_3, both in its hydrolysis and its oxidizing reactions, is understandable when it is recognized that SO_3 is isoelectronic and isostructural with BF_3. The much greater effective charge on $S(VI)$ and the greater polarizability of O are expected to greatly facilitate multicenter π bonding in SO_3 relative to BF_3, hence its lower reaction rate. In the gas phase, SO_3 consists of monomeric planar molecules, but in the condensed phase it gives a number of polymorphs in which polymerization has taken place by coordination of one oxygen of each SO_3 molecule to the acidic sulfur of its neighbor to give the unit

$$\left(\begin{array}{c} O \\ \parallel \\ -S-O- \\ \parallel \\ O \end{array} \right)_n$$

The so-called α-SO_3 (melting point, 17 °C) is a six-membered ring ($n = 3$). The acidity of SO_3 is manifest in its formation of adducts with tertiary amines and through its reaction with basic oxides to give sulfates. The sulfate ion $[SO_4]^{2-}$ is isoelectronic and isostructural with $[BF_4]^-$.

Selenium and tellurium dioxides are white solids obtained by oxidation of the elements with nitric acid. Selenium dioxide has an infinite-chain structure in which selenium is coordinated to three oxygens, two shared and one unshared. Tellurium dioxide is a three-dimensional structure with each tellurium coordinated to a distorted tetrahedron of oxygens. Selenium dioxide dissolves readily in water to give selenous acid, but TeO_2 is only slightly soluble. Both react with alkalis to give the selenite and tellurite salts. Both are easily reduced to the elements and oxidized by strong oxidizing agents to selenates and tellurates. Selenium and tellurium trioxides are obtained as solids by dehydration of H_2SeO_4 and H_6TeO_6, respectively. Selenium trioxide is polymorphic, but the various forms contain polymerized SeO_4 tetrahedra. One form of TeO_3 has the ReO_3 structure (Chapter 4) in which the Te is octahedrally coordinated to oxygen.

The trioxides are strong oxidizing agents and undergo fairly easy thermal decomposition to the dioxides and oxygen. Although selenium trioxide and selenic acid parallel SO_3 and H_2SO_4 in their structures and reactivity, telluric acid is actually $Te(OH)_6$ and is only a relatively weak acid. The "tellurates" usually contain the ions $[TeO(OH)_5]^-$ and $[TeO_2(OH)_4]^{2-}$, and the $[TeO_4]^{2-}$ ion is thus far unknown.

Sulfur forms a group of fluorides, remarkable for their variety and chemical properties. There are two compounds of formula S_2F_2 (F—SS—F and S=SF_2) and one each of formulas SF_4, S_2F_{10}, and SF_6. All these compounds are gases. Sulfur difluoride has been observed but not isolated as a pure compound.

Sulfur tetrafluoride is made by reaction of SCl_2 with NaF in acetonitrile. In the gas phase, it has the pseudotrigonal-bipyramidal structure predicted by VSEPR theory. It is quite reactive, being a useful fluorinating reagent and rapidly hydrolyzed by water, according to Equation 10.51:

$$SF_4 + H_2O \longrightarrow \quad \begin{matrix} F \\ \diagdown \\ \diagup \\ F \end{matrix} S{=}O + 2HF \xrightarrow{\text{H}_2\text{O}} SO_2 \qquad (10.51)$$

In contrast to SF_4, SF_6 and S_2F_{10} are extremely unreactive and reminiscent of CF_4 in their properties. Sulfur hexafluoride is the archetypal twelve-electron octahedral molecule, in which the six σ-bonding orbitals are just completely filled. It is likely that the molecule gains further stability from p_π—d_π bonding between the fluorine and the $3d$ orbitals on the sulfur. This is one of the cases where d-orbital participation is favored because the d orbitals are greatly contracted under the influence of the high effective positive charge on the sulfur. Figure 10.16 shows the overlaps involved in p_π—d_π bonding. The effect of this type of overlap is to make three of the nonbonding p orbitals of fluorine more bonding and three of

Figure 10.16
p_π—d_π bonding in an octahedral complex. Similar overlaps occur in the xz and yz planes.

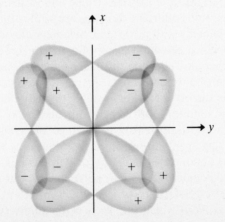

the nonbonding *d* orbitals of sulfur antibonding. We will encounter this type of bonding again in the hexafluorides of Group VIB (6).

The extreme unreactivity, nontoxicity, and good dielectric properties of SF_6 have led to its use as an electrical insulator in large transformers. Although S_2F_{10} is similarly unreactive, it is for some reason highly toxic to humans. Selenium and tellurium also form tetrafluorides and hexafluorides with properties similar to the sulfur analogs, but with generally enhanced reactivity. Whereas SF_6 and SeF_6 show no acid properties, TeF_6 forms the $[TeF_8]^{2-}$ anion on reaction with CsF. Again, the tendency of large atoms is to give compounds with high coordination numbers.

The chlorides of sulfur are much less stable and more reactive than are the fluorides. Compounds of the type S_2Cl_2, SCl_2, and SCl_4 are known. All are strongly oxidizing, easily hydrolyzed, and generally disagreeable compounds. The tetrachlorides of Se and Te are more stable than is SCl_4, and they are the only known binary halides of Se and Te, other than the fluorides. In the presence of alkali chlorides in concentrated hydrochloric acid, both $SeCl_4$ and $TeCl_4$ yield complex octahedral anions $[MCl_6]^{2-}$. Like the $[SbX_6]^{3-}$ ions discussed earlier, the undistorted octahedral structure of these ions indicates location of the nonbonding electron pair in a spherically symmetrical orbital of largely *s* character.

10.6 Interhalogen and Noble Gas Compounds

Interhalogen and noble gas compounds comprise a relatively limited family of highly reactive and unstable molecules whose primary importance is their role in testing chemical bonding theory. At first it may seem rather strange to treat the chemistry of the halogens and the noble gases, two groups that represent the extremes of chemical activity and inertia, in the same section. The superficial differences between the halogens and the noble gases are much reduced, however, if we focus our attention on the comparison of halide ions (particularly F^-) with the isoelectronic noble gas atoms and the noble gas compounds with halogen atoms or the halogens in their higher positive-oxidation states.

Noble gases are exceptional in their reluctance to either gain or lose an electron. Halide ions—because of their excess negative charge, relative to the isoelectronic noble gas atoms—have both a lower ionization energy and a lesser electron affinity. On the other hand, noble gas cations have greater electron affinities and greater ionization energies than do isoelectronic halogen atoms. From such considerations, it is obvious that inert gases should be less reactive than are halide ions, but their compounds should show even higher reactivity than the halogens. The big question remaining is: Are there any chemically significant conditions under which noble gases can be persuaded to yield electrons sufficiently to produce stable compounds? The answer is definitely, yes! (The same question can be asked of halogen atoms, which have ionization energies comparable to those of the inert gases[2].)

Another obvious point of similarity between halogen and noble gas compounds is the characteristically large number of electrons that must be accommodated in the valence shell. For a noble gas atom bonded to any number of other atoms, the octet rule must be exceeded; for a halogen atom to be bonded to more than one other atom, the same must be true. It is a curious historical fact that the mythical inertia of a closed shell did much to diminish the energy expended in the search for noble gas compounds, long after numerous examples of superoctet valence shells were known, particularly among interhalogen compounds.

We may roughly classify the interhalogen compounds into two categories: those in oxidation state zero (the binary analogs of the elementary diatomics) and those in which one of the halogens is in a formally positive-oxidation state. Heterodiatomic halogens are generally formed readily on mixing the required pair of halogens in a 1:1 ratio. The bond energies are always higher in the heteropolar molecules than are the average bond energies of the two constituents and in some cases higher than either. It is this factor that drives the reactions. All heterodiatomics are more or less stable under ambient conditions except for BrF, which spontaneously disproportionates to BrF_3 and Br_2. The bonding in the halogen diatomics can be attributed to a single σ bond, formed by overlap of p orbitals. In the heterodiatomics, the principal new features are the poorer orbital overlaps that are possible between atoms of widely different principal quantum number (n), the polarity arising from the difference in electronegativity, the contribution of ionic terms to increase bond energy, and the relief in interelectronic repulsion in the fluorides, relative to difluorine.

Dihalogens (except for F_2) usually react by dissociation into atoms or by heterolytic dissociation under the influence of an attacking reagent. Thus, reaction of Cl_2 with hydroxide may be viewed as displacement of Cl^- from Cl_2 by OH^-:

$$Cl_2 + OH^- \rightarrow HOCl + Cl^- \tag{10.52}$$

The tendency to undergo heterolytic fission increases on descending the group, and the I_2 molecule can actually be cleaved to two stable species:

$$I_2 + 2C_5H_5N + AgNO_3 \rightarrow [(C_5H_5N)_2I]NO_3 + AgI \tag{10.53}$$

---

[2] One may wonder why we do not consider the possibility of the inert gas *acquiring* electrons. Because ionization always requires input of energy and because the electron affinities of noble gases are highly endothermic, there is little hope that compounds could be spontaneously formed by chemical transfer of an electron to a noble gas. The best chance for compound formation would be for the noble gas to lose an electron to an atom with a highly exothermic electron affinity.

Table 10.1 *Interhalogen Compounds and Some Properties*

Compound	(bp) [mp] °C	Single X—F Bond Energy (kJ mol^{-1})
ClF_3	(12)	175
BrF_3	(126)	200
IF_3	[−28]	270
ICl_3	[64, sub.]	—
ClF_5	(−14)	140
BrF_5	(41)	185
IF_5	(100)	265
IF_7	[5, sub.]	230

The increased homolytic bond energies of the heterodiatomic halogens decrease the tendency toward homolytic reactions, but the increased polarity increases the tendency toward heterolytic reactions. Thus, ICl is a much better electrophilic iodinating agent than is I_2 and unlike I_2 even iodinates aromatic compounds.

Table 10.1 lists the interhalogen compounds in which one of the halogen atoms may be assigned a positive oxidation state. As may be expected, the general trends reflect the increasing difficulty of withdrawing electrons from the central atom on ascending the group and with increasing oxidation state. As might be anticipated, all known stable compounds are fully electron-paired, and the series IF, IF_3, IF_5, and IF_7 give us a homogeneous sequence of molecules exemplifying all possible odd-coordination numbers and their associated geometries (there is no known nine-coordinate neutral binary molecule, although many nine-coordinate complexes are known). We may complement this series with some of the fluorides discussed earlier and with the xenon fluorides to be discussed below, thereby completing the primary family of molecular structures and electron configurations for main-group elements (Figure 10.17). In each of the cases shown, the experimental evidence indicates that the molecule adopts the structure predicted by VSEPR theory. A valence–bond description of the molecules with more than eight-valence electrons requires the inclusion of *d* orbitals in the hybridization scheme. An MO scheme without the participation of *d* orbitals requires location of electrons in antibonding orbitals, and therefore bond order of less than one, for molecules with more than eight-valence electrons. The mixing of empty *d* orbitals into the scheme can lower the energy of the antibonding electrons (make them less antibonding) and thereby increase the bond strength.

The mystique of the chemical inertness of the closed shell was such as to render the first synthesis of an inert gas compound by Bartlett in 1962 an event of historical importance to chemistry. It has been pointed out on

C.N.	2				3				4			
	Molecule	Symmetry	VE	BO	Molecule	Symmetry	VE	BO	Molecule	Symmetry	VE	BO
	BeF_2	$D_{\infty h}$	4	2	BF_3	D_{3h}	6	3	CF_4	T_d	8	4
	(CF_2)	C_{2v}	6	2	NF_3	C_{3v}	8	3	SF_4	D_{4h}	10	4
	OF_2	C_{2v}	8	2	IF_3	D_{3h}	10	3	XeF_4	D_{4h}	12	4
	XeF_2	$D_{\infty h}$	10	2								

C.N.	5				6				7			
	PF_5	D_{3h}	10	5	SF_6	O_h	12	6				
	IF_5	C_{4v}	12	5	XeF_6	?	14	6	IF_7	D_{5h}	14	7

Figure 10.17
The family of binary molecular structures as exemplified by the fluorides.
(VE = number of electrons;
BO = number of σ-bonding orbitals.)

many occasions since that discovery, that there was no sound theoretical reason for anyone to think that inert gas compounds should not exist. Nevertheless, it is indisputable that the twin barriers of the closed-shell myth and the general difficulty of doing chemistry with elemental fluorine were sufficiently formidable to hold off this last great advance to a very late date indeed. The discovery of the noble gas compounds effectively completed the "classical period" of inorganic chemistry—the period in which the natural elements were discovered and their broad reactivity patterns established. The pioneering reaction whereby Bartlett synthesized the first noble gas compound $Xe^+[PtF_6]^-$ by reaction of Xe with the extremely powerful oxidant PtF_6 is now of little more than archival interest. There very quickly followed reports of syntheses of fluorides of krypton and xenon and of xenon oxides. The period since those early discoveries has seen a continuous trickle of new compounds, mixed oxohalides, inert gas–nitrogen compounds, halo complex ions, and oxoacids.

We confine our discussion here to a brief description of the fluorides and oxides of xenon. Although some chlorides are known, they are extremely unstable. The three fluorides of xenon are all made by direct reaction between the elements. The difluoride is obtained by UV photolysis of a mixture of the elements under ambient conditions; the tetrafluoride is produced by use of heat or electric discharge on a mixture of the elements; the hexafluoride results from a reaction at high pressure and temperature and with a large excess of fluorine. All xenon fluorides are colorless solids under ambient conditions. They are all strong oxidizing and fluorinating agents and are hydrolyzed by water, according to Equations 10.54–10.56:

$$2XeF_2 + 2H_2O \xrightarrow{\text{Slow}} 2Xe + 4HF + O_2 \tag{10.54}$$

$$6XeF_4 + 12H_2O \xrightarrow{\text{Fast}} 2XeO_3 + 4Xe + 24HF + 3O_2 \tag{10.55}$$

$$XeF_6 + 3H_2O \xrightarrow{\text{Fast}} XeO_3 + 6HF \tag{10.56}$$

Despite their great reactivity, the fluorides, unlike the oxides, are thermodynamically stable with respect to the elements. The difluoride and tetrafluoride are linear and square-planar molecules, respectively, and appear to retain their molecular form in the solid. The structure of XeF_6 in the vapor is not known, but it is known *not* to be regularly octahedral. A number of solid-state forms are known, all based on polymerization of units that are best described as $[XeF_5]^+F^-$. The $[XeF_5]^+$ ion assumes a square-pyramidal coordination, as predicted by VSEPR theory; $[XeF_5]^+$ is isoelectronic with IF_5.

Both XeF_2 and XeF_6 lose a fluoride ion to strong fluoride acceptors such as SbF_5. The products from XeF_2 contain the XeF^+ and the $[Xe_2F_3]^+$ ions, but the XeF_6 compounds are almost exclusively salts of $[XeF_5]^+$. The hexafluoride also behaves as a fluoride acceptor and with alkali metal fluorides yields complexes of the type $MXeF_7$ and M_2XeF_8. The latter are the most stable xenon compounds known.

Oxides of xenon are dangerously explosive in the solid state. Trioxide, obtained from the reaction shown in Equation 10.55, is isoelectronic with SbF_3 and $[IO_3]^-$, and it presumably has a trigonal-pyramidal structure like the latter compounds. It is weakly acidic in water and with alkali metal hydroxides gives xenates $MHXeO_4$, which slowly decompose in aqueous solution to produce perxenates, formally $[XeO_6]^{4-}$, but not likely to be fully ionized. It is by treating these xenates with concentrated sulfuric acid that XeO_4 is obtained.

BIBLIOGRAPHY

Suggested Reading

Bailar, J. C., Jr., et al., eds. *Comprehensive Inorganic Chemistry*. Oxford, England: Pergamon Press, 1973. A comprehensive reference text to the chemistry of inorganic compounds.

Greenwood, N. N., and A. Earnshaw. *Chemistry of the Elements*. New York: Pergamon Press, 1984. An excellent encyclopedic coverage of descriptive and structural chemistry of inorganic compounds.

Wulfsberg, G. *Principles of Descriptive Inorganic Chemistry*. Pacific Grove, Calif.: Brooks/Cole, 1984.

PROBLEMS

10.1 Identify two important ways in which hydrogen differs from the other elements in its bonding properties.

10.2   Using the orbital overlap diagram in Figure 10.15, draw an orbital level diagram showing the frontier orbitals of SF_6. Explain the unusual stability of this molecule.

10.3   The products of burning Li, Na, and K in O_2 are Li_2O, Na_2O_2, and KO_2, respectively. How do you explain this difference in behavior?

10.4   Write balanced equations for the dissolution of BeO in concentrated aqueous HF and in concentrated aqueous NaOH.

10.5   Using the oxides of a group of your choice, illustrate the changes in chemistry that typically occur on descending a group.

10.6   On the basis of inductive effect, the Lewis acidity of the boron halides is expected to be $BF_3 > BCl_3 > BBr_3$. Experimentally, the opposite is observed. Explain this apparent anomaly.

10.7 N_2O_4 dissociates easily into two molecules of NO_2, but N_2O_5 has a tendency to autoionize. Discuss this difference in terms of the electronic structures of the reactants and products.

10.8   Why can phosphorus(III) oxide be obtained as a molecular crystal of discrete P_4O_6 molecules, but boron(III) oxide cannot?

10.9   Dissolving SO_3 in concentrated sulfuric acid gives a family of linear polysulfuric acids of general formula

$$HO-\left(\begin{array}{c} O \\ \| \\ S \\ \| \\ O \end{array}-O\right)_n H$$

The acidity (pK_a) of these acids increases with increasing n. Provide an explanation for this behavior.

10.10   Both PCl_3 and BCl_3 form a trimethylamine adduct. What would you expect the structures of these adducts to be?

10.11   Use the VSEPR theory to predict the structures of each of the species depicted in Equation 10.41.

10.12   Write balanced equations for the synthesis of ClO_2, $POCl_3$, and CH_3SiCl_3, in which the initial source of chlorine is Cl_2.

10.13   Predict the geometries of all the compounds in Table 10.1.

10.14   Give one use for each of the following compounds:
   a. $SiCl_4$
   b. BF_3
   c. MgO
   d. Al_2O_3
   e. $NaClO_3$
   f. SF_6

10.15 Give one example of each of the following:
  a. A solid oxide in which a main-group atom is four-coordinate
  b. A liquid main-group halide in which the main-group atom is three-coordinate
  c. A Lewis acid–base complex in which a main-group fluoride coordinates to two additional fluorides
  d. A basic oxide from Group VA (15)

10.16 Compare the chemical and physical properties of the hydrides of Li and F.

10.17 Explain each of the following:
  a. The B—F bond energy in BF_3 is 646 kJ mol^{-1}, but the N—F bond energy in NF_3 is only 280 kJ mol^{-1}.
  b. The bond order of $CO_2 > NO_2 > O_3 > ClO_2$.
  c. The B—F bond is much stronger than the B—H bond, but the O—H bond is much stronger than the O—F bond.
  d. The melting points of BF_3 and AlF_3 differ by almost 1200 °C, but the melting points of CF_4 and SiF_4 differ by only about 100 °C.

10.18 Use the oxides of Group IVA (14) to illustrate how coordination number and tendency to polymerize vary on descending a group.

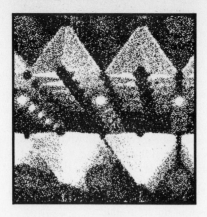

11

Halides and Oxides of Transition Elements

11.1 Some General Considerations

A number of factors leads to significant differences between the properties of the simple binary compounds of the transition groups and those of the analogous main-group elements with the same numbers of valence electrons. Perhaps the most striking difference is the tendency of transition elements to give stable compounds of many of the oxidation states from $+2$ to N (where N is the group number), whereas the main-group elements tend to exhibit a more limited number of oxidation states, most commonly N alone or N and $(N - 2)$. The underlying reason for this difference is

Table 11.1 *Ionization Energies and Radii of Some* d- *and* p-*Group Elements*

Element	Ionization Energy (MJ mol^{-1})[a]					Radius (pm)[b]
	I	II	III	IV	V	
Si	0.787(*0.790*)	1.577(*1.655*)	3.232(*1.124*)	4.356		40
Ti	0.659(*0.651*)	1.310(*1.343*)	2.653(*1.522*)	4.175		56
Ge	0.762(*0.775*)	1.537(*1.765*)	3.302(*1.108*)	4.410		53
Zr	0.660(*0.607*)	1.267(*0.951*)	2.218(*1.095*)	3.313		73
P	1.012(*0.891*)	1.903(*1.009*)	2.912(*2.045*)	4.957(*1.317*)	6.274	31
V	0.650(*0.764*)	1.414(*1.414*)	2.828(*1.679*)	4.507(*1.782*)	6.299	49.5
As	0.944(*0.854*)	1.798(*0.938*)	2.736(*2.101*)	4.837(*1.206*)	6.043	47.5
Nb	0.664(*0.718*)	1.382(*1.034*)	2.416(*1.274*)	3.690(*1.187*)	4.877	62

[a] The figures in parentheses are the increments on going from lower to higher oxidation state.
[b] Radii of highest oxidation states in tetrahedral coordination.

evident from the ionization-energy data shown in Table 11.1. The transition metal species have a greater tendency to lose electrons than do their main-group analogs, except for the last valence electron for the groups beyond Group IVB (4). However, the trends in the *increments* in the ionization energies for the first members of transition groups, compared to their main-group analogs, are quite different, reflecting the different behaviors of *nsnp* and of $ns(n-1)d$ valence shells.

The first two electrons to be removed from a transition atom are effectively the *ns* electrons. Ionization of the second electron is so easy that there is no tendency to give compounds in the oxidation state +1. The main-group elements lose their *np* electrons fairly easily, as reflected in the relatively small increments in ionization energy, but there is a distinct step in the increment for loss of the first *ns* electron. This step is important in stabilizing the $(N-2)$ oxidation state. For the 3*d* shell, the ionization increments tend to be larger than for the 3*p* and 4*p* shells, and it is these slightly larger increments (which result from the much greater sensitivity of *d*-orbital energy to effective nuclear charge) that lend stability to the intermediate oxidation states.

It would be naïve to assume that we could explain all differences between the main-group and transition elements with a single parameter. From Table 11.1, it is evident that radius effects are also important. Indeed, for the 4*d* and 5*d* elements, radius effects are perhaps more important than are ionization-energy effects.

Up to and including Group VIIB (7), the chemistry of the simple binary transition compounds is dominated by ionization energy and radius effects. The group oxidation states of all elements in these groups are known, and the trends between the stabilities and properties of the different oxidation states are fairly systematic.

Beyond Group VIIB (7), it is impossible to reach the group oxidation states of any first-row elements, and there is a dramatic change in periodic behavior for these elements. With the exceptions of Ru and Os, for both of which the group oxidation state is known and whose chemistry follows fairly systematically from that of Tc and Re, the same is true of all heavier elements beyond Group VIIB (7).

Another factor that has a great influence on periodic trends in transition chemistry is the crystal field stabilization energy. Although occasionally referring to this phenomenon in this chapter, we leave its detailed treatment to Chapter 13.

A final general feature that tends to distinguish the behavior of *d*-block elements is the tendency for transition compounds in low oxidation states to exhibit metallic or pseudometallic properties. This tendency is because *d* orbitals, under the influence of a low effective nuclear charge, are spatially quite extended relative to *s* and *p* orbitals and can overlap directly when the metal atoms are separated by small main-group atoms.

11.2 Oxides of Transition Elements

All $3d$ transition elements from Ti to Zn form a monoxide, either by direct combination of the elements or by reduction of a higher oxide by the metal. Most of these monoxides have the NaCl structure and are basic. With the exception of TiO, they all dissolve in aqueous mineral acids to give stable salts or complexes of M^{2+} ions. The Ti^{2+} ion liberates hydrogen from aqueous acid, and so dissolution of TiO in aqueous acid gives Ti^{3+} and hydrogen.

Monoxides of Ti and V show quasi-metallic behavior—for example, high reflectivity and electronic conductivity—and it may be assumed that although the relationship between the metal and oxygen is largely ionic the metal-ion lattice is fully connected through metal–metal bonds. Whereas CrO is marginal, the monoxides from Mn to Zn are typical ionic insulators (or more precisely, semiconductors).

All elements from Ti to Fe give stable M_2O_3 oxides with corundum-type structures. These oxides are all ionic and predominantly basic. The increase in effective nuclear charge in Ti_2O_3 and V_2O_3 contracts the $3d$ orbitals, relative to Ti(II) and V(II), to the point that they no longer overlap to give strong metal–metal bonds. In air the M_2O_3 oxide is the most stable for Cr, Mn, and Fe.

Mixed oxidation–state species $M^{II}M^{III}O_4$ are formed by Mn, Fe, and Co. Although of little chemical significance, this small family provides some interesting challenges in the interpretation of physical properties. For example, Fe_3O_4, which occurs abundantly in nature as the mineral magnetite, is ferromagnetic, a relatively rare property in common binary compounds. In Chapter 13, we will return to the intriguing question as to why the Mn and Co compounds adopt a *spinel* structure (all M^{3+} ions in O_h sites), while Fe_3O_4 adopts the *inverse spinel* structure (half of the Fe^{3+} ions in O_h sites and half in T_d sites).

The elements Ti, V, Cr, and Mn give MO_2 oxides of rutile or distorted rutile structures. Although we have not mentioned color, it is worth noting that of all the transition oxides so far mentioned (with the exception of ZnO) TiO_2 is the first to be colorless. The color of transition compounds is the subject matter of Section 13.2, and until then we will defer the theoretical explanation of why most transition compounds are colored. The whiteness of TiO_2—coupled with a number of other characteristics to do with particle size, shape, and its very low toxicity—have made TiO_2 the most widely used inorganic pigment in the modern paint industry. Both TiO_2 and VO_2 are weakly basic and dissolve in strong mineral acids to give complicated mixtures of polymeric and monomeric Ti(IV) and V(IV) (or $V^{IV}{=}O$) complexes. The Cr(IV) and Mn(IV) oxidation states are unstable or highly reactive in aqueous solution; thus, aqueous dissolution of the dioxides never (in the case of Cr) or rarely (in the case of Mn) occurs with the formation of stable M(IV) species. This series of oxides exhibits remarkable variation in electronic properties (described in more detail in Chapter 13). Like magnetite, CrO_2 is ferromagnetic and is particularly useful for the manufacture of high-quality recording tape.

With TiO_2 we reached the ultimate oxidation state of titanium. In fact, this is the most stable state for titanium, TiO_2 exhibiting neither significant acid–base nor redox properties, whereas Ti_2O_3 is quite strongly reducing. The ultimate oxides of the next three elements (V_2O_5, CrO_3, and Mn_2O_7) exhibit progressively stronger oxidizing and acidic properties. We see here again the appearance of valence saturation of oxygen with increasing effective positive charge on the central metal atom; Mn_2O_7 is a blackish purple liquid that decomposes exothermically (often violently) to MnO_2 and oxygen and reacts violently with organic material. It is an oxo-bridged dimer like I_2O_7, and the bridging oxygen is subject to extreme polarization stresses, probably resulting in a tendency for heterolytic splitting to give $[MnO_3]^+$ and $[MnO_4]^-$ (permanganate). The permanganate ion, although still thermodynamically very unstable, shows much greater kinetic stability than does Mn_2O_7, for reasons akin to those already discussed for main-group MF_4 and $[MO_4]^{n-}$ species.

Chromium trioxide is a red crystalline, water-soluble compound produced by treatment of chromate(VI) salts with concentrated sulfuric acid. It is a powerful oxidizing agent but less violently reactive than Mn_2O_7. In the solid, CrO_3 has a layer-lattice structure in which the chromium atoms are tetrahedrally coordinated to oxygen. The covalency of CrO_3 is suggested by its low melting point (197 °C). The trioxide gives acid solutions in water, and dissolution in alkali gives chromates M_2CrO_4. At lower pH, protonation of $[CrO_4]^{2-}$ occurs, followed by condensation to $[Cr_2O_7]^{2-}$, the dichromate ion, isostructural and isoelectronic with Mn_2O_7. The reluctance of Cr(VI) and Mn(VII) to expand their coordination spheres beyond four is undoubtedly a size problem and contrasts with the behavior of the heavier members of these groups.

Vanadium pentoxide varies from orange-red in large crystals (melting point 650 °C) to yellow powders because particle size varies depending on the mode of preparation. It can be made by direct combination of the elements or by precipitation from solutions of vanadate(V) salts. Its structure consists of zigzag, double chains of edge-sharing VO_5 trigonal bipyramids. There are two vanadium–oxygen bond lengths characteristic of V—O—V and V=O bonds. Vanadium pentoxide is amphoteric and dissolves in both strong alkali and strong acid solutions to give a variety of oxo-bridged polyvanadate anions. The relatively strong oxidizing power of the V(V) species is illustrated by the liberation of Cl_2 on dissolution of V_2O_5 in concentrated HCl:

$$V_2O_5 + 6HCl \rightarrow 2VOCl_2 + Cl_2 + 3H_2O \qquad (11.1)$$

The highest oxidation–state oxides of the congeners of Groups IVB–VIIB (4–7) tend to be less acidic and less oxidizing than those of the first members of the group. Zirconium and hafnium dioxides are chemically inert, white materials whose main distinction from TiO_2 is that they do not exhibit the rutile structure. One crystal form of ZrO_2 places the metal in seven coordination to oxide. The high melting point and chemical inertness of ZrO_2 lead to its use as a ceramic and as a high-temperature, solid-state oxide ion conductor. The lower oxides of Zr and Hf are poorly characterized and of little importance.

Pentoxides of Nb and Ta are also white, chemically inert compounds. Unless freshly precipitated, they are resistant to attack in aqueous medium by virtually all reagents except HF, which dissolves them in the form of fluoro- and oxofluorocomplexes. They can also be reacted by fusion with strongly basic oxides to give polyniobate and tantalate salts. Niobium also forms a well-characterized NbO_2, which has a distorted rutile structure in which pairs of niobium atoms are joined through metal–metal bonds, and NbO, which is metallic.

The simple oxides of Mo and W are MoO_3, WO_3, Mo_2O_5 and MoO_2 and WO_2. Trioxides are easily formed by combustion of the metals in an abundant oxygen supply. Both oxides occur with the metal in octahedral coordination, but they have different structures (MoO_3 has a layer-lattice, and WO_3 has a distorted ReO_3 structure). They are weakly oxidizing and weakly acid. Although simple molybdates and tungstates $[MoO_4]^{2-}$ and $[WO_4]^{2-}$ exist in basic solution, at lower pH they polymerize to form a wide variety of homopolyanions and heteropolyanions (see Chapter 16 for some examples). An important difference between Cr(VI) and Mo(VI)/ W(VI) is the ease with which the latter pair expand their coordination sphere to six by hydration (Equation 11.2):

$$[MoO_4]^{2-} + 2H_2O \longrightarrow \left[\begin{array}{c} O \\ \| \ \ \ \ OH \\ HO-Mo-OH \\ HO \ \ \| \\ O \end{array} \right]^{2-} \qquad (11.2)$$

Both trioxides undergo easy reduction by either electropositive metals or organic materials. The products are usually nonstoichiometric and are treated further in Chapter 19. For now, we limit mention of this type of reactivity to the fact that careful heating of Mo and MoO_3 to 750 °C in the right stoichiometric ratio leads to formation of Mo_2O_5. This violet compound dissolves in acids to give Mo(V) complexes. The dioxides of Mo and W have similar properties to NbO_2.

The only oxide of technetium of any importance is Tc_2O_7 (melting point 119.5 °C), which can be distilled from acidified $[TcO_4]^-$ solutions. Because of the element's scarcity and radioactivity, technetium chemistry had not been widely explored, but this situation has changed in recent years because of the wide application of ^{198}Tc in medical radiography.

Rhenium is one of the last discovered and rarest of the elements, but its extremely rich and varied chemistry has been fairly extensively explored. Its well-established oxides are Re_2O_7, ReO_3, Re_2O_5, and ReO_2. The heptoxide (melting point 200 °C), like Tc_2O_7, is much more stable than is Mn_2O_7 because it can be distilled from the reaction of concentrated sulfuric acid with an alkali metal perrhenate. Both Tc_2O_7 and Re_2O_7 give acidic solutions in water.

Table 11.2 summarizes the remaining important oxides of Groups VIIIB–IIB (8–12). It can be seen that the rutile structured dioxides continue to appear consistently, reflecting the particular stability of the structure. The volatile covalent oxides RuO_4 (melting point 25 °C) and OsO_4 (melting point 40 °C) represent the highest oxidation state achieved by the transition elements. Both tetroxides can be distilled from compounds of Ru and Os, following attack by strong oxidants such as $NaBiO_3$. They are both fairly strong oxidants and are used for the selective oxidation of organic compounds despite their high cost. They both oxidize aqueous halides (Cl^-, Br^-, and I^-) to give the halogen and lower oxidation–state complexes of the metal. Neither is a particularly strong acid. Osmium

Table 11.2 *Oxides of the Platinum Group and Groups IB and IIB (11 and 12)*

Compound	Structure	Color	Mode of Preparation
RuO_2	Rutile	Dark blue	Metal in O_2 at 1200°C
RuO_4	T_d molecules	Orange	Ruthenium compound + $NaBiO_3$ or $NaOCl$
OsO_2	Rutile	Copper	$OsO_4 + O_5$ at 1200°C
OsO_4	T_d molecules	Colorless	Metal in O_2 at 800°C
Rh_2O_3	Corundum	Brown	ppt from Rh^{III} salts
RhO_2	Rutile	Black	Freshly pptd $Rh_2O_3 + O_2$ at 750°C
Ir_2O_3	Corundum	Brown	ppt from Ir^{III} salts
IrO_2	Rutile	Black	Metal + O_2 at 1200°C
PdO		Black	ppt from Pd^{II} salts
PtO_2		Brown	ppt from Pt^{IV} complexes and dehydrate
CuO	NaCl (distorted)	Black	Decomposition of $Cu(OH)_2$
Cu_2O	Cuprite[a]	Red-yellow	Thermal decomposition of CuO
AgO	(Actually $Ag^IAg^{III}O_2$)	Black	Oxidation of alkaline Ag_2O
Ag_2O	Cuprite	Brown	ppt from Ag^I salts
Au_2O_3		Brown	ppt from Au^{III} complexes and dehydrate
ZnO	Wurzite	White	Metal in O_2
CdO	NaCl	(Varies due to defects)	Metal in O_2
HgO	Zigzag chains	Red	Metal in O_2

[a] The cuprite structure is an "antiquartz" structure, with the metal being two-coordinate and the O four-coordinate.

tetroxide dissolves in strong alkali to give osmates in which the osmium remains in oxidation state VIII but expands its coordination sphere to six. Ruthenium tetroxide dissolves in aqueous alkalis with liberation of oxygen, to give green perruthenate salts $MRuO_4$ and eventually orange ruthenates M_2RuO_4. The ease with which osmium and ruthenium give the highly toxic and volatile tetroxides precludes many of the high-temperature and electrical contact applications for which the other noble metals are widely used.

Except for RuO_4 and OsO_4, the oxides shown in Table 11.2 are more or less basic in character. Compared to the oxides of the earlier groups, they are thermodynamically less stable with respect to dissociation into the elements. This instability increases with increasing group number and on descending the groups. In terms of the number of oxides and their stoichiometries, Group IB and IIB (11 and 12) oxides punctuate the end of transition metal behavior and a return to main-group behavior.

11.3 Transition Metal Halides

All $3d$ elements form dihalides with all halogens, with the exceptions of TiF_2, CuI_2, and MX_2, where M = Ag and Au and X = Cl, Br, and I. The instability of TiF_2 is probably due to easy disproportionation to Ti and TiF_3, whereas the high oxidizing power of M^{2+} ions of Group IB (11) accounts for instability of the others. Anhydrous dihalides can generally be synthesized by reaction of the metal with hydrogen halide or by dehydration of hydrated salts with a covalent halogen compound, for example, $SOCl_2$. Thermal dehydration normally gives some hydrolysis and a product contaminated with oxides, but pyrolytic dehydration in a stream of hydrogen halide usually gives the anhydrous dihalide.

The difluorides commonly have rutile structures, the dichlorides $CdCl_2$ structures, and the diodides CdI_2 structures. Dibromides have either $CdCl_2$ or CdI_2 structures or both. Dihalides are all ionic and typically dissolve in water to give aquo-complexes or mixed aquo-halo-complexes. The solutions of Ti(II), V(II), and Cr(II) are very strongly reducing. They react extremely rapidly with O_2, and Ti(II) even rapidly reduces the water to liberate hydrogen. Solutions of Fe(II) undergo slow oxidation in air, but in acid or neutral solution Mn(II), Co(II), Ni(II), and Cu(II) are quite stable to oxygen.

Copper(II) halides are moderate oxidizing agents due to the Cu^I/Cu^{II} couple. In water, where the potential is largely that of the aquo-complexes, there is not a great deal of difference between the different halides, but in nonaqueous media, the oxidizing (or more exactly, the halogenating) power increases in the sequence $CuF_2 \ll CuCl_2 < CuBr_2$. In fact, in nonaqueous media, the equilibrium shown in Equation 11.3 lies substantially to the right:

$$CuBr_2 \rightleftharpoons CuBr + \tfrac{1}{2}Br_2 \qquad\qquad\qquad (11.3)$$

All trihalides of all elements from Ti to Cr are known. Mn(III) and Co(III) ions are too oxidizing to coexist with any halide ion except F^- under ambient conditions, whereas Ti(III) and V(III) are moderately strongly reducing. Chromium(III) is fairly stable toward either oxidation or reduction. There is a marked tendency toward decreasing ionic character on passing from left to right across the period and from the fluorides to the heavier halides. Ferric chloride and bromide show essentially covalent molecular properties such as low melting points and solubility in donor organic solvents, but in the solid they have layer lattices based on hexagonal halide-ion packing. As with all trends across the transition periods, the changes are not monotonic and depend strongly on crystal field stabilization energy effects.

Many trihalides can be prepared by direct combination of the elements. In those cases where direct combination gives a higher oxidation state, trihalides can be produced by either thermal dissociation, disproportionation of the higher halide, by reduction—for example, $TiCl_3$ can be produced by reduction of $TiCl_4$ with H_2 at high temperature—or under the influence of an electric discharge. Another method with some generality for trichlorides is heating the oxide in a stream of CCl_4 vapor, as shown in Equation 11.4:

$$Cr_2O_3 + 3CCl_4 \xrightarrow{600\,°C} 2CrCl_3 + 3COCl_2 \qquad \textbf{(11.4)}$$

All these trihalides adopt structures in which the metal is six-coordinate, either octahedral or distorted octahedral. Many of the lattices are complicated; however, several including $TiCl_3$, VCl_3, and $CrCl_3$ adopt layer lattices similar to $CdCl_2$, but with one-third of the cations of the M^{3+} layers systematically missing.

Titanium tetrahalides are easily obtained by direct combination. The tetrachloride is produced in the metallurgical extraction of the metal from its ores by chlorination of the mass produced by fusion of the ore (either rutile, TiO_2, or ilmenite $FeTiO_3$) with carbon and is isolated by fractional distillation. Titanium tetrafluoride is a fairly high-melting (sublimes at 284 °C) crystalline solid, whereas other halides are volatile molecular species with tetrahedral structures. The chemical properties of the titanium tetrahalides are similar to those of the silicon and germanium halides. Besides its importance in extractive metallurgy, $TiCl_4$ is important in the manufacture of Ziegler–Natta catalysts (see Chapter 23), although the high efficiency of the modern versions of these catalysts leads to only a modest consumption of $TiCl_4$ by the polyalkenes industry.

Whereas titanium tetrahalides are fairly unreactive in redox and halogenation chemistry, vanadium tetrahalides are quite reactive. The physical properties of VF_4, VCl_4, and VBr_4 are quite similar to their titanium counterparts, but VCl_4 and VBr_4 (dark brown fuming liquids) dissociate spontaneously under ambient conditions to VX_3 and X_2. They also tend to halogenate organic material. Beyond vanadium the first transition–period elements do not typically form tetrahalides, with the particular exceptions noted below.

The only stable pentahalide of vanadium is VF_5. Chemically, it resembles a Group VA (15) pentafluoride, being readily hydrolyzed and a strong Lewis acid. In the solid state, VF_5 probably exists (as do most other pentafluorides) as square tetramers with linear or slightly bent bridges along the square edges (compare with the SbF_5 structure, Chapter 10). In the vapor phase, it has a trigonal-bipyramidal monomeric structure with D_{3h} symmetry. Beyond V the only higher halides with any stability are the fluorides of chromium, CrF_4 (green solid, sublimes at 100 °C) and CrF_5 (red solid, melting point 30 °C).

When we move to the second and third transition periods, we find the same trends in stabilities as were described for the oxides. In general, high oxidation states become less oxidizing, more stable, and less acidic, whereas low oxidation states become more reducing. It is necessary to qualify the latter statement because metal–metal bonding becomes so important in the lower oxidation states of the lower members of the groups that a comparison with those of the first members is in many cases not reasonable or possible.

Zirconium tetrafluoride has square antiprismatic, eight-coordination in the solid state, whereas the other tetrahalides of Zr and Hf have structures based on zigzag chains of octahedral MX_6 units, sharing edges. In the vapor, $ZrCl_4$ is molecular and tetrahedral.

All trihalides of Zr and Hf have been prepared, and their chemistries have been well explored. Several phases ZrX_n (X = Cl, Br, and I) and $HfCl_n$ ($2 < n < 3$) have been claimed but little structural information is available. It is very likely that at least some of these materials contain clusters analogous to those found with the elements in the succeeding groups. Monochlorides of Zr and Hf are rather interesting metallic compounds resembling graphite in appearance but dissolving in water with evolution of hydrogen.

In stark contrast to the $3d$ elements for which the simple MX_2 and MX_3 halides are highly typical, the $4d$ and $5d$ elements of Groups VB–VIIB (5–7) have virtually no simple binary halides containing the M^{2+} and M^{3+} ions (exceptions to this generality are MoF_3 and $MoCl_3$). Instead, all the compounds MX_n ($n \leq 3$) contain metal–metal bonded clusters. The compounds are usually prepared by reduction of higher halides, either with the same metal as the compound to be prepared or with an electropositive metal such as Na, Mg, or Al. The preparations involve high-temperature, solid-state reactions and require very careful control of conditions. The early history of this subject is strewn with confusion due to the difficulties involved in defining optimum reaction conditions and isolating and identifying products. An exhaustive discussion of the descriptive chemistry of binary halide cluster compounds is beyond the scope of this book, but Table 11.3 gives some examples of the major types. The first example Nb_3Cl_8 is a purely solid-state compound based on partial occupation of the octahedral holes in a HCP lattice of halide ions by the metal. In the *core-type nomenclature*, the subscripts $x/2$ and $x/3$ indicate that x halides are shared between two and three cluster units, respectively.

In the case of the remaining cluster types, it is possible to dissolve the cluster core intact in both water and in some cases nonaqueous solvents. The breakdown of the solids involves the replacement of halogen bridges by electron-pair donors, for example, H_2O or additional halide ions.

The basic structures of the M_6X_8 and M_6X_{12} clusters can be approximated by relating them to a cubic framework. In M_6X_8 the octahedron of

Table 11.3 *Some Examples of Binary Halide Clusters*

Compound	Cluster Geometry	Core Type
Nb_3Cl_8	Triangular	$Nb_3Cl_3Cl_{6/2}Cl_{6/3}$
Ta_6Br_{14}	Octahedral	$[Ta_6Br_{12}]Br_{4/2}$
Ta_6Br_{15}	Octahedral	$[Ta_6Br_{12}]Br_{6/2}$
Nb_6I_{11}	Octahedral	$[Nb_6I_8]I_{6/2}$
Mo_6Cl_{12}	Octahedral	$[Mo_6Cl_8]Cl_2Cl_{4/2}$
Re_3Br_9	Triangular	$Re_3Br_6Br_{6/2}$

metal atoms are placed at the face centers of a cube, whereas the eight X groups are placed at the eight corners, just beyond the limits of the cube (Figure 11.1a). The M_6X_{12} unit is similar, but the twelve X units are located at the edge centers, again just outside the cube limits (Figure 11.1b). In Re_3X_9 the metal-atom cluster is an equilateral triangle of Re atoms, each with a complement of five halogen atoms and two Re atoms in its coordination sphere. Two of the five halides bridge between clusters, two within the cluster, and the fifth is nonbridging (Figure 11.1c). When

Figure 11.1
Skeletal structures of M_6X_8 and M_6X_{12} clusters.

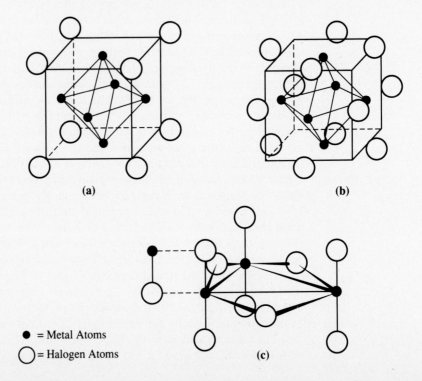

(a)

(b)

● = Metal Atoms

◯ = Halogen Atoms

(c)

Re_3X_9 is treated with aqueous hydrogen halide and the halide of a large cation—for example, Cs^+—salts of the anion $[Re_3X_{12}]^{3-}$ can be crystallized out. In this anion, the intracluster bridges are preserved, and there are three terminal halides on each Re.

The binary clusters and the soluble complexes derived from them have not found any great practical use, but an understanding of their synthesis and an explanation of their bonding have been important developments in fundamental inorganic chemistry. These developments have helped greatly in the rational search for interesting new solid-state materials (see for example, the superconducting Chevrel phases, Chapter 18).

With the exception of TaF_4, all tetrahalides of Nb and Ta are known. Niobium tetrafluoride differs from the other tetrahalides in that the octahedrally coordinated Nb(IV) is symmetrically located in its octahedron, indicating the absence of the Nb—Nb interaction evident in the other tetrahalides. (This lack of interaction is also evident from the paramagnetism of NbF_4, which shows that the single d electron remains unpaired.) Niobium tetrafluoride is isostructural with SnF_4 (see Figure 10.9). The same structure is common to other transition tetrafluorides. In all other tetrahalides of Nb and Ta (and those of Mo, W, Tc, and Re), structures are adopted where the metal is octahedrally coordinated, but occupation of octahedra is such that the metal atoms can pair by mutual displacement toward each other and away from the center of the octahedron. (This statement is approximate because the octahedra are distorted and not really centrosymmetrical.) The $NbCl_4$ structure is based on sheets of $NbCl_6$ octahedra like NbF_4, but the occupation of the octahedra is changed to allow Nb_2 pairing. The NbI_4 structure is based on chains of NbI_6 octahedra with all octahedra occupied but with alternating displacements of the Nb atoms (to the left and to the right) to permit pairing. The $NbCl_4$ and NbI_4 structures are also common in the other tetrahalides of Groups VIB and VIIB (6 and 7).

The tetrahalides cannot in general be made by direct halogenation but are usually produced by reduction of the pentahalides or hexahalides. Table 11.4 shows some representative examples. They are generally strong Lewis acids and undergo facile hydrolysis to monomeric or polymeric halo-oxo-complexes. The detailed nature of these hydrolysates has not been elucidated for most cases.

Pentafluorides of all second and third members of Groups VB–VIIIB (5–8) are known. They are made either by direct fluorination (Nb, Ta, Ru, Os) or by reduction of the hexafluoride (Mo, W, Te, Re). Except for Ru_4F_{20}, which has bent bridges, they all exist as square tetramers in the solid state, with octahedral coordination completed by M—F—M bridges (see SbF_5 above). As might be expected, they are all very strong Lewis acids and subject to hydrolysis to fluoro-oxo-complexes by pure water. The pentachlorides of Nb, Ta, Mo, W, and Re are stable and formed by direct

Table 11.4 *Some Representative Syntheses of Transition Tetrahalides*

Compound	Method of Synthesis	Structure
$NbCl_4$	$NbCl_5 \xrightarrow[450°C]{Nb} NbCl_4$	$NbCl_4$
$MoCl_4$	$MoO_3 \xrightarrow[450°C]{H_2} MoO_2 \xrightarrow[400°C]{CCl_4} MoCl_4$	NbI_4
WCl_4	$WCl_6 \xrightarrow[400°C]{Al} WCl_4$	NbI_4
$ReBr_4$	$HReO_4 + HBr \xrightarrow[Evaporation]{Slow} ReBr_4$	NbI_4
RuF_4	$10RuF_5 + I_2 \rightarrow 10RuF_4 + 2IF_5$	ZrF_4

combination of the elements. All are isostructural, the metals completing an octahedral coordination by dimerization through two bent M—Cl—M bridges:

The pentabromide of W is known, but no pentaiodides have been prepared. Like the pentafluorides, the other pentahalides are strong Lewis acids and form many complexes with two-electron donors. Of particular interest are the halo-complexes, which in the case of fluoride can exhibit high coordination numbers. Table 11.5 shows some examples of halo-complexes and their syntheses. As usual, an important distinction between the fluoro-complexes and the heavier halide complexes is the compatibility of the former with water.

Hexafluorides of Mo, W, Tc, and Re are the normal products of direct fluorination of these metals under mild conditions. They are remarkable for their relative lack of reactivity, considering the high formal charge of the metal. This lack of reactivity, although not quite so spectacular as that of SF_6, probably has its origins in the same type of p_π—d_π bonding (see Chapter 10). The slightly greater reactivity, particularly regarding hydrolysis, of the transition hexafluorides is no doubt related to the difficulty of overlapping $2p$ fluorine orbitals with $4d$ and $5d$ metal orbitals in the presence of the large inner-core electrons on the metal. Although the hexafluorides of all platinum-group metals except Pd are known, they tend to be extremely reactive, particularly with respect to reduction. Hexafluo-

Table 11.5 *Some Examples of Heavy Transition Metal Halo-Complexes*

Compound	Coordination[a]	Synthesis
Cu_2ZrF_8	*SA*	—
Na_3HfF_7	*CTP*	$HfF_4 + NaF(aq)$
Li_2ZrF_6	O_h	$ZrF_4 + LiF(aq)$
K_2NbF_7	*CTP*	$NbF_5 + KF(aq)$
K_3TaF_8	*SA*	$TaF_5 + KF(aq)$
$NaTaCl_6$	O_h	$TaCl_5 + NaCl$ (fusion)
Cs_3MoCl_6	O_h	Electrolysis of MoO_3 in concentrated HCl
K_2WF_8	?	$W(CO)_6 + IF_5 + KI$
K_3WF_8	?	$W(CO)_6 + IF_5 + KI$
K_2ReCl_6	O_h	$KReO_4 + HCl$ (concentrated) $+ KI$
K_2ReF_8	*SA*	$KF + ReF_6$

[a] *SA* = square antiprism; *CTP* = capped trigonal prism.

rides of Ir and Pt are so strongly oxidizing that they combine directly with O_2 and Xe by electron abstraction to give the $[IrF_6]^-$ and $[PtF_6]^-$ anions (besides polymeric derivatives of these ions) and O_2^+ or Xe^+. The increasing oxidizing power that develops on moving beyond Group VIB (6) may be attributed to the rise in the sixth-ionization energy on crossing the period. A general decline in stability would also be expected because the additional electrons that are present in the MF_6 molecules beyond Group VIB (6) must be accommodated in orbitals that are π antibonding in character.

Hexachlorides and hexabromides of W and Re are known. Tungsten hexachloride is used as a catalyst in conjunction with aluminum alkyls for the metathesis of alkenes (see Chapter 23). All hexahalides are monomeric molecules in all phases. The nd^0 species MoF_6 and WF_6 are predicted and found to be perfectly octahedral. Some others may be slightly distorted (such distortion is predicted by VSEPR theory and by the Jahn–Teller theorem; see Chapter 13).

Both Re and Os give heptafluorides. Little is known of their chemistries, and we will not mention more than that ReF_7 can be made by direct combination, it is a volatile solid, and it is isostructural with IF_7.

The quartet of elements Ru, Os, Rh, and Ir give a full complement of trihalides. Most of them are layer-lattice compounds, and the chlorides are extremely difficult to solubilize other than by fusion with alkali metal halides to form the hexahalometallate(III) complexes. This behavior contrasts with that of the so-called hydrate trihalides $RhCl_3 \cdot nH_2O$ and

$RuCl_3 \cdot nH_2O$. These compounds are useful starting materials for preparing a wide range of Rh and Ru compounds because of their ready solubility in water and a number of polar organic solvents. The true composition of these "compounds" is not known but is certainly not reflected in their formulas. Osmium also forms all of the tetrahalides, except the iodide.

The Pd and Pt dyad is characterized by oxidation states II and IV in binary compounds, with Pd(IV) being rather unstable. All dihalides are known, and PdF_2, a violet paramagnetic compound with a rutile structure, is the only known high-spin palladium compound (see Chapter 14). Apart from PdF_2, dihalides of Pd and Pt have the metal in a characteristic square-planar coordination. In $PdCl_2$ there is an infinite chain structure based on squares sharing common edges. One form of $PtCl_2$ has the same "open" structure, but another is based on a "closed" Pt_6Cl_{12} unit, which is superficially very similar to the M_6X_{12} cluster illustrated in Figure 11.1b, but both MO calculations and structural parameters indicate that there is no Pt—Pt bonding in Pt_6Cl_{12} and the unit is held together entirely by Pt—Cl—Pt bridges. The Pt_6Cl_{12} compound is readily soluble in organic solvents, even hydrocarbons such as benzene. Pd and Pt dihalides are slightly soluble in water and some hydroxylic organic species such as carboxylic acids. They are often used as starting materials in synthesis.

Except for PdF_4, palladium does not form stable tetrahalides, but all tetrahalides of platinum are known. Platinum tetrachloride is deliquescent and very soluble in water. In its hydrated form, known commercially as chloroplatinic acid, it is widely used in the preparation of platinum compounds and of platinum-based catalysts.

The halides of Group IB (11), like the oxides, reflect the intermediate character of the elements between transition and main-group behavior. Apart from the existence of almost all monohalides (except CuF and AuF), this group is more notable for its unperiodic rather than periodic properties. The nonexistence of CuF and AuF is related to the instability of Cu^+ and Au^+ ions with respect to disproportionation into the metal and Cu^{2+} and Au^{3+} ions. The other Cu(I) halides can be produced by reduction of Cu(II) by copper metal or other mild reducing agents even in the presence of water because Cu(I) halides are very insoluble in water. They dissolve in concentrated aqueous hydrogen halides to give $[CuX_2]^-$ complex ions, which are not susceptible to disproportionation, and the white insoluble halide can be reprecipitated by dilution. The disproportionation equilibrium (Equation 11.5) lies almost completely to the left in water under ambient conditions, but at 150 °C the equilibrium lies substantially to the right:

$$Cu^\circ + Cu^{2+}(aq) \rightleftharpoons 2Cu^+(aq) \qquad \textbf{(11.5)}$$

The Ag^+ ion does not disproportionate, and although the halides can be produced by direct combination, the chloride bromide and iodide are most easily produced by precipitation from aqueous solution. The fluoride may be produced by dissolving Ag_2CO_3 in dilute hydrofluoric acid. Unlike other halides, it is soluble in water. This solubility trend, along with the general difference between the solubilities of the Group IA and IB (1 and 11) halides, is attributable to an additional nonionic contribution to the lattice energy of the solid Group IB (11) halides. This energy difference is evident from a comparison of experimental and Born–Haber cycle lattice energies and increases dramatically on going from F to I. Because the solution energy for the Ag^+ ion remains constant, the free energy of solution is less favorable on going from the fluoride to the iodide. The insolubility of the heavier silver halides makes AgF (and other salts with nonpolarizable anions, for example, $[ClO_4]^-$, $[NO_3]^-$, $[BF_4]^-$) useful for exchanging other halides with fluoride (or other nonpolarizable groups).

Gold(I) halides are not compatible with water and are best made by thermal decomposition of gold(III) halides. The chloride, bromide, and iodide are the only simple binary compounds of gold(I) known.

The fluoride, chloride, and bromide of copper(II) are the products of direct combination of the elements. Their colors—white, brown-black (depending on state of division), and black, respectively—are a manifestation of the decreasing energy of the electronic transition corresponding to $Cu^{II}X^- \rightarrow Cu^IX^{\cdot}$ (see Chapter 13). The low energy for this transition in $CuBr_2$ and CuI_2 leads to the easy thermal decomposition of the former and the nonexistence of the latter (an almost parallel behavior is seen in the trihalides of iron). Despite its mild oxidizing properties, the Cu(II) state is the most commonly encountered oxidation state for copper.

Binary halides of Cu(II) have structures in which the Cu(II) is located in a highly distorted octahedral environment, due to Jahn–Teller effects (see Chapter 13). The distortion in these cases is always observed to be in the form of an elongation along one axis and a compression along the other two. The fluoride has a distorted rutile structure, whereas the chloride and bromide have highly distorted $CdCl_2$ structures. The structures of $CuCl_2$ and $CuBr_2$ can also be viewed as chains of square-planar coordinated Cu(II) (Cu—Cl, 230 pm), stacked in a staggered way so that the Cl in one chain occupies an axial coordination site of a Cu(II) in an adjacent chain (Cu—Cl, 295 pm). The stacked chains form the layers of the distorted $CdCl_2$ structure.

Stable compounds of Ag(II) and Au(II), unlike those of Cu(II), are rare and nonexistent, respectively. One of the few known such compounds of silver is AgF_2, produced by reaction of F_2 with the metal at about 250 °C. It is a good fluorinating agent and is instantly reduced by water. Its structure is complex, with highly distorted AgF_6 octahedra.

The inorganic chemistry of gold is mostly that of Au(III), and both the chloride and fluoride are prepared by reaction of the metal with halogen or with BrF_3 at moderate temperature. In both compounds, Au(III) exhibits the typical square-planar coordination of a $5d^8$ ion; the chloride being bridged dimers, Au_2Cl_6; and the fluoride consisting of infinite helical chains with linkage through cis Au—F—Au bridges. Both trihalides are moderate Lewis acids and form the $[AuCl_4]^-$ and $[AuF_4]^-$ ions very easily. Unlike the fluorides of the lighter transition elements, AuF_3 is rapidly hydrolyzed by water. The only higher oxidation–state compound of Group IB (11) is AuF_5. This compound resembles PtF_6 in its reactivity, as may be concluded from its synthesis by the following reaction:

$$Au + 6F_2 + 2O_2 \longrightarrow 2[O_2]^+[AuF_6]^- \xrightarrow{150\,°C} 2AuF_5 + 2O_2 + F_2$$

$$\textbf{(11.6)}$$

Dihalides of Group IIB (12) are easily synthesized by direct combination or in most cases by thermal dehydration of hydrated halides in a stream of gaseous hydrogen halide. Fluorides of Zn(II) and Cd(II) are highly ionic, and both behave very much like heavier Group IIA (2) halides, particularly with regard to their poor solubility in water. The zinc compound has a rutile structure, whereas CdF_2 has a fluorite structure, reflecting the large radius of Cd(II). Other dihalides of Zn and Cd show more covalent character than do their Group IIA (2) analogs, both from lattice-energy point of view and from chemistry. They tend to be very soluble in water and many coordinating organic solvents. Zinc chloride like the other dihalides of Zn and Cd is a mild Lewis acid, with a number of applications as a Friedel–Crafts catalyst in organic chemistry. An important difference between the heavier halides of Groups IIA and IIB (2 and 12) in aqueous solution is the tendency of the latter to exist as mixtures of aquo- and halo-complexes, whereas the former give uniquely the hydrated ions.

Mercury(II) fluoride is quasi-ionic with a fluorite structure. It is unusual for a simple ionic fluoride in that it undergoes complete hydrolysis in water. Other mercury(II) halides are molecular both in the condensed and dissolved state. They are linear, and the solids are easily sublimed and dissolve in donor organic solvents. When mercury(II) halides are reacted with Hg metal, they are reduced to mercury(I) halides, which in fact are compounds of the metal–metal bonded Hg_2^{2+} species.

11.4   *Oxides and Halides of Group IIIB (3): Lanthanides and Actinides*

In their chemical, as opposed to physicochemical, properties, the binary compounds of Group IIIB (3) and the lanthanides follow a pattern more typical of the main-group elements than do transition elements. The chemistry of these elements is almost entirely limited to oxidation state III, and the small periodic trends are largely related to size factors and to a lesser extent ionization energies. The elements themselves are quite electropositive and so show many chemical similarities to the lower members of Group IIIA (13). They react readily with both oxygen and halogens.

Scandium is very similar to aluminum. Its oxide Sc_2O_3 is amphoteric like that of aluminum and the trichloride (isostructural with $AlCl_3$) is readily hydrolyzed by water. The trifluoride has a ReO_3 structure.

In Group IIIB (3), there is no lanthanide contraction; thus, the ionic radius increases continuously on progressing from Sc to Y to La. This leads to the curious fact that, whereas La^{3+} behaves very much like a lighter lanthanide, Y^{3+} behaves very much like a heavier lanthanide because the lanthanide contraction reduces the radii of the later lanthanides to a value very close to that of Y^{3+}. Trioxides of Y, La, and the lanthanides are basic, the larger radius cations giving more strongly basic oxides than those of smaller radius. Similarly, trihalides are quite saltlike in their behavior, but the tendency toward hydrolysis increases with decreasing radius.

A number of exceptions occur to the general rule that lanthanides give compounds uniquely in oxidation state III. Oxidation state IV is commonly encountered with Ce. Burning cerium in air produces CeO_2 and high-temperature fluorination of CeF_3 gives CeF_4. The Ce^{4+} ion is thermodynamically unstable with respect to oxidation of water, and so the tetrafluoride undergoes rapid reduction and hydrolysis in water. The dioxide is soluble in concentrated sulfuric acid to give ceric sulfate, a useful oxidant in analytical and organic chemistry. Other lanthanide(IV) compounds are known but are very unstable.

Oxidation state II occurs with Eu, Sm, and Yb. Dihalides can be produced by reduction of the trihalides with an electropositive metal (calcium is most frequently used). All lanthanide(II) compounds are strong reducing agents. The Eu^{2+}/Eu^{3+} couple resembles Cr^{2+}/Cr^{3+}; the others are more strongly reducing and are even oxidized by water. Unlike trihalides, which tend to have unusual structures and often exhibit high coordination numbers, dihalides tend to have simple structures, for example, SmF_2 and EuF_2 (fluorite) and TmI_2 and YbI_2 (CdI_2).

It should be evident that the tendency of lanthanides to give the IV or II oxidation state is strongly linked to the creation of a full, half-full, or empty $4f$ shell.

We conclude the discussion of the periodic behavior of the oxides and halides of the elements with a very brief mention of the actinide elements. Although their scarcity and dangerous radioactivity makes study of most of these elements difficult, a strong imperative has been provided by military and industrial applications of nuclear fission. The chemistry of the actinides is more complicated than that of the lanthanides because of the tendency of the actinides up to Am to exhibit several valence states. Beyond the half-filled shell, oxidation state III becomes dominant, and the chemical behavior more closely parallels that of the lanthanides. (We must note, however, that this observation is probably colored because our knowledge of the chemistry of these elements decreases dramatically with increasing atomic number.)

Actinium exhibits only oxidation state III, and it behaves consistently with its character as a congener of La. With Th, Pa, and U, the ultimate oxidation states are the most important, although oxidation state IV is also important for U. Thorium behaves like a congener of Hf, giving a very stable, high-melting oxide ThO_2 (fluorite structure), a tetrafluoride, and a tetrachloride. The absence of lower oxidation states sets it apart from either Hf or Ce.

Uranium is somewhat representative of the general features of the actinides, and its binary compounds are commercially important. Yellow uranium trioxide is obtained by roasting the hydrated oxides that precipitate from U(VI) solution under basic conditions. It redissolves in acids to give complexes that contain the linear "uranyl" moiety $[O{=}U{=}O]^{2+}$. Severe heating of UO_3 converts it to a black mixed U^{IV}/U^{VI} oxide U_3O_8, which can be further reduced by heating in hydrogen to black UO_2. Uranium dioxide can be transformed by heating in a stream of HF to UF_4 and a stream of CCl_4 to UCl_4. Heating the involatile tetrafluoride in fluorine gives the volatile (boiling point 60°C) UF_6. Despite its high reactivity with respect to hydrolysis and fluorination of both organic and inorganic material, this compound has proven to be the best derivative of uranium for enrichment of ^{235}U by fractional distillation. The trifluoride, pentafluoride, and pentachloride of U are also known. Pentahalides have the same structures as their transition analogs. The chemistry of the uranium halides is strongly reminiscent of those of Mo and W.

BIBLIOGRAPHY

Suggested Reading

Brown, D. *Halides of the Lanthanides and Actinides*. New York: Wiley-Interscience, 1969.

Colton, R., and J. H. Canterford. *Halides of the First Row Transition Metals*. New York: Wiley-Interscience, 1969.

Colton, R., and J. H. Canterford. *Halides of the Second and Third Row Transition Metals*. New York: Wiley-Interscience, 1969.

Also see reading list for Chapter 10.

PROBLEMS

11.1   It is common for the oxides of transition metals to undergo the following reaction:

$$MO_n + nF_2 \rightarrow MF_{2n} + \tfrac{n}{2}O_2$$

What energy factor is mainly responsible for the exothermicity of these reactions?

11.2   The nonexistence of TiF_2 has been attributed to the ease with which it disproportionates to TiF_3 and Ti. Why doesn't this problem arise with $TiCl_2$? (HINT: Think about the structures and the radius ratios of the various species involved.)

11.3   Using the oxides of Mn, illustrate how oxidation state influences the acid–base properties and the redox properties of a compound.

11.4   Draw structures of the following compounds:
   a. Nb_3Cl_8
   b. Mo_6Cl_{12}
   c. Ta_6Cl_{14}

11.5   Give one use for each of the following compounds:
   a. TiO_2 **c.** ZrO_2
   b. CrO_2 **d.** OsO_4

11.6   Give the most commonly adopted structures for the following classes of transition compounds:
   a. MF_4 **d.** MO_2
   b. MF_5 **e.** MO_3
   c. MCl_4 **f.** MO

11.7   How would you synthesize each of the following?
   a. $FeCl_2$ **c.** RuO_4
   b. CrO_3 **d.** $ReBr_4$

11.8   Draw the structure of AuF_3. In what way do the structures of AuF_3 and $AuCl_3$ resemble those of NbF_5 and $NbCl_5$?

11.9   Why does the chemistry of Y^{3+} resemble that of the lighter lanthanides, whereas that of La^{3+} resembles that of the heavier lanthanides, rather than the reverse?

11.10   Describe two ways in which V_2O_5 and P_2O_5 are similar to each other and two ways in which they differ.

PART FOUR

Complex Compounds: Coordination Chemistry

12

Werner's Classical Coordination Theory

12.1   Historical Development of Coordination Chemistry

Coordination chemistry has its roots in two major theoretical advances that were well separated in time. The first was the formulation of the coordination theory of Werner; the second was the formulation of the concept of the dative covalent bond, initially by Lewis, with later modifications by Langmuir and Sidgwick. The historical importance of these developments to the theories of chemical bonding was mentioned briefly in Chapter 2. Werner's coordination theory was based on studies of the chemistry of the transition elements, whereas the theories of Lewis, Langmuir, and Sidgwick extended to all the elements. However, because the early theories of the two-electron bond were strongly tied to ideas of inert gas shells, the dative-bond concept fitted more easily into main-group chemistry than into transition metal chemistry, where frequent exceptions to the inert gas rule were encountered.

Although we will only touch on its central ideas, Werner's theory was intensely phenomenological, that is, it was based on a staggering number of experimental observations. He also suffered constant rebuttal by an astute and aggressive adversary by the name of Jørgensen. In the long run, Werner's theory benefited from Jørgensen's persistent attacks and the need to build up unassailable experimental proof of all its details.

A major dilemma facing chemists in the latter half of the nineteenth century may be exemplified by reference to a simple case. Atoms were believed to combine in simple ratios or simple multiples of those ratios, as established by Dalton's laws. Also, atoms seemed to exhibit a limited number of valencies (for the main-group elements, one or two). However, it became increasingly evident that there were many disturbing exceptions to these rules. For example, although cobalt was known to exhibit only two common valencies in its binary compounds (two and three), the compound $CoCl_2$ showed a remarkable tendency to combine with other molecules such as ammonia to give products of constant stoichiometry in which it was difficult to assign a rational valency to the constituent atoms. Thus, with ammonia, $CoCl_2$ gave a series of compounds $CoCl_2 \cdot (NH_3)_n$ in which $n = 1–6$. An early explanation of this behavior was based on the reasonable assumption that nitrogen can exhibit a valency of four (as in ammonium chloride) and that the ammonia molecules were forming chains, thus:

$$Cl_2CoNH_3 \cdot NH_3 \cdot NH_3 \cdot NH_3$$

and so on. This hypothesis could also accommodate the observation that the compound where $n = 4$ existed in two different forms with quite different properties. Thus, another structure with the same stoichiometry was proposed but with different connectivity from that shown above, for example:

$$\begin{array}{c} Cl \\ | \\ H_3NCoNH_3 \cdot NH_3 \cdot NH_3 \\ | \\ Cl \end{array}$$

On the basis of this model, a third isomer would be expected, but no one succeeded in making one. Such negative evidence was not fatal to the model because negative evidence of this kind was easily explained by assuming that the missing compound is unstable with respect to the other forms. For a while, this appealingly simple explanation, the so-called catenation theory of coordination, became the prevailing orthodoxy. However, as Werner accumulated more and more compounds, he not only became convinced that the catenation theory was incorrect but also devised tests that unequivocally proved that he was right. Furthermore, he pro-

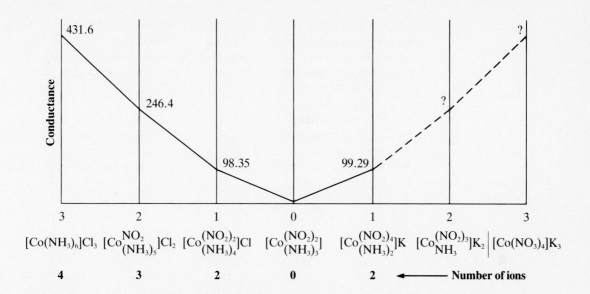

Figure 12.1
Variation of conductivity for a family of cobalt compounds as originally published by Werner.

duced and tested a new model that has stood the test of time to the present day, including corroboration by crystallographic structure determinations, which followed many years after Werner's formulation of his theory.

The first key hypothesis that Werner proposed and then established experimentally beyond reasonable doubt was that besides the usual valence atoms may exhibit a secondary combining tendency, which he called *coordination*. Thus, an atom is surrounded by a constant number of atoms or groups. This number, the *coordination number*, may be different from the conventional valence of the atom. The most common coordination number is six, although coordination by four groups was also recognized in Werner's time. The proof of six coordination was provided by Werner in a series of classical experiments, one of which is illustrated by the results in Figure 12.1. In this example, Werner used the new ionic conductance theory to determine the total number of ions in each compound. Combining the conductimetric evidence with chemical evidence such as precipitation of the chloride ions with $AgNO_3$, Werner concluded that cobalt is always associated with six groups, that is, it has a coordination number of six.

The second major theoretical concept that Werner established by elegant experiment was that of fixed geometry. This idea had already been established in organic chemistry by the demonstration of optical activity in tetrahedral carbon compounds by van't Hoff and LeBel. The same principle was applied to coordination compounds by Werner, who recognized that they could be rendered asymmetric by the use of groups that simultaneously occupied two coordination positions. (In modern parlance, we call such groups *bidentate*—"two toothed." The modern name for a coordinated group is a *ligand*—a group that has been "tied down.") The first such compound isolated by Werner is shown below. The bidentate ligand in this case was ethylene diamine, and the cationic character of the complex made resolution in the form of a tartrate salt easy.

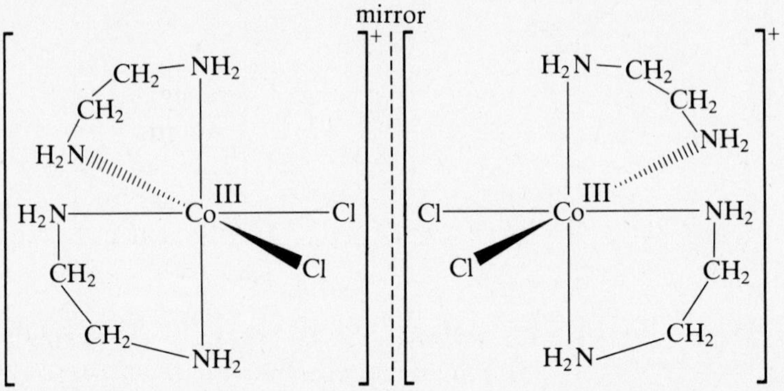

Opponents of Werner's ideas refused to accept the conclusions he drew from this experiment, preferring to believe that the optical activity was in some way a function of the organic part of the compound and not related to the geometry of the coordination sphere of Co(III). One of Werner's crowning achievements was the refutation of this opposition by the successful synthesis and resolution of the entirely inorganic complex shown below:

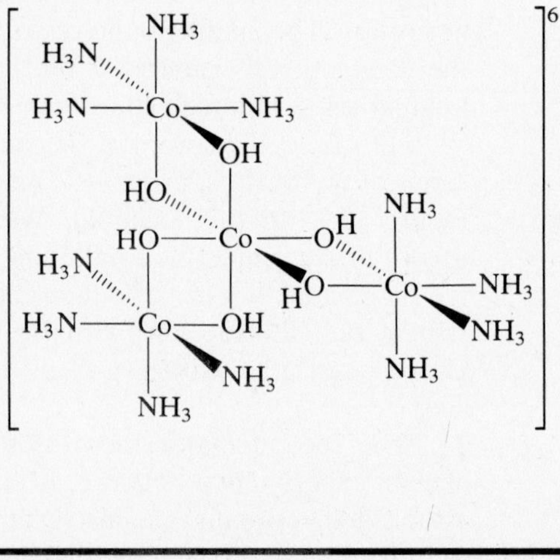

12.2 Types of Ligands

We will adopt the convention that when a ligand is removed from the complex it takes both electrons of the bond with it. Under this convention, there are those ligands that acquire a negative charge on dissociation and those that remain neutral. For example, if we consider one of the classical ammine platinum complexes studied by Werner, we can imagine the removal of either an NH_3 or a Cl^- ligand:

$$Pt(NH_3)_2Cl_4 \longrightarrow \begin{cases} [Pt(NH_3)_2Cl_3]^+ + Cl^- \\ \\ Pt(NH_3)Cl_4 + NH_3 \end{cases} \qquad \textbf{(12.1)}$$

For obvious reasons, those ligands that acquire a negative charge by dissociation under this convention are classified as *anionic ligands*. It should be noted that in reality the bonds involved may be very covalent. This is particularly true of the transition metal to hydrogen bond, which is commonly highly covalent, but under our convention we will always treat it as hydridic unless we are specifically concerned with the polarity of the bond.

Anionic ligands include all the simple halide and halogenoid anions encountered in elementary inorganic chemistry. Singly charged perfluoro-anions or peroxoanions derived from highly polarizing main-group ions tend to be very poor ligands, for example, $[PF_6]^-$, $[BF_4]^-$, and $[ClO_4]^-$. More highly charged ions such as $[SO_4]^{2-}$ and $[PO_4]^{3-}$ are more strongly coordinating.

Essentially, all products of deprotonation of main-group hydrides and their organic derivatives can function as anionic ligands. Some representative anionic ligands are the following:

Cl^- (chloride)	NH^{2-} (imide)
$[CNO]^-$ (cyanate)	R^- (alkide)
OH^- (hydroxide)	RO^- (alkoxide)
CN^- (cyanide)	$[RCOO]^-$ (carboxylate)
$[NO_2]^-$ (nitrite)	RS^- (alkyl sulfide)
$[NH_2]^-$ (amide)	$[R_2P]^-$ (dialkylphosphide)
O^{2-} (oxide)	H^- (hydride)
$[CO_3]^{2-}$ (carbonate)	

Neutral ligands include in principle all those molecules that formally possess one or more lone pairs of electrons. These molecules are found among the simple neutral compounds of Groups VA–VIIA (15–17). In situations where the electronegativity of the ligating atom is very high—for example, hydrogen halides or alkyl halides—or the effective electronegativity of the ligating atom is enhanced by bonding to very electronegative atoms—for example, NF_3—the molecule may be a very poor ligand even though it possesses lone pairs. On the other hand, there are some notable cases where neutral molecules that do not possess formal lone pairs—for example, H_2—are nevertheless coordinated (see Chapter 23). Some commonly encountered neutral ligands are the following:

H_2O	NO	RNH_2
NH_3	R_2S	RCN
CO	$R_2S=O$	RNC

12.3 The Chelate Effect and Chelating Ligands

All the ligands described above contain only one atom through which the ligand binds to the central atom. There is, however, nothing to stop a ligand having a number of different atoms, all of which can bind to the central atom. These atoms may be neutral, anionic, or a mixture of both types. Those ligands endowed with a single ligating atom are called *uni-* or *monodentate* (Latin: "single-toothed"), whereas those with more than one ligating atom are called bi-, tri-, tetra-, and so on or polydentate. The process of binding by more than one ligating atom is called *chelation* (*chelos*: Greek for claw). It is found that ligands with certain separations between ligating atoms bind unusually strongly to metal ions. It is usually the case that the formation of five- or six-membered rings leads to particularly high stability.

The affinity of a metal ion for a ligand is usually measured in terms of a formation constant, which is the equilibrium constant for an equilibrium of the kind

$$ML_n + mL' \xrightleftharpoons{K} ML_{n-m}L'_m + mL \tag{12.2}$$

or

$$ML_n + mL' \xrightleftharpoons{K} ML_nL'_m \tag{12.3}$$

The major part of the data in the literature applies to systems where L is water, but of course the principle is perfectly general. Table 12.1 lists the formation constants for a number of ligands in such a way that a comparison can be made between the ligating power of monodentate and multidentate ligands carrying the same functional groups.

The chelate effect is the result of both enthalpic and entropic contributions. The enthalpic effect is because the electrostatic repulsion, resulting from placing two negatively charged ligand atoms in close proximity, has already been largely overcome if the ligand atoms are close together in the same molecule. The entropy effect arises because, when one end of the ligand is already attached to the metal center, there is no further sacrifice of translational entropy accompanying the attachment of the subsequent

Table 12.1 *Formation Constants for Some Mono- and Bifunctional Complexes from Aqueous Ions*

Metal Ion (with $6H_2O$ Ligands)	Complexing Ligand	log (Formation Constant)[a]
Fe^{2+}	Acetate	2.1
Co^{2+}	Acetate	1.9
Cu^{2+}	Acetate	3.3
Fe^{2+}	$^-OOCCH_2COO^-$	2.8
Co^{2+}	$^-OOCCH_2COO^-$	3.7
Cu^{2+}	$^-OOCCH_2COO^-$	5.5
Co^{2+}	$^-OOCCOO^-$	4.7
Cu^{2+}	$^-OOCCOO^-$	6.3
Fe^{2+}	Pyridine	0.7
Co^{2+}	Pyridine	1.5
Cu^{2+}	Pyridine	4.3
Fe^{2+}	2,2'-Bipyridyl	4.4
Co^{2+}	2,2'-Bipyridyl	5.7
Cu^{2+}	2,2'-Bipyridyl	6.3

[a] The formation constants are for substitution of two water molecules by two mono- or one bidentate ligand.

Table 12.2 *Some Chelating Ligands and Their Applications*

Name	Structure	Application
EDTA	$\begin{array}{c} HOOCCH_2 \qquad\qquad CH_2COOH \\ {>}NCH_2CH_2N{<} \\ HOOCCH_2 \qquad\qquad CH_2COOH \end{array}$	Sequestering metal ions, analytical determinations, detoxification
Chiraphos	$Ph_2PCH(CH_3)CH(CH_3)PPh_2$	Chiral catalytic synthesis
Pennicillamine	$HSC(CH_3)_2CH(NH_2)COOH$	Detoxification, Wilson's disease therapy
Citric acid	$\begin{array}{c} CH_2COOH \\ | \\ HOCCOOH \\ | \\ CH_2COOH \end{array}$	Sequestering trace metals in aqueous systems
Acetylacetone	$CH_3C(O)CH_2C(O)CH_3$	Metal complexes soluble in nonpolar organic solvents

ligating atoms because the number of particles in the system does not change. The sacrifice of configurational entropy is small for the formation of rings with six or less members but increases rapidly with increasing ring size with a concomitant disappearance of the chelate effect. Table 12.2 lists some important chelating ligands and their uses.

A special feature of chelating ligands, referred to above in our discussion of Werner's coordination theory, is their ability to confer chirality on metal centers. Although the same result can be achieved with an octahedral complex by attaching six different substituents, this is a formidable synthetic challenge. With chelating ligands, it is possible to achieve chirality with a single ligand type by forming a complex with three bidentate ligands. If the bidentate ligand is symmetrical, the complex has the D_3 symmetry of a three-bladed propeller.

12.4 Naming Coordination Compounds

The generally accepted rules for naming chemical compounds are formulated by a committee of the International Union of Pure and Applied Chemistry (IUPAC). The rules outlined below are taken from the most recent version of the IUPAC rules. Unfortunately, these like all rules get changed from time to time, and you may note in your reading of inorganic chemistry texts that there have been a number of fairly significant changes in the nomenclature rules over the past twenty-five years.

Ligands

The names of neutral ligands are usually unchanged in naming a coordination compound containing them, with the following important exceptions: H_2O becomes *aquo*; NH_3 becomes *ammine*; CO becomes *carbonyl*; NO becomes *nitrosyl*.

The names of coordinated anions always end in -o. Free anions, which end in -ate, or -ite, change to -ato or -ito. For those that end in -ide—for example, all of the halides and the anions derived from O_2, O, and CN^-—in some cases replace -ide with -o and in others -e with -o. For example:

Free Anion	Coordinated Anion
Chloride Cl^-	Chloro
Hydroxide OH^-	Hydroxo
Superoxide O_2^-	Superoxo
Cyanide CN^-	Cyano

Free Anion	Coordinated Anion
Cyanate $[CNO]^-$	Cyanato
Carbonate $[CO_3]^{2-}$	Carbonato
Methoxide $[CH_3O]^-$	Methoxo
Nitrate $[NO_3]^-$	Nitrato
Nitrite $[NO_2]^-$	Nitrito, or nitro
Nitride N^{3-}	Nitrido
Azide N_3^-	Azido
Amide $[NH_2]^-$	Amido

The hydride ligand is most commonly referred to as hydrido, but the use of hydro is not uncommon.

Complexes

Ligands are named first in alphabetical order, followed by the name of the central atom. This is the reverse order of that used in writing the formula of a complex—for example, $PtCl(N_3)(H_2O)(NH_3)$, ammineaquoazidochloroplatinum(II). Note that the oxidation number of the central atom is included at the end of the name as a parenthetical Roman numeral.

There is no ending modification in neutral or cationic complexes, but in anionic complexes the name of the metal is modified to an -ate ending, for example, molybdate, ferrate, platinate, cuprate. As with simple ionic compounds, ionic complexes are named with the cation first, followed by the anion, as in the following examples:

Sodium hexacyanoferrate(III)	$Na_3[Fe(CN)_6]$
Lithium hexachloroplatinate(IV)	$Li_2[PtCl_6]$
Hexamminecobalt(III) tartrate	$[Co(NH_3)_6]_2[C_4H_6O_6]_3$

Multipliers

As a general rule, a complex contains several of the same ligand attached to the central atom. The number of a given ligand is indicated by the appropriate Greek prefix di-, tri-, tetra-, penta-, hexa-, hepta-, and so on if the ligand is monatomic or one of the neutral ligands with a special name noted above. Polyatomic ligands are parenthetical, and their number is indicated by a prefix outside the parentheses. The prefixes are bis (2), tris (3), tetrakis (4), pentakis (5), and so on, for example:

Potassium tris(oxalato)chromate(III)	$K_3[Cr(C_2O_4)_3]$
Bis(acetato)diamminepalladium(II)	$Pd(NH_3)_2(C_2H_3O_2)_2$
Lithium tetrakis(methylisocyanide)cobaltate(-I)	$Li[Co(NCCH_3)_4]$

Locators

Certain structural features of molecules are indicated by prefixes attached to the name. These prefixes are *italicized* and separated from the compound or ligand name by a hyphen. In certain geometries—for example, square or octahedral but not tetrahedral—there may be inequivalent locations for a ligand to occupy to give geometrical isomers. The only important case of this kind is where a ligand may occupy an adjacent site (*cis*), or a diagonally opposite site (*trans*), with respect to a particular ligand, for example:

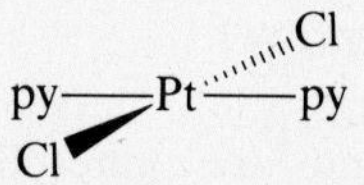

trans-dichlorobis(pyridine)platinum(II)

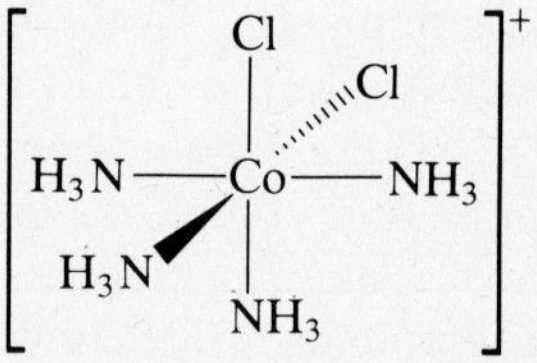

cis-tetraamminedichlorocobalt(III)

Another common form of geometrical isomerism in octahedral compounds is found when there are three ligands the same and the remaining three are different. The three identical ligands may occupy the corners of a common triangular face of the octahedron (*faci*al) or they may occupy three consecutive corners of a square plane of the octahedron (*meridio*nal), for example:

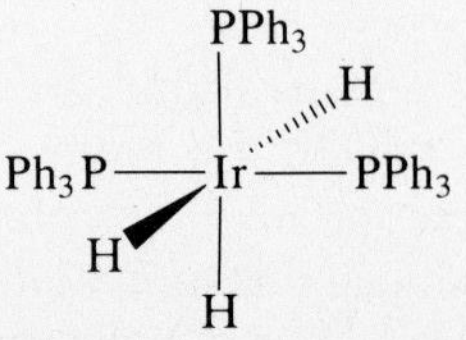

mer-trihydridotris(triphenylphosphine)iridium(III)

For more complicated molecules, it is necessary to use a numbering system to unequivocally name a geometrical isomer. The numbering system used for the octahedron is shown below:

Note that in this system, the mer isomer shown above is named 1,2,6-trihydrido-3,4,5-tris(triphenylphosphine)iridium(III).

Bridging Ligands

Ligands with more than one lone pair of electrons frequently bond to more than one metal center at the same time. In this situation the ligand is said to be *bridging*, and it becomes necessary to indicate both the fact that it is bridging and the number of metal atoms to which it is bridging. This is achieved by preceding the name of the ligand with the Greek letter *mu* (μ) to indicate that the ligand is bridging together with a superscript n to indicate the number of metal atoms to which the ligand is attached (when $n = 2$, it is usually omitted). Figure 12.2 shows a number of examples.

Figure 12.2
Examples of complexes containing bridging ligands.

$[(NC)_5Co-N\equiv C-Fe(CN)_5]^{6-}$

Pentacyanocobalt(II)-μ-cyanopentacyanoferrate(III)

$[(NH_3)_5Co-O-O-Co(NH_3)_5]^{4+}$

μ-peroxobis[pentaamminecobalt(III)]

Bis[dicarbonyl-μ-chlororhodium(I)]

Bis[η^5-cyclopentadienylcarbonyl(μ-carbonyl)iron]

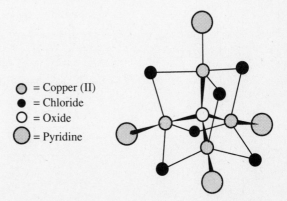

○ = Copper (II)
● = Chloride
○ = Oxide
◉ = Pyridine

Hexa-μ-chloro-μ^4-oxotetrakis[pyridinecopper(II)]

Unsaturated Ligands Bonded Through One or Several Atoms

It is often the case that an unsaturated ligand can bond in several different modes, involving the attachment of different numbers of ligand atoms to the central atom. In this case, the name of the ligand is preceded by the Greek letter *eta* with a superscript n to indicate the number of contiguous ligating atoms. An alternative method is to precede the *eta* with a numerical designation of the string of ligating atoms. The examples in Figure 12.3 illustrate this nomenclature.

Figure 12.3
Examples of the use of the *eta* nomenclature.

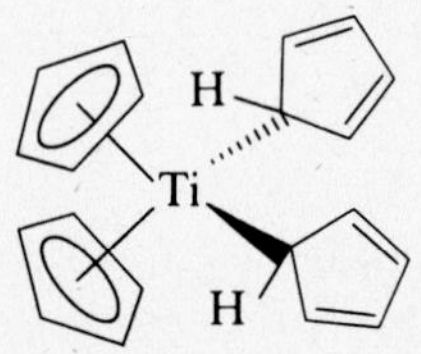

Bis(η^5-cyclopentadienyl)bis(η^1-cyclopentadienyl)titanium(IV)

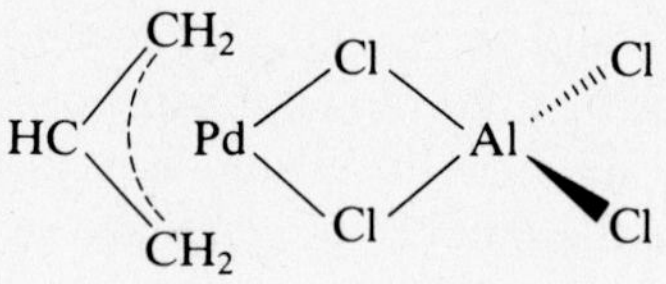

η^3-allylpalladium(II)di-μ-chlorodichloroaluminum(III)

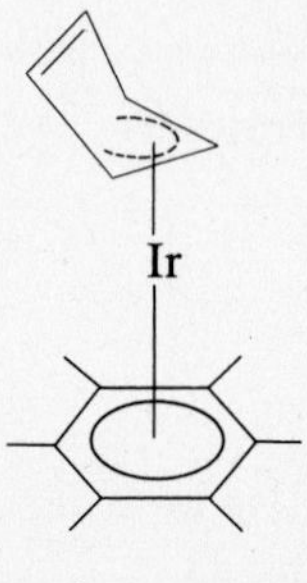

η^3-cyclopentadienyl(η^6-hexamethylbenzene)iridium(I)

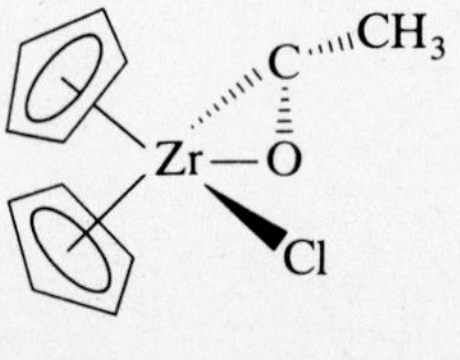

η^2-acetylchlorobis(η^5-cyclopentadienyl)zirconium(IV)

Clusters

Cluster molecules are three-dimensional assemblies of metal atoms in which the metals are attached directly to each other, without intervening bridging ligands. The naming of such compounds, especially when they contain more than one kind of metal or have irregular geometries, can be very complicated. In the case of homometallic, regular clusters, use can be made of the geometry of the metal assembly to facilitate naming. This is achieved by preceding the name of the metal by an italicized designator derived from the name of the geometric form, for example, *triangulo*, *quadro*, *tetrahedro*, *hexahedro*, *octahedro*, and so on. The examples in Figure 12.4 illustrate this method.

Figure 12.4
Names of some polymetallic clusters.

Triangulo-dodecacarbonyltriruthenium(0)

Tetrahedro-nonacarbonyl-μ-tricarbonyltetracobalt(0)

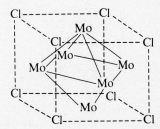

Octahedro-octa-μ³-chlorohexamolybdate

12.5 Geometries of Coordination Compounds

At the time Werner formulated his coordination theory and for the subsequent 100 years, the vast majority of coordination compounds whose structures were known with a reasonable degree of certainty were either tetrahedral, octahedral, or square-planar. When rapid structure determination by X-ray diffraction arrived in the 1960s, the number of known coordination numbers and geometries increased rapidly. The complexity of this area of coordination chemistry was also increased with the development of NMR methods, which allow the study of dynamic transformations between different topological states of a stereochemically nonrigid molecule.

Linear Coordination

This type of coordination is not very common and is largely confined to species with a closed d-shell configuration. Linear coordination can sometimes be forced by using very bulky ligands.

Complexes of Cu(I), Ag(I), Au(I), and Hg(II) commonly have linear structures, for example, $[CuCl_2]^-$, $[Ag(NH_3)_2]^+$, $Au(CH_3)(PEt_3)$, and $Hg(CH_3)_2$.

Sterically demanding phosphines such as tricyclohexylphosphine (Cy_3P) allow the isolation of two-coordinate complexes of the zero-valent nickel-group elements, for example, $Pt(PCy_3)_2$. Even the analogous dicarbonyls can be observed spectroscopically at very low temperatures by cocondensing CO with metal atoms in an inert gas matrix.

Trigonal Coordination

Trigonal coordination occurs with the same species as mentioned in the preceding paragraph but also has a much wider occurrence. Figure 12.5 shows some examples of main-group compounds with three coordination.

Some very bulky ligands that have been used to achieve low coordination numbers are $[N(SiMe_3)_2]^-$, $[CH_2C(SiMe_3)]^-$, and 2,6-ditert-butylphenoxide.

Four Coordination

This coordination number is very common and is encountered in three ideal geometries: tetrahedral, tetragonal (square) planar, and the product of removal of one vertex from a trigonal-bipyramid (sawhorse).

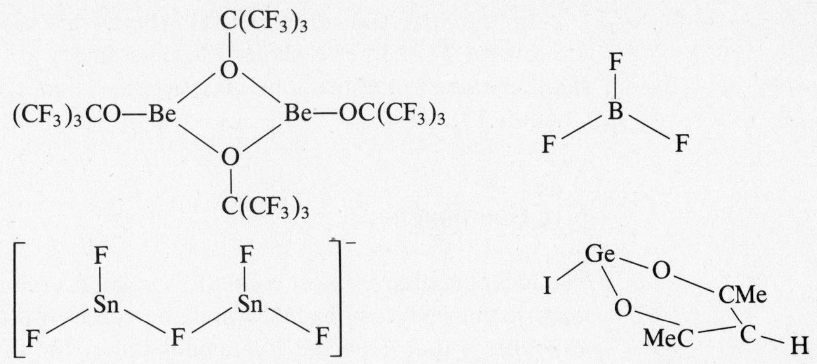

Figure 12.5
Some main-group
three-coordinate
compounds.

Tetrahedral coordination is the most common form encountered in the compounds of the first members of the main groups. In the transition groups, it is encountered frequently in d^0 compounds, particularly in the oxo-species—for example, $[CrO_4]^{2-}$, $[MnO_4]^-$, and OsO_4—and in d^{10} species—for example, $Ni(CO)_4$, $[Co(CO)_4]^-$, and $Pt(PEt_3)_4$.

Ionic radius plays an important role in determining whether a complex adopts tetrahedral or some higher coordination. However, the problem is more complicated than in the case of closed-shell, main-group ionic compounds because the crystal field stabilization energy, which is dependent on ionic radius among a number of other factors (see Chapter 13), can also strongly influence the choice of coordination geometry. Thus, apart from the d^0 and d^{10} cases mentioned, tetrahedral complexes are restricted almost entirely to the first transition period. They are more common for higher oxidation states and for elements in the second half of the period where the effect of d-orbital contraction is stronger.

The halides MX_4 ($M = Ti$; $X = Cl$, Br, and I. $M = V$; $X = Cl$, and Br) are tetrahedral in the liquid and gaseous states, but most other Ti and V compounds exhibit higher coordination numbers. Many complexes of Mn(II), Fe(II), Co(II), and Ni(II) exhibit tetrahedral geometry. This form of coordination is virtually unknown in Cr(III) and Co(III) compounds because of the strong favoring of octahedral geometry by the crystal field stabilization energy for d^3 and d^6 ions (see Chapter 13).

Square-planar geometry is most commonly encountered in complexes of ions with d^8 configuration, for example, M(III), M = Cu, Ag, and Au; M(II), M = Ni, Pd, and Pt; M(I), M = Co, Rh, and Ir. The appearance of square-planar geometry is a clear indication that simple coulombic and steric effects, which favor tetrahedral geometry, are being overridden by covalent forces. The following are typical square-planar complexes: $[AuCl_4]^-$, $[PdCl_4]^{2-}$, and $RhCl(CO)(PPh_3)_2$.

In the first transition period where various factors that steer the ion toward T_d or square-planar (D_{4h}) geometry are more or less balanced, small chemical modifications may tip the balance in either direction (see Chapter 13).

Five Coordination

The ideal geometries most commonly encountered in five coordination are the trigonal-bipyramid (TBP) and the square-pyramid (SQP). The TBP structure is that expected for ligands under the influence of purely electrostatic forces, but the energy is only marginally lower than that of the SQP. Besides the relatively small energy difference between the two extreme structures, the activation energy for the transformation of one into the other is also quite small. This gives rise to two important consequences: first, stereochemical nonrigidity; second, structural variability.

Recall from Chapter 7 that a good example of stereochemical nonrigidity is the $Fe(CO)_5$ molecule. In the solid state, a crystal-structure determination shows $Fe(CO)_5$ to be a TBP molecule. In solution, two ^{13}C-NMR resonances are expected for such a geometry, corresponding to the axial and equatorial CO groups. The fact that only a single resonance is observed down to the lowest achievable temperatures is attributed to a rapid interconversion between two TBP structures, probably through a SQP intermediate (Figure 12.6), a process that leads to interchange of the axial and equatorial positions. The PF_5 molecule behaves in much the same way and only exhibits one ^{19}F-NMR resonance even though it is known to be TBP. The mechanism for group interchange shown in Figure 12.6 is called *pseudorotation*.

Structural irregularity is the rule rather than the exception in five-coordinate complexes. Only a few compounds in which all of the ligands are the same exhibit ideal geometry (all vertices and angles of the same

Figure 12.6
Interconversion of axial and equatorial positions in $Fe(CO)_5$.

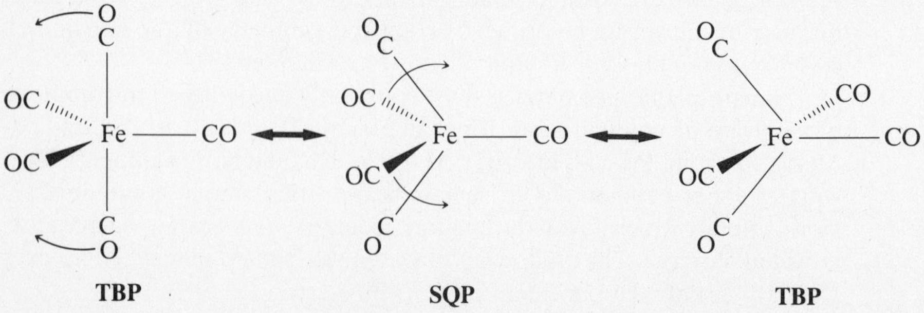

TBP   SQP   TBP

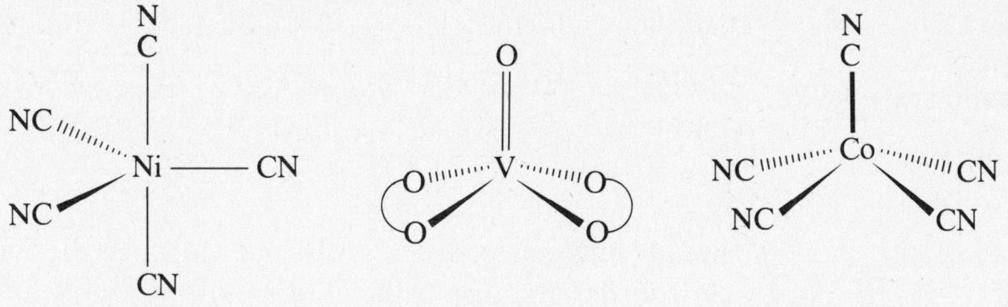

Figure 12.7
Some examples of
five-coordinate
complexes.

$[Ni(CN)_5]^{3-}$ $VO(acac)_2$ $[Co(CN)_5]^{3-}$

kind are equal). When more than one ligand is present, the coordination
geometry is always irregular. Figure 12.7 shows some examples of five-
coordinate structures.

Six Coordination

Six coordination with octahedral geometry is the most widely distributed
and encountered of all coordination numbers. It is found among the
compounds of all elements except H, B, C, N, O, the lighter halogens, and
the lighter noble gases. The octahedron is the most stable configuration for
six mutually repelling charges of the same sign. It is also the geometry that
permits the maximum occupancy of bonding or nonbonding orbitals for a
number of important electronic configurations (see Chapter 3).

A second possible geometry for six coordination, the trigonal prism, is
only rarely encountered and usually with larger, polarizable ligand atoms
such as S, Se, or As. The earliest examples of such compounds were the
dithiolene complexes of Mo and W and some of the neighboring elements.
Figure 12.8 shows the structure of such a compound.

Other examples of trigonal prismatic coordination, which we have

Figure 12.8
Structure of
$[Re(S_2C_2Ph_2)_3]$.

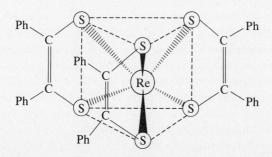

Figure 12.9
Examples of octahedral complexes.

Cationic	$[Cr(NH_3)_6]^{3+}$	$[Ru(dipy)_3]^{2+}$	$[Mn(CO)_6]^+$
Neutral	$IrCl_3(PMe_3)_3$	$[NbCl_4(py)]_2$	$Cr(CO)_3(NH_3)_3$
Anionic	$[Co(NO_2)_6]^{3-}$	$[Fe(CN)_6]^{4-}$	$[V(CO)_6]^-$

already encountered, are NiAs and MoS_2. In the former, the nickel atoms are coordinated to a trigonal prism of As atoms; in the latter, each Mo is surrounded by a trigonal prism of S atoms with the edges connecting the trigonal faces formed by the S—S bonds of S_2^{2-} ions (see Chapter 4).

The octahedron is a trigonal antiprism and is transformed into a trigonal prism by the rotation of two opposite trigonal faces by 60° relative to each other (see Figure 4.6). If the bond lengths are kept constant, the interligand distance diminishes in the course of this operation and the coulombic repulsion between ligands increases. For this reason, the trigonal prismatic structure is only encountered in situations where covalent bonding is important.

Almost every conceivable type of coordination compound is exemplified by octahedral complexes. Figure 12.9 lists some examples that illustrate some of the possible categories.

Seven Coordination

Coordination numbers greater than six are commonly encountered with large metal atoms (heavier members of the p and d groups, lanthanides, and actinides), with small ligands (H, F, or CN) and with cyclic polyene ligands. In the case of coordination number seven, three geometries are relatively common. The highest symmetry (D_{5h}) structure is the pentagonal-bipyramid, encountered in the alkali metal salts of $[ZrF_7]^{3-}$ and $[HfF_7]^{3-}$.

One form of ZrO_2, found in the mineral baddeleyite, has the zirconium coordinated to a distorted octahedron of O, with a seventh O capping a trigonal face of the octahedron. The alkali metal–ion salts of the fluorocomplexes $[NbF_7]^{2-}$ and $[TaF_7]^{2-}$ have a similar trigonally capped octahedral structure in the solid state (Figure 12.10). In solution all of these fluoro-complexes show only a single fluorine resonance in their ^{19}F-NMR spectra, indicating stereochemical nonrigidity (see fluxionality of four- and five-coordinate complexes described above). Such fluxionality is due to the relatively small energy differences between the different geometries and is a common feature of many compounds with high coordination numbers.

A third geometry encountered with seven coordination is the tetragonally capped trigonal prism such as found in the ion $[NbOF_6]^{3-}$ (Figure 12.10).

Figure 12.10
Structures of some
seven-coordinate
complexes.

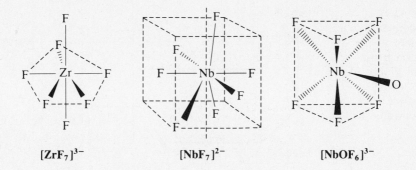

$[ZrF_7]^{3-}$ $[NbF_7]^{2-}$ $[NbOF_6]^{3-}$

A number of polyhydride complexes of the heavier transition metals adopt seven-coordinate structures. Figure 12.11 shows a few examples.

High coordination numbers are often encountered in complexes where the metal is coordinated to a conjugated polyene, polyenyl, or polyenonium ligand. In such complexes, it is customary to assign the number of coordination sites occupied by the ligand as being equal to the number of π-electron pairs in the π-bonding orbitals of the ligand. Examples of these kinds of ligands are cyclopentadienyl and hexamethylbenzene, as illustrated in Figure 12.3. Each of these ligands is considered to occupy three-coordination sites. Thus, the complex $CpWH(CO)_3$ is seven-coordinate.

Eight Coordination

The commonest geometries for eight coordination are the cubic antiprism and the dodecahedron. On the same grounds that the octahedron is more stable than is the trigonal prism, the cubic antiprism is more stable than is the cube. Rare examples of nearly ideal cubic coordination are found in the fluoroactinide salts $Na_3[An\ F_8]$ (where An = Pa, U, and Np). The cubic antiprism is derived from the cube by rotation of two opposite faces with

Figure 12.11
Examples of
seven-coordinate
hydrides.

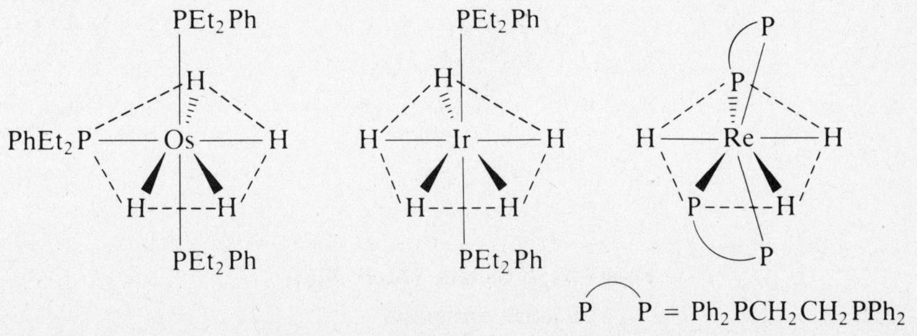

$$PEt_2Ph \qquad PEt_2Ph$$

$$PhEt_2P \longrightarrow Os \longrightarrow H \qquad H \longrightarrow Ir \longrightarrow H \qquad H \longrightarrow Re \longrightarrow H$$

$$PEt_2Ph \qquad PEt_2Ph$$

$$P \qquad P = Ph_2PCH_2CH_2PPh_2$$

Figure 12.12
Transformation of a cube into a dodecahedron. (NOTE: Vertices *A* and *B* are not equivalent in the dodecahedron.)

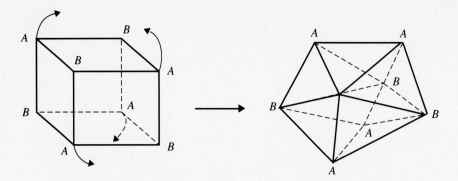

respect to each other by 45° (see Figure 4.6). The cube is transformed into a dodecahedron by the operations shown in Figure 12.12.

The difference in energy between the square antiprism and the dodecahedron is small, and it is not possible to predict which of these two structures will be preferred. Figure 12.13 illustrates some structures that are adopted in the solid state.

A number of cyclopentadienyl complexes of the earlier transition groups exhibit coordination eight according to the rules described for seven coordination. Many bis(cyclopentadienyl)metal(IV) compounds where the metal is a member of the group of elements Ti, Zr, Hf, V, Nb, Ta, Mo, and W have this kind of coordination. Figure 12.14 shows several examples.

Nine Coordination and Beyond

Nine coordination is the ultimate for *d*-block elements using covalent bonding because it represents the full use of all valence orbitals:

Figure 12.13
Examples of eight-coordinate complexes.

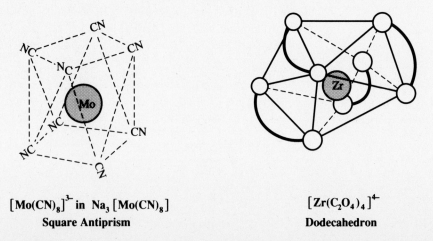

$[Mo(CN)_8]^{3-}$ in $Na_3[Mo(CN)_8]$
Square Antiprism

$[Zr(C_2O_4)_4]^{4-}$
Dodecahedron

Figure 12.14
Examples of high coordination number metallocene complexes of early transition period elements.

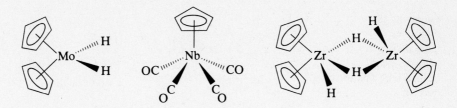

$ns + [3 \times np] + [5 \times (n - 1)d]$. Nonorthogonal overlap of these metal orbitals with nine ligand orbitals gives nine bonding and nine antibonding MOs. A beautifully symmetrical example of nine coordination is provided by the nonahydridorhenate ion. The geometry of this ion has been shown by neutron diffraction to be a trigonal prism with each of the tetragonal faces capped with additional hydride ligands:

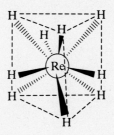

From solution ^{1}H-NMR measurements, the $[ReH_9]^{2-}$ ion is known to be stereochemically nonrigid in solution.

Nine coordination is quite rare among d-block elements but widely encountered among the complexes of the lanthanides and actinides. All lanthanides form nine-coordinate aquo-ions $[Ln(H_2O)_9]^{3+}$ with the same geometry as $[ReH_9]^{2-}$. Theoretical studies show that the f orbitals of the actinides and lanthanides do not participate in covalent bonding but that the lowest unoccupied d orbitals do. Thus, in situations where covalency is important, one would expect that nine would be the limiting coordination number. However, many of the complexes involving small, nonpolarizable main-group ligand atoms have high ionic character and exhibit coordination numbers greater than nine. Figure 12.15 gives some examples.

Figure 12.15
Examples of
lanthanide and
actinide complexes
of high coordination
number.

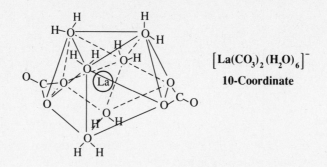

$\left[La(CO_3)_2 (H_2O)_6 \right]^-$

10-Coordinate

$Th(NO_3)_4 \cdot 3H_2O$

11-Coordinate

$\left[Ce(NO_3)_6 \right]^{2-}$

12-Coordinate

12.6   *Isomerism in Coordination Compounds*

In previous sections, we have already referred to geometrical isomerism in coordination compounds. A number of other kinds of isomerism are encountered, and some are briefly described next. In the characterization of new coordination compounds, the possible existence of isomers should always be remembered.

Ionization Isomerism

Often in ionic complexes, the anion can act as a ligand. If the cation has anionic ligands that are different from the counterion, the possibility exists that they may interchange and give rise to two chemically different species. Some examples are the following:

$[Pt(en)_2Cl_2]Br_2$ and $[Pt(en)_2Br_2]Cl_2$

$[Co(NH_3)_4(SCN)Br]Cl$, $[Co(NH_3)_4(SCN)Cl]Br$, and $[Co(NH_3)_4BrCl]SCN$

Coordination Isomerism

When the anion and cation of an ionic compound are complex, ligands may be exchanged between the two to produce chemically different species, thus:

$$[Co(NH_3)_6][Cr(CN)_6] \text{ and } [Cr(NH_3)_6][Co(CN)_6]$$

$$[Ni(phen)_3][Co(SCN)_4] \text{ and } [Co(phen)_3][Ni(SCN)_4]$$

Linkage Isomerism

Some polyatomic ligands possess a number of chemically different coordination sites and may yield isomers by the selective use of one or the other. Such ligands are also called *ambidentate ligands*. The most commonly studied compounds of this type are complexes of $[NO_2]^-$, $[CNO]^-$, $[CNS]^-$ and to a lesser extent CN^-. Figure 12.16 shows some examples of this type of isomerism.

Linkage isomerism is important to take into account when studying the binding of metal complexes to biopolymers such as proteins and nucleic acids. For example, because certain platinum(II) complexes have found widespread use as chemotherapeutic agents against some forms of cancer and it is believed that the key to this activity is the selective binding of the metal to the bases of the nucleotides of the malignant cells, much work has been done to elucidate the nature of the binding of metal ions to nucleotides. Figure 12.17 shows some examples of compounds that have been prepared and isolated.

Polymerization Isomerism

This term is applied to compounds that have the same empirical formula but different multiples of a given molecular weight, for example, $Pt(NH_3)_2Cl_2$, $[Pt(NH_3)_4][PtCl_4]$, and $[Pt(NH_3)_3Cl]_2[PtCl_4]$.

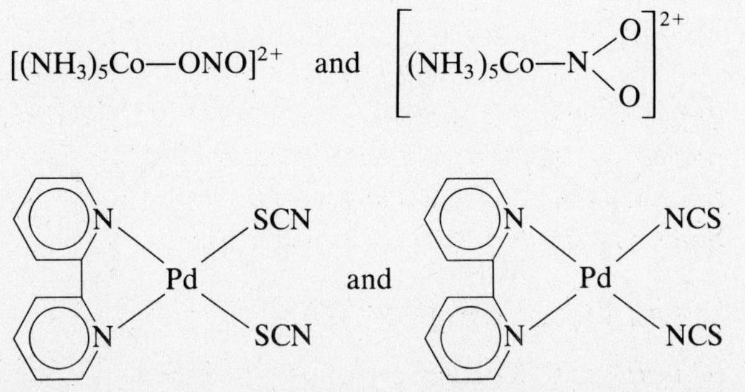

Figure 12.16
Examples of linkage isomerism.

$$[(NH_3)_5Co-ONO]^{2+} \text{ and } \left[(NH_3)_5Co-N\begin{array}{c}O\\O\end{array}\right]^{2+}$$

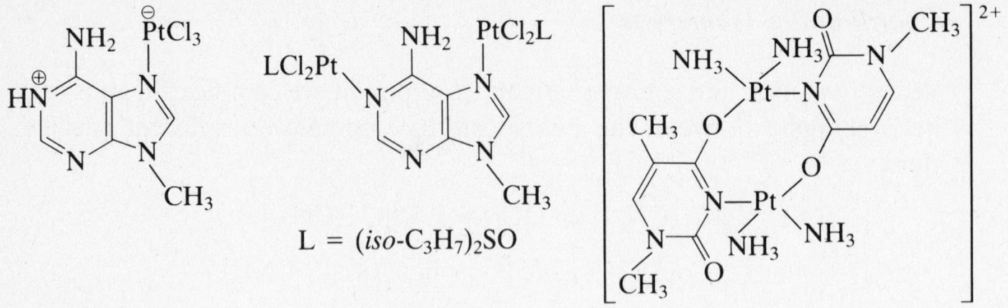

Figure 12.17
Some modes of coordination of Pt(II) to nucleotide base derivatives.

Conformational Isomerism

These isomers only differ by rotations about metal–ligand bonds. Usually, the energy differences and the activation barriers to interconversion of such isomers are small, and their separation is difficult or impossible. An example of this kind of isomerism is provided by ferrocene (see Chapter 22). In practice the ferrocene molecule can have two extreme conformations: one with two cyclopentadienyl rings eclipsed and the other with rings staggered. Early interpretations of X-ray diffraction data concluded that the molecule had an inversion center of symmetry and was therefore staggered. This was surprising because the isomorphous ruthenocene was found to be eclipsed. More recent studies on ferrocene using electron diffraction and neutron diffraction have shown that the molecule is eclipsed in its equilibrium state; a reinterpretaion of the X-ray data, in terms of a disordered, almost eclipsed structure, is now in accord with the other results.

The compound dicarbonyl (dichloromethylsilyl) (η^5-cyclopentadienyl) iron(II) gives twice the expected number of ν(co) bands in the infrared. This has been interpreted to signify the existence of the two conformers shown in Figure 12.18. If the interpretation is correct, the interconversion of the isomers is very rapid in solution because they cannot be detected by NMR spectroscopy, which has as a much longer time constant than IR spectroscopy.

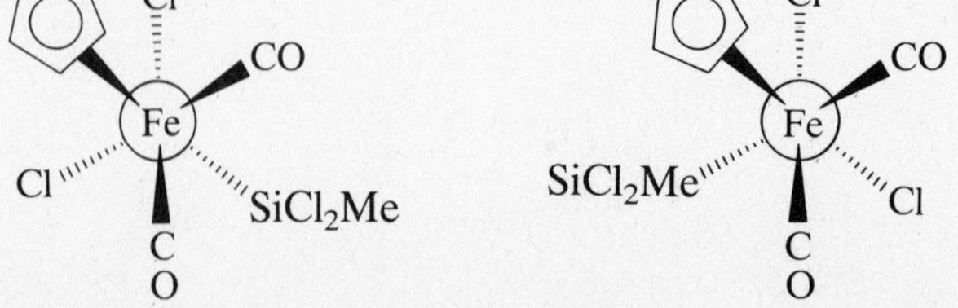

Figure 12.18
Conformational isomerism about the Fe—Si bond in CpFe(CO)$_2$SiCl$_2$Me.

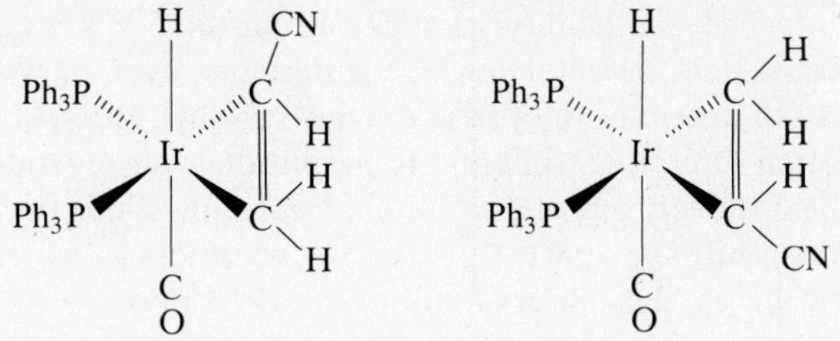

Figure 12.19
Two conformational
isomers of an
iridium alkene
complex.

The rotation of unsymmetrically substituted alkenic ligands can give rise to conformational isomers. The two isomers shown in Figure 12.19 can be detected by NMR spectroscopy (two different Ir—H resonances) at room temperature, but they interconvert too rapidly to be isolated.

Stereoisomerism

Molecules devoid of a plane of symmetry, an improper rotation axis, or a center of symmetry exhibit the property of stereoisomerism. The best known example of this behavior is the existence of enantiomeric forms of carbon compounds in which the carbon carries four different substituents. In transition metal chemistry, classical tetrahedral complexes with four different ligands are relatively rare, and those that exist are too labile to be resolvable. A special class of compounds, which may be considered to be tetrahedral, are the *piano-stool* type, in which the η-^{5}cyclopentadienyl ligand is considered to occupy a single coordination position. Figure 12.20 shows some examples of complexes of this type.

In the case of octahedral geometry, it is the bis- and tris-chelate complexes that have been most widely synthesized and resolved. However, the number of known asymmetric octahedral complexes with only monodentate ligands has increased dramatically during the recent, explosive growth of coordination chemistry.

Figure 12.20
Examples of chiral
piano-stool
complexes.

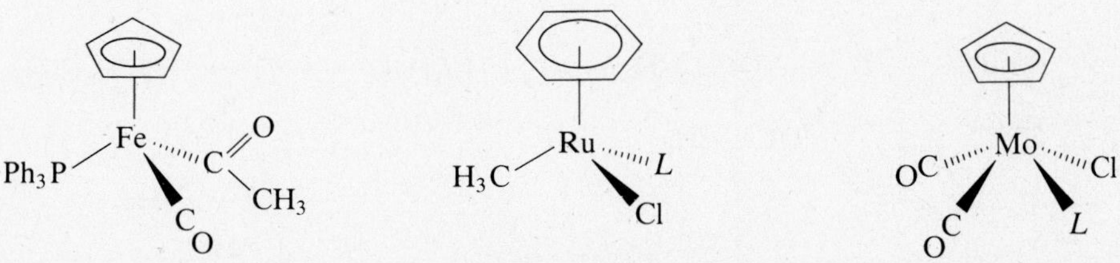

$L = Ph_2PN(H)CH(CH_3)Ph$ $L = Ph_2PN(CH_3)CH(CH_3)Ph$

The availability of high-resolution NMR spectroscopy has greatly facilitated the detection of enantiomers, even in systems where resolution is not possible. In optically active solvents, the spectra of enantiomers are often sufficiently different to permit their observation. The same effect can be achieved with the addition of optically active paramagnetic compounds that shift the spectra of the two enantiomers in different ways. A third probe is the observation of *diastereotopy* in groups attached to the stereocenter. This phenomenon depends on the fact that groups that are related by a plane of symmetry are NMR equivalent. Thus, the protons of a CH_2 group are equivalent in the NMR, provided they can be reflected through a molecular plane of symmetry. If the CH_2 group does not lie on a molecular plane of symmetry, the two hydrogens are no longer equivalent. For example, the compound $Ti(acac)_2(OCH_2CH_3)_2$ can in principle exist as either the cis or trans isomer, as shown in Figure 12.21. In the trans isomer, both CH_2 groups of the ethoxy ligands and the methyl groups of the acac ligands all lie on a molecular plane of symmetry, and their protons are therefore equivalent. In the cis isomer, the molecule has no plane of symmetry, and so neither the two CH_2 protons of a given ethoxy ligand nor the two methyl groups of a given acac ligand are equivalent. The NMR observation of two resonances of equal intensity for the CH_2 group of the ethoxy ligand and for the methyl groups of the acac ligands confirms that they are attached to a stereocenter and that the complex exists almost entirely in the cis form in solution.

Miscellaneous

Polytopal isomerism refers to molecules that differ only in topology or geometrical form. For example, the pentacyanonickelate(II) ion

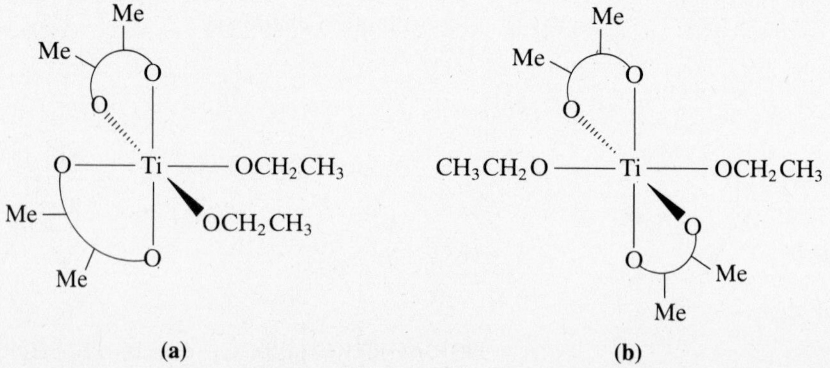

Figure 12.21
Cis (**a**) and trans (**b**) isomers of $Ti(acac)(OCH_2CH_3)_2$.

(a)   (b)

(Figure 12.7) exists in both SQP and TBP forms. Another case is found among the bisphosphine(or phosphite)nickel dihalides. $Ni(PPh_3)_2Cl_2$ has been observed only as a pseudotetrahedral molecule, $Ni(OPPh_3)_2Cl_2$ has been observed only as a square-planar molecule, and the compound $Ni(Ph_2PCH_2CH_2CH_2PPh_2)_2Cl_2$ exists in both tetrahedral and square-planer forms.

Spin-state isomerism is unique to transition metal complexes and is used to describe situations where a compound can exist in two different spin states under the same experimental conditions. (For a discussion of the spin states of transition compounds, see Chapter 14.) This phenomenon is encountered most importantly among the heme proteins, in which Fe(III) has been observed to exist in both high and low spin states in the same coordination environment.

Ligand isomerism arises when the ligand itself can exist in different isomeric forms—for example, in $V(OCH_2CH_2CH_3)_4$ and $V[OCH(CH_3)_2]_4$ or tricarbonyl(η^6-2,4-dimethylbenzene)chromium(0) and tricarbonyl(η^6-2,3-dimethylbenzene)chromium(0).

BIBLIOGRAPHY

Suggested Reading

Gould, R. F., ed. *Advances in Chemistry Series*, No. 62. In Kauffmann, G. B., ed. *Werner Centennial*. Washington, D. C.: American Chemical Society, 1967.

PROBLEMS

12.1 The two key concepts of Werner's coordination theory are coordination number and fixed geometry. Outline the two experimental strategies that Werner used to establish these two tenets.

12.2 Give two examples of coordination complexes to illustrate each of the following terms:
a. A tridentate ligand
b. A trans complex
c. A singly bridging ligand
d. An η^1-cyclopentadienyl complex

12.3   Draw the structures of the following compounds:
     a. Triangulo-tri-μ-chloro-hexachlorotris(pyridine)trirhenium
     b. 1,2-η^3-allyl-3,4,6-triiodo-5-trifluoromethylplatinate(IV)
     c. Bis(η^8-cyclooctatetraene)uranium
     d. (η^5-cyclopentadienyl)dicarbonylhydrido[hydridobis(triphenylphos-phine)platinum]rhenium

12.4   Give two examples each of complexes with the following properties:
     a. Pentagonal-bipyramidal geometry
     b. Chirality
     c. A metal cluster with a threefold symmetry axis
     d. Two chemically different cyclopentadiene ligands

12.5   Give the correct IUPAC names for the following compounds:
     a. $Cs[AuCl_4]$
     d.

     b. $Pt(NH_3)(H_2O)Cl_2$
     e.

     c. $Rh(acac)(C_2H_4)_2$
     f.

12.6   Identify the coordination geometry of the following complexes:
     a. Pentacarbonylosmium(0)
     b. Bis(2,4-pentanedionato)oxovanadium(IV)
     c. Nonohydridorhenate(VII)
     d. Octacyanomolybdate(V)
     e. Tris(tricyclohexylphosphine)platinum(0)
     f. Tetraaminecarbonatocobalt(III) chloride

12.7   Identify the type of isomerism exhibited by each of the following:
     a. $Pt(NH_3)_2Cl_2$
     b. $Rh(acac)_3$
     c. $[Co(NH_3)_5(NO_2)]^{2+}$
     d. $CpFe(CO)_2SiCl_2CH_3$
     e. $CpFe(CO)_2 (C_3H_7)$
     f. $[Ni(CN)_5]^{3-}$

12.8 Give one example of each of the following:
   a. A chiral complex containing no carbon
   b. An eight-coordinate complex with dodecahedral geometry
   c. A hexadentate ligand
   d. A tetrahedral metal cluster compound

12.9 What are the two d-electron configurations that characteristically give tetrahedral coordination with oxide and carbonyl ligands, respectively?

12.10 Why is nine coordination the highest expected for complexes of d-block elements?

13

Crystal Field Theory

13.1   Bonding in Coordination Compounds

The fundamental principles of bonding developed in Chapter 3 apply equally well to all types of chemical compounds. However, just as certain approximations may be more appropriate in some situations than in others in the case of binary compounds, so different bonding models may be appropriate to describe different extremes in coordination compounds. For example, just as an ionic description may serve adequately to discuss the binary compounds of magnesium but be less than adequate for a description of beryllium compounds, so an ionic model applies well to the complex ion$[CoCl_4]^{2-}$ but is useless in describing the $[Co(CO)_4]^-$ ion. As in the case of simple compounds, the ionic view is retained where possible because of its greater computational simplicity. Borderline situations can be accommodated by adjustments analogous to the introduction of the concept of polarization in the ionic model; however, if the covalency becomes too high, a model must be used that explicitly takes account of orbital mixing.

We begin our discussion of bonding models in coordination compounds with a theory that was developed particularly to describe the behavior of transition metal complexes. This model, known as the *crystal field theory* (*CFT*), is essentially an ionic description that takes explicit account of the influence of the coulombic fields of the ligands on the energies of the metal-ion *d* electrons.

13.2 Crystal Field Theory

The compounds of the transition elements are distinct from those of the main groups by virtue of one very obvious property and another more subtle property. The obvious property is color. Colored transition compounds are the rule and colorless ones the exception, whereas the reverse is true of compounds that contain only main-group elements. The more subtle property is the magnetic activity of transition compounds, which is a reflection of their tendency to possess unpaired electrons.

The novelty of these properties attracted the attention of physicists in the 1930s. The eventual outcome of this interest was the formulation of CFT by Bethe and van Vleck, who developed the theory to explain the properties of simple solid-state compounds of the transition elements but did not extend the theory to compounds in solution. The theory was not adopted by chemists until twenty years later to explain the physical properties of coordination compounds. One reason for the difficulty that chemists had in accepting the CFT model was that it used an essentially ionic model and the ideas of Pauling's VB approach had already become so entrenched in their thinking that they found it difficult to abandon a covalent picture of the coordinate bond.

The essential idea of the CFT model is that the coordination sphere of anions or ligands around a metal ion in a crystal or complex behaves as a set of point negative charges and that these charges interact repulsively with the electrons on the central metal cation. The result of this repulsive interaction is that the degeneracy of the orbitals on the metal ion is altered (this is the classical Stark effect). For example, if we place two equal negative charges on the z and $-z$ axes of a set of p orbitals, a test electron in the p_z orbital will feel a greater coulombic repulsion than will a test electron in the p_x and p_y orbitals. All of the orbital set will be raised in energy, but the p_z will be raised more than the other two, as shown in Figure 13.1.

This lifting of the degeneracy is because the p_z orbital does not belong to the same symmetry representation as do the other two in the linear field.

Figure 13.1
Effect of two point
charges on the *x* axis
on the energies of
the *p*-orbital set.

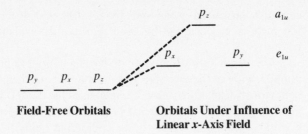

Field-Free Orbitals   **Orbitals Under Influence of**
                                 Linear *x*-Axis Field

The quantitative treatment of this splitting of the energy levels is quite simple if one assumes ionic interaction (that is, no mixing of the metal and ligand orbitals) and if the orbital set is split into only two new energy states. In this case, only a single parameter is required to describe the difference in energy between the two orbital sets.

It may seem strange to be talking about a theory in which the orbital energies are all being raised, but one must remember that we have ignored the large attractive coulombic interaction between the excess positive charge on the cation and the negative charges on the anions. We do this because the properties we wish to explain have little to do with the interionic attraction but are strongly related to the nonspherical part of the interelectronic repulsion described for the case of the *p* orbitals in a linear field. Because only the nonspherical part of the interaction is of interest to us, we will adopt the convention of using energy-level diagrams in which the field-free orbitals are placed at the energy they would have if exposed to the spherically symmetrical part of the field. Under this convention, the diagram in Figure 13.1 becomes that shown in Figure 13.2.

Although our example illustrates adequately the principles of the CFT model, it is of little relevance to our interest in transition complexes. For these cases, it is the *d*-orbital set that is important, and the fields are generally of lower symmetry than $D_{\infty h}$ for a linear molecule. In fact, because the cubic fields are both extremely common and fairly simple to

Figure 13.2
Energy-level
diagram showing the
effect of the
nonspherical
component of a
linear CF on the
p-orbital set.

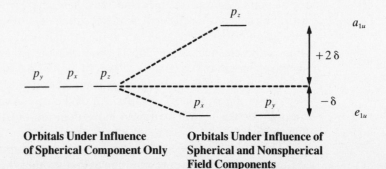

Orbitals Under Influence   **Orbitals Under Influence of**
of Spherical Component Only   **Spherical and Nonspherical**
                                  Field Components

Figure 13.3
Interaction of the
d-orbital set with an
octahedral ligand
set.

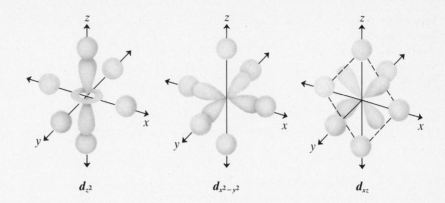

d_{z^2} $d_{x^2-y^2}$ d_{xz}

visualize, we will treat them in some detail. The groups of cubic symmetry are the cube, the tetrahedron, and the octahedron.

We begin our discussion with the octahedron because it is the easiest to visualize. Figure 13.3 shows a set of *d* orbitals circumscribed by an octahedral set of ligands. By inspection it is obvious that the coulombic interactions between the ligands and those orbitals that lie along the cartesian coordinates are much stronger than are the interactions with the orbitals that are directed between the axes. Thus, the orbitals that we label $d_{x^2-y^2}$ and d_{z^2} are more destabilized than those that we label d_{xy}, d_{xz}, and d_{yz}. The higher-energy pair of orbitals belong to a set whose symmetry is labeled t_{2g}. Figure 13.4 shows a schematic energy-level diagram for the interaction of an octahedral set of charges with the *d*-orbital set. The cube is not a common geometry in coordination compounds, but we use it as a starting point because it is easier to visualize the interaction of a cubic set of charges than a tetrahedral set. Once the interactions in cubic symmetry are understood, it is easy to progress to the tetrahedron. Figure 13.5 shows each of the *d*-orbitals separately interacting with a set of ligands at the corners of a cube.

Figure 13.4
Energy-level
diagram for the
effect of an
octahedral field on
the *d*-orbital set.

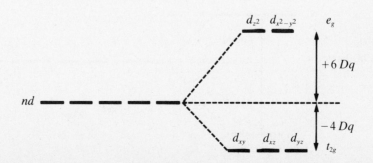

Figure 13.5
Interaction of the
d-orbital set with a
cubic array of
ligands.

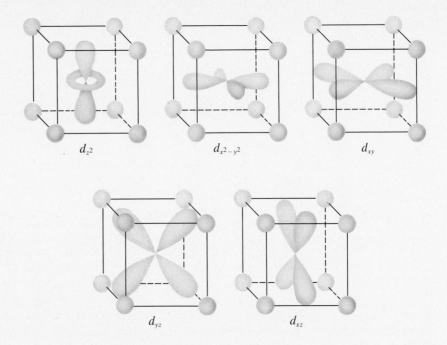

d_{z^2} $d_{x^2-y^2}$ d_{xy}

d_{yz} d_{xz}

From this figure, it is evident that the three orbitals we identify as d_{xy}, d_{xz}, and d_{yz}, which are directed toward the edge centers of the cube, are on the average closer to the ligands than are the $d_{x^2-y^2}$ and d_{z^2} orbitals, which are directed toward the face centers. We therefore conclude that it is the e_g set that is relatively more stable than the t_{2g} set in this case. Figure 13.6a gives a schematic energy-level diagram for the cubic field.

To extend the model to the T_d case, we simply remove every alternate ligand from the cubic example. Because each ligand has an identical interaction with the *d*-orbital set, we reduce the total coulomb field by one-half without modifying the geometry of the interaction or the degeneracy relative to the cubic case. Figure 13.6b illustrates the splitting of the

Figure 13.6
Energy-level
diagram for the
influence of (a) cubic
and (b) tetrahedral
fields on the
d-orbital set.

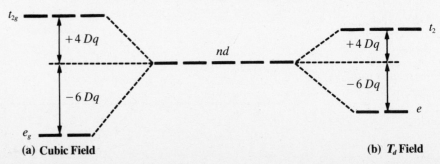

t_{2g} t_2

$+4\,Dq$ *nd* $+4\,Dq$

$-6\,Dq$ $-6\,Dq$

e_g e

(a) Cubic Field   **(b) T_d Field**

d-orbital set by a T_d field. Because the tetrahedron has no center of symmetry, the g subscript is dropped.

The energy difference between the two sets of energy levels in a CF is called *crystal field splitting*, variously labeled $10\,Dq$, or Δ. The quantity Dq, or $\frac{1}{10}\,\Delta$, is called the *crystal field parameter*. The energies of the t and e levels are measured relative to the center of gravity of the orbital set. This center of gravity is $4\,Dq$ above the t_{2g} set and $6\,Dq$ below the e_g set in an octahedral field or $6\,Dq$ above the e set and $4\,Dq$ below the t set in a tetrahedral field. Thus, for a half-filled or filled shell there is no gain or loss of electronic energy on application or removal of the CF. The magnitude of Dq is a function of (a) the identity of the central metal ion, (b) the charge on the central metal ion, (c) the position of the element in its group, and (d) the identity of the ligand. There is no simple rule relating Dq to the particular element. As a rough approximation, the value of Dq is about $1000\ \text{cm}^{-1}$ for $2+$ ions of the $3d$ elements and $2000\ \text{cm}^{-1}$ for $3+$ ions. The value increases by about 50% on descending from the $3d$ to the $4d$ element and another 50% on going from the $4d$ to the $5d$ element.

In applying the CFT model to explaining the electronic configurations of transition metal compounds, there are two extreme situations that must be considered. First, if the magnitude of Dq is less than the interelectronic repulsion energy, the electron occupancy of the d orbitals follows the Aufbau principle in exactly the same way as in the field-free ion (see Chapter 2). Second, if the CF splitting energy exceeds the spin-pairing energy, electrons prefer to pair in the t_{2g} levels before occupying the upper levels.

In practice CF splitting in simple T_d complexes virtually never exceeds the spin-pairing energy, but in O_h complexes it commonly does. Cases where spin pairing occurs in a lower-energy set of crystal field–split orbitals are referred to as *low spin*, and those where the filling is the same as in the field-free ion are called *high spin*. Low-spin complexes are the rule in d^6 cases, due in part to the particularly large gain in energy on going from high to low spin in this case, as shown in Figure 13.7. We emphasize that

Figure 13.7
Change in CFSE with spin pairing for the d^6 case in an O_h field.

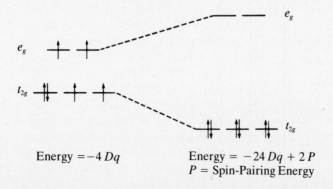

Energy $= -4\,Dq$

Energy $= -24\,Dq + 2\,P$
$P = $ Spin-Pairing Energy

the energy change is only negative if the value of 10 Dq is larger than the spin-pairing energy P.

The low-spin state is also the rule for ions of the second and third transition periods, reflecting the higher Dq values for those periods mentioned above. Orbital occupancies for the d^n configurations for the high- and low-spin states in an O_h field are the following:

	High-Spin	Low-Spin
d^1	$t_{2g}^1 e_g^0$	Same
d^2	$t_{2g}^2 e_g^0$	Same
d^3	$t_{2g}^3 e_g^0$	Same
d^4	$t_{2g}^3 e_g^1$	$t_{2g}^4 e_g^0$
d^5	$t_{2g}^3 e_g^2$	$t_{2g}^5 e_g^0$
d^6	$t_{2g}^4 e_g^2$	$t_{2g}^6 e_g^0$
d^7	$t_{2g}^5 e_g^2$	$t_{2g}^6 e_g^1$
d^8	$t_{2g}^6 e_g^2$	Same
d^9	$t_{2g}^6 e_g^3$	Same

We defer further discussion of the effect of ligands on Dq until we have dealt with the spectra of transition metal ions.

Colors of Transition Metal Ions

One characteristic of the compounds of the transition elements that sets them apart from the elements of the main group is their color. As a rough rule, most compounds of the transition elements are colored, whereas most of those of the main groups are not. It is in providing a simple explanation of these colors that CFT has one of its greatest successes.

Although it was not explicitly stated, the d-orbital set, which was used in the discussion of CFT, are one-electron orbitals. The test electron used to calculate such an orbital set can thus have two different energies in a d-orbital set under the influence of a cubic CF. In the case of an O_h field, the lower-energy state is triply degenerate, and we label this energy state T_{2g}. The upper-energy state is doubly degenerate and given the label E_g. The electronic structure of these two states is illustrated in Figure 13.8. Note that we use lower-case labels when referring to one-electron orbitals and upper-case when referring to atomic energy states. These states are fully analogous to the terms of field-free atoms, described in Chapter 2.

Figure 13.8
Electronic structure of the T_{2g} and E_g terms of d^1 in an O_h field.

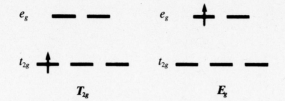

The ground term of a d^1 ion is ^2D, and under the influence of a cubic field this fivefold degenerate term is split into the T_{2g} and E_g terms. The separation of the two CF terms is $10\,Dq$, and the magnitude of this energy increases with increasing CF strength. These effects can be summarized in a simple energy diagram, known as an Orgel diagram. Figure 13.9 shows such a diagram for a d^1 ion. The Orgel diagram for a d^1 ion in a T_d field is obtained by extrapolating the one for the O_h field back through the origin, as shown in Figure 13.9.

The CF splitting for $3d$ ions is generally in the range of 5000–25,000 cm^{-1}, and this corresponds to the energy of visible light. The Ti^{3+} ion is an excellent example of the d^1 case and the visible absorption (or electronic) spectrum of the $[Ti(H_2O)_6]^{3+}$ ion is illustrated in Figure 13.10. This spectrum was measured on a solution at room temperature, and the CF band is very broad. The large bandwidth is due to the vibrations of the metal–ligand bonds inducing a fluctuation in the value of Dq, which in turn induces a fluctuation in the value of the transition energy. The spectra of solids measured at very low temperatures exhibit much narrower bands with finer structure.

The color of a compound that is absorbing light in the visible region is the complement of the absorbed light. In the Ti^{3+} ion, the violet color results from the absorption of most of the green and yellow light and the transmission of blue and some red. The intensity of the absorption bands is

Figure 13.9
Orgel diagram for a d^1 ion in O_h and T_d fields.

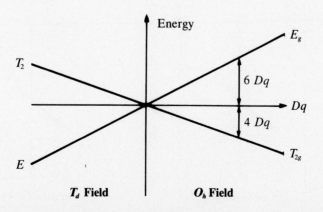

Figure 13.10
Visible absorption
spectrum of the
aqueous
$[Ti(H_2O)_6]^{3+}$ ion.

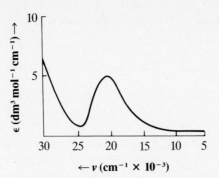

relatively weak. The absorption of electromagnetic radiation requires the coupling of the oscillating electric field of the light with an oscillating electric field of the absorber. This oscillating field is only possible if the electric dipole of the excited state is different from that of the ground state. This condition leads to *selection rules*, which define the criteria determining whether an excitation can take place. For our present purposes, there are two selection rules that govern electronic transitions in atoms: The *Laporte rule* states that transitions between AOs of the same kind are forbidden; the *spin rule* states that only transitions between states with the same spin are allowed.

According to the Laporte rule, transitions between d orbitals in an isolated atom or ion are strictly forbidden, in conformity with experimental observation; the same should be true for a perfectly octahedral complex, but in reality perfectly octahedral environments are distorted by vibrations about the metal-to-ligand bonds. These vibrations lower the symmetry of the ligand field, which in turn distorts the shape of the electron cloud of the central atom. In particular, it is the perturbation of the d orbitals by unsymmetrical ligand vibrations such as that shown in Figure 13.11, that

Figure 13.11
Unsymmetrical and
symmetrical
vibrations of an O_h
molecule.

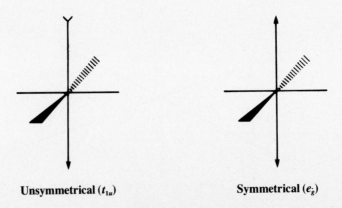

Unsymmetrical (t_{1u})

Symmetrical (e_g)

render the *d-d* transitions of transition metal complexes slightly allowed. It can be seen from Figure 13.11 that the unsymmetrical vibration introduces an oscillating electric dipole, whereas the symmetric vibration does not.

If all ligands are the same kind, octahedral complexes typically have a molar extinction coefficient (or molar absorptivity) ϵ of about 10 dm^3 mol^{-1} cm^{-1}. If there is more than one kind of ligand and the ligands differ considerably in their field strength, the extinction coefficients can be larger, particularly in the case where there is a unique pair of ligands in a cis relationship to give the complex a large permanent dipole. In such cases — for example, *cis*-$[(en)_2Cr(CN)_2]^+$ — the extinction coefficient can be as high as 100. Also, the extinction coefficients for T_d complexes, which are not centrosymmetrical, tend to be higher than those of O_h complexes. However, it should be remembered that it is the integrated area under the peak that is a true measure of the transition probability rather than the peak height at a particular wavelength.

The Orgel diagram in Figure 13.9 can also be applied to three other *d*-electron configurations, d^4, d^6, and d^9. An examination of a box diagram (for an introduction to box diagrams, see Chapter 2) quickly shows that each of these configurations also has a *D*-term ground state:

	m_l					
	-2	-1	0	1	2	
d^4		↑	↑	↑	↑	L = 2; S = 2; ^5D
d^6	↑	↑	↑	↑	↓↑	L = 2; S = 2; ^5D
d^9	↑	↓↑	↓↑	↓↑	↓↑	L = 2; S = $\frac{1}{2}$; ^2D

An examination of the CF energies of the ground and first excited states reveals that the d^4 configuration in an O_h field has a CF energy in the ground state of $-6\,Dq$ and in the first excited state of $+4\,Dq$. This is the same as we found for d^1 in a T_d field. A similar treatment of d^4 in a T_d field reveals that the energies parallel those for the d^1 case in an O_h field. Figure 13.12 illustrates the one-electron energy diagrams and model calculations of the energies of the states.

If we examine the d^6 case in the same way, it turns out that the energies of the ground and first excited states in an O_h field are $-4\,Dq$ and $+6\,Dq$, respectively (see Figure 13.12). That is, the states behave analogously to the d^1 configuration. A parallel result is obtained for d^6 in a T_d field.

Finally, in d^9 the ground state is at $-6\,Dq$, and the first excited state is at $+4\,Dq$ in an O_h field and at $-4\,Dq$ and $+6\,Dq$ in a T_d field. All these results can be summarized in the single diagram illustrated in Figure 13.13.

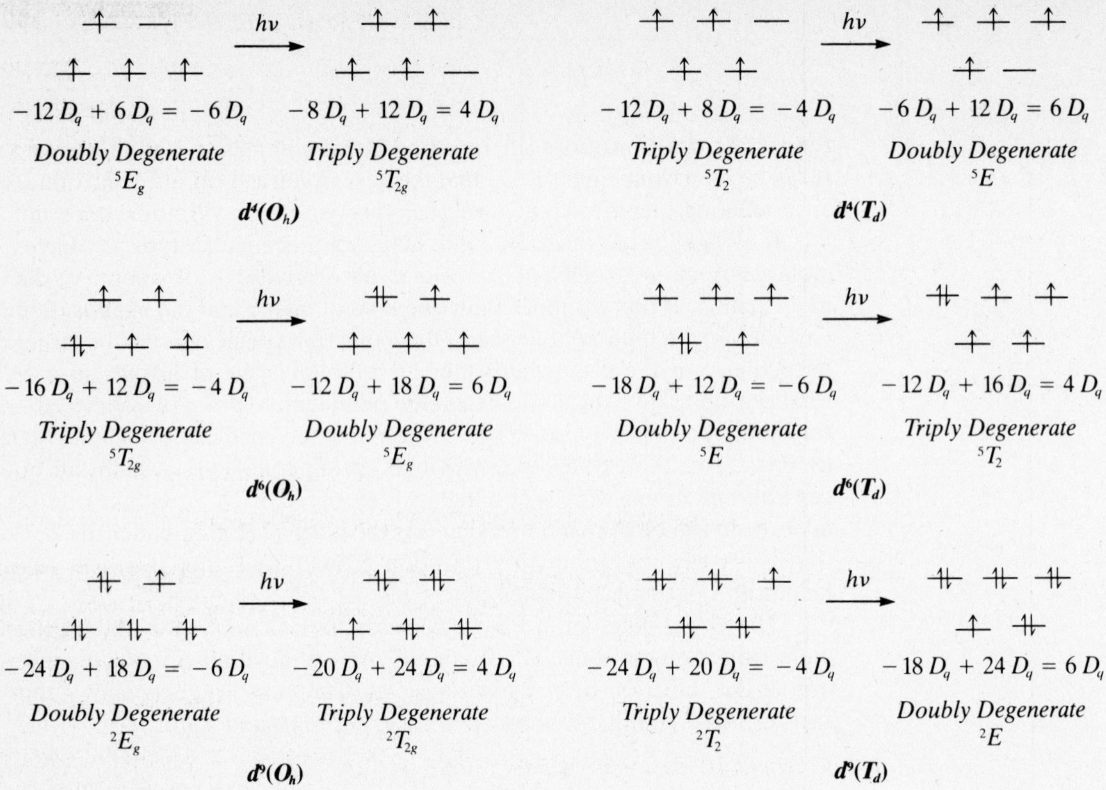

$-12\,D_q + 6\,D_q = -6\,D_q$

Doubly Degenerate
^5E_g

hv

$-8\,D_q + 12\,D_q = 4\,D_q$

Triply Degenerate
$^5T_{2g}$

$d^4(O_h)$

$-12\,D_q + 8\,D_q = -4\,D_q$

Triply Degenerate
$^5T_2$

hv

$-6\,D_q + 12\,D_q = 6\,D_q$

Doubly Degenerate
^5E

$d^4(T_d)$

$-16\,D_q + 12\,D_q = -4\,D_q$

Triply Degenerate
$^5T_{2g}$

hv

$-12\,D_q + 18\,D_q = 6\,D_q$

Doubly Degenerate
^5E_g

$d^6(O_h)$

$-18\,D_q + 12\,D_q = -6\,D_q$

Doubly Degenerate
^5E

hv

$-12\,D_q + 16\,D_q = 4\,D_q$

Triply Degenerate
$^5T_2$

$d^6(T_d)$

$-24\,D_q + 18\,D_q = -6\,D_q$

Doubly Degenerate
^2E_g

hv

$-20\,D_q + 24\,D_q = 4\,D_q$

Triply Degenerate
$^2T_{2g}$

$d^9(O_h)$

$-24\,D_q + 20\,D_q = -4\,D_q$

Triply Degenerate
$^2T_2$

hv

$-18\,D_q + 24\,D_q = 6\,D_q$

Doubly Degenerate
^2E

$d^9(T_d)$

Figure 13.12 Energy-level diagrams for the ground and excited states of the *D*-term configurations.

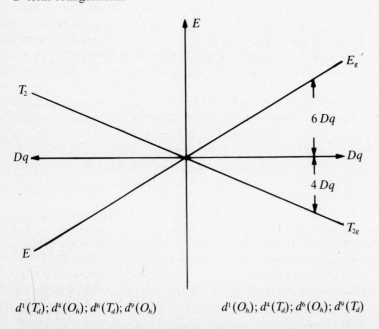

E

E_g'

T_2

$6\,Dq$

Dq

Dq

$4\,Dq$

T_{2g}

E

$d^1(T_d);\ d^4(O_h);\ d^6(T_d);\ d^9(O_h)$ $d^1(O_h);\ d^4(T_d);\ d^6(O_h);\ d^9(T_d)$

Figure 13.13 Orgel diagram for ground and excited states of the *D*-term configurations.

Ions with F-Ground Terms

All the configurations d^2, d^3, d^7, and d^8 have an F-ground term, as illustrated in the following box diagram:

$$m_l$$

	-2	-1	0	1	2	
d^2				↑	↑	L = 3; S = 1; ^3F
d^3			↑	↑	↑	L = 3; S = $\frac{3}{2}$; ^4F
d^7	↑	↑	↑	↓↑	↓↑	L = 3; S = $\frac{3}{2}$; ^4F
d^8	↑	↑	↓↑	↓↑	↓↑	L = 3; S = 1; ^3F

Besides the F-ground term, there is also a P term of the same spin multiplicity at higher energy. We cannot ignore this term for two reasons. First, the separation between the F and P term is of the same order as the CF splitting. Second, transitions between terms of the same spin multiplicity are allowed, whereas those between terms of different multiplicity are forbidden. This is the spin selection rule mentioned previously.

It is easy to see from a visual examination of a set of d orbitals occupied by two electrons that there are two states of different energy. If the two occupied orbitals are in the same plane, as in Figure 13.14a, the coulombic repulsion between them is clearly greater than if they are in different planes, as in Figure 13.14b. The configurations in Figure 13.14a are the three components of the ^3P term, whereas those of Figure 13.14b are three of the seven components of the ^3F term. The ^3P term is higher in energy than is the ^3F term, as would be expected from Hund's second rule.

A further level of complication in the F-term ions is that there are two excited states rather than the single excited states of the D-term ions. Figure 13.15 illustrates the ground and the two excited states for the d^2 case. The ground state is triply degenerate, it has the term symbol T_{1g}, and its CF energy is $-8\ Dq$. From the one-electron energy-level diagram, the first excited state appears to be sixfold degenerate: a triply degenerate t_{2g} level and a doubly degenerate e_g level. However, this is an illusion that arises because one-electron levels do not take into account the effects of interelectronic repulsion. In fact, these six components arise from two different terms, ^3F and ^3P, which we discussed above, and they belong to two triply degenerate CF states: $^3T_{2g}(F)$ and $^3T_{1g}(P)$. Both states appear to have a CF energy of $+2\ Dq$, but once again the one-electron energy levels are deceptive. Because $T_{1g}(F)$ and $T_{1g}(P)$ states have the same symmetry, they can interact and do so at low CF strengths in such a way that the energy of the $T_{1g}(P)$ is lowered by $2\ Dq$ and that of $T_{1g}(F)$ is raised by an equal amount. As the CF strength increases and the separation of the two

Figure 13.14
Orbital occupancies
for permuting two
electrons in two
separate *d* orbitals.

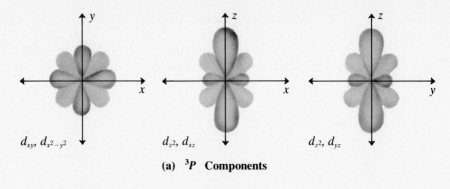

(a) ^3P **Components**

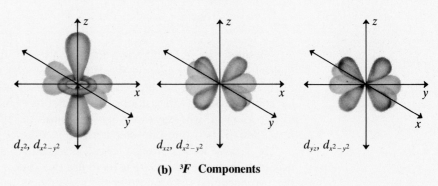

(b) ^3F **Components**

Figure 13.15
Four energy states
for a d^2 ion in an O_h
field.

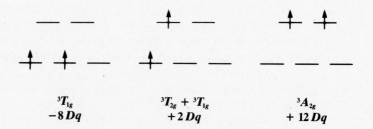

$^3T_{1g}$ $^3T_{2g} + {}^3T_{1g}$ $^3A_{2g}$
$-8\,Dq$ $+2\,Dq$ $+12\,Dq$

states becomes larger, this configuration interaction (CI) diminishes, and the CF energies asymptotically approach those predicted by the one-electron energy-level diagram.

The second excited state, in which both electrons are promoted to the e_g level, is nondegenerate. It is at an energy of between 18–20 Dq above the ground state, depending on the amount of CI between the $T_{1g}(F)$ and $T_{1g}(P)$ levels, and it has the symmetry symbol A_{2g}. Figure 13.16 shows an Orgel diagram summarizing these conclusions.

Analyzing the d^3 case reveals that the ordering of the energy levels is reversed relative to the d^2 case. Thus, from Figure 13.17 it can be seen that

Figure 13.16
Orgel diagram for a
d^2 ion in an O_h field.

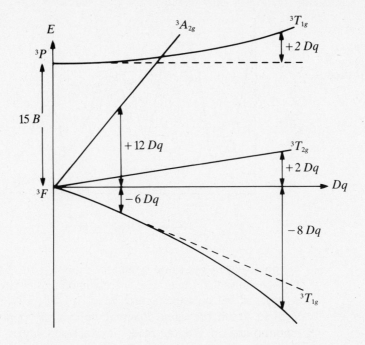

the ground state is nondegenerate ($^4A_{2g}$), the first excited state contains six components [$^4T_{2g}$ and $^4T_{1g}(P)$] and the second excited state is triply degenerate [$^4T_{1g}(F)$]. The Orgel diagram in this case is produced by back extrapolation of the d^2 diagram, as shown in Figure 13.17. A slight difference in the behavior of the d^3 relative to the d^2 case is because the two states of T_{1g} symmetry converge for the d^3 case and consequently the degree of CI between these states increases progressively with increasing CF strength. They are also subject to the *no-crossing rule*, which states that

Figure 13.17
Orgel diagram for
ions with F-term
ground states in O_h
and T_d fields.

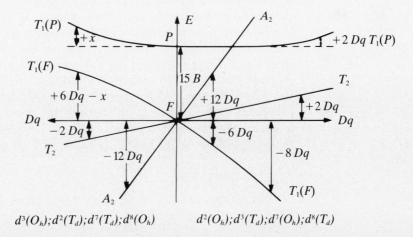

$d^3(O_h); d^2(T_d); d^7(T_d); d^8(O_h)$ $d^2(O_h); d^3(T_d); d^7(O_h); d^8(T_d)$

Figure 13.18
Electronic
absorption spectra
of aqueous solutions
of $[V(H_2O)_6]^{2+}$ and
$[V(H_2O)_6]^{3+}$.

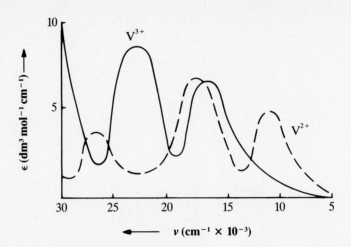

energy levels of the same symmetry cannot cross. Extending the analysis
further shows that the d^7 case is accommodated by the same diagram as d^2,
whereas d^8 behaves analogously to d^3. It is also easy to show that ions in a
T_d field fit on the same diagram but on the opposite side from their
counterparts in the O_h field, in exact analogy to the case of the D-
term ions.

From the Orgel diagram it can be seen that d^2 ions should have spectra
with three peaks, except in the region where the A_{2g} and $T_{1g}(P)$ levels
cross. Relatively few d^2 ions in regular ligand environments are known.
The V^{3+} ion is easy to obtain, either in binary compounds or in complexes.
Figure 13.18 shows the UV–visible spectrum of the $[V(H_2O)_6]^{3+}$ ion. The
first peak (between 8–10 Dq energy) occurs at 17,241 cm^{-1}, which affords
a Dq value of between 2155–1724 cm^{-1}, depending on the degree of CI.
Using this value of Dq, we can predict that the second transition should be
between 40,000–30,000 cm^{-1} (18–20 Dq), if it is from $^3T_{1g}$ to $^3A_{2g}$. In fact,
the second peak is seen at much lower energy, 24,000 cm^{-1}. This anomaly
is resolved if it is recognized that the transition we are observing experimentally is not the $^3T_{1g}$ to $^3A_{2g}$ transition but the $^3T_{1g}(F)$ to $^3T_{1g}(P)$
transition because the CF of the 3+ ion is so high that the $^3T_{2g}$ state has
crossed over the $^3T_{1g}(P)$ state. In the field-free ion, the separation between
the ^3F and ^3P terms is 15 B, where B is the atomic interelectronic repulsion
parameter (see Chapter 2). If we use the value of B, determined by
analyzing the gas-phase spectrum of the V^{3+} ion (860 cm^{-1}), to calculate
the energy of the $^2T_{1g}(F)$ to $^3T_{1g}(P)$ transition we get

Minimum value $6\,Dq + 15\,B = 23{,}115$ cm^{-1}

Maximum value $8\,Dq + 15\,B + 2\,Dq = 27{,}200$ cm^{-1}

This result suggests that the weak-field approximation is better than the strong-field approximation in the $[V(H_2O)_6]^{3+}$ ion. This better agreement is unfortunately illusory because in fact the value of B does not remain the same in the complex as it is in the gaseous ion. In general, the value of B is smaller in complexes than in the free ions, and the disparity tends to increase with increasing charge on the ion. This reduction in the interelectronic repulsion parameter on formation of a compound is one of the indicators of the inadequacy of a purely ionic model. One interpretation of the reduction in B is that the spreading of d electrons over delocalized MOs reduces the interelectronic repulsion. We will encounter a method for the evaluation of B in the complex (B') below.

Despite the limitations of our analysis of the V^{3+} system, the agreement between the observed and predicted energy of the second band is a considerable advance over the situation we encountered with the D-term ions. In that case, we had only one datum and one parameter to determine. When you have only the same number of observable quantities as you have parameters in your theory, the theory cannot be tested. For the $[V(H_2O)_6]^{3+}$ ion, we have two observables and one parameter to be determined. We use the first observable to fix the parameter (Dq) and then predict the second observable with considerable success, given the crudeness of the model. An unfortunate feature of the V^{3+} case is the obscuring of the third d-d transition by the very intense *charge transfer band* below $30,000$ cm^{-1}. Discussion of the nature of these bands will be deferred until Section 13.6.

The d^3 configuration is represented by the left-hand-side of the Orgel diagram in Figure 13.17. Figure 13.18 shows the spectrum of the $[V(H_2O)_6]^{2+}$ ion. All three d-d transitions are evident in this spectrum because the lower value of Dq for a 2+ ion shifts all of the d-d transitions to lower energy, while the charge-transfer band moves to higher energy with decreasing charge on the metal. Because the first and second energy levels are unaffected by CI, the first transition gives us 10 Dq precisely. Note that the Dq value for the 2+ ion is roughly half the value found for the 3+ ion. If we recognize that the CI between the two T_{1g} states leads to equal but opposite changes in their energies (this is required by the principle of conservation of energy), we can do some interesting calculations using the three available transition energies. To begin, assume that $T_{1g}(F)$ is lowered by an amount x and $T_{1g}(P)$ is raised by this amount. From the Orgel diagram for d^3, it is easy to show that

Energy of first transition $\nu_1 = 10\ Dq$

Energy of second transition $\nu_2 = 18\ Dq - x$

Energy of third transition $\nu_3 = 12\ Dq + 15B' + x$

where B' is the Racah parameter of the complex. Therefore,

$$15B' = \nu_2 + \nu_3 - 3\nu_1$$

From the peak energies of the spectrum in Figure 13.18, we can calculate the value of B':

$$B' = \frac{17{,}587 + 27{,}397 - (3 \times 11{,}764)}{15} = 664 \text{ cm}^{-1}$$

because

$$x = 18\,Dq - \nu_2 = 3311 \text{ cm}^{-1}$$

We now have values for the three unknown parameters, but again we had to use three data to get the parameters, and the quality of our theory does not seem to have advanced beyond the level of the V^{3+} case. The situation is not so bad as it seems because x is in fact a function of Dq and B'.

Our calculation above has given a value for B' (664 cm^{-1}), which we can compare with the free-ion value B (755 cm^{-1}). Here we see quantitative evidence for a reduction in the Racah parameter in the complex ion.

The d^5 Case

The d^5 configuration is unique in that the ground state is nondegenerate and of unique spin multiplicity (ground term ^6S). Because any movement of electrons within the d-orbital set must lead to a change of spin, excitations of the ^6S state are spin-forbidden. Because of spin–orbit coupling the spin-selection rule is not totally operative; instead of transitions being completely forbidden, their intensities are greatly reduced relative to those of the spin-allowed transitions of the other d configurations. Thus, complexes of Mn^{2+}, the most accessible example of a d^5 ion, appear to be almost colorless. Very concentrated solutions of the $[Mn(H_2O)_6]^{2+}$ ion appear very faint pink. The spectrum of this solution exhibits many more lines than do those of the other d configurations. This is because the d^5 configuration gives rise to an unusually large number of terms, and the spin-forbidden transitions in the other configurations are obscured by the more intense spin-allowed transitions. Figure 13.19 shows the spectrum of the hexaaquomanganese(II) ion. Figure 13.20 illustrates an energy-level diagram showing the origins of the various transitions.

Note particularly that the sharp peaks in the spectrum correspond to transitions to upper levels that are parallel to the ground state. Neither the ground state nor the parallel upper states vary in energy with varying CF. The reason why is easily seen from an examination of the influence of CF on the total energy of the states, as illustrated in Figure 13.21.

Figure 13.19
Electronic
absorption spectrum
of the $[Mn(H_2O)_6]^{2+}$
ion.

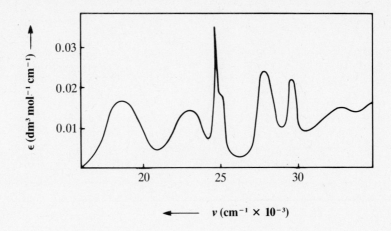

Figure 13.20
Energy-level
diagram showing the
effect of CF on a d^5
ion.

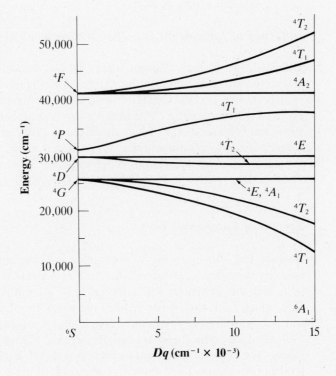

Figure 13.21
Some states of the
d^5 configuration
whose energy does
not vary with CF.

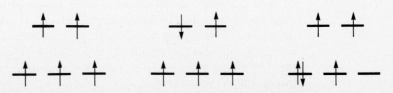

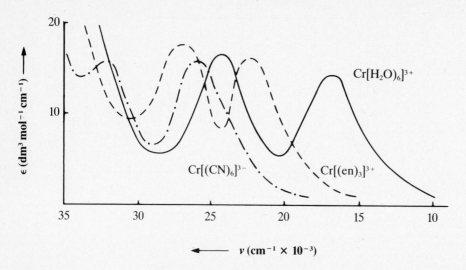

Figure 13.22
Electronic
absorption spectra
of some
chromium(III)
complexes in
aqueous solution.

Influence of Ligands on Dq: The Spectrochemical Series

Different ligands give different values for Dq; hence, the positions of the d-d bands are dependent on the nature of the ligand. Figure 13.22 shows the spectra of several Cr(III) complexes. The ligands can be placed in order of their increasing CF, and this order is independent of the nature of the metal. The sequence of ligands placed in order of their CF strength is called the *spectrochemical series*. A series that includes some of the more frequently encountered ligands is the following: $I^- < Cl^- < F^- < OH^- < H_2O < NH_3 < RS^- < R_3P < CN^- < CO \sim CH_3^- \sim H^-$.

Spectra of Lanthanide Ions

Because the $4f$ orbitals in the lanthanide ions are shielded from outside fields by the filled $5s$ and $5p$ shells, the CF is a relatively minor perturbation, and the spectra of the compounds of a lanthanide element do not differ much from each other or from the spectrum of the gaseous ion. The transitions, which are essentially between free-ion terms, are quite weak in comparison to typical d-d transitions because the vibronic coupling that relaxes the Laporte rule in transition complexes is almost absent in the lanthanides. Figure 13.23 shows the spectrum of the Nd^{3+} ion ($4f^3$). It can be seen that the visible spectrum is an image of the term energy-level diagram for the ion. The spectrum of $[Ti(H_2O)_6]^{3+}$ is shown superimposed on the Nd^{3+} spectrum to illustrate the considerable difference in the integrated intensity of the d-d band relative to the f-f band. It is this difference that highlights the very low transition probability for the f-f transitions.

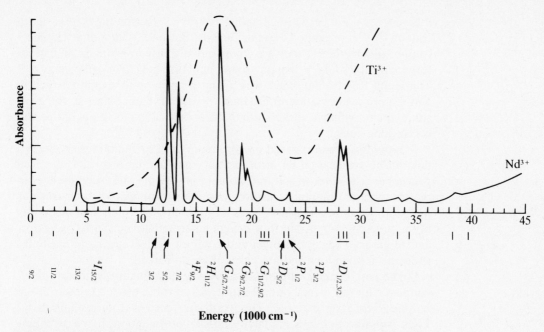

Figure 13.23 Electronic absorption spectrum of the Nd^{3+} ion with the spectrum of $[Ti(H_2O)_6]^{3+}$ to illustrate narrower lines in the lanthanide spectrum.

13.3 *Gemstones, Lasers and Such*

One attractive feature of compounds of the transition elements is their color, and nowhere is this more true than in their role as the source of color in precious and semiprecious stones. In modern times, much of the mystery and to some degree the value of these beautiful objects have been removed by our understanding of the source of their colors and our ability to synthesize many of the gemstones in the laboratory. Most gems are mixed-oxide crystals (silicates, aluminosilicates, borosilicates, or quartz) with small amounts of transition ions substituting for main-group ions in the lattice. The dimensions and geometry of the site are determined largely by the host lattice, thus, the metal ion may be in a coordination state that is difficult to achieve in isolated complex ions.

Emeralds are mainly beryl, a beryllium aluminosilicate. The best stones, brilliant green in color, have some of the aluminum host ions replaced by Cr^{3+} ions. The coordination sites are distorted octahedral, and the Cr—O distance is slightly longer than in the $[Cr(H_2O)_6]^{3+}$ ion. In a ruby, some of the aluminum ions of corundum (Al_2O_3) are replaced by Cr^{3+}. The coordination sites are again distorted octahedral, but the Cr—O distance is shorter than in the hexaaquo ion. Hence, the remarkable difference in color between emeralds and rubies, despite the fact that both contain Cr^{3+} coordinated to six oxide ions. Because of the simple chemical composition of rubies, they are comparatively easy to make in the laboratory, and rubies larger than any found in nature have been manufactured. This is not true for emeralds, which have a much more complex structure and require manipulation of the highly toxic beryllium oxide. In the gemstone, the beryllium is locked into the very inert aluminosilicate lattice and is less dangerous.

Sapphires are also based on corundum, and splendid blue monocrystals can be synthesized by incorporating small amounts of Ti^{3+}. Like all gemstones, the color of natural sapphire is variable due to differences in the amounts of different transition ions. The semiprecious stones such as the garnets, jades, and tourmalines lack the durability and pristine quality of the precious stones, but they exhibit an amazingly rich variety of colors and textures. Their more open and complex lattices are more easily invaded by foreign ions and are more prone to defects.

The easy manufacture of large-size and high-quality synthetic rubies, coupled with the special optical properties of ruby, has led to their use as an important laser material. The word *laser* derives from the initials for Light Amplification by Stimulated Emission of Radiation. In the preceding sections on the spectra of transition ions, we repeatedly referred to the excitation of ions that give rise to absorption bands. We did not mention that these excited states must return to the ground state if the entire assembly of ions is not to end up in the excited state. In reality, an excited state that has the same spin as the ground state will have a very short lifetime (about 1 μs–1 ns) before dropping spontaneously to the ground state. Because both the excitation and the emission are essentially random processes with respect to time, there is no rational relationship between the ages of the photons emerging from the system. In wave terms, we say that the phase of the light is highly dispersed or *incoherent*.

Spontaneous deexcitation is not the only way in which an excited state can return to the ground state. Another route is by *stimulated emission*. In this process, a photon of precisely the energy corresponding to the deexcitation process can interact with the excited state and induce the emission of a second photon. The resulting pair of photons are in phase or *coherent*. Such coherent radiation is not produced under most normal circumstances, either because the excessively short lifetime of the excited state precludes its interception by an appropriate photon or because its concentration is too small. The special criterion for the observation of coherent emission of this kind is that the population of excited-state species must exceed the population of ground-state species, a condition referred to as *population inversion*. It is possible to achieve this population inversion in ruby because of the fortunate coincidence of a number of factors. These factors can be described in terms of the energy-level diagram for the Cr^{3+}

Figure 13.24
Energy-level
diagram showing the
transitions involved
in the ruby laser.

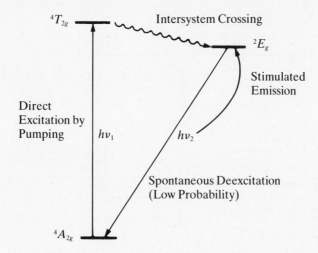

$^4T_{2g}$

Intersystem Crossing

^2E_g

Stimulated
Emission

Direct
Excitation by
Pumping $h\nu_1$ $h\nu_2$

Spontaneous Deexcitation
(Low Probability)

$^4A_{2g}$

ion shown in Figure 13.24. The critical feature of this set of energy levels is
the location of the ^2E state just below the $^4T_{2g}$ state. Normally, transitions
between the quartet states and the doublet state would not be very
probable because of the spin-selection rule. There is, however, a third
route to deexcitation. This route does not involve the emission of a photon
and is not subject to the spin-selection rule. It involves radiationless
transfer of energy to the surrounding medium or, in the present case, the
lattice. The criterion for this process, known as *intersystem crossing*, is that
the deexcitation energy must be virtually equal to the energy of one of the
vibrational modes of the lattice. This criterion is met in the case of ruby,
and there is a fairly efficient relaxation of the $^4T_{1g}$ excited state to the ^2E_g
state. In most other Cr^{3+} complexes with different CF splittings, this
condition is not met.

 Once the ion arrives at the ^2E_g state, it is constrained to stay in the
excited doublet state for a relatively long time because it is spin-forbidden
to decay spontaneously to the ground state. Its lifetime is long enough to
permit a population inversion and to allow stimulated emission to occur.
The population inversion is achieved in practice by *pumping* the crystal
with an intense flash of light, rich in the energy corresponding to the $^4A_{2g}$
to $^4T_{2g}$ transition. Rapid intersystem crossing then populates the ^2E_g state.
The slow spontaneous deexcitation of this state furnishes the photons to
initiate the rapid stimulated emission that returns the doublet ions to the
ground state. The ruby crystal is in the form of a long rod with reflecting
ends. The pulse of light arising from stimulated emission sweeps back and
forth between the reflecting ends of the rod, picking up more and more
photons as it goes. Photons that are not traveling parallel to the long axis of
the rod are quickly lost through the side walls, and so the pulse is nearly
perfectly collimated and coherent. After appropriate amplification, the

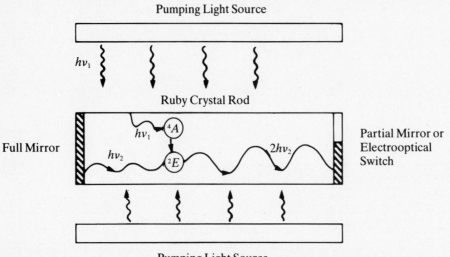

Pumping Light Source

$h\nu_1$

Ruby Crystal Rod

Full Mirror

$h\nu_1$ — ^4A

$h\nu_2$ — ^2E

$2h\nu_2$

Partial Mirror or
Electrooptical
Switch

Pumping Light Source

Figure 13.25
Schematic
representation of a
ruby laser.

pulse is eventually released from the end of the crystal by an optical or electrooptical shutter. Figure 13.25 shows a schematic illustration of the working of a ruby laser.

Laser light has many useful properties. Unlike incoherent light, there is no theoretical limit to the focusing of coherent light; thus, the focusing of laser pulses, which already have unusually high-power density, can produce extremely high concentrations of energy. This property has led to the use of lasers for the surgical reattachment of detached retinas, precision welding of refractory metals, and thermally induced nuclear fusion. Coherent light also shows little tendency to diverge, so it is possible to send a pulse of light to the moon, to reflect off a mirror, and to return to earth with relatively little increase in the cross-section of the beam. This property has also been used for a variety of other ranging and surveying devices such as artillery targeting, the measurement of large terrestrial distances to precisions of 1 in 10^{10}, and the amplitude of resonant vibrations of the moon. A third important application of coherent light is in telecommunications. A coherent light beam can be modulated to carry dozens of independent signals simultaneously, and many long-distance telephone calls are already transmitted in this way.

Lasers based on *nd* ions tend to be costly; their size and power are limited because they must be single-crystal materials. To a large degree, the ruby laser has been superceded by rare-earth glass lasers. Because the *f-f* transitions are not sensitive to the environment of the ion, it is not necessary to incorporate them into a highly regular lattice. In addition, the abundance of transitions available in 4*f* ions (see Figure 13.23) makes it possible to have lasing action between an excited state and a lower excited level. Because the lower excited level is not normally populated, it is very easy to achieve the condition of population inversion for such transitions. Giant lasers based on rare-earth glasses have been extensively used in research on controlled nuclear fusion where the light pulse serves the dual function of compressing the sample of fusible material (for example, LiD) and of heating it to temperatures of 10^6–10^7 K.

13.4   Some Structural and Thermodynamic Effects of Crystal Fields

Because the effect of the CF is to lower the energy of some of the d orbitals and to raise the energy of others, the occupation of these orbitals by electrons gives rise to a stabilization energy relative to the energy the system would have if the CF were spherically symmetrical. If we consider the case of an O_h field, each electron in a t_{2g} level contributes a stabilization energy of 4 Dq, while each election in an e_g level destabilizes the system by 6 Dq. The sum of such one-electron energies, the crystal field stabilization energy (CFSE), is easily computed for each d^n configuration. Figure 13.26 summarizes the results for T_d and O_h fields.

The double-humped curve for the CFSE for high-spin complexes parallels the behavior of many physical and thermodynamic parameters of compounds containing d^n ions. As an example, Figure 13.27 shows the variation of the heat of formation of the MCl_2 compounds with atomic number for the first-row transition dihalides. The deviations of the experimental points from a smooth curve drawn through the values for the spherically symmetrical ions Ca^{2+}, Mn^{2+}, and Zn^{2+} correspond roughly to the values of the CFSE calculated using spectroscopic values for Dq. Other

Figure 13.26
CFSE for the d^n configurations in T_d (+) and in O_h fields ($\times$ = high-spin; ($\square$ = low-spin).

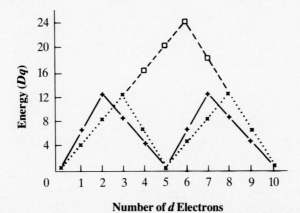

Figure 13.27
Variation of the
heat of formation
with atomic number
for the $3d$
dichlorides (MCl_2).

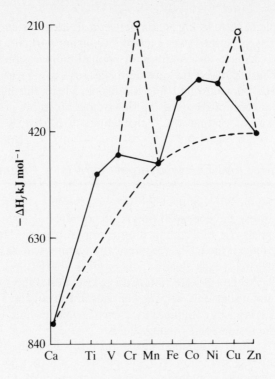

experimental quantities that behave in a similar way are heats of hydration and ligation and heats of vaporization of the metals. As a general rule, the CFSE is only one of a number of energy factors involved in a particular process. Therefore, strong correlation with CFSE is to be expected only in situations where all other factors other than CFSE remain essentially constant across the comparison series. This is clearly the case for the several examples cited above, but it is not the case for others such as reduction potential comparisons (see Chapter 20). In comparing the reduction potentials for the reactions

$$M^{3+} + e^- \rightarrow M^{2+} \tag{13.1}$$

the irregularities in the third ionization energies, which have values of the order MJ mol^{-1}, overwhelm the irregularities in CFSE, which have values of the order of 0.1 MJ mol^{-1}.

Although there is no obvious correlation between CFSE and reduction potential across a period, the reduction potential change for a given ion on varying the ligands usually does correlate well, provided there is no significant change in geometry of the complexes. In the following redox equilibria, it is clear that the potentials follow the spectrochemical series:

$$[Co(H_2O)_6]^{3+} + e^- \rightleftharpoons [Co(H_2O)_6]^{2+} \qquad E° = 1.84 \text{ V} \qquad \textbf{(13.2)}$$

$$[Co(NH_3)_6]^{3+} + e^- \rightleftharpoons [Co(NH_3)_6]^{2+} \qquad E° = 0.10 \text{ V} \qquad \textbf{(13.3)}$$

$$[Co(CN)_6]^{3-} + e^- \rightleftharpoons [Co(CN)_5]^{3-} + CN^- \quad E° = -0.83 \text{ V} \quad \textbf{(13.4)}$$

A further example of the influence of CFSE on thermodynamic properties is the variation of stability constants for a series of complexes of several metals with the same ligand. If we consider the general equilibrium

$$[M(H_2O)_6]^{n+} + 6L \xrightarrow{K} [ML_6]^{n+} + 6H_2O \qquad \textbf{(13.5)}$$

it is usually the case that the formation constant K increases for the series of metal ions $Mn^{2+} < Fe^{2+} < Co^{2+} < Ni^{2+} < Cu^{2+} > Zn^{2+}$ if $Dq(L) > Dq(H_2O)$. This order, known as the Irving–Williams order after the workers who first noted the phenomenon, is with the exception of the Cu^{2+} ion the order of increasing CFSE.

Even, when CFSE correlations work reasonably well, serious discrepancies always occur for those ions that have degeneracy in the e_g levels (for example, high-spin d^4 and d^9). Thus, the experimental points for $CrCl_2$ and $CuCl_2$ in Figure 13.27 are much higher than expected on the basis of CFSE. These anomalies are due to a phenomenon known as the Jahn–Teller effect. This effect is always present in nonlinear molecules with orbitally degenerate ground states but is particularly pronounced in O_h compounds with E_g ground states. The Jahn–Teller effect manifests itself as a distortion of the molecule from the state of highest symmetry, which in turn results in a lifting of the degeneracy of the electronic state and in a lowering of the energy of the system. The effect is illustrated in Figure 13.28 for a d^4 ion in an O_h field. If the ligands along the z axis of a perfectly octahedral complex are moved away from the metal ion, the energy of the

Figure 13.28
Effects of axial distortions on the energy of the d^4 configuration. The splitting of the levels is exaggerated.

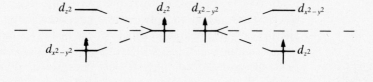

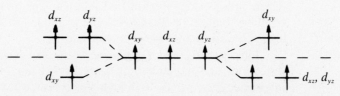

Compression Along z Axis   **Extension Along z Axis**

d_{z^2} orbital decreases. To satisfy the requirement of energy conservation, the ligands in the xy plane must move closer to the metal ion with a corresponding increase in energy of the $d_{x^2-y^2}$ orbital. In the high-spin d^4 ion, the lone e_g electron, which is located in the d_{z^2} orbital, is lowered in energy. A compression along the z axis and extension in the xy plane would give a similar result. It is impossible on the basis of the simple argument given here to predict which of these distortions will occur, but experiments indicate that it is usually the extension along the z axis that occurs in d^4 and d^9 systems.

A beautiful example of the Jahn–Teller effect in a d^4 ion is furnished by the mixed-valence chromium fluoride Cr_2F_5. This compound consists of an approximately CCP lattice of fluoride ions with two-fifths of the octahedral holes occupied symmetrically by Cr^{2+} and Cr^{3+} ions. Whereas Cr^{3+} ions, which have a nondegenerate A_{1g} ground state, reside in perfectly octahedral holes, the sites occupied by Cr^{2+} ions are severely distorted, with two colinear fluorides at 260 pm and the other four coplanar fluorides at 200 pm. Virtually all Cr^{2+} complexes whose structures have been determined exhibit similar distortions.

Because of its oxidative stability and its tremendous importance in metalloproteins and the catalysis of oxidation reactions, the Cu^{2+} ion has received far more attention than has Cr^{2+}. Besides Jahn–Teller effects, which operate strongly in six-coordinate complexes and moderately in four-coordinate complexes, the small CFSE of d^9 and the relatively small radius of the Cu^{2+} ion lead to little difference in energy between four and six coordination. These factors combine to give Cu^{2+} the most variable stereochemistry of any transition ion. Indeed, it is almost the case that no two compounds of Cu(II) have exactly the same geometry.

Distortion along the z axis leads to stabilization of d^4 and d^9 ions. In the case of compression, it is easy to see that beyond a certain point the electron clouds of the central metal and the ligand will begin repelling each other and preventing further compression. In an extension along an axis, it is not so easy to predict what the end result will be. In some cases, the electrostatic attraction between the effective charge of the central metal and the negative charge or dipole of the ligands is large enough to oppose complete removal of ligands. In others, there is effectively complete ligand removal, and the complex is best considered to be square-planar. Of course, there are also many cases where it is difficult to decide whether distant groups along the fourfold axis of an apparently square molecule are bonding. Figure 13.29 shows some examples of copper complexes that exhibit a variety of different geometries.

Ions with T_{2g} and T_{1g} ground states also should exhibit Jahn–Teller distortion, and a number of cases have been reported. A simple and interesting case is FeF_2. In a high-spin d^6 ion, it is possible to predict that a compression along the z axis will be more stabilizing than will an extension,

Figure 13.29
Geometrical
variability of Cu(II)
complexes. Bond
distances are in pm.

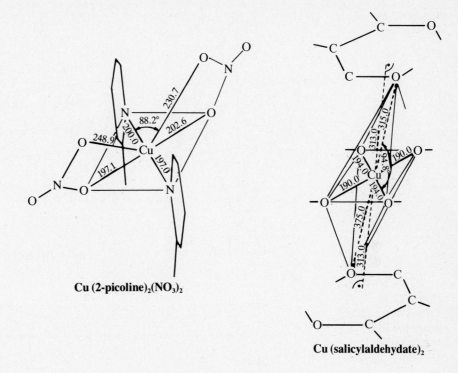

Cu (2-picoline)$_2$(NO$_3$)$_2$

Cu (salicylaldehydate)$_2$

**Two isomers of
Cu (4-picoline-N-oxide)$_2$Cl$_2$**

Green Isomer

Yellow Isomer
(Continued)

Figure 13-29
(*Continued*)

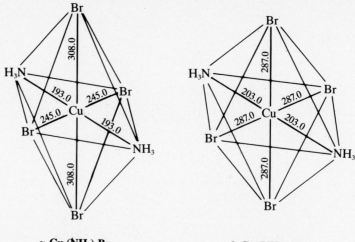

α-**Cu (NH₃)₂Br₂**   β-**Cu (NH₃)₂Br₂**

Figure 13.30
Effects of tetragonal
distortion of the d^6
configuration.
O_h field.

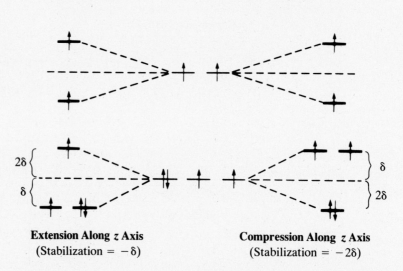

Extension Along *z* Axis   **Compression Along *z* Axis**
(Stabilization = −δ)   (Stabilization = −2δ)

as illustrated in Figure 13.30. The crystal structure of FeF_2 shows the iron
to be surrounded by a tetragonally distorted set of six fluorides, two at a
distance of 199 pm and four at a distance of 212 pm.

The d^8 configuration in O_h fields does not gain any stabilization
through small tetragonal distortions. However, if a sufficiently large distortion occurs, the energy difference between the d_{z^2} and $d_{x^2-y^2}$ orbitals can be
greater than the spin-pairing energy, thus inducing both formerly e_g electrons to occupy the same orbital. As shown in Figure 13.31, this leads to a
preference for square-planar over O_h geometry. Weak field d^8 complexes

Figure 13.31
Effect of tetragonal distortion on the d^8 configuration.

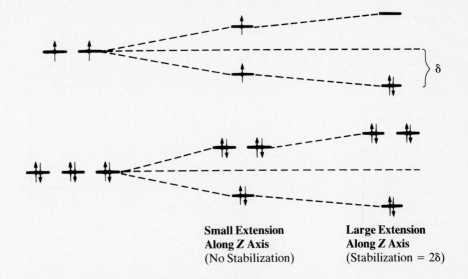

Small Extension Along Z Axis (No Stabilization)

Large Extension Along Z Axis (Stabilization = 2δ)

such as those of Ni^{2+} with most simple ligands tend to be octahedral. In strong-field situations such as $[Ni(CN)_4]^{2-}$ or most of the compounds of Pd(II) and Pt(II) square-planar coordination is encountered. PdF_2 is one of the rare compounds of Pd(II) in which the Pd(II) is octahedrally coordinated.

A final topic on the subject of structural effects of CFSE is the so-called *site-preference problem*. As was explained in Chapter 4, many ionic structures can be approximated by assuming that one of the ions forms a close-packed structure and the other ion occupies either the O_h or T_d holes in the lattice. Which type of hole is occupied is determined to a large degree by the radius ratio of the ions. However, in borderline cases it is difficult to predict correctly which type of hole the cation will occupy. One factor that can influence which kind of occupancy is adopted by transition ions is the CFSE.

The spinels are a family of mixed-metal oxides, isostructural with the mineral spinel $MgAl_2O_4$. The structure may be approximated by a CCP lattice of oxides in which one-eighth of the tetrahedral holes are occupied by the divalent ion and one-half of the octahedral holes are occupied by the trivalent ions. This structure is rather difficult to visualize, but it is related to a NaCl structure in which alternate rows of cations have been removed. This gives a lattice with the formal composition AX_2 (or A_2X_4). The additional cations (B) of the spinel structure are then included in tetrahedral holes of this lattice, such that each anion is tetrahedrally coordinated to three A and one B cation. Figure 13.32 shows a schematic illustration of the relationship between the NaCl and spinel structures.

Figure 13.32
Schematic representation of the spinel structure. Cations A and cation vacancies X form a FCC lattice, which with the FCC lattice of anions complete the NaCl structure. One of the tetrahedral cavities of the unit cell shown is occupied by the second cation B.

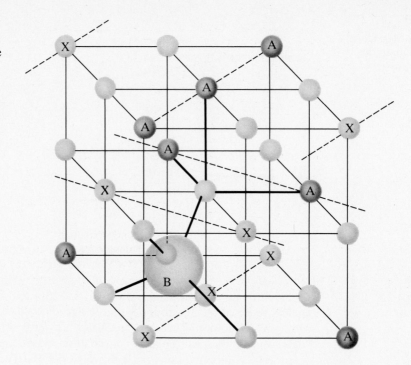

Occasionally, compounds of the same stoichiometry adopt an *inverse spinel* structure in which the divalent ions displace half of the trivalent ions from their octahedral sites into tetrahedral sites. This behavior is found in the nickel spinel $NiAl_2O_4$, a fact that is undoubtedly related to the maximum difference between CFSE (O_h) and CFSE (T_d), which occurs at d^8. Table 13.1 shows the differences in CFSE for O_h and T_d fields (Δ CFSE) for

Table 13.1	*CFSE for the Divalent and Trivalent Ions of Mn to Cu (Dq)*				
Ion	Mn(II)	Fe(II)	Co(II)	Ni(II)	Cu(II)
CFSE (O_h)	0	4	8	12	6
CFSE (T_d)	0	$2\frac{2}{3}$	$5\frac{1}{3}$	$3\frac{5}{9}$	$1\frac{7}{9}$
$(\Delta$CFSE$)$ (O_h)	0	$1\frac{1}{3}$	$2\frac{2}{3}$	$8\frac{4}{9}$	$4\frac{2}{9}$
Ion	Mn(III)	Fe(III)	Co(III)	Ni(III)	Cu(III)
CFSE (O_h)	6	0	4	8	12
CFSE (T_d)	$1\frac{7}{9}$	0	$2\frac{2}{3}$	$5\frac{1}{3}$	$3\frac{5}{9}$
$(\Delta$CFSE$)$ (O_h)	$4\frac{2}{9}$	0	$1\frac{1}{3}$	$2\frac{2}{3}$	$8\frac{4}{9}$

the divalent and trivalent ions from Mn to Cu. It should be remembered that $Dq(T_d)$ is about $\frac{4}{9} Dq(O_h)$ for the same ligands and internuclear separations. In Table 13.1, we have normalized the $Dq(T_d)$ to $Dq(O_h)$ to take account of this fact in calculating the CFSE.

At the simple level developed here, CFSE cannot be used to predict reliably whether a particular compound will adopt a normal or an inverse spinel structure; however, it can be used to predict which of a series of compounds is most likely to adopt the inverse structure. For example, one of the series of mixed-metal oxides Mn_3O_4, Fe_3O_4, and Co_3O_4 has the inverse, whereas the other two have the normal spinel structure. (These compounds all are of the type $M^{II}M_2^{III}O_4$.) Which one has the inverse structure? If we consult Table 13.1, it can be seen that Mn(II) has no CFSE in either site, while Mn(III) strongly prefers the O_h site; Fe(III) has no preference, while Fe(II) is more stable in O_h; if we take into account that Dq for a tripositive ion is about double that for a dipositive ion, the Co(II) and Co(III) both have the same preference for the O_h site. Thus, only in the case of Fe_3O_4 is there a driving force displacing the M(II) into an O_h site that is stronger than the driving force resisting displacement of M(III) out of the O_h site; this is indeed the compound with the inverse spinel structure.

13.5   Some Kinetic Consequences of Crystal Field Stabilization Energy

The ability of transition metal complexes to undergo ligand-substitution reactions is strongly affected by CFSE. The subject of substitution mechanisms in transition metal complexes is complicated and largely beyond the scope of this text. However, some of the generalizations are fundamentally important and any discussion of CFT would be incomplete without some mention of them.

A number of different mechanisms have been identified for substituting ligands in O_h complexes. To a first approximation, these can be identified as associative (increasing coordination number from six to seven) and dissociative (decreasing coordination number from six to five). In both cases, there are profound changes in CFSE on the passage of the reactant to the transition state, and the influence of CFSE is present in both the initial state and in the transition state. As a general rule, CFSE is particularly large in O_h fields, and any disruption of O_h coordination leads to a sacrifice of CFSE. Also, the larger is the CFSE in the initial O_h complex, the larger is the sacrifice on going to a transition state of lower symmetry. Thus, one might expect the ease of substitution to parallel the magnitude of the CFSE. This expectation is upheld by the data for one of the simplest substitution processes, the substitution of a water molecule in a hexaaquo ion by another water molecule. In general such reactions are extremely fast

Table 13.2	*Exchange Rates of Ligand for Solvent Water in Hexaaquo Complexes of 3d Ions*			
Metal Ion	Rate (s^{-1})	d^n	CFSE (Dq)	CFAE[a]
Cr^{2+}	7×10^9	4	6	-1.1
Mn^{2+}	3×10^7	5	0	0
Fe^{2+}	3×10^6	6	4	-0.6
Co^{2+}	1×10^6	7	8	-1.1
Ni^{2+}	3×10^4	8	12	2.0
Cu^{2+}	8×10^9	9	6	-3.1
Fe^{3+}	3×10^3	5	0	0
Cr^{3+}	3×10^{-6}	3	12	2.0

[a] Crystal field activation energy: the difference between CFSE in the O_h reactant molecule and in the transition state, assuming a dissociative mechanism with a SQP transition state.

and cannot be measured by simple sampling techniques. The data listed in Table 13.2 were obtained by measuring the proton NMR–line broadening of water resulting from the rapid exchange of coordinated water with solvent. Even though all the reactions have extremely high rates, their rates do fall into the familiar order of CFSE, with $[Cr(H_2O)_6]^{2+}$ and $[Cu(H_2O)_6]^{2+}$ the usual exceptions.

Particularly dramatic cases of substitution inertia due to CFSE are the Cr^{3+} and Co^{3+} ions. Here, the conjunction of maxima in CFSE with high effective positive charge on the ion leads to the classic cases of first-row complexes that are sufficiently inert to substitution to survive dissolution in water and recovery without significant substitution of the ligands. It is this stability that allowed Werner and his contemporaries to prepare so many different complexes of these ions.

Because of the generally higher CFSEs of low-spin complexes, they also tend to be more substitution-inert. Thus, the hexacyanometallate complexes of Cr(III), Fe(III), and Co(III) can be dissolved in and recovered from cyanide-free water.

13.6   Breakdown of the Crystal Field Theory

Like the ionic model, the CFT can be modified and adapted to deal with moderate deviations from the notion that metal and ligands can be represented as discrete point charges or dipoles. At some point, however, these adjustments become patently ridiculous. Perhaps the most extreme examples (the analog of methane in the ionic bonding case) are the cases involving metals in oxidation state zero, where there is in fact no charge on the metal to hold the molecule together by pure coulombic attraction.

Even in the case of highly ionic compounds, the CFT is fatally flawed. CF–type calculations based on more realistic models, taking account of the explicit shapes and degrees of interpenetration of electronic charge clouds, can even give the opposite result to the point-charge approximation with respect to the lifting of the degeneracy of the d orbitals. Fortunately, the point-charge CF model gives the same predictions regarding ordering and degeneracy of frontier orbitals as does the MO theory. What we have described as the e and t sets of orbitals in cubic-crystal fields are metal AOs whose degeneracy has been reduced by Stark splitting in the coulombic field of the ligands. In MO theory (Chapter 3), the frontier orbitals also are of e and t symmetry in cubic fields. However, the energy difference between e and t is attributed to covalent interaction. In the O_h case, if only σ bonding is involved, the t_{2g} set is nonbonding and still essentially atomic d orbitals in character. The e_g set is antibonding in character and, if the ligands are highly electronegative, located mainly on the metal. If π bonding is involved, both e and t sets are molecular. Thus, it is evident that the descriptions of the frontier orbitals in the two models are very similar, differing only in the cause of the splitting. This is fortunate because it allows most of what we have developed using the CFT to be transferred into MO terms. Nevertheless, the computational simplicity of CFT is lost because in rigorous covalent treatments the spitting cannot be reduced to a simple single-parameter quantity as in CFT. As a result, the CFT formalism is still extensively used in describing spectroscopic and other properties of transition metal complexes. However, it is important to remember that this useful descriptive framework has little fundamental theoretical justification.

A large body of experimental evidence supports the view that the metal d electrons are delocalized and have lost their purely atomic character. Coupling of metal-electron spin with the nuclear spins of ligand nuclei can be observed by both ESR and NMR spectroscopies. Figure 13.33 shows the ESR spectrum of the complex $Cp_2Ti(SiH_2Ph)(THF)$. The unpaired electron of Ti(III) is coupled to the nuclear spins $(\frac{1}{2})$ of the two H atoms of the silyl ligand. Such coupling can occur only if the electron spin has a finite density at the H nucleus. At the same time, hyperfine coupling to the Ti isotopes with spins of $\frac{5}{2}$ (^{47}Ti, 7% abundance) and $\frac{7}{2}$ (^{49}Ti, 6% abundance) is also evident, proving that the electron also has a finite density on the Ti. The odd electron has little density on the protons of the Cp ligands, and the coupling is too small to be resolved.

MO theory also gives a natural explanation of the so-called charge-transfer bands observed in the spectra of all coordination compounds.

Figure 13.33
ESR spectrum of
$Cp_2Ti(SiH_2Ph)(THF)$.

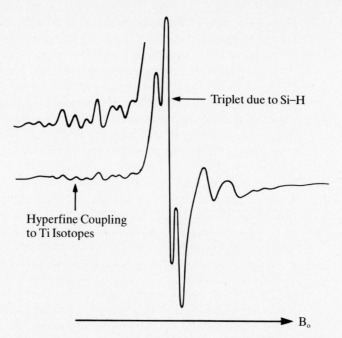

Triplet due to Si–H

Hyperfine Coupling
to Ti Isotopes

B_o

These bands are due to transitions between lower bonding orbitals and the empty frontier orbitals or between filled frontier orbitals and higher antibonding orbitals; they have no place in CFT. In compounds where there is a large electronegativity difference between metal and ligand atoms, the MOs are relatively localized on the atoms—for example, σ-bonding orbitals are concentrated on highly electronegative ligands, antibonding orbitals on the metal. In this case, excitation of an electron from a bonding to an antibonding orbital involves a transfer of charge from the ligand to the metal, hence the name charge transfer. Other transitions can involve excitation from an orbital concentrated on the metal to a high-energy empty orbital on the ligand. These excitations are not subject to the Laporte selection rule and are highly allowed, with molar extinction coefficients from 10^4–10^5 $dm^3\ mol^{-1}\ cm^{-1}$.

Finally, the existence of compounds such as the binary carbonyls, involving a ligand with no formal charge and almost devoid of a molecular dipole, makes no sense at all in terms of CFT but finds a natural explanation in MO theory. In particular, it is the possibility of multiple covalent bonding that not only explains the existence of the binary carbonyls but also that CO is one of the strongest field ligands. In forming bonds to transition metals, the unoccupied, high-energy π^a orbitals of the CO overlap with filled metal frontier orbitals (see Chapter 22). The resulting π-bonding orbitals are primarily metal in character and are at lower energy than they would be in the absence of π bonding. This additional lowering

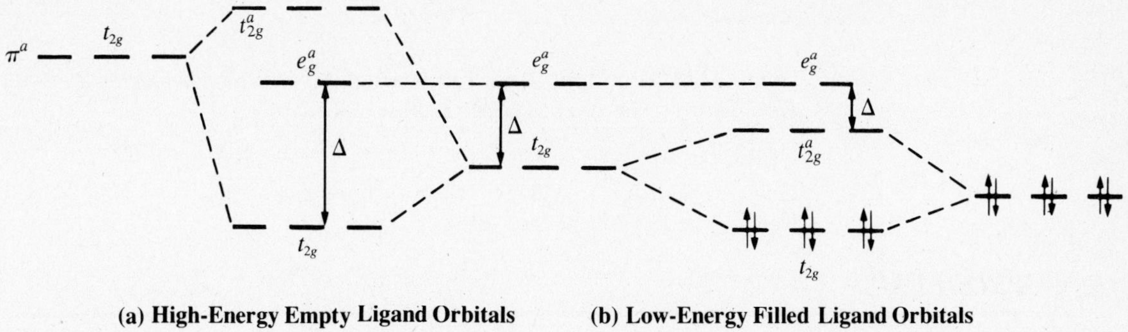

(a) High-Energy Empty Ligand Orbitals   **(b) Low-Energy Filled Ligand Orbitals**

Figure 13.34
Bonding overlaps in
metal-to-heavier-
main-group bonds.

of the metal frontier orbitals results in an apparently large Dq if carbonyl compounds are treated in CFT terms. A similar effect occurs in ligands with heavier main-group elements. In these cases, it is low-lying empty d orbitals of π symmetry that participate in π bonding with the metal. Figure 13.34 shows the schematic representation of d_π—p_π bonding between a transition metal and CO and the d_π—d_π to a heavier main-group atom.

If the orbitals on the ligand with π symmetry are lower in energy than are the metal orbitals and filled, the result of π overlap is to make the metal orbitals π antibonding in character (raises their energy). In O_h complexes, this has the effect of reducing the separation between t_{2g} and e_g levels and making the ligand appear to have an unusually low CF. This effect is apparent in small second-period atoms with completed octets, particularly F^-, which gives smaller values of Dq than expected (see spectrochemical series, Section 13.2).

BIBLIOGRAPHY

General Texts

Cotton, F. A., and G. Wilkinson. *Advanced Inorganic Chemistry*, 5th. ed. New York: Wiley, 1988.

Huheey, J. E. *Inorganic Chemistry: Principles of Structure and Reactivity*. New York: Harper & Row, 1983.

Jolly, W. L. *Modern Inorganic Chemistry*. New York: McGraw-Hill, 1984.

Moeller, T. *Inorganic Chemistry: A Modern Introduction*. New York: Wiley, 1982.

Porterfield, W. W. *Inorganic Chemistry: A Unified Approach*. Reading, Mass.: Addison-Wesley, 1984.

Suggested Reading

Dunn, T. M., McClure, R. S., and Pearson, R. G. *Some Aspects of Crystal Field Theory*. New York: Harper & Row, 1965.

Figgis, B. N. *Introduction to Ligand Fields*. New York: Wiley, 1966.

Lewis, J., and R. G. Wilkins. *Modern Coordination Chemistry*. New York: Interscience Publishers, Inc., 1960.

PROBLEMS

13.1   With the aid of one-electron energy levels, deduce the effect of an octahedral CF on the quartet terms of the Cr^{3+} ion. Express your conclusions in the form of an Orgel diagram.

13.2   A tetrahedral complex of Co(II) gives three absorption bands with extinction coefficients of about 20 dm^3 mol^{-1} cm^{-1} in the visible region. The lowest-energy band is 6000 cm^{-1}. Predict the energies of the other two bands. [The free-ion value for B for Co(II) is 971 cm^{-1}.]

13.3   A first-row transition metal dissolved in dilute sulfuric acid to give 400 cm^3 of H_2 per gram of metal. The resulting solution had a spectrum in the visible region consisting of a single broad band centered at 10,000 cm^{-1}. Explain these observations.

13.4   Fluorination of a mixture of KCl and an unknown metal led to the formation of a hexafluoro complex that gave spectrum (**a**) in the solid state. Dissolution of the complex in boiling dilute perchloric acid resulted in the slow evolution of oxygen. After oxygen evolution ceased, the spectrum of the solution was as shown in (**b**). Suggest an identity for the metal and the species responsible for the spectra. Explain as many of the features of the spectra as possible. [Assume a value of the free-ion Racah parameter B of 1071 cm^{-1} for your discussion of the spectrum (**b**).]

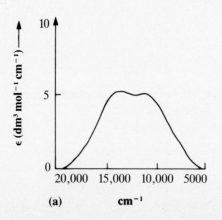

(a)   cm^{-1}

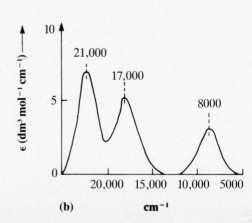

(b)   cm^{-1}

13.5   Suggest an explanation for each of the following observations:

   a. The optical absorption bands due to d-d transitions in transition metal compounds are broad.

   b. The optical absorption bands of lanthanide compounds are much narrower than those of transition complexes.

   c. High-spin Mn(II) compounds are only very faintly colored compared to most other first-row transition compounds.

   d. The maximum molar absorbivity for $[CoCl_4]^{2-}$ in the visible region is an order of magnitude greater than that of $[Co(H_2O)_6]^{2+}$.

13.6   Evaluate the correctness of the following statements:

   a. In an octahedral field, it takes less energy to promote a t_{2g}^2 ion to the $t_{2g}(d_{xy})\, e_g(d_z^2)$ state than to the $t_{2g}(d_{xy})\, e_g(d_{x^2-y^2})$ state.

   b. A set of p orbitals remains degenerate in an octahedral CF.

   c. The Mn^{4+} ion in $K_2[Mn(CN)_6]$ is in a low-spin state.

   d. Transition metal complexes are weakly colored because the excitation energies for d-d transitions are small.

13.7   Of the compounds $FeAl_2O_4$, $CoAl_2O_4$, and $NiAl_2O_4$, which is the most likely to have an inverse spinel structure?

13.8   The hydration energies of Ca^{2+}, Mn^{2+}, and Zn^{2+} were plotted versus atomic number. A smooth curve passing through these points gave a value of 3 MJ mol^{-1} for the hydration energy of Ni^{2+}. If the $^3A_{2g} \rightarrow {}^3T_{2g}$ transition of the $[Ni(H_2O)_6]^{2+}$ ion occurs at 8600 cm^{-1}, estimate the true hydration energy of the Ni^{2+} ion.

13.9   Crystalline ferrous fluoride has a distorted rutile structure. In rutile, all oxide ions are equidistant from the Ti ion, but in ferrous fluoride two of the fluoride ions are much closer to the Fe than the other four. How can this distortion be explained?

13.10   Jahn–Teller distortions arising from degeneracy in nonbonding orbitals are small; those arising from degeneracy in antibonding orbitals are large. Discuss the equilibrium structure of the E_g and T_{2g} states of $[Cr(H_2O)_6]^{2+}$ in terms of this statement.

13.11   A very high proportion of the experiments performed by Werner to establish the coordination theory were done with Cr(III) and Co(III) complexes. Why?

13.12   List the factors that are responsible for the highly variable coordination geometries of Cu(II) complexes.

13.13   On the basis of a point-charge model such as the CFT, the fluoride ion should exert a particularly strong CF, and the CO molecule should exert hardly any field at all. Experimental measurements lead to the opposite conclusion. How is this paradox resolved?

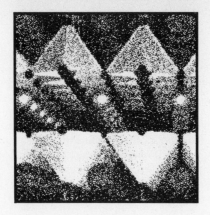

14

Magnetic Properties of d- and f-Block Compounds

14.1 Theoretical Basis of Magnetochemistry

The presence of unpaired electrons in molecules gives rise to magnetic properties that are very useful in determining electronic structure. Because of this, magnetochemistry has played an important role in the development of understanding of the electronic structure of transition compounds.

Accelerating charges—for example, electrons—generate a magnetic field perpendicular to their direction of motion. An electron pair in a molecule has no net magnetic moment because in occupying the same orbital they must have equal and opposite angular momenta; thus, their individual magnetic fields are equal and opposite. Applying an external magnetic field to such an electron pair opposes the motion of one of the electrons and accelerates the other. Thus, an applied magnetic field induces a net magnetic dipole in the electron pair. It can easily be shown that the direction of this dipole is such as to oppose the applied field.

In the case of an unpaired electron, a permanent magnetic field exists because there is no second electron of equal and opposite momentum to cancel it. Thus, molecules with unpaired electrons possess permanent magnetic dipoles. Applying an external magnetic field to such a dipole tends to orient it to become aligned with the field and to increase the field strength in its vicinity. The dipole associated with an unpaired electron is an order of magnitude greater than the induced dipole arising from the effect of an external field on an electron pair.

The combined effects of the induced magnetism, *diamagnetism*, and the permanant magnetism of unpaired electrons, *paramagnetism*, result in the experience of a force by matter when placed in a magnetic field gradient. Opposing the applied field by the diamagnetic effect results in a repulsion of the sample by the applied field (the sample moves in the direction of lower applied field strength), and the paramagnetic effect results in an attraction of the sample to regions of higher field. The total force per unit quantity of material experienced by the sample is proportional to the strength of the applied field and to the field gradient. This may be expressed mathematically in the form of Equation 14.1:

$$\frac{dF}{dm} = H \cdot \chi \cdot \frac{dH}{dx} \tag{14.1}$$

where F = force experienced by sample, m = mass of sample, H = applied field, x = distance, and χ = specific susceptibility. As is usual in chemistry, we use the mole as the unit of mass, and the susceptibility is then referred to as the molar susceptibility. The absolute measurement of susceptibility is laborious. Fortunately, several experimental methods are available to measure the susceptibility of experimental samples relative to readily available standards. Most methods are based on the integration of Equation 14.1, which gives

$$F = \frac{A}{2} \cdot \chi \cdot (H^2 - H_0^2) \tag{14.2}$$

where A = the cross-sectional area of the sample and H and H_0 = the upper and lower limits of field strength experienced by the sample. Thus, by weighing a standard and an unknown under identical conditions of geometry and field, the unknowns can be factored out of Equation 14.2 and the susceptibility of the unknown determined.

The diamagnetic susceptibilities of atoms do not change much with changing molecular environment and can be found in tables. The molecular diamagnetic susceptibility is simply the sum of the atomic susceptibilities, and the total susceptibility is the sum of the paramagnetic and diamagnetic contributions. The paramagnetic susceptibility is therefore obtained from the simple subtraction:

$$\chi_{para} = \chi_{tot} - \chi_{dia} \tag{14.3}$$

The key to understanding the relationship between the macroscopic susceptibility and the microscopic electronic structure of atoms resulted from the empirical observation of Pierre Curie that the magnetic susceptibility of many materials varies inversely with the absolute temperature:

$$\chi_{para} = C/T \tag{14.4}$$

where C = the Curie constant. This equation was eventually shown to result from the competition between the ordering of molecular magnetic moments in the applied magnetic field and the randomization of orientation by thermal motion. The energy of a typical laboratory magnetic field is about 1 cm^{-1}, whereas the thermal energy at room temperature is about 100 cm^{-1}. From a theoretical treatment of such a model, Equation 14.5 was deduced:

$$\chi_{para} = N \cdot \frac{\mu^2}{3\,kT} \tag{14.5}$$

where μ = molecular magnetic dipole moment. By substitution of Equation 14.4 into Equation 14.5 we obtain

$$C = N \cdot \frac{\mu^2}{3\,k} \tag{14.6}$$

Thus, the Curie constant gives us the connection between the experimentally measured χ and the molecular property μ.

The magnetic dipole moment is produced by all the various motions of the unpaired electrons. For our purposes, the most important motions are the spinning of the electron about its own axis and the movement of the electron about the nucleus. In cases where there is no orbital degeneracy in the ground state, the electrons cannot move independently of each other, and there is no orbital contribution to the magnetic moment (we say that the orbital contribution is *quenched*). Such is the case with high-spin d^5 ions and with all configurations that have an A ground state in a ligand field, such as d^3 and d^8 in an O_h field and d^2 and d^7 in a T_d field. In these cases, the magnetic moment is given accurately by the so-called *spin-only* Equation 14.7:

$$\mu_s = 2\sqrt{S(S+1)} \tag{14.7}$$

$S = n/2$, where n is the number of unpaired electrons. The moments given by Equations 14.7–14.9 are in units of Bohr magnetons (*BM*).

If the ground state of the molecule is orbitally degenerate, there is an orbital contribution to the total moment. Two extreme cases are possible. The first is where the spin-orbit coupling energy is much smaller than the available thermal energy. In this situation, all possible J states are equally occupied, and the moment is given by Equation 14.8:

$$\mu_{L+S} = \sqrt{4S(S + 1) + L(L + 1)} \tag{14.8}$$

In the event that the spin–orbit coupling energy is greater than the thermal energy, the ground state is uniquely that with the lowest (less than half-filled shell) or highest (more than half-filled shell) J value. In this second case, the moment is given by Equation 14.9:

$$\mu_J = g\sqrt{J(J + 1)} \tag{14.9}$$

In Equations 14.7–14.9, the coefficient g is the *gyromagnetic ratio* and is given by Equation 14.10:

$$g = 1 + \frac{J(J + 1) + S(S + 1) - L(L + 1)}{2J(J + 1)} \tag{14.10}$$

It is easily seen that for the case $L = 0$ (spin-only case), $g = 2$ and Equation 14.10 reduces to Equation 14.7.

14.2 Magnetic Properties of 3d *Compounds*

All the above equations for μ allow the calculation of the number of unpaired electrons in the molecule because the various many-electron quantum numbers are a function of the number of unpaired electrons. For example, because $S = n/2$, Equation 14.7 yields the values for μ_s shown in Table 14.1.

It is clear that the ions with A ground states give excellent agreement with the spin-only moments. The less than half-filled configurations with degenerate ground states also give good agreement, but this is fortuitous. Besides the ground-state orbital contribution to the moment, there is a further contribution from the mixing of excited-state orbital contributions that is mediated by spin–orbit coupling. This contribution is taken into account by the expression

$$\mu_{\text{obs}} = \mu_s\left(1 - \frac{a\lambda}{Dq}\right) \tag{14.11}$$

where $a = $ a constant, $\lambda = $ the spin–orbit coupling constant, and $Dq = $ CF splitting. Because the sign of λ is positive for less than half-filled

Table 14.1 *Spin-Only and Experimentally Measured Moments for Some Complexes of 3d Ions*

Metal Complex	n	$\mu_s(BM)$	$\mu_{obs}(BM)$
$[Ti(H_2O)_6]^{3+}$	1	1.73	1.75
$[V(H_2O)_6]^{3+}$	2	2.83	2.80
$[Cr(H_2O)_6]^{3+}$	3	3.87	3.88
$[Mn(H_2O)_6]^{3+}$	4	4.90	4.93
$[Fe(H_2O)_6]^{3+}$	5	5.92	5.40
$[CoF_6]^{3-}$	4	4.90	4.26
$[Co(H_2O)_6]^{2+}$	3	3.87	4.85
$[CoCl_4]^{2-}$	3	3.87	4.59
$[Ni(H_2O)_6]^{2+}$	2	2.83	2.83
$[NiCl_4]^{2-}$	2	2.83	3.80
$[Cu(H_2O)_6]^{2+}$	1	1.73	1.75

shells and negative for more than half-filled shells, the excited-state contribution cancels the ground-state orbital contribution for less than half-filled shells and augments it for more than half-filled shells. It is also evident from Equation 14.11 that the excited-state correction term becomes less important for large values of Dq. This explains why the positive deviation from μ_s for more than half-filled shells is greater for T_d than for O_h complexes.

The magnetic properties of complexes can be very useful in establishing details of electronic structure and stereochemistry. This usefulness can be illustrated through consideration of the magnetic properties of some nickel complexes. Take, for example the complex $[Ni(Py)_4][ClO_4]_2$. Under slightly different preparative conditions, two complexes of this composition can be isolated. One is yellow and diamagnetic, and the other is blue with a magnetic moment of 2.9 BM. The blue complex can be assigned an O_h structure with some confidence. The yellow complex is most likely the square-planar species in which the perchlorate ions are not coordinating, but five-coordinate geometries with a single coordinating perchlorate ligand cannot be excluded on the basis of magnetic data alone. The complex $NiCl_2(PPh_3)_2$ is diamagnetic and square-planar. The analogous complex with the bulky tricyclohexylphosphine ligand has a magnetic moment of 3.5 BM, indicating a tetrahedral structure. In intermediate cases with moderately bulky phosphines, the square-planar and tetrahedral forms are in equilibrium in solution.

Magnetic measurements have also been important in the study of the spin states of metalloproteins. This topic is discussed in more detail in Chapter 24.

14.3   Magnetic Behavior of Lanthanide Compounds

In first-row d-configuration ions, spin–orbit coupling can be viewed as a small perturbation on CF splitting. For lanthanides the reverse is true due to the shielding of the $4f$ electrons from the CF by outer s and p shells and to the larger spin–orbit coupling in heavier ions. Consequently, moments of the lanthanide ions are best predicted by Equation 14.9. The predictions of Equation 14.9 for each of the lanthanide 3+ ions are compared with the experimentally observed values in Table 14.2. The agreement is generally excellent except for Sm and Eu. This discrepancy arises because the separations between the ground and excited J states are less than kT and there is a substantial thermally induced population of the excited J states. The terms of Eu(III) $(4f^6)$ are $^7F_0$, $^7F_1$, $^7F_2$, and so on. Because S and L are both 3, they cancel out in Equation 14.11, and the g value for all terms is 1.5. Using Equation 14.11, the values for the first three excited-state moments are $J = 1$, $\mu = 2.1$; $J = 2$, $\mu = 3.7$; $J = 3$, $\mu = 5.1$. It is evident from these figures that the $J = 3$ state must be substantially populated at room temperature to explain the experimentally observed moments in Table 14.2.

14.4   Cooperative Interactions

In all of the foregoing discussion, we have implicitly assumed that the magnetic moments of individual ions interact with externally applied fields but that they do not interact with each other. For first-row ions surrounded by diamagnetic, saturated ligands or for dilute solutions of metal complexes, either in liquid solvents or isomorphously substituted into a diamagnetic lattice, this is generally a good approximation. In such cases, the

Table 14.2   *Observed and Calculated (μ_J) Magnetic Moments for Some Lanthanide Ions*

Ion	$4f^n$	Ground Term	g	μ_J	μ_{exp}
Ce^{3+}	1	$^3F_{5/2}$	6/7	2.54	2.5
Nd^{3+}	3	$^4I_{9/2}$	8/11	3.62	3.6
Sm^{3+}	5	$^6H_{5/2}$	2/7	0.86	1.6
Eu^{3+}	6	$^7F_0$	1	0.00	3.4
Gd^{3+}	7	$^8S_{7/2}$	2	7.94	8.0
Dy^{3+}	9	$^6H_{15/2}$	4/3	10.63	10.6
Er^{3+}	11	$^4I_{15/2}$	6/5	9.59	9.5
Yb^{3+}	13	$^2F_{7/2}$	8/7	4.54	4.5

Figure 14.1
Schematic
two-dimensional
representation of
ferromagnetic and
anti-ferromagnetic
spin arrays.

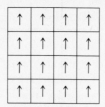

Ferromagnet

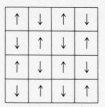

Antiferromagnet

system is said to be *magnetically dilute*, which means that whatever coupling may occur between the paramagnetic ions is small compared to the thermal energy (kT) available for disrupting the coupling.

Binary compounds and polynuclear complexes that have paramagnetic metal centers brought into close proximity or bridged by simple ligand bridges rarely meet the criterion of magnetic dilution. In these cases, the paramagnetic centers couple with the fields of their neighbors in much the same way that bar magnets become coupled at short approaches or a compass needle is coupled to the earth's magnetic field. When these macroscopic magnets interact, they only do so in a single predictable way. That is, the north poles are always as close as possible to south poles and vice versa. At the atomic level, things are more complicated because the atomic paramagnets may align themselves in either of two ways, with their spins parallel (ferromagnetic coupling) or with their spins opposed (antiferromagnetic coupling); it is impossible to easily predict which kind of coupling will occur. However, antiferromagnetic coupling is by far the more common mode. Figure 14.1 shows a schematic, two-dimensional representation of the two kinds of coupling. The coupling energy for two interacting spin systems can be expressed in the form of Equation 14.12:

$$E = J \cdot S_1 \cdot S_2 \qquad\qquad \textbf{(14.12)}$$

where J = a coupling constant and S_1 and S_2 = the two spins that are coupled. When J is positive, the lower energy is obtained when the spins have opposite signs (antiferromagnetic); when J is negative, the lower energy results when the spins have the same sign (ferromagnetic).

The spin ordering illustrated in Figure 14.1 only occurs if the coupling energy is greater than the available thermal energy. When the thermal energy exceeds the coupling energy, the spins can tumble and assume the random orientation relative to each other that is characteristic of paramagnetic behavior. Thus, for both ferromagnetic and antiferromagnetic systems, there is a temperature above which the cooperative behavior disappears and the system obeys Curie's law. Figure 14.2 illustrates the way in which the susceptibility changes with temperature for the various kinds of magnetic behavior. The transition temperature for change from

Figure 14.2
Variation of
susceptibility with
temperature for a
ferromagnet, an
antiferromagnet, and
a paramagnet.

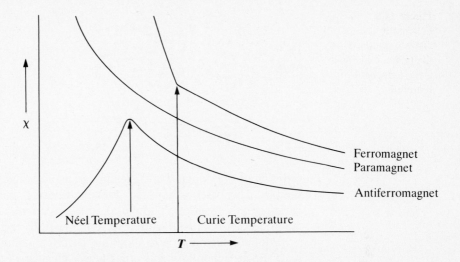

the ferromagnetic to the paramagnetic state is called the *Curie temperature* and that for the antiferromagnetic to paramagnetic is called the *Néel temperature*.

Antiferromagnetism hides the microscopic magnetic moments of the individual atoms or ions, and on a macroscopic scale the material appears to be diamagnetic. Thus in MnO, which has the NaCl structure, each Mn(II) is in a ^6S state, but the unit cell contains equal numbers of ions of $S = +\frac{5}{2}$ and $-\frac{5}{2}$ and there is no net magnetic moment. Even though the material is magnetically inactive, the presence of high-spin Mn(II) ions can be demonstrated by neutron-scattering experiments.

Antiferromagnetic behavior of the kind exhibited by MnO is very common among the structurally simple, binary compounds of the transition elements. As a rough rule, the Néel temperatures of oxides are well above room temperature whereas those of halides are lower, often well below room temperature.

Ferromagnetism is a much more useful phenomenon than is antiferromagnetism, but it is more rarely encountered. It occurs in the metals Fe, Co, and Ni and their alloys, and several of the lanthanide metals are ferromagnetic. The dramatic manifestation of ferromagnetism is the exhibition of a giant permanent magnetic dipole in ferromagnetic materials. As a general rule, this dipole is not observably present in an unmagnetized sample because the ordering of spins does not extend uniformly through the whole sample. Instead, there are macroscopic *domains* in which the spins are all aligned in the same direction, but a crystal normally contains many such domains randomly oriented, as shown in Figure 14.3. The average of the moment of these domains sums to 0. Applying an external force such as an applied magnetic field or a mechanical stress can lead to

Figure 14.3
Schematic
representation of
magnetic domains in
a ferromagnetic
material.

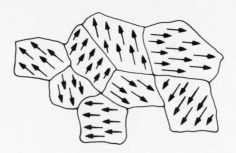

ordering of the domains and the appearance of a net giant moment in the sample. The ease with which domain ordering and disordering occurs varies greatly from one alloy to another, and a whole range of magnetic materials can be manufactured to satisfy various technological demands.

Metallic ferromagnetic alloys are used for making magnets for large-scale use in generators and other heavy electrical equipment. Such uses require massive metal, and the problem of oxidative corrosion is not serious. This is not the case in situations where the magnetic material must be very finely divided. Finely divided iron, for example, undergoes fairly rapid oxidation in moist air to the antiferromagnetic FeO and Fe_2O_3. Fortunately, there are several metal oxides that are ferromagnetic, and these are not subject to problems of further oxidation. The most commonly used are Fe_3O_4 and CrO_2. These are the materials that are used in the manufacture of high-fidelity magnetic storage devices such as magnetic tapes and disks. In these devices, information is stored in the form of magnetized and unmagnetized regions. To get high-density data storage, the ferromagnetic particles must be very small and therefore resistant to attack by ambient chemical species such as O_2 and H_2O.

It is impossible to explain the origin of ferromagnetism in any simple way, nor can the magnetic behavior of a material be easily predicted. The problem is dramatically illustrated by comparing the magnetic and electronic properties of the isostructural neighbors of Fe_3O_4 (inverse spinel structure) and CrO_2 (distorted rutile structure). Table 14.3 lists the relevant properties.

An interesting approach to the synthesis of ferromagnetic materials has recently been reported. This approach takes advantage of the easily achieved antiferromagnetic coupling, rather than trying to achieve the poorly understood ferromagnetic coupling. The underlying principle is to allow chains of two alternating ions of differing spin to align antiferromagnetically. For example, a chain of Cu^{2+} and Mn^{2+} would give alternating and opposed spins of, for example, $+\frac{1}{2}$ and $-\frac{5}{2}$, as shown in Figure 14.4a. If such chains are stacked in two dimensions, the resulting structure can be either antiferromagnetically or ferromagnetically coupled, as illustrated in Figures 14.4b and c. The successful alignment of the chains in the form

Table 14.3 *Some Magnetic and Electronic Properties of Some Binary Oxides*

Compound	Mn_3O_4	Fe_3O_4	Co_3O_4	TiO_2	VO_2	CrO_2	MnO_2
Structure	S^a	I	S	R	R	R	R
Electronic property	sem	met	sem	ins	sem	met	sem
Magnetic property	A	F	A	D	A	F	A

[a] S = spinel; I = inverse spinel; R = rutile; sem = semiconductor; met = metal; ins = insulator; A = antiferromagnet; F = ferromagnet; D = diamagnetic.

Figure 14.4
Alignment of spins in an antiferromagnetically coupled alternating chain of Cu^{2+} and Mn^{2+} ions.
(a) One-dimensional ferromagnet,
(b) two-dimensional antiferromagnet, and
(c) two-dimensional ferromagnet.

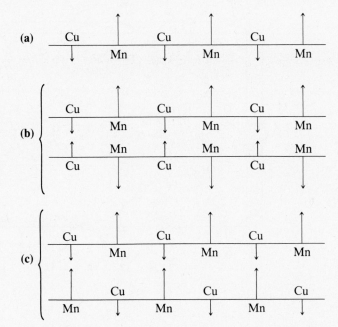

shown in Figure 14.4c is a remarkable example of architectural coordination chemistry. The ligand used to achieve this alignment is 2-hydroxy-1,-3-bis(oxamate). Figure 14.5 shows the structure of this ligand. In the crystal, the ligand links the Cu^{2+} and Mn^{2+} ions into infinite, parallel polymeric chains. Figure 14.6 shows the structure of the unit cell.

Although the compound shown in Figure 14.6 only becomes ferromagnetic at about 4.6 K, its synthesis represents a major achievement in the application of systematic molecular architecture to a problem in materials science, and this result opens the way to further advances toward the achievement of practical molecular ferromagnets.

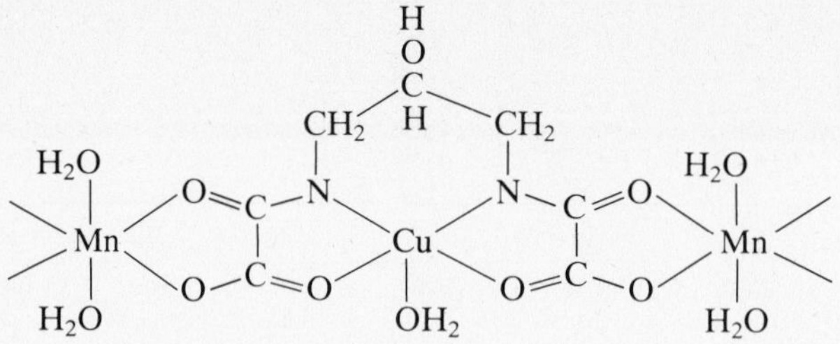

Figure 14.5
The
2-hydroxy-1,3-
propylene-
bis(oxamate) ligand.

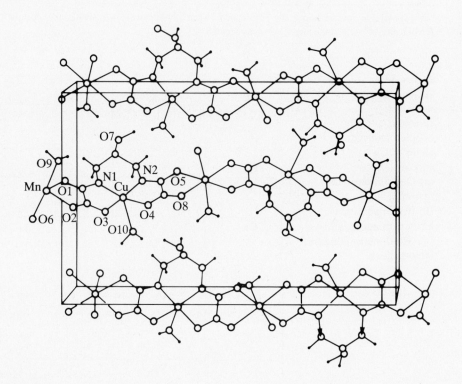

Figure 14.6
Crystal structure of
the ferromagnetic
Cu^{2+}/Mn^{2+}
oxamato-complex.
Reprinted with
permission from
Y. Pei, J. Am.
Chem. Soc. 108
(1986): 7428.

BIBLIOGRAPHY

General Texts

Cotton, F. A., and G. Wilkinson. *Advanced Inorganic Chemistry*, 5th ed. New York: Wiley, 1988.

Huheey, J. E. *Inorganic Chemistry: Principles of Structure and Reactivity*. New York: Harper & Row, 1983.

Jolly, W. L. *Modern Inorganic Chemistry*. New York: McGraw-Hill, 1984.

Moeller, T. *Inorganic Chemistry: A Modern Introduction*. New York: Wiley, 1982.

Porterfield, W. W. *Inorganic Chemistry: A Unified Approach*. Reading, Mass.: Addison-Wesley, 1984.

Suggested Reading

Dunn, T. M., R. S. McClure, and R. G. Pearson. *Some Aspects of Crystal Field Theory*. New York: Harper & Row, 1965.

Figgis, B. N. *Introduction to Ligand Fields*. New York: Wiley, 1966.

Lewis, J., and R. G. Wilkins. *Modern Coordination Chemistry*. New York: Interscience Publishers, Inc., 1960.

PROBLEMS

14.1. List the four main types of magnetic behavior. Briefly describe the characteristics of each type of behavior.

14.2. Write the equations for the magnetic moment of an atom in which
   a. Only the electron spin contributes to the moment.
   b. Both spin and orbital moments are present, but spin-orbit coupling is negligible.
   c. Spin-orbit coupling is strong.

14.3. Explain the following observations:
   a. $K_4[Mo(CN)_8]$ is diamagnetic.
   b. $NiCl_2(OPPh_3)_2$ has a magnetic moment of 3.9 *BM*.
   c. $Na_4[Co(NO_2)_6]$ has a magnetic moment of 1.8 *BM*.
   d. $K_3[Fe(CN)_6]$ has a magnetic moment of 2.4 *BM*.

14.4. A transition metal chloride, MCl_3, was found to have a molar paramagnetic susceptibility (after the usual corrections) of $\chi_M = 1.4 \times 10^{-2}$ at 300 K. Reaction of the chloride with aqueous cesium cyanide yielded a complex $Cs[M(CN)_6]$ with a corrected $\chi_M = 1.25 \times 10^{-3}$ at 300 K. Suggest an electronic configuration for M^{3+} and explain the change in susceptibility.

14.5. Explain the following:
   a. Mn(II) is the only divalent ion of the first transition period whose magnetic moment is correctly predicted by the formula $\mu = g\sqrt{J(J + 1)}$.
   b. Gd(III) is the only trivalent lanthanide ion whose magnetic moment is correctly predicted by the spin-only formula.
   c. The $[Cl_5MoOMoCl_5]^{2-}$ ion has no measurable paramagnetic moment.
   d. Potassium tetracyanonickelate(II) is diamagnetic.

14.6. Room-temperature experimental data gave the magnetic moments listed in the following table for three lanthanide ions:

Ion	No. of 4f Electrons	Ground Term	μ_{exp} (BM)
Nd^{3+}	3	$^4I_{9/2}$	3.6
Eu^{3+}	6	$^7F_0$	3.4
Dy^{3+}	9	$^6H_{15/2}$	10.63

Assuming strong spin-orbit coupling, what are the theoretical moments for these ions? For which ion is there a significant population of excited J states at room temperature?

14.7. A $3d$ transition metal dissolved in dilute perchloric acid under nitrogen gave, after evaporation, a compound $M(ClO_4)_2 \cdot 6H_2O$ with a corrected molar paramagnetic susceptibility of 1.04×10^{-2} at 300 K. After the compound was redissolved in dilute perchloric acid in air and the solution reevaporated, a new compound $M(ClO_4)_3 \cdot 6H_2O$ was recovered. This compound had a corrected molar susceptibility of 0.62×10^{-2} at 300 K. Determine the electronic structures of M in its two oxidation states and suggest an identity for M.

PART FIVE

Complex Compounds: Rings, Chains, Cages, and Clusters

15

Inorganic Polymers

15.1 The Bridge Between Small and Infinite Molecules

Much of chemistry deals with relatively small molecules of well-defined molecular weight and structure. An equally large domain of chemistry deals with substances that are essentially infinite aggregates of atoms or ions. The latter, which include metals, ionic solids, and infinite covalent solids, exist only in the solid state and can be dissolved or volatilized only by breaking down the infinite structure to small aggregates. Usually, this results in a profound change in the properties of the material.

In between the domains of discrete molecular compounds and infinite solids, there is a third world in which molecules are large but far from infinite in size and usually polydisperse (containing molecules of differing molecular weights, but having the same empirical formula). In some cases, the high molecular weight material may have chemical properties that are essentially the same as those of the small analog—for example, polyethylene is chemically very similar to low molecular weight paraffins.

Figure 15.1
Schematic illustration of the similarity between long chains and large rings.

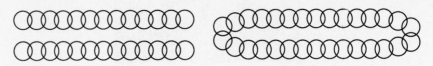

Two Parallel Chains   **A Large Ring**

In other cases such as metals, the properties of a bulk metal can be very different from those of a cluster of twenty or thirty atoms, and it is still not clear how big a cluster must be before it has fully developed metallic properties.

In this chapter, we review a number of topics that relate to the subject of inorganic macromolecules. The agglomeration of atoms into macromolecules can occur in one, two, or three dimensions. In the one-dimensional case, we include all molecules that are built up from linear, unbranched chains. This definition includes both chains and isolated rings, even though the latter are strictly speaking two-dimensional objects. This definition is excusable on the grounds that the properties of linear chains and isolated rings are generally very similar, especially when the size of the molecule is very large. Indeed, it becomes extremely difficult to distinguish between them at high degrees of polymerization. The only real distinguishing feature between long chains and large rings is the presence of chemically distinct chain ends in the former. At very high molecular weights, the chain ends become very dilute and difficult to detect. These factors should be evident from the schematic illustration in Figure 15.1. An important property shared between chains and isolated rings is that they can be dispersed in molecular form, either by dissolving in a solvent or if the molecular weight is not too high by vaporization.

To form isolated chains or rings, a polymerizing species must have only two coordination positions free. For example, sulfur in elemental sulfur always has a coordination number of two but exists as a wide variety of allotropes, which may be either linear chains or rings of different sizes. Figure 15.2 shows some examples of known polysulfur rings. It is possible that plastic sulfur, produced by rapidly quenching molten sulfur at 200 °C to room temperature, contains rings up to much larger sizes, but because there is no known physical method for detecting their presence, we do not know for sure. For reasons that are not understood, the thermodynamically stable form of sulfur is the eight-membered ring, and all other forms revert rather easily to octasulfur at room temperature. As will be seen below, the equilibration of rings with chains is an important general method of making inorganic polymers.

The equilibrium between rings and chains is influenced by both enthalpic and entropic factors. In the free energy of the reaction $\Delta G = \Delta H - T\Delta S$, a larger translational entropy favors many small molecules over a

Figure 15.2
Structures of some
polysulfur rings in
the solid state.

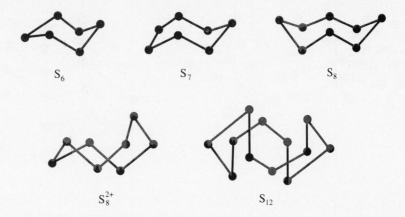

few large molecules, and the negative contribution of the entropy term
becomes greater with increasing temperature. The enthalpy term for ring-
opening polymerization is mainly derived from the release of ring strain.
For small rings (six to eight atoms or less), this strain energy can be large
enough to drive the reaction in the direction of polymer formation.

When branching occurs on isolated rings or chains, it is possible to
build up two- and three-dimensional infinite structures. Infinite sheets are
produced by linking together isolated rings or chains. Figure 15.3 illustrates this process. Such structures require a coordination number of three
at the branch point. The elements of Group VA (15) provide simple
examples of such structures. Phosphorus, arsenic, and antimony form
puckered sheets in one of their allotropic forms, as illustrated in Figure
15.4. Because these forms can be obtained as single crystals, their structures have been determined unequivocally by X-ray crystallography. The
formation of sheets requires that the conformations about the bonds in the
structure be highly regular. For example, black arsenic is made up of
interconnected cyclohexanelike rings in which all of the ring bonds assume

Figure 15.3
Formation of infinite
sheets by cross-
linking chains
or rings.

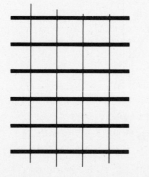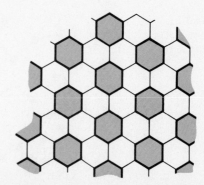

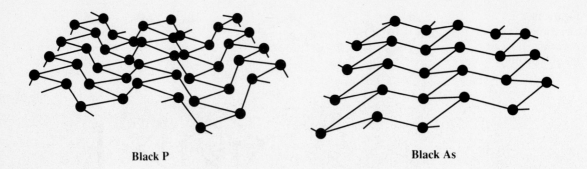

Black P   Black As

Figure 15.4
Ring conformations
and inter-ring
stereochemistries in
the black allotropes
of P and As.

a trans conformation, as shown in Figure 15.4. All ring interconnections are through equatorial bonds in the arsenic and antimony structures, but in black phosphorus the interconnections are [trans,cis,cis], as shown in Figure 15.4.

If the conformations about the bonds are not fixed with respect to repetitivity or angle, the closing of a ring becomes a problem in statistics, both with regard to ring size and to the orientation of the ring in space. Under such conditions, a random three-dimensional network results, as shown schematically in Figure 15.5. Such a network has no order and is an amorphous glass (see Chapter 19). This is the three-dimensional analog of the randomly coiled chain of plastic sulfur.

Group VA (15) elements can be assembled into rings or chains by blocking one of the coordination positions with an organic group:

$$4(CF_3)AsCl_2 + 4Hg \rightarrow [(CF_3)As]_4 + 4HgCl_2 \qquad (15.1)$$

The interconnection of four-coordinate, tetrahedral atoms can only occur in one way if the exocyclic bonds of the six-membered rings are constrained to be in an all trans arrangement. This is the diamond structure (Figure 18.4). Another way of visualizing the diamond structure is as sheets of carbon atoms, isostructural with the black arsenic sheets, with

Figure 15.5
Schematic
illustration of the
random
interconnection of
three-coordinate
atoms.

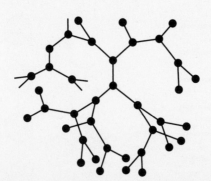

Figure 15.6
The diamond structure as axially interconnected sheets of cyclohexanelike rings.

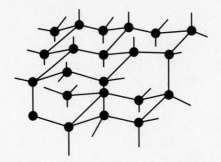

the sheets linked to similar sheets above and below through axial bonds (Figure 15.6).

Diamond cannot be prepared in an amorphous form because it is thermodynamically unstable relative to graphite under ambient conditions. Graphite has an infinite sheet structure made up of planar six-membered rings (Figure 15.7). Part of the extraordinary stability of graphite may be attributed to the large resonance energy associated with its infinite π system. There is some evidence that moderately stable structures consisting of clusters of carbon atoms can exist in the gas phase, but these species are not subunits of a diamond-type lattice. A particularly stable entity of this kind that has been identified is C_{60}, which may have a truncated icosahedral structure (Figure 15.8). If this structure is correct, the molecule is closer to being a graphite-type subunit folded back on itself to form a sphere than to being a diamond-type subunit.

Silicon can be produced in a metastable amorphous form. In this material, the silicon is still believed to be mostly in tetrahedral coordination, but the interconnections between tetrahedra are not systematically with a trans conformation. The resulting amorphous material has many unpaired electrons due to "dangling" bonds and is consequently very

Figure 15.7
Structure of graphite.

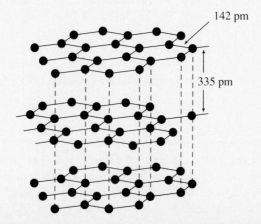

142 pm

335 pm

Figure 15.8
Truncated
icosahedral structure
proposed for C_{60}.
◯ = C atoms in
front hemisphere;
● = C atoms in rear
hemisphere.

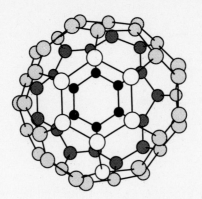

chemically reactive. A material of approximate composition [SiH], produced by passing a microwave discharge through SiH_4, is believed to have a similar structure, but the dangling bonds are capped with hydrogen atoms.

In principle, capping one of the coordination sites of a tetrahedral Group IVA (14) atom with a substituent gives rise to the possibility of producing a sheet structure, isostructural with the Group VA (15) sheets described above. Such structures have never been reported for a number of reasons. In the case of carbon, the difficulty of removing defects from such a structure during its synthesis presents a serious problem because of the high C—C bond energy. [This is a much less severe problem in the case of the heavier congeners of Group VA (15) where the bonds can be fairly easily broken and remade under relatively mild conditions.] Another problem is the strong tendency of such a structure to convert to graphite under high-temperature conditions. The absence of a graphite analog for the heavier congeners of Group IVA (14) and the weaker bonds between these elements suggest the possibility of making puckered-sheet structures, but they have not yet been reported.

In the same way that the polymer of a difunctional atom can close on itself to form a ring, the polymers of atoms of higher functionality can close on themselves to form cluster or cage compounds. The simplest examples of this behavior are the tetrahedral molecules P_4, As_4, and Sb_4. Group IVA (14) analogs, which would include tetrahedrane $(CH)_4$, are not known, but larger structures such as cubane $(CH)_8$ have been made.

15.2 Polyboranes

Whereas the polymers built of Group IVA (14) atoms readily form open structures such as rings and chains, analogous compounds of boron have a strong tendency to form cage structures. The underlying reason for this difference has already been touched on in our earlier discussion of diborane (Chapter 9). We concluded that hydrogen-bridge bonding can be used by boron to raise the energy of its unused nonbonding orbital with an accompanying stabilization of some of its filled bonding orbitals. On the other hand, ethane would lose energy if it adopted a hydrogen-bridged structure. In the higher boron hydrides, boron takes advantage of multicenter bonding to both hydrogen and other boron atoms to achieve the same end.

Structures of Polyboranes

All known boron hydrides can be considered as cage compounds based on deltahedral structures. A deltahedron is a regular three-dimensional object, all of whose facets are equilateral triangles. Figure 15.9 shows a number of the more symmetrical deltahedra. Besides the deltahedra themselves, Figure 15.9 also shows the objects produced by removing one and two adjacent vertices from the parent deltahedra. These three series of objects provide the frameworks for the structures of most known boranes. The three families of structures shown in Figure 15.9 are referred to as the *closo* (closed), *nido* (nestlike), and the *arachno* (spiderlike) structures.

   The closoboranes are always encountered as the dianions, $[B_nH_n]^{2-}$, which are known for $n = 6-12$. Table 15.1 lists some representative boron hydrides of the three structural classes. Figure 15.10 shows the structures of the three compounds $[B_6H_6]^{2-}$, B_5H_9, and B_4H_{10}. It can be seen from the molecular formulas that the structural progression shown in Figure 15.10 involves not only the removal of boron atoms from the parent deltahedron but also an increase in the ratio of hydrogen to boron.

   In naming the boron hydrides, neutral molecules are always called boranes, and anions are called hydroborates. The number of boron atoms in a neutral species is designated by a Greek prefix, and the number of hydrogen atoms is designated with a parenthetical Arabic numeral at the end. For example, B_5H_9 is called pentaborane(9). In the names of anionic species, both the numbers of boron and hydrogen atoms are designated with Greek prefixes, and the number of negative charges is designated in parentheses at the end. For example, $[B_6H_6]^{2-}$ is called hexahydrohexaborate(2−).

Electronic Structure and Bonding in Polyboranes

Early bonding concepts were focused on the principle that the completion of an electron octet about a *p*-group atom is an essential feature of bonding. Although this simple idea works remarkably well for atoms with four or more valence electrons, it is impossible to construct compounds

Figure 15.9
Idealized deltahedra and framework structures for some members of the three families of boranes.

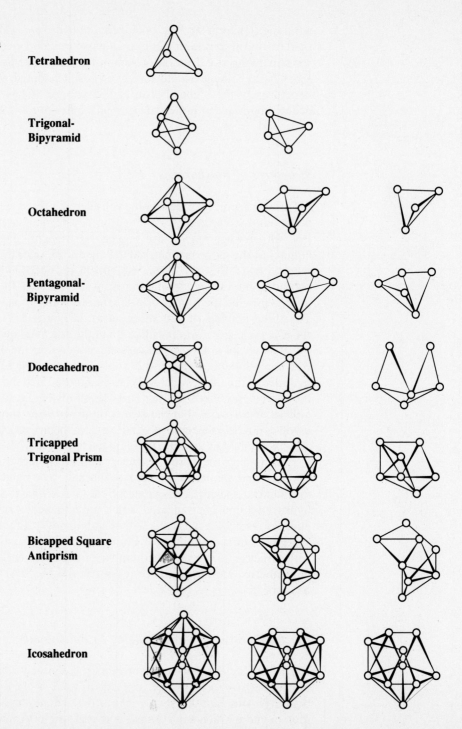

Tetrahedron

Trigonal-Bipyramid

Octahedron

Pentagonal-Bipyramid

Dodecahedron

Tricapped Trigonal Prism

Bicapped Square Antiprism

Icosahedron

Table 15.1	*Some Boron Hydrides of the Three Major Structural Classes*	
Closo	Nido	Arachno
$[B_6H_6]^{2-}$	B_2H_6	B_4H_{10}
$[B_8H_8]^{2-}$	B_5H_9	B_5H_{11}
$[B_{12}H_{12}]^{2-}$	B_6H_{10}	B_6H_{12}
	B_8H_{12}	B_8H_{14}

Figure 15.10 Structures of closohexaborate, nidopentaborane(9), and arachnotetraborane(10).

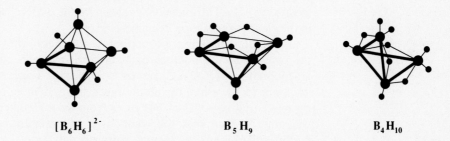

$[B_6H_6]^{2-}$　　　　B_5H_9　　　　B_4H_{10}

containing only boron and one-electron donor atoms that fully satisfy the octet rule. In addition, the classical concepts of octets and localized tetrahedral bonding provide no clue as to why the boranes adopt deltahedral structures. A clearer insight into this problem was furnished by applying MO methods to the problem; Lipscomb received the Nobel Prize in 1976 for his pioneering work on the theoretical description of bonding in the boranes.

Before examining the bonding in boranes, it is useful to look at some qualitative features of the structures and at the way in which the electrons can be counted and partitioned in the molecule. There are three distinct types of bonding interaction in these molecules: (a) two-center interactions between B and H, of which there will be at least one per B atom in the molecule (terminal B—H bonds); (b) three-center interactions involving two B atoms and one H atom (H bridges), and (c) multicenter interactions involving only B atoms. The terminal B—H bonds are conventional, and we can partition them out from a discussion of the bonding within the cluster. These bonds use one electron and one orbital from the boron and one electron and one orbital from the hydrogen. The remaining three-valence orbitals and two electrons from each boron atom are considered to be used in the framework bonds. The bridging H atoms are considered to contribute one orbital and one electron to the framework bonds. Any excess negative charges on the species are considered to contribute to the framework electron count. Thus, we can partition the units of any borane in the following manner:

$$[B_aH_b]^{c-} \rightarrow [(BH)_aH_{b-a}]^{c-}$$

From this partioning we can see that there will be $(a + (b - a)/2 + c/2)$ pairs of electrons in the framework, and it is expected that an ideal structure provides just that number of bonding orbitals.

The generic formula for a closoborane is $[B_nH_n]^{2-}$ — that is, in the partitioned formula above, $a = b$, and so the number of framework electrons is $(a + c/2)$ or $(n + 1)$. Although the proof goes beyond the scope of this book, it can be shown from symmetry arguments that the number of bonding orbitals associated with a regular deltahedron with n vertices is always $(n + 1)$. The underlying reason for the preference for such geometry in the closoboranes then becomes evident.

A similar treatment of nidoboranes, generic formula B_nH_{n+4}, yields the result that the number of framework electron pairs is $(n + 2)$; for arachnoboranes (B_nH_{n+6}), it is $(n + 3)$. So, for a series such as that shown in Figure 15.10, where the nidoboranes and arachnoboranes are generated by sequential removal of BH units from an n-membered closoborane, it can now be seen that the compounds are related by having the same number of framework electron pairs, namely $(n + 1)$. Detailed MO calculations also show that the number of bonding orbitals remains constant for such a series, although there is a tendency for the HOMOs to become less bonding and for their electrons to assume some of the character of basic lone pairs. This analysis of the relationship between structure, framework electron pairs, and bonding MOs was first suggested by Wade.

The high symmetry of $[B_6H_6]^{2-}$ and the absence of H bridges greatly simplifies the problem of constructing linear combinations of AOs to describe the framework bonding. We will describe the bonding in more detail for this case because the principles involved are generally applicable to more complicated structures. In the case of $[B_6H_6]^{2-}$, each boron is contributing three-valence orbitals to the framework; therefore, there are eighteen framework MOs. Also, each boron contributes two electrons, and with the two negative charges, this yields seven electron pairs.

To construct the MOs, we choose a coordinate system such that each boron atom has a pair of p orbitals perpendicular to the body diagonal of the octahedron and a third p orbital directed along the body diagonal. To simplify matters, we hybridize the s and p orbitals and use one hybrid to form the terminal B—H bond and the other, which projects into the interior of the molecule, to form bonds with the other boron atoms. This hybrid orbital and the two perpendicular p orbitals constitute the three-valence orbitals involved in framework bonding. Figure 15.11 shows the eighteen linear combinations of the framework orbitals. The combinations are arranged in order of decreasing bonding character; note that they

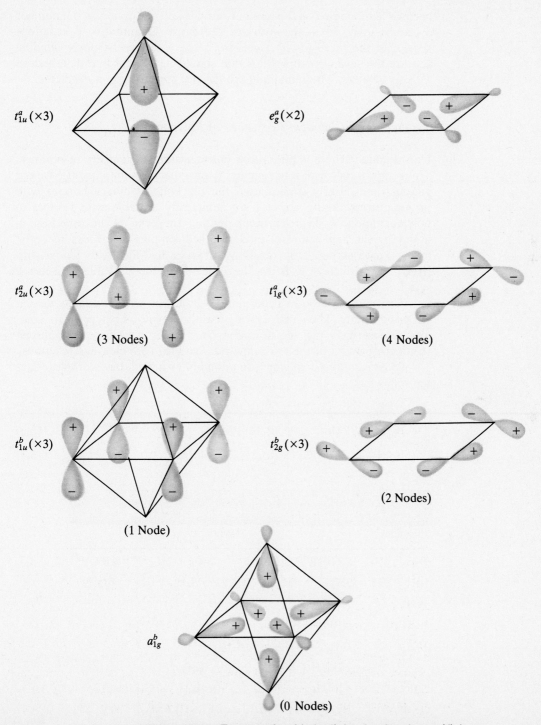

Figure 15.11 Framework orbitals of the closohexaborate(6) ion.

become less bonding as the number of nodes increases and as the number of atoms involved in the overlaps decreases. Quantitative calculations show that the three lowest levels ($a_{1g} + t_{2g} + t_{1u}$) are strongly bonding and are the seven bonding MOs that should be present in a deltahedron with six vertices. The remaining orbitals are antibonding in character.

Synthesis and Chemical Properties of Polyboranes

Unfortunately, there is little order in the synthetic chemistry of boranes. Three approaches with some generality have been commonly employed at various times. Early studies used the reactions of binary borides with protonic acids. Many boranes are sufficiently unstable with respect to polymerization or disproportionation that they can be decomposed to others by the application of moderate heat or pressure. More recently, many syntheses based on reactions of lower hydrides with bases, with $[BH_4]^-$ or with ionic hydrides, have been developed. Table 15.2 lists a selection of syntheses.

The neutral boron hydrides tend to be Lewis acids and many exhibit Brønsted acidity. Thus, with Lewis bases three types of reactions are commonly observed: (a) the formation of Lewis acid–base complexes, (b) deprotonation of the boron hydride, and (c) base cleavage reactions.

Some reactions involving formation of Lewis acid–base complexes are the following:

$$B_2H_6 + 2Et_2O \quad \rightarrow \quad 2Et_2OBH_3 \qquad\qquad (15.2)$$

$$B_5H_9 + 2PMe_3 \quad \rightarrow \quad B_5H_9[PMe_3]_2 \qquad\qquad (15.3)$$

$$B_{10}H_{14} + 2MeCN \rightarrow B_{10}H_{14}[NCMe]_2 \qquad\qquad (15.4)$$

Table 15.2 *Some Reactions for the Synthesis of Higher Boranes*

Product	Synthesis
B_2H_6	$2NaBH_4 + 2H_2PO_3F \rightarrow B_2H_6 + 2H_2 + 2NaHPO_3F$
B_4H_{10}	$2MgB_2 + 4Mg + 12HCl \rightarrow B_4H_{10} + H_2 + 6MgCl_2$
B_5H_9	$5B_2H_6 \xrightarrow{250\ ^\circ C} 2B_5H_9 + 6H_2$
$[B_6H_6]^{2-}$	$2[BH_4]^- + 2B_2H_6 \xrightarrow{100\ ^\circ C} [B_6H_6]^{2-} + 7H_2$
B_5H_{11}	$B_4H_{10} + KH \rightarrow K[B_4H_9] + H_2$
	$2K[B_4H_9] + B_2H_6 \xrightarrow[\text{ether}]{-35\ ^\circ C} 2K[B_5H_{12}]$
	$K[B_5H_{12}] + HCl(l) \xrightarrow{-110\ ^\circ C} B_5H_{11} + H_2 + KCl$
$[B_{12}H_{12}]^{2-}$	$B_{10}H_{14} + 2Et_3NBH_3 \rightarrow [Et_3NH]_2B_{12}H_{12} + 3H_2$

Figure 15.12
Molecular structure
of $B_5H_9[PMe_3]_2$.

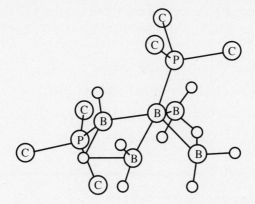

Figure 15.12 shows the structure of the product of the reaction in Equation 15.3. This compound is interesting in that it is a member of the very rare class of *hyphoboranes*, which have $n + 4$ framework electron pairs [the two phosphines each donate an electron pair into the framework of the *nidopentaborane(10)*]. Although the two phosphines occupy chemically different positions, at the apex and at a base corner of the square pyramid, dynamic ^{11}B-NMR measurements in solution indicate that there is a fluxional process that moves the basal phosphine between the four corners of the pyramid base but does not interchange the apical and basal phosphines.

Proton abstraction from boranes usually occurs at the bridging hydrogen positions. The four bridging hydrogens of decaborane(14) can be selectively deuterated with D_2O, as shown in Figure 15.13. The order of acidity of some boron hydrides is the following:

Nido
$$B_5H_9 < B_6H_{10} < B_{10}H_{14} < B_{16}H_{20}$$

Arachno
$$B_4H_{10} < B_5H_{11} < B_6H_{12}$$

Figure 15.13
Structure of
$B_{10}H_{10}D_4$. The
bridging D atoms
are shown in black.

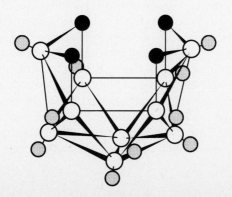

Base cleavage may occur homolytically to give neutral products or heterolytically to give ionic products. Some examples are given in Equations 15.5–15.8:

$$B_2H_6 + 2Me_3N \rightarrow 2Me_3NBH_3 \qquad \textbf{(15.5)}$$

$$B_2H_6 + 2NH_3 \rightarrow [(H_3N)_2BH_2]^+ \, [BH_4]^- \qquad \textbf{(15.6)}$$

$$B_5H_{11} + 2CO \rightarrow H_3BCO + B_4H_8(CO) \qquad \textbf{(15.7)}$$

$$B_6H_{12} + Me_3P \rightarrow Me_3PBH_3 + B_5H_9 \qquad \textbf{(15.8)}$$

Carboranes and Metallaboranes

A large number of heteroboranes in which one or more boron atoms have been replaced by other atoms are known. Over half of the elements in the periodic table have been incorporated into borane structures, and there is no reason to believe that the number will not eventually include most of the elements. The most widely studied examples of this type of compound are the carboranes. The availability of the carboranes stems from the facile reactions of acetylene with the boranes and their much greater stability to oxidation and hydrolysis makes them much easier to study than the boranes themselves.

Some typical carborane syntheses are shown in Equations 15.9 and 15.10:

$$2B_8H_{12} + MeC\equiv CMe \rightarrow 2[Me_2C]B_7H_9 + B_2H_6 \qquad \textbf{(15.9)}$$

$$B_{10}H_{14} + HC\equiv CH \rightarrow C_2B_{10}H_{12} + 2H_2 \qquad \textbf{(15.10)}$$

The initial (kinetically determined) product of such reactions usually has the two carbon atoms adjacent to each other. High temperatures result in migration of the carbon atoms to positions more distant from each other to give thermodynamically more stable products. Figure 15.14 shows the structures of some carboranes.

Because the CH group contributes three electrons to the carborane cage (compared to the two electrons contributed by BH), the neutral closo-$C_2B_nH_{n+2}$ carboranes are isoelectronic with the closo-$[B_nH_n]^{2-}$ anions.

An extraordinary variety of metallaboranes and metallacarboranes have been synthesized. Reactions are usually carried out with metal halides or organometallics under basic conditions. There are usually several major products from any given reaction, and these can be separated by column chromatography. Some metallaborane and organometallaborane syntheses are shown in Equations 15.11–15.17. The nidodianion $[C_2B_9H_{11}]^{2-}$ is electronically very similar to the cyclopentadienyl anion. It is easily synthe-

Figure 15.14
Framework
structures of some
carborane
molecules. The
carbon atoms are
black.

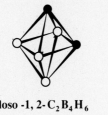

closo -1, 2- $C_2B_4H_6$

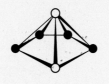

nido -3, 4, 5, 6- $C_4B_2H_6$

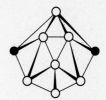

arachno -1, 3- $C_2B_7H_{13}$

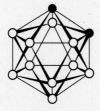

closo -1, 2-$C_2B_{10}H_{12}$
(ortho)

closo -1, 7- $C_2B_{10}H_{12}$
(meta)

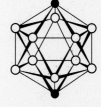

closo -1, 12- $C_2B_{10}H_{12}$
(para)

sised by degradation of closo-$C_2B_{10}H_{12}$ as shown in Equations 15.16 and 15.17:

$$NaB_5H_8 + CoCl_2 + NaC_5H_5 \xrightarrow{\text{THF, } -20\,°C} 2\text{-}[Co(C_5H_5)B_4H_8] \quad \textbf{(15.11)}$$

$$2\text{-}[Co(C_5H_5)B_4H_8] \xrightarrow{\text{Pyrolysis, } 200\,°C} 1\text{-}[Co(C_5H_5)B_4H_8] \quad \textbf{(15.12)}$$

$$B_{10}H_{14} + 2ZnEt_2 \xrightarrow{\text{Et}_2O} [B_{10}H_{12}Zn(Et_2O)_2] + 2C_2H_6$$
$$\textbf{(15.13)}$$

$$[B_{10}H_{12}Zn(Et_2O)_2] \xrightarrow{\text{H}_2O/\text{Me}_4NCl} [Me_4N]_2[Zn(B_{10}H_{12})_2]$$
$$\textbf{(15.14)}$$

$$2[C_2B_9H_{11}]^{2-} + FeCl_2 \xrightarrow{\text{THF}} [Fe(C_2B_9H_{11})_2]^{2-} + 2Cl^-$$
$$\textbf{(15.15)}$$

$$1,2\text{-}C_2B_{10}H_{12} + EtO^- \xrightarrow{\text{EtOH/85}\,°C} [7,8\text{-}C_2B_9H_{12}]^- \quad \textbf{(15.16)}$$

$$[7,8\text{-}C_2B_9H_{12}]^- + NaH \longrightarrow [7,8\text{-}C_2B_9H_{11}]^{2-} + H_2 + Na^+$$
$$\textbf{(15.17)}$$

The structures of the products of the reactions in Equations 15.12–15.15 are shown in Figure 15.15. In the product of the reaction in Equation 15.11, cobalt occupies the apical (1)-position of the SQP structure.

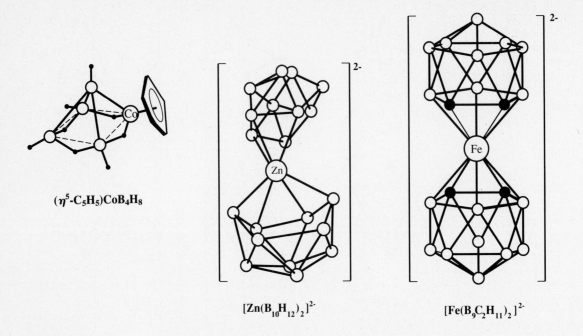

$(\eta^5\text{-}C_5H_5)CoB_4H_8$

$[Zn(B_{10}H_{12})_2]^{2-}$

$[Fe(B_9C_2H_{11})_2]^{2-}$

Figure 15.15
Structures of
$1\text{-}[Co(C_5H_5)B_4H_8]$,
$[Zn(B_{10}H_{12})_2]^{2-}$, and
$[Fe(C_2B_9H_{11})_2]^{2-}$.

The pyrolysis step of Equation 15.12 leads to a migration of cobalt to a corner of the square base (2-position). Co(III) of the $[CpCo]^{2+}$ unit is analogous to a CH unit and contributes three of its six d electrons to bonding to the Cp^- and three to the framework. The boron fragment can be viewed as the dianion of arachnotetraborane(10). The resulting complex is isoelectronic and isostructural with nidopentaborane(9). The Fe(II) of the carborane complex functions in a very similar manner.

The Elusive Linear Polyboranes

All known polyboranes have cage structures, and no linear chains have been reported. This is not surprising in the case of the binary hydrides because of their tendency to form intramolecular hydrogen bridges and B—B bonds. On the other hand, we might expect polymerization of Lewis-base adducts of BH_3 to form stable polymeric chains of the kind

$$\begin{array}{ccccc} L & L & L & L & L \\ | & | & | & | & | \\ -B\!-\!\!&\!\!B\!-\!\!&\!\!B\!-\!\!&\!\!B\!-\!\!&\!\!B\!- \\ | & | & | & | & | \\ H & H & H & H & H \end{array}$$

(where L = a Lewis base) because such a molecule is isoelectronic with a polycarbon chain. A serious problem with such a molecule is the steric hindrance that arises from having a substituent on every atom of the chain.

Even in the case of carbon-based polymers, they are not usually stable if there is a substituent on every backbone atom. The excessive build-up of charge on the boron resulting from the coordinate bonds also reduces stability. It thus seems unlikely that such polymers will ever be synthesized, but there is nothing more guaranteed to stimulate an effort to do something than to state that it cannot be done!

15.3 Polysilylenes

The rapid development of carbon-based polymer chemistry in the middle of the twentieth century was heavily dependent on the fact that carbon forms metastable multiple bonds to itself. Ethylene and a large number of its substituted derivatives can be easily synthesized and handled. They are, however, unstable with respect to polymerization and can be easily transformed into polymers in the presence of appropriate catalysts or initiators. Because oxygen can initiate the polymerization of monomers such as styrene and methylmethacrylate, the adventitious discovery of their polymers was almost inevitable. The near absence of silicon analogs of vinyl monomers required a more devious path to the discovery of useful high molecular weight polysilylenes. The path was made more difficult because the earliest attempts to make such polymers produced materials that were intractible because of their high crystallinity. These difficulties, combined with the extreme reactivity of unsubstituted polysilylenes, helped develop the myth that high polymers based on a silicon backbone were intrinsically unstable due to the weakness of the Si—Si bond. This is now known to be untrue, and polymers with chains of hundreds of thousands of silicon atoms have been prepared.

Unsubstituted silicon analogs of the linear paraffins have been fully characterized up to about Si_8, and a variety of unsubstituted ring compounds have also been made. These compounds are all spontaneously explosive in the presence of oxygen and show a strong tendency to disproportionate to SiH_4 and polymeric solids of unknown composition.

The substituted polysilylenes are much less oxygen sensitive than those without organic substituents and, provided they are not highly crystalline (such as polydimethylsilylene, for example), dissolve in many organic solvents. They can be made by two routes. The classical route, the most successful so far for achieving high molecular weight polymers, uses the reduction of a chlorosilane with an active metal, as shown in Equation 15.18:

$$RR'SiCl_2 + 2Na \xrightarrow[120\,°C]{Toluene} \begin{array}{c} R \\ | \\ \!-\!\!\left[Si\right]_n\!\!- \\ | \\ R' \end{array} + 2NaCl \qquad (15.18)$$

Figure 15.16
Molecular structure
of $[(CH_3)_2Si]_{16}$.
● = Si
⬤=CH₃

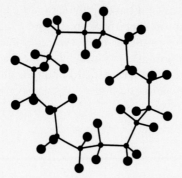

The product of such a reaction is a complicated mixture of rings and chains. The ratios of the products can be varied over a limited range by varying the reaction conditions. For the case of R = R′ = Me, rings of all sizes from $n = 4\text{--}35$ have been prepared. Figure 15.16 shows the structure of $(Me_2Si)_{16}$, determined by X-ray crystallography. A remarkable difference between these cyclopolysilanes and analogous carbon-ring systems is that the bonding and antibonding σ orbitals are highly delocalized in the silicon compounds. Strong electron donors such as alkali metals can transfer an electron to the permethylcyclosilane ring to give a radical anion in which the electron is equally distributed over all of the silicon atoms. Figure 15.17 shows the ESR spectrum of the $[SiMe_2]_5^{\tau}$ radical anion. The large number of lines in this spectrum are the central part of the thirty-one-line pattern arising from the interaction of the unpaired electron with the thirty equivalent protons in the molecule. At higher resolution, the lines due to hyperfine interactions with spin $\frac{1}{2}$ ^{29}Si nuclei can also be detected. This behavior is more characteristic of aromatic hydrocarbon rings than of aliphatic rings. The strong delocalization of the σ bonds in the chain and the relatively low-energy difference between the bonding and antibonding

Figure 15.17
ESR spectrum of
the $[SiMe_2]_5^{\tau}$ radical
anion. Reprinted
with permission
from R. West and
E. Carberry,
Science, 189 (1975):
179.

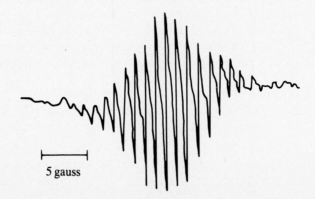

5 gauss

manifolds are responsible for many of the interesting electronic properties of polysilanes, such as photoconductivity, low-energy UV absorption, and exceptional charge-carrying ability.

A second recently discovered method for the synthesis of polysilanes is the catalytic dehydrocoupling of silicon hydrides. The reaction shown in Equation 15.19 is catalyzed by organotitanium and organozirconium complexes such as bis(cyclopentadienyl)dimethyltitanium and zirconium:

$$n\text{RSiH}_3 \xrightarrow{\text{Catalyst}} \text{H}\!-\!\!\begin{smallmatrix} \text{R} \\ | \\ \text{Si} \\ | \\ \text{H} \end{smallmatrix}\!\!\Big]_n\!\!-\!\text{H} + (n-1)\text{H}_2 \qquad\qquad \textbf{(15.19)}$$

The renaissance of interest in polysilanes came about when it was demonstrated that fibers of poly(dimethylsilylene) can be pyrolyzed to give silicon carbide fibers. Subsequently, monolithic objects of silicon carbide have been made by pyrolysis of polysilanes. Some of the reactions that are believed to take place in this transformation are shown in Equations 15.20 and 15.21:

$$-\!\!\big[\text{SiMe}_2\big]_n\!\!- \xrightarrow{400\,°\text{C}} -\!\!\big[\text{Si(H)(Me)CH}_2\big]_n\!\!- \qquad\qquad \textbf{(15.20)}$$

$$-\!\!\big[\text{Si(H)(Me)CH}_2\big]_n\!\!- \xrightarrow{>800\,°\text{C}} n\text{SiC} + n\text{CH}_4 + n\text{H}_2 \qquad\qquad \textbf{(15.21)}$$

This kind of transformation of an organic-type polymer into an inorganic ceramic or refractory material has opened up a promising new technology for production of finished objects in such materials. The useful properties—hardness, high melting temperature, and chemical inertness—of materials such as SiC, BN, Si_3N_4, and WC present enormous problems in fabrication. Traditionally, they have been manufactured as powders, and objects are made by cementing and sintering the powder into the required form. Such methods require many steps and the consumption of much energy. In addition, the ultimate properties of the final product are severely limited by the presence of voids and other defects in the structure. The possibility that objects could be made from a meltable polymer, which can then be transformed under relatively mild conditions to a refractory, is an attractive one. This is currently a very active research area. Besides rings and chains, a few cage polysilanes have also been isolated and characterized by, including the trifunctional reactant methyltrichlorosilane in reaction in Equation 15.18 which leads to the formation of cage compounds. Figure 15.18 shows one of these.

As a final word on polysilanes, it is recognized that these compounds are essentially molecular analogs of elemental silicon. There are great hopes that developments will lead to their use as a lower-cost alternative to elemental silicon in applications such as solar collectors.

Figure 15.18
The structure of
tetradecamethyl-
bicyclo[2.2.2]-
octasilaoctane.

= CH$_3$

= Si

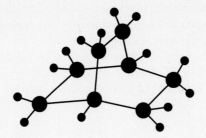

BIBLIOGRAPHY

Suggested Reading

Chivers, T. "Chemistry of Homonuclear Sulfur Species." *Chem. Soc. Rev. 2* (1973): 233.

Muetterties, E. L., ed. *Boron Hydride Chemistry*. New York: Academic Press, 1975.

Ray, N. H. *Inorganic Polymers*. New York: Academic Press, 1978.

West, R. "Organopolysilanes." *J. Organomet. Chem.* (1986): 300, 327.

Zeldin, M., K. J. Wynne, and H. R. Allcock, eds. *Inorganic and Organometallic Polymers*. ACS Symposium Series, Vol. 360. Washington, D. C.: American Chemical Society, 1988, Chapters 2–9.

PROBLEMS

15.1 Suggest a reason why phosphorus can be transformed into a crystalline allotrope with a puckered-sheet structure, but an isostructural compound based on the CH$_3$C— unit has not been made.

15.2 Outline the reasons why polystyrene (CH$_2$CHPh)$_n$ has been known and widely used for decades, but (SiH$_2$SiHPh)$_n$ has yet to be made.

15.3 How many framework electron pairs does each of the following species have?
a. B$_5$H$_{11}$
b. [B$_9$H$_9$]$^{2-}$
c. C$_2$B$_{10}$H$_{12}$
d. P$_4$
Describe the structure of each of these species.

15.4 Construct an MO energy-level diagram for the hypothetical [B$_3$H$_3$]$^{2-}$ ion. Show the orbital overlaps that give rise to each MO. How many framework electron pairs does this species have and in which orbitals are they likely to be accommodated in the ground-state ion? What influence will the charge have on the stability of this ion compared to $[B_6H_6]^{2-}$?

15.5 Starting only from elements, how would you synthesize B_5H_9?

15.6 Predict the outcome of the following reactions:

  a. $B_4H_{10} + 2[BH_4]^- \xrightarrow{100\,°C}$

  b. $5B_2H_6 + 2PMe_3 \xrightarrow{250\,°C}$

  c. $B_2H_6 + 2NH_3 \longrightarrow$

  d. $B_{10}H_{14} + CH_3C\equiv CH \longrightarrow$

15.7 Give one example each of an element with the following structures:
  a. Isolated ring molecules
  b. Planar sheets
  c. Puckered sheets
  d. Isolated deltahedral molecules

15.8 What do you understand by the following terms?
  a. A carborane
  b. A nido structure
  c. A deltahedron
  d. A preceramic polymer

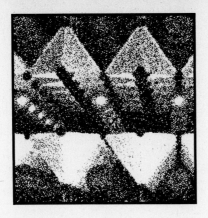

16

Heteropolar Inorganic Polymers

In Chapter 15, we dealt with polymers that contain only a single type of atom in the backbone. With the exception of carbon, all such polymers exhibit a high chemical reactivity, and in many cases the backbone bonds are rather weak. It is these factors that retarded the growth of interest in such polymers but that more recently have been shown to be highly advantageous in certain applications. This pattern is completely reversed in the cases of polymers where the backbone consists of an alternating chain of heavier *p*-group atoms and either C, N, or O. The bonds between such pairs of elements are among the strongest known, and as long as they are not too ionic, they are resistant to chemical attack. The archetype of this kind of polymer is the siloxane family, which includes the silicate minerals and the synthetic silicone polymers. As the name suggests, this family is based on chains of alternating silicon and oxygen atoms. It is in the siloxane family where we find the most varied and elaborate polymer architecture; they are considered in detail later. We begin, however, with nitrogen-containing polymers.

16.1 Ring and Chain Molecules Containing Nitrogen

Nitrogen forms only very unstable bonds with itself or with the elements of Groups VIA and VIIA (16 and 17) because of the large interelectronic repulsions that build up when such atoms are close enough to each other to achieve effective overlaps of their valence orbitals. Nitrogen does form strong bonds to other atoms in its vicinity, particularly B, P, and S. In each case, there is a rich chemistry of polymeric species in the form of both rings and chains.

Boron–nitrogen rings exist as two families, which may be likened to the aromatic and the aliphatic ring systems of organic chemistry. The aromatic analog borazine is readily made by the reactions shown in Equations 16.1 and 16.2:

$$3NH_4Cl + 3BCl_3 \rightarrow B_3N_3Cl_3H_3 + 9HCl \qquad (16.1)$$

$$4B_3N_3Cl_3H_3 + 3NaBH_4 \rightarrow 4B_3N_3H_6 + 3NaBCl_4 \qquad (16.2)$$

Although there are many superficial similarities between $B_3N_3H_6$ and benzene, pushing this similarity too far is not justified. Although the π-electron system in borazine is delocalized, the bonding orbitals are concentrated on the more electronegative nitrogen, making it much more basic than the carbon atoms of benzene. The antibonding π orbitals are concentrated on the boron atoms, making them more acidic than the carbons of benzene. One consequence of this polarity is that borazine undergoes addition reactions much more easily than does benzene and is much more reluctant to undergo substitution. Thus, Br_2 readily adds to borazine but substitutes into the benzene ring under the influence of appropriate catalysts, as shown in Equations 16.3 and 16.4:

$$+ \; Br_2 \; \longrightarrow \qquad\qquad\qquad + \; HBr \qquad (16.4)$$

A number of four-membered and eight-membered borazine rings have been reported, but so far no linear polymers.

Borazanes, the analogs of saturated hydrocarbons, are made as a mixture of rings by the reaction shown in Equation 16.5:

$$B_2H_6 + NaNH_2 \;\rightarrow\; 1/n[BH_2NH_2]_n + NaBH_4 \qquad\qquad (16.5)$$

From this mixture, the main isolated product is the surprisingly stable ten-membered ring ($n = 5$), cyclopentaborazane. The unstable four-membered ring and the stable six-membered ring have also been isolated.

Borazines and borazanes parallel their carbon analogs in the same way that boron nitride has two forms that are isostructural with diamond and graphite. Borazine can be used as a pyrolytic precursor to hexagonal boron nitride (the graphite analog) if it is pyrolyzed under an atmosphere of ammonia.

Although a number of phosphorus nitrides have been reported in the literature, relatively little is known about them. A white compound of the formula P_3N_5 is the best known. This compound has been used in the electronics industry for doping silicon. Its structure does not appear to be known.

Many ring and chain compounds based on the P—N bond are known. Polyphosphazenes, based on alternating P(V)/N(III) chains, are electronically similar to borazines. Thus, in a [BRNR'] unit, each B and N contributes one electron to forming the bond to the substituent. This leaves two electrons from B and four electrons from N to contribute to ring bonding. For an n-membered ring, this gives $3n$ ring electrons, $2n$ of which are used to make σ bonds and n of which are available for π bonding. The monomer unit of a polyphosphazene is [PX$_2$N]. In this unit, the phosphorus uses two electrons for bonding to X and contributes three to ring bonding. The N contributes three to ring bonding, and the other two reside in a nonbonding orbital (lone pair). This formal similarity in the bonding for borazines and phosphazenes is probably only superficial because the most recent calculations indicate that the $3d$ orbitals of phosphorus are significantly involved in the bonding in phosphazenes.

Reaction of PCl$_5$ with NH$_4$Cl in a chlorinated hydrocarbon solvent yields a mixture of rings and chains according to Equation 16.6:

$$n\text{PCl}_5 + n\text{NH}_4\text{Cl} \xrightarrow{\;120-150\,°C\;} [\text{NPCl}_2]_n + 4n\text{HCl} \qquad (16.6)$$

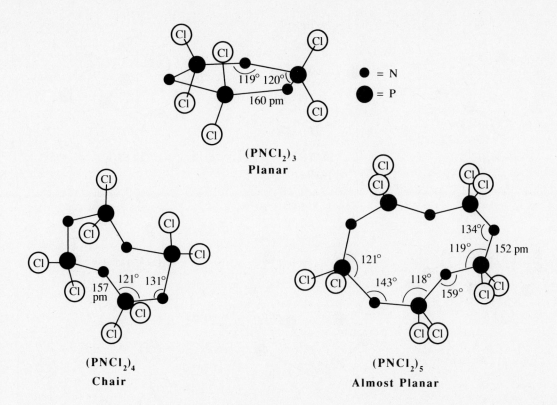

(PNCl₂)₃
Planar

● = N

● = P

(PNCl₂)₄
Chair

(PNCl₂)₅
Almost Planar

Figure 16.1
Structures of
$[NPCl_2]_n$, n = 3–5.

The linear polymer thus obtained has an average degree of polymerization (n) of about 15,000. Rings with n = 3–8 have been isolated as pure compounds and have been structurally characterized. Figure 16.1 shows the structures of the n = 3–5 compounds. As in the case of the borazines, there is substantial evidence in favor of delocalized electrons in the cyclo-polyphosphazenes. For example, all P—N bonds are of equal length (about 158 pm). This value is close to that encountered in P═N bonds and much shorter than the length expected for a single P—N bond (about 177 pm). A large number of derivative polyphosphazenes in which the Cl is replaced by some other substituent are known. Some exemplary substituents are EtO, PhO, CF_3CH_2O, Et_2N, and Ph. This substitution may be achieved by reaction of the dichlorophosphazene with an appropriate nucleophile, for example, liquid NH_3, Me_2NH, PhONa, or PhLi.

The chloro-compounds have been known for 150 years, but they found little practical use because of their hydrolytic instability. Replacement of the chlorine by other substituents such as those listed above produces materials of remarkable chemical stability, which have already found use in a number of applications. Although substituted, high molecular weight polyphosphazenes can be prepared by functionalization of linear polymers produced by the reaction in Equation 16.6, a better route is

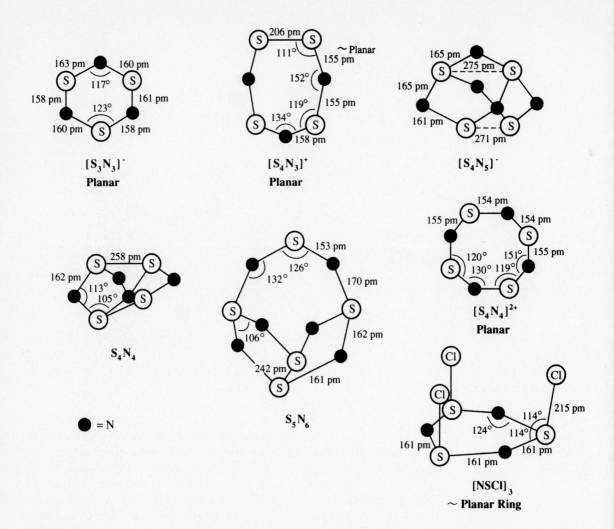

Figure 16.2
Structures of some sulfur nitride species. Bond distances are in pm.

the thermal ring–opening polymerization of the appropriately substituted cyclotriphosphazene. Polymers with outstanding flexibility at very low temperatures, water repellency, flame resistance, solvent resistance, and biocompatibility have been produced.

The chemistry of the S—N bond is rich and varied. Figure 16.2 shows the structures of a variety of ring compounds containing S—N bonds. The analogy between the sulfur nitrides and the cyclosulfurs is evident. In a homopolysulfur, two of the electrons from each sulfur are used to form S—S—S bonds, and the remaining four-valence electrons are in essentially nonbonding orbitals. Although these nonbonding orbitals can overlap nonorthogonally, there is nothing to be gained because they are both filled. Replacement of S by N results in a reduction of one in the skeletal electron

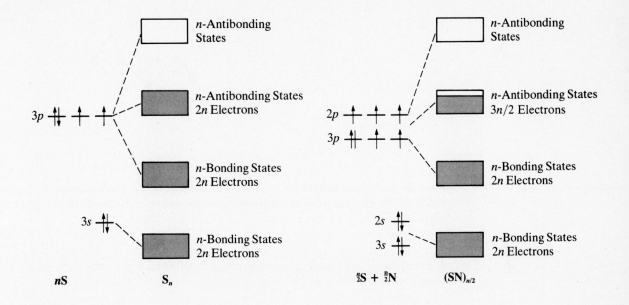

Figure 16.3
Schematic illustration of the difference in bonding between an S_n and an $[SN]_{n/2}$ polymer.

count, which in turn allows an increase in the total bond order because half of the orbitals that were filled and nonbonding in the homopolymer are now only half-filled. This is illustrated schematically in Figure 16.3. It is evident from this argument that neutral $[SN]_n$ molecules with odd n are not likely to be very stable. A number of odd-n $[SN]_n$ cations and anions are known, however. The sulfur nitrides are notably less stable than the borazines and phosphazines, and many of them are dangerously explosive. This character can be attributed to the necessity of accommodating excess electrons in compounds of the later main groups in antibonding orbitals.

The planar $[S_3N_3]^-$ ion is of particular interest. In this anion, each of the nitrogens has a lone pair, two σ bonds to S, and contributes a single electron to the π system. Each sulfur forms two σ bonds to N, has one lone pair, and contributes two electrons to the π system. Thus, the S and N contribute a total of nine π electrons that together with the extra electron of the anion charge results in a total of ten electrons to be accommodated in the six π MOs. Because only three of these π orbitals are bonding, four of these electrons must go into antibonding orbitals. This ion obeys the Hückel rule for aromaticity—namely, the molecule must have $(4n + 2)$ π electrons to be aromatic. Six-electron π systems (that is, $n = 1$) are common in carbon chemistry, but the ten-electron systems ($n = 2$) are much rarer because they require either high negative charges or unusually large rings. Because S and N both contribute more electrons than does C, the cylcothiazenes can easily achieve the ten-electron count, even in relatively small rings. Besides the $[S_3N_3]^-$ ion described, $[S_4N_4]^{2+}$ and $[S_4N_3]^+$

Figure 16.4
(a) and **(b)** Two of the resonance structures of S_4N_4 and **(c)** one of the structures of S_8^{2+}.

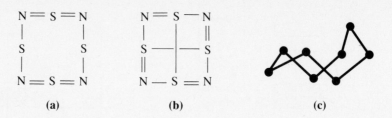

(a) (b) (c)

are also known examples of ten-electron π systems. It is difficult to describe many electron-excess compounds with localized two-electron bonds without resorting to large numbers of resonance structures. Figure 16.4 shows two of the resonance structures for S_4N_4. The actual structure of S_4N_4, a boat with unusually short distances between the transannular sulfur atoms, indicates a considerable contribution from the structure Figure 16.4 (b). This S—S bonding is also confirmed by MO calculations. The sulfur cation S_8^{2+}, produced by the reaction shown in Equation 16.7

$$S_8 + 3AsF_5 \rightarrow [S_8][AsF_6]_2 + AsF_3 \tag{16.7}$$

has a half-boat, half-chair configuration with only one short transannular S—S interaction (Figure 16.4c). The isoelectronic S_6N_2 ring has not yet been synthesized, but it is expected to have the same type of structure.

A number of reactions used to synthesize sulfur nitrides are shown in Equations 16.8–16.12. In general, the stoichiometries shown in these reactions are only approximate.

$$24SCl_2 + 64NH_3 \longrightarrow 4S_4N_4 + S_8 + 48NH_4Cl \tag{16.8}$$

$$S_4N_4 \xrightarrow[200\ °C]{Ag} 2S_2N_2 \tag{16.9}$$

$$3S_4N_4 + 6Cl_2 \longrightarrow 4S_3N_3Cl_3 \tag{16.10}$$

$$2NH_3 + S_4N_4 \longrightarrow NH_4[S_4N_5] + H_2 \tag{16.11}$$

$$10[S_4N_5]^- + 5Br_2 \longrightarrow 8S_5N_6 + N_2 + 10Br^- \tag{16.12}$$

As mentioned, sulfur nitrides are much less thermodynamically stable than are the nitrogen compounds of phosphorus and boron. If subjected to shock or rapid heating, S_2N_2 and S_4N_4 have a tendency to explode. Controlled heating of these compounds can yield polythiazul $\{SN\}_n$. We will mention this remarkable, bronze-colored, fibrous, metallic polymer again in Chapter 18. Solid-state polymerization of crystals of S_2N_2 give a crystalline polymer whose structure has been determined by X-ray crystallography. Unlike the helical structures of S, Se, and Te, the polythiazyl chain is almost planar (Figure 16.5). The electrical conductivity of the fibers is about 50 times greater in the direction perpendicular to the fiber axis at room temperature and about 1000 times greater at 40 K. The metallic

Figure 16.5
Structure of
polythiazyl.

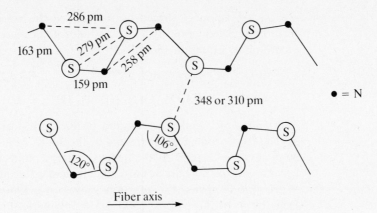

conduction arises from a band that is σ antibonding in character. In S_n, this band is completely filled, giving rise to dielectric behavior, but in SN_n the band is only partly filled (see Figure 16.3).

16.2 Oxyanions and Polyoxyanions

Some General Considerations

The small size and high polarizing power of the elements C to F result in their oxides and oxyanions being simple molecules. The electron demand of the central atom is so great that the oxygen electrons are too tightly bound to be available for intermolecular interactions (we refer to this reduction in the availability of electrons as *saturation*). The small sizes of C, N, O, and F restrict the coordination numbers of their oxides to low values, three or less. Boron is a borderline case. In its oxygen compounds, there is enough residual electron density on the oxygen for it to donate intermolecularly, and the boron can exist in either three or four coordination, thus allowing the formation of three-dimensional structures. Therefore, although B_2O_3 could have the structure O=B—O—B=O, it exists under normal conditions as a polymer (see Chapter 10).

In the third period, there is a dramatic change in behavior. Due to the larger sizes of the atoms from Al to P, the ability to saturate the oxygen electrons is greatly diminished, and coordination numbers of four or higher are the rule. The oxides of these elements are therefore polymeric as described in Chapter 10. The same is true of the oxyanions. Although there is no doubt that aluminum forms a family of polymeric oxyanions in aqueous solution, very little is known of their structures. We will concentrate our attention on silicon and phosphorus.

An oxyanion is formed by reaction of the oxide ion with an oxide according to Equation 16.13 or by deprotonation of a hydroxo-species according to Equation 16.14:

$$MO_x + yO^{2-} \rightleftharpoons [MO_{x+y}]^{2y-} \tag{16.13}$$

$$M(OH)_x \rightleftharpoons [M(O)_y(OH)_{x-y}]^{y-} + yH^+ \tag{16.14}$$

The values of x and y are limited by the ability of the central atom to expand its coordination sphere and by the ability of the product to sustain a large negative charge. As the negative charge builds up, the oxyanion can become a stronger base toward the element oxide than is the oxide ion. When this happens, the reaction shown in Equation 16.15 can occur. Repetitions of this process can lead to polymeric products.

$$MO_x + [MO_{x+y}]^{2y-} \rightleftharpoons [O_xM—O—MO_{x+y-1}]^{2y-} \tag{16.15}$$

In aqueous solution, the acidities and basicities of the reactants and products, relative to that of water and of the hydroxide ion, must be considered. For example, of the isostructural and isoelectronic species $[MO_3(OH)]^{n-}$, where M = Al(III), Si(IV), P(V), S(VI), and Cl(VII), perchloric acid is completely dissociated to perchlorate ion when dissolved in water, the sulphate species is substantially but far from completely dissociated to sulfate, and the monohydrogenphosphate ion is almost completely protonated to dihydrogenphosphate and phosphoric acid. The Si and Al species have not been detected in aqueous solution, but we can be reasonably sure that they would be almost completely protonated.

A final reaction that we must consider with respect to the polymerization of hydroxides and oxyanions is the condensation process depicted in Equation 16.16:

$$M{-}OH + H{-}OM \rightleftharpoons M—O—M + H_2O \tag{16.16}$$

This is a reaction of enormous generality throughout the periodic table, and the equilibrium may range from totally to the left to totally to the right, depending on the particular system. In pure sulfuric acid, the equilibrium (Equation 16.17) leads to a concentration of about 0.009 molal dimer. Pure phosphoric acid contains the dimer at a concentration of about 0.3 molal at equilibrium (Equation 16.18). Neither orthosilicic acid nor aluminum hydroxide exist as the monomer under normal conditions.

$$2H_2SO_4 \rightleftharpoons [(HO)O_2S—O—SO_3]^- + H_3O^+ \tag{16.17}$$

$$3H_3PO_4 \rightleftharpoons [(HO)O_2P—O—PO_3(OH)]^{2-} + [H_4PO_4]^+ + H_3O^+ \tag{16.18}$$

Besides the decrease in thermodynamic stability of the condensed species on going from Si to S, there is also an accompanying increase in kinetic activity of the oxo-bridge toward hydrolysis. This accounts for the evolutionary choice in living systems of pyrophosphate linkage for energy storage. Pyrophosphate linkage is thermodynamically unstable with respect to hydrolysis, so energy is liberated on hydrolysis, but it is kinetically stable enough to survive relatively long periods in water in the absence of catalysts. Thus, adenosine triphosphate is produced in the energy production and storage apparatus of living cells and is subsequently hydrolyzed under the influence of enzymes to furnish the energy for a wide range of biochemical processes. The equation for this fundamental life process is given in Equation 16.19:

$$
\text{(ATP)} + \text{H}_2\text{O} \longrightarrow
$$

$$
\text{(ADP)} + [\text{HPO}_4]^{2-} \quad \textbf{(16.19)}
$$

The pyrosilicate linkage is thermodynamically stable but is also very kinetically inert to hydrolysis under physiological conditions. The pyrosulfate linkage is very energy-rich, but it is very unstable kinetically in the presence of water (half-life is only a few minutes). On the other hand, the great stability of the pyrosilicate linkage explains why most terrestrial silicon is found in the form of a bewilderingly varied family of polymeric siloxanes.

Polyphosphates

Thermal dehydration of phosphate salts containing P—OH groups provides a general method for the production of linear polyphosphates. These

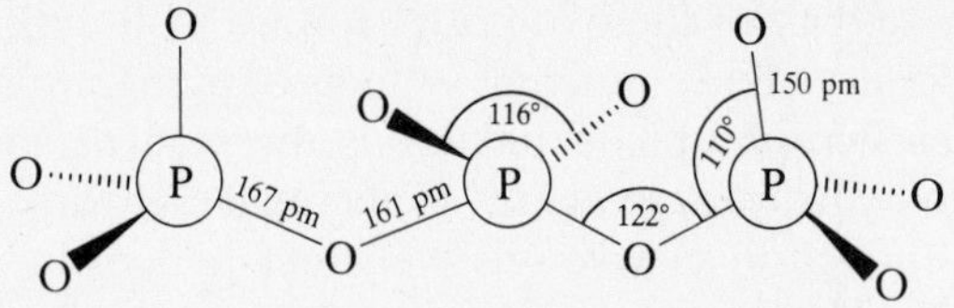

Figure 16.6
Structure of the tripolyphosphate ion.

compounds consist of strings of phosphate tetrahedra joined at their corners by P—O—P bridges. Diphosphates can be prepared by controlled heating of alkali metal dihydrogenphosphates or hydrogenphosphates:

$$2M_2HPO_4 \rightarrow M_4P_2O_7 + H_2O \tag{16.20}$$

In polyphosphates the P—O—P angle is quite variable (about 130–160°). The bridging P—O bonds (about 160–170 pm) are longer than the terminal P—O bonds (about 150 pm).

Sodium tripolyphosphate is made on a large scale industrially by the carefully controlled heating of a stoichiometric mixture of sodium hydrogen and dihydrogen phosphates (Equation 16.21):

$$NaH_2PO_4 + 2Na_2HPO_4 \rightarrow Na_5P_3O_{10} + 2H_2O \tag{16.21}$$

The product has been widely used as a "builder" in synthetic detergents. The builder acts as a chelator to keep polypositive ions in solution. These products may contain as much as 25–45% tripolyphosphate. It serves to solubilize Ca and Mg ions by chelation, thus preventing the formation of insoluble soap scums when laundry is done in hard water. Because of the harmful effects of sewage phosphate in stimulating the undesirable growth of algae in natural waters, the use of tripolyphosphate in detergents has declined greatly in recent years, particularly in North America. Figure 16.6 shows the structural details of the tripolyphosphate ion.

Linear polyphosphates such as those discussed have the generic formula $[P_nO_{(3n+1)}]^{(n+2)-}$. Pure compounds have been isolated with ions of n up to 10, but beyond 10 the problem of separation becomes excessively difficult. Higher molecular weight materials are glassy mixtures, and the polymetaphosphates have an essentially infinite chain that approaches the generic formula $[PO_3]^-$. A wide variety of structural forms have been identified for linear polyphosphates in the solid state, depending on the nature of the cation(s). This structural variety comes from the essentially unlimited number of variations in the conformational relationships between the linked tetrahedra of the chain. Figure 16.7 shows some of these variations that result in helical backbone arrangements of different pitch.

Fusion of sodium dihydrogen phosphate at 625 °C, followed by rapid quenching of the melt, produces a glassy solid known as Graham's salt. This is a mixture of mostly linear polymetaphosphates, but it also contains

Figure 16.7
Some different
helical arrangements
of the
polymetaphosphate
backbone.

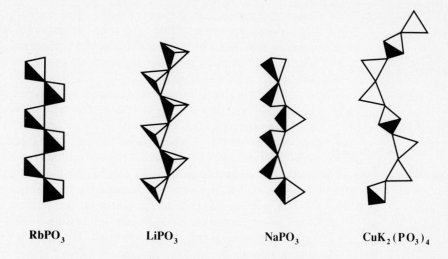

RbPO$_3$ LiPO$_3$ NaPO$_3$ CuK$_2$(PO$_3$)$_4$

small amounts of cyclometaphosphates, which can be separated by chromatography of solutions of Graham's salt. Cyclometaphosphates with n from 3–8 have been identified in this way. The $n = 3$ and 4 compounds are more easily made by controlled hydrolysis of P$_4$O$_{10}$. Figure 16.8 shows the relationship between the structure of P$_4$O$_{10}$ and those of the $n = 3$ and 4 ions.

Polysilicates

Finite Silicates The larger size and lower effective nuclear charge of silicon, compared to phosphorus, makes the oxygens of the [SiO$_4$]$^{4-}$ ion

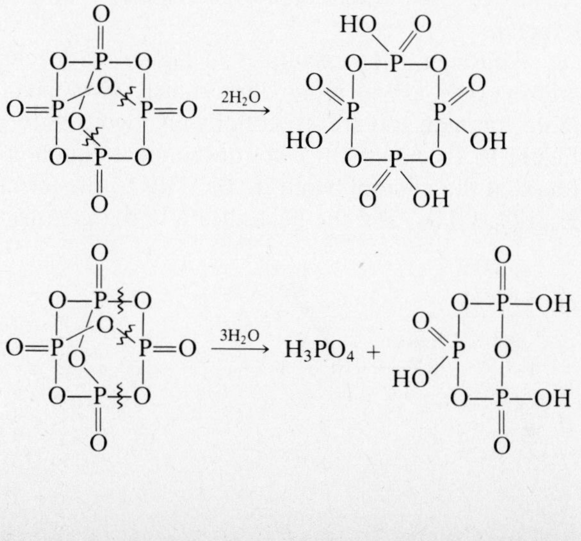

Figure 16.8
Schematic
illustration of
hydrolysis of P$_4$O$_{10}$
to give
cyclometatriphos-
phoric and
cyclometatetraphos-
phoric acids.

much more basic than those of $[PO_4]^{3-}$. Silicon differs from phosphorus in that it readily forms anything up to four-bridge bonds, whereas phosphorus forms a maximum of three under ambient conditions. This is because the larger and less polarizing silicon does not saturate the oxygen electrons as effectively as does phosphorus.

Silicon is almost exclusively tetrahedral in oxygen coordination. There are a few examples of simple salts of the orthosilicate ion $[SiO_4]^{4-}$, but they are relatively rare. The gemstone zircon appears to be such a case, but the ionicity of the bonds between the highly charged Zr^{4+} and $[SiO_4]^{4-}$ ions must be very low. This compound is probably more sensibly viewed as a covalent mixed oxide of Si(IV) and Zr(IV). The synthetic orthosilicates Na_4SiO_4 and K_4SiO_4 are, on the other hand, highly ionic materials.

A number of disilicates are also known but are rare. The series of lanthanide disilicates $Ln_2Si_2O_7$ are interesting in that they illustrate the great flexibility of the Si—O—Si bond. We have already referred to this phenomenon in relation to the P—O—P bond. The flexibility of M—O—M bonds in $3p$ relative to $2p$ compounds is often explained in terms of the participation of d orbitals in p_π—d_π bonding, which gives added stability to the linear bonding arrangement. Most rigorous theoretical calculations suggest that this is probably not true. Greater flexibility is not all that surprising as one goes to bonds with greater ionicity and the stringent directional demands of covalent forces are relaxed. The mineral thortveitite $Sc_2Si_2O_7$ has a linear Si—O—Si bond. The synthetic lanthanide disilicates referred to above show a progressive decrease in the bond angle from 180° to 133° on traversing $4f$ elements. Besides the change in Si—O—Si bond angle, the coordination number of the lanthanide ion increases from six to eight on traversing the series. The Si—O—Si bond angle responds to crystal forces that change with the changing structure. In these compounds, the two terminal SiO_3 moieties are staggered with respect to each other. Figure 16.9 shows a schematic illustration of this structure.

Although a few examples of higher, short-chain, linear silicates are known, they are of minor importance. Cyclometasilicates $[SiO_3]_n^{2n-}$ are more common and are structurally analogous to the cyclometaphosphates. Figure 16.10 shows structures of the more common rings. The trisilicate is found in the mineral benitoite $BaTiSi_3O_9$, the tetrasilicate in the synthetic $K_4Si_4O_8(OH)_4$, and the hexasilicate in the gemstone beryl.

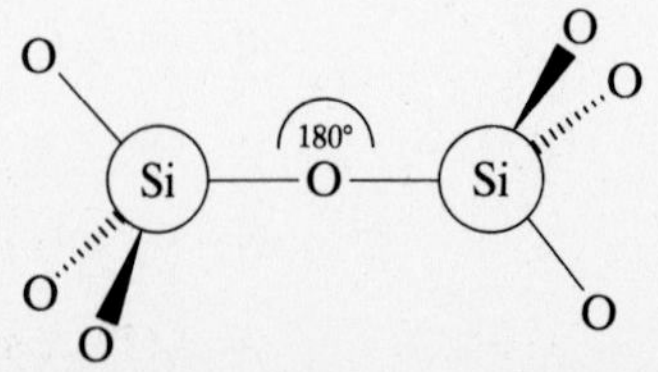

Figure 16.9
Schematic illustration of the structure of the disilicate ion in $Sc_2Si_2O_7$.

Figure 16.10
Some cyclometasilicate rings.

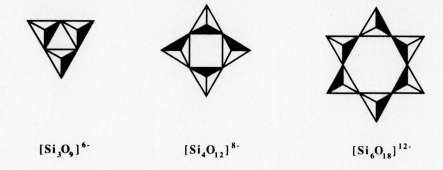

$[Si_3O_9]^{6-}$ $\qquad$ $[Si_4O_{12}]^{8-}$ $\qquad$ $[Si_6O_{18}]^{12-}$

Infinite Silicates: Chains and Sheets Infinite-chain metasilicates $[SiO_3^{2-}]_n$ are very common in nature. Like the polymetaphosphates, great diversity is possible through variation in the conformation of the chain and by variation of the cation(s). Sodium metasilicate has a twofold repeat unit, $Ca_3Si_3O_9$ (which has an infinite-chain structure despite its empirical formula) has a threefold helical repeat unit, $(Ca,Fe)_7Si_7O_{27}$ has a sevenfold repeat unit, and so on. Figure 16.11 illustrates these structures schematically.

Silicons in metasilicate ions are coordinated to two bridging oxygens and two terminal oxygens. Protonation of the terminal oxygens can give Si—OH groups, which can undergo condensation to increase the dimensional complexity of the structure. Formation of one more bridging oxide per silicon leads to either closed or open structures. The most closed

Figure 16.11
Some metasilicate chains with different conformational motifs.

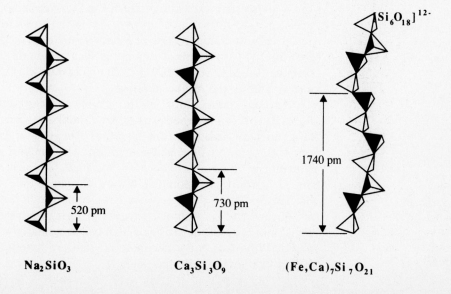

$[Si_6O_{18}]^{12-}$

520 pm

730 pm

1740 pm

Na_2SiO_3 $\qquad$ $Ca_3Si_3O_9$ $\qquad$ $(Fe,Ca)_7Si_7O_{21}$

Figure 16.12
Schematic illustration of the double-chain silicate structure.

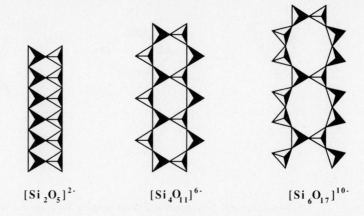

$[Si_2O_5]^{2-}$ $\qquad$ $[Si_4O_{11}]^{6-}$ $\qquad$ $[Si_6O_{17}]^{10-}$

structures are the double-chain, or ladder, polysilicates (Figure 16.12). It should be evident that these structures result from the fusion of chains with different helical repeat units.

The molecular structure of the infinite double-chain silicates expresses itself in the morphology of the bulk material. These are the compounds that constitute the group of fibrous minerals known as asbestos. Asbestos has been used in the manufacture of fireproof textiles for hundreds and perhaps thousands of years. More recently it has been used in the manufacture of fireproof construction materials such as roofing panels, ceiling tiles, and pipes. In motor vehicle manufacture, it is used in clutch and brake linings. Unfortunately, exposure to asbestos has been shown to lead to a dangerous lung disease, asbestosis, which can eventually lead to lung cancer. There is presently a vigorous search for replacements for asbestos to avoid excessive exposure of miners and the general public to its harmful effects.

The most common class of asbestos minerals, the amphiboles, adopt the $[Si_4O_{11}]^{6-}$ structure. Tremolite is such a mineral, with the empirical formula $[Ca_2Mg_5(Si_4O_{11})(OH)_2]$. The hydroxide ions in this structure are coordinated to the magnesium ions.

Further condensation of the double-chain structures leads to the formation of infinite-sheet structures. These two-dimensional structures are even more varied than are the chains. The structural variables are again ring size, conformation, and counterion. The simplest and most common structure is based on cyclohexasilicate units in which all of the terminal Si—O$^-$ groups are pointing toward the same side of the sheet (Figure 16.13). This type of network has a composition $[Si_2O_5]^{2-}$ (each of the six silicons and six terminal oxygens of a ring is shared with three other rings, and each of the six bridging oxygens is shared with two rings). We will describe two important minerals that are based on the binding of a layer of cations to this type of sheet.

Figure 16.13
Structure of the
$[Si_2O_5]^{2-}$ planar
network.

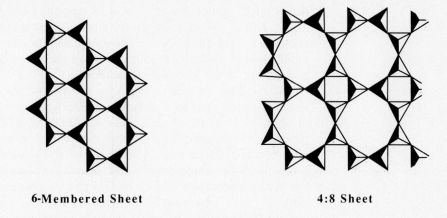

6-Membered Sheet　　　　　　　　**4:8 Sheet**

     Chrysotile is an important fibrous mineral that has the idealized formula $Mg_3(OH)_4(Si_2O_5)$. In crystalline $Mg(OH)_2$, the magnesium ion is octahedrally coordinated. The structure of chrysotile can be approximated by simply replacing two of the OH ligands of the $Mg(OH)_2$ structure with two of the terminal oxide ions of the $[Si_2O_5]^{2-}$ unit. Each Si—O$^-$ unit is shared by three magnesium ions (one Mg ion bridging every edge of the hexagon). Figure 16.14 shows the structure schematically. You may well be wondering why a fibrous mineral would result from something with a sheetlike structure. The answer is that the interoxygen separation in the

Figure 16.14
Idealized
representation of the
formation of the
chrysotile structure.

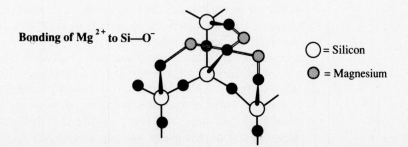

MgO$_6$ Octahedra

SiO$_4$ Tetrahedra

● = Oxygen

• = Hydrogen

Bonding of Mg^{2+} to Si—O$^-$

○ = Silicon

◉ = Magnesium

flat $[Si_2O_5]^{2-}$ network is a bit shorter than the ideal for coordination to Mg^{2+}. This results in an expansive stress on one side of the sheet, which causes it to curl. In chrysotile, the sheets curl up into tubes of essentially infinite length, and it is the morphology of these secondary structures that is expressed in the macroscopic material.

If the Mg^{2+} ion is replaced by the smaller Al^{3+} ion, there is no expansive stress, and the sheets remain flat. Kaolinite, a major constituent of China clay, has the idealized formula $Al_2(OH)_4(Si_2O_5)$ and is similar in structure to chrysotile. The planar structure with relatively weak interactions between adjacent sheets in the crystal gives a lubricious material that is capable of absorbing considerable amounts of water between the sheets by hydrogen bonding to the ligand OH groups. These are the properties of clays that make them useful in ceramics and indispensable in retaining moisture in the soil. The potter uses the plasticity of water-laden clay to form it into objects and then transforms the object into a dimensionally stable artifact by driving off the water at high temperature.

More of the residual hydroxide ligands in the chrysotile or kaolin can be substituted by the terminal Si—O$^-$ of a second silicate sheet to form a sandwich of metal ions between two silicate sheets. The resulting materials, such as talc $Mg_3(OH)_2(Si_2O_5)$, retain the lubricity associated with weakly interacting lamellas but lose the ability to absorb large amounts of water. Indeed, these materials tend to be hydrophobic because the covalent siloxane sheets are relatively nonpolar and all ions are inaccessible in the interior of the sandwich.

An important extension of the sandwich structure arises when some of the silicon atoms of the network are replaced by aluminum atoms. Recall from the discussion of the diagonal relationship between the chemistries of the second- and third-period elements, aluminum and silicon are roughly the same size and have many other similarities. Thus, Al(III) readily replaces Si(IV) in silicate structures to give the so-called aluminosilicates. This replacement leaves an excess of negative charge on the structure, and more cations must be added to balance the charge. Mica minerals, of which phlogopite $KMg_3(OH)_2Si_3AlO_{10}$ is an example, have this type of structure. Potassium ions bind the sandwiches more strongly than they are bound in the talcs, but cleavage along the potassium planes is still relatively easy. Micas are not lubricious, but they can be cleaved to thinner and thinner sections, almost down to the molecular level. Micas are used extensively in electrical applications where their excellent insulating properties and high-temperature resistance are particularly valuable. Vermiculite is a mica-type mineral where the intersandwich ions are hydrated, for example, $[Mg(H_2O)_6]^{2+}$. Although the native material is hard and dense, when it is heated the water of hydration is driven off as steam, causing the structure to puff up like popcorn. The resulting light, soft material is extremely useful as a noninflammable packing and a soil conditioner.

Three-Dimensional Silicate Structures   In principle it is possible to construct a wide variety of three-dimensional structures based on silicon bonded to three covalent-bridging oxygens and one terminal oxide. Such structures are rarely encountered, either in nature or as a result of synthesis. On the other hand, the three-dimensional aluminosilicates are the most abundant minerals, constituting over 65% of the earth's crust. Most abundant by far of the aluminosilicates are the feldspars, which may be envisaged as the products of isomorphously replacing the Si in silica with Al and balancing the charge with various cations. A discussion of the extremely complex structures and composition of the feldspars goes beyond the scope of this text. We will simply mention that most feldspars can be categorized ideally as phases resulting from the ternary system $NaAlSi_3O_8/KAlSi_3O_8/CaAl_2Si_2O_8$. Orthoclases are rich in the potassium component, anorthoclases are rich in the sodium component, and the plagioclases are rich in the calcium component.

A fascinating class of three-dimensional polysilicates is the zeolites. In these compounds, the SiO_4/AlO_4 tetrahedra are interconnected to form highly symmetrical structures containing channels and large cavities. A common basic unit of these structures is the cuboctahedron, formed by connecting squares of tetrahedra (Figure 16.15). The cuboctahedra may be interconnected either through their square faces or through their hexagonal faces (Figure 16.15). In the Linde zeolite A, a cubic cavity of about 1100 pm in diameter is formed by connection through the square faces in three dimensions, and these cavities are interconnected by octagonal holes with a diameter of about 420 pm. The central cavity is connected through six-membered rings, diameter of about 200 pm, to the cuboctahedra whose internal diameters are about 660 pm. The cubes may be strung together *ad infinitum* linearly in two-dimensional sheets or in three dimensions. This open structure allows the easy passage of small molecules and ions, and this property is the basis of the variety of practical applications of the zeolites. As manufactured, Linde zeolite A is in the form of the sodium salt, idealized formula $[Na_{12}(Al_{12}Si_{12}O_{48})] \cdot 27H_2O$, and sodium ions are located in the cavities of the structure. Sodium ions are easily replaced by ions of higher charge by an ion-exchange process; thus, zeolite can be used to soften hard water by extracting Ca^{2+} and Mg^{2+} ions. Once zeolite is fully saturated with these ions, sodium salt can be regenerated by washing with a concentrated solution of NaCl. By virtue of the mass-action effect, sodium replaces the more tightly bound dipositive ions.

Neutral molecules can also be selectively removed from gas streams by zeolites, and in this application they are known as *molecular sieves*. Much of the formula water can be driven off under vacuum at elevated temperatures, and the resulting material, having a very high affinity for water, can be used as a drying agent for gases and organic solvents.

A very large-scale use of zeolites is as catalyst supports. The zeolite

Figure 16.15
Schematic
illustration of how
the zeolite structure
is built up from
cyclotetrasilicate
units arranged in
the form of a
cubocathedron.

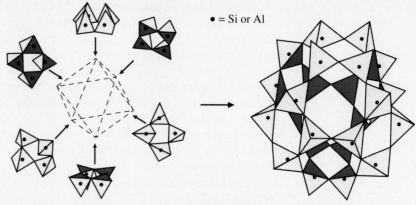

• = Si or Al

Connections of Tetrahedra

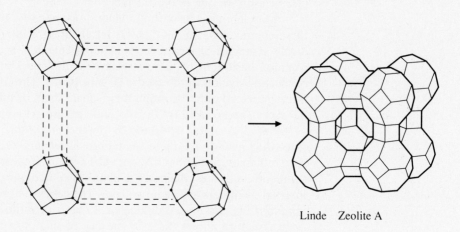

Linde   Zeolite A

Connections of Cuboctahedra

structure confers two novel properties when used as a catalyst support. The first is the ability of the cavities to support catalyst particles in a very highly dispersed state. The second is the ability of the channels and cavities to restrict the size of reactant molecules that can gain access to the catalyst and to restrict the size of molecules produced in the catalytic reaction.

Although some zeolites occur naturally, they are sufficiently rare that the commercial developments referred to would not have been possible if they had only been available from natural sources. Synthetic zeolites are made by hydrothermal recrystallization of aqueous mixed–alkaline aluminosilicate gels.

A third group of framework aluminosilicates are the ultramarines. These substances are based on the cuboctahedral unit with alternating aluminum and silicon, but they are less open in structure than are zeolites.

They contain no water, but they do contain anions other than the alumino-silicate entity. Sodalite has the idealized formula $[Na_8Cl_2(Al_6Si_6O_{24})]$ and is colorless. The mineral lapis lazuli has some of the chlorides of sodalite replaced by sulfide, and the beautiful blue color of this mineral has encouraged its use in decoration and jewelery since ancient times. Complete replacement of the chlorides by disulfide produces the mineral ultramarine. This intensely blue-colored stone has been used in powder form since antiquity as a pigment for paint and ceramics. Synthetic ultramarines may be made by reacting kaolin, sulfur, and sodium carbonate at high temperature. Variation in the sulfur content allows the production of green, red, and violet ultramarines as well as the blue type. The origin of the colors of these materials was only recently established to be varying proportions of sulfur radical anions ($S_2^{\cdot-}$, $S_3^{\cdot-}$, blue; $S_4^{\cdot-}$, red). These species are also generated on dissolution of alkali polysulfides in donor solvents such as dimethylformamide and their spectroscopic properties (UV/vis, IR, resonance Raman and ESR) have been thoroughly characterized. A comparison of these species in solution with those in the ultramarines allows confident assignment of the species present in the latter.

16.3   *Organopolysiloxanes: The Silicones*

The silicones are a family of polymers with an $[SiR_2O]_n$ backbone and have found wide application and considerable commercial success. The building blocks of commercial silicones are the units $(CH_3)_3SiO$—, abbreviated as

M; —$OSi(CH_3)_2O$—, abbreviated as D, and $(CH_3)Si{\overset{\displaystyle O-}{\underset{\displaystyle O-}{\diagdown}}}O$—, abbreviated

as T. Each species has a separate function in determining the molecular dimensions and shape of a silicone polymer. The M units (*M*onofunctional) are used as end groups to limit the molecular weight and to remove reactive OH groups from the ends of the chains. The D units (*D*ifunctional) provide the main polymer backbone, and the T units (*T*rifunctional) can be used for introducing cross-links. Figure 16.16 shows a schematic representation of a polymer containing each of these units.

Silicones are made by the hydrolysis of methylchlorosilanes. This hydrolysis gives rise to reactive silanols, which condense to form siloxanes. These reactions are shown in Equations 16.22 and 16.23:

$$(CH_3)_2SiCl_2 + 2H_2O \rightarrow (CH_3)_2Si(OH)_2 + 2HCl \qquad \textbf{(16.22)}$$

$$n(CH_3)_2Si(OH)_2 \rightarrow [Si(CH_3)_2O]_n + nH_2O \qquad \textbf{(16.23)}$$

The product of such an aqueous hydrolysis is a complex mixture of rings and chains. Because the components of the mixture can be equilibrated in

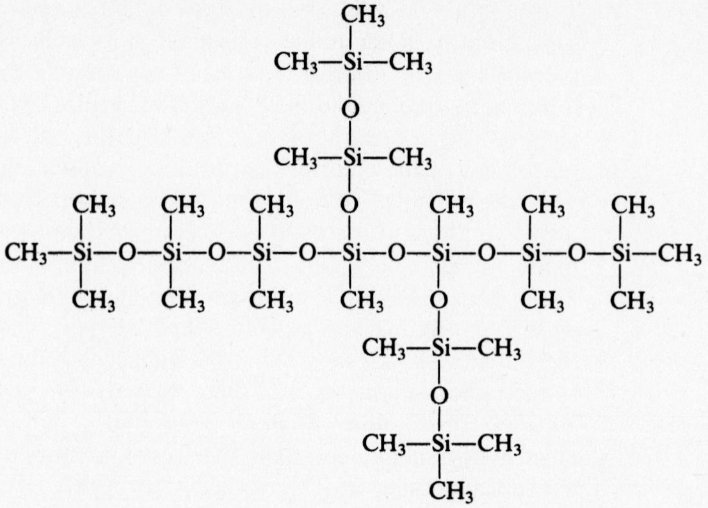

Figure 16.16
Schematic
representation of a
silicone polymer
containing *M*, *D*,
and *T* units.

the presence of basic or acidic catalysts, the initial hydrolysate can be converted by distillation, more or less completely, into the cyclic tetramer (D_4). This tetramer can then be converted, in a well-controlled manner, to high polymer by ring-opening polymerization under the influence of a base or acid catalyst. The molecular weight is governed by the ratio of monomer to catalyst as is evident from Equation 16.24:

$$x[(CH_3)_2SiO]_4 + y[R_3SiO]^- \rightarrow yR_3SiO[(CH_3)_2SiO]_{4x/y}^- \qquad (16.24)$$

Such techniques allow the synthesis of polymers with molecular weights of about 10^6.

The crystalline melting point of polydimethylsiloxane is well below room temperature, and the linear polymers are viscous liquids. These gums can be converted into tough elastomers (rubbery materials) by cross-linking, using a variety of chemical reactions. The curing reaction for silicone bathtub seal is shown in Equations 16.25 and 16.26. The acetic acid liberated in this reaction is responsible for the characteristic vinegary smell of silicone caulking compounds. The water necessary for the curing reaction comes from atmospheric humidity.

$$-\overset{|}{\underset{|}{Si}}-O-C(=O)CH_3 + H_2O \longrightarrow$$

$$-\overset{|}{\underset{|}{Si}}-OH + CH_3COOH \qquad (16.25)$$

$$2 -\overset{|}{\underset{|}{Si}}-OH \longrightarrow -\overset{|}{\underset{|}{Si}}-O-\overset{|}{\underset{|}{Si}}- + H_2O \qquad (16.26)$$

Polydimethylsiloxanes can also be cross-linked by free-radical reactions, as shown in Equations 16.27–16.29:

$$PhC(=O)OO(O=)CPh \longrightarrow 2PhC(=O)O^{\cdot} \qquad (16.27)$$

$$
\underset{\underset{CH_3}{|}}{\overset{\overset{CH_3}{|}}{-(O-Si)_n}} + PhC(=O)O^{\cdot} \longrightarrow \underset{\underset{CH_3}{|}}{\overset{\overset{CH_2^{\cdot}}{|}}{-(O-Si)_n}} + PhC(=O)OH
$$

$$(16.28)$$

$$
2 \underset{\underset{CH_3}{|}}{\overset{\overset{CH_2^{\cdot}}{|}}{-(O-Si)_n}} \longrightarrow CH_3 \underset{|}{\overset{\overset{|}{O}}{Si}}CH_2CH_2 \underset{|}{\overset{\overset{|}{O}}{Si}}CH_3 \qquad (16.29)
$$

One feature of polysiloxanes that has enabled them to fill a wide variety of niches in the synthetic polymer market is the ease with which their properties can be slightly modified by substitution of different organic groups at the silicon. Their relatively high cost of production, however, restricts their use to applications where their exceptional properties lead to savings or the achievement of otherwise inaccessible performance. Modifications to the organic substituents can be made at the monomer level, for example, dimethyldichlorosilane can be partly or wholly replaced by diphenyldichlorosilane, methylvinyldichlorosilane, or methyldichlorosilane in the reaction in Equation 16.22. On the other hand, other functions can be introduced onto the backbone of an already formed polymer. A particularly useful reaction for achieving the latter is the catalyzed hydrosilation (or hydrosilylation) reaction (see Chapter 24). An example of hydrosilation is shown in Equation 16.30:

$$
\underset{\underset{CH_3}{|}}{\overset{\overset{CH_3}{|}}{-(O-Si)_n}} \ \underset{\underset{H}{|}}{\overset{\overset{CH_3}{|}}{(O-Si)}} \ \underset{\underset{CH_3}{|}}{\overset{\overset{CH_3}{|}}{(O-Si)_m}} + CH_2=CHCN
$$

$$\downarrow \text{CuCl/Amine catalyst}$$

$$(16.30)$$

$$
\underset{\underset{CH_3}{|}}{\overset{\overset{CH_3}{|}}{-(O-Si)_n}} \ \underset{\underset{\underset{\underset{CN}{|}}{\underset{CH_2}{|}}}{CH_2}}{\overset{\overset{CH_3}{|}}{(O-Si)}} \ \underset{\underset{CH_3}{|}}{\overset{\overset{CH_3}{|}}{(O-Si)_m}}
$$

The prepolymer in this case would be produced by cohydrolysis of $(CH_3)_2SiCl_2$ and $(CH_3)SiHCl_2$. In most applications, only small amounts of nonmethyl substituents are needed to produce the modification of properties for a new application.

The silicones are exceptionally biocompatible and have found wide-range application in medicine. Because of the unusually high diffusion rate of oxygen through silicone membranes, they can be used as temporary skin substitutes to protect severely burned areas from infection during the healing process. Silicone fluids are used as lubricants in applications where their high temperature or radiation stability offsets their relatively high cost compared to organic oils. By far the largest volume of silicone products is used in the construction industry, in the form of elastic caulking compounds and water-repellent sealants.

16.4   *Some Transition Metal Polyoxyanions*

Isopolyanions

In their higher oxidation states, the metals of the earlier transition groups have chemistries that bear some formal resemblance to the main-group elements with the same characteristic valencies. One such resemblance is in the tendency of the oxyanions of these transition elements to exist as polymeric species. The chemistry of Group IVB (4) oxyanions has not been studied very much, and we will not discuss this group beyond saying that a compound of formula $Ti(O)SO_4 \cdot H_2O$ appears to consist of long chains of Ti—O—Ti—O with six coordination at the Ti being completed by sulfate and water molecules.

The oxyanion chemistry of Group VB (5) is much more elaborate, and aqueous solutions of V(V) contain many species whose concentrations vary greatly with pH and total vanadium concentration. In strongly alkaline solution, the dominant species is the tetrahedral orthovanadate ion $[VO_4]^{3-}$, and salts of this ion are easily prepared. Increasing concentration and lower pH favor polymerization. In the range of pH 8–14, the divanadate species, in which the vanadium is still tetrahedrally coordinated but dimerized through a V—O—V bridge, becomes important. Between pH 4–8, a variety of polymeric species appear; at the lower end of this range, decavanadate species, ideally $[V_{10}O_{28}]^{6-}$, are prevalent. Figure 16.17 shows the decavanadate structure. It consists of ten fused, highly distorted, octahedral VO_6 units of three different kinds. Four of the vanadiums have a terminal, four μ^2, and one μ^6 oxides; four vanadiums have a terminal, two μ^2, two μ^3, and one μ^6 oxides; two of the vanadiums have two μ^2, two μ^3, and two μ^6 oxides. The solid-state structure of the decavanadate

Figure 16.17
Structure of the
decavanadate ion
$[V_{10}O_{28}]^{6-}$.

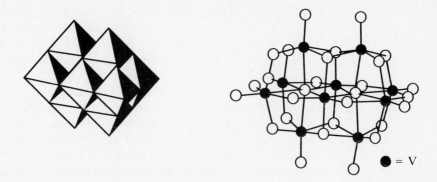

● = V

ion has been determined by X-ray analysis, and the nonequivalence of the
vanadium ions in this structure has greatly aided its detection and assay in
solution using ^{51}V-NMR spectroscopy.

A variety of metavanadates, similar in composition and structure to
the metaphosphates and metasilicates, have been prepared and studied.
Anhydrous metavanadates such as NH_4VO_3 contain chains of corner-
linked VO_4 tetrahedra. In the hydrated versions, the vanadium in these
compounds adopts trigonal-pyramidal coordination, and the chains are
formed by edge sharing.

Niobium(V) and tantalum(V) do not give as wide a variety of species
as does V(V). The best known polyanions of these elements are prepared
by fusion of Nb_2O_5 or Ta_2O_5 with alkali hydroxides or carbonates.
Aqueous solutions of the products of such reactions yield the $[M_6O_{19}]^{8-}$
ion, whose structure has been determined crystallographically and is shown
in Figure 16.18. The structure consists of an octahedron of equivalent MO_6
octahedra all sharing a common vertex at the center of the framework and
an edge with each of the four nearest neighbors.

Chromium(VI) is very small and highly polarizing. Its coordination
number in oxyanions is limited to four, and it shows little inclination to

Figure 16.18
Two representations
of the $[M_6O_{19}]^{8-}$ ion
(M = Nb and Ta).

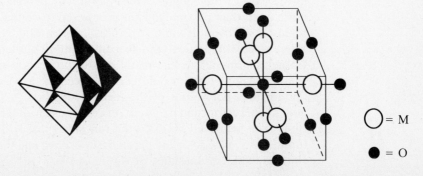

○ = M

● = O

polymerize. In aqueous solution, only the chromate $[CrO_4]^{2-}$ and dichromate $[Cr_2O_7]^{2-}$ ions are present at low and high pH, respectively. An extraordinary change occurs on descending the group, and Mo and W have the most varied and complicated polyoxyanion chemistry of all the transition elements.

At high pH, the ions $[MO_4]^{2-}$ are dominant, and the corresponding orthomolybdates and tungstates can be crystallized out. As the pH is lowered, protonation of the oxide ligands occurs. This protonation triggers two processes, expansion of the coordination sphere of the metal and condensation polymerization through the resulting OH groups. The first major species produced as the pH goes below 6 in aqueous molybdate solutions is heptamolybdate or paramolybdate $[Mo_7O_{24}]^{6-}$. Further lowering of the pH causes further polymerization as shown in Equations 16.31–16.33:

$$7[MoO_4]^{2-} + 8H^+ \rightarrow [Mo_7O_{24}]^{6-} + 4H_2O \qquad \textbf{(16.31)}$$

$$8[MoO_4]^{2-} + 12H^+ \rightarrow [Mo_8O_{26}]^{4-} + 6H_2O \qquad \textbf{(16.32)}$$

$$36[MoO_4]^{2-} + 64H^+ \rightarrow [Mo_{36}O_{112}]^{8-} + 32H_2O \qquad \textbf{(16.33)}$$

The Mo_{36} species shown in Equation 16.33 is the largest transition metal isopolyanion currently known. There are many other species including $[Mo_6O_{19}]^{2-}$, isostructural with the niobium and tantalum ions shown in Figure 16.18. Figure 16.19 shows the structures of the three products of Equations 16.31–16.33. The Mo_{36} entity is unique among the isopolymolybdates in having seven-coordinate, pentagonal-bipyramid, molybdenum ions.

Although there are many superficial similarities between the polymolybdates and the polytungstates, the differences are also very great. From the viewpoint of remembering details, it is unfortunate that there are very few structural identities in the two series. A feature of the polytungstates is their tendency to give more open frameworks, somewhat reminiscent of the zeolite structural units described previously. In the polytungstates, the building blocks are octahedral rather than tetrahedral, as in the case of the zeolites. One reason why polytungstate structures are more open than polymolybdates is believed to be the effect of intercation repulsions. Polymolybdate structures are rich in edge-sharing octahedra, whereas polytungstates tend to contain more corner-sharing octahedra. Corner sharing allows the metal ions to be farther apart than does edge sharing. Another major difference between the Mo and W systems is that the aqueous polyanions of the former equilibrate rapidly (less than one hour), whereas it can take up to several weeks for thermodynamic equilibrium to be achieved between the tungsten species. Two important polytungstates are the so-called paratungstate-*Z* ion and the metatungstate ion. Both are dodecatungstates of formulas $[W_{12}O_{42}]^{12-}$ and $[H_2W_{12}O_{40}]^{6-}$,

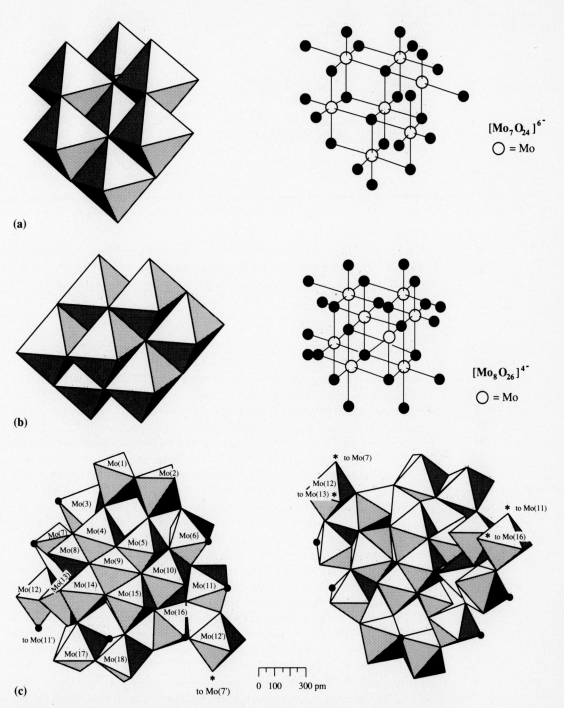

Figure 16.19 Schematic representations of the structures of some isopolymolybdates: (a) paramolybdate, (b) octamolybdate, (c) two halves of $[Mo_6O_{112}(H_2O)_{16}]^{8-}$. Reprinted with permission from I. Paulat-Böschen. *J. Chem. Soc. Chem. Commun.* (1979): 780.

Figure 16.20
Structures of
(a) paratungstate-Z
and **(b)**
metatungstate.

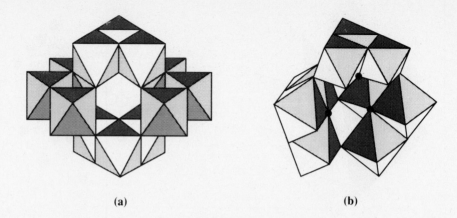

(a)   (b)

Figure 16.20
Structures of
(a) paratungstate-Z
and **(b)**
metatungstate.

respectively. Figure 16.20 shows the structures of these two ions. Paratungstate-Z can be crystallized from tungstate solutions at pH of about 6 over a period of weeks. Metatungstate is obtained by slow crystallization of tungstate solutions at pH of about 4.

Heteropolyanions

As mentioned above, isopolyanion structures can be relatively open. The cavities in such structures can be occupied by another main-group or transition element. Historically, the heteropolyanions of Mo and W have been known the longest and have been studied most intensively. The largest class to have been structurally characterized is based on the metatungstate structure. The cavity at the center of this structure has a tetrahedron of triply bridging oxygens directed toward its center, and it can contain a wide range of small atoms, for example, P(V), As(V), Si(IV), B(III), and Ti(IV). The heteropolymolybdates of this series are isostructural with the tungstates, even though the analogous isometamolybdate structure is not known. An interesting example of how the cavity can impose an unusual coordination on the central atom is $[Co^{III}W_{12}O_{40}]^{5-}$, which contains the extremely rare, high-spin, tetrahedral Co(III).

If the heteroatom is too large to be accommodated in the meta polyanion cavity, the heteropolyanion can be obtained with a structure in which the cavity provides octahedral coordination for the guest atom. A number of examples are known where the guest ion sits at the center of the octahedral cavity formed by a hexagon of edge-sharing MO_6 octahedra (Figure 16.21). The ions $[M^{n+}Mo_6O_{24}]^{(12-n)-}$ —where M = Te(VI), I(VII), and Co(III)—have this type of structure.

Despite many years of work, the synthesis of isopolyanions and hetero-

Figure 16.21 Schematic illustration of the structure of the 1:6 octahedral heteropolymolybdates.

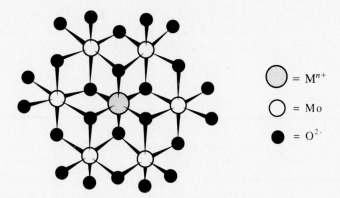

$\bigcirc$ = M^{n+}

$\bigcirc$ = Mo

$\bullet$ = O^{2-}

polyanions remains very much an empirical art. The syntheses are usually arrived at by crystallization of promising aqueous solutions, single compounds, or mixtures and by adjusting pH, temperature, and concentration in the hope of obtaining interesting new species. Problems also are often encountered in determining structures due to poor quality of crystals or difficulty in locating hydrogen atoms that can allow an unequivocal distinction between O, OH, or H_2O.

BIBLIOGRAPHY

Suggested Reading

Allcock, H. R. "Poly(organophosphazenes)—Unusual New High Polymers." *Angew. Chem. Int. Ed. Eng.* 16 (1977): 147.

Chivers, T. "Synthetic Methods and Structure–Reactivity Relationships in Electron-Rich Sulphur–Nitrogen Rings and Cages." *Chem. Rev.* 85 (1985): 341.

Griffith, E. J. "The Chemistry and Physical Properties of Condensed Phosphates." *Pure Appl. Chem.* 44 (1975): 173.

Haiduc, I., and D. B. Sowerby. *The Chemistry of Inorganic Homo- and Heterocycles.* New York: Academic Press, 1987.

Roesky, H. W. "Structure and Bonding in Cyclic S—N Compounds." *Angew. Chem. Int. Ed. Eng.* 18 (1979): 91.

Tytoko, K.-H., and O. Glemser. "Isopolymolybdates and Tungstates." *Adv. Inorg. Chem. Radiochem.* 19 (1976): 239.

Wells, A. F. *Structural Inorganic Chemistry*, 5th ed. London: Oxford University Press, 1984, Chapter 23.

Zeldin, M., K. J. Wynne, and H. R. Allcock, eds. *Inorganic and Organometallic Polymers.* ACS Symposium Series, Vol. 360. Washington D.C.: American Chemical Society, 1988, Chapters 19–25.

PROBLEMS

16.1 How would you synthesize the following compounds?
  a. $B_3H_3N_3(CH_3)_3$ **c.** S_4N_4
  b. $N_3P_3Cl_6$ **d.** $Na_5P_3O_{10}$

16.2 Outline the factors that favor the P—O—P linkage over other main-group acid anhydrides as the preferred biochemical energy-storage system.

16.3 List four applications of inorganic polymers. Identify one specific polymer with each use.

16.4 Draw a diagram of P_4O_{10} on paper. Cut your diagram with scissors in such a way as to convert it into cyclometatetraphosphate.

16.5 What factors are important in determining the following?
  a. Carbon only gives monocarbonates, but silicon gives a huge number of polymeric silicates.
  b. Sulfur nitrides tend to be thermodynamically unstable, but the nitrides of silicon and phosphorus are stable.
  c. Even though chrysotile has a sheet structure, it is a fibrous mineral.
  d. Borazines undergo addition reactions much more easily than does benzene.

16.6 Draw structures for the following species:
  a. $[PNCl_2]_3$ **c.** $[Si_2O_7]^{6-}$
  b. $[S_4N_5]^-$ **d.** $[Ta_6O_{19}]^{8-}$

16.7 Explain the following experimental observations:
  a. A colorless compound of formula $Na_8Cl_2Al_6Si_6O_{24}$ became highly colored when boiled with sodium polysulfide solution.
  b. A colorless crystal of S_2N_2 developed a deep bronze color after storing for several months at room temperature under argon.
  c. A sample of $NaNO_3$ dissolved easily in all proportions in water to give a mobile liquid. A sample of $NaPO_3$ only dissolved in water with great difficulty to give a highly viscous solution.
  d. A bottle of dimethyldichlorosilane, which had stood for several months under the fume hood, had a badly corroded cap with a large, gummy incrustation that was soluble in organic solvents. A bottle of phenyltrichlorosilane standing next to it was in a similar condition, but the incrustation around the cap was extremely hard and insoluble in organic solvents.

16.8 Compounds containing $[SO_4]^{2-}$ and $[PO_4]^{3-}$ ions are very common, but those containing the $[SiO_4]^{4-}$ ion are very rare. Why?

16.9 Compare the electronic structures of the following:
   a. $[HBN(CH_3)]_3$
   b. $[(CH_3)_2SiN(CH_3)]_3$
   c. $[(CH_3)_2PN]_3$
   d. $[S_3N_3]^-$

16.10 Complete the following equations:
   a. $3S_4N_4 + 6Cl_2 \rightarrow$
   b. $P_4O_{10} + 2H_2O \rightarrow$
   c. $7Na_2MoO_4 + 4H_2SO_4 \rightarrow$
   d. $B_2H_6 + NaNH_2 \rightarrow$

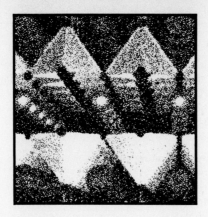

17

Metal Cluster Compounds

The special properties that we associate with metals, such as high electrical conductivity, high optical reflectivity, malleability, and ductility, are essentially a property of the macroscopic solid state. A question that has fascinated chemists and physicists for many years is: How many metal atoms must be combined before we begin to observe these typical metal properties?

A similar situation exists in the field of heterogeneous catalysis, in which the later transition metals have played an important role. It has been known since the earliest studies that the metal component of a catalyst is more efficiently used if it is finely divided; a variety of methodologies have been developed to produce and maintain metals in the finely divided state. For example, the tendency of small metal particles to fuse is suppressed if they are supported on metal oxide surfaces, and it is possible to produce platinum metal microcrystallites smaller than 5000 pm supported on silica–alumina. Although such particles are very small, they still contain more than 1000 atoms, of which only about 600 are at the surface. A question that has been asked by researchers in catalysis is, How small can a metal particle be and still retain its catalytic properties?

In this section, we describe some transition metal clusters that help, at least partly, to answer the questions posed above. We have encountered the phenomenon of metal-bonded clusters in the description of the halides of Groups VB and VIB (5 and 6) in Chapter 11; in Chapter 22, we discuss the simple metal carbonyls. In recent years, the range and size of carbonyl clusters have been greatly increased, and rational synthetic methods for their production are beginning to be devised.

As a consequence of the extremely strong bonds that osmium forms to itself and to CO, osmium is one of the most intensively studied of the elements that form metal carbonyl clusters. Because of these strong bonds, osmium carbonyls have a lesser tendency to decompose to the metal and CO than do many of the other carbonyls. Figure 17.1 shows some of the remarkable range of high nuclearity clusters that can be isolated from pyrolysis reactions of triosmium dodecacarbonyl. In this figure, we also include some hydrido- and carbido-clusters that are formed under pyrolysis conditions, as a result of the presence of water or the disproportionation of two molecules of CO into C and CO_2, respectively.

Aqueous alkali is also a useful reducing medium for metal carbonyls because of the low oxidation potential of the hydride ion produced in the reaction

$$CO + OH^- \rightarrow CO_2 + H^- \qquad (17.1)$$

Figure 17.2 shows some syntheses of high nuclearity clusters, using reduction of lower carbonyls.

The structures of transition metal clusters are frequently based on the same deltahedra as the polyboranes. Figure 17.3 shows the structures of some deltahedral clusters. These can often be rationalized with some

Figure 17.1
Some products of the pyrolysis of $Os_3(CO)_{12}$.

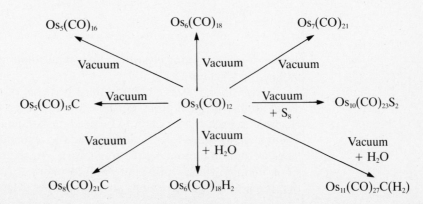

Figure 17.2
Synthesis of some
high nuclearity
clusters by reduction
of simple carbonyls.

$$[Ni_6(CO)_{12}]^{2-} \xleftarrow{\text{NaBH}_4} Ni(CO)_4 \xrightarrow{\text{Na/THF}} [Ni_{12}(CO)_{21}]^{4-}$$

$$Rh_4(CO)_{12} \xrightarrow{\text{OH}^-/\text{Pr}^i\text{OH/Heat}} [Rh_{22}(CO)_{37}]^{4-}$$

$$[PtCl_6]^{2-} \xrightarrow{\text{CO/NaOH/MeOH}} [Pt_9(CO)_{18}]^{2-} \xrightarrow{\text{MeCN/Heat}} [Pt_{19}(CO)_{22}]^{4-}$$

Figure 17.3
Some deltahedral
metal clusters.

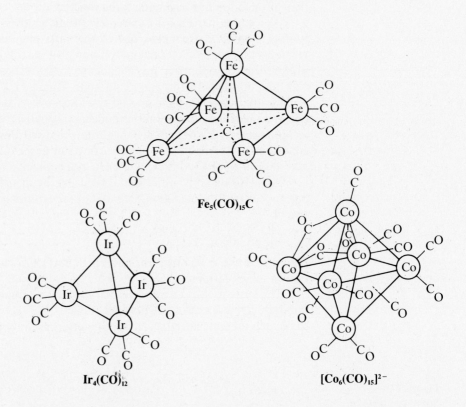

$Fe_5(CO)_{15}C$

$Ir_4(CO)_{12}$

$[Co_6(CO)_{15}]^{2-}$

degree of success, using electron- and orbital-counting rules such as those outlined in Chapter 15.

In transition clusters, account must be taken of the *d* orbitals on the metal atoms. In main-group clusters, we attributed one orbital and one electron on each cluster atom to be used for bonding to a terminal atom. The remaining three-valence orbitals and the remaining cluster-atom electrons were used in bonding the framework. An analogous model for transition metals, which works quite well for many smaller clusters, assumes that three of the metal orbitals are used to bind external ligands, three are

Table 17.1   *Framework Electron Contributions for Some Metal Carbonyl Fragments*

Metal	Cluster unit		
	$M(CO)_2$	$M(CO)_3$	$M(CO)_4$
Cr, Mo, W	−2	0	2
Mn, Tc, Re	−1	1	3
Fe, Ru, Os	0	2	4
Co, Rh, Ir	1	3	5
Ni, Pd, Pt	2	4	6

nonbonding, and the remaining three are used in forming framework orbitals. Filling the ligand bonding and the metal nonbonding orbitals requires twelve electrons. If the metal has v-valence electrons and terminal ligands contribute x electrons, the number of electrons that the metal can contribute to framework bonding is $(v + x - 12)$. Table 17.1 lists the numbers of framework electrons for some commonly encountered transition metal carbonyl fragments.

The terminal carbonyls are assumed to contribute two electrons to the metal-fragment electron count. We can use these numbers to help predict the structures of some clusters. For example, to predict the structure of the $Rh_6(CO)_{16}$ molecule, we start with the assumption that there are six $Rh(CO)_2$ units and four nonterminal CO groups (each of which contributes two electrons to the framework). The framework electron count is

$6 \times Rh(CO)_2$	6
$4 \times CO$ (nonterminal)	8
Total	$14e^-$

These seven framework electron pairs, binding a six-atom cluster, would lead to a closo-structure with six vertices, that is, an octahedron as in the case of $[B_6H_6]^{2-}$. The four nonterminal CO groups are actually triply bridging on alternate trigonal faces of the octahedron. The electron count for $[Ni_6(CO)_{12}]^{2-}$ is the same as for the Rh_6 custer, but it has six terminal and six bridging carbonyl groups. The $[M_6(CO)_{18}]^{2-}$ clusters (M = Ru and Os) are also isoelectronic with the Ni_6 and Rh_6 compounds and contain an octahedral metal unit.

Unfortunately, there are many exceptions to these simple electron-counting rules, and they are virtually useless for very large clusters. Figure 17.4 shows some clusters that do not obviously obey the rules. The

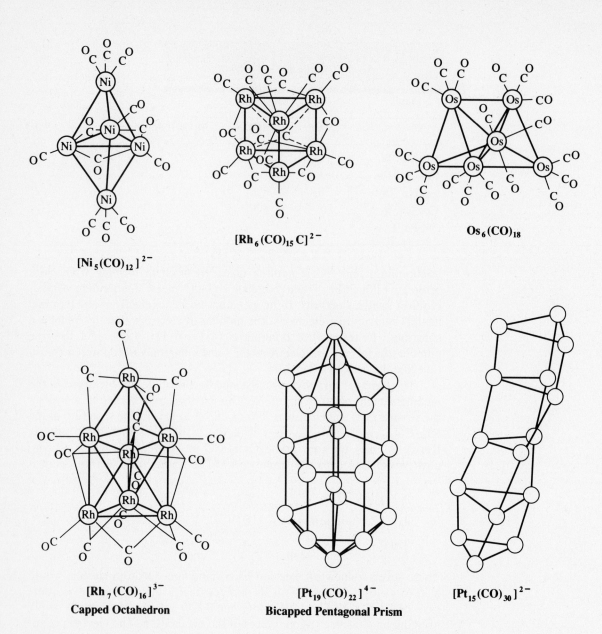

$[Ni_5(CO)_{12}]^{2-}$

$[Rh_6(CO)_{15}C]^{2-}$

$Os_6(CO)_{18}$

$[Rh_7(CO)_{16}]^{3-}$
Capped Octahedron

$[Pt_{19}(CO)_{22}]^{4-}$
Bicapped Pentagonal Prism

$[Pt_{15}(CO)_{30}]^{2-}$

Figure 17.4 Some clusters that do not obey the simple electron-counting rules. Only the metal skeletons are shown for the Pt_{19} and Pt_{15} clusters.

$[Rh_7(CO)_{16}]^{3-}$ ion represents a case where the rules almost fit. The electron count for this ion gives seven framework bonding pairs, which in turn suggests the possibility of a closo-octahedral structure, but in fact we need a structure with seven vertices. The experimentally determined structure is a capped octahedron, which is not a regular deltahedron. In this type of capping, the additional metal atom is able to make maximum use of its three unused AOs without modifying the number of deltahedral skeletal bonding MOs. Cases of this kind, where the number of framework electron pairs equals the number of skeletal atoms and where the structure is a capped deltahedron, are common. The $Os_6(CO)_{18}$ structure (Figure 17.4) consists of a bicapped tetrahedron.

Many high nuclearity clusters are built up by polymerization of simpler units—see $[Pt_{15}(CO)_{30}]^{2-}$ and $[Pt_{19}(CO)_{22}]^{4-}$ in Figure 17.4—or contain one or more metal atoms within the molecular polyhedron. These structures are not easily explained by the simple electron-counting rules. Finally, there are numerous relatively simple structures that do not obey the rules. These are exemplified in Figure 17.4 by the ions $[Ni_5(CO)_{12}]^{2-}$ and $[Rh_6(CO)_{15}C]^{2-}$. The nickel complex, being a regular TBP seems to represent a closo-structure and should have six framework pairs. In fact, it has eight and ought to be a more irregular arachno-structure. The Rh cluster has a framework electron pair count of nine, suggesting an irregular arachno-structure rather than the regular trigonal prismatic structure actually observed. However, the prismatic structure, which is not a deltahedron, involves nine metal–metal bonds, thus allowing full use of the nine electron pairs in an electron-precise structure.

A number of the clusters described above contain heteroatoms. Such compounds furnish interesting models for the unit cells of interstitial compounds of the transition elements. Hydrides are frequently produced in the reductive syntheses of the type illustrated in Figure 17.2, and the anionic species can often be protonated to give hydrides. Hydrogen usually ends up in a bridging or interstitial, rather than a terminal, position.

Hydride-containing clusters present a particularly interesting structural challenge because the hydrogen atom is particularly difficult to locate by X-ray methods when it is located close to a large number of electron-rich atoms. It is sometimes possible to define the location of the hydride on the basis of unusually long bond lengths in certain parts of the cluster framework. However, it usually requires the use of costly and inaccessible neutron diffraction to definitively locate the hydrogen. Multinuclear NMR studies can be most helpful in establishing structure, particularly in solution. Simple chemical shift data are not of much use because they cover an enormous range for structurally similar compounds. For example, in both $[Co_6(CO_{15}H]^-$ and $[Ni_{12}(CO)_{21}H]^{3-}$, the hydrogen occupies an octahedral interstitial position, but the ^{1}H-NMR chemical shifts are +23.2 and −24 ppm, respectively! Both compounds can be considered as models for the

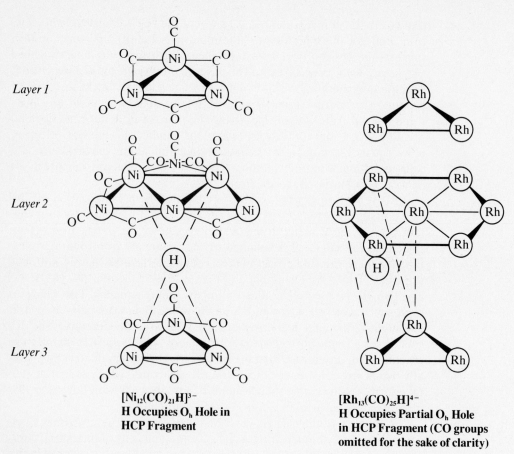

Layer 1

Layer 2

Layer 3

$[Ni_{12}(CO)_{21}H]^{3-}$
**H Occupies O_h Hole in
HCP Fragment**

$[Rh_{13}(CO)_{25}H]^{4-}$
**H Occupies Partial O_h Hole
in HCP Fragment (CO groups
omitted for the sake of clarity)**

Figure 17.5
Exploded views of
some clusters that
show relationship to
HCP structures.

hydride sites in PdH_x, in which the hydride phase has a distorted rock-salt structure. The nickel complex is particularly interesting because the cluster is a distorted fragment of a HCP lattice. Figure 17.5 shows an exploded version of this structure, which shows its relationship to the HCP packing.

Because ^{103}Rh has a 100% natural abundance and spin $= \frac{1}{2}$, multi-nuclear NMR is particularly useful for rhodium clusters. Although $[Rh_6(CO)_{15}H]^-$ has not thus far been available in crystals suitable for diffraction studies, its structure has been unequivocally assigned by a combination of ^1H-, ^1H-$\{^{103}Rh\}$-, ^{13}C-, ^{13}C-$\{^{103}Rh\}$- and ^{103}Rh-NMR spectroscopies.

Despite the enormous advances in the last decade in the synthesis, structural characterization, and reactivity of metal clusters, relatively little progress has been made in answering the fundamental questions posed at the beginning of this section. All clusters described above have covalent molecular properties and shown no sign of beginning to exhibit metallic behavior. This is true of both physical phenomena related to band structure and of catalytic behavior. In fact, it has proved to be extraordinarily difficult to demonstrate conclusively that any catalytic behavior of these compounds is not due to their breakdown to mononuclear species. In fact, for most systems this usually turns out to be the case.

Despite the lack of progress in answering the key questions related to the solid versus dispersed states of metals, the basic foundation of acquired knowledge leaves little doubt that the resolution of these questions cannot be far off. In one of those surprising coincidences (which seem to occur with greater than random frequency when you do research and read the current literature), the day after the first paragraph of this section was written, a report of the UV/vis spectroscopic behavior of $(Ph_3P)_{12}Au_{18}Ag_{20}Cl_{14}$ appeared in the latest issue of the *Journal of the American Chemical Society*. The metal atoms in this cluster, shown in Figure 17.6, consist of three Au_6Ag_6 dodecahedra, each sharing adjacent Au vertices with its two neighbors. Two additional Ag atoms lie along the perpendicular to the Au triangle formed by the shared dodecahedral vertices, and there is an Au atom at the center of each dodecahedron. The UV/vis spectrum of the compound exhibits an electronic absorption band at 495 nm. This band is at the same energy as the so-called *plasmon frequency* of Au/Ag alloy particles of the same stoichiometry as the

Figure 17.6
Arrangement of
metal atoms in
$(Ph_3P)_{12}Au_{18}Ag_{20}Cl_{14}$.

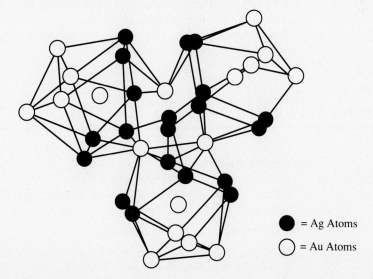

● = Ag Atoms

○ = Au Atoms

Figure 17.7
UV/vis spectra
of **(a)**
$(Ph_3P)_{12}Au_{18}Ag_{20}Cl_{14}$,
(b) colloidal gold in
aqueous solution,
and **(c)** colloidal
$Au_{0.22}Ag_{0.78}$.
Reproduced with
permission from
B. K. Teo et. al.
*J. Amer. Chem.
Soc.* 109 (1987):
3494.

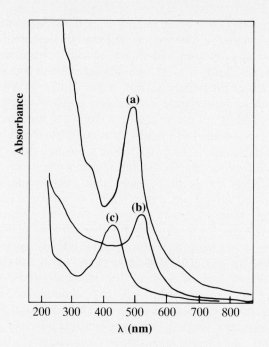

cluster but of much larger size (20–40 nm). This plasmon frequency is due
to the resonant interaction of light with the highly polarizable conduction-
band electrons of the metal and is what gives rise to the characteristic
colors of gold and copper and their alloys. Figure 17.7 shows the spectra of
the cluster and the colloidal alloy. The authors conclude that the apparent
observation of the plasmon absorption is a "strong indication that the
cluster possesses the free electron quasi-band structure behavior characteristic of bulk metal."

BIBLIOGRAPHY

Suggested Reading

Cotton, F. A., and M. H. Chisholm. "Bonds Between Metal Atoms," Special
Report. *Chem. Eng. News* (June 1982): 40.

Johnston, B. F. G., ed. *Transition Metal Clusters.* New York: Wiley, 1980.

Vargas, M. D., and J. N. Nicholls. "High Nuclearity Metal Carbonyl Clusters."
Adv. Inorg. Chem. Radiochem. 30 (1986): 123.

PROBLEMS

17.1 On the basis of framework electron count, predict the structures of the following species:
  a. $Fe_2Co_2(CO)_{12}$ **c.** $Fe_5(CO)_{15}C$
  b. $Ni_5Os(CO)_{14}$ **d.** $Pt_7(CO)_{16}$

17.2 The structure of $[Ni_5(CO)_{12}]^{2-}$ can be described as a capped nido-cluster. Analyze the electron count and explain how the latter structural designation is justified.

17.3 Give a synthesis for each of the following compounds:
  a. $Na_2Pt_9(CO)_{18}$
  b. $Os_8(CO)_{21}C$
  Suggest sources for the electrons needed to reduce the platinum to produce (a) and the carbon atom needed to produce (b).

17.4 Ti, Zr, and Hf do not form extended metal cluster compounds. Consult Table 17.1 and try to explain the absence of metal-cluster compounds in Group IVB (4).

17.5 Define the two classes of close packing of spheres. Give an example of a metal cluster in which the metal atoms occupy the positions characteristic of a close-packed structure.

PART SIX

Solid-State Chemistry

18

Ordered Solids

For most of its modern history, chemistry has had an overwhelming bias toward molecular phenomena and chemical theory; experiment and general thinking reflect this fact. Perhaps this is nowhere more evident than in the domain of solid-state science, which despite its enormous technological importance and the clear role of chemistry in its development is rarely accorded more than a casual and brief recognition by chemistry texts. In large measure, this is because of the numerous difficulties encountered in dealing with the solid state; these difficulties often seem to yield more fruit to the arcane inventor than to the basic scientist. The tools available for probing the structure of solid materials are much more limited than those available for elucidating the structures of molecules. It is only recently that modern computers have made crystal-structure determination by diffraction methods a relatively routine operation (see Chapter 4). Impressive though it may be, the contribution of crystal-structure determination is still limited because the interesting features of many solids reside in their disorder or in defects rather than their ordered crystalline structure. It is an interesting commentary on the symbiotic relationship between science and technology that the modern generation of storage batteries is well understood technically because it has been developed largely from established, basic solid-state science. On the other hand, the detailed mechanisms of the lead-acid battery or the simple flashlight battery, which were invented a century ago and then improved empirically, are still very poorly understood because the electrode materials are highly defective, essentially amorphous solids.

The theoretical treatment of bonding in solids has also presented many problems for the chemist. When quantum mechanics was first being applied to problems in chemical bonding, the electronic structure of solids was a much more urgent problem in physics than in chemistry. Physicists needed a theoretical framework to deal with the optical, electronic, magnetic, and acoustic properties of solids, and they rapidly developed a model based on the scattering of standing electron waves by the constituent atoms of a lattice. The results of this theory (the band theory of solids) bore some similarities to those of the atomic and molecular theories, in that the electrons were found to have certain allowed energies whereas other energies were forbidden. On the other hand, the physicists preferred to deal with the momenta of electrons, rather than their energies. This fact, combined with the lack of any clear connection between the band theory and the then prevalent (in chemistry) valence bond theory, deterred most chemists from becoming generally interested in the bonding properties of solids. Band theory also deterred physicists from becoming seriously interested in amorphous solids because all its elegance and computational simplicity was based on the properties of an ordered lattice. Out of this grew the prejudice that all the fascinating properties of crystals, which band theory so elegantly explained, must be peculiar to crystals because band theory apparently could not be applied to amorphous materials. Fortunately, this prejudice did not endure for long; experiments and inventions soon began to undermine it. It is now known that the differences between crystalline and amorphous materials are quite subtle, rather than dramatic, and that the theories used both by physicists and chemists can make valuable contributions to understanding the structure of solids.

In large measure, the rapprochement between the chemistry and physics of the solid state is due to the necessity of using more localized bonding models for the description of amorphous solids. This has encouraged physicists to adopt theoretical models that are more familiar to chemists and encouraged chemists to embark more vigorously on the synthesis and characterization of new solid materials.

In this chapter, we discuss bonding in solids entirely on the basis of localized and delocalized AOs and MOs. We also confine our discussion almost entirely to materials that only exist as solids and lose their identity when dissolved or vaporized.

It would be inappropriate to end a general introduction to solid-state chemistry without drawing attention to the enormous contributions solid-state science has made to modern technology. The production, transmission, storage, and utilization of electrical energy has always been heavily dependent on the special properties of solids. The new demands of the so-called energy crisis, which will be with us for at least the next several decades, have greatly accelerated the search for new materials to help produce and use electrical energy more cheaply and efficiently. The great revolution in information handling that has occurred over the past half-century has also grown out of developments in solid-state science. In computers, there are microelectronics, magnetic-storage disks and tapes, and various display devices; all are based on solids with special properties. Photography and electrophotography (xerography) have revolutionized our leisure and business activities. In none of these areas is there a belief that invention and discovery has run its course. The opportunities and needs for development of new solid materials are very great, and as compositions of greater complexity are explored, the chemistry will become more demanding.

18.1 *Bonding in Solids*

Bonding in Metals and Semiconductors

In Chapter 3, we discussed the interactions of small groups of hydrogen atoms, in different geometrical arrangements, and the resulting formation of delocalized MOs. In principle this hypothetical exercise can be continued *ad infinitum* to produce a macroscopic lattice of interacting hydrogen atoms. This in turn gives rise to a manifold of an infinite number of bonding levels and a manifold of an infinite number of antibonding levels. The first important questions we must ask are: How wide are these manifolds and how far apart are they? In fact, there is no single answer to these questions, and the hypothetical hydrogen lattice is a good illustrative case because it does not exist under any conditions that have thus far been achieved in the laboratory. There is a general expectation that it will be realized some day under conditions of extremely high pressure, which will force the hydrogen atoms into closer proximity than is allowed under ambient conditions. It is this internuclear separation that is the critical parameter determining the energy characteristics of the bonding and antibonding manifolds (henceforth, we will use the more general names of *valence band* and *conduction band*).

If we compress H_2 gas below its critical temperature, it eventually liquefies and then solidifies to a molecular solid in which the H_2 molecules are only very weakly coupled to each other. In this solid, all bonding levels have almost the same energy, all antibonding levels have almost the same energy, and the difference between the energies is more or less the energy characteristic of the H_2 molecule in the gas phase. Applying further pressure forces the H_2 molecules closer together and, as the intermolecular separation falls below the equilibrium van der Waals limit, coupling between molecules becomes stronger and intermolecular orbital overlaps increase. The effect of these overlaps is to spread the range of energies in the valence and conduction bands and to reduce the separation between them. Figure 18.1 shows an approximate representation of this behavior. In reality the energy changes are not linear with distance, and for the H_2 case, there is not a single separation coordinate. Figure 18.1 illustrates that for internuclear separation values greater than the value x, the valence band and the conduction band are separated by a "forbidden zone," which is called the *band gap*. In addition the valence band is completely filled with electrons, and the conduction band is completely empty. Below the critical separation x, the two bands merge to form a continuum of energy states that is now only half-filled. This state corresponds to the formation of a metal because the characteristic high electrical and thermal conductivities and the high optical reflectivity–absorbtivity of metals is a direct consequence of this kind of partially filled band of energy states.

Electron pairs have very poor mobility, and a filled band is analogous to an hourglass that is completely filled with sand; no matter how many times we turn it, there will be no transfer of sand from one lobe to the other. If, on the other hand, one of the lobes is only partially filled, turning

Figure 18.1
Schematic representation of the transformation of a molecular H_2 lattice to a metallic H lattice.

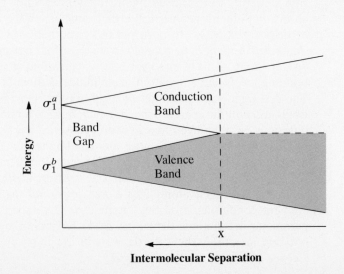

the hourglass causes a flow of sand between the lobes. In a partially filled band, electrons are more spatially mobile because they can move into unfilled levels within the band. Thus, electrons can move easily down an electrical potential gradient and thereby carry current. Local heating increases electron energy; these "hot" electrons move easily through the metal, carrying excess kinetic energy that gets transferred to the lattice as the electrons scatter from the constituent atoms. The increased lattice vibration is detected as a rise in temperature; the rate at which the excitation moves away from the source is the thermal conductivity. The strong absorption and reflection of light throughout the visible region is because the bandwidth is typically larger than the visible-spectrum width; thus, there is a continuum of transitions between filled and empty states that covers the energy range from zero to the bandwidth.

The way in which the band structure of H_2 is shown in Figure 18.1 is unconventional; it is more normal to correlate the bands of the solid with the AOs of isolated atoms, rather than of diatomic molecules. However, the conclusion is the same in both cases. Figure 18.2 shows the more usual representation of the relationship of band structure to internuclear separation. In this type of representation, we see the changes in energy states as we compress an infinitely expanded lattice of atoms to small internuclear separations. For different ns^1 atoms, the diagrams are quantitatively different but qualitatively similar. Alkali metals show the reverse behavior to H_2 in that they exist as solid metals under ambient conditions and must be evaporated at elevated temperature to produce the diatomic molecules. In hydrogen the lack of inner-shell repulsion allows the formation of an unusually short and strong bond in the diatomic molecule. The relatively expanded inner shells of the alkali metals lead to rather long, weak bonds in the diatomic molecules. The high coordination number of a close packed solid requires a considerable depletion of the electron density in the internuclear regions and a corresponding increase in the internuclear repulsion term. This effect is more severe for H than for alkali metals because the internuclear separation is shorter for metallic H, while the effective nuclear charge is more or less constant. Thus, relative to alkali metals, the diatomic molecule is much more stable, and the metal is much less stable. Alkali metals have weak bonds in both states, which explains the softness, low melting points, and high volatility of alkali metals.

In the discussion of the bonding in diatomic molecules (Chapter 3), it was pointed out that the Group IIA (2) elements should not form diatomic molecules because such molecules would inevitably have equal numbers of occupied bonding and antibonding orbitals. In principle this same conclusion might appear to be valid for the metal where the bands analogous to those shown in Figure 18.2 would have to accommodate twice as many electrons and there would be equal occupancy of bonding and antibonding states. Even if such a solid were stable, we would expect the material to be

Figure 18.2
Band-structure diagram for an ns^1 metal.

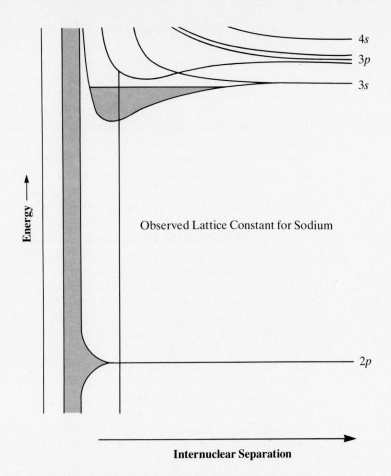

an insulator, not a metal, because the bands are completely full and all electrons are paired. These conclusions do not conform to experimental observations (all ns^2np^0 elements are stable metals) because we have ignored the presence of the np orbitals. Configuration interaction (CI) between some of the empty np states and the filled ns states leads to a lowering of the energy of the latter at the expense of the former. In addition, the lower edge of the np band (the more strongly bonding np states) overlaps the upper part (the more strongly antibonding ns states) of the ns band, thus furnishing the partially occupied continuum necessary for metallic behavior. Figure 18.3 shows the band structure of ns^2np^0 metals.

The increasing electronegativity and decreasing radii of the later groups reduces the tendency toward metallic bonding by favoring the localization of electrons and disfavoring the high coordination numbers normally associated with the close-packed metallic structures. Thus, boron has a number of allotropes, all nonmetallic, and most of them built from

Figure 18.3
Schematic
representation of the
band structure of
ns^2p^0 metals.

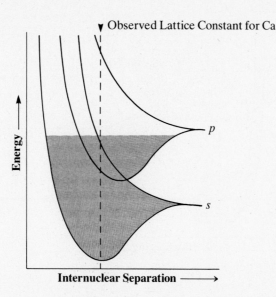

Observed Lattice Constant for Ca

Energy

p

s

Internuclear Separation ⟶

basic units of regular polyhedra of boron atoms, for example, octahedra and icosahedra. The remaining members of Group IIIA (13) are metals.

Group IVA (14) provides the most comprehensive example of the trends in electronic structure of solids within a group. Carbon, silicon, and germanium all exist in the so-called diamond structure, in which each atom is surrounded by a regular tetrahedron of neighboring atoms (Figure 18.4). This structure is analogous to the zinc blende structure, but the Group IVA (14) atom occupies both the zinc and the sulfide sites. Because the local symmetry is different from that in the close-packed ns^2np^0 metals, the band structure is slightly different from that shown in Figure 18.3. The way in which the diamond bands are formed is easily seen by reference to the MOs of methane and is illustrated in Figure 18.5. The only difference is that the ligand orbitals are no longer hydrogen $1s$ orbitals but appropriate

Figure 18.4
Diamond structure.

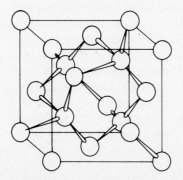

Figure 18.5
Band structure of
the Group IVA (14)
diamond lattice.
Adapted with
permission from
J. B. Hannay, *Solid
State Chemistry*.
Englewood Cliffs,
N.J: Prentice Hall,
1967, p. 36.

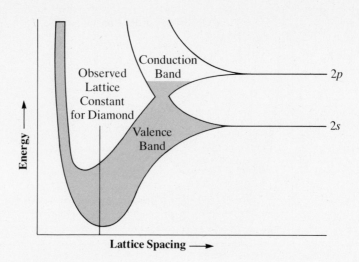

combinations of carbon $2s$ and $2p$ orbitals. The bonding a_1 and t_2 sets meld at an intermediate internuclear separation to give a valence band that is three-quarters p states and one-quarter s states. The antibonding a_1 and t_2 sets furnish a conduction band with the same proportions of s and p states as the valence band. Because the four-valence orbitals of each atom are, on the average, equally divided between bonding and antibonding states and because each atom contributes four electrons, the valence band is completely filled, and the conduction band is empty. This is the state of affairs that exists in diamond and thus accounts for its transparency, hardness, and insulating character. The lack of color of pure diamond also indicates that the band gap is of greater energy than is the highest-energy visible light.

On descending the group, the band gap diminishes in the same way that the energy separation between bonding and antibonding levels of simple molecules decreases for heavier atoms. This diminished band gap leads to the appearance of the properties associated with *intrinsic semiconduction*. These properties arise when electrons from the completely filled valence band can be thermally excited to the conduction band in sufficient numbers to give appreciable electrical conductivity. The usual convention is to designate the departure of an electron from the valence band as creating a positive hole in the valence band; in the charge-carrying process, the positive holes in the valence band migrate in the opposite direction to the electrons in the conduction band. Silicon and germanium are both intrinsic semiconductors, and one of the allotropes of tin (gray tin) has a diamond lattice but shows metallic conductance. Lead only exists in a close-packed lattice and is metallic.

Figure 18.6
Correlation of the band gap with electronegativity for Group IVA (14) elements and IIB–VIA (12–16) and IIIA–VA (13–15) compounds. χ_A and χ_B are the electronegativities of the constituent atoms.

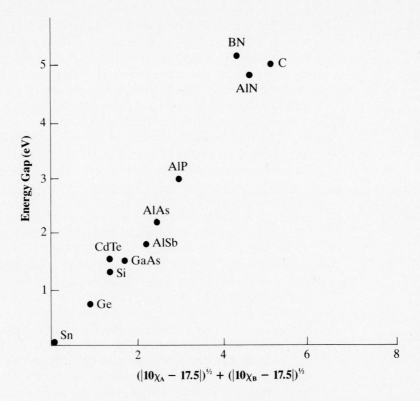

Besides their increased electrical conductivity, good intrinsic semiconductors are opaque and highly reflective because the band gap is less than the energy of much of the visible spectrum. Photoexcitation of electrons also increases electrical conductivity, and semiconductors therefore also exhibit *photoelectric conduction*.

The number of isoelectronic and isostructural analogs of C, Si, Ge, and Sn is quite large. Most of the binary IIB/VIA (12/16)(for example, CdS) and IIIA/VA (13/15)(for example, GaAs) compounds can be made with zinc blende structures and therefore have band structures qualitatively similar to those of Group IVA (14) elements. The band gap correlates with the difference in electronegativity (more ionic compounds tend to have larger band gaps) and with the sum of the electronegativities (two electronegative elements—for example, B and N—give a large band gap). These two tendencies can be combined in the electronegativity expression plotted in Figure 18.6 to show the correlation between band gap and electronegativity (χ). Cubic boron nitride (diamond structure) can be made by combination of the elements under ultrahigh-pressure synthesis conditions, of the kind used for making synthetic diamonds, and it bears a remarkable resemblance to diamond in all of its properties. Gallium arsenide is widely used in the manufacture of light-emitting diodes, and cadmium sulfide is a useful photoconductor. Both compounds can be made by relatively cheap and conventional chemical routes. Gallium arsenide is made by the copyrolysis of trimethylgallane and trimethylarsine. Cadmium sulfide is made by reaction of a cadmium salt with H_2S in aqueous solution followed by crystallization from the melt.

The more stable allotrope of carbon is graphite. Graphite is the example par excellence of a layer lattice and, from the viewpoint of band structure, is essentially two-dimensional. The layers are infinite sheets of trigonally coordinated carbon atoms in which the in-plane σ bonding uses two $2p$ and $2s$ orbitals, while the third $2p$ orbital, perpendicular to the plane, gives rise to a π band. The structure of graphite is shown in Figure 15.7. In graphite the bonding and antibonding π bands overlap, giving rise to the near-metallic conductivity and high optical opacity of graphite. The conductivity in the plane of the layers is very much greater than that in the direction perpendicular to the plane. The combination of cheapness, chemical inertness, good electronic conductivity, and lubricity makes graphite enormously useful in maintaining electrical contact under conditions of sliding contact—for example, brushes of electric motors and generators—and as an electrode material. The other members of Group IVA (14) do not exist in the graphite structure, although a class of silicon hydride materials (SiH_n; $n < 1$) are believed to contain extended graphitelike units. Hexagonal boron nitride, although isostructural with graphite, is a colorless insulator.

Nitrogen exists only as the diatomic molecule under ambient conditions. The middle elements of Group VA (15)—phosphorus, arsenic, and antimony—all exist as M_4 molecules in the vapor. Molecular solids, built up from tetrahedral M_4 molecules, are produced by condensation of the vapor. The most stable allotropes of these three elements consist of extended sheets in which the atoms are trigonally coordinated with bond angles between 95–100°. The latter bond angles exclude the possibility of planarity, and the sheets are in fact corrugated (see Figure 15.4). Despite this nonplanarity of the sheets, black phosphorus resembles graphite in its electronic and optical properties. This implies that there must be overlap of bands arising from either higher-energy, unoccupied AOs [see Group IIA (2) metals] or from σ-antibonding states [see Group IVA (14) metals]. On descending the group, the interatomic spacing within the sheets increases, as expected, but the interplane separation remains almost constant, reflecting stronger and stronger bonding interaction between sheets. These trends in bond lengths parallel the increasing metallic character on descending the group. In fact, bismuth has a slightly distorted close-packed structure and exhibits metallic behavior. Although the sheet structures are the thermodynamically most stable allotropes, their production from M_4 requires high temperature and pressure.

As in Group VA (15), oxygen, the first element of Group VIA (16), exists only as gaseous molecular species under ambient conditions. With the lower chalcogens [elements of Group VIA (16)], the trend to produce extended structures of lower and lower dimensionality [Group IVA (14), three-dimensional; Group VA (15), two-dimensional] continues, and the dominant structural units for Group VIA (16) solids are rings and chains. The most stable forms of sulfur are crystals made up of cyclic S_8 molecules with a crown structure. A number of other cyclics, with ring sizes from six to twenty atoms, have also been prepared by reactions of hydrogen polysulfides with polysulfur dichlorides:

$$H_2S_x + S_yCl_2 \rightarrow S_{x+y} + 2HCl \tag{18.1}$$

The S—S bond is subject to fairly easy thermal rupture, and all S_n rings undergo reactions on modest heating. The most interesting is the polymerization to long chains, which begins in the melt above 130 °C, and yields polymer of ever-increasing molecular weight up to a maximum of about 10 million at about 300 °C. The increasing molecular weight is manifest by a dramatic rise in viscosity from a mobile yellow liquid at 130 °C to a dark amber molasses at 300 °C. At temperatures above 300 °C, the viscosity declines again due to the $T\Delta S$-driven breakdown of the polymer chains. Provided the temperature changes are not too rapid, these phenomena are quite reversible, but rapid quenching from 300 °C to 0 °C or lower produces a metastable plastic, or fibrous form of sulfur, in which the long polymeric chains are preserved. The slow reversion of polymeric sulfur to S_8 under ambient conditions has precluded large-scale use of plastic sulfur as a structural material despite extensive research efforts and occasional claims of stable compositions. All forms of elemental sulfur are insulators with a large band gap under normal pressure conditions.

Whereas sulfur is more stable as small rings, but can give a metastable polymer, selenium and tellurium are most stable in the form of the long-chain polymers. Moreover, the chain polymers of Se and Te readily crystallize. The crystal structure is based on packing threefold helical chains (Figure 18.7). Selenium and tellurium form a continuous range of mixed compounds Se_xTe_{1-x}. The bond lengths in Se and Te show the same trends as noted for P, As, and Sb, in that the intrachain bond lengths increase substantially on going from Se (237 pm) to Te (284 pm), while the interchain distances (344 and 347 pm, respectively) remain almost constant. Polonium, which is metallic, has almost symmetrical octahedral coordination. In the dark, selenium is a fairly poor semiconductor but becomes highly conducting when irradiated with visible light. Its behavior is quite

Figure 18.7
Structure of Se and
Te.

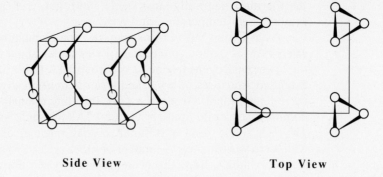

Side View   **Top View**

different from the photoconducting semiconductors of the diamond type. We return to this question in our discussion of xerography.

An interesting binary compound that is closely related to the long-chain chalcogens is polymeric $(SN)_x$, made from careful thermal polymerization of the eight-membered ring compounds S_4N_4 or S_2N_2 (see Chapter 15), as a fibrous crystalline material with a metallic, coppery sheen. Surprisingly, the polymer has essentially metallic conductance along the fiber axis. This conductivity can be understood in terms of the same band structure as a polychalcogen, but because N has one less electron than has S, only partial filling of the valence band occurs (see page 469).

Another interesting binary system, which has features in common with polysulfur and graphite, is poly(acetylene) of general formula $(CH{=}CH)_n$. Pure polyacetylene is a semiconductor, but its conductivity can be increased to metallic levels by inclusion of either oxidants (for example, I_2 and AsF_5), which deplete the valence band of electrons, or strong reductants (for example, alkali metals), which inject electrons into the conduction band.

In Group VIIA (17), the electronegativity of even the heavier members is so high that all members exist only in the molecular state. All transition elements, lanthanides, and actinides are metals. Their band structures are complicated by the involvement of d and f orbitals, but the same principles that govern the simple s- and p-band metals are operative. The transition metals tend to be much stronger and higher melting than are main-group metals because the extra orbitals and electrons can lead to higher bond orders.

Imperfections in Solids

The concept of stoichiometry applies very well to molecules. In molecular compounds, there may be impurities, but because the coupling between molecules is weak, the presence of impurities does not disturb the system generally, and any effects are more or less linearly dependent on concentration. Because the concentration of impurities can be reduced to very low proportionate levels, their effects can be essentially eliminated. This cannot be said of solids. First, the general rigidity of solids slows the migration of defects and impurities to a degree that makes their removal very difficult. Second, the strong coupling of local electronic and vibrational effects to the entire lattice greatly amplifies the effects of their presence. Consequently, solids exhibit a much wider structural variability than do simple molecules.

In this section, we consider some effects of the presence of minor defects in a crystal lattice. Although crystals can exhibit defects involving whole planes or rows of atoms, we limit our discussion to *point defects*, where there is an imperfection at a single lattice point, or interstice. There are three basic types of point defect: an *interstitial defect*, where an atom occupies a normally unoccupied hole in the lattice; a *vacancy*, where there is a hole at a normally occupied lattice site; and a *foreign-atom substitution*, where a lattice site is occupied by a chemically different atom from the normal occupant. The rate at which crystallization occurs at the atomic level is very rapid; even under the most careful conditions, some of each of these kinds of defects are trapped in the lattice. Here, we deal with those situations where the crystal contains small numbers of such defects. Later, we describe systems where defects are possible in such numbers as to introduce large deviations from stoichiometry.

One of the more dramatic examples of defect influences at low levels is the influence of foreign-atom substitution on semiconductor behavior. The lattices of all diamond–zinc blende structure semiconductors can accommodate appreciable amounts of substitution by atoms from neighboring groups in the periodic table—for example, silicon can be replaced by Al/Ga or by P/As. Figure 18.8 schematically shows the replacement of silicon by aluminum and by phosphorus. In the schematic format, the presence of an aluminum atom, with one less electron than silicon, creates a hole in the normal electron distribution, and the presence of the phosphorus atom creates an excess electron. Because the orbital overlap between host atom and foreign atom is different from host–host overlaps, the

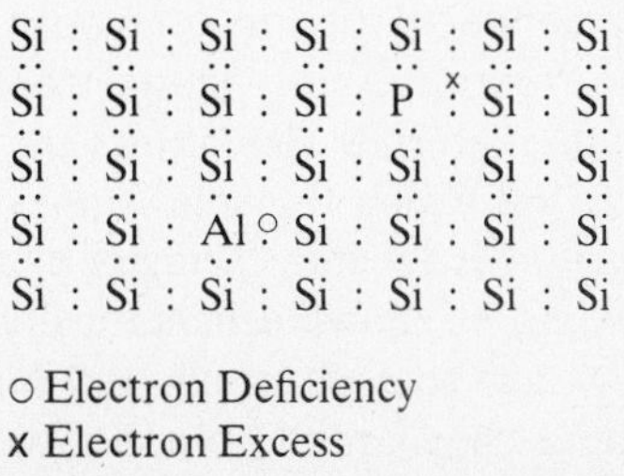

Figure 18.8
Schematic representation of Al and P impurities in a Si lattice.

Si : Si : Si : Si : Si : Si : Si
Si : Si : Si : Si : P $\overset{x}{:}$ Si : Si
Si : Si : Si : Si : Si : Si : Si
Si : Si : Al $\circ$ Si : Si : Si : Si
Si : Si : Si : Si : Si : Si : Si

$\circ$ Electron Deficiency
x Electron Excess

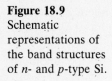

Figure 18.9
Schematic
representations of
the band structures
of *n*- and *p*-type Si.

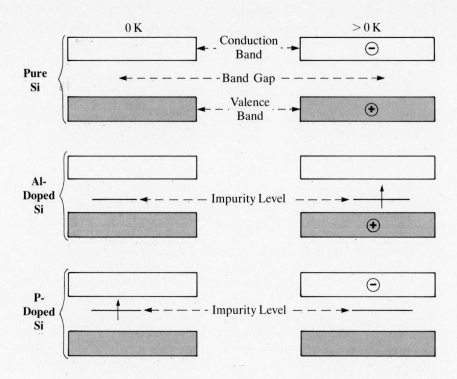

impurity energy levels in these cases do not lie in the conduction or valence band of the host but in the band gap, and they are quite localized. The most important result of these impurity levels is that they provide much lower-energy processes for creating positive holes in the valence band or for injecting electrons into the conduction band. Figure 18.9 shows schematically the way in which this happens. Those impurities that have an electron deficiency and allow easier creation of *positive* holes are called *p* type, and those that have an electron excess and inject *negative* charge carriers are called *n* type. Semiconductors that have essentially no impurities and operate by excitation across the natural band gap are called *intrinsic semiconductors*, whereas those that have their effective band gap reduced by impurities are known as *extrinsic semiconductors*.

The controlled incorporation of *n*- and *p*-type impurities allows the achievement of higher conductivities in otherwise poor semiconductors and allows much greater reproducibility in electronic behavior. The combination of *n*- and *p*-type materials also allows the fabrication of the different electronic devices on which the solid-state electronic revolution is based. The simplest example is the interfacing of *n*- and *p*-type materials to produce a rectifying diode, as shown schematically in Figure 18.10. The principle of such a device is that electrons and holes can migrate from where they are to where they aren't, but not vice versa. By suitably

Figure 18.10
Schematic representation of an *n-p* diode.

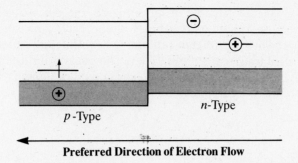

Preferred Direction of Electron Flow

arranging the band gaps under current-flow conditions, the positive holes and negative electrons can be made to recombine with high efficiency at the junction with emission of visible light. This is the basis of the light-emitting diode (LED). The reverse process, creation of positive hole–electron pairs by absorption of light, can cause current to flow and is the basis of photovoltaic conversion.

Another important class of extrinsic semiconductors can be generated by the introduction of mixed oxidation states into ionic or pseudoionic compounds by the use of *controlled valency*. This approach allows some of the ions in the lattice to be in higher or lower oxidation states without disrupting the lattice. This is achieved by replacing some of the host ions with redox-resistant ions of higher or lower charge. For example, semiconducting nickel oxide materials can be made by reacting a mixture of NiO and Li_2O with oxygen at high temperature, according to Equation 18.2:

$$\frac{x}{2} Li_2O + NiO + \frac{x}{4} O_2 \rightarrow Li_x Ni^{III}_x Ni^{II}_{1-x} O_{1+x} \tag{18.2}$$

Figure 18.11 shows a schematic representation of such a doped nickel oxide. In such a lattice, the Ni^{3+} ion is an electron-deficient (*p*-type) impurity. Often, if not usually, compounds of this type can exist over a wide range of stoichiometries and can exhibit a range of properties from insulator to metal.

Figure 18.11
Schematic representation of a $Li_x Ni^{III}_x Ni^{II}_{1-x} O_{1+x}$ lattice, showing charge compensation.

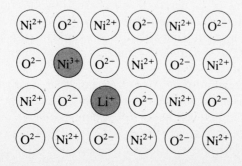

In the example of nickel oxide, the defect ions occupy lattice sites because the interstitial holes of the NaCl-type lattice are too small to accommodate foreign cations. A very different situation exists in the so-called tungsten bronzes. These compounds are usually made by reacting WO_3 with electropositive metals in a range of molar ratios from 0 to 1, as shown in Equation 18.3:

$$xM + W^{VI}O_3 \rightarrow M^I_x W^V_x W^{VI}_{1-x} O_3 \qquad \textbf{(18.3)}$$

In the lower range of $x \sim 0.3$, the bronzes are dark blue and semiconducting; the higher ranges of x (> 0.3) give a metallic sheen and high electronic conductivity. At the highest values of x, conductivity falls again, and the compounds are yellow. It is apparently not possible to increase x above about 0.95 without causing phase separation.

The WO_3 (distorted ReO_3 structure) unit cell may be represented as a cube of W^{VI} ions, with oxide ions at the centers of each cube edge. This structure is the same as the perovskite structure (ABO_3), except that all the large cations, which normally occupy the large cavity inside the cube, are missing. It is in these cavities that the alien ions of the nonstoichiometric bronzes are accommodated and the bronzes with large x values have distorted perovskite structures. Obviously, the extreme changes of composition in these compounds lead to quite dramatic changes in the band structure. However, in general terms it can be seen that for small values of x (W^V impurity in W^{VI} host) we are dealing with an n-type semiconductor. As the W^V content increases, what was an empty conduction band of WO_3 becomes so heavily populated with electrons as to confer metallic conductivity. For large values of x (W^{VI} impurity in W^V host), the band is almost filled, and the material is best viewed as a p-type semiconductor.

We have discussed the important effects that defects can have on the *electronic conductance* of solids. Another effect of great technological importance is the way in which defects can be exploited to increase the *electrolytic* (ion) *conductance* of solids. A good example of how this can be achieved in a crystalline solid is the phenomenon of *charge compensation*. Like controlled valency, charge compensation is achieved by incorporating foreign ions of higher or lower charge into the host lattice. Instead of compensating for the impurity by oxidation or reduction of a host ion, however, the charge imbalance is countered by creating a lattice vacancy at a host-ion site. Figure 18.12 illustrates this phenomenon for $CdCl_2$-doped AgCl.

The creation of this type of vacancy can greatly increase the diffusion rate of ions through the lattice. In simple lattices, the effect is not dramatic at temperatures well below the melting point because it still takes much work to squeeze an ion through the small gaps between lattice sites. Even so, the effect is useful, and a CaO-doped ZrO_2 can act as an excellent

Figure 18.12
Creation of cation
vacancies in AgCl
by CdCl$_2$ doping.

Ag$^+$	Cl$^-$	Ag$^+$	Cl$^-$	Ag$^+$	Cl$^-$
Cl$^-$	Ag$^+$	Cl$^-$	Ag$^+$	Cl$^-$	Ag$^+$
Ag$^+$	Cl$^-$		Cl$^-$	Ag$^+$	Cl$^-$
Cl$^-$	**Cd^{2+}**	Cl$^-$	Ag$^+$	Cl$^-$	Ag$^+$
Ag$^+$	Cl$^-$	Ag$^+$	Cl$^-$	Ag$^+$	Cl$^-$

oxide-ion conductor and oxygen electrode in high-temperature fuel-cell applications. Compounds with more complicated structures such as Na$_x$AlO$_2$ compounds, which contain wide-diameter channels through the structure, can exhibit solid-state ionic conductivites as high as 0.1 M aqueous solution of strong electrolyte.

Interstitial Compounds, Alloys, and Superclusters

In our discussion of the oxides and halides of the transition elements of the earlier groups, a number of simple binary compounds such as TiO and NbC with typical metallic properties were mentioned. These compounds are members of a much larger group that includes binary compounds of transition elements with the lighter elements of Groups IIIA–VA (13–15). At one time, this class of compounds was known as *interstitial compounds* because the structures of many of the simpler ones seemed to be derived from placement of the heteroatoms into the interstitial holes of the metal lattice. Thus, many of the carbides, MC, have a rock-salt structure where the C atoms occupy the octahedral interstices of a CCP metal lattice. On the other hand, the borides almost never have such simple structures and usually consist of cage clusters of B atoms linked together through BMB bonds. Table 18.1 lists some examples of typical interstitial compounds.

All interstitial compounds are produced by direct combination of the elements at high temperature. The phase diagrams can be very complex, involving several compounds of different stoichiometry as well as a true solution phase of the heteroatom in the metal. This complexity, together

Table 18.1 *Some Interstitial Compounds and Their Structures*

Rock Salt						NiAs	
TiC	TiN	TiO	VC	VN	VO	CrC	CrN
ZrC	ZrN	ZrO	NbC	NbN	NbO	MoC	MoN
HfC	HfN	—	HfC	—	—	WC	WN

with the chemical inertness and refractory properties of the interstitial compounds, has deterred chemists from a serious evaluation of their chemical properties. On the other hand, their immense importance in metallurgy has stimulated a great deal of research into their phase behavior and their physical properties.

As if the multiplicity of phases were not enough, the problem of determining the exact composition of an interstitial phase is further complicated by the tendency of the pure phases to exhibit nonstoichiometric behavior. The TiO system represents a fairly simple case. The compound is prepared by heating the metal in oxygen. But by varying the pressure of O_2 in equilibrium with the solid, a continuous range of homogeneous phases TiO_n ($0.6 < n < 1.35$) can be produced. All have the NaCl structure, but for $n < 1$ there are oxide-ion vacancies and for $n > 1$ there are titanium-ion vacancies. At very low O_2 pressures, the $TiO_{0.6}$ phase is in equilibrium with metallic titanium. At high O_2 pressures, Ti_2O_3 and TiO_2 phases separate out. The most recent studies on the structure of TiO suggest that it is best considered as made up of fused $Ti_6O_{12/2}$ clusters, analogous to the Mo_6X_{12} compounds described in Chapter 11. This view probably applies to many of the interstitial compounds.

In many respects, the interstitial compounds are a bridge between the alloys (combinations of two more or less electropositive metals) and the saltlike binary compounds (combinations of an electropositive and an electronegative element), in the same way that the semiconductors bridge metallic and ionic behavior. For example, a correlation such as shown in Figure 18.6 could be envisaged for all of the rock-salt structure, binary compounds of the transition elements (correlations across transition periods almost never give straight lines though!). Figure 18.13 illustrates another way of showing the continuum of properties between the various types of solid-state bonding. A further variable, which can be treated in such triangular relationships, is oxidation state.

Most metals crystallize in either one of the close-packed structures or the body-centered cubic (BCC) structure. Metals with similar radii, similar electronegativity, and the same crystal structure in the pure state form a continuous series of solid solutions in which the lattice sites are randomly occupied by the two metals. In certain cases, with very slow and careful crystallization, ordered lattices may result (for example, Cu/Ag and Ti/Zr). As the differences between the radii and/or the electronegativities of the two metals increase, the mutual solubility limits become much narrower, and phases of more or less fixed stoichiometry appear. Such phases may or may not be ordered, but the tendency toward ordering increases with the discrepancies in radius and electronegativity. When these discrepancies are extreme, they result in interstitial or ionic compounds.

In highly symmetrical structures such as the NaCl structure, there is a unique metal–metal distance, and metal–metal bonding is the same be-

Figure 18.13
Triangular relationships between properties of elements and compounds. (*SeBr$_2$ has not yet been prepared.)

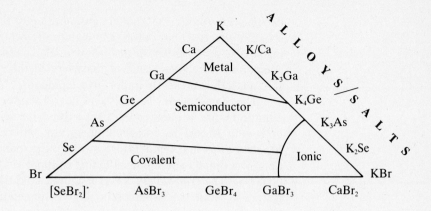

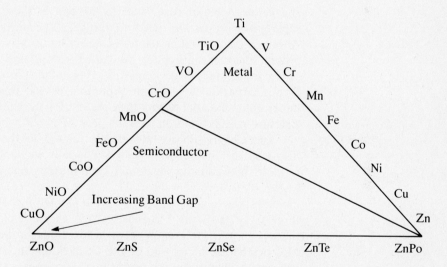

tween all nearest neighbors. This is not the case for structures of lower symmetry. Even the NiAs structure (see Chapter 4) results in each metal atom having two other metal atoms in particularly close proximity. Thus, in NiAs itself there are two nickel atoms at 252 pm, only slightly farther away than the six arsenic atoms at 243 pm. This close contact gives rise to rows of metal–metal bonded nickel atoms, running through the crystal in a direction perpendicular to the close-packed planes of arsenic atoms and giving rise to the metallic conductivity of NiAs. Many binary compounds of transition elements with heavy *p*-block elements have the NiAs structure and show high electrical conductivity along the hexagonal axis.

The different types of intermetallic interaction in binary compounds are quite numerous, but for ternary compounds the number of possibilities becomes enormous. A detailed consideration of this subject is beyond the scope of this text, but before leaving the subject of electronically conducting solids, we look briefly at one particularly interesting class of ternary compounds that gives some of the flavor of the more complicated solid-state materials presently under investigation.

The compounds, known as Chevrel-Phases, have attracted much attention since a number of them exhibit interesting superconductivity, both with regard to critical temperature and critical field. The *critical temperature* is that temperature at which the material undergoes a transition from metallic or semiconducting to superconducting (zero resistance) behavior. The *critical field* is that magnetic field that is required to cause a transition from the superconducting to the metallic or semiconducting state. The enormous technological advances that could flow from the availability of cheap zero-resistance conductors has stimulated a lengthy search for new superconducting materials with higher critical temperatures and fields. A major cost factor in the best of the current generation of superconductors is the need to refrigerate materials to 20 K or less. Although the general principles of superconductivity are well understood, present theories have poor predictive power, and the search for new superconductors is largely empirical. The interest aroused by the Chevrel-Phases is partly due to their promising critical properties, but equally important is the opportunity they present for the study of the influence of structure modification on physical properties.

The Chevrel-Phases are closely related to the $[Mo_6X_8]^{4+}$ clusters discussed in Chapter 11. They are usually made by high-temperature reaction between the elements. One subclass has the general formula $M_2Mo_6X_6$ (where M = In or Tl; X = S, Se, or Te). In the molybdenum dihalides, the octahedral clusters are separated by halide bridges along all three C_4 axes of the Mo_6 octahedron. In the chalcogenides $M_xMo_6X_8$, there is a direct interaction between Mo atoms at the corners of adjacent Mo_6 octahedra, along one of the C_4 axes of the octahedron, and chalcogenide bridging along the other two axes. Thus, the $M_xMo_6X_8$ structure contains linear chains of Mo_6X_8 octahedra, linked through weak Mo—Mo bonds (the intercluster distance is 20–30% longer than the intracluster Mo—Mo distance); the chains of Mo_6X_8 are stacked in such a way as to form chains of cubic cavities surrounded by bridging X ligands. It is in these cubic cavities that the second metal-ion M is located; they are empty in the binary Mo_6X_8 compounds. In the $M_2Mo_6X_6$ structure (Figure 18.14), the Mo_6 clusters form chains by the sharing of opposed trigonal faces.

The structural parameters and band filling can be varied widely by changing M, x, and X (even mixed chalcogenides have been made). Further families of the same type can be made by replacing chalcogenide with halide, thereby changing the oxidation state of the cluster (for example,

Figure 18.14
Structure of the
Chevrel compound
$TlMo_6Se_6$.
Reprinted with
permission from
M. Potel,
R. Chevrel, and
M. Sargent. Acta
Cryst. B36 (1980):
1545.

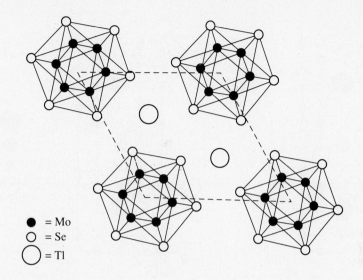

● = Mo
○ = Se
◯ = Tl

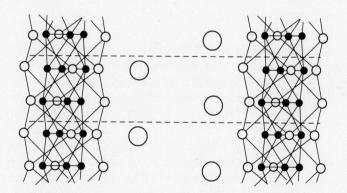

$Mo_6S_6Br_2$) or by substituting Mo by another metal in the cluster (for example, $Mo_4Re_2Te_8$ and Mo_5RuTe_8). These variations alone lead to infinite variety because many of the types mentioned can exist in continuous ranges of solid solution. Because other families based on Mo_6X_{11} or $Mo_{12}X_{14}$ clusters or even on infinite chains of Mo_6 octahedra sharing two opposite trigonal faces have been prepared and characterized, the enormity of the task of searching for superconductors by empirical means becomes evident.

Recently, a new class of mixed-oxide superconductors with critical temperatures above the boiling point of liquid nitrogen has been discovered. The greatly reduced cost and ease of handling of liquid nitrogen, as opposed to the liquid helium that has been necessary in all superconducting

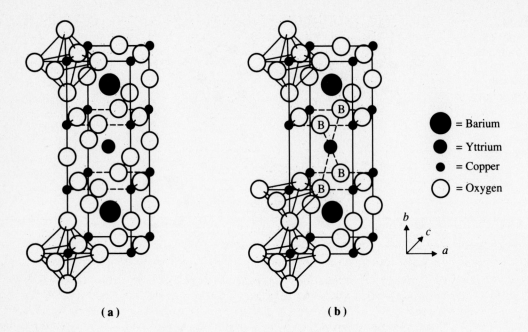

= Barium

= Yttrium

= Copper

= Oxygen

(a)   (b)

Figure 18.15
Structure of
$YBa_2Cu_3O_7$.
(a) The idealized
structure of the
hypothetical
compound
$YBa_2Cu_3O_9$ and
(b) the idealized
structure of
$YBa_2Cu_3O_7$.

applications up to this time, are expected to revolutionize a number of technologies such as power transmission and storage, magnetic levitation in transport, and magnetic resonance imaging.

The new high-temperature superconductors have approximate compositions $YBa_2Cu_3O_y$ ($y \simeq 7$) and $(La_{1-x}M_x)_2CuO_4$ (La = a tripositive lanthanide ion; M = Sr or Ba). All these materials exist as a single phase over a fairly wide range of composition, but the appearance of superconductivity is very dependent on composition. The excess oxygen, over that required if these were Cu(II) compounds, balances the charge of Cu(III) ions in the lattice. The yttrium compound, known popularly as the 1-2-3 HT superconductor (because of the ratios of the metal ions), has a zero–field critical temperature of 96 K.

The structure of the 1-2-3 HT superconductor is related to that of perovskite (see Figure 4.12). An ideal perovskite would require a composition $YBa_2Cu_3O_9$, and the structure would be as that shown in Figure 18.15a. Such a structure would require at least one of the cations to be in an unachievably high oxidation state; hence, it cannot exist in practice. In the actual 1-2-3 structure, where $y \approx 7$, the copper is present as $Cu(II)_2Cu(III)$, and the charge balance is achieved by removing a horizontal layer of oxides from the central plane of the unit in Figure 18.15a [the plane containing the Y(III) ions]. The result is shown in Figure 18.15b. In this unit, the uppermost and lowest layers of Cu ions are octahedrally coordinated, as in the perovskite structure, but the two inner layers of copper ions are coordinated to a square pyramid of oxides. Stacking of such units produces a layer lattice in which identical layers of perovskite-like $Ba_2Cu_3O_7$ are separated by layers of Y(III). A distortion of four of the eight oxide ions (labeled B in Figure 18.15b) toward the yttrium cation segregates the SQP–coordinated copper ions into chains along the *c* axis of the crystal. The superconductivity is presently believed to be intimately connected with these chains, and their destruction by, for example, over-heating the sample leads to destruction of superconductivity. At the time of writing, new copper oxide–based HT superconductors containing Bi and Tl, rather than Y or a lanthanide, have been reported. The T_c of these materials is around 120 K, and the copper ions are arranged in sheets rather than chains.

18.2   *Storage Batteries: Old and New*

The storage battery in its commonest form is an electrochemical redox cell. The two electrodes, one of which is capable of oxidizing the other, are made of materials that give up their chemical potential by passage of ions across an electrolyte separating the electrodes and by passage of electrons through an external circuit. Virtually all practical batteries presently use solid electrodes and liquid (or paste) electrolytes. In the course of the charge and discharge cycles of a battery, profound chemical changes take place in the electrodes, and proper functioning of the device requires the efficient transport of electrons and often ions into and out of the solid electrodes. If the battery is to be rechargeable, these processes must be both chemically and physically reversible—that is, the electrodes must return to their original chemical and physical condition after recharging.

The classical batteries most commonly in use are the MnO_2/Zn (disposable flashlight) and the lead-acid (rechargeable automotive) batteries. In the MnO_2/Zn battery, the dissolution of the Zn anode occurs from the surface and does not involve bulk solid-state transport. The cathode, which is a very highly divided, hydrated form of MnO_2, is converted during discharge into a hydrated oxide of Mn(III), [MnO(OH)] as shown in Equations 18.4 and 18.5:

Cathode
$$2MnO_2 + 2H^+ + 2e^- \rightarrow 2MnO(OH) \tag{18.4}$$

Anode
$$Zn + 2OH^- \rightarrow ZnO + H_2O + 2e^- \tag{18.5}$$

This change requires the efficient transport of both protons and electrons into the cathode material. The electron transport is assisted because throughout most of the discharge cycle the material is a mixed-valency compound (see doped NiO and tungsten bronzes discussed earlier). The ion transport is facilitated by the high surface area and highly defective nature of the solid. These characteristics are common to many battery electrode materials.

In the lead-acid battery, both the anode and cathode are subject to ion and electron transport during the charge–discharge cycles, according to Equations 18.6 and 18.7:

Cathode
$$PbO_2 + [SO_4]^{2-} + 4H^+ + 2e^- \rightarrow PbSO_4 + 2H_2O \qquad \text{(18.6)}$$

Anode
$$Pb + [SO_4]^{2-} \rightarrow PbSO_4 + 2e^- \qquad \text{(18.7)}$$

Again, it is necessary for the electrode materials to be highly divided and highly defective to sustain the necessary ion currents. Both electrodes are prepared in the form of pastes, which are the subject of much proprietary art. The paste must protect the active electrode material from *sintering* (becoming more perfectly crystalline) and from undergoing mechanical degradation. The mechanical degradation occurs because the density of the electrode material is not the same in the oxidized and reduced states. The volume changes occurring during charge–discharge therefore give rise to cyclic stress and eventually to mechanical failure of the electrode (crumbling, flaking, and so on). For this reason, the lifetime of any rechargeable battery system is maximized by using as little of the capacity of the battery as possible on any given charge–discharge cycle.

The MnO_2/Zn battery is not rechargeable because following dissolution of the zinc, it is difficult to redeposit it again in the form of a coherent sheet. In addition, MnO_2 undergoes severe sintering during charge–discharge. A rechargeable flashlight-type battery, based on electrode reactions shown in Equations 18.8 and 18.9, has found widespread application, but the high cost of such cells has as yet prevented their displacing the MnO_2/Zn cell.

$$2NiO(OH) + 2e^- \rightarrow 2NiO + 2OH^- \qquad \text{(18.8)}$$

$$Cd + 2OH^- \rightarrow CdO + H_2O + 2e^- \qquad \text{(18.9)}$$

In the presence of an alkaline electrolyte, the CdO remains insoluble, and the charge–discharge cycle involves a solid-state transformation of metal to oxide (and vice versa), rather than electrodissolution and replating. Figure 18.16 shows schematic illustrations of the structures of the standard flashlight and lead-acid batteries.

In all the systems described above, a detailed description of the electrode-material structure is precluded because the materials owe their favorable properties to a highly dispersed and defective character. Such

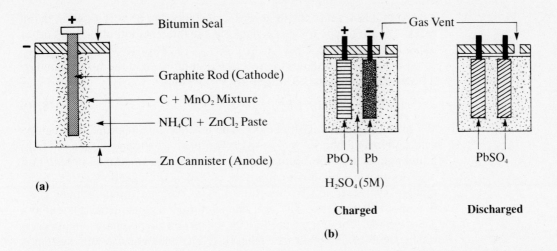

(a)

(b)

Figure 18.16
Schematic
illustrations of the
construction of
(a) standard
flashlight and
(b) lead–acid
batteries.

structures do not yield easily to analysis by conventional methods; even today, our fundamental understanding of them is poor. Because none of the classical batteries even approaches an ideal energy-storage device and because large-scale use of solar energy requires efficient and cheap storage, the search for new battery systems is still very active. Some of the newer approaches to battery design draw heavily on the systematic knowledge of solid-state structure-property relations that has developed in the past two decades, and the search has become less empirical with the passage of time. The ideal storage battery should have at least the following characteristics:

1. The number of charge–discharge cycles, without change in performance, should be unlimited.

2. The weight per mole of electrons stored should be minimal.

3. The cell potential should be as high as possible and reasonably constant.

4. The resistive losses within the battery should be minimal.

5. The battery materials should be cheap, nontoxic, and nonnoxious.

One of the more promising lines of research to achieving higher-efficiency batteries is the use of *intercalation compounds*. The phenomenon of intercalation involves the incorporation of guest atoms, molecules, or ions into the interstices of a layered host lattice. Some of the more promising substances in terms of battery applications are the transition metal dichalcogenides. The majority of these compounds have either CdI_2 or MoS_2 (NiAs) lattices (see Chapter 4). The migration of ions and basic neutral molecules into the intersandwich planes can be very rapid in many cases. The prototype for this behavior is the intercalation of Li^+ and other alkali metal ions into TiS_2.

Equation 18.10 shows the reaction of interest, and the formal similarity between Equations 18.10 and 18.3 is obvious.

$$x Li + TiS_2 \rightarrow Li_x TiS_2 \tag{18.10}$$

We return to this similarity below. The reaction in Equation 18.10 can be achieved in a number of ways. An early preparation used solutions of Li in liquid ammonia as the source of lithium. Although ammonia intercalation also occurs, the ammonia can easily be expelled by gentle heating. More modern syntheses involve reaction of TiS_2 with an organolithium compound or electrolysis of a lithium salt solution in a nonaqueous solvent at a TiS_2 cathode. The intercalate $Li_x TiS_2$ exists as a single-phase, highly crystalline solid for all values of x from 0–1. The analogous sodium system is more complicated.

Like the tungsten bronzes, the metal chalcogenide intercalation compounds seem to be highly ionic, and the intercalated metal atom gives up its electron to the lowest unoccupied bands of the host lattice. The difference between the bronzes and the intercalates is quantitative, rather than qualitative, in that the latter exhibit very much higher guest-diffusion coefficients in two dimensions than do the former. In the bronzes, diffusion of a guest atom occurs at an equal rate in all three dimensions. The host crystal undergoes expansion in a direction perpendicular to the intercalation planes as the guest is intercalated. In $Li_x TiS_2$, the change in composition from $x = 0$ to $x = 1$ gives an expansion of about 10%. The oxidation potential of TiS_2 is about two volts higher than that of lithium (depending on the solvent), and a promising lithium battery based on the half-cell reactions can be constructed:

$$Li \rightarrow Li^+ + e^- \tag{18.11}$$

$$x Li^+ + TiS_2 + xe^- \rightarrow Li_x TiS_2 \tag{18.12}$$

Because lithium reduces water, a nonaqueous electrolyte must be used. Irreversible electrochemical decomposition of the solvent and mechanical stability of the electrode, under the large number of charge-discharge cycles required, are two of the major problems that have so far prevented practical applications in transportation and large-scale utility-storage operations.

18.3   *A Brief Note on Corrosion and Tarnishing*

The chemical degradation of metals and alloys is an extremely complicated phenomenon and can only be treated with justice through an interdisciplinary approach that includes, besides inorganic chemistry, a number of areas of physical chemistry, physics, and chemical engineering. Along with the degradation of organic polymers and the weathering of inorganic structural materials, metallic corrosion is, on one hand, a key factor in sustaining the producer–consumer economy that characterizes our civilization, and on the other, enormously costly in energy and resources. This does not mean that great advances in the inhibition of corrosion will stop the production–consumption cycle; a more likely result is that it will cut the costs of producing goods that wear out just as quickly! Hence, a great deal of research and invention is dedicated to finding ways to reduce corrosion.

We do not attempt to do more than draw attention to the relevance of much of the foregoing solid-state chemistry to the general question of the chemical degradation of metals. This degradation may take place with a metal exposed to a gaseous environment, a liquid environment (especially water), or even in contact with another solid. Although the physical processes for these different media may be very different, we do not draw distinctions between them. Because, in general, the crystal lattice of a metal is impermeable to attacking reagents, the chemical attack normally takes place at the bulk surface or along grain boundaries of polycrystalline metals. An exception to this rule is the hydrogen embrittlement of hydride-forming metals such as titanium where diffusion of hydrogen into the bulk crystal can be quite rapid.

Unless special precautions are taken and special techniques are used, metal surfaces usually get covered with a layer of a derivative of the metal, most commonly the oxide. Once an oxide (or some other coherent film) covers the metal surface, the rate-controlling processes for further oxidation become either transport processes (molecules, atoms, ions, or electrons) or the rate of oxygen reaction at the *oxide* surface. If the oxide has a very different density from the metal, has poor mechanical strength, or has poor adhesion to the metal, it may fracture or detach; further oxidation can occur by the rather easy passage of oxygen through cracks to the clean metal surface. Oxides that are highly ionic (low metal-ion charge) usually fail on all three of these physical factors and are likely to be readily soluble in neutral water or dilute acids. Similarly, highly covalent oxides (high metal-ion charge) also afford poor protection on mechanical grounds because of volatility and of solubility in water or dilute bases. An interesting example of the volatility problem is furnished by osmium, which despite its relative nobility cannot be used in many applications because of the ease with which it forms the volatile and highly toxic OsO_4. Trioxides of molybdenum and tungsten also volatilize at sufficiently low temperatures that filaments made of these metals can only operate successfully when sealed in an evacuated envelope.

Of the main-group metals, aluminum is best protected by its surface oxide. In large part, this is due to the great physical strength of Al_2O_3 and resistance to chemical attack. Bulk Al_2O_3, in the form of corundum, is commonly used as an abrasive, with a hardness not greatly inferior to diamond. Although freshly precipitated aluminum hydroxide, a highly defective nonstoichiometric substance, dissolves readily in both aqueous acids and bases, monocrystalline Al_2O_3 can only be dissolved with great difficulty.

A number of precious and semiprecious stones (rubies, emeralds, and sapphires) are largely Al_2O_3; the great physical and chemical durability of macrocrystalline alumina endows on these stones the ability to survive mechanical and chemical attack on a geological time scale.

The transition metals of the earlier groups exhibit great resistance to corrosion. Titanium is widely used in chemical plant construction, usually as an alloy with other metals because of its resistance to a broad range of corrosive media. Although this is partly attributable to the good physical properties of TiO_2, another important factor is the pronounced ability of Ti (and other neighboring metals) to form boundary layers of intermediate levels of oxidation (for example, $TiO_{0.6}$ to $TiO_{1.35}$), in addition to other more highly oxidized phases. The more gradual density gradient at the Ti/TiO_2 boundary reduces the mechanical stress and increases the adhesion of the oxide layer. This feature is common to the other corrosion-resistant transition metals of Groups IVB–VIB (4–6). In addition, the oxides of corrosion-resistant metals tend to have very compact structures with little interstitial space for the transport of reactive species across the protective layer. They are also relatively weak acids or bases.

Zirconium has found widespread application in nuclear reactor systems. Although more costly than titanium, it has superior properties under intense radiation fluxes, particularly a low-neutron capture cross-section. Tantalum is the most corrosion-resistant of the nonnoble metals; despite its low abundance and fairly high cost, it is used in medical engineering applications such as the fabrication of gauzes to reinforce the abdominal wall following serious hernia surgery. Chromium is protected by Cr_2O_3, which is isostructural with Al_2O_3, but it is also stabilized by a substantial CFSE (see Chapter 13). The noble metals, most notably platinum and gold, resist corrosion by virtue of their intrinsic lack of chemical reactivity, rather than through formation of protective oxide films.

We conclude our discussion of oxide-film protection of metals with a brief mention of corrosion-resistant alloys. Often, an alloy shows considerably greater resistance to corrosion than do its major constituents in the pure state. Such alloys can give rise to ternary (or more complicated) oxide layers such as spinels and perovskites (see Chapter 4) at the surface. Because these ternary oxides often have structures based on the filling of interstices of more open structures by a second cation, they tend to be dense and impermeable. Also, because one of the components is frequently more basic and the other more acidic, the mixed oxides have a lower affinity for external acids or bases.

Halogens and halide ions tend to be particularly corrosive to metals. Fluorine and fluorides are a rather special case. Fluoride forms very stable complexes with most of the metals that form very strong, coherent oxide films, and fluorine tends to give volatile, higher oxidation state fluorides. The most resistant metal for containing F_2 is nickel.

The heavier halogens and chalcogens are especially corrosive. The layer-lattice compounds produced by these elements have poor mechanical qualities; even if they produce coherent surface layers, the very properties that make them good candidates for battery electrodes make them ineffective barriers to the migration of electrons, atoms, or ions. A notable engineering disaster occurred at Glace Bay, Canada, when the builders of a heavy-water plant failed to recognize the corrosion problems inherent in the H_2S/H_2O equilibration process (see Chapter 9) and in the use of seawater as a coolant. After brief use, most of the plumbing had to be replaced at a cost of many millions of dollars. A number of promising catalytic processes that use iodine or bromine cocatalysts have become costly failures because of corrosion problems. Such problems have, however, also been surmounted—for example, the Monsanto process for carbonylation of methanol (see Chapter 23) uses an iodine-resistant Ni/Mo alloy (Hastelloy B-2) for the plant. The considerable art involved in making the alloy provides a very effective barrier to the unauthorized exploitation of the process.

BIBLIOGRAPHY

Suggested Reading

Donahue, J. *The Structure of the Elements*. New York: Wiley, 1974.

Greenwood, N. N. *Ionic Crystals, Lattice Defects, Nonstoichiometry*. London: Butterworths, 1968.

Hannay, N. B. *Solid State Chemistry*. Englewood Cliffs, N.J.: Prentice-Hall, 1967.

Hume-Rothery, W. *Electrons, Atoms, Metals, Alloys*. London: Metal Industry, 1955.

"Photovoltaic Cells." Special Report. *Chem. & Eng. News* (7 July 1986): 34.

PROBLEMS

18.1   With the aid of a suitable example in each case, explain what is meant by the following:
   a. Valence band
   b. Conduction band

    c. Band gap
    d. Insulator
    e. Intrinsic semiconductor
    f. Intercalation compound
    g. Photoelectric conduction
    h. Point defect
    i. Controlled valency
    j. Charge compensation
    k. Interstitial compound
    l. Sintering

18.2 Give five physical properties associated with metallic bonding.

18.3 If a unit cell is the simplest repeating, three-dimensional unit in any crystal structure, how many NaCl molecules are there per sodium chloride unit cell?

18.4 Titanium monoxide (TiO) is isomorphic with NaCl (that is, has the same crystal structure). If the unit-cell length of TiO is 424 pm, what is its density?

18.5 Why is metallic hydrogen but not metallic helium likely to exist?

18.6 Describe the effect of traces of Se on the electronic structure of gallium arsenide (GaAs).

18.7 Which of the following would be p-type semiconductors?
    a. Silicon doped with As
    b. AlP doped with S
    c. ZnO doped with Zn
    d. NiO doped with CuO

18.8 Comment on the validity of the following statements:
    a. Band gap increases with increasing electronegativity.
    b. Ge is an intrinsic semiconductor.
    c. Cubic boron nitride is metallic.
    d. $TiO_{0.6}$ has a NaCl structure with O^{2-} vacancies.

18.9 Describe the structures of the following compounds in terms of close-packed lattices:
    a. Diamond   **e.** TiH_2 (fluorite structure)
    b. VC   **f.** $YBa_2Cu_3O_7$
    c. GaAs   **g.** Cubic BN
    d. $CaTiO_3$

18.10 Describe the physical and chemical processes occurring in the MnO_2 electrode of a dry cell during discharge.

18.11 Molecular S_2N_2 is square with alternating S and N atoms. Crystals of S_2N_2 undergo solid-state polymerization on warming to give a fibrous crystalline polymer $(SN)_x$ in which helical chains of alternating S and N atoms pack in manner similar to the structure of crystalline Se. Explain the observation that $(SN)_x$ is a metallic conductor.

18.12 Freshly precipitated metal oxides are usually easily soluble in dilute acids or bases but become very difficult to dissolve if they are dried for extended periods at high temperatures. What chemical and physical processes are responsible for this behavior?

18.13 Tantalum and platinum are two of the more corrosion-resistant metals. Describe the properties that are responsible for this behavior in each case.

18.14 Oxidation of a metal surface requires that the metal and oxygen come together across the oxide layer. Suggest some mechanisms whereby this may happen.

18.15 Give three properties in which elemental sulfur differs from selenium.

18.16 Place the following compounds in order of increasing ability to conduct electrons.
a. MgSe   **c.** SN
b. NaCl   **d.** InTe

19

Amorphous Solids

19.1 *General Considerations and Definitions*

The constraints of thermodynamics make perfectly ordered systems impossible to achieve. On the other hand, geometric factors tend to make ordered systems more stable than those that are highly disordered. Therefore, unless special effects are operating, most simple solids tend to exist as imperfect crystals.

A solid that exhibits no evidence of crystallinity is called *amorphous*. The most common criterion for establishing noncrystallinity is the failure to observe a diffraction-pattern (X-ray, electron, or neutron) characteristic of an identifiable unit cell. Even amorphous materials diffract because of characteristic repeating properties not related to high-level order. However, they give a small number of diffuse halos rather than the large number of sharp rings or spots observed with crystalline materials. The lack of long-range order may be due to either extremely small particle size or high-level disorder in a homogeneous solid. The former situation results in an *amorphous powder* and the latter in a *glass*. The boundary between these two states and between the amorphous and crystalline states is not a sharp one because, in principle, both a glass and a crystal can be divided indefinitely to give an amorphous powder and an amorphous powder can be *sintered* (subjected to heat and pressure to cause fusion at unstable surface sites but not in the bulk solid). In practice these transformations may not be easy to achieve because the act of division puts enough energy into the system to cause local heating and fusion of very small, metastable particles.

The growth of a crystal involves uniting atoms or molecules in very large numbers and in very specific spatial relationships to each other. A combination of these spatial constraints and because it takes time for the material to diffuse to the right location places an upper limit on the rate at which crystals can grow. Because small, geometrically simple species can diffuse rapidly and easily find the right orientation, they can form crystals with incredible speed. The metals and simple spherical ions are extreme examples of this type of behavior. Molecules that have complicated shapes and that are bound by fairly strong intermolecular forces even in the liquid state tend to crystallize much more slowly. The sugars provide an excellent example in organic chemistry, and an inorganic analog would be a small cyclic phosphoric acid:

$$
\begin{array}{c}
\text{O}\diagdown\quad\;\;\text{O}\diagdown\;\;\text{P}\diagup\!\!\diagup\text{O}\diagdown\;\;\text{O} \\
\text{O}\!=\!\text{P}\!-\!\text{O}\!-\!\!\overset{\displaystyle}{\underset{\displaystyle}{}}\!\!\text{P}\!=\!\text{O} \\
\text{HO}\diagup\qquad\text{OH}\;\;\big|\;\; \\
\text{OH}
\end{array}
$$

Such molecules form very stable supersaturated solutions (syrups) in which the molecules are randomly oriented but polymerized through hydrogen bonding. It is particularly difficult to detach such a molecule from its network in the liquid state and attach it in the correct orientation to the growing crystal. The hydrogen bonding also slows the diffusion of the molecules in the liquid phase. Despite these difficulties, using good *nucleating agents* (tiny foreign particles around which crystal growth is initiated) and having a great deal of patience eventually will be rewarded with well-formed macrocrystals. Emil Fisher, the founder of modern carbohydrate chemistry, was reputed to owe much of his success in crystallizing particularly obstinate sugars to the quality of the dandruff from his beard!

Most liquids exhibit an increase in viscosity when they are cooled as a result of the slowing of molecular motions. If a liquid is cooled in a manner that precludes crystallization (for example, by very rapid cooling), a temperature eventually will be reached at which there is a sudden, very

Figure 19.1
A typical volume versus temperature curve for a material undergoing a liquid to glass transition.

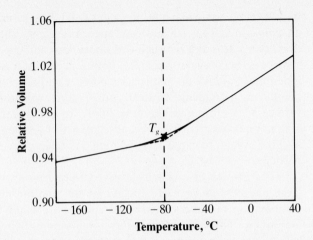

dramatic rise in viscosity. The change does not occur with a complete discontinuity in properties such as happens with a first-order transition like the melting of a crystal, but it is quite dramatic anyway, as shown in Figure 19.1. We will not go into the theoretical explanations for this phenomenon but will limit our discussion to pointing out that the viscosity below this transition temperature becomes so high as almost to stop diffusion and reduce the crystal-growth rate to zero. The critical temperature at which this dramatic rise in viscosity occurs is called the *glass-transition temperature* (T_g). With very complicated molecules, especially high polymers, it is easy to cool the liquid below T_g without crystallization; with spherical atoms, molecules, or ions, however, it is very difficult because of the extremely high cooling rates required.

19.2 Glassy Metals

Glassy metals have been recognized since 1930, but their wide investigation and commercial exploitation is a recent phenomenon. The key development in their history was the invention of the so-called splat-cooling technique: Globules of molten metal are projected at high velocity at a cold surface where they undergo cooling rates on the order of 1 million degrees per second. Even using such techniques, it is rarely possible to produce a glass of pure metal, and even those cases in which it is possible give glasses that crystallize quite rapidly well below T_g.

The second important step in producing useful metallic glasses was the discovery that they can be greatly stabilized by alloying with main-group elements. For example, a commercial composition for use in transformer cores has the approximate composition: Fe (40%), Ni (40%), P (14%), and B (6%). The precise manner in which such alloying stabilizes the glass is not known, but it is likely that the formation of fairly large, nonspherical clusters in the melt is one major factor. The intrinsically greater difficulty of forming pure crystals from complex mixtures is undoubtedly another.

Because so many of the physical and chemical properties of conventional metals are strongly linked to the existence and properties of *grain boundaries* (the surfaces at which crystallites of the various phases meet), it is not unexpected that homogeneous glassy metals, which are totally lacking in grain boundaries, would have dramatically different properties. The grain boundaries of conventional metals are particularly susceptible to chemical attack and represent an important site for the initiation and penetration of corrosion into the metal. Glassy metals show remarkable resistance to corrosion. They also exhibit very high tensile strength, combined with unusually high plasticity. For example, a metallic glass fiber can retain its tensile fracture strength (three times that of stainless steel) while undergoing a 50% elongation. Most strong, hard, conventional metals are quite brittle and cannot survive anything like a 50% strain. A third very useful property of some glassy metals is their extreme ease of magnetization and demagnetization. Their ability to adjust their magnetization to the ac field in a transformer reduces power losses due to heating and thereby saves energy and reduces the demands for heat dissipation. A 1978 estimate indicated that this single application of glassy metals could result in annual power savings of $100 million in the United States alone and represented a market of about 200 million pounds of material per year.

The technology of glassy metals is still in its infancy, and almost nothing is known of their chemistry; they represent, however, a fascinating example of the most simple type of inorganic glass.

19.3   *Chalcogenide Glasses*

Whereas the metals represent the most difficult kind of material to transform to the glassy state, the chalcogenides below oxygen very easily form glasses when they are in the form of long-chain polymers. We have already described the polymeric allotropes of these elements in Chapter 18. The glass-transition temperature of highly polymerized sulfur is normally below ambient temperature (-30 °C), and it exhibits rubbery behavior at room temperature. This has been shown to be an artifact arising from the plasticizing effect of residual S_8 molecules, and removal of these by extraction with CS_2 gives a polymer with $T_g = 75$ °C, which is a tough resinous solid at room temperature. The easy depolymerization of plastic sulfur and its failure to exhibit interesting electrical properties are responsible for its place in the closet marked "Scientific Curiosities."

Figure 19.2
Introduction of
mobility into
chalcogenide
chains by a
bond-interchange
mechanism.
● = Free radical
chain end.

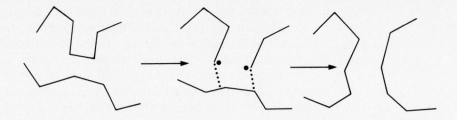

Unlike polysulfur, which does not crystallize as such, the linear allo-tropes of selenium and tellurium and of the continuous series of Se_xTe_y copolymers can exist in both crystalline and glassy forms. Rapid, conventional cooling of selenium melts readily produces a glass with a T_g of 31 °C, which quickly devitrifies above 70 °C. Small amounts of tellurium catalyze devitrification, perhaps because the weaker Te—Se bond undergoes more rapidly the chain-breaking and -making reaction that constitutes one of the mechanisms for molecular mobility in the solid state (Figure 19.2).

Although pure chalcogenide glasses possess the same interesting electronic and optical properties as the crystalline materials, they are of little practical value because of their ease of devitrification. However, introduction of other elements with a characteristic valency of greater than two— for example, the heavier elements of Groups IIIA–VA (13–15)—induces cross-linking of the otherwise linear polymers and greatly impedes the crystallization process. Thus, besides being resistant to devitrification, these types of glasses provide a family of materials with an extraordinary range of electronic and optical properties. At one extreme, there are simple binary combinations such as $(AsX_{3/2})_n$ (X = S, Se, Te; n ~50–90), which have been widely used in the manufacture of IR–transparent windows for scientific, military, and aerospace applications. Figure 19.3 shows the proposed structure for a typical AsX_n glass.

Obviously, the numbers of three, four, five, and so on component systems become very great. Such multicomponent systems have, however,

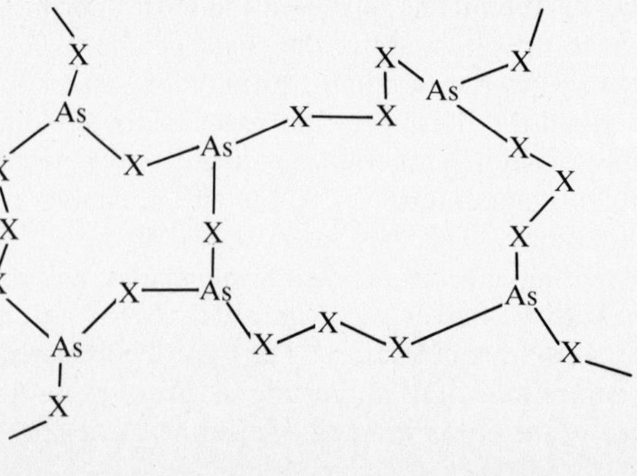

Figure 19.3
Schematic structure
of an $AsX_{3/2}$ glass.

assumed great technological importance because they form the basis of the vast industry of *electrophotography*, of which the Xerox machine is the best known example. We will describe the process of electrophotography in some detail for two reasons. First, it is an excellent illustration of the application of the properties of chalcogenide glasses; second, the vast majority of people, including highly educated and mature scientists, are completely ignorant of the workings of this extraordinary invention that has so transformed the efficiency of dissemination of the written word.

One of the keys to the invention is that certain chalcogenide glasses (based particularly on Se) are very good photoconducting switches—that is, in the dark they are insulators, but in the light they are good electronic conductors. This property allows a glass film to maintain a static electric charge on its surface in the dark, which can be conducted away upon illumination. Thus, an optical image can be recorded on the surface of the glass in the form of electric-charge density: light areas, low electric charge; dark areas, high electric charge. This image is "developed" by dusting the charged surface with a powder carrying the opposite charge to the surface. In practice this powder consists of an organic polymer containing a black pigment and is known as the *toner*. The toner (and the image) then is transferred from the chalcogenide glass to a sheet of charged paper and fixed to the paper by heating it above the glass-transition temperature of the organic polymer. Figure 19.4 shows a schematic representation of the electrophotographic process. In practice the chalcogenide glass and the metal substrate are in the form of a drum that rotates synchronously with the image scanner. This roller mode simplifies the engineering requirements for a rapid copying device. The detailed composition of the glasses used in commercial devices are generally proprietary and are optimized for photoelectric properties, mechanical strength, abrasion resistance, resistance to oxidative and other chemical degradation, and many other factors.

Like most major inventions, electrophotography was not dependent on a basic understanding of its physical principles for its discovery. This was a fortunate circumstance because the basic mechanism of photo-conduction of chalcogenide glasses is still not fully understood. However, a more profound understanding of the fundamentals has aided and accelerated the refinement and improvement of the basic device.

It is evident from certain obvious differences in their electronic behavior, particularly the relative indifference of chalcogenide glasses to impurities and the absence of unpaired electrons, that there is a fundamental difference between the conduction mechanism in chalcogenide glasses and that in the Group IVA (14) semiconductors. The currently accepted model, using selenium as an example, is as follows: It is believed that glassy selenium contains, besides simple runs of linear chains, regions where selenium atoms in different chains are in close enough contact to give rise to cross-linking (Figure 19.5a). These trivalent cross-links,

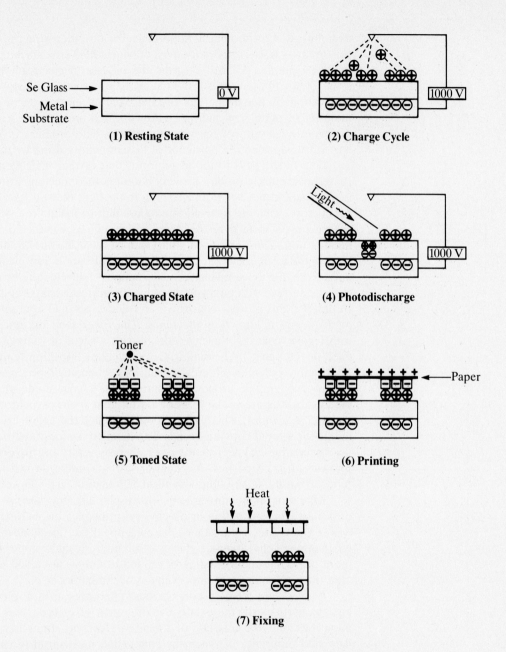

Figure 19.4 Schematic illustration of the electrophotographic process.

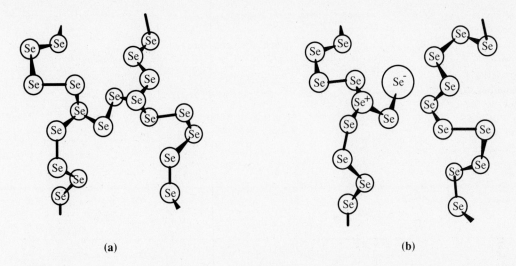

(a)   **(b)**

Figure 19.5
Structure of **(a)** the cross-links and **(b)** the valence alternation pairs in vitreous selenium.

however, are unstable with respect to disproportionation into two much more stable species, a selenide ion (—Se$^-$) and a selenonium ion ($\overset{\mid}{\underset{\mid}{-\text{Se}^+}}$) (Figure 19.5b). Such ion pairs are referred to as *valence alternation pairs*.

It is one of the significant differences between the first members of the main groups and the heavier members that the latter give abnormal valence states that are more stable than analogous states of the former. Figure 19.6 shows some of the bonding states available to selenium, together with a simple MO energy diagram for each state and an indication of the relative stability of the species. In this figure, it is assumed that only the 4p orbitals of Se are used to form bonds and the 4s remains nonbonding (a helpful and reasonable simplification but not essential to the argument). A comparison between Figures 19.5 and 19.6 reveals that the formation of a neutral cross-link leads to one of the pairs of nonbonding electrons on the interacting selenium atoms becoming more bonding, but another pair becomes strongly antibonding. On the other hand, disproportionation results in one nonbonding pair becoming bonding (on Se$^+$) and one bonding pair becoming nonbonding.

The introduction of Se$^+$ and Se$^-$ into Se^0 is obviously analogous to the mixed oxidation–state oxide systems discussed in Chapter 18, except we are now faced with a system that simultaneously generates equal numbers of *n*- and *p*-type impurities. The filled nonbonding levels of Se$^-$ and the empty antibonding levels of Se$^+$ are both in the band gap between the valence and conduction bands arising from the bonding and antibonding orbital manifolds of the linear runs of Se^0. The variation of coordination number of the Se in the three valence states permits the unusual result that there are three sequential oxidation states present in the solid, none of

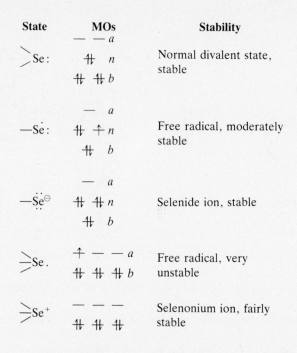

Figure 19.6
Some valence states of selenium. (The $4s$ electrons are omitted for clarity.) The total number of MOs is equal to the number of p orbitals on selenium plus an orbital from each bonding atom.

State	MOs	Stability
Se:	a, n, b	Normal divalent state, stable
—Se:	a, n, b	Free radical, moderately stable
—Se$^{\ominus}$	a, n, b	Selenide ion, stable
Se.	a, b	Free radical, very unstable
Se^{+}		Selenonium ion, fairly stable

which has unpaired electrons. This state of affairs is much more likely to occur in a glass, where the symmetry of the local environment of an atom is varied, than in a crystal, where the lattice imposes severe constraints on the symmetry of the local environment.

The photoconduction of glassy selenium therefore can be attributed to excitation of electrons from the valence band to the antibonding levels of Se^{+} and to promotion of electrons from Se^{-} to the conduction band. The photoconductivity of chemically pure selenium is not sensitive to the presence of small amounts of foreign impurities, as are the Group IVA (14) intrinsic semiconductors, because the "valence impurities" (Se^{+} and Se^{-}) are already present in high concentration.

19.4 Oxide Glasses

The vitreous oxide mixtures based on silica, commonly known as glass, were first used in the glazing of ceramic ornaments about 14,000 years ago in Egypt. Transparent, bulk glass similar to the forms used today appeared in the eastern Mediterranean region about 4000 years ago. A detailed recipe for making such glass was found in Mesopotamia on a clay tablet at least 3600 years old!

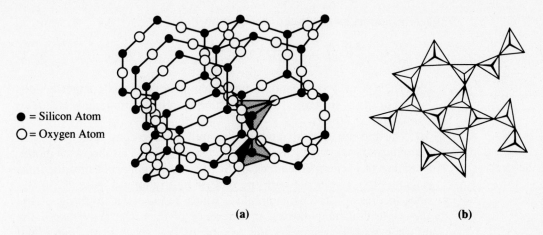

● = Silicon Atom

○ = Oxygen Atom

(a)

(b)

Figure 19.7
Structures of
(a) crystalline and
(b) glassy quartz.

The oxides of virtually all the elements in Groups IIIA–VA (13–15) can be readily produced in the form of a glass, particularly when they are mixed with other oxides. We will concentrate our attention on the silica-based glasses because they are the most widely used and the principles that apply to silicate glasses are generally applicable to other oxide systems.

Silica readily forms a glass when the melt is cooled; in fact, crystallization from the melt is very difficult to achieve. A clue to why this is so is because silica does not melt to a mobile liquid but to an extremely viscous liquid, indicating that the melt consists of very large molecules. This is not surprising when it is recalled that solid SiO_2 consists of an infinite, three-dimensional, covalently bonded structure and the melting process requires the breaking of very strong covalent chemical bonds rather than the weaker physical bonds that usually bind the crystals of small molecules. Thus, like all highly polymerized compounds, silica already has a high viscosity around the crystalline melting point, and the rate of crystallization is expected to be slow.

The ease of glass formation by silica and silicates is further enhanced by the ability of the Si—O—Si bond system to accommodate a highly disordered structure because of its great flexibility. Thus, although each silicon is coordinated to a regular tetrahedron of oxygens, the angles between the tetrahedra (which share common corners in the supermolecule) can vary over a considerable range. In the crystalline forms of silica, the Si—O—Si angles are constant; in the melt and in the glass, however, they span a range of values (Figure 19.7).

When silica, which is the anhydride of the family of silicic acids $SiO_{2-x}(OH)_{2x}$ (where $O < x < 2$), is reacted with highly basic oxides such as those of Groups IA and IIA (1 and 2), the covalent Si—O—Si bonds are cleaved to give terminal Si—O$^-$ ions, associated with the heterocations, as illustrated in Equation 19.1:

$$SiO_2 + \frac{x}{n}MO_n \rightarrow SiO_{2+x}M_{x/n} \tag{19.1}$$

where n = 1 or 2. The random introduction of these ionic imperfections into the silica structure further inhibits crystallization and facilitates vitrification.

The introduction of ionic groups into silica greatly increases chain mobility and lowers the glass-transition temperature. This increased mobility is due to nucleophilic attack by Si—O$^-$ on neighboring siloxane links (Figure 19.8). (Compare with the process illustrated in Figure 19.2.) Besides increasing the chain mobility, which results in a lowering of the glass-transition temperature, the presence of ionic charges expands the polymer network, increases the amount of free space in the glass, and lowers its density. The lowered glass-transition temperature was a great advantage in the early days of glass technology because the temperature at which quartz becomes easily workable is rather high (greater than 1000 °C) and beyond the practical reach of earlier generations. The increased free volume of the glass is a disadvantage because it results in a high thermal expansion coefficient. Below T_g the glass is extremely brittle, and the forces produced by shrinkage on cooling can easily shatter it. Glasses with high densities, but still containing ionic groups, can be formulated by including other covalent oxides in the mixture. The most successful formulations are based on the addition of B_2O_3. These borosilicate glasses, of which Pyrex is perhaps the most renowned, show remarkable resistance to temperature shock (as does quartz itself), yet they still have a glass-transition temperature low enough to permit easy manipulation in the molten state. The replacement of Si in quartz by boron leaves a positive-charge deficiency, which is balanced by added cations. This is the same, in principle, as the replacement of silicon by aluminum in the aluminosilicate minerals (see Chapter 16).

The traditional method for making glass involves the fusion of the ingredients at high temperature (about 800 °C). This is a very energy-

Figure 19.8
Anion transfer in
ionic polysiloxanes.

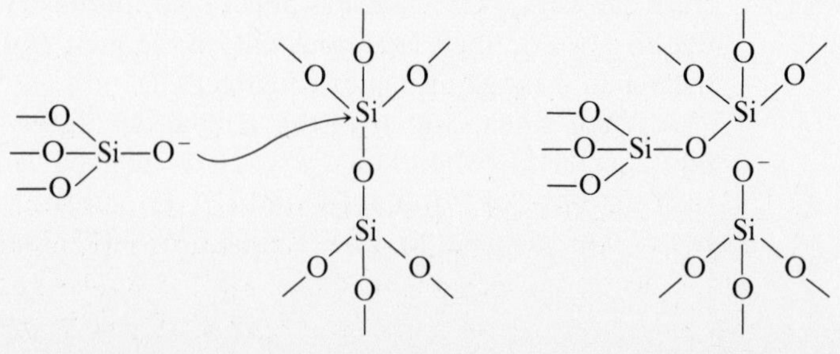

intensive procedure. Recently, attention has been directed to new methods for producing glasses at much lower temperatures. A very promising approach, which has already found some major applications, is the so-called sol–gel method. The sol–gel method uses the controlled hydrolysis of organic derivatives such as metal alkoxides to produce oxides in a glassy state under more or less ambient conditions. A typical reaction is shown in Equation 19.2:

$$Si(OC_2H_5)_4 + 2H_2O \xrightarrow{C_2H_5OH} SiO_2 + 4C_2H_5OH \qquad (19.2)$$

Under the proper conditions, such a hydrolysis leads to a clear viscous fluid, *sol*. This sol can then be allowed to dry by solvent evaporation to give a rigid but still very brittle *gel*. This gel has a very open structure, but it can be densified by relatively mild heating (300–400 °C) to give a glass. The fluid character of the sol lends itself to applications such as surface coatings; the sol–gel technique is used to apply nonreflecting, nonfogging, and nonchargeable coatings to mirrors, automobile windshields, and television screens. High-quality optical components can also be made by this technology. Glasses with many different components can be made easily by the sol–gel process. For example, incorporating small amounts of palladium compounds provides a tinted-glass coating, which has been widely used in the tinted glazing of many urban high-rise buildings.

19.5   Amorphous Carbon

Carbon exists in a number of different forms, each with its own unique set of properties. The properties of graphite—a metallic, opaque, lubricious, chemically inert material—and diamond—a hard, transparent insulator—have already been discussed (Chapter 18). The two other technologically important forms of carbon are *carbon black* and *glassy carbon*. Carbon black results from the combustion of high carbon–content organic material (pitch, aromatic compounds). The resulting smoke from such combustion consists of submicron particles that show no crystallographic order but that are believed to consist of a random network of small graphitelike structures. Carbon black is fairly inert chemically, has an open structure, and can be obtained with extremely high surface areas. Because of the high strength of the C—C bonds, carbon black resists sintering and finds extensive use as a catalyst support and as a filler for rubber used in automobile tires. It is very cheap and greatly increases the strength and durability of rubber.

Structurally, glassy carbon is closely related to carbon black. It differs mainly in being monolithic, rather than consisting of microscopic particles. Glassy carbon is made by the controlled pyrolysis of organic polymers such as poly(acrylonitrile). The required object is preformed as the organic polymer and then carefully heated to transform the organic material into carbon with the same form as the original organic object. Glassy carbon is electronically conducting and extremely resistant to electrochemical and chemical corrosion. It also has high mechanical strength. These properties have resulted in its common use as an electrode material.

Glassy carbon is easily obtained in the form of fibers, again by the pyrolysis of prespun organic fibers. These fibers have extremely high tensile strengths and survive up to very high temperatures. They also have a very low density compared to metals. Their combination of lightness and high strength makes them exceptionally valuable as reinforcing agents for plastics and other glassy matrices. Carbon fiber–reinforced materials have found their way into such diverse applications as tennis rackets, ultralight airframes, and turbine blades for jet engines.

BIBLIOGRAPHY

Suggested Reading

Brinker, J. J., ed. *Better Ceramics Through Chemistry*. Symposium Proceedings of the Materials Research Society, Vol 3, 1986.

Brinker, J. J., "A Chemical Route to Advanced Ceramics." *Science* 233 (1986): 25.

Hench, L. L., and D. R. Ulrich, eds. *Science of Ceramic Chemical Processing*. New York: Wiley, 1986.

Roy, R. "Ceramics by the Solution Sol–Gel Route." *Science* 238 (1987): 1664.

Wyl, W. A., and E. Chastner-Marloe. *The Constitution of Glasses: A Dynamic Interpretation*. New York: Wiley-Interscience, 1962.

PROBLEMS

19.1 Explain what is meant by the following terms:
  a. An amorphous material   **c.** T_g
  b. A glass   **d.** A borosilicate

19.2 Describe the processes that occur when rhombic sulfur (S_8) is heated slowly from room temperature to $200\,°C$ and then rapidly quenched to room temperature again. Why is the final product a rubber rather than a glassy material?

19.3 Speculate on the structural details of the material resulting from the following reaction (HINT: Consult Chapter 16):

$$SiO_2 + 0.1Na_2O \rightarrow Na_{0.2}SiO_{2.1}$$

19.4 Identify four properties of glass that make it an attractive material for household use.

19.5 What are the properties of glassy carbon that make it an attractive material? Identify two applications that take advantage of these properties.

PART SEVEN

Solution Chemistry

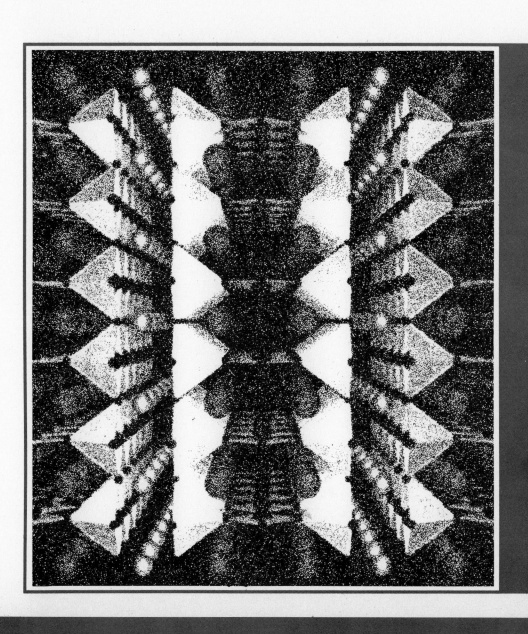

20

Aqueous and Nonaqueous Solutions

20.1   *Aqueous Solutions*

Most of us are already familiar with water as a solvent. Oceans cover about two-thirds of the surface of our planet, and water plays a vital role in many important biological processes, including our own survival. In this chapter, we examine several properties of water that make it an ideal solvent for ionic species such as Na^+ and Cl^-. Besides this aqueous solution chemistry, we consider briefly some aspects of inorganic chemistry in nonaqueous solvents such as liquid NH_3 and anhydrous H_2SO_4. Many important chemical, biochemical, and electrochemical processes that affect our environment take place in solution, and a few of these are described here. We begin by examining the properties of pure water, which are largely shaped by hydrogen bonding. This phenomenon in water and other polar solvents is crucial to understanding solvation properties.

The Hydrogen Bond

Melting and boiling points tend to increase with increasing molecular weight. There is therefore something strange about the fact that water is a relatively high boiling-point liquid, while all of its Group VIA (16) congeners (H_2S and H_2Se, for example) are gases (Figure 20.1). This trend of anomalously high boiling (and melting) points for the MH_n compounds of N, O, and F is quite general and was one of the major pieces of experimental evidence that led to the concept of the hydrogen bond (Figure 20.2). Because the idea of the hydrogen bond was conceived at a time when the two-center, two-electron bond reigned supreme, there is an entrenched prejudice that the hydrogen bond is somehow "unusual" because it involves the simultaneous interaction of a hydrogen atom with two other atoms.

The hydrogen bond is encountered among compounds with the most polar M—H bonds, and early attempts at obtaining a quantitative description of the hydrogen bond were preoccupied with dipole–dipole interactions. These models were mathematically tractable and did not conflict with the two-center, two-electron bond concept, but they often

Figure 20.1
Comparison of the boiling points of H_2X and MH_4 molecules.

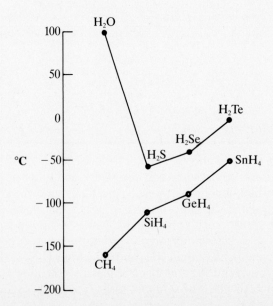

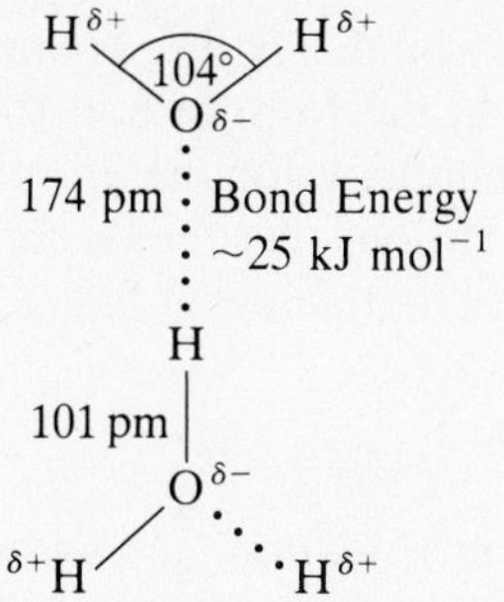

Figure 20.2
Schematic drawing
of a typical
hydrogen-bond
interaction for H_2O
molecules. The data
given are from a
neutron-diffraction
study of deuterated
ice I_h at 100 K.

$H^{\delta+}$ 104° $H^{\delta+}$
$O^{\delta-}$

174 pm : Bond Energy
: $\sim$25 kJ mol^{-1}

H

101 pm |

$O^{\delta-}$

$\delta+$H $\cdots$ H$^{\delta+}$

gave poor results. Despite this, it is undeniable that a term involving coulombic attraction will be very important in whatever model is ultimately successful.

A great deal of structural information is now available on hydrogen-bonded compounds, and the common observation of interatomic separations of less than the sum of van der Waals radii confirms the intervention of strong covalent effects in many cases. Table 20.1 shows some data on calculated and observed interatomic separations. The strongest known hydrogen bond occurs in the linear $[HF_2]^-$ ion, which unlike most cases of hydrogen bonding has the hydrogen atom located midway between the two fluorines:

$$[F \overset{113\ pm}{\rule{1.5cm}{0.4pt}} H \overset{113\ pm}{\cdots} F]^-.$$

It is likely that this system represents the greatest degree of covalency in a hydrogen bond, and there is now some justification in viewing the bonding system as a three-center, four-electron interaction analogous to that of BeH_2. Two of the four electrons occupy a strongly σ-bonding orbital and

Table 20.1 *Comparison of Observed and Calculated Interatomic Distances (pm) in Some Hydrogen Bonds (A—H $\cdots$ B)*[a]

Hydrogen-Bond Type	H $\cdots$ B (Observed)	H $\cdots$ B (Calculated)[b]
O—H $\cdots$ O	170	260
O—H $\cdots$ N	190	270
N—H $\cdots$ N	220	270
N—H $\cdots$ O	200	260
O—H $\cdots$ Cl	220	300

[a] Data taken from W. C. Hamilton and J. C. Ibers. *Hydrogen Bonding in Solids*. New York: Benjamin, 1968.
[b] Based on van der Waals radii.

the other two a nonbonding orbital. The F—H bond energy of the $[HF_2]^-$ ion, which is only available in the form of nonvolatile ionic salts, is very difficult to determine. The enthalpy for the gas-phase process (Equation 20.1) has been estimated by various theoretical and experimental methods to lie between 150–260 kJ mol^{-1} (the higher figure is probably closer to the truth).

$$HF(g) + F^-(g) \rightarrow [HF_2]^-(g) + \Delta H^\circ \tag{20.1}$$

Even if the ΔH° value for Equation 20.1 were known accurately, it is not easy to compare it with anything. (Should we compare it to the dissociation of HF into H$^\bullet$ and F$^\bullet$, 560 kJ mol^{-1}, or into H$^+$ and F$^-$, 1000 kJ mol^{-1}?) The best we can say in qualitative terms is that for a H—F bond with a bond order of $\frac{1}{2}$ and with the increased interelectronic repulsion arising from a formal negative charge we would expect a bond energy of something less than half the bond energy of molecular HF.

Most other hydrogen bonds are neither symmetrical nor anywhere near as strong as the $[HF_2]^-$ bond. There is no fixable lower limit on hydrogen-bond strength; the weaker ones simply merge into the weak polar and dispersion force interactions. For example, carbon is not usually considered to form hydrogen bonds, but the linear chain structure of HCN contains C—H $\cdots$ N hydrogen bonds and the acetone–CHCl$_3$ complex contains the C—H $\cdots$ O linkage.

A hydrogen bond is still probably best treated as the result of a simple dipole–dipole interaction. The situation is akin to that of the wide range of "ionic solids," which in fact are substantially covalent. For most inorganic chemists, excessive soul searching over the precise character of the bonds in LiH or SiO$_2$ is considered futile. That the former reacts as though it were a hydride salt while the latter reacts as though it were covalent is perhaps more important.

The hydrogen bond plays an extremely important role in biochemical structures. Its relative weakness permits reversibility in reactions involving its formation and a greater subtlety of interaction than is possible with normal covalent bonds. Hydrogen bonds are of paramount importance in determining the secondary structure of proteins and in binding substrates to enzymes, receptors, and carriers (Figure 20.3).

The special properties of water, particularly its liquid range and its conductivity for ions (especially the proton), are highly important for the sustenance of life. The exceptionally high conductivity of aqueous protons is attributed to the facile transfer of the proton from a hydronium ion to neighboring water molecules, through formation of hydrogen bonds (Equation 20.2):

$$\begin{array}{c} H \\ \diagdown \\ H \diagup \end{array} O^+ \!\!-\!\! H + \begin{array}{c} H \\ \diagup \\ \diagdown H \end{array} O \longrightarrow \begin{array}{c} H \\ \diagdown \\ H \diagup \end{array} O^+ \!\!-\!\! H \cdots \begin{array}{c} H \\ \diagup \\ \diagdown H \end{array} O \tag{20.2}$$

Figure 20.3
Possible hydrogen-bonding interactions in the zinc-containing enzyme, carbonic anhydrase.

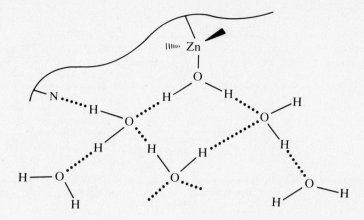

The base strength of H_2O and the bond strength of $[H_3O]^+$ seem to have just the optimum values for maximum rate of transit of the proton across the hydrogen bond. That the hydrogen-bond network extends in three dimensions in water is also an important factor. Because there is only one hydrogen atom in HF, it can only form linear polymers via hydrogen bonding.

Most of the molecules in liquid water are believed to be fully hydrogen-bonded at any given instant, but there is also a small concentration of free OH groups. The disorder in liquid water is allowed by the extremely rapid equilibrium between the free and hydrogen-bonded OH groups.

The structures of both liquid water and ice have been the subject of much attention since the turn of the century. Nine different structural forms of ice have been identified, chiefly by neutron-diffraction studies of the deuterated species, the particular form depending critically on the external temperature and pressure. The lattices are held together by rigid hydrogen bonding. The most common ice structure (ice I_h) is formed at 101.3 kPa (1 atm) pressure. This structure has open hexagonal channels (Figure 20.4), which account for the familar hexagonal appearance of snowflake crystals and ice formations on window panes during winter. This structure is closely similar to wurtzite (hexagonal ZnS, Figure 3.25). Each oxygen atom occupies a Zn or S position of the wurzite lattice, and the hydrogens are colinearly and unsymmetrically located between each pair of oxygens. Because the O—O distance is insensitive to the location of the H atom and because the energy for displacing the proton from one equilibrium position to the other is not large, the ice I_h structure is highly susceptible to disorder.

The maximum density of liquid water occurs at 4 °C, and this phenomenon is attributed to the persistence of the ice I_h hexagonal channels at this temperature, with additional water molecules produced by the melting

Figure 20.4
Structure of ice I
with almost linear
O—H · · · O bonds.
Adapted with
permission from
R. E. Dickerson and
I. Geis. *Chemistry,
Matter, and the
Universe.* Menlo
Park, Calif.:
Benjamin/Cummings,
1976.

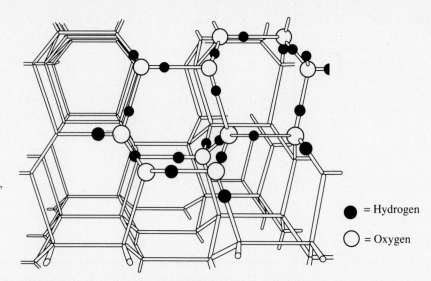

● = Hydrogen

○ = Oxygen

ice being trapped in the channels. Extensive hydrogen bonding exists in liquid water as well, even close to its boiling point (compare with Et_2O in Table 20.2).

There is an obvious relationship between the stronger hydrogen bonds and the bridging covalent bonds encountered in borane and polyborane chemistry. This type of bridging is also very prevalent in the transition metal hydrides where, although a hydrogen coordination number of two is most common, coordination numbers up to six have been encountered. In the boranes and metal hydride complexes, we are dealing with much more

Table 20.2   *Some Physical Properties of Water Compared with Those of Diethylether*

Property	H_2O	Et_2O
Boiling point (°C, 1 atm)	100.00	34.51
Melting point (°C)	0.00	−116.2
Dielectric constant (ϵ_0)	81.1 (18 °C)	4.34
Density (kg dm^{-3})	1.00 (3.98 °C)	0.736 (0 °C)
Dipole moment of gas (μ, debye)	1.85	1.15
Refractive index (25 °C)	1.333	1.352
Surface tension (erg cm^{-2})	71.97 (25 °C)	17.01 (20 °C)
Viscosity (ρ, 25 °C, centipoise)	0.959	0.222

SOURCE: *Handbook of Chemistry and Physics*, 52nd ed. Cleveland: The Chemical Rubber Co., 1971.

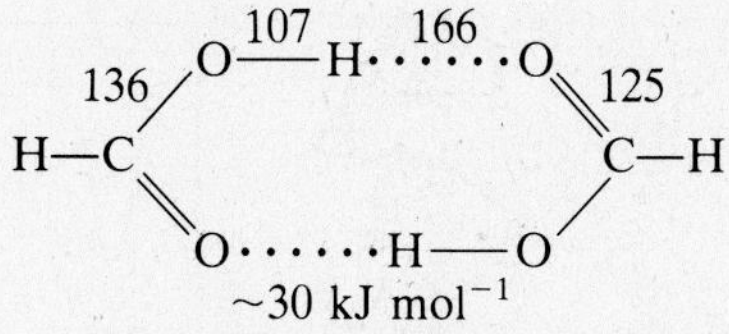

Figure 20.5
Hydrogen-bonded
dimers in formic
acid. Bond distances
are in picometers.

basic systems (in the sense that the electrons are less tightly bound rather than in terms of numbers) than those involved in classical hydrogen bonds, and the polarities of the component atoms are probably reversed.

Hydrogen bonding in water and ice is a direct consequence of the electronegativity difference (1.3 on the Allred–Rochow scale) between oxygen (3.5) and hydrogen (2.2), leading to polar $O^{\delta-}$—$H^{\delta+}$ bonds. Hydrogen bonds have much lower bond energies (about 25 kJ mol^{-1}) than normal covalent bonds (about 450 kJ mol^{-1}); this is why the O—H $\cdots$ O system is unsymmetrical with two different O—H bond lengths. Intermolecular hydrogen bonds are also formed in carboxylic acids—for example, formic acid (HCO$_2$H)—with dimers being produced (Figure 20.5).

As mentioned, hydrogen bonding is also important for several other highly electronegative elements, mainly N, F, and Cl, for example, in NH$_3$, HF, and *o*-chlorophenol. The last compound undergoes both *inter*molecular and *intra*molecular hydrogen bonding:

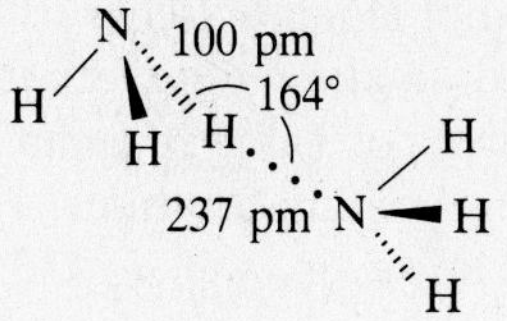

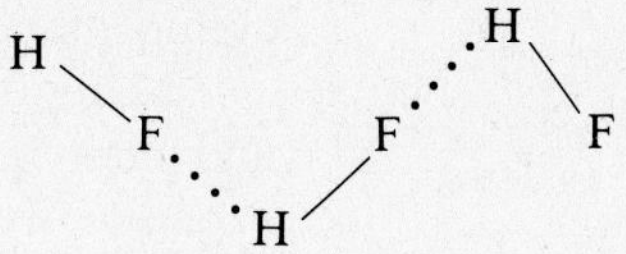

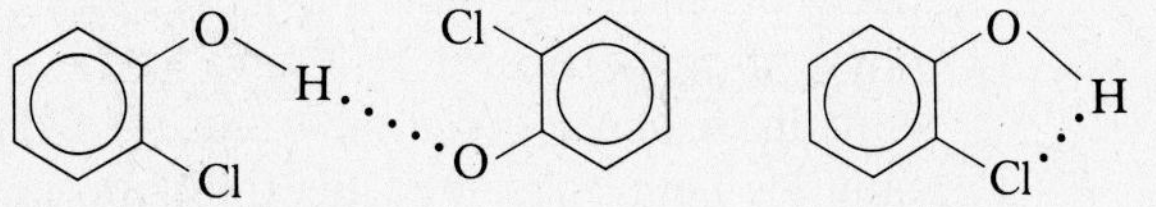

Inter   **Intra**

The meta and para isomers of chlorophenol can only undergo intermolecular hydrogen bonding; consequently, they appear to have much higher molecular weights than the ortho isomer. This difference is reflected in the higher boiling points of the meta and para compounds:

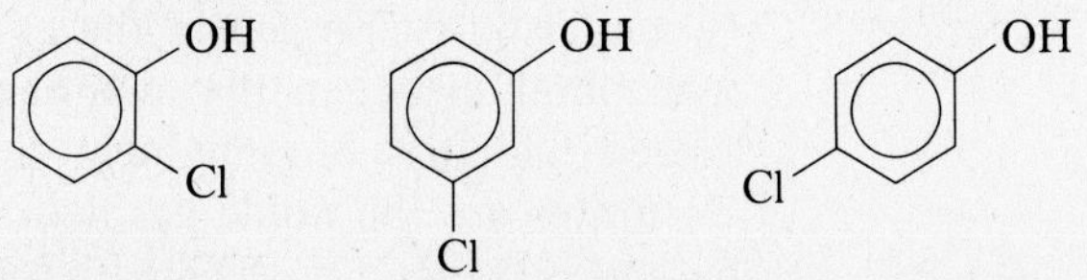

ortho, 174.9 °C   **meta, 214.0 °C**   **para, 219.8 °C**

Figure 20.6
IR spectra of
cis-cyclopentane-1,2-
diol in CCl$_4$
solution. On
dilution, the
OH–stretching band
at about 3500 cm^{-1}
due to the
intermolecular
hydrogen bond
disappears.

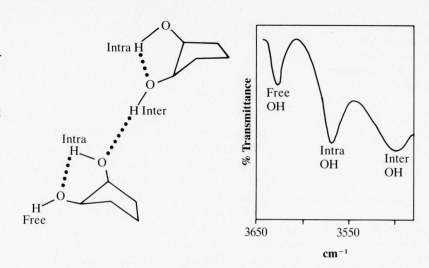

Whenever there is a possibility of a system having both intramolecular and intermolecular hydrogen bonding in solution, the intramolecular hydrogen bonding occurs first in dilute solution; as the concentration increases, the intermolecular hydrogen bonding increases. This is nicely demonstrated by IR spectroscopy in the O—H–stretching region for *cis*-cyclopentane-1,2-diol (Figure 20.6).

Many of the dramatic differences in the physical properties of water and diethylether given in Table 20.2 can be rationalized by the presence of extensive hydrogen bonding in water. Hydrogen bonding is not the only factor concerned, however, and secondary effects due to permanent and induced dipole interactions also play a role. Finally, it should be mentioned that plots similar to that shown in Figure 20.1 can be made for several other physical properties of H$_2$X and MH$_4$—for example, melting points and enthalpies of vaporization—and for MH$_3$ and HX, with similar discontinuities being observed for the hydrogen-bonded species NH$_3$ and HF.

Hydrogen bonding also plays an essential role in the stabilization of some curious compounds known as clathrates. The name is derived from the Greek word *klethra* ("cage" or "trap"). For instance, an ammoniacal solution of nickel cyanide Ni(CN)$_2$ is perfectly stable for long periods of time, but when benzene is added, crystals of the clathrate Ni(CN)$_2$ · NH$_3$ · C$_6$H$_6$ immediately begin to form (Figure 20.7). The C$_6$H$_6$ molecule is trapped in the cavity made by the intermolecular hydrogen bonds between the ammonia groups. The benzene burns when a lighted match is brought near the crystals. Another important class of clathrates are the noble gas hydrates such as Xe(H$_2$O)$_6$, in which the noble gas is trapped in the cavity formed by the hydrogen-bonded water molecules.

Figure 20.7
Structure of
$Ni(CN)_2 \cdot NH_3 \cdot C_6H_6$.
The dashed lines
represent
intermolecular
hydrogen bonds
between the NH_3
groups.

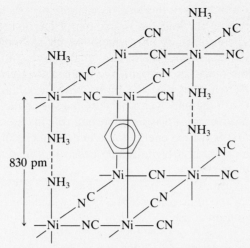

For a compound to dissolve in water, enough energy must be provided externally to break down the hydrogen bonds in water. This energy is obtained chiefly from the *aquation* (the specific term of *solvation* of water) of the species present, such as the Na^+ and Cl^- ions of sodium chloride (Equation 20.3):

$$NaCl(s) \xrightarrow{\text{H}_2\text{O}} Na^+(aq) + Cl^-(aq) \qquad\qquad \textbf{(20.3)}$$

It should be remembered, however, that we are proceeding from a highly ordered system to a disordered one, and so entropy effects also play a role in the dissolution of a compound.

In the case of ionic compounds, the solvation energy is so great that it suffices to overcome the hydrogen bonding of the water and the lattice energy of the salt. Aquation is the direct result of electrostatic attraction between the ions and the water molecules; this interaction is the subject of the next section.

Solvation and Dielectric Properties

One of the most important physical properties of water listed in Table 20.2 is its dielectric constant (ε_0). In general the energy of electrostatic attraction between two ions of opposite charge (such as Na^+ and Cl^-) is given by Equation 20.4:

$$E = -\left(\frac{q^+ q^-}{4\pi d \epsilon_0}\right) \qquad\qquad \textbf{(20.4)}$$

where q^+ and q^- = the electric charges on the two ions, d = the distance between the ions, and ε_0 = the dielectric constant of the medium. Dielectric constants are primarily the result of dipole–dipole interactions, and they increase with increasing temperature. From Equation 20.4, the greater the dielectric constant, the weaker the electrostatic attraction between the two ions and the more easily a solution is formed. This is obviously not the only factor because the specific solvation energies of the ions themselves must play an important role in determining whether solvation occurs. Solvation energies are determined by dipole–ion interactions, and an increase in solvent dipole leads to an increase in both ε_0 and the solvation energy.

An alternative way of looking at aquation is in terms of the enthalpy of solution of the ionic compound ΔH°_{soln}, which is given by the Born expression (Equation 20.5):

$$\Delta H^{\circ}_{soln} = \frac{q^+ q^-}{2d} \left(1 - \frac{1}{\epsilon_0} \right) \tag{20.5}$$

An increase in ε_0 results in an increase in ΔH°_{soln} as well; the more easily dissolution should take place provided that the aquation energies of the ions are greater than the lattice energy (U_0) of the crystal. As an example, let us take crystalline sodium chloride NaCl(s) in water, for which the following thermodynamic cycle (Equation 20.6) can be generated (D = dissociation energy, I = ionization energy, and A = electron affinity):

$$
\begin{array}{lll}
Na(s) \xrightarrow{\Delta H^{\circ}_{sub}} Na(g) & Cl(g) \xleftarrow{D/2} \tfrac{1}{2}Cl_2(g) \\
\quad \big\downarrow I & \quad \big\downarrow A \\
Na^+Cl^-(s) \xleftarrow{U_0} Na^+(g) \;+\; Cl^-(g) & \tag{20.6} \\
\quad \big\downarrow \Delta H^{\circ}_{aq}(Na^+) \quad \big\downarrow \Delta H^{\circ}_{aq}(Cl^-) \\
\xrightarrow{\Delta H^{\circ}_{soln}} Na^+(aq) \;+\; Cl^-(aq)
\end{array}
$$

that is,

$$\Delta H^{\circ}_{soln} + U_0 = \Delta H^{\circ}_{aq}(Na^+) + \Delta H^{\circ}_{aq}(Cl^-) \tag{20.7}$$

or

$$\Delta H^{\circ}_{soln} = \Delta H^{\circ}_{aq}(Na^+) + \Delta H^{\circ}_{aq}(Cl^-) - U_0 \tag{20.8}$$

There are many nonaqueous solvents that have large dielectric constants—for example, at 20°C, CH_3COCH_3 (20.7), CH_3CN (36), CH_3CONH_2 (60), and $CH_3CON(CH_3)_2$ (179). Consequently, these sol-

Table 20.3 *Some Physical Data for the Hydration of the Alkali Metal Cations*

Cation	Ionic Radius (pm)	Estimated Radius of Hydrated Cation (pm)	Relative Ionic Mobility	Estimated Hydration Number
Li^+	60	340	1.0	25
Na^+	95	280	1.3	17
K^+	133	230	1.9	11
Rb^+	148	230	2.0	10
Cs^+	169	230	2.0	10

vents might well be expected to be useful for the dissolution of ionic compounds, and this is indeed true.

A particularly striking illustration of the effect of solvation of ions comes from the values for the ionic mobilities of the alkali metal cations in aqueous solution (Table 20.3). Surprisingly, the supposedly smaller Li^+ ion moves much more slowly than does the bulky Cs^+ ion. This situation results from the different *hydration numbers* of the ions. The hydration number refers to the total number of H_2O molecules in strong contact with the metal cation. The Li^+ ion apparently has a much larger hydration number than does Cs^+, such that the "effective atomic weight" of Li^+ is much greater than that of Cs^+. Therefore, the now heavier and larger $Li^+(aq)$ ion would be expected to move appreciably slower than would $Cs^+(aq)$, as is observed experimentally.

The exact nature of the coordination of water molecules around cations and anions is not fully resolved. It is generally accepted, however, that there is more than one coordination sphere. On the basis of ^1H- and ^{17}O-NMR measurements, there appear to be six water molecules arranged octahedrally in the first coordination sphere of most metal cations:

The polar water molecules are attracted toward the central M^{n+} ion and are held in place mainly by electrostatic effects. But, because of the polarity of the $O^{\delta-}$—$H^{\delta+}$ bonds, it is possible for water molecules in a second coordination sphere to hydrogen bond to those in the first:

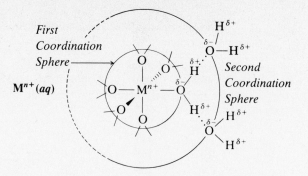

Anions are also aquated through hydrogen bonding:

In the case of transition metals, the H_2O molecules in the first coordination sphere interact so strongly with the d orbitals of the central metal ion that crystal field splitting of the d orbitals takes place (for example, into t_{2g} and e_g sets of orbitals for O_h symmetry). The energy difference between the orbital sets usually lies in the visible range, and so solutions of transition metal ions in solution are often highly colored—for example, $[Ti(H_2O)_6]^{3+}$ (violet) and $[Fe(H_2O)_6]^{2+}$ (green) (see Chapter 13).

Many common inorganic salts crystallize with water of crystallization as an intimate part of the unit cell, for example, $CuSO_4 \cdot 5H_2O$ and $Na_2SO_4 \cdot 12H_2O$ (Glauber's salt). Once again, it is the electrostatic interactions between the cations and the H_2O molecules that contribute greatly to the stabilization of the crystal lattice as a whole. For $CuSO_4 \cdot 5H_2O$, the first coordination sphere consists of four H_2O molecules and two $[SO_4]^{2-}$ ions (Figure 20.8). The fifth water molecule is located some distance away from the Cu^{2+} ion, but it is hydrogen bonded to $[SO_4]^{2-}$ ions, as are the four H_2O molecules.

It is also common to find other solvents present as discrete molecules in the unit cell of many compounds, especially in organometallic chem-

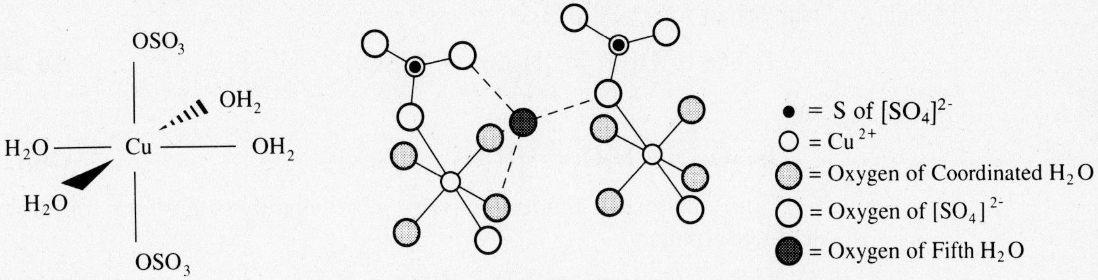

Figure 20.8
Structure of
$CuSO_4 \cdot 5H_2O$.

istry where potentially coordinating solvents are often used in synthetic procedures, for example, $Ru_3(CO)_8(Ph_2PCH_2PPh_2)_2 \cdot 2Me_2CO$ and $Br(CO)_3Mn[Ph_2PCH_2CH_2P(Ph)CH_2CH_2PPh_2]Cr(CO)_5 \cdot CH_2Cl_2$.

At high temperatures and pressures near its critical point, water becomes almost like an organic liquid because of the diminished hydrogen bonding present, and hydrocarbons even dissolve in it. This property is important in geology and accounts for the large quantities of water that are mixed with oil in the earth's crust.

Acids and Bases

This topic is one of the most fundamental in all of inorganic chemistry, and many different but often interrelated theories and definitions have been put forward over the past sixty or so years. Only three of these approaches will be considered here: Brønsted–Lowry (1923), Lewis (1923), and Pearson (1963).

Brønsted–Lowry Acids and Bases The Brønsted-Lowry definition is especially important when considering conventional acids such as CH_3CO_2H and H_2SO_4. For instance, the equilibrium shown in Equation 20.9 is established upon mixing acetic acid (CH_3CO_2H) with water:

$$CH_3CO_2H + H_2O \rightleftharpoons [H_3O]^+ + [CH_3CO_2]^- \qquad (20.9)$$

| Proton | Proton | Proton | Proton |
| Donor | Acceptor | Donor | Acceptor |

Acetic acid loses a proton to a water molecule leading to the subsequent formation of the hydronium ion $[H_3O]^+$, which is itself a proton donor and is extensively hydrogen bonded: $[H_5O_2]^+$, $[H_7O_3]^+$, and so on. Similarly, $[CH_3CO_2]^-$ is a proton acceptor. The interrelated species $CH_3CO_2H/[CH_3CO_2]^-$ and $H_2O/[H_3O]^+$ are referred to as *conjugate acid–base pairs*, where a proton donor is defined as a Brønsted–Lowry acid and a proton acceptor is a Brønsted–Lowry base. A similar equilibrium is established for H_2SO_4/H_2O:

$$H_2SO_4 + H_2O \rightleftharpoons [H_3O]^+ + [HSO_4]^- \tag{20.10}$$

and

$$[HSO_4]^- + H_2O \rightleftharpoons [H_3O]^+ + [SO_4]^{2-} \tag{20.11}$$

Some materials *autoionize* (self-ionize) slightly to produce conjugate acid–base pairs:

$$H_2O + H_2O \rightleftharpoons [H_3O]^+ + OH^- \tag{20.12}$$

$$NH_3 + NH_3 \rightleftharpoons [NH_4]^+ + [NH_2]^- \tag{20.13}$$

Liquid   Liquid   (K_a = about 10^{30})

All that autoionization means is that certain species can function simultaneously as both a Brønsted–Lowry acid and base, that is, both giving and receiving a proton in the process.

Not all Brønsted–Lowry acids and bases are as simple as those mentioned. There are some exotic examples in organometallic chemistry:

$$Ru_3(CO)_8(dppm)_2 + CF_3CO_2H \rightarrow$$

$$[HRu_3(CO)_8(dppm)_2]^+ + [CF_3CO_2]^- \tag{20.14}$$

where dppm = $Ph_2PCH_2PPh_2$.

The triangular triruthenium(0) complex effectively acts as a Brønsted–Lowry base by accepting a proton from trifluoroacetic acid, which ultimately finishes as a hydrogen atom bridging two Ru atoms. The H atom has been located in the cationic metal cluster complex by X-ray diffraction of the more easily crystallized $[PF_6]^-$ salt (Figure 20.9). The lengthening of the unique Ru—Ru bond by about 13 pm and the Ru(2)—Ru(1)—Ru(3) angle [63.03(2)°] directly reflects the presence of the H atom.

Lewis Acids and Bases   The Lewis definition of acids and bases is truly the central principle of modern coordination chemistry. Rather than speaking of proton donors and acceptors, Lewis proposed *electron-pair* donors and acceptors. A molecule that can give up one or more electron pairs is a *Lewis base*, whereas a molecule that can accept one or more electron pairs is a *Lewis acid*. Consider the following simple addition reaction:

$$F_3B(g) + NH_3(g) \rightarrow F_3BNH_3(s) \tag{20.15}$$

Boron trifluoride is an electron-deficient species that needs two additional electrons to achieve a stable octet; NH_3 has an electron pair available for donation. The two species combine almost instantaneously to produce a white, solid addition product. In coordination chemistry, it is

Figure 20.9
Perspective
drawing of
$[HRu_3(CO)_8(dppm)_2]^+$
with phenyl
hydrogen atoms
omitted for the sake
of clarity. Bond
distances are in
picometers. Adapted
with permission
from J.-J. Bonnet,
N. Lugan,
F. Mansilla, and I. S.
Butler, unpublished
work.

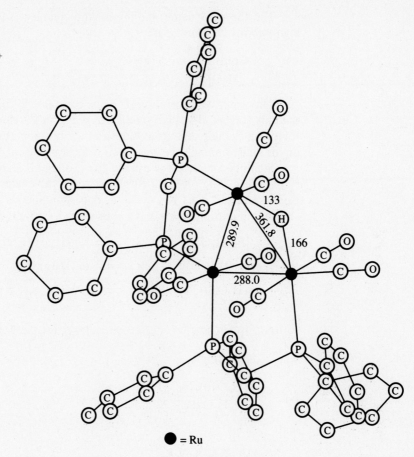

● = Ru

usually the metal ion that acts as the Lewis acid accepting electron pairs
from a variety of ligands. For instance, transition metal ions are believed to
be hydrated in aqueous solution with normally six H_2O molecules coordinated to the metal ion in the first coordination sphere, as in $[Fe(H_2O)_6]^{2+}$:

$$Fe^{2+} + 6H_2O \rightarrow [Fe(H_2O)_6]^{2+} \qquad (20.16)$$

Lewis   Lewis
Acid   Base

There is a wide range of Lewis basicities for ligands; that is, some
ligands are clearly better donors than others. A good example of a strong
Lewis base toward transition metal ions is CN^-. This ligand displaces all six
of the water molecules (weaker Lewis bases) in $[Fe(H_2O)_6]^{2+}$ to form the
more thermodynamically stable hexacyanoferrate(II) anion:

$$[Fe(H_2O)_6]^{2+} + 6CN^- \rightarrow [Fe(CN)_6]^{4-} + 6H_2O \qquad (20.17)$$

The thermodynamic stability of the complex is indicated by its very high *stability constant* (β):

$$\beta = \frac{\text{Constant}}{[H_2O]^6} = \frac{[Fe(CN)_6^{4-}]}{[Fe(H_2O)_6^{2+}][CN^-]^6} \simeq 10^{40} \tag{20.18}$$

The water concentration $[H_2O]$ is constant (55.5 M) because water is the bulk solvent. The high stability of $[Fe(CN)_6]^{4-}$ can lead to problems in removing cyanide ions from industrial waste—for example, in silver plating or gold winning—because iron is always present in such processes.

Stability constants are usually given as $\log_{10}\beta$, that is, 40 for the Fe(II) system here. The replacement of six water molecules by $6CN^-$ shown in Equation 20.17 does not occur in one spontaneous reaction. There are several equilibrium steps, each with its own *stepwise stability constant*:

$$[Fe(H_2O)_6]^{2+} + CN^- \xrightleftharpoons{K_1} [Fe(H_2O)_5(CN)]^+ + H_2O \tag{20.19}$$

$$[Fe(H_2O)_5(CN)]^+ + CN^- \xrightleftharpoons{K_2} Fe(H_2O)_4(CN)_2 + H_2O \tag{20.20}$$

$$Fe(H_2O)_4(CN)_2 + CN^- \xrightleftharpoons{K_3} [Fe(H_2O)_3(CN)_3]^- + H_2O \tag{20.21}$$

$$\vdots$$

$$[Fe(H_2O)(CN)_5]^{3-} + CN^- \xrightleftharpoons{K_6} [Fe(CN)_6]^{4-} + H_2O \tag{20.22}$$

where, for example,

$$K_1 = \frac{[Fe(H_2O)_5(CN)^+][H_2O]}{[Fe(H_2O)_6^{2+}][CN^-]} \tag{20.23}$$

and

$$\beta = K_1 K_2 K_3 K_4 K_5 K_6 \tag{20.24}$$

Table 20.4 gives some typical stability constant values. Notice the strikingly increased stability associated with chelation, due to the so-called *chelate effect*. Chelation is favored on entropy grounds because substitution of monodentate ligands by chelating ones results in an increase in the number of species present:

$$[M(H_2O)_6]^{n+} + 6NH_3 \rightarrow [M(NH_3)_6]^{n+} + 6H_2O \tag{20.25}$$

Seven Species   Seven Species

$$[M(H_2O)_6]^{n+} + 3en \rightarrow [M(en)_3]^{n+} + 6H_2O \tag{20.26}$$

Four Species   Seven Species

In general it is safe to conclude that transition metal ions are Lewis acids and that the ligands coordinated to them are Lewis bases. Neverthe-

Table 20.4 *Selected Stability Constants in Aqueous Solution*[a]

Ligand	Equilibrium Expression	$\log_{10}\beta$					
		Co^{2+}	Co^{3+}	Ni^{2+}	Cu^{2+}	Zn^{2+}	Ca^{2+}
NH_3	$\dfrac{[M(NH_3)_6]}{[M][NH_3]_6}$	4.4	34.4	8.3			
	$\dfrac{[M(NH_3)_4]}{[M][NH_3]^4}$	5.1		7.7	11.8	8.9	
en^b	$\dfrac{[M(en)_3]}{[M][en]^3}$	14.1	48.7	18.1			
	$\dfrac{[M(en)_2]}{[M][en]^2}$	10.8		13.8	20.0		
$edta^c$	$\dfrac{[M(edta)]}{[M][edta]}$	16.3	41.4	18.5	18.7	16.4	10.6

[a] For concentrations in mol dm^{-3}. Adapted with permission from T. Moeller. *Inorganic Chemistry. A Modern Introduction*. New York; Wiley, 1982, Table 11.1, p. 706. M = cation concerned.
[b] en = ethylenediamine, $H_2N—CH_2CH_2—NH_2$, bidentate ligand.
[c] edta = ethylenediaminetetraacetate anion $(^-O_2CCH_2)_2N—CH_2CH_2—N(CH_2CO_2^-)_2$, hexadentate ligand.

less, there are some notable exceptions. For example, Ir(I) in Vaska's compound *trans*-$IrCl(CO)(PPh_3)_2$ exhibits *metal basicity*; that is, it acts as a Lewis base under certain conditions:

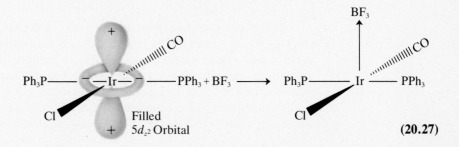

$$(20.27)$$

A lone pair of valence electrons in the square-planar Ir(I) complex in Equation 20.27 is localized on the $5d_{z^2}$ orbital and is available for donation to the Lewis acid BF_3.

Another unusual example of normal Lewis acid–base behavior involves the weak interaction of the bridging carbonyl groups in $Fe_2(CO)_9$

with various Lewis acids such as BF_3:

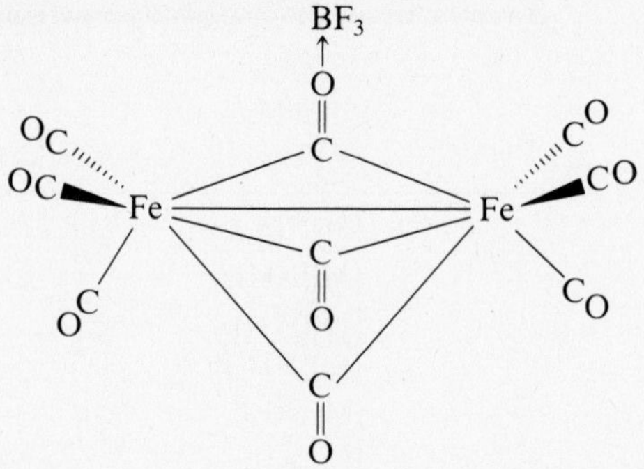

Pearson Acids and Bases The final classification of acids and bases is more recent in origin. By combining experimental data from several different approaches, Pearson evolved the Principle of Hard and Soft Acids and Bases (HSAB) about twenty years ago. A hard base is associated with the small donor atom in a Lewis base that holds on extremely tightly to its valence electrons (for example, F^- and the N atom in NH_3). A soft base has the opposite behavior—it is associated with large, easily polarizible Lewis donor atoms (for example, I^- and the P atom in PPh_3; see Fajan's rules, page 115).

In terms of chemical reactivity, hard acids interact more readily with hard bases, whereas soft acids prefer to react with soft bases. For example, Al^{3+} is a typical hard acid and forms $[AlF_6]^{3-}$ with F^- ions very easily. On the other hand, the soft acid Hg^{2+} reacts more readily with I^- than with F^- to yield HgI_2. Clearly, the definitions given represent the extremes, and there are many situations in which species are best termed borderline (for example, Fe^{2+} and Br^-).

Table 20.5 gives an abbreviated version of the table of hard and soft acids originally presented by Pearson. As an illustration of the use of this table, small, highly charged cations tend to be hard acids and so do not easily form metal carbonyl complexes with the soft base CO. As you will see later, one of the striking features about metal carbonyls and the closely related metal phosphine complexes is that the metals are most often in low oxidation states—that is, they behave as typical soft acids.

20.2 *Inorganic Chemistry in Nonaqueous Solvents*

Thus far, we have restricted our discussions to inorganic reactions in aqueous solution. There are many other solvent systems available, for example, liquid NH_3 and H_2SO_4. In this section, some nonaqueous solution chemistry is described briefly, initially that of the hydrides (or *protic*

Table 20.5 *Some Selected Examples of Hard and Soft Acids and Bases*

Hard Acids	Hard Bases
H^+, M^+ [Group IA (1) elements]	NH_3, RNH_2
M^{2+} [Group IIA (2) elements]	H_2O, ROH, RO^-, R_2O, OH^-, O^{2-}
M^{3+} (rare earths)	$[CH_3CO_2]^-$, $[PO_4]^{3-}$, $[SO_4]^{2-}$, $[ClO_4]^-$, $[CO_3]^{2-}$
Cr^{3+}, Cr^{6+}, Mn^{2+}, Mn^{7+}, Fe^{3+}, Co^{3+}	F^-
M^{4+} [Group IVA (14) elements]	
M^{3+} [Group IIIA (13) elements]	
N^{3+}, As^{3+}	
Cl^{3+}, Cl^{7+}, I^{5+}, I^{7+}	

Borderline Acids	Borderline Bases
Fe^{2+}, Co^{2+}, Ni^{2+}, Cu^{2+}, Zn^{2+}	C_5H_5N, N_2, N_3^-
Rh^{3+}, Ir^{3+}	$[SO_3]^{2-}$, $[NO_2]^-$
Sn^{2+}, Pb^{2+}	Br^-
Sb^{3+}, Bi^{3+}	

Soft Acids	Soft Bases
Pd^{2+}, Pt^{2+}, Pt^{4+}	H^-
M^+ [Group IB (11) elements]	CO, C_2H_4, CN^-, CNR
BH_3, GaX_3 (X = Cl, Br, I)	SCN^-, PR_3, $P(OR)_3$, AsR_3
CO, PR_3 (R = alkyl, aryl)	R_2S, RS^-
RX^+ (X = O, S, Se, Te)	I^-
Cl, Br, I	
M^0	

solvents) HF, NH_3, and H_2SO_4, and then that of a few selected organic solvents. A few important physical properties of protic solvents are compared in Table 20.6. There are many examples of nonaqueous solvents not containing hydrogen—for example, SO_2 and $POCl_3$—but these are not considered here.

Protic Solvents

Liquid Hydrogen Fluoride (HF)   The dielectric constant of HF (83.6 at 0 °C) is similar to that of water (81.1 at 18 °C), and so HF should be a good solvent for ionic substances. In concentrated solution, HF is strongly hydrogen-bonded, owing to the formation of the very stable $[HF_2]^-$ species

Table 20.6   *Some Physical Properties of HF, NH$_3$, and H$_2$SO$_4$*[a]

Property	HF	NH$_3$	H$_2$SO$_4$
Melting point (°C)	−89.4	−77.7	10.4
Boiling point (°C)	19.5	−33.4	270 (dec.)
Density (kg dm^{-3})	0.991(20 °C)	0.683(−33.4 °C)	1.83(25 °C)
Dielectric constant (ε_0)	83.6(0 °C)	22(−34 °C)	101(25 °C)
Viscosity (ρ, centipoise)	0.256(0 °C)	0.254(−33.5 °C)	24.5(25 °C)
Specific conductivity (ohm^{-1} cm^{-1})	1.4×10^{-5} (−15 °C)	1×10^{-11} (−33.4 °C)	1.04×10^{-2} (25 °C)

[a] Data from T. Moeller. *Inorganic Chemistry. A Modern Introduction*. New York: Wiley, 1982, Table 10.1, p. 620.

by autoionization:

$$3HF \rightleftharpoons [H_2F]^+ + [HF_2]^- \tag{20.28}$$

Therefore, HF is a very strong Brønsted–Lowry acid under these conditions. Hydrogen fluoride is probably the most toxic and reactive chemical that most students will ever encounter in the laboratory. A major difficulty in working with liquid HF is the problem of the container itself because HF reacts directly with all kinds of silicate and borate glasses. In fact, both liquid and gaseous HF are used industrially to etch glassware—for example, the graduations on thermometer and burette scales and the floral designs on glassware. The glass object is dipped into hot wax; when the wax has solidified, small marks are cut into the wax coating before the object is exposed to the HF medium. After the etching process is complete, the wax coating is washed off. A major use of gaseous HF is to "frost" light bulbs through direct reaction with the glass surface. The reactions occurring during the etching of a silicate glass may be approximated as

$$[SiO_4]^{4-} + 8HF \rightarrow SiF_4 + 4F^- + 4H_2O \tag{20.29}$$

$$SiF_4 + 4H_2O \rightarrow Si(OH)_4 + 4HF \tag{20.30}$$

$$Si(OH)_4 \rightarrow SiO_2 + 2H_2O \tag{20.31}$$

Nowadays, liquid HF can be easily handled in Teflon (polytetrafluoroethylene) and Kel-F (polychlorotrifluoroethylene) plastic containers. Also, unreactive metal alloys such as Monel (Ni/Cu) are used for vacuum lines when working with liquid or gaseous HF.

Many simple inorganic salts are soluble in liquid HF:

$$LiF + HF \rightarrow Li^+ + [HF_2]^- \tag{20.32}$$

$$BaF_2 + 2HF \rightarrow Ba^{2+} + 2[HF_2]^- \tag{20.33}$$

In general, solubilities *increase* on descending a specific group:

$$LiF \quad < NaF < KF < RbF < \quad CsF$$
$$(0.01 \text{ g/g HF}) \qquad\qquad\qquad (0.2 \text{ g/g HF})$$

This behavior parallels the solubilities of the Group IA(1) hydroxides in water.

Solubility *decreases* with increasing cation charge:

$$NaF \quad \gg \quad MgF_2$$
$$(0.03 \text{ g/g HF}) \quad (2.5 \times 10^{-3} \text{ g/g HF})$$

Some inorganic and organic compounds are completely miscible with liquid HF in all proportions:

$$SbF_5 + HF \rightarrow H^+ + [SbF_6]^- \tag{20.34}$$
Fluoride
Acceptor

$$ClF_3 + HF \rightarrow [ClF_2]^+ + [HF_2]^- \tag{20.35}$$
Fluoride
Donor

$$CH_3OH + HF \rightarrow [CH_3OH_2]^+ + F^- \tag{20.36}$$

$$CH_3CO_2H + HF \rightarrow [CH_3CO_2H_2]^+ + F^- \tag{20.37}$$

Note that methanol and acetic acid in Equations 20.36 and 20.37 function as Brønsted–Lowry bases and not as acids as they usually do in aqueous solution.

Other simple inorganic salts undergo complete solvolysis in liquid HF:

$$NaCN + HF \rightarrow HCN(g) + Na^+ + F^- \tag{20.38}$$

There is widespread use of liquid HF in anodic oxidation processes in the electrochemical industry for the production of fluorinated organic compounds. For instance, trifluoroacetic acid (CF_3COOH) is sometimes made by the electrolysis of CH_3COOH in liquid HF using a steel cathode and a nickel anode.

Finally, despite the high reactivity of liquid HF, certain biologically active materials such as insulin, ribonuclease, and vitamin B_{12} can be recovered from the solvent unchanged.

Liquid Ammonia (NH₃) This solvent is the most thoroughly studied of all nonaqueous solvents. Many compounds are more soluble in liquid NH_3 than in water despite the fact that its dielectric constant (22 at $-34\,°C$) is less than that of water. However, salts of highly charged cations are generally insoluble. If the salt has a large, easily polarizable anion, then the salt is usually more soluble than those salts that do not have such an anion. These observations are illustrated by the following solubility data:

$$BaCl_2 \quad \text{Insoluble}$$

$$\underset{\text{(1 g/g NH}_3)}{NH_4Cl} \quad < \quad NH_4Br \quad < \quad \underset{\text{(3.7 g/g NH}_3)}{NH_4I}$$

Organic materials are often more soluble in liquid NH_3 than in water—for example, aliphatic alcohols are miscible in all proportions, esters are very soluble, and ethers are only slightly soluble.

Ammonia undergoes autoionization in a similar manner to water:

$$NH_3 + NH_3 \rightleftharpoons [NH_4]^+ + [NH_2]^- \tag{20.39}$$

There are many examples of transition metal ammine complexes that are directly analogous to aquo-complexes—that is, the metal ions are solvated, usually giving a coordination number of six: $[Co(NH_3)_6]^{3+}$ and $[Co(H_2O)_6]^{3+}$; $[Cr(NH_3)_6]^{2+}$ and $[Cr(H_2O)_6]^{2+}$. Moreover, these ammine complexes are more thermodynamically stable than are the corresponding aquo-complexes.

Two typical reactions in liquid NH_3 are *ammonolysis* and *metathesis*:

$$SO_2Cl_2 + 4NH_3 \rightarrow SO_2(NH_2)_2 + 2NH_4Cl \tag{20.40}$$

$$NH_4Br + KNH_2 \rightarrow KBr + 2NH_3(g) \tag{20.41}$$

One of the most fascinating areas involving liquid NH_3 has been its ability to dissolve certain metals, notably the alkali metals. The resulting solutions are blue and extremely good electrical conductors. When the solutions are very concentrated, they are a bronze color and almost as good conductors as the free metals themselves. The blue color results from the tail of an extremely intense absorption band at about 1500 nm tailing into the visible region. The solutions are stable for long periods if they are maintained at low temperatures. Gradual decomposition affords hydrogen gas:

$$2Na/NH_3 \rightarrow 2NaNH_2 + H_2(g) \tag{20.42}$$

Magnetic and ESR studies have shown that the solutions contain unpaired electrons, especially when dilute. It is believed that the alkali metal atoms are ionized in liquid NH_3 solution, giving solvated electrons as well:

$$\text{Na} \xrightarrow{\text{NH}_3} \text{Na}^+(\text{NH}_3)_n + e^-(\text{NH}_3)_m \qquad (20.43)$$

(where n and m are unknown).

When the solutions, referred to as *metal electrides*, are concentrated by evaporation, the electrical conductivity increases, presumably due to the closer association of the solvated metal ions and the solvated electrons tending to make the cations more metal-like. Complete evaporation, surprisingly, affords the free metals unchanged. Recently, the first example of a crystalline metal electride Cs$^+$ (crown ether) $\cdot$ e^- has been reported. Crown ethers are large organic coordinating solvents, which we will discuss shortly.

Metal electride solutions are excellent reducing agents:

$$\text{O}_2 + [\text{Li}^+/e^-] \rightarrow \text{LiO}_2 \qquad (20.44)$$

followed by

$$\text{LiO}_2 + [\text{Li}^+/e^-] \rightarrow \text{Li}_2\text{O}_2 \qquad (20.45)$$

$$2\text{NH}_4\text{S} + 2[\text{Li}^+/e^-] \rightarrow 2\text{Li}_2\text{S} + 2\text{NH}_3(g) + \text{H}_2(g) \qquad (20.46)$$

$$\text{Fe(CO)}_5 + 2[\text{Na}^+/e^-] \rightarrow \text{Na}_2[\text{Fe(CO)}_4] + \text{CO}(g) \qquad (20.47)$$

$$\text{Ge}_2\text{H}_6 + 2[\text{Na}^+/e^-] \rightarrow 2\text{NaGeH}_3 \qquad (20.48)$$

Anhydrous Sulfuric Acid (H_2SO_4)   Sulfuric acid has a higher dielectric constant (101 at 25 °C) than water and is an excellent solvent for ionic compounds. Because it is highly corrosive and extremely viscous, it is certainly not a pleasant solvent with which to work, and there have been many serious accidents. Like HF and NH$_3$, sulfuric acid autoionizes but much more extensively, as illustrated by the larger specific conductance (Table 20.6). The hydrogen sulfate anion [HSO$_4$]$^-$ and the sulfuric acidium cation [H$_3$SO$_4$]$^+$ are produced:

$$\text{H}_2\text{SO}_4 + \text{H}_2\text{SO}_4 \rightleftharpoons [\text{H}_3\text{SO}_4]^+ + [\text{HSO}_4]^- \qquad (20.49)$$

Some of the reactions that occur in anhydrous H$_2$SO$_4$ are shown below; compounds behaving as Brønsted bases give [HSO$_4$]$^-$, whereas those acting as Brønsted acids yield [H$_3$SO$_4$]$^+$.

$$\text{H}_2\text{SO}_4 + \text{H}_2\text{O} \rightarrow [\text{H}_3\text{O}]^+ + [\text{HSO}_4]^- \qquad (20.50)$$

$$\text{H}_2\text{SO}_4 + [\text{SO}_4]^{2-} \rightarrow 2[\text{HSO}_4]^- \qquad (20.51)$$

$$3\text{H}_2\text{SO}_4 + [\text{PO}_4]^{3-} \rightarrow \text{H}_3\text{PO}_4 + 3[\text{HSO}_4]^- \qquad (20.52)$$

$$\text{H}_2\text{SO}_4 + \text{CH}_3\text{CO}_2\text{H} \rightarrow [\text{CH}_3\text{C(OH)}_2]^+ + [\text{HSO}_4]^- \qquad (20.53)$$

Perchloric acid ($HClO_4$), normally a very strong acid, behaves as a very weak acid in H_2SO_4:

$$H_2SO_4 + HClO_4 \rightarrow [H_3SO_4]^+ + [ClO_4]^- \tag{20.54}$$

The strong acid $HB(HSO_4)_4$ is produced by dissolving various boron compounds in H_2SO_4:

$$B_2O_3 + 9H_2SO_4 \rightarrow 3[H_3O]^+ + 2[B(HSO_4)_4]^- + [HSO_4]^- \tag{20.55}$$

Sulfur trioxide (SO_3) dissolves in H_2SO_4 to produce pyrosulfuric acid $H_2S_2O_7$:

$$H_2SO_4 + SO_3 \rightarrow H_2S_2O_7 \tag{20.56}$$

Sulfuric acid is another extensively hydrogen-bonded system:

$$
\begin{array}{ccc}
\overset{\displaystyle O}{\underset{\displaystyle\|}{}} & & \overset{\displaystyle O}{\underset{\displaystyle\|}{}} \\
O{=}S{-}O{-}H\cdots O{=}S{-}O{-}H \\
\underset{\displaystyle O}{|} & & \underset{\displaystyle O}{|} \\
\underset{\displaystyle H}{|} & & \underset{\displaystyle H}{|}
\end{array}
$$

Because of its high dielectric constant, ionic compounds dissolve fairly readily in H_2SO_4, often with chemical reactions taking place. For example, sulfates dissolve easily but are recovered as bisulfates as indicated above. Many typical organic compounds (for example, aldehydes, ketones, and ethers) behave as strong Brønsted bases in H_2SO_4:

$$H_2SO_4 + CH_3{-}O{-}CH_3 \rightarrow [CH_3{-}(OH){-}CH_3]^+ + [HSO_4]^- \tag{20.57}$$

Organic Solvents

Tetrahydrofuran (C_4H_8O) In organometallic chemistry, there are now numerous examples of reactions being performed in this solvent, which coordinates weakly through an oxygen lone pair:

= THF

Tetrahydrofuran (THF) can be easily dried and deoxygenated by distillation (boiling point 65.4 °C) from sodium strips in a nitrogen or argon atmosphere. Also, because it is so volatile, it can be removed under reduced pressure from reaction mixtures. It is probably the most commonly used solvent for organometallic reactions for these reasons. Sometimes, the solvent is actually involved in the formation of an active intermediate species:

$$Cr(CO)_6 + THF \xrightarrow{h\nu} Cr(CO)_5(THF) + CO \qquad \textbf{(20.58)}$$

$$Cr(CO)_5(THF) + PPh_3 \longrightarrow Cr(CO)_5(PPh_3) + THF \qquad \textbf{(20.59)}$$

Crown Ethers and Cryptands In recent years, considerable interest has been generated in macrocyclic ligands. As the name implies, these are large (macro) closed-ring ligands, sometimes with a strong resemblance to the porphyrin ligands present in biologically important systems such as hemoglobin. They have more than one coordinating center, usually O, N, S, or P atoms.

Cyclic ethers comprise one class of macrocyclic ligands that has particularly excited the imagination of chemists because cyclic ethers have the ability to complex the alkali metal cations. Some of these cyclic ethers adopt a crownlike configuration—hence the name "crown ethers." By coordinating the alkali metals, it has been possible to solubilize them in ethers. An example of a monocyclic crown ether is dibenzo-{18}-crown-6, where "crown-6" indicates six oxygen-coordination sites and the "{18}" gives the number of atoms in the macrocyclic ligand that adopts a crown configuration. The size of the hole in this crown ether is about 310 pm, and a K^+ ion, which has a diameter of about 305 pm, fits snugly in the hole. It is possible to design crown ethers to coordinate other Group IA(1) elements selectively—for example, crown-4, Li^+; crown-5, Na^+.

Crown-6   Crown-4

The extension of crown ethers into three-dimensional structures has led to the synthesis of an important class of macrocyclic ligands known as *cryptands* (*Gk.* "hidden"). The resulting complexes are called *cryptates*; metal cations, especially alkali metal ones, can be trapped inside the three-dimensional cavities of these cagelike organic systems. An example of a cryptand ligand is 2,2,2-crypt; this ligand coordinates Na^+ more strongly than does a crown-5 ligand.

Moreover, the scientific community was startled to learn that the resulting $[(2,2,2\text{-crypt})Na]^+$ cation stabilizes the formation of lustrous golden crystals containing the Na^- ion. The presence of the Na^- anion was substantiated by both X-ray diffraction and ^{23}Na-NMR spectroscopy.

Two current important uses of alkali metal crown-ether complexes and cryptates are the preparation of potassium permanganate ($KMnO_4$) and potassium superoxide (KO_2) solutions as more easily controlled reagents for specific organic oxidations.

20.3   Electrochemistry

Oxidation and Reduction

Many important metallurgical processes take place in solution, and these processes more often than not involve oxidation–reduction (redox) reactions. For example, in industrial copper plating of expensive cookware, aqueous solutions of Cu^{2+} are reduced to copper metal (Cu^0). During a redox reaction, electrons are transferred from one chemical species to another: *gain of electrons (reduction), loss of electrons (oxidation).* You are already familiar with several common redox reactions: Your pocket calculator is almost certainly powered by a nickel–cadmium battery. Also, one of the classic experiments in a junior home chemistry set involves the precipitation of copper metal from an aqueous solution of copper sulfate by the addition of powdered metallic iron filings.

This last reaction has more pertinent industrial uses; it is the basis of the industrial recovery of copper from metallic waste and is one of the vital steps in the tertiary treatment of sewage for the removal of unwanted Cu^{2+} ions.

$$CuSO_4(aq) + Fe(s) \rightarrow FeSO_4(aq) + Cu(s) \qquad (20.60)$$
$$\underset{\substack{\text{Scrap} \\ \text{Iron}}}{}$$

The reaction can be described in terms of two half-reactions:

$$\text{Fe}(s) \rightleftharpoons \text{Fe}^{2+}(aq) + 2e^- \qquad (oxidation) \quad \textbf{(20.61)}$$

$$\text{Cu}^{2+}(aq) + 2e^- \rightleftharpoons \text{Cu}(s) \qquad (reduction) \quad \textbf{(20.62)}$$

$$\text{Cu}^{2+}(aq) + \text{Fe}(s) \rightleftharpoons \text{Fe}^{2+}(aq) + \text{Cu}(s) \qquad \textbf{(20.63)}$$

The $[\text{SO}_4]^{2-}$ ions are *spectactor ions* and are not involved in the actual redox process. The thermodynamic driving force for the overall reaction is the *standard electrode* or *reduction potential* (E^0). This term must be defined, and to do this we need a reference point. Over the years, the standard hydrogen electrode has been arbitrarily chosen as the reference potential:

$$\text{H}^+(aq) + e^- \rightleftharpoons 1/2\text{H}_2(g) \qquad E^0 = 0.0 \text{ V} \qquad \textbf{(20.64)}$$

By definition, this particular system has a standard reduction potential of zero ($E^0 = 0.0$ V) for an aqueous solution of H^+ ions with unit activity ($a_{\text{H}^+} = 1.0$) in equilibrium with gaseous H_2 at 101.3 kPa (1 atm) pressure. Under the conditions specified, this potential is independent of temperature. Electrolytic cells can be designed such that the standard reduction potential of the electrode is either positive or negative with respect to the standard hydrogen electrode:

$$\text{Pd}^{2+}(aq) + 2e^- \rightleftharpoons \text{Pd}(s) \qquad E^0 = +0.987 \text{ V} \qquad \textbf{(20.65)}$$

$$\text{Rb}^+(aq) + e^- \rightleftharpoons \text{Rb}(s) \qquad E^0 = -2.925 \text{ V} \qquad \textbf{(20.66)}$$

For the sign convention used here (the so-called American convention), E^0 is positive for the $\text{Pd}^{2+}(aq)/\text{Pd}(s)$ couple because Pd^{2+} gains electrons (that is, is reduced) more easily than does $\text{H}^+(aq)$. Similarly, E^0 is negative for the $\text{Rb}^+(aq)/\text{Rb}(s)$ couple because Rb loses electrons (that is, is oxidized) more readily than does molecular hydrogen under the defined standard conditions.

One of the most important relations for electrode processes is

$$\Delta G^0 = -n\mathscr{F}E^0 \qquad \textbf{(20.67)}$$

where ΔG^0 = the Gibbs free energy for a process (ΔG^0 negative implies spontaneous reaction, and ΔG^0 positive implies no reaction under the conditions specified), n = number of moles of electrons transferred, $\mathscr{F}$ = the Faraday constant (9.65×10^4 C mol^{-1}), and E^0 = the standard reduction potential for the process.

Consider the copper–iron example again. Table 20.7 is an abbreviated form of the extensive tables of standard reduction potentials available in

Table 20.7   *Some Selected Standard Reduction Potentials and the Half-Reactions Involved in Acid Solution at 25 °C and 101 kPa (1 atm) Pressure*[a]

Half-Reaction[b]	$E°$ (V)
$F_2 + 2e^- \rightleftharpoons 2F^-$	2.866
$Ag^{2+} + e^- \rightleftharpoons Ag^+$	1.980
$H_2O_2 + 2H^+ + 2e^- \rightleftharpoons 2H_2O$	1.776
$PbO_2 + 4H^+ + [SO_4]^{2-} + 2e^- \rightleftharpoons PbSO_4 + 2H_2O$	1.691
$[MnO_4]^- + 4H^+ + 3e^- \rightleftharpoons MnO_2 + 2H_2O$	1.679
$[MnO_4]^- + 8H^+ + 5e^- \rightleftharpoons Mn^{2+} + 4H_2O$	1.507
$PbO_2 + 4H^+ + 2e^- \rightleftharpoons Pb^{2+} + 2H_2O$	1.455
$Cl_2 + 2e^- \rightleftharpoons 2Cl^-$	1.358
$[Cr_2O_7]^{2-} + 14H^+ + 6e^- \rightleftharpoons 2Cr^{3+} + 7H_2O$	1.232
$O_2 + 4H^+ + 4e^- \rightleftharpoons 2H_2O$	1.229
$MnO_2 + 4H^+ + 2e^- \rightleftharpoons Mn^{2+} + 2H_2O$	1.224
$Br_2(l) + 2e^- \rightleftharpoons 2Br^-$	1.066
$[AuCl_4]^- + 3e^- \rightleftharpoons Au + 4Cl^-$	1.002
$[NO_3]^- + 4H^+ + 3e^- \rightleftharpoons NO + 2H_2O$	0.957
$2Hg^{2+} + 2e^- \rightleftharpoons Hg_2^{2+}$	0.920
$Ag^+ + e^- \rightleftharpoons Ag$	0.800
$Hg_2^{2+} + 2e^- \rightleftharpoons 2Hg$	0.800
$Fe^{3+} + e^- \rightleftharpoons Fe^{2+}$	0.771
$O_2 + 2H^+ + 2e^- \rightleftharpoons H_2O_2$	0.695
$[MnO_4]^- + e^- \rightleftharpoons [MnO_4]^{2-}$	0.558
$I_2 + 2e^- \rightleftharpoons 2I^-$	0.536
$Cu^+ + e^- \rightleftharpoons Cu$	0.521
$Cu^{2+} + 2e^- \rightleftharpoons Cu$	0.342
$Hg_2Cl_2 + 2e^- \rightleftharpoons 2Hg + 2Cl^-$	0.268
$AgCl + e^- \rightleftharpoons Ag + Cl^-$	0.222
$[SO_4]^{2-} + 4H^+ + 2e^- \rightleftharpoons H_2SO_3 + H_2O$	0.172
$Cu^{2+} + e^- \rightleftharpoons Cu^+$	0.153
$2H^+ + 2e^- \rightleftharpoons H_2$	0.000
$Pb^{2+} + 2e^- \rightleftharpoons Pb$	-0.126
$Sn^{2+} + 2e^- \rightleftharpoons Sn$	-0.138
$Ni^{2+} + 2e^- \rightleftharpoons Ni$	-0.257
$PbSO_4 + 2e^- \rightleftharpoons Pb + [SO_4]^{2-}$	-0.359
$Cd^{2+} + 2e^- \rightleftharpoons Cd$	-0.403
$Cr^{3+} + e^- \rightleftharpoons Cr^{2+}$	-0.407
$Fe^{2+} + 2e^- \rightleftharpoons Fe$	-0.447
$Ga^{3+} + 3e^- \rightleftharpoons Ga$	-0.560
$U^{4+} + e^- \rightleftharpoons U^{3+}$	-0.608

Table 20.7 (Continued)	
Half-Reaction[b]	$E°$ (V)
$Cr^{3+} + 3e^- \rightleftharpoons Cr$	-0.744
$Zn^{2+} + 2e^- \rightleftharpoons Zn$	-0.762
$Cr^{2+} + 2e^- \rightarrow Cr$	-0.913
$Mn^{2+} + 2e^- \rightleftharpoons Mn$	-1.185
$Al^{3+} + 3e^- \rightleftharpoons Al$	-1.660
$H_2 + 2e^- \rightleftharpoons 2H^-$	-2.230
$Mg^{2+} + 2e^- \rightleftharpoons Mg$	-2.372
$La^{3+} + 3e^- \rightleftharpoons La$	-2.522
$Na^+ + e^- \rightleftharpoons Na$	-2.710
$Ca^{2+} + 2e^- \rightleftharpoons Ca$	-2.868
$Ba^{2+} + 2e^- \rightleftharpoons Ba$	-2.912
$K^+ + e^- \rightleftharpoons K$	-2.931
$Li^+ + e^- \rightleftharpoons Li$	-3.040

[a] Data from *Handbook of Chemistry and Physics*, 66th ed. Boca Raton, Fla.: CRC Press, 1985.
[b] These half-reactions refer (unless otherwise specified) to aqueous solution, e.g., $F_2(g) + 2e^- \rightarrow 2F^-(aq)$.

the literature. From this table, we have

$$Fe^{2+}(aq) + 2e^- \rightleftharpoons Fe(s) \qquad E^0 = -0.440 \text{ V} \qquad \textbf{(20.68)}$$

$$Cu^{2+}(aq) + 2e^- \rightleftharpoons Cu(s) \qquad E^0 = +0.337 \text{ V} \qquad \textbf{(20.69)}$$

If we reverse the equation for the $Fe^{2+}(aq)/Fe(s)$ couple, we obtain the original half-reactions and the overall reaction

$$Fe(s) \rightleftharpoons Fe^{2+}(aq) + 2e^- \qquad E^0 = +0.440 \text{ V} \textbf{(20.70)}$$

$$Cu^{2+}(aq) + 2e^- \rightleftharpoons Cu(s) \qquad E^0 = +0.337 \text{ V} \textbf{(20.71)}$$

$$Cu^{2+}(aq) + Fe(s) \rightleftharpoons Fe^{2+}(aq) + Cu(s) \qquad E^0 = +0.777 \text{ V} \textbf{(20.72)}$$

Because E^0 for the reaction is positive, it follows immediately from Equation 20.67 that ΔG^0 is less than zero, and so the reaction proceeds spontaneously as written:

$$\Delta G^0 = -(2 \text{ mol } e^-)\mathscr{F}(+0.777 \text{ V}) = \text{A Negative Number} \qquad \textbf{(20.73)}$$

Obviously, standard reduction potentials can be extremely useful in predicting chemical reactivity. But note that E^0 values are *intensive properties*—that is, they are independent of the stoichiometry of the reaction.

If you double the overall reaction above to get

$$2Cu^{2+}(aq) + 2Fe(s) \rightleftharpoons 2Fe^{2+}(aq) + 2Cu(s) \qquad \textbf{(20.74)}$$

then $E^0 = +0.777$ V and not 2×0.777 V $= 1.554$ V.

Let us do two more examples. First, suppose that you were interested in knowing whether Cu(I) is thermodynamically stable with respect to *disproportionation* into Cu(II) and Cu(0) in aqueous solution; that is, does the following reaction occur?

$$2Cu^{+}(aq) \rightarrow Cu^{2+}(aq) + Cu(s) \qquad \textbf{(20.75)}$$

This overall equation can be derived from the summation of the two following half-reactions:

$$Cu^{+}(aq) \rightleftharpoons Cu^{2+}(aq) + e^{-} \qquad E^0 = -0.153 \text{ V} \qquad \textbf{(20.76)}$$

$$Cu^{+}(aq) + e^{-} \rightleftharpoons Cu(s) \qquad E^0 = +0.521 \text{ V} \qquad \textbf{(20.77)}$$

$$2Cu^{+}(aq) \rightleftharpoons Cu^{2+}(aq) + Cu(s) \quad E^0 = +0.368 \text{ V} \qquad \textbf{(20.78)}$$

Again, substitution of the E^0 value ($+0.368$ V) for the proposed reaction into Equation 20.67 indicates that ΔG^0 is negative, and so Cu(I) should indeed disproportionate quite easily into Cu(II) and Cu(0) at 25 °C. However, at higher temperature, the equilibrium shifts to the left and 30% of the dissolved copper is Cu(I) at 160°C in 0.85M H_2SO_4. There is a similar disproportionation of Ag(I) into Ag(II) and Ag(0).

The second example is the lead-acid storage battery used so extensively throughout the world by the automobile industry. The two important half-reactions are

Oxidation at Anode
$$Pb(s) + [SO_4]^{2-}(aq) \rightleftharpoons PbSO_4(s) + 2e^{-}$$
$$E^0 = +0.36 \text{ V} \quad \textbf{(20.79)}$$

Reduction at Cathode
$$PbO_2(s) + 4H^{+}(aq) + [SO_4]^{2-}(aq) + 2e^{-} \rightleftharpoons PbSO_4(s) + 2H_2O$$
$$E^0 = +1.69 \text{ V} \quad \textbf{(20.80)}$$

$$Pb(s) + PbO_2(s) + [2SO_4]^{2-}(aq) + 4H^{+}(aq) \rightleftharpoons 2PbSO_4(s) + 2H_2O$$
$$E^0 = +2.05 \text{ V} \quad \textbf{(20.81)}$$

The cell potential is positive, and so the reaction is spontaneous; the discharge reaction proceeds from left to right as written. Six of these cells are joined in series to give the conventional 12-V car battery. Note that the reduction potential for the reverse reaction is negative (-2.05 V), and so it is not spontaneous. This is why your battery must be recharged from an external source such as a battery charger when it has run down.

Standard reduction potentials strictly imply unit activities for the species concerned, but most people usually prefer to work with concentrations. In this case, the cell potential varies with changes in concentration according to Equation 20.82 first proposed by Nernst in 1881 and known universally as the *Nernst equation*:

$$E = E^0 - \left(\frac{RT\ln Q}{n}\right) \tag{20.82}$$

$$= E^0 - \left(\frac{2.303RT\log_{10}Q}{n}\right) \tag{20.83}$$

where E = the actual cell potential, E^0 = the standard reduction potential, n = the number of moles of e^- transferred, and Q = the ratio of the concentrations of the products to reactants, each raised to its proper stoichiometric coefficient. For a typical reaction such as

$$aA + bB \rightleftharpoons cC + dD \tag{20.84}$$

Q is the equilibrium constant K_{eq}

$$Q = \frac{[A]^a[B]^b}{[C]^c[D]^d} = K_{eq} \tag{20.85}$$

For room temperature (298 K) conditions, Equation 20.83 reduces to

$$E = E^0 - \left(\frac{0.05916}{n}\log_{10}Q\right) \tag{20.86}$$

Notice that an increase in the concentration of the products leads to a decrease in the cell potential.

One important use of the Nernst equation in real-life situations is in calculating solubility products (K_{sp}). These are notoriously difficult values to obtain accurately for salts that are only slightly soluble in water. For example, AgCl is of crucial importance to the photographic industry and its K_{sp} value is vital in evaluating the developing process. How can we use Equation 20.83 to calculate $K_{sp}(AgCl)$ at room temperature? For a saturated solution

$$AgCl(s) \rightleftharpoons Ag^+ + Cl^- \qquad E^0 = ? \tag{20.87}$$

And from Table 20.7, we have

$$Ag^+(aq) + e^- \rightleftharpoons Ag(s) \qquad E^0 = +0.7991 \text{ V} \tag{20.88}$$

$$AgCl(s) + e^- \rightleftharpoons Ag(s) + Cl^-(aq) \qquad E^0 = +0.2222 \text{ V} \tag{20.89}$$

Reversing the first half-reaction and adding it to the second, we find that

$$AgCl(s) \rightleftharpoons Ag^+(aq) + Cl^-(aq) \qquad E^0 = -0.5769 \text{ V} \qquad \textbf{(20.90)}$$

From the Nernst equation (Equation 20.83), we obtain

$$E = -0.5769 - \left(\frac{0.05916}{1} \log_{10} \frac{[Ag^+][Cl^-]}{[AgCl(s)]}\right) \qquad \textbf{(20.91)}$$

$$= -0.5769 - 0.05916 \log_{10}[Ag^+][Cl^-] \qquad \textbf{(20.92)}$$

because $n = 1$ and $[AgCl(s)] = 1$ for our purposes.

At equilibrium, $\Delta G = 0 = n\mathscr{F}E$. So, $E = 0$ and

$$\log_{10}[Ag^+][Cl^-] = -\left(\frac{0.05769}{0.05916}\right) = -9.75 \qquad \textbf{(20.93)}$$

Finally, we obtain $K_{sp}(AgCl) = [Ag^+][Cl^-] = 1.78 \times 10^{-10}$.

As an example of the dramatic effect of concentration on the reduction potential, consider the following half-reaction that is important to the ecological balance in rivers and streams.

$$O_2(g) + 4H^+(aq) + 4e^- \rightleftharpoons 2H_2O(l) \qquad E^0 = +1.23 \text{ V} \qquad \textbf{(20.94)}$$

Relatively pure rainwater has a pH of 5.6, whereas acid rain resulting from SO_2 industrial emissions that eventually finish up as H_2SO_4 aerosols in the upper atmosphere can have a pH of 2! What are the reduction potentials under these two different pH conditions if $[O_2(g)] = 1.50 \times 10^{-3} \text{ mol dm}^{-3}$ at room temperature?

$$E = E^0 - \left(\frac{0.05916}{n} \log_{10} \frac{[H_2O(l)]^2}{[O_2(g)][H^+(aq)]^4}\right) \qquad \textbf{(20.95)}$$

but $n = 4$ and $[H_2O(l)] = 1$. So,

$$E = 1.23 + \left(\frac{0.05916}{4} \log_{10} [O_2(g)][H^+(aq)]^4\right) \text{ V} \qquad \textbf{(20.96)}$$

$$= 1.23 + 0.01479\log_{10}[1.50 \times 10^{-3}][H^+(aq)]^4 \text{ V} \qquad \textbf{(20.97)}$$

Also,

$$pH = -\log_{10}[H^+(aq)] = 5.6 \qquad \textbf{(20.98)}$$

Therefore, $E = +0.90$ V. Similarly, for acid-rain conditions (pH = 2), $E = +1.07$ V. This value is sufficiently different from the normal one for the redox chemistry of the waterways concerned to be sharply affected. The deleterious effects of acid rain are already being felt worldwide. More will be said about this topic later.

Metal cations are often complexed by ligands present in aqueous solution. Such complexations also affect the reduction potential of a system.

Consider a silver electrode in an aqueous silver cyanide (AgCN) solution. This is the typical cell used for electroplating tableware. The following equilibrium is established:

$$AgCN(s) \overset{H_2O}{\rightleftharpoons} Ag^+(aq) + CN^-(aq) \tag{20.99}$$

$$\Big\Updownarrow {\scriptstyle 2CN^-}$$

$$[Ag(CN)_2]^-$$

The equilibria lie to the right, and so virtually all Ag present is complexed as the dicyanoargentate(I) species. This means that there is a large concentration of the complex ion; thus, E is much lower than the standard reduction potential.

$$Ag^+(aq) + e^- \rightleftharpoons Ag(s) \qquad E^0 = +0.7991 \text{ V} \tag{20.100}$$

But the activity of Ag(s) is unity, and $n = 1$; therefore, at room temperature, we have

$$E = 0.7991 + (0.05916\log_{10}[Ag^+]) \text{ V} \tag{20.101}$$

Remember $\log_{10}[Ag^+]$ is negative because $[Ag^+]$ is very small.

Reduction potentials play an important part in explaining some perplexing coordination chemistry. For example, why do most known cobalt complexes contain Co(III) despite the fact that Co(III) should, in principle, be easily reduced to Co(II) because of the favorable reduction potential?

$$Co^{3+}(aq) + e^- \rightleftharpoons Co^{2+}(aq) \qquad E^0 = +1.808 \text{ V} \tag{20.102}$$

In the presence of strongly complexing ligands such as CN^-, Co(II) is oxidized even by water to yield the stable $[Co(CN)_6]^{3-}$ complex. Again, from the Nernst equation, we find that at room temperature

$$E = +1.808 - \left(0.05916\log_{10}\frac{[Co^{2+}]}{[Co^{3+}]}\right) \text{ V} \tag{20.103}$$

The concentrations of both Co^{2+} and Co^{3+} are drastically reduced by complex formation, but more for Co^{3+} than for Co^{2+}. Therefore, E is much less than E^0, making reduction of Co(III) to Co(II) unfavorable.

Metallurgical Applications

Electrochemical processes play a major role in extracting several industrially important metals. All metals exist in their minerals in one or more oxidized forms, and reduction to the metal is the basis of their extraction. The values of the standard reduction potentials for the most common oxidation states of metals can be taken as a rough guide of the comparative ease with which the reductions can be performed. For those metals with high negative E^0 values [for example, K(I), -2.92 V; Ca(II), -2.87 V; Mg(II), -2.36 V; Al(III), -1.66 V], electrolytic reduction is necessary. For other metals with less negative or positive E^0 values, chemical reducing agents are employed, typically carbon, hydrogen, or aluminum. Those metals that have fairly high positive E^0 values—for example, the coinage metals, copper, silver, and gold—are often found free in nature, especially gold [E^0Au(I) $= +1.69$ V].

To illustrate the practical applications of electrolytic reduction, the extractions of aluminum and copper are discussed.

Aluminum Extraction The method used to extract Al is referred to as the Hall–Héroult procedure and has been known since the late 1860s. Initially, the Bayer process is used to purify the alumina (Al_2O_3) necessary for the Hall–Héroult electrolytic reduction step. Aluminum is the most abundant metal found in the earth's crust (about 8%, 30–40 km depth). One of the chief minerals of Al is bauxite, a mixture of about 50–70% Al_2O_3, Fe_2O_3, and SiO_2. Bauxite occurs widely in the tropics and subtropics. In the Bayer process, bauxite is treated with a strong caustic soda solution (30% NaOH) under about 800 kPa (8 atm) pressure at about 190 °C.

$$\text{Bauxite} \longrightarrow [AlO_2]^-(aq) + [SiO_3]_n^{2n-}(aq) + Fe_2O_3 \qquad \textbf{(20.104)}$$

$$\Big\downarrow CO_2$$

$$Al_2O_3 \cdot 3H_2O(s)$$

Aluminum and many of its compounds are *amphoteric*—that is, they dissolve in acids to form Al^{3+} salts and in alkalis to form aluminates. Aluminates formally contain the $[AlO_2]^-$ ion, but little is known of the detailed structures of such species. The SiO_2 present (sand) is converted into soluble polymeric silicates $[SiO_3]_n^{2n-}$. Examples of these ions are commonly found in various silicate minerals (see page 475): benitoite $BaTlSi_3O_9$, which contains the cyclic $[Si_3O_9]^{6-}$ ion, and in pyroxenes such as enstatite ($MgSiO_3$) with the polymeric $[SiO_3]_n^{2n-}$ ion. The insoluble Fe_2O_3 is allowed to settle to the bottom of the reaction vessel, and then CO_2 gas is passed into the solution. The hydrated alumina ($Al_2O_3 \cdot 3H_2O$) crystallizes out on addition of seed crystals of the pure compound. Before this hydrate can be used in the Hall–Héroult electrolytic cell, however, it must be dehydrated by heating it to about 1200 °C. The resulting pure alumina is mixed with sufficient cryolite (Na_3AlF_6, another naturally occurring aluminum mineral) in the electrolytic cell so that the melting point of the nonaqueous electrolyte is about 970 °C, a convenient temperature in terms of energy consumption and reaction efficiency.

Because the aluminum industry uses so much electrical energy, it is necessary to have a relatively cheap energy source nearby, usually hy-

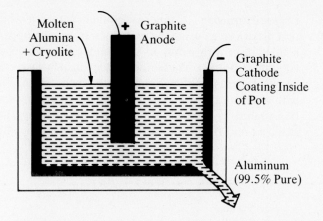

Figure 20.10
Schematic representation of a typical electrolytic pot used in the production of aluminum metal. Typical operating conditions: 970 °C, 5.0 V, 100,000 A.

droelectric power. One of the largest aluminum refineries in the world is located in Arvida, Quebec, Canada, and there are literally hundreds of electrolytic cells (or pots as they are called) linked in series. The choice of Arvida for the plant was based in part on the presence of abundant hydroelectric power obtained by harnessing the Saguenay River and in part because of an easily accessible deep-water port (Port Arthur) suitable for the delivery of bauxite and the subsequent shipping of the pure aluminum.

Figure 20.10 shows a schematic drawing of a typical pot used in aluminum manufacture. One of the factors contributing to the cost of Al production is the cost of the graphite electrodes, which are gradually combusted during the electrolysis of the molten nonaqueous electrolyte and must be replaced at regular intervals. The molten aluminum is drained from the pots and cast into ingots, rolled into sheets, or extruded into wire, depending on its ultimate fabrication use. The aircraft industry is one of the major users of aluminum because of the light weight and strength of the metal. Although Al metal should be easily oxidized $[E^0(\text{Al}) = +1.66 \text{ V}]$, it does not corrode very easily because when freshly prepared its surface quickly becomes coated with a thin protective film of Al_2O_3.

The important reactions occurring during the actual electrolysis are

Oxidation at the Anode **(20.105)**
$$3O^{2-}(l) \rightarrow {}^3\!/_2 O_2(g) + 6e^-$$

Reduction at the Cathode **(20.106)**
$$2Al^{3+}(l) + 6e^- \rightarrow 2Al(l)$$

$${}^3\!/_2 O_2(g) + {}^3\!/_2 C(s) \rightarrow {}^3\!/_2 CO_2(g) \quad \textbf{(20.107)}$$

The last reaction is quite exothermic and provides sufficient heat to bring about some thermal reduction of alumina as well as contributing to the depletion of the graphite anodes.

$$Al_2O_3(s) + {}^3\!/_2 C(s) \rightarrow 2Al(l) + {}^3\!/_2 CO_2(g) \quad \textbf{(20.108)}$$

Copper Extraction   Copper is another metal for which electrolytic reduction plays a role during the extraction process. Copper occurs principally in combination with sulfur in such minerals as chalcocite (Cu_2S) and chalcopyrite ($CuFeS_2$). These minerals are usually low-grade quality, and so the copper fraction must be concentrated by flotation. The minerals are mixed with water and a surfactant in a special apparatus in which the mixture can be agitated by a stream of air. The copper mineral particles are partially oxidized and rise to the surface in the froth concentrate:

$$2Cu_2S + 3O_2 \rightarrow 2Cu_2O + 2SO_2 \tag{20.109}$$

$$2FeS + 3O_2 \rightarrow 2FeO + 2SO_2 \tag{20.110}$$

Smelting Step

$$Cu_2S + 2Cu_2O \rightarrow 6Cu + SO_2 \tag{20.111}$$

The copper metal formed by smelting the ore concentrate is termed *blister copper*, and this is refined electrolytically to produce the purity of copper necessary (99.5%) for the electrical industry where it is used in enormous quantities. The blister copper acts as the anode of the electrolytic cell, whereas a pure copper rod is used as the cathode. The electrolyte is an aqueous solution of $CuSO_4/H_2SO_4$. The impurities in the blister copper either go into solution or drop to the bottom of the cell. The copper from the anode enters the solution as Cu^{2+} ions, where they are discharged at the cathode as pure copper:

Oxidation at the Anode
$$Cu(s) \rightarrow Cu^{2+}(aq) + 2e^- \tag{20.112}$$
Impure

Reduction at the Cathode
$$Cu^{2+}(aq) + 2e^- \rightarrow Cu(s) \tag{20.113}$$
Pure

The residue at the bottom of the cell often contains recoverable amounts of the other coinage metals (Ag, Au) and the platinum-group metals (Pd, Pt).

Chlorine Manufacture

Chlorine (Cl_2) is the most industrially important member of the halogens. It occurs mainly as Cl^- in seawater (about 55% of all dissolved species) and as NaCl in dried-up salt lakes and massive underground deposits from the evaporation of fossil seas. Chlorine gas is obtained commercially by electrolysis of brine (NaCl) solutions (chloralkali process). One main type of cell used is the diaphragm cell, which is equipped with an asbestos di-aphragm to keep apart the reactive gases (Cl_2 and H_2) produced during the electrolysis. The electrodes are graphite (anode) and iron (cathode), and the cells are operated typically at 3.8 V, 50,000 A, and 100 °C. The impor-tant electrolytic reactions are

Oxidation at the Anode

$$Cl^-(aq) \rightarrow \tfrac{1}{2}Cl_2(g) + e^-$$ **(20.114)**

Reduction at the Cathode

$$H^+(aq) + e^- \rightarrow \tfrac{1}{2}H_2(g)$$ **(20.115)**

During the electrolysis, H^+ ions from H_2O are being consumed, thus leading to a build-up in concentration of OH^- and formation of NaOH.

There is another electrolytic cell, which uses large quantities of mer-cury as an electrode, still being used in Cl_2 production. However, owing to pollution problems associated with mercury, this type of cell has fallen into disfavor and is being phased out.

One major use of Cl_2 is in the production of numerous chlorine-containing compounds, especially chlorinated solvents such as carbon tet-rachloride (CCl_4) and chloroform ($CHCl_3$) for use in organic synthesis, by the dry-cleaning industry, and for industrial degreasing. Other important derivatives include the perchlorates ($MClO_4$) and chlorine dioxide (ClO_2), which find considerable use in the pulp and paper industry. Many North American pulp mills are equipped with on-site electrolytic cells for chlorate $[ClO_3]^-$ production, which is then used to manufacture ClO_2:

$$2NaClO_3 + SO_2 + H_2SO_4 \rightarrow 2ClO_2 + 2NaHSO_4$$ **(20.116)**

Photogalvanic Energy Production

There are many electrical devices that are powered by photochemically induced electrochemical (photogalvanic) processes. Much of the interest in this area resulted from the U.S. space program and the need for solar-powered cells for use on communications satellites and space vehicles such as the Space Shuttle. Silicon solar cells are the most important, and the operation of these is based on the coupling of *n*- and *p*-type semiconduc-tors. Basically, solar energy falls on the cell causing the photoejection of an electron from the *n*-type semiconductor (for example, Si/As), which then moves across the *pn* junction into the *p*-type semiconductor (for example, Si/In) where it occupies a previously vacant positive hole. This flow of electrons produces an electrical current that can be used to power vari-ous components, for instance, the life-support system in a spacecraft. Fig-ure 20.11 illustrates the process schematically.

Figure 20.11
Basic design of a
solar cell.

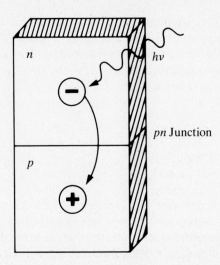

Although solar cells of this type can convert light into electricity, the intermittent character of the light source (at least at the surface of the earth) presents a problem of storing the excess energy produced during the day, for use during the night. One approach to solving this dilemma is to couple the photoelectric conversion into a fuel-producing electrochemical process. A promising direction, which is attracting much current attention, is to split water into $H_2(g)$ and $O_2(g)$. Chlorophyll *a* in plants performs the water-splitting process in nature, and this fundamental reaction is the very basis of life on our planet. Solar energy (in the laboratory, visible light from powerful lamps or lasers) is allowed to fall onto an *n*-type semiconducting photoanode such as TiO_2 and $SrTiO_3$. Electron promotion occurs from the valence band into the conduction band of the semiconductor where electrons then become available to perform some electrochemical process in solution.

Considerable research effort has recently gone into studies of the photochemical decomposition of water, whereby $H_2(g)$ is liberated at the cathode and $O_2(g)$ at the anode. The available systems are far from being economical, in large part due to their low efficiency and the highly corrosive electrochemical environment. Even when these problems are overcome, hydrogen-gas storage is difficult. Suggestions have been put forward to store $H_2(g)$ in the form of interstitial hydrides (for example, TiH_2), as discussed earlier.

$$H_2O(aq) \rightarrow H^+(aq) + OH^-(aq) \tag{20.117}$$

Oxidation at the Anode
$$OH^-(aq) \rightarrow \tfrac{1}{2}O_2(g) + 2e^- + H^+(aq) \tag{20.118}$$

Reduction at the Cathode
$$2H^+(aq) + 2e^- \rightarrow H_2(g) \tag{20.119}$$

Transition metal complexes that undergo photoredox reactions—for example, $[Ru(bipy)_3]^{2+}$ (bipy = bipyridyl)—are also being actively studied in the search for alternative energy sources.

Photographic Silver Recovery

Silver salts are consumed in large quantities by the film industry, especially in the film used by radiography departments in hospitals throughout the world. The value of silver has fluctuated greatly over the past few years (as has the value of gold), but the usual price levels make it worthwhile to reclaim the residual silver from the compounds in discarded photographic and X-ray films, and in exhausted developing solutions.

The silver salts in these films are usually silver halides (AgCl, AgBr) because of their photosensitivity. Thin plastic films are coated with gelatin containing the silver salts. On exposure to light or X rays, the salts in the areas affected are more sensitive to reduction to silver metal in the gelatin emulsions by the developer (hydroquinone). The unsensitized salts can be leached out by complexation (fixation) with sodium thiosulfate (hypo, $Na_2S_2O_3$) to give the silver complex $Na_3[Ag(S_2O_3)_2]$. The silver metal in the emulsion in the exposed part of the film remains there, and the result is the negatives that we are all accustomed to seeing. The silver compounds, both in the silver thiosulfate solutions and in the thousands of old negatives and X-ray films, are reclaimed in two steps: (a) burning the film and (b) reducing the Ag^+ ions left in the ash either by heating with cheap electropositive metals such as steel wool or electrochemically using a pure silver metal cathode. These are both inexpensive reclamation procedures, and very little equipment is necessary.

$$Fe(s) + 2Ag^+ \rightarrow Fe^{2+} + 2Ag(s) \qquad \textbf{(20.120)}$$

$$Ag^+(s) + e^- \rightarrow Ag(s) \qquad \textbf{(20.121)}$$

20.4   Inorganic Chemistry and the Environment

Because inorganic chemistry covers every element except carbon and ninety of these occur distributed in the earth's crust, oceans, and atmosphere, it is an understatement to say that inorganic chemistry plays an essential role in our environment. The subject encompasses so many diverse fields that it is only possible to present a cursory view of how inorganic chemistry is important in such diverse areas as manganese nodules in ocean seabeds around the world and acid rain in the atmosphere. It is hoped that you will become a little more aware just why our environment *must* be protected for the future of our planet. The foolish mistakes we make today may be with us forever.

Pyrometallurgy/Acid Rain

Many of the extraction processes for metals involve roasting sulfide minerals such as millerite (NiS), molybdenite (MoS_2), and pentlandite (Co_2S_3) in air to produce metal oxides, which are fairly easy to reduce to the metals. Unfortunately, one of the major by-products of these *pyrometallurgic processes* (Latin *pyro*, "heat") is SO_2. This volatile gas is carried into the atmosphere, and so begins the first step in our problems with air pollution and acid rain in particular.

Sulfur dioxide is quite toxic, and there are numerous accounts throughout recorded history when burning coal has produced sufficient SO_2 in the atmosphere to cause death, especially in elderly persons who often have respiratory problems. About 700 years ago in England, King Edward I issued a proclamation to the effect that if you were caught burning coal in the city of London you were liable to have your head cut off! Condensation of water vapor on the smoke and other pollution particles leads to the formation of smog (a contraction of the words *smoke* and *fog*). The infamous London smogs in England were principally the result of coal-burning fires, leading to much SO_2 and smoke in the atmosphere. It was dreadful smogs, such as that in December 1952 when 4000 persons died, that finally led the British Government to pass the Clean Air Act and to make London a smokeless zone. It should be emphasized, however, that not all the blame can be attributed to coal because automobiles burning high sulfur–content fuels also contribute to SO_2 production. Volcanoes are other sources of atmospheric SO_2 through the reduction of molten-metal sulfates in the lava.

During the extraction of lead from galena (PbS), the following process is performed:

$$2PbS + 3O_2 \rightarrow 2PbO + 2SO_2 \tag{20.122}$$

What happens to the SO_2 after it goes up the smokestack into the atmosphere? Reactions in the upper atmosphere are complicated by the presence of sunlight, but it seems fairly clear that SO_2 is either converted into SO_3 and ultimately H_2SO_4, or it forms sulfates [$(NH_4)_2SO_4$ and NH_4HSO_4] with the ammonia in the atmosphere, produced by the decay of organic matter. These materials become encapsulated in water to produce aerosols that eventually return to the earth's surface in the form of acid rain. One important reaction of SO_2 in the upper atmosphere is with ozone O_3 to produce SO_3. Some typical upper atmosphere reactions are

$$2SO_2 + 2H_2O + O_2 \rightarrow 2H_2SO_4 \tag{20.123}$$

$$(NH_4)_2SO_4 + 2H_2O \rightarrow 2NH_3 + [2H_3O]^+ + [SO_4]^{2-} \tag{20.124}$$

$$SO_2 + O_3 \rightarrow SO_3 + O_2 \tag{20.125}$$

Often the effects of acid rain are felt some distance from their original source because atmospheric winds cause the rain clouds to drift quite long distances. For instance, acid rain from the Ruhr valley, the industrial heartland of West Germany, is believed to be killing large areas of the Black Forest. The death of the forests is not due directly, however, to the acidity of the rain but to the leaching of metals from the soil and rocks. Even normally harmless aluminum may be a major culprit in the killing of evergreens. It is possible that certain British industries may be the cause of fish dying in lakes in Scandinavia, several hundreds of miles away. There is currently considerable debate between the U.S. and Canadian governments about the effects and causes of acid rain in the two countries.

Many of the world's famous monuments and statues are made from marble ($CaCO_3$). Acid rain reacts with the marble to yield more soluble $CaSO_4$, which is gradually washed away, leaving the priceless monuments and statues badly pitted. Because of industrial development and relatively poor pollution controls, our heritage of many centuries is in danger of being destroyed within a couple of generations.

How do we control the amount of SO_2 in the atmosphere? Many countries have already introduced rigorous air-pollution controls, and the petroleum industry is endeavoring to reduce the amount of sulfur present in crude oil before conversion into gasoline. Sulfur dioxide is being removed from stack gases by passage through fluidized beds of $CaCO_3$ and basic slag. These well-intentioned efforts have to produce results soon, hopefully before it is too late for our environment.

A new problem in this connection has recently surfaced. Nuclear power plants, especially in the United States, are under serious political attack because of the potential dangers from a failure of safeguards presently in place—witness the repercussions from the Three-Mile Island and Chernobyl incidents. One obvious way to reduce the amount of energy being supplied by nuclear power in the United States is to use alternative energy sources, but the cheapest currently available is strip-mined coal. However, much of the coal in the United States contains relatively large amounts of sulfur, that is, the precursor to the unwanted atmospheric pollutant SO_2! The way out of the this dilemma has not been realized, although major efforts are being made to provide supplementary alternate energy sources such as hydroelectric power (in Canada especially), wind power, solar energy, tidal power (already realized on the River Rance near St. Malo in France), and even wood as used by our forefathers. Of course, these alternate energy sources also have their own attendant problems. For instance, woodburning stoves and fireplaces sound particularly romantic, but there are numerous carcinogens produced that are released into the atmosphere from the chimneys when wood is burned. It is your generation that ultimately will make the decisions about the earth's future, and you should be prepared to make these decisions with a full knowledge of the consequences.

As a final comment, there are some agricultural *benefits* of acid rain. Sulfur is an essential element in plant growth, and it has been found that it is not necessary to add sulfur fertilizers in those industrial areas where the acid rain fall is quite high.

Hydrometallurgy

Apart from the atmospheric pollution problems associated with metal extraction by pyrometallurgical processes, there is also the question of energy costs. Heat must be provided somehow, and this is costly. Consequently, industry is continually seeking new refinement methods, and one of these that has been used for gold and copper extraction for some time is *hydrometallurgy*. In this approach, the extraction is performed in aqueous solution, which greatly reduces the pollution problems.

Traditionally, gold in low-grade ores, such as those remaining in the Comstock Lode in Nevada following the aftermath of the initial 1870s gold rush, is extracted by treatment of the crushed ore with potassium cyanide solution. The complexed gold is precipitated by addition of powdered zinc metal to the aqueous solution of the dicyanoaurate(I) complex:

$$4Au + 8KCN + O_2 + 2H_2O \rightarrow 4K[Au(CN)_2] + 4KOH \quad \textbf{(20.126)}$$

$$2K[Au(CN)_2] + Zn \rightarrow K_2[Zn(CN)_4] + 2Au \quad \textbf{(20.127)}$$

Of course, disposal of the spent cyanide solutions containing the various cyano-compounds presents its own set of environmental hazards.

Copper is sometimes extracted hydrometallurgically from aqueous solutions of Cu^{2+} ions using chelating agents, and the pure metal is obtained by reduction of the complexed Cu^{2+} ions by H_2 gas.

Natural Leaching of Exposed Toxic Substances

Rain is an extremely important part of any environmental cycle. Unless the water that evaporated from the surface of the earth was continually being replenished, the earth as we now know it would not remain the same for very long. As important as water is to us, however, it still causes problems: It is such an excellent solvent that it gradually leaches out all kinds of unwanted materials from the earth's crust. Many of these materials are toxic to humans.

One very common leaching problem arises from disused mines. As water penetrates the depths of these mines, sulfur is gradually leached out of the mine tailings (unwanted ores) and other minerals still there. Ultimately, there is a dangerous build-up in concentration of sulfuric acid in the mine water. The dilute acid solution in turn starts to dissolve out other unwanted species because H_2SO_4 is a much better solvent for many naturally occurring minerals than is water. A typical reaction that produces H_2SO_4 in mine waters is shown in Equation 20.128:

$$4FeS_2 + 14H_2O + 15O_2 \rightarrow 4Fe(OH)_3 + 8H_2SO_4 \qquad \textbf{(20.128)}$$

Iron
Pyrites

Although this reaction can have harmful effects, it is also probably the reaction that produced the massive hematite (Fe_2O_3) ore deposits in northern Minnesota and at the Quebec/Labrador border.

Other dangerous materials can be brought into play in the ecosphere by the leaching action of water. Serious problems are just beginning to surface because of the gradual leaching taking place in refuse dumps. Our society has so many unwanted things that we need such dumps. Some of them are fairly old, and their precise locations have been forgotten. Over the years, all kinds of noxious materials—for example, dioxins in the Love Canal—have found their way into the water table. Among the unwanted elements are arsenic, cadmium, mercury, and lead. Once these chemicals get into the water system in abnormal concentrations, they not only affect aquatic and wildlife but often also people, inevitably with deleterious effects.

Arsenic is fairly widespread in nature as minerals such as As_2S_3 and is associated with its congener, phosphorus, in phosphate rocks, for example, $Ca_5(PO_4)_3OH$. These phosphate minerals are crushed and often used directly as fertilizers, thereby unwittingly putting arsenic compounds in contact with foodstuffs. The element is believed to enter the water table as the As(V) oxyanion, arsenate $[AsO_4]^{3-}$. Bacteria in the soil reduce the As(V) to As(III), subsequently forming, for example, the highly toxic dimethylarsenic(III) cation $[As(CH_3)_2]^+$.

Cadmium is another ubitiquous element, always associated with zinc unless very pure. Zinc is a crucial element in various biological processes that involve zinc-containing enzymes. The chemistry of Cd is very similar to that of Zn; moreover, Cd actually replaces Zn in many of its compounds. Cadmium has been implicated in hypertension. Water-soluble Zn salts are sometimes added to flour as preservatives, and inevitably some Cd is present as well. People who eat too many pastries and breadstuffs are often susceptible to hypertension; the replacement of Zn by Cd in the enzyme carboxypeptidase may be one of the factors that leads to hypertension. Cadmium-containing carboxypeptidase can no longer perform its required biological function because the "fine tuning" of the enzyme has been subtly interfered with. Severe Cd poisoning seriously weakens the internal structure of bone, leading to Itai-Itai disease—the extremely painful, physical collapse of limbs.

Mercury has become particularly notorious over the past twenty years with the discovery that under aqueous conditions certain bacteria convert the metal itself into the water-soluble methylmercury(II) cation $[HgCH_3]^+$. This soluble mercury unfortunately is concentrated in fish and so enters our food chain. Because of the relatively rare occurrence of mercury, natural Hg pollution is limited. It is occasionally encountered in the vicinity of volcanic and geothermal features. There have been several cases of serious industrial Hg pollution throughout the world that have resulted in widespread Hg poisoning and even deaths. The Hg that was lost from the chemical plants eventually found its way into nearby water supplies (lakes, rivers, and so forth), resulting in heavily contaminated fish, for example, in Minimata in Japan.

Mercury poisoning causes severe brain damage, depression, and in-sanity—the same effects occurring upon prolonged exposure to the vapor of the metal as well. Rumor has it that the Flying Dutchman, the ship that was found several centuries ago floating in the Atlantic Ocean with all sails hoisted but no crew in sight, was carrying in its cargo some barrels full of the liquid Hg metal. If this were indeed true, then the concentration of Hg vapor in the air might well have been sufficient to drive the crew mad enough to jump ship! Another related story concerns the Mad Hatter in Lewis Carroll's *Alice in Wonderland*—hats used to be treated with aqueous solutions of Hg mercury compounds to inhibit the growth of fungi and mildew! Mercury is still being used by industry, but its use has greatly diminished; there are such strict controls being enforced in many countries that hopefully Hg pollution will cease to be an ecological problem.

Atmospheric lead pollution resulting from adding tetraethyllead(IV) $(C_2H_5)_4Pb$ to gasoline in order to increase the combustion efficiency of automobile engines is a well-known environmental problem. Lead addi-tives in gasoline have almost been outlawed worldwide. Lead is toxic, and so its industrial use is carefully monitored. The airborne-lead pollutants are lead halides ($PbCl_2$, $PbBr_2$, $PbClBr$), resulting from the high-temperature reactions of $(C_2H_5)_4Pb$ with ethylene dichloride and dibromide (other fuel additives):

$$(C_2H_5)_4Pb + C_2H_4Cl_2 \rightarrow PbCl_2 + \text{Hydrocarbons} \qquad (20.129)$$

These lead halides eventually finish up in seas, lakes, and other water systems either as the halides themselves or as coordination complexes such as $[PbCl]^+$. Lead poisonings have apparently resulted from the leaching of Pb out of vessels made from it. This is especially true if acidic solutions are involved. Possibly, the reason why Nero was fiddling while Rome burned might be linked to the lead drinking goblets used for wine at that time! Another major source of lead pollution is the careless disposal of auto-mobile batteries.

As soon as the petroleum industry realized the problems with Pb pollution, there was an intensive effort to find other antiknock reagents to replace $(C_2H_5)_4Pb$. One of the promising compounds was $(\eta^5\text{-}C_5H_4CH_3)\text{-}Mn(CO)_3$. The discovery of this compound and many others like it led to the fantastic development of organometallic chemistry over the past twenty years. Organometallic chemistry is discussed in some detail in Chapter 22.

Manganese Nodules

Manganese is a vital element both to humans and in photosynthesis. Without the extensive redox chemistry of manganese, no photosynthesis would occur. The element occurs chiefly as manganite MnO_2, usually as the result of volcanic activity. The preponderance of manganese nodules (MnO_2) on the ocean seabeds is believed to owe its origin to undersea volcanoes. These nodules are so rich in manganese that mining them will no doubt become a serious venture in the future.

BIBLIOGRAPHY

Suggested Reading

Cotton, F. A., and G. Wilkinson. *Advanced Inorganic Chemistry*, 5th ed. New York: Wiley, 1988.

Douglas, B. E., D. H. McDaniel, and J. Alexander. *Concepts and Models of Inorganic Chemistry*, 2d ed. New York: Wiley, 1983.

Fergusson, J. E. *Inorganic Chemistry and the Earth.* New York: Pergamon Press, 1982

Huheey, J. E. *Inorganic Chemistry*, 3rd ed. New York: Harper & Row, 1983.

Jolly, W. L. *The Principles of Inorganic Chemistry*, 2nd ed. New York: McGraw-Hill, 1984.

Moeller, T. *Inorganic Chemistry: A Modern Introduction.* New York: Wiley, 1982.

Porterfield, W. W. *Inorganic Chemistry. A Unified Approach.* Reading, Mass.: Addison-Wesley, 1984.

Purcell, K. F., and J. C. Kotz. *Inorganic Chemistry.* Philadelphia: Saunders, 1977.

Sanderson, R. T. *Inorganic Chemistry.* New York: Van Nostrand-Reinhold, 1967.

Sharpe, A. G. *Inorganic Chemistry*. New York: Longman, 1981.

Sebera, D. K. *Electronic Structure and Chemical Bonding*. Waltham, Mass.: Blaisdell, 1964.

PROBLEMS

20.1   The molecular weights of methanol (CH_3OH) and dimethylether (CH_3OCH_3) are quite similar, 32.04 and 46.07, respectively. Explain why their dielectric constants (32.63 and 5.02 at room temperature) and boiling points (64.7 and $-23.7\,°C$) are so radically different.

20.2   Which of the following chlorides would you classify as ionic? Which of them would you expect to dissolve fairly easily in polar solvents?
   a. CsCl   **e.** $AlCl_3$
   b. CuCl   **f.** CCl_4
   c. $BaCl_2$ **g.** SCl_4
   d. $MgCl_2$

20.3   Sketch structures for dimeric C_2H_5COOH and tetrameric CH_3OH.

20.4   Explain the difference between intermolecular and intramolecular hydrogen bonding. What types of hydrogen bonding would you predict for the following molecules?

1,5-dihydroxynapthalene, $N\overset{..}{H}_3$, $C_6H_5(CH_2)_4OH$

= 17β-estradiol

20.5   Describe some of the physical evidence for the presence of hydrogen bonding. Are there any spectroscopic methods for the distinction between intermolecular and intramolecular hydrogen bonding?

20.6   Arrange the following molecules in order of decreasing boiling point:
   a. Rn   **e.** RaO
   b. H_2S **f.** $GeCl_4$
   c. NaF   **g.** SeO_2
   d. NaI

20.7   Explain, with the aid of a suitable example, what is meant by a clathrate. Why do you think it is impossible to form a clathrate with H_2O?

20.8   What are the half-reactions involved in a nickel–cadmium calculator battery if the overall discharge reaction is

$$2Ni(OH)_3(s) + Cd(s) \rightarrow 2Ni(OH)_2(s) + CdO(s) + H_2O(l)$$

20.9 Predict the voltage of a proposed electrolytic cell for which the overall reaction is

$$H_3AsO_4 + 2H^+ + 2I^- \rightarrow HAsO_2 + I_2 + 2H_2O$$

given that $E^0 = +0.56$ V for $H_3AsO_4 + 2H^+ + 2e^- \rightarrow HAsO_2 + 2H_2O$.

20.10 Would you expect the following reactions to proceed as written?
a. $F_2 + 2Br^- \rightarrow Br_2 + 2F^-$
b. $2Cr + 3Cu^{2+} \rightarrow 2Cr^{3+} + 3Cu$
c. $Fe + 2Fe^{3+} \rightarrow 3Fe^{2+}$

20.11 Calculate the voltage for the following reaction when the concentrations of the products are both 0.1M:

$$2Al + 3I_2 \rightarrow 2Al^{3+} + 6I^-$$

20.12 In your opinion, what are the most important factors involved in inorganic chemistry in
a. Nonaqueous solvent systems
b. Environmental chemistry

21

Mechanisms of Inorganic Reactions in Solution

Many of the thousands of known inorganic reactions take place in solution. At first sight, the task to categorize the multitude of different reactions would seem to be impossible. Fortunately, however, there are several main reaction types; this chapter introduces you to these different types and discusses their possible mechanisms. It is often very difficult to pin down a specific mechanism precisely, especially when proof of an important intermediate is lacking. Because the vast majority of kinetic studies has been concerned with the transition metals, most of the examples presented are drawn from this field rather than from main-group chemistry.

21.1 *Main Reaction Types*

There are six principal types of reaction that we consider in this chapter: substitution, oxidation–reduction (or redox), addition–dissociation, oxidative–addition/reductive–elimination, free radical, and insertion.

Substitution

A substitution reaction involves the replacement of one ligand in a complex by another external ligand, without any change in either the oxidation state or the coordination number (CN) of the central atom:

$$[PtCl_4]^{2-} + NH_3 \rightarrow [Pt(NH_3)Cl_3]^- + Cl^- \tag{21.1}$$

Pt(II)   Pt(II)
CN = 4   CN = 4

$$Cr(CO)_6 + py \rightarrow Cr(CO)_5(py) + CO \tag{21.2}$$

Cr(0)   Cr(0)
CN = 6   CN = 6

Oxidation–Reduction (Redox)

This class of reaction is identified by changes in the oxidation state of the central atom; there may or may not be a change in the coordination number. Another way of regarding this class of reaction is *electron transfer* (Equation 21.3), *atom transfer* (Equation 21.4), or *ligand transfer* (Equation 21.5):

$$2[V(H_2O)_6]^{2+} + 2H^+ \rightarrow 2[V(H_2O)_6]^{3+} + H_2 \tag{21.3}$$

V(II)   H(I)   V(III)   H(0)
CN = 6   CN = 6

$$Na_2SO_3 + HClO \rightarrow Na_2SO_4 + HCl \tag{21.4}$$

S(IV)   Cl(I)   S(VI)   Cl(−I)
CN = 3   CN = 4

$$[Co(NH_3)_5Cl]^{2+} + [Cr(H_2O)_6]^{2+} + 5[H_3O]^+ \rightarrow \tag{21.5}$$

Co(III)   Cr(II)
CN = 6   CN = 6

$$[Co(H_2O)_6]^{2+} + [Cr(H_2O)_5Cl]^{2+} + 5[NH_4]^+$$

Co(II)   Cr(III)
CN = 6   CN = 6

Addition–Dissociation

Electron-deficient and other molecules with available vacant orbitals readily form complexes through the *addition* of extra ligands (Equations 21.6 and 21.7). The reverse process is *dissociation* (Equation 21.8). There are only changes in the coordination number of the central atom.

$$BF_3 + F^- \rightarrow [BF_4]^- \tag{21.6}$$

B(III) B(III)
CN = 3 CN = 4

$$TiCl_4 + 2Cl^- \rightarrow [TiCl_6]^{2-} \tag{21.7}$$

Ti (IV) Cl (−I) Ti (IV)
Cl (−I) Cl (−I)
CN = 4 CN = 6

$$Cr(CO)_5(CS) \rightarrow Cr(CO)_4(CS) + CO \tag{21.8}$$

Cr(0) Cr(0)
CN = 6 CN = 5

Oxidative–Addition/Reductive–Elimination

These types of reactions involve changes in both the oxidation number and the coordination number of the central atom. Oxidative–addition (Equations 21.9–21.11) is the reverse of reductive-elimination (Equations 21.12 and 21.13).

$$2[Co(CN)_5]^{3-} + H_2 \rightarrow 2[Co(CN)_5H]^{3-} \tag{21.9}$$

Co(II) H(0) Co(III)
CN = 5 H(−I)
CN = 6

$$IrH(CO)(PPh_3)_3 + Cl_2 \rightarrow IrH(CO)Cl_2(PPh_3)_2 + PPh_3 \tag{21.10}$$

Ir(I) Cl(0) Ir(III)
CN = 5 Cl(−I)
CN = 6

$$PCl_3 + Cl_2 \rightarrow PCl_5 \tag{21.11}$$

P(III) Cl(0) P(V)
CN = 3 Cl(−I)
CN = 5

$$IrCl(CO)(PPh_3)_2O_2 \rightarrow \textit{trans-}IrCl(CO)(PPh_3)_2 + O_2 \tag{21.12}$$

Ir(III) Ir(I) O(0)
O(−I) CN = 4
CN = 6

$$[PdCl_6]^{2-} \rightarrow [PdCl_4]^{2-} + Cl_2 \qquad (21.13)$$

Pd(IV)	Pd(II)	Cl(0)
Cl(−I)	Cl(−I)	
CN = 6	CN = 4	

Free Radical

This type of reaction involves paramagnetic species and, sometimes, photochemically activated species:

$$2[Cr(H_2O)_6]^{2+} + CHCl_3 \rightarrow$$

$$[Cr(H_2O)_5Cl]^+ + [Cr(H_2O)_5CHCl_2]^+ + 2H_2O \qquad (21.14)$$

This reaction necessitates the homolytic fission of a C—Cl bond in $CHCl_3$:

$$CHCl_3 \rightarrow CHCl_2^{\cdot} + Cl^{\cdot} \qquad (21.15)$$

Insertion

This class of reaction involves placement of a group between the central atom and another ligand already bonded to the central atom, without any changes in oxidation state or coordination number taking place:

$$CH_3Mn(CO)_5 + PPh_3 \rightarrow (CH_3CO)Mn(CO)_4(PPh_3) \qquad (21.16)$$

Mn(I)	Mn(I)
CN = 6	CN = 6

In concluding this brief introduction, it should be mentioned that reactions of the same type often proceed by completely different mechanisms. In many instances, the exact mechanism is still unknown, and a great deal of current research focuses on determining reaction rates and mechanisms.

21.2 Substitution Reactions

Inert and Labile Complexes

In 1952 Taube suggested a classification of transition metal complexes based on the rapidity of their substitution reactions. He proposed that complexes with a half-life ($t_{1/2}$) less than one minute at room temperature (for 0.1 M solutions) be termed *labile*, whereas those complexes with $t_{1/2}$

greater than one minute should be called *inert*. It must be clearly under-stood that these labels refer to only the rates of reaction and not to thermodynamic stability. Many examples of kinetically inert complexes are thermodynamically unstable. For example, the famous cationic Werner complex $[Co(NH_3)_6]^{3+}$ decomposes slowly in acid solution (about 6M HCl) over a period of several days at room temperature:

$$[Co(NH_3)_6]^{3+} + 6[H_3O]^+ \rightarrow [Co(H_2O)_6]^{3+} + 6[NH_4]^+ \qquad \textbf{(21.17)}$$

The thermodynamic driving force behind this reaction is the very strong Brønsted–Lowry base behavior of the six NH_3 ligands that leads to the formation of the $[NH_4]^+$ ions. The equilibrium constant for the reaction is very large ($\sim 10^{30}$), further illustrating the thermodynamic instability of $[Co(NH_3)_6]^{3+}$ with respect to acid hydrolysis. An example of a complex with the opposite reactivity behavior is $[Ni(CN)_4]^{2-}$. This complex is kinetically labile but thermodynamically stable; it undergoes rapid exchange with isotopically labeled cyanide ($^{14}CN^-$):

$$[Ni(CN)_4]^{2-} + 4^{14}CN^- \rightarrow [Ni(^{14}CN)_4]^{2-} + 4CN^- \qquad \textbf{(21.18)}$$

The $[Cu(NH_3)_4]^{2+}$ complex is another good example of a compound that is kinetically labile but thermodynamically stable, except in the presence of strong acid:

$$[Cu(NH_3)_4]^{2+} + 4HCl \rightarrow [CuCl_4]^{2-} + 4[NH_4]^+ \qquad \textbf{(21.19)}$$

Addition of concentrated HCl to the solution of the tetraamminecop-per(II) cationic complex results in an almost instantaneous color change from dark blue to green. This tetraammine-complex is of particular historical significance because its aqueous solutions dissolve cellulose (paper, rags, and the like). When these solutions are squirted through very fine holes into either an acid or an alkali bath, the cellulose reprecipitates as threads of artificial silk, known industrially as Rayon.

An alternative way of considering kinetically labile and inert complexes is in terms of the method necessary to determine the reaction rates in the first place. Kinetic rates of substitution vary over about eighteen orders of magnitude (10^{-9} s^{-1} for Cr^{3+} to 10^9 s^{-1} for Li^+, Na^+, and so on). For inert complexes, all the usual conventional kinetic techniques can be employed, for example, UV–visible, IR, and NMR spectroscopy and electrical conductivity measurements. For labile complexes, specialized techniques such as temperature- and pressure-jump and stopped-flow methods are required. In the first two procedures, the equilibrium of the system is perturbed by extremely rapid changes in temperature or pressure, and the rate of return to equilibrium is measured. In the case of the stopped-flow measurements, the two reacting solutions are mixed as quickly as possible (around 10^{-3} s^{-1}), directly in the optical path of a

Table 21.1 *Correlation of Taube's Predictions with Experimental Results*

d Electronic Configuration (O_h Symmetry)	Example of Metal Ion	Typical Octahedral Complex	Taube Classification	Approximate $t_{1/2}$ (s)
$d^0(t_{2g})^0$	Ti(IV)	$[TiCl_6]^{2-}$	Labile	—
$d^1(t_{2g})^1$	Ti(III)	$[Ti(H_2O)_6]^{3+}$	Labile	—
$d^2(t_{2g})^2$	V(III)	$[V(phen)_3]^{3+}$	Labile	—
$d^3(t_{2g})^3$	V(II)	$[V(H_2O)]^{2+}$	Inert	7×10^{-5}
$d^4(t_{2g})^3(e_g)^1$	Cr(II)	$[Cr(H_2O)_6]^{2+}$	Labile	3×10^{-9}
$d^4(t_{2g})^4$	Cr(II)	$[Cr(CN)_6]^{4-}$	Inert	Very large
$d^5(t_{2g})^3(e_g)^2$	Mn(II)	$[Mn(H_2O)_6]^{2+}$	Labile	1×10^{-7}
$d^5(t_{2g})^5$	Mn(II)	$[Mn(CN)_6]^{4-}$	Inert	Very large
$d^6(t_{2g})^4(e_g)^2$	Fe(II)	$[Fe(H_2O)_6]^{2+}$	Labile	7×10^{-7}
$d^6(t_{2g})^6$	Fe(II)	$[Fe(CN)_6]^{4-}$	Inert	Very large
$d^7(t_{2g})^6(e_g)^1$	Co(II)	$[Co(H_2O)_6]^{2+}$	Labile	1×10^{-6}
$d^8(t_{2g})^6(e_g)^2$	Ni(II)	$[Ni(H_2O)_6]^{2+}$	Labile	4×10^{-5}
$d^9(t_{2g})^6(e_g)^3$	Cu(II)	$[Cu(H_2O)_6]^{2+}$	Labile[a]	2×10^{-9}
$d^{10}(t_{2g})^6(e_g)^4$	Ga(III)	$[Ga(H_2O)_6]^{3+}$	Labile	$<7 \times 10^{-5}$

[a] This complex is not strictly octahedral because of Jahn–Teller effects; two of the water molecules are located farther away from the Cu^{2+} ion than are the other four.

UV–visible spectrometer, and then the change in optical density at a fixed wavelength is monitored using a fast-response electronic oscilloscope.

Taube's classification has a sound theoretical basis. The substitution-reaction rates measured for octahedral transition metal ML_6 complexes correlate well with the ground-state d^n electronic configurations of metal ions. Basically, if a metal ion has d electrons located in its antibonding e_g orbitals, ligand replacement is anticipated to be fairly easy because of the relatively weak M—L bonds. Additionally, if the metal ion has fewer than three electrons in the t_{2g} level, the entering ligands should not experience very large electrostatic repulsions, and substitution reactions for these complexes should be quite rapid. Any other configuration leads to large electrostatic repulsions; thus, the remaining possibilities should give inert complexes. Table 21.1 summarizes these predictions.

Octahedral Complexes

Substitution Reactions There are several mechanistic pathways possible for the substitution of a ligand L in an octahedral ML_6 complex by another ligand L'. These pathways could involve complete or only partial bond making or bond breaking in the transition state. For the purposes of our discussion, however, we consider only the two extreme situations: complete ligand dissociation (D mechanism) and complete ligand association (A mechanism).[1]

Complete Ligand Dissociation   The dissociative, or D, mechanism involves total bond breaking in the transition state. The older nomenclature of S_N1 would fall into this classification. The rate is dependent on only the concentration of the substrate (starting complex), that is, *first order*:

$$ML_6 \xrightarrow[k_1]{\text{Slow}} ML_5 + L \quad \text{Rate} = -\frac{d[ML_6]}{dt} = k[ML_6] \qquad \textbf{(21.20)}$$

$$ML_5 + L' \xrightarrow{\text{Fast}} ML_5L'$$

The five-coordinate intermediate (ML_5) could be either a trigonal-bipyramid or a square-pyramid:

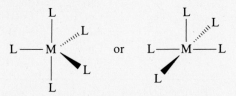

This mechanism is a limiting case of the following more general one

$$ML_6 \underset{k_{-1}}{\overset{k_1}{\rightleftharpoons}} ML_5 + L$$

$$ML_5 + L' \xrightarrow{k_2} ML_5L' \qquad \textbf{(21.21)}$$

for which the rate expression using the steady-state approximation is

$$\text{Rate} = \frac{k_1k_2[ML_6][L']}{k_{-1} + k_2[L']}$$

When $[L']$ is very large, $k_2[L']$ is much greater than k_{-1}; thus, this rate law reduces to

$$\text{Rate} = k_1[ML_6][L'] = k[ML_6] \text{ (first order)}$$

Complete Ligand Association   The associative, or A, mechanism involves complete bond making in the transition state. This class of reaction would include the familiar S_N2 category of organic chemistry. The reaction rate is dependent on the concentrations of both the substrate and the entering

---

[1] For a full discussion of the nomenclature used when discussing reaction mechanisms, see C. H. Langford and H. B. Gray. *Ligand Substitution Processes*. Menlo Park, Calif.: Benjamin/Cummings, 1966.

ligand, that is, *second order*:

$$ML_6 + L' \xrightarrow[k]{\text{Slow}} ML_6L' \quad \text{Rate} = -\frac{d[ML_6]}{dt} = k[ML_6][L']$$

$$ML_6L' \xrightarrow{\text{Fast}} ML_5L' + L$$

$$(21.22)$$

The seven-coordinate ML_6L' intermediate is most probably a pentagonal-bipyramid with L' positioned either in the pentagonal plane or in one of the axial positions:

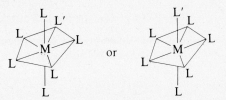

Note that the slow, first steps in both the D and A mechanisms are considered to be the rate-determining steps.

The classic work of Eigen's group in West Germany on the exchange rates of water molecules in the first coordination sphere of numerous octahedral $[M(H_2O)_6]^{n+}$ complexes is an excellent illustration of the approach used in rationalizing a mechanism. It was partly Eigen's development of the experimental techniques mentioned earlier for studying very fast reactions that eventually led to him sharing the 1967 Nobel Prize in Chemistry with Norrish and Porter (U.K.). Some of the results obtained for the water-exchange reactions using isotopically labeled water are summarized below:

1. The rates of exchange for the alkali metal cations decrease in the order

$$Cs^+ > Rb^+ > K^+ > Na^+ > Li^+$$
$$6 \times 10^9 \quad 4 \times 10^9 \quad 2 \times 10^9 \quad 9 \times 10^8 \quad 7 \times 10^8 \text{ s}^{-1}$$

2. The exchange rates for the alkaline earth metal cations decrease as

$$Ba^{2+} > Sr^{2+} > Ca^{2+} > Mg^{2+} > Be^{2+}$$
$$4 \times 10^9 \quad 8 \times 10^8 \quad 5 \times 10^8 \quad 1 \times 10^5 \quad 1 \times 10^2 \text{ s}^{-1}$$

3. For metal cations of similar size in the two series above (for example, Rb^+ and Sr^{2+}), the one with the lower charge exchanges faster.

These three observations are in accord with a D mechanism for the substitution reactions rather than an A mechanism. In both series, the larger the central metal ion (M), the faster the exchange reaction because the $M-OH_2$ bond is weaker. Moreover, for metal cations of similar size, the one with the lower charge exchanges faster because the $M^{n+} \cdots O^{\delta-}H_2$ electrostatic interactions are weaker in this case.

Numerous examples of substitution reactions apparently satisfy both extreme types of reaction kinetics. The observation of a certain rate law, however, does not prove definitively either mechanism because almost all the kinetic studies have been performed in coordinating solvents (H_2O, CH_3OH, and so on), and these molecules can also function as ligands. For example, consider the following *aquation reaction*:

$$trans\text{-}[Co(en)_2Cl_2]^+ + H_2O \rightarrow trans\text{-}[Co(en)_2Cl(H_2O)]^{2+} + Cl^-$$

$$\textbf{(21.23)}$$

$$\text{Rate} = k_{obsd}[trans\text{-}Co(en)_2Cl_2^+]$$

The observed rate law suggests a D mechanism, but because the H_2O ligand is also the bulk solvent, it could also participate in an A pathway:

$$trans\text{-}[Co(en)_2Cl_2]^+ + H_2O \xrightarrow[k_2]{\text{Slow}} [Co(en)_2Cl_2(H_2O)]^+ \qquad \textbf{(21.24)}$$

$$\Big\downarrow \text{Fast}$$

$$[Co(en)_2Cl(H_2O)]^{2+} + Cl^-$$

Therefore, the true rate law is more likely to be

$$\text{Rate} = \{k_1 + k_2[H_2O]\}[trans\text{-}Co(en)_2Cl_2^+]$$

$$= k_{obsd}[trans\text{-}Co(en)_2Cl_2^+]$$

where $k_{obsd} = k_1 + k_2[H_2O]$ = a constant because $[H_2O] \approx 55M$; this will be unchanged in the reaction. This kind of combined mechanism is often observed and is referred to as a *mixed-order reaction*. For instance, the substitution reactions of the Group VIA (6) metal hexacarbonyls $M(CO)_6$ (M = Cr, Mo, W) in noncoordinating solvents such as CCl_4 and C_2Cl_4 have been extensively investigated in recent years, and it is clear that a mixed-order mechanism is operative:

$$\text{D Pathway} \begin{cases} M(CO)_6 \xrightarrow[k_1]{\text{Slow}} M(CO)_5 + CO \\ M(CO)_5 + L \xrightarrow{\text{Fast}} M(CO)_5L \end{cases} \qquad \textbf{(21.25)}$$

$$\text{A Pathway} \begin{cases} M(CO)_6 + L \xrightarrow[k_2]{\text{Slow}} M(CO)_6L \\ M(CO)_6L \xrightarrow{\text{Fast}} M(CO)_5L + CO \end{cases} \qquad \textbf{(21.26)}$$

$$\text{Rate} = (k_1 + k_2[L])[M(CO)_6]$$

These reactions were performed with [L] much greater than $[M(CO)_6]$, that is, under *pseudo-first-order conditions*, such that

$$\text{Rate} = k_{obsd}[M(CO)_6]$$

Figure 21.1
Typical plot of observed kinetic data for a metal hexacarbonyl substitution reaction. These data were obtained by monitoring the rate of disappearance of the single $\nu(CO)$ peak in the IR spectrum (see inset) due to the t_{1u} mode of the metal hexacarbonyl.

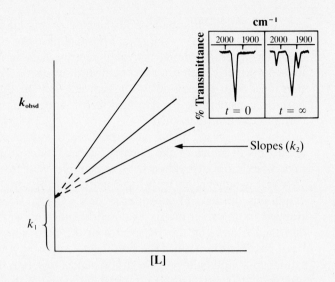

and so

$$k_{obsd} = k_1 + k_2[L]$$

The values of the rate constants k_1 and k_2 can be obtained from plots like those shown in Figure 21.1. Moreover, by measuring the reaction kinetics over a reasonably wide temperature range (usually 20–40 °C), it is possible to determine the enthalpy ($\Delta H^{\ddagger}$) and entropy ($\Delta S^{\ddagger}$) of activation for each reaction. Table 21.2 gives some typical activation parameters.

The entropy of activation ($\Delta S^{\ddagger}$) is a particularly useful parameter in providing support for a proposed mechanism. When $\Delta S^{\ddagger}$ is large and negative, the activated intermediate complex is entropically unfavorable, as it would be in an associative (A) mechanism involving the formation of a

Table 21.2   *Typical Activation Parameters for the CO Substitution Reactions of $M(CO)_6$ Complexes*[a,b]

	D Pathway		A Pathway	
Complex	$\Delta H^{\ddagger}$ (kJ mol^{-1})	$\Delta S^{\ddagger}$ (e.u.)	$\Delta H^{\ddagger}$ (kJ mol^{-1})	$\Delta S^{\ddagger}$ (e.u.)
$Cr(CO)_6$	167	23	109	−15
$Mo(CO)_6$	139	7	92	−15
$W(CO)_6$	167	14	121	−7

[a] Data taken from J. R. Graham and R. J. Angelici. *Inorg. Chem.* 6 (1967): 2082.
[b] With P(n-Bu$_3$).

seven-coordinate intermediate in the case of an octahedral substrate. On the other hand, if $\Delta S^{\ddagger}$ is large and positive, a dissociative (D) mechanism involving a less hindered activated intermediate is indicated, that is, a five-coordinate complex for an octahedral substrate. The entropy of activation must be used cautiously for such mechanistic distinctions, however, because considerable overlap has been found in the values observed for certain D and A reactions.

In a few cases, it has been possible to obtain detailed information about the structure of the reactive intermediates from low-temperature, matrix-isolation experiments. For example, the five-coordinate species $Cr(CO)_5$ is believed to be formed under both thermal and photochemical conditions during CO substitution of $Cr(CO)_6$. By irradiating $Cr(CO)_6$ trapped in an argon matrix at liquid-helium temperatures with UV light and then examining the IR spectra of the matrix-isolated species, it has been possible to establish the presence of square-pyramidal $Cr(CO)_5$ (C_{4v} symmetry). Upon allowing the argon matrix to soften slightly, the $Cr(CO)_5$ intermediate reacts with CO trapped in the matrix to regenerate $Cr(CO)_6$. A similar photochemical/matrix-isolation study for $Cr(CO)_5(CS)$ has shown that there are two geometrical isomers (C_{4v} and C_s symmetries) for the coordinatively unsaturated $Cr(CO)_4(CS)$ intermediate:

C_{4v} symmetry C_s symmetry

A considerable amount of work has been performed on the kinetics and mechanisms of the substitution reactions of metal carbonyl complexes, especially the simple $M(CO)_nL_{6-n}$ derivatives, where L is a monodentate ligand and $n = 3, 4,$ or 5. Often, these substitution reactions are stereospecific. For example, the equimolar reaction of $W(CO)_5(CS)$ with ^{13}CO produces only *trans*-$W(CO)_4(^{13}CO)(CS)$, whereas that of $Mn(CO)_5Br$ with 1 mole of PPh_3 yields exclusively *cis*-$Mn(CO)_4(PPh_3)Br$. These different stereochemical preferences have been rationalized on the basis of the kinetic stabilities of the five-coordinate intermediates formed. For $W(CO)_5(CS)$, the more stable intermediate appears to be whereas for $Mn(CO)_5Br$, it is

(assuming square-pyramidal geometry).

For the thiocarbonyl complex, the CS group is considered to be a better π-acceptor ligand than is CO, whereas Br^- is a poorer π-acceptor than is CO.

The base hydrolyses of cobalt(III) ammine complexes apparently follow a second-order pathway:

$$[Co(NH_3)_5Cl]^{2+} + OH^- \rightarrow [Co(NH_3)_5(OH)]^{2+} + Cl^- \qquad \textbf{(21.27)}$$

$$\text{Rate} = k[Co(NH_3)_5Cl^{2+}][OH^-]$$

The OH^- ion is not a particularly good ligand in any other substitution process, however, especially in the absence of an ionizable hydrogen—for example, in $[Co(py)_4Cl_2]^+$, for which there is no marked dependence on OH^- concentration. To bring the kinetics for the base hydrolyses into line with more rational behavior for the OH^- ion, an alternative mechanism has been proposed. This mechanism is referred to as the *substitution nucleophilic unimolecular conjugate base mechanism* (S_N1CB):

$$[Co(NH_3)_5Cl]^{2+} + OH^- \underset{}{\overset{\text{Fast, } K}{\rightleftharpoons}} [Co(NH_3)_4(NH_2)Cl]^+ + H_2O$$

$$\textbf{(21.28)}$$

$$[Co(NH_3)_4(NH_2)Cl]^+ \xrightarrow[k_1]{\text{Slow}} [Co(NH_3)_4(NH_2)]^{2+} + Cl^-$$

$$\textbf{(21.29)}$$

$$[Co(NH_3)_4(NH_2)]^{2+} + H_2O \xrightarrow{\text{Fast}} [Co(NH_3)_5(OH)]^{2+} \qquad \textbf{(21.30)}$$

The first step involves an NH_3 group acting as a Brønsted–Lowry acid in giving a proton to the OH^- to form H_2O. The conjugate pairs are NH_3/NH_2^- and OH^-/H_2O.

$$K = \frac{[Co(NH_3)_4(NH_2)Cl^+][H_2O]}{[Co(NH_3)_5Cl^{2+}][OH^-]}$$

The second step is rate-determining, and a conventional D mechanism is assumed to take place:

$$\text{Rate} = k_1 [Co(NH_3)_5Cl^{2+}] = \frac{k_1K[Co(NH_3)_5Cl^{2+}][OH^-]}{[H_2O]}$$

$$= k[Co(NH_3)_5Cl^{2+}][OH^-]$$

because $[H_2O]$ is constant. Therefore,

$$k = \frac{k_1 K}{[H_2O]}$$

As a general rule, the substitution rates of complexes increase with decreasing effective nuclear charge on the central metal atom; thus, k_1 is much faster than direct substitution into the starting complex.

Isomerization Reactions   Isomerizations of octahedral complexes can be treated as a subcategory of octahedral substitution reactions. The most common form of isomerization is cis-trans isomerization in ML_4X_2 complexes. For example, the following reversible isomerizations take place in aqueous solution:

$$trans\text{-}[Co(en)_2Cl_2]^+ \xrightleftharpoons[HCl]{NH_3} cis\text{-}[Co(en)_2Cl_2]^+ \qquad (21.31)$$

The mechanism of these reactions appears to involve a D process in which a five-coordinate intermediate (probably with trigonal-bipyramidal geometry) is implicated:

$$\text{Rate} = k_1[trans\text{-}Co(en)_2Cl_2^+]$$

Several isomerization reactions have been studied kinetically for octahedral organometallic systems:

Once again, a D mechanism has been proposed for this isomerization, the necessary five-coordinate intermediate being $Mn(CO)_3LBr$. Isomerization has also been observed during the replacement of the arene rings in $(\eta^6\text{-Arene})Cr(CO)_2(CSe)$ complexes with monodentate ligands L, for example, tertiary phosphines and phosphites. The mer-I structure of the $Cr(CO)_2L_3(CSe)$ product from the reaction of methylbenzoate derivative with $(MeO)_3P$ has been established by ^{31}P-NMR spectroscopy and X-ray diffraction:

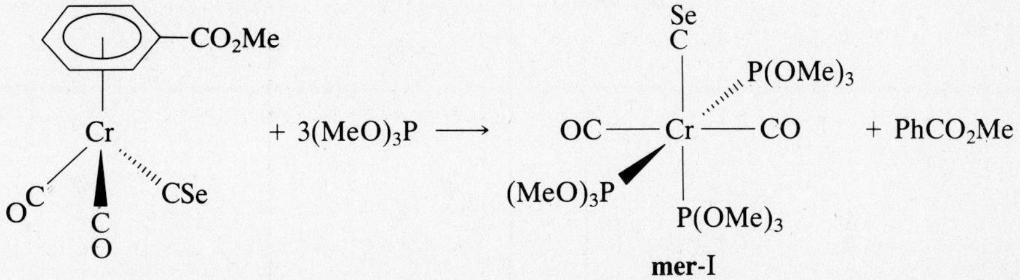

The other two possible isomers for the product are fac (three phosphite ligands mutually cis to one another) and mer-II (CSe trans to CO). The most likely reaction product is the fac isomer, but this apparently isomerizes to the mer-I isomer in order to mimimize steric crowding due to the presence of the three tertiary phosphite ligands. There is some ^{31}P-NMR evidence for a small amount of the mer-II isomer. Moreover, it is believed that the mer-I $\rightleftharpoons$ mer-II isomerization most probably takes place by an intramolecular trigonal twist mechanism:

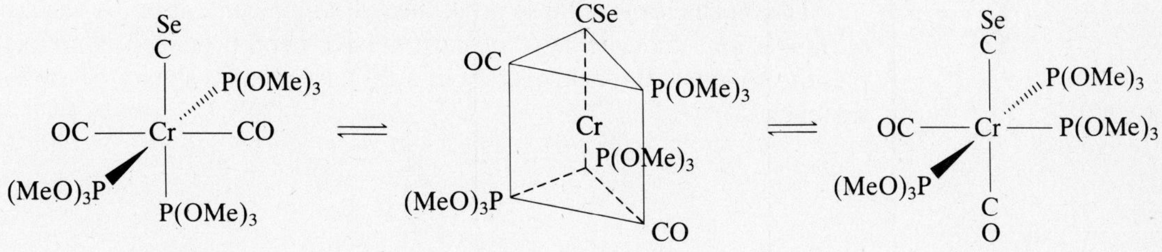

or

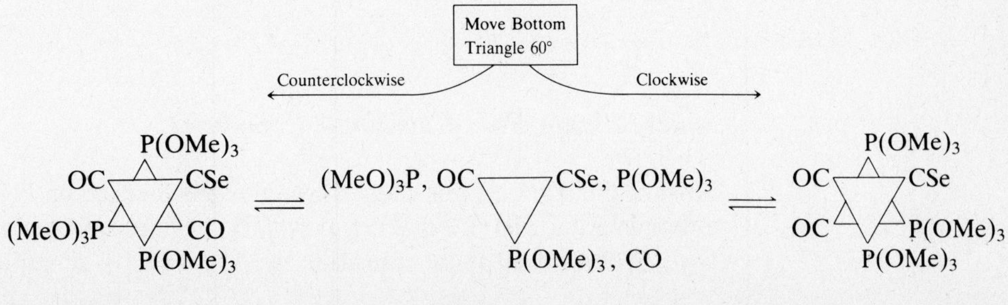

Racemization Reactions   This type of reaction for octahedral complexes is another important subclass of octahedral substitution reactions. Often, when a specific octahedral, optical isomer has been successfully isolated and then left to stand in solution, it is converted into the other optical isomer (mirror image) until eventually a $1:1$ *racemic mixture* is produced. Unlike the individual optical isomers, this racemic mixture does not rotate plane-polarized light at all. Tris(chelate) complexes such as $[Co(en)_3]^{3+}$ and the closely similar compound $[Co(en)_2Cl_2]^+$ undergo racemizations fairly easily:

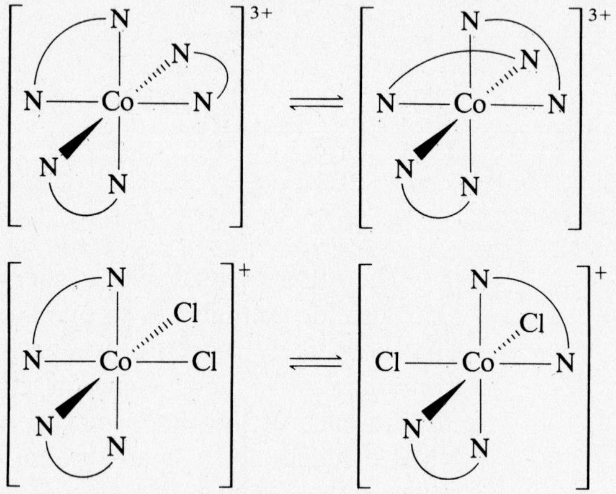

$$N \frown N = H_2N—CH_2—CH_2—NH_2 = en$$

The mechanisms of these processes are still in contention. A variety of *intramolecular* and *intermolecular twists* have been proposed. One example, the trigonal (Bailar) twist about a threefold axis is shown below for a tris(chelate):

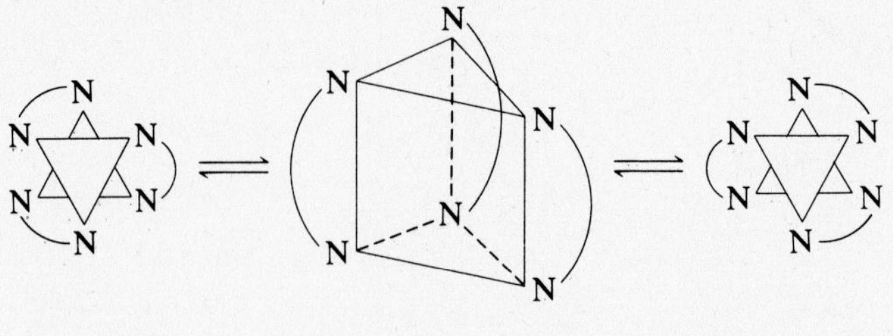

Square-Planar Four-Coordinate Complexes

Most of the work on this class of complex has focused on Pt(II) chemistry, although Rh(I), Ir(I), Pd(II), and Au(III) have received substantial attention. Note that all these transition metal ions have d^8 ground-state electronic configurations, leaving the $(n - 1)d$, ns, np_x, and np_y orbitals available for square-planar bond formation. A typical substitution reaction of Pt(II) is

$$[Pt(NH_3)_4]^{2+} + Cl^- \xrightarrow{H_2O} [Pt(NH_3)_3Cl]^+ + NH_3 \qquad \textbf{(21.32)}$$

Under pseudo-first-order conditions, the observed rate law for this and all other similar square-planar substitution reactions is

$$\text{Rate} = k_{obsd}[Pt(NH_3)_4^{2+}]$$

where $k_{obsd} = k_1 + k_2[Cl^-]$.

The k_1 pathway appears to be solvent-dependent, suggesting that there is an associative mechanism operative in addition to the second-order k_2 pathway. When the reactions are performed in a noncoordinating solvent such as CCl_4, the k_1 pathway disappears. Figure 21.2 shows some typical results graphically.

Some spectroscopic evidence has been obtained for the existence of five-coordinate intermediates of Rh(I) and Ni(II). The geometry of the intermediates during the reactions is still open to question, however, because both square-pyramidal and trigonal-bipyramidal complexes are now well documented. The energy differences between these two

Figure 21.2
Typical plots of the variation of k_{obsd} for a four-coordinate, square-planar Pt(II) complex undergoing substitution by various ligands L′ in a coordinating solvent (MeOH). Adapted with permission from U. Belluco, L. Cattalini, F. Basolo, R. G. Pearson, and U. Turco. *J. Am. Chem. Soc.* 87 (1965): 241.

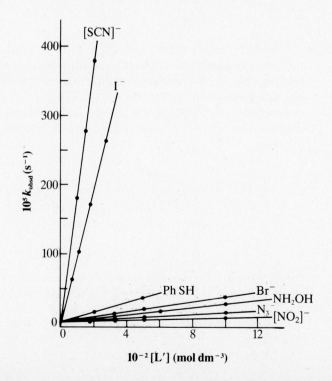

geometries can be quite small. For instance, the Cr(III)/Ni(II) salt $[Cr(en)_3][Ni(CN)_5] \cdot 5H_2O$ crystallizes with two molecules per unit cell, one in which the Ni(II) complex anion is square-pyramidal and the other in which it is trigonal-bipyramidal.

The overall associative (A) mechanism now accepted for the substitution reactions of a typical square-planar ML_4 complex in a coordinating solvent (S) is as follows:

$$\text{Overall Reaction} \quad ML_4 + L' \xrightarrow{\text{S}} ML_3L' + L \quad (21.33)$$

k_1' Pathway Where
$k_1 = k_1'$ [S]
(Associative)
$$\begin{cases} ML_4 + S \xrightarrow[k_1']{\text{Slow}} ML_4S \\ ML_4S \xrightarrow{\text{Fast}} ML_3S + L \\ ML_3S + L' \longrightarrow ML_3L' + S \end{cases}$$

k_2 Pathway
(Associative)
$$\begin{cases} ML_4 + L' \xrightarrow[k_2]{\text{Slow}} ML_4L' \\ ML_4L' \xrightarrow{\text{Fast}} ML_3L' + L \end{cases}$$

Therefore,

$$k_{\text{obsd}} = k_1'[S] + k_2[L']$$
$$= k_1 + k_2[L']$$

Because square-planar transition metal complexes with d^8 ground-state electronic configurations are coordinatively unsaturated (sixteen valence electrons), it is not too surprising that there is a driving force to form eighteen-electron, five-coordinate species. In fact, in electron-rich d^8 complexes, the five-coordinate state is the ground state, and four-coordinate complexes are only produced as unstable intermediates:

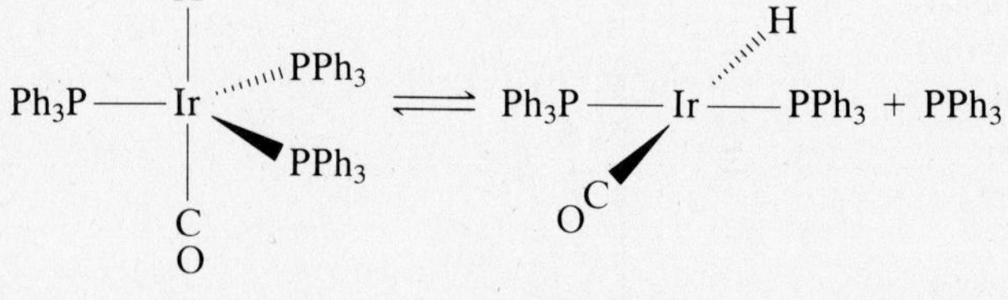

One of the most important results in terms of synthetic chemistry from the exhaustive studies on the substitution reactions of Pt(II) complexes has been the discovery of the *trans effect*. The pioneering work in this field was by Chernaev (U.S.S.R.) and Chatt, Ducanson, and Venanzi (U.K.) in the early 1950s. Their studies resulted in the first examples of stereospecific substitution.

As mentioned earlier, square-planar complexes such as $Pt(NH_3)_2Cl_2$ can exist in two isomeric forms:

$$\text{Cl} \longrightarrow \overset{\overset{\displaystyle NH_3}{|}}{Pt} \longrightarrow NH_3 \qquad H_3N \longrightarrow \overset{\overset{\displaystyle Cl}{|}}{Pt} \longrightarrow NH_3$$

cis (C_{2v} symmetry)   **trans (D_{4h} symmetry)**

The stereochemistry dramatically affects the chemical reactivity of the two compounds. For instance, as mentioned earlier in this book, only the cis isomer is effective as a therapeutic drug in the treatment of cancer. The two isomers may be synthesized as follows:

$$[PtCl_4]^{2-} \xrightarrow[-Cl^-]{+NH_3} [Pt(NH_3)Cl_3]^- \xrightarrow[-Cl^-]{+NH_3} \textit{cis-}Pt(NH_3)_2Cl_2 \qquad \textbf{(21.34)}$$

Sole Product

$$[Pt(NH_3)_4]^{2+} \xrightarrow[-NH_3]{+Cl^-} [Pt(NH_3)_3Cl]^+ \xrightarrow[-NH_3]{+Cl^-} \textit{trans-}Pt(NH_3)_2Cl_2 \qquad \textbf{(21.35)}$$

Sole Product

The two different synthetic routes yield only one geometrical isomer in each case. The geometrical preferences are the result of the relative metal–ligand bond strengths, especially in the last substitution step. The structures of the starting materials in these two substitutions are

(a)   **(b)**

where the shaded groups trans to the chloro-groups are the ones being replaced. The trans effect is the name given to the labilization (easier replacement) of ligands trans to a specific group in a square-planar complex. By studying a wide variety of complexes, an order of decreasing of relative trans-directing ability has been established. An abbreviated version of this series is: $CO \sim CN^- \sim C_2H_4 > PR_3$ (R = alkyl, aryl) $> H^- > Me^- > Ph^- \sim [NO_2]^- \sim I^- \sim [SCN]^- > Br^- > Cl^- > py > NH_3 > F^- \sim OH^- > H_2O$. Because the trans effect of Cl^- is greater than that of NH_3, it therefore follows that *trans-*$Pt(NH_3)_2Cl_2$ is produced in situation **(b)**.

Imagine that there is a need to synthesize *cis-*chloroiodobis-(pyridine)platinum(II), *cis-*$PtCl(I)(py)_2$ from $[PtCl_4]^{2-}$, I^-, and pyridine (py). To do this, the relative trans-directing abilities of the I^- and py ligands must be taken into account. The actual sequence of reactions would be as follows:

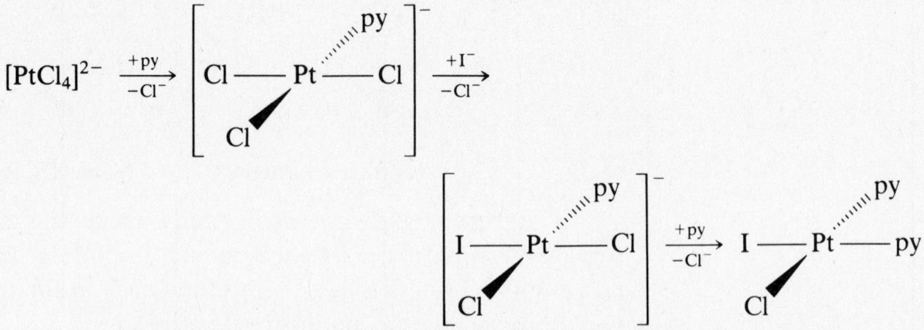

$$[PtCl_4]^{2-} \xrightarrow[-Cl^-]{+py} \left[\begin{array}{c} py \\ Cl - Pt - Cl \\ Cl \end{array} \right]^- \xrightarrow[-Cl^-]{+I^-}$$

$$\left[\begin{array}{c} py \\ I - Pt - Cl \\ Cl \end{array} \right]^- \xrightarrow[-Cl^-]{+py} \begin{array}{c} py \\ I - Pt - py \\ Cl \end{array}$$

The first substitution step has no stereochemical consequence; the second takes place because of the greater bond strength of the Pt—py bond compared to that of Pt—Cl. The final step follows the trans-effect order in that I^- is a better trans-directing ligand than is Cl^-.

Using similar reasoning, it has been possible to synthesize all three isomers of square-planar $Pt(NH_3)(CH_3NH_2)(NO_2)Cl$:

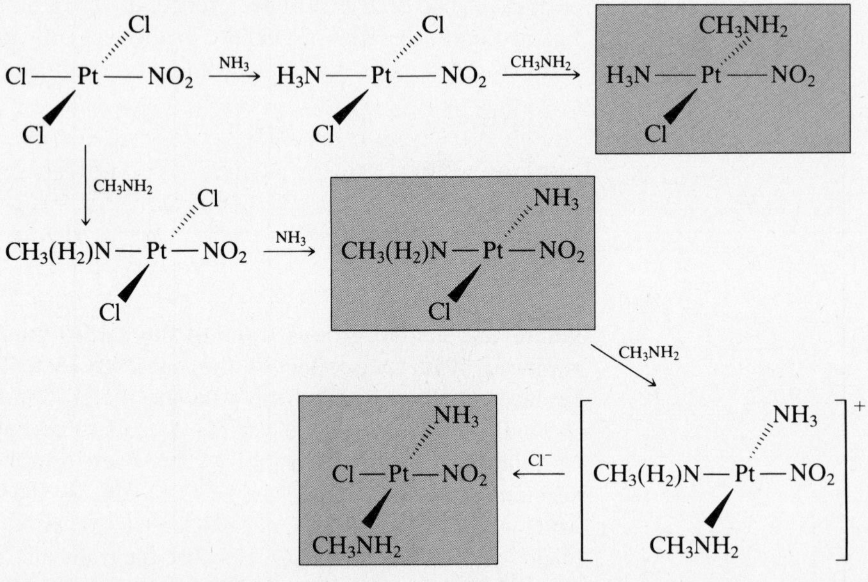

$$\begin{array}{ccc}
\text{Cl} & & \text{Cl} \\
Cl - Pt - NO_2 & \xrightarrow{NH_3} & H_3N - Pt - NO_2 \\
\text{Cl} & & \text{Cl}
\end{array} \xrightarrow{CH_3NH_2} \begin{array}{c} CH_3NH_2 \\ H_3N - Pt - NO_2 \\ Cl \end{array}$$

$$\downarrow CH_3NH_2$$

$$\begin{array}{cc}
\text{Cl} & \\
CH_3(H_2)N - Pt - NO_2 & \xrightarrow{NH_3} \\
\text{Cl} &
\end{array} \begin{array}{c} NH_3 \\ CH_3(H_2)N - Pt - NO_2 \\ Cl \end{array}$$

$$\searrow CH_3NH_2$$

$$\begin{array}{c} NH_3 \\ Cl - Pt - NO_2 \\ CH_3NH_2 \end{array} \xleftarrow{Cl^-} \left[\begin{array}{c} NH_3 \\ CH_3(H_2)N - Pt - NO_2 \\ CH_3NH_2 \end{array} \right]^+$$

As another example of Pt(II) geometrical isomerism, the three isomers of $Pt(CO)(Ph)(PMePh_2)Cl$ have also been synthesized as shown below. These isomers are readily distinguished by their ^{13}C- and ^{31}P-NMR parameters, especially the Pt—C, P—C, and Pt—P coupling constants (Table 21.3).

Table 21.3 *NMR Parameters for the Pt(CO)(Ph)(PMePh$_2$)Cl Isomers*[a]

Isomer	δ(CO) (ppm)	J(Pt, CO) (Hz)	J(P, CO) (Hz)	δ(P) (ppm)	J(Pt, P) (Hz)
I	162.1	1947	8	6.8	1402
II	177.4	906	6	−1.7	3920
III	173.3	1427	158	5.7	3481

[a] From G. K. Anderson and R. J. Cross. *Acc. Chem. Res.* 17 (1984): 67.

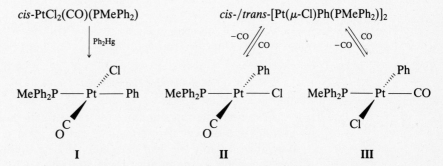

The theoretical explanation for the trans effect is still being debated. Two main theories have been promulgated; both are probably operative to a greater or lesser degree at any one time depending on the system in question. First is the *polarization theory*, which is useful in accounting for the high placement of H$^-$ and Me$^-$ in the series. Second is the *π-activation theory*, which explains the extremely strong trans-directing ability of the π-acceptor ligands, CO, CN$^-$, alkenes, and so on.

The polarization theory is a σ effect based on the interaction of induced dipoles. For example, in the trans H—Pt—Cl interaction, the hydride H$^-$ is very polarizable and thus reduces the effective positive charge on the central Pt(II). This in turn lowers the electrostatic part of the Pt—Cl interaction, rendering the Pt—Cl bond more susceptible to chemical attack. In general, the bulkier the trans-directing ligand, the more easily its electrons are polarized, and such ligands appear high in the trans-effect series, for example, I$^-$ > Br$^-$ > Cl$^-$ > F$^-$.

The π-bonding theory necessitates that the trans-directing ligand is capable of accepting π electrons from the central metal through an interconnecting framework of orbitals. The drawing below illustrates the interaction of the π-acceptor orbitals on ligands, such as CO and C$_2$H$_4$, with a transition metal that is believed to result in a weakening of the trans metal–ligand bond. The π bonding results in a reduction of electron density in the interligand region, making nucleophilic attack easier. Furthermore, such a π-bonding interaction is thought to stabilize the five-coordinate intermediate (or activated complex) that is implicated in the associative mechanism, whereby L is replaced by the incoming ligand L'.

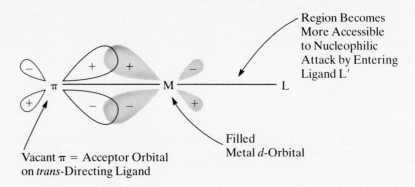

Vacant π = Acceptor Orbital on *trans*-Directing Ligand

Region Becomes More Accessible to Nucleophilic Attack by Entering Ligand L'

Filled Metal d-Orbital

π-Bonding Interaction

The two different situations refer to ground-state weakening (σ polarization) and activated intermediate stabilization (π theory). Moreover, they are probably the two extremes, and no doubt both effects play an important role in the relative trans-directing ability of ligands.

Tetrahedral Four-Coordinate Complexes

The substitution reactions of essentially all tetrahedral complexes in solution are extremely fast and have been little investigated. The reactions of the organometallic systems $Co(CO)_3(NO)$ and $Fe(CO)_2(NO)_2$ have proven more amenable to study and have been thoroughly studied. For example, $Co(CO)_3(NO)$ undergoes reaction with various monodentate ligands (L) in nonaqueous solvents to form $Co(CO)_2(NO)L$ without any evidence of NO substitution:

$$Co(CO)_3(NO) + L \rightarrow Co(CO)_2(NO)L + CO \qquad (21.36)$$

Second-order kinetics were observed, and there was clearly no solvent involvement. The reactions were examined under pseudo-first-order conditions and the corresponding rate law is

$$Rate = k_{obsd}[Co(CO)_3(NO)]$$

where $k_{obsd} = k_1[L]$.

The A mechanism, which was proposed to satisfy this rate law, necessitates initial production of the five-coordinate $Co(CO)_3(NO)L$

intermediate:

$$Co(CO)_3(NO) + L \xrightarrow[k_1]{Slow} Co(CO)_3(NO)L \tag{21.37}$$

$$Co(CO)_3(NO)L \xrightarrow{Fast} Co(CO)_2(NO)L + CO \tag{21.38}$$

The activation parameters for the CO substitution of the cobalt–nitrosyl complex in nitromethane solution when L = PPh$_3$ are $\Delta H^{\ddagger} = 88$ kJ mol^{-1} and $\Delta S^{\ddagger} = -13$ e.u. The negative entropy value is supportive of the proposed A mechanism. A similar associative mechanism appears to be operative for the related tetrahedral Fe(CO)$_2$(NO)$_2$ complex.

21.3 Oxidation–Reduction Reactions

Reactions involving electron transfer from one metal-ion center to another are termed *oxidation–reduction*, or *redox*, *reactions*; they play an extremely important role in many different fields of chemistry. Before discussing the proposed mechanisms of selected redox reactions, however, we consider Taube's classification of electron-transfer mechanisms involving two metal complexes: *inner-sphere* and *outer-sphere*. For his remarkable work on the mechanisms of inorganic reactions, particularly electron-transfer reactions, Taube was awarded the 1983 Nobel Prize in Chemistry.

An inner-sphere reaction is one in which the two metal ions are considered to be bonded together at the time of actual electron transfer (that is, in the transition state) through a bridging ligand:

$$M\text{———————}L\text{———————}M'$$

The outer-sphere electron transfer mechanism is one in which the first coordination spheres of the two metal ions remain intact in the transition state or any reaction intermediate. The electron hops from one metal-ion center to the other, without the intermediacy of a bridging ligand. We consider this class of reactions first.

Outer-Sphere Electron-Transfer Reactions of Octahedral Complexes

Two fundamental conditions must be satisfied for this type of reaction to occur:

1. The rate of the reaction must depend on the concentrations of the two complexes [A] and [B] because both complexes are present in the activated complex when electron transfer must occur.

    Rate = k[A][B]

2. The rate of electron transfer between complexes A and B must be significantly faster than the rates of ligand substitution for either complex.

An excellent example of an outer-sphere electron-transfer reaction taking place in aqueous solution is shown below:

$$[Fe(CN)_6]^{4-} + [IrCl_6]^{2-} \longrightarrow [Fe(CN)_6]^{3-} + [IrCl_6]^{3-} \qquad \text{(21.39)}$$

Fe(II)   Ir(IV)   Fe(III)   Ir(III)

The electron-transfer rate for this reaction at 25 °C is about $4 \times 10^5 \, \text{mol}^{-1} \, \text{dm}^3 \, \text{s}^{-1}$, while the aquation (water substitution) reactions of the two complexes proceed at about $10^3 \, \text{s}^{-1}$.

Another classic example where the two requisite conditions for an outer-sphere mechanism are met is

$$[Co(NH_3)_5Cl]^{2+} + [Ru(NH_3)_6]^{2+} \longrightarrow [Co(NH_3)_5Cl]^{+} + [Ru(NH_3)_6]^{3+}$$

Co(III)   Ru(II)   Co(II)   Ru(III)

$$\text{(21.40)}$$

Both of the above two reactions involve real chemical changes in the nature of the species concerned. Other simpler electron-transfer reactions do not result in any apparent chemical change:

$$[Fe(CN)_6]^{4-} + [Fe(CN)_6]^{3-} \longrightarrow [Fe(CN)_6]^{3-} + [Fe(CN)_6]^{4-}$$

Fe(II)   Fe(III)   Fe(III)   Fe(II)

$$\text{(21.41)}$$

To monitor this last type of reaction, isotopic-labeling experiments (^{57}Fe) are necessary to identify the products. For the other reactions, an examination of the changes in spectroscopic properties such as UV–visible spectra can be used.

In an outer-sphere redox reaction, an electron must cross over somehow from one metal-ion system to the other. The current theory considers that the amplitudes of the metal–ligand vibrational modes play a crucial role. At any given instant, it is possible to imagine that the potential-energy surfaces associated with metal–ligand bond stretching in the two complexes are in close enough proximity for electron hopping to take place. This situation is shown schematically below for a homonuclear electron transfer, such as in the last reaction (Equation 21.41). In the case of a heteronuclear exchange, the potential well for the products is deeper than that for the reactants.

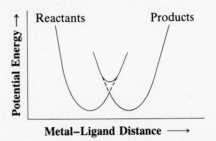

When the metal–ligand bond distances are similar in the two complexes, electron transfer is quite feasible; moreover, the transfer should take place quickly. Conversely, for complexes of markedly different sizes, electron transfer should be appreciably more difficult, and the observed reaction rates are expected to be significantly slower. These predictions are vindicated by the experimental data for some selected outer-sphere reactions, as shown in Table 21.4.

Usually, reactions involving chemical changes are quite fast, but other considerations come into play as well. For instance, for those reactions that involve removal of an electron from an antibonding orbital (e_g in O_h

Table 21.4 *Some Typical Rate Constants for Outer-Sphere Electron-Transfer Reactions*[a]

Reactants	Products	Approximate Second-Order Rate Constants ($mol^{-1}\ dm^3\ s^{-1}$, 25 °C)
$[Fe(CN)_6]^{4-}$, $[Fe(CN)_6]^{3-}$	$[Fe(CN)_6]^{3-}$, $[Fe(CN)_6]^{4-}$	7×10^2
$[W(CN)_8]^{4-}$, $[IrCl_6]^{2-}$	$[W(CN)_8]^{3-}$, $[IrCl_6]^{3-}$	6×10^7
$[IrCl_6]^{3-}$, $[IrCl_6]^{2-}$	$[IrCl_6]^{2-}$, $[IrCl_6]^{3-}$	1×10^3
$[Co(NH_3)_6]^{2+}$, $[Co(NH_3)_6]^{3+}$	$[Co(NH_3)_6]^{3+}$, $[Co(NH_3)_6]^{2+}$	1×10^{-9}
$[V(H_2O)_6]^{2+}$, $[V(H_2O)_6]^{3+}$	$[V(H_2O)_6]^{3+}$, $[V(H_2O)_6]^{2+}$	1×10^{-2}

[a] From F. Basolo and R. G. Pearson. *Mechanisms of Inorganic Reactions*, 2nd ed. New York: Wiley, 1967.

symmetry systems), there subsequently will be fairly substantial reorganization in the metal–ligand bond lengths; this type of complex generally reacts the most slowly. Reactions involving transfer of electrons between t_{2g} orbitals (O_h symmetry), on the other hand, can be quite fast because there is little difference in bond length anticipated. The symmetry arrangement of the electrons in the orbitals is also a factor. Consider the reaction between the low-spin complexes $[Fe(CN)_6]^{4-}$ and $[Fe(CN)_6]^{3-}$:

$$
\begin{array}{cccc}
\text{Fe(II)} + \text{Fe(III)} & \rightarrow & \text{Fe(III)} + \text{Fe(II)} \\
d^6 \quad\quad d^5 & & d^5 \quad\quad d^6 \\
(t_{2g})^6 \quad (t_{2g})^5 & & (t_{2g})^5 \quad (t_{2g})^6
\end{array}
\qquad \textbf{(21.42)}
$$

This reaction is *electronically symmetrical* and does not involve any changes in e_g occupancy. Consequently, it is reasonably fast ($k \sim 7 \times 10^2$ mol^{-1} dm^3 s^{-1}). For the following reaction, however, there are changes in e_g occupancy; thus, the reaction is substantially slower ($k \sim 2 \times 10^{-5}$ mol^{-1} dm^3 s^{-1}).

$$
\begin{array}{cccc}
\text{Cr(II)} + \text{Cr(III)} & \rightarrow & \text{Cr(III)} + \text{Cr(II)} \\
d^4 \quad\quad d^3 & & d^3 \quad\quad d^4 \\
(t_{2g})^3(e_g)^1 \quad (t_{2g})^3 & & (t_{2g})^3 \quad (t_{2g})^3(e_g)^1
\end{array}
\qquad \textbf{(21.43)}
$$

Inner-Sphere Electron-Transfer Reactions of Octahedral Complexes

The elegant work of Taube's research group on Co(III)/Cr(II) complexes in acid solution provided the first evidence for electron transfers taking place through bridging groups:

$$
[Co(NH_3)_5Cl]^{2+} + [Cr(H_2O)_6]^{2+} \xrightarrow[\text{H}_2\text{O}]{\text{H}^+}
$$
$$
[Co(H_2O)_6]^{2+} + [Cr(H_2O)_5Cl]^{2+} + 5NH_4^+ \qquad \textbf{(21.44)}
$$

During electron transfer, the Cl^- group is transferred from Co(III) to Cr(II), while Co(III) is reduced to Co(II) and Cr(II) is oxidized to Cr(III). The proposed intermediate prior to electron transfer in the reaction is

$$
[(NH_3)_5Co \text{---} Cl \text{---} Cr(H_2O)_5]^{4+}
$$
$$
\text{Co(III)} \qquad\qquad\qquad \text{Cr(II)}
$$

A stable analog of such an intermediate is shown in the following reaction:

$$
[Fe(CN)_6]^{3-} + [Co(CN)_5]^{3-} \rightarrow [(CN)_5Fe\text{—}CN\text{—}Co(CN)_5]^{6-} \qquad \textbf{(21.45)}
$$

The following mechanism has been substantiated by radiochemical tracer studies using ^{36}Cl:

$$[Cr(H_2O)_6]^{2+} \longrightarrow [Cr(H_2O)_5]^{2+} + H_2O$$

$$[Co(NH_3)_5Cl]^{2+} + [Cr(H_2O)_5]^{2+} \longrightarrow [(NH_3)Co^{III}---Cl---Cr^{II}(H_2O)_5]^{4+}$$

Immediately Before e^- Transfer

$$[(NH_3)_5Co^{II}---Cl---Cr^{III}(H_2O)_5]^{4+}$$

Immediately After e^- Transfer

$$[Co(NH_3)_5]^{2+} + [Cr(H_2O)_5Cl]^{2+}$$

H^+, H_2O

$$[Co(H_2O)_6]^{2+} + 5NH_4^+ \qquad\qquad \textbf{(21.46)}$$

This type of reaction has been studied for many $[Co(NH_3)_5X]^{2+}$ complexes, and the effect of varying X has been investigated. For instance, when X = halide, the rates of reaction decrease in the order $I^- > Br^- > Cl^- > F^-$. Because electron transfer is presumably easier when the bridging ligand is highly polarizable, the observed order for the halides provides strong support for the inner-sphere mechanism. It should be emphasized that this process is also known as *ligand atom transfer* and is analogous to free radical abstraction reactions:

$$C^{III}H_3 + CH_4 \rightarrow H_3C---H---CH_3 \rightarrow CH_4 + CH_3^{\cdot}$$

21.4 *Fluxional Molecules*

With the advent of high-field, multinuclear NMR instrumentation equipped with variable-temperature accessories, it is becoming increasingly clear that many molecules are not really stereochemically rigid. There are numerous examples in the literature for various different coordination numbers. Probably, the simplest to visualize is the inversion of the pyramidal three-coordinate ammonia molecule:

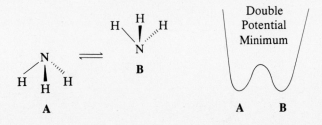

This inversion may be likened to the behavior of an umbrella when it is turned inside out on a wet, windy day. It is believed that all pyramidal molecules undergo inversion; whether the inversion process can be detected ultimately depends on the barrier height in the double-potential minimum that describes the two states involved. Vibrational spectroscopy is useful for the detection of inversion in ammonia. There is an a_1 vibrational mode of the C_{3v} symmetry NH_3 molecule (ν_2), which closely matches the presumed inversion process. The vibrational frequency for this mode is split into a doublet at 968.08 and 931.58 cm^{-1} in the IR spectrum of gaseous NH_3. This 37 cm^{-1} splitting provides evidence of the two inverted states for NH_3:

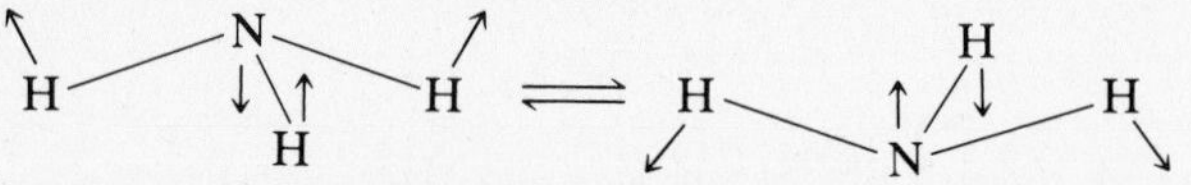

Fluxionality is tremendously important for five-coordinate, trigonal-bipyramidal molecules, especially in AB_5 systems amenable to NMR studies like PF_5 (^{19}F and ^{31}P) and $Fe(CO)_5$ (^{13}C and ^{17}O). From the NMR studies on these two molecules, the conclusion is that the five ligands exchange between the nonequivalent axial and equatorial sites in the trigonal bipyramids through an intramolecular pathway. The most prominent mechanism for the exchange has been termed the *Berry mechanism*, or *pseudorotation*:

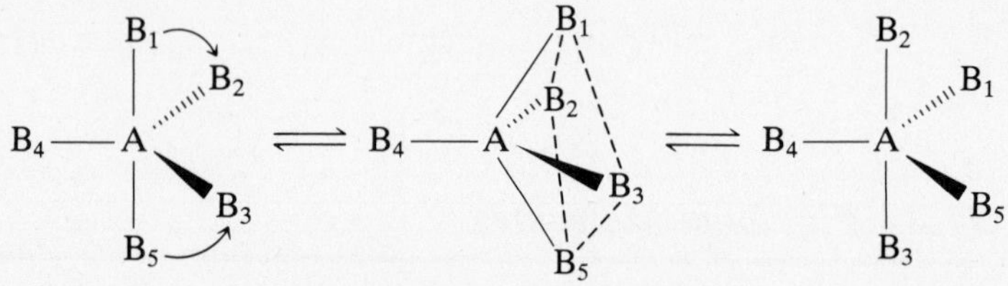

The classic work on fluxional organometallic molecules was concerned with the iron and ruthenium complexes of cyclooctatatraene $(C_8H_8)M(CO)_3$. On the basis of low-temperature ^1H-NMR studies, it was concluded that above the coalescence temperature of the NMR signals, the exchange mechanism involves a 1, 2 shift, as shown below:

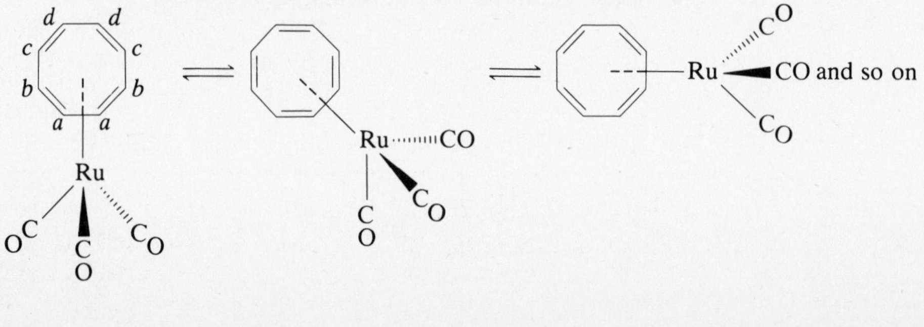

In the case of the iron complex, it was not possible to freeze out the limiting structures, but for ruthenium the limiting structures could be identified at about $-150\,°C$. As expected, four different ^{1}H-NMR resonances corresponding to the four different sets of equivalent protons (a, b, c, d) were observed. Upon raising the temperature, the four signals collapsed into one because of the fluxional motion of the cyclooctatetraene ring.

Other examples of fluxionality include $Rh_4(CO)_{12}$ and $(\eta^5\text{-}C_5H_5)$-$(\eta^1\text{-}C_5H_5)Fe(CO)_2$. The possible exchange mechanisms are shown below:

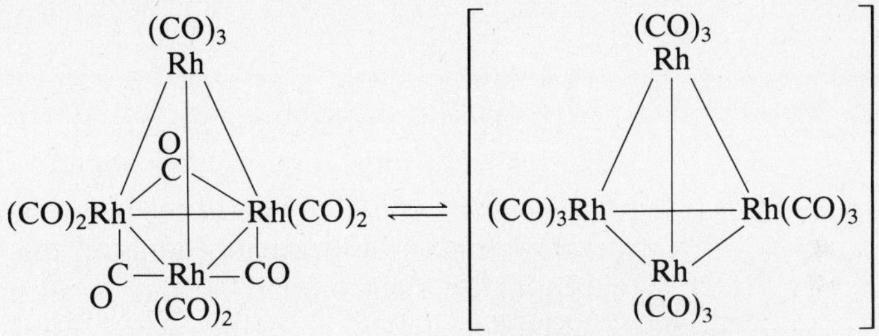

Different Terminal CO Groups Become Bridging, Thereby Permuting the CO Groups

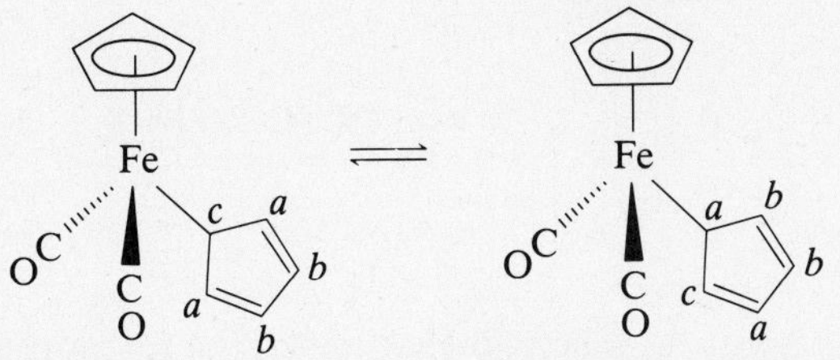

1,2-Shift of η^1-C_5H_5 Ring

In each of the above molecules, there are more than two chemically equivalent configurations, and as we have already noted, the molecules are termed *fluxional*. If the configurations are chemically *inequivalent*—for example, square-planar and tetrahedral $Ni[P(C_2H_5)_3]_2Cl_2$—then the molecules are termed *stereochemically nonrigid*, not fluxional. Interconversions of geometrical isomers are the most common examples of stereochemical nonrigidity, for example, *fac*- and *mer*-$Mo(CO)_3L_3$.

21.5 Oxidative–Addition/Reductive–Elimination Reactions

These classes of reaction became especially well known in the 1960s following the discovery of Vaska's compound *trans*-$IrCl(CO)(PPh_3)_2$. This compound is remarkable in that it oxidatively adds O_2 and many other simple molecules such as HCl and Cl_2, often reversibly. In its reversible reaction with O_2, Vaska's compound bears a chemical resemblance to hemoglobin; for this reason, the mechanistics of oxidative–addition reactions have been widely studied. With HCl the following reaction takes place:

$$trans\text{-}IrCl(CO)(PPh_3)_2 + HCl \longrightarrow$$

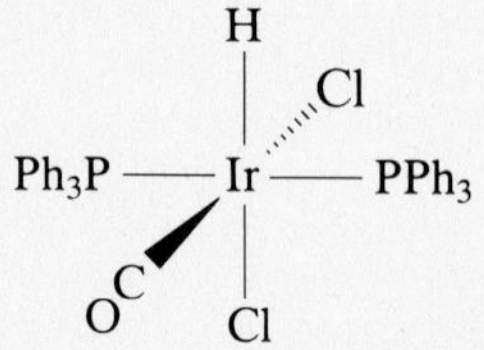

The Ir(I) is formally oxidized to Ir(III), and the coordination number increases from four in the square-planar substrate to six in the octahedral product. The oxidative–addition may be crudely described as the Ir(I) species acting as a Lewis base (donating one electron pair) and a Lewis acid (accepting an electron pair) at the same time.

Mechanistically speaking, these rather simple addition reactions are often complicated. For example, consider the following reaction of Vaska's compound with molecular hydrogen:

$$trans\text{-}IrCl(CO)(PPh_3)_2 + H_2 \rightleftharpoons$$

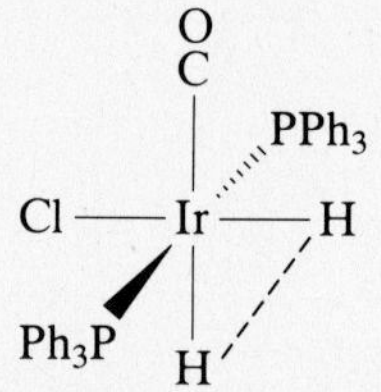

The two H atoms of the H_2 molecule finish up cis to each other in the product. It has been suggested that the pair of electrons from the Ir(I) substrate enter into the σ-antibonding orbital on the H_2 molecule. Such a situation would result in a reduction in the H—H bond order, and so the transition state in the reaction involves a three-center attack. The kinetics are in accord with such a concerted mechanism, and the rate law is first-order in both the starting complex and H_2:

$$Rate = k[trans\text{-}IrCl(CO)(PPh_3)_2][H_2]$$

Vaska's compound and its square-planar analogs with other halides and halogenoid ligands usually react with hydrogen to give a single octahedral isomer in which the tertiary phosphine ligands are trans to one another. Other types of square-planar compounds, however, afford mixtures of two octahedral isomers:

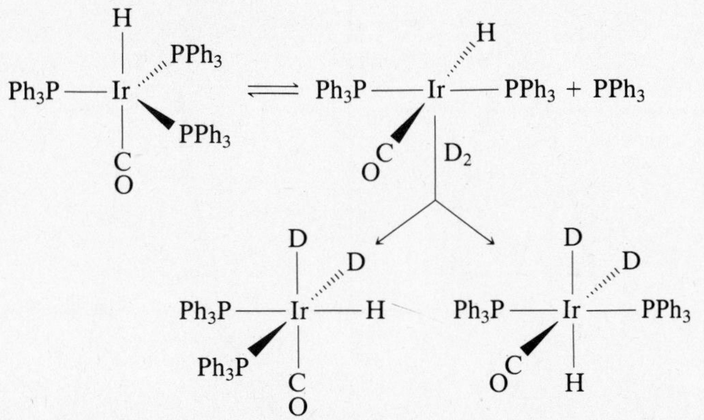

In the early 1980s, the scientific community was startled, and even skeptical, to learn that it was possible to prepare $W(CO)_3[P(Pr\text{-}i)_3]_2(H_2)$, where Pr-$i$ = isopropyl, in which the H_2 molecule remains intact and is bonded sideways-on. The H—H distance was determined from low-temperature X-ray diffraction to be 75 ± 16 pm. The sideways-on bonding of H_2 was subsequently verified by a ^{1}H-NMR study of the analogous HD complex. The resonance for the H atom of the coordinated HD appears as a $1:1:1$ triplet due to coupling with the D atom ($I = 1$). The J(HD) value for the coordinated HD molecule is 33.5 Hz, comparable to the value for HD gas of 43.2 Hz. The isolation of this molecular hydrogen complex was particularly interesting because it provided support for the proposed intermediates for oxidative–addition reactions involving molecular hydrogen.

Over thirty molecular H_2 complexes have now been reported in which the H—H bond remains essentially intact. A particularly simple example is $Cr(CO)_5(H_2)$, which is produced by the UV photolysis of $Cr(CO)_6$ dissolved in liquid xenon doped with H_2 at about 200 K. This compound and its HD and D_2 analogs are readily identified by the respective H—H, H—D, and D—D stretching vibrations in their IR spectra (Figure 21.3). As expected, the increase in reduced mass in going from H_2 to D_2 leads to a decrease in energy of the associated stretching vibration.

21.6 *Free-Radical Reactions*

Inorganic chemists are now becoming aware that free-radical reactions may be more widespread in solution than originally thought. The following three examples suffice to give you an idea of what kinds of reaction can take place.

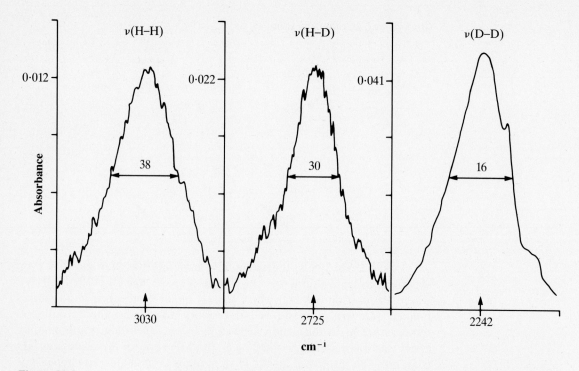

Figure 21.3
IR spectra of $Cr(CO)_5(H_2)$ and those of its deuterium-substituted analogs in liquid xenon at -70 °C. Note the shift in the $\nu(H-H)$ band position upon deuterium substitution. Reproduced with permission from M. Poliakoff, University of Nottingham, U.K., 1965.

Benzyl bromide ($PhCH_2Br$) reacts with Cr^{2+} ions in aqueous solution to produce the organochromium(III) cation according to the following two-step free-radical mechanism:

$$PhCH_2Br + Cr^{2+} \xrightarrow{\text{H}_2\text{O}} [CrBr]^{2+} + PhCH_2^{\bullet} \qquad \begin{array}{l}\text{Inner-Sphere}\\\text{Electron Transfer}\end{array}$$

$$PhCH_2^{\bullet} + Cr^{2+} \xrightarrow{\text{Fast}} [CrCH_2Ph]^{2+} \qquad \textbf{(21.47)}$$

Overall Reaction

$$PhCH_2Br + 2Cr^{2+} \longrightarrow [CrBr]^{2+} + [CrCH_2Ph]^{2+} \qquad \textbf{(21.48)}$$

The final product is believed to be hydrated $[(H_2O)_5CrCH_2Ph]^{2+}$. A similar mechanism has been proposed for the reaction of m,m'-$(BrCH_2C_6H_4)_2O$ with Cr^{2+} in 1 : 1 acetone–water solution. In this case, the reaction is

$$m,m'\text{-}(BrCH_2C_6H_4)_2O + 4Cr^{2+} \rightarrow$$
$$2[CrBr]^{2+} + [m,m'\text{-}(CrCH_2C_6H_4)_2O]^{4+} \qquad \textbf{(21.49)}$$

Under pseudo-first-order conditions with the bromide being the limiting reagent, the rate law is

$$\text{Rate} = k_{\text{obsd}}[Cr^{2+}]$$

in agreement with the type of mechanism given above.

The above free-radical reactions are essentially one-electron, oxidative additions. It is also believed that certain two-electron, oxidative–addition reactions proceed by sequential one-electron steps:

$$Pt(PPh_3)_3 + CH_3I \rightarrow PtI(PPh_3)_3 + CH_3^{\cdot} \tag{21.50}$$

$$PtI(PPh_3)_3 + CH_3^{\cdot} \rightarrow PtI(CH_3)(PPh_3)_2 + PPh_3 \tag{21.51}$$

Many metal carbonyl–bridged dimers such as $[CpMo(CO)_3]_2$ and $M_2(CO)_{10}$ (M = Mn, Re) can be cleaved homolytically under photochemical conditions to yield radicals:

$$(OC)_5M—M(CO)_5 \xrightarrow{h\nu} 2M(CO)_5^{\cdot}$$

These extremely reactive radicals can participate in numerous types of reactions including ligand substitution and atom transfer:

$$Mn(CO)_5^{\cdot} + PPh_3 \rightarrow Mn(CO)_4(PPh_3)^{\cdot} + CO \tag{21.52}$$

$$Re(CO)_5^{\cdot} + CCl_4 \rightarrow Re(CO)_5Cl + CCl_3^{\cdot} \tag{21.53}$$

Kinetic studies on these reactions have been completed recently. For example, flash photolysis of $Mn_2(CO)_{10}$ in the presence of organic halides (RX = CCl_4, $CHBr_3$, and so on) yields $Mn(CO)_5X$ exclusively. Bimolecular atom-transfer constants were determined by monitoring the decay of the visible- and IR-absorption bands of the radicals. The proposed mechanism is

$$Mn_2(CO)_{10} \xrightarrow[\text{Fast}]{h\nu} 2Mn(CO)_5^{\cdot}$$

$$Mn(CO)_5^{\cdot} + RX \xrightarrow{\text{Slow}} Mn(CO)_5(RX)^{\cdot}$$

$$Mn(CO)_5(RX)^{\cdot} \xrightarrow{\text{Fast}} Mn(CO)_5X + R^{\cdot}$$

One major research area for many years has been the study of coenzyme B_{12} and its catalytic behavior in certain enzymatic reactions (see Chapter 24). It now appears that these reactions proceed through free-radical mechanisms. Coenzyme B_{12} has a cobalt ion bonded to four nitrogen atoms in a square plane—the nitrogen atoms are part of a large macrocyclic ligand called *corrin*. The corrin also provides a fifth-coordination ligand, which bonds to the cobalt at the bottom of an octahedron:

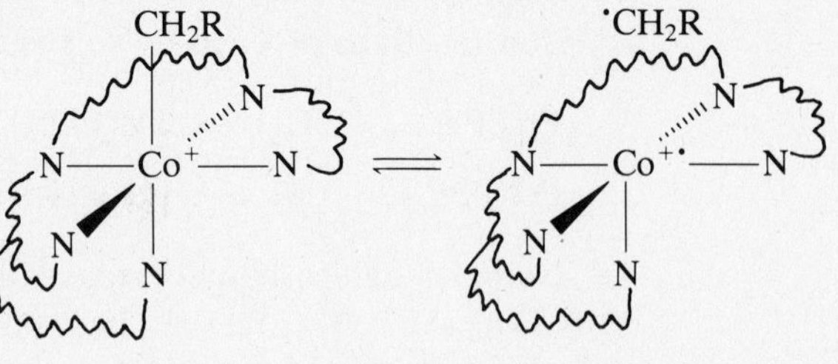

Note that coenzyme B_{12} differs from vitamin B_{12} in that the CH_2R grouping is replaced by CN^- in the vitamin. Vitamin B_{12} is used in the treatment of pernicious anemia and as a component of many multivitamins.

21.7 Insertion Reactions

The most famous insertion reactions are probably those involving organometallic complexes in which a carbonyl group inserts into a metal–alkyl bond with the formation of a metal–acyl bond (the other ligands have been omitted for the sake of clarity):

$$(OC)M—R + CO \rightarrow (OC)M—\underset{\underset{O}{\|}}{C}—R \qquad (21.54)$$

Similar reactions occur when the entering ligand is not CO as well, for example, PPh_3:

$$(OC)M—R + L \rightarrow LM—\underset{\underset{O}{\|}}{C}—R \qquad (21.55)$$

These reactions have been studied kinetically because of the direct relation between them and the industrial hydroformylation of alkenes (addition of HCHO in the form of $CO + H_2$) in the presence of cobalt carbonyl catalysts. More is said about hydroformylation in Chapter 23, but the important insertion step is the following when the starting alkene is ethylene:

$$C_2H_5Co(CO)_4 + CO \rightarrow C_2H_5—\underset{\underset{O}{\|}}{C}—Co(CO)_4 \qquad (21.56)$$

Among the complexes that have been studied kinetically are $CH_3Mn(CO)_5$, $CpFe(CO)_2CH_3$, and $CpMo(CO)_3CH_3$. There has been a long debate about whether the apparent insertion step is actually CO insertion or migration of an alkyl group:

$$
\begin{array}{cc}
\underset{\text{CO Insertion}}{
\begin{matrix}
\text{O} \\
\| \\
\text{C} \\
\text{M} - \text{R}
\end{matrix}} &
\underset{\text{R Migration}}{
\begin{matrix}
\text{O} \\
\| \\
\text{C} \\
\text{M} - \text{R}
\end{matrix}}
\end{array}
$$

CO Insertion   **R Migration**

The kinetics of these reactions have been widely investigated—for example, the iron complex $CpFe(CO)_2CH_3$ in tetrahydrofuran (THF) as a solvent. The final mechanism proposed involves a solvent-assisted methyl migration:

This mechanism can be written in shorthand form as

$$
A \underset{k_2}{\overset{k_1}{\rightleftharpoons}} B \underset{k_4}{\overset{k_3}{\rightleftharpoons}} C
$$

Applying the steady-state approximation to the concentration of the intermediate B, we obtain

$$
-\frac{d[A]}{dt} = k_1[A] - k_2\left(\frac{k_1[A] + k_4[C]}{k_2 + k_3[L]}\right)
$$

If $k_2 \simeq k_3$, there is competition for the intermediate B, and the observed rate constant depends on [L], as is observed when L is a tertiary phosphine or phosphite. Moreover, because the reactions go to completion, k_4 can be ignored and the complete equation reduces to

$$
-\frac{d[A]}{dt} = \left(\frac{k_1 k_3[L]}{k_2 + k_3[L]}\right)[A] = k_{\text{obsd}}[A]
$$

that is,

$$
\frac{1}{k_{\text{obsd}}} = \frac{k_2}{k_1 k_3}\left(\frac{1}{[L]}\right) + \frac{1}{k_1}
$$

Support for the proposed mechanism is because the plots of $1/k_{\text{obsd}}$ versus $1/[L]$ are indeed linear (Figure 21.4). It should be mentioned,

Figure 21.4
Plot of $1/k_{obsd}$ versus $1/[PPh_3]$ for the reaction of $CpFe(CO)_2CH_3$ with PPh_3 in tetrahydrofuran at 50.7 °C. Reproduced with permission from I. S. Butler, F. Basolo, and R. G. Pearson. *Inorg. Chem.* 6 (1967): 2074.

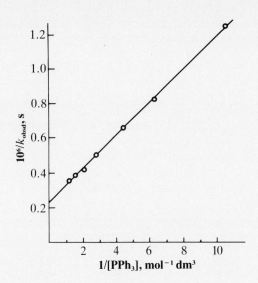

however, that although solvent-assisted alkyl migration is the generally accepted mechanism for CO insertion reactions, it is probably not that simple.

BIBLIOGRAPHY

Suggested Reading

Atwood, J. D. *Inorganic and Organometallic Reaction Mechanisms*. Pacific Grove, Calif.: Brooks/Cole, 1985.

Basolo, F., and R. G. Pearson. *Mechanisms of Inorganic Reactions: A Study of Metal Complexes in Solution*, 2nd ed. New York: Wiley 1967.

Heck, R. F. *Organotransition Metal Chemistry: A Mechanistic Approach*. New York: Academic Press, 1974.

Langford, C. H., and H. B. Gray. *Ligand Substitution Processes*. Menlo Park, Calif.: Benjamin/Cummings, 1965.

Purcell, K. F., and J. C. Kotz. *Inorganic Chemistry*. Philadelphia: Saunders, 1977.

Tobe, M. L. *Inorganic Reaction Mechanisms*. London: Nelson, 1972.

Wilkins, R. G. *The Study of Kinetics and Mechanisms of Reactions of Transition Metal Complexes*. Boston: Allyn and Bacon, 1974.

PROBLEMS

21.1   Give an example (not already in the chapter) for each of the six main types of chemical reaction presented in Section 21.1.

21.2   Explain carefully what is meant by each of the following terms:
- **a.** Coordination number
- **b.** Oxidation state
- **c.** Inert and labile complexes
- **d.** Aquation
- **e.** Base hydrolysis
- **f.** Pseudo-first-order conditions
- **g.** Fluxional molecules
- **h.** Stereochemical nonrigidity

21.3   Write essays describing the kinetics and mechanisms associated with the substitution reactions of
- **a.** Octahedral complexes
- **b.** Four-coordinate complexes

21.4   For what aspects of kinetic studies are the following men famous?
- **a.** Taube
- **b.** Eigen
- **c.** Chernaev
- **d.** Chatt

21.5   Give an example for each of the following reaction types:
- **a.** Isomerization
- **b.** Racemization
- **c.** S_N1CB
- **d.** Reductive–elimination

21.6   Propose mechanisms for the following reactions:
- **a.** Cl^- substitution in $[(\eta^2\text{-}C_2H_4)PtCl_3]^-$ by Br^- in methanol to give *trans*-$[(\eta^2\text{-}C_2H_4)PtCl_2Br]^-$.
- **b.** Arene substitution in $(\eta^6\text{-}C_6H_6)Mo(CO)_3$ by PPh_3 to give *fac*-$Mo(CO)_3(PPh_3)_3$.

21.7   Write the most likely mechanisms for the following reactions bearing in mind your knowledge of the established mechanisms of closely related reactions. (NOTE: L = monodentate ligand; A—A = bidentate chelating ligand.)
- **a.** $Fe(CO)_2(NO)_2 + L \rightarrow Fe(CO)(NO)_2L + CO$
- **b.** $CH_3Co(CO)_4 + L \xrightarrow{\text{THF}} (CH_3CO)Co(CO)_3L$
- **c.** *cis*-$[Mn(CO)_4Br_2]^- + L \rightarrow$ *cis*-$Mn(CO)_4LBr + Br^-$
- **d.** *cis*-$Re(CO)_4LCl + A\text{—}A \rightarrow$ *fac*-$Re(CO)_3(A\text{—}A)Cl + CO + L$

21.8   Account for the following observations:
- **a.** The base hydrolysis of $[Co(NH_3)_4Cl_2]^+$ obeys a second-order rate expression.
- **b.** The product formed upon successive treatment of $K_2[PdBr_4]$ with sodium nitrite, ammonia, and trimethylamine is 1-ammine-2-chloro-4-(trimethylamine)nitrito-N-palladium(II).

21.9   Given that $\Delta H^\ddagger = 80$ kJ mol^{-1} and $\Delta S^\ddagger = -13$ e.u., suggest a plausible mechanism for the following CO–substitution reaction:

$$Mn(CO)_4(NO) + L \rightarrow Mn(CO)_3(NO)L + CO$$

[For further information, see H. Wawersik and F. Basolo. *J. Amer. Chem. Soc.* 89 (1967): 4626.]

21.10   Kinetic data have been obtained for the replacement of *cis*-cyclooctene in $CpMn(CO)_2(C_8H_{14})$ and $CpMn(CO)(CS)(C_8H_{14})$ by PPh_3. The observed reaction rates are first-order in substrate and independent of the concentration of PPh_3. The measured activation parameters are shown below. Propose mechanisms for these substitution reactions.

Compound	$\Delta H^{\ddagger}$ (kJ mol^{-1})	$\Delta S^{\ddagger}$ (e.u.)
$CpMn(CO)_2(C_8H_{14})$[a]	146	28
$CpMn(CO)(CS)(C_8H_{14})$[b]	128	20

[a] From R. J. Angelici and W. Loewen. *Inorg. Chem.* 6 (1967): 682.
[b] From I. S. Butler and A. E. Fenster. *Inorganica Chim. Acta* 7 (1973): 79.

PART EIGHT

Advanced Topics in Coordination Chemistry

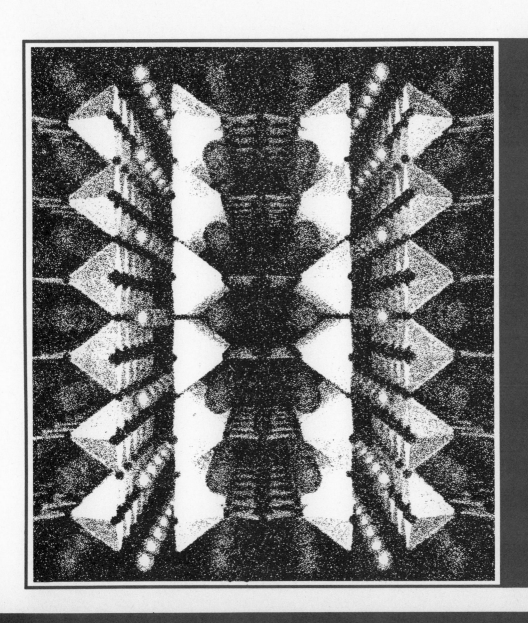

22

Organometallic Chemistry

Organometallic chemistry is an important subdiscipline that bridges the fields of organic and inorganic chemistry. As the name implies, compounds that fall into this classification contain both a metal and an organic component. By some people's definition, organometallic compounds are required to have at least one metal–carbon bond, and the metal can be either a main-group element or a transition metal. Compounds containing organic moieties linked to the metal through an atom other than carbon—that is, oxygen, sulfur, nitrogen, phosphorus, and so on—are related to organometallic compounds under the broader term of *metal–organic chemistry*. Metal alkyls, metal carbonyls, and metal–arene compounds are therefore all classified as organometallics. In addition, organic compounds of elements such as silicon and germanium are also usually considered organometallics. Figure 22.1 gives some examples of typical organometallic compounds.

The earliest reports of organometallic compounds are believed to date back to the discoveries of the platinum–ethylene salt $K[PtCl_3(C_2H_4)]$

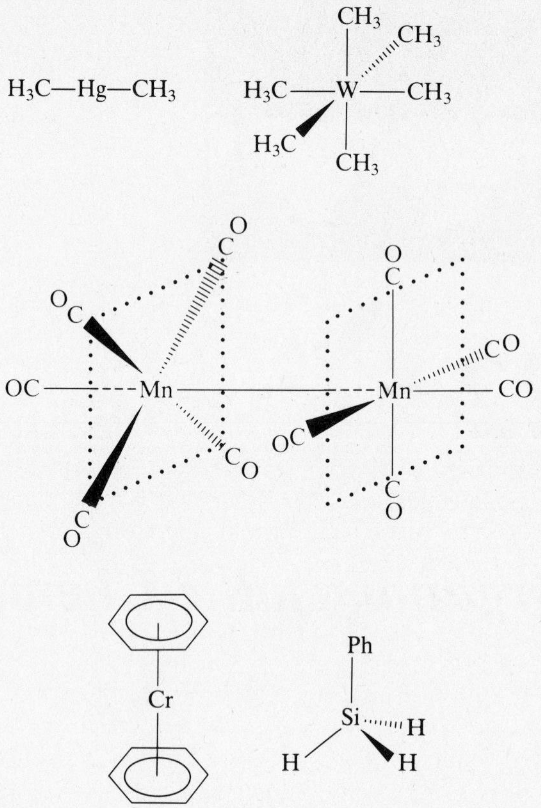

Figure 22.1
Some representative
examples of
organometallic
compounds.

(Zeise 1827) and diethylzinc(II) $(C_2H_5)_2Zn$ (Frankland 1852), although the true nature of the platinum anion was not discovered until the early 1950s when its structure was determined by X-ray diffraction. Organometallic chemistry is a well-established field, which has been recognized by the award of four Nobel Prizes in Chemistry:

1912 V. Grignard (France) for his discovery of organomagnesium reagents useful in organic synthesis

1963 K. Ziegler (West Germany) and G. Natta (Italy) for their work on the chemistry and technology of high polymers, especially the synthesis of isotactic polypropylene using organometallic catalysts

1973 E. O. Fischer (West Germany) and G. Wilkinson (United Kingdom) for their independent discoveries of organometallic "sandwich" compounds

1984 K. Fukui (Japan) and R. Hoffmann (United States) for their research into the theoretical basis of chemical reactions, including organometallic reactions

There are several scientific journals devoted solely to organometallic chemistry, including *Organometallics, Journal of Organometallic Chemistry*, and *Applied Organometallic Chemistry*. Organometallic compounds are finding widespread use as reagents or catalysts in organic synthesis—for example, the syntheses of *l*-DOPA (used in the treatment of Parkinson's disease) and *d,l*-estrone, using soluble, chiral organorhodium compounds and $CpCo(CO)_2$ ($Cp = \eta^5$-C_5H_5), respectively, as catalysts. Large-scale industrial applications of organometallics are also well known—for example, the rhodium-catalyzed carbonylation of CH_3OH to CH_3COOH (see Chapter 23).

This chapter introduces you to the principal areas of organometallic chemistry. Most of the earlier research in this field was directed toward the synthesis of new compounds. Today, much more emphasis is being placed on the potential applications of organometallic compounds, especially as homogeneous catalysts, as reagents in organic synthesis and even as markers in molecular biology. We begin with a brief introduction to the metal–carbon bond and then discuss main-group and transition metal organometallic compounds.

22.1 The Metal–Carbon Bond

Metal–carbon bonds may involve two or more electrons. Table 22.1 illustrates some of the various types of known metal–carbon bonds. Note the wide choice in the number of electrons that can be donated by the organic ligands. Table 22.2 lists the number of electrons donated by the more common ligands encountered in organometallic chemistry. The main-group elements mainly form two-electron covalent bonds with alkyl and aryl ligands, as in $(CH_3)_2Hg$.

Electron counting is a useful starting point for the systematization of the compositions and structures of organometallic compounds. For example, many atoms in main-group compounds, like $(CH_3)_4Si$, attain valence-shell electronic structures corresponding to those of the noble gases, namely, eight-valence electrons (Lewis octet rule). A parallel situation exists for organotransition metal compounds—the *eighteen-electron rule*, also known as the *effective atomic number rule*, whereby the central metal atom in a complex surrounds itself with sufficient ligands to achieve the same atomic number as the next noble gas. This means that the number of valence electrons originating with the central metal atom *plus* the number of electrons donated by the surrounding ligands must total eighteen because the valence orbitals are ns ($2e^-$), np ($3 \times 2 = 6e^-$), and $(n-1)d$ ($5 \times 2 = 10e^-$), where n is the principal quantum number.

Organotransition metal compounds are usually diamagnetic and have stoichiometries such that the number of valence electrons from the central metal, together with the electrons donated by the various ligands, does

Table 22.1 *Some Selected Types of Metal–Carbon Bonding*

Bond Type	Number of Electrons Donated	Eta Notation[a]
M—CH$_3$	2 from [CH$_3$]$^-$	η^1
M=CR$_2$	2 from carbene ligand	η^1
M ← ‖ (CR$_2$=CR$_2$)	2 from alkene ligand	η^2
M≡CR	3 from carbyne ligand	η^1
(allyl structure)	4 from allyl anion ligand	η^3
(diene structure)	4 from diene ligand	η^4
(cyclopentadienyl–M)	6 from cyclopentadienyl ligand, [C$_5$H$_5$]$^-$ (Cp$^-$)	η^5
(arene–M)	6 from arene ligand	η^6

[a] In general, η^n, where n = number of carbon atoms considered to be bonded to metal atom M.

Table 22.2 *Number of Electrons Donated to a Transition Metal by Some Typical Ligands Found in Organometallic Compounds*

Ligand	Number of Electrons Donated
H$^-$, [alkyl]$^-$ ([CH$_3$]$^-$, [C$_2$H$_5$]$^-$, etc.), [aryl]$^-$ ([C$_6$H$_5$]$^-$, [C$_6$H$_5$CH$_2$]$^-$, etc.), CO, CS, CSe, PR$_3$, P(OR)$_3$, SbR$_3$, F$^-$, Cl$^-$, Br$^-$, I$^-$, carbenes (=CR$_2$), alkenes	2
NO, NS, carbynes (≡CR)	3
Dienes, [π-allyl]$^-$	4
[C$_5$H$_5$]$^-$, [C$_4$H$_4$]$^{2-}$, [C$_7$H$_7$]$^+$, arenes (C$_6$H$_6$, C$_6$H$_5$CO$_2$CH$_3$, etc.) [C$_3$H$_3$]$^{3-}$	6
Tetraenes (C$_8$H$_8$)	8
[C$_8$H$_8$]$^{2-}$	10

total eighteen—that is, the eighteen-electron rule is obeyed. As an example, consider tetracarbonylnickel(0) $Ni(CO)_4$:

Ni(0):	[Ar] $4s^2 3d^8$	
Ni(0) in $Ni(CO)_4$:	[Ar] $3d^{10}$	10 $\Big\}$ $18e^-$
Four terminal CO groups		8
providing two electrons each		

Some other examples are pentacarbonyl(methyl)rhenium(I), CH_3-$Re(CO)_5$; bis(η^5-cyclopentadienyl)ruthenium(II), $(\eta^5\text{-}C_5H_5)_2Ru$; tricarbonyl ($\eta^6$-benzene)tungsten(0), $(\eta^6\text{-}C_6H_6)W(CO)_3$; and nonacarbonyl-bis[iron(0)], $Fe_2(CO)_9$:

$CH_3Re(CO)_5$	Re(I)	6
	$[CH_3]^-$ at	
	two electrons	2 $\Big\}$ $18e^-$
	Five terminal CO	
	groups at two electrons each	10
Cp_2Ru	Ru(II)	6 $\Big\}$ $18e^-$
	$2Cp^-$ at six electrons each	12
$(\eta^6\text{-}C_6H_6)W(CO)_3$	W(0)	6
	C_6H_6 at six electrons	6 $\Big\}$ $18e^-$
	Three terminal CO	
	groups at two electrons each	6
$Fe_2(CO)_9$	Fe(0)	8
	Three terminal CO	
	groups at two electrons each	6
	Three bridging CO	$\Big\}$ $18e^-$
	groups giving one electron each	3
	One Fe—Fe bond giving one	
	electron	1

The eighteen-electron rule can be used to predict the structures of new compounds. For instance, both ruthenium and osmium form metal carbonyl compounds with the empirical formula $M(CO)_4$. Mass spectroscopy indicates that the compounds are trimers of stoichiometry $M_3(CO)_{12}$, while their IR spectra show that there are only terminal CO groups present. Therefore, the most logical assumption is that each metal atom has four terminal CO groups attached to it. This means that we have eight-valence electrons for each metal(0) atom plus eight electrons for the four terminal CO groups, leading to a total of sixteen electrons—that is, two electrons short of the requisite eighteen. This situation can be most easily remedied by including two single M—M bonds as well. The final predicted structure that satisfies all these conditions is the triangular D_{3h} symmetry one shown below; this has been verified by X-ray diffraction and vibrational spectroscopy:

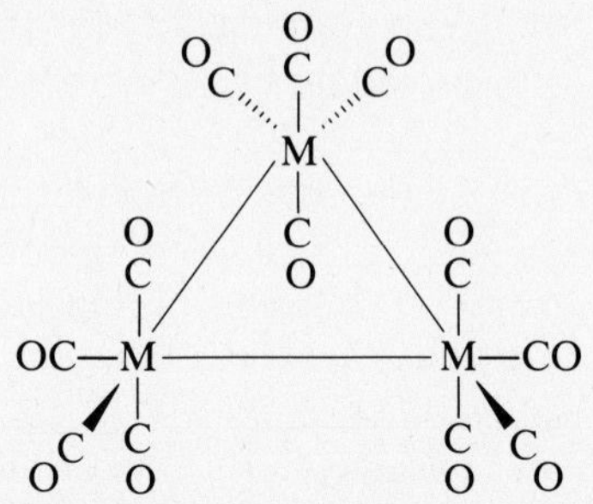

In some cases, multiple metal–metal bonds can also be predicted. For example, rhodium(I) forms a dimeric complex of stoichiometry $[Cp^*Rh(CO)]_2$ $[Cp^* = \eta^5\text{-}C_5(CH_3)_5]$, the IR spectrum of which indicates the presence of bridging but not terminal CO groups. Electron counting yields sixteen electrons: eight electrons for Rh(I), six electrons for Cp^{*-}, and two electrons for two bridging CO groups. To satisfy the eighteen-electron rule, there must therefore be a Rh≡Rh bond:

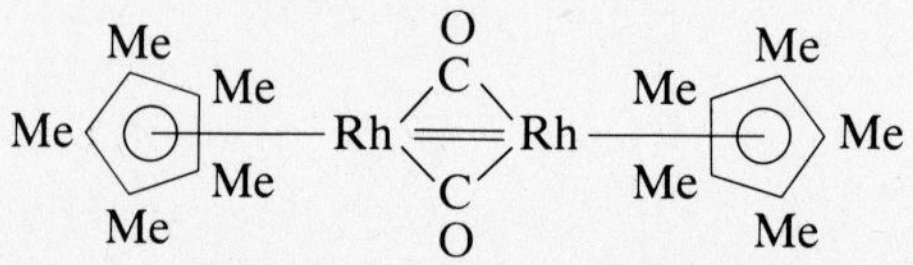

As with any rule, there are always exceptions, especially with the early transition metals, for example, $V(CO)_6$ (seventeen electrons), Cp_2V (fifteen electrons), Cp_2HfCl_2 (sixteen electrons), $(CH_3)_6W$ (twelve electrons), and Cp_2Ni (twenty electrons). Most square-planar complexes do not obey the eighteen-electron rule—for example, Vaska's compound *trans*-$IrCl(CO)(PPh_3)_2$ (sixteen electrons), due to the special crystal field stabilization afforded d^8 metal ions in this geometry. Low electron counts can also result from the presence of strong π-donor interaction with ligands or to steric limitations on the number of ligands, for example, Cp_2TiCl_2 (sixteen electrons).

In conclusion, the eighteen-electron rule is a good rule of valency provided the coordination sphere of the central metal is dominated by ligands like CO (π-acceptor ligands, see Section 22.4). It is a very poor rule, however, if the dominant ligands are π-donor ligands such as OH^- and halides (F^-, Cl^-, Br^-, I^-) or σ-bonding ligands like alkyl groups (even if they are strong-field ligands like $[CH_3]^-$).

22.2 *Main-Group Organometallics*

Because the important electronic subshells of main-group elements are the *s*- and *p*-valence orbitals, it is obvious why they can never satisfy the eighteen-electron rule. Two types of organometallic compounds are formed: ionic and covalently bonded. Group IA (1) metal alkyls provide good examples of both extremes.

Lithium alkyls are soluble in hydrocarbons and other nonpolar solvents indicative of covalent character. They react violently in air and water, often igniting spontaneously. Other Group IA (1) metal alkyls are appreciably more ionic than are the lithium compounds. They are not particularly soluble in hydrocarbon solvents, but they are still extremely reactive toward air and moisture. These ionic organometallic compounds can be prepared either by treatment of $(CH_3)_2Hg$ with the appropriate alkali metal or by the reaction of an alkyl halide with an alkali metal:

$$(CH_3)_2Hg + 2K \rightarrow 2[CH_3]^-K^+ + Hg \qquad (22.1)$$

$$C_2H_5Cl + 2K \rightarrow [C_2H_5]^-K^+ + KCl \qquad (22.2)$$

Their structures are typical of ionic compounds; for example, $[CH_3]^-K^+$ crystallizes in the nickel–arsenide (NiAs) lattice. Lithium alkyls, on the other hand, have strikingly different structures than the other metal alkyls; because of their greater covalent character, they form tetrameric or octahedral clusters, for example, $(CH_3Li)_4$ and $(C_2H_5Li)_6$. Figure 22.2 shows the structure of $(CH_3Li)_4$.

Clusters or polymers are also found for Groups IIA (2) and IIIA (13) metal alkyls. Monomeric precursors for Groups IA–IIIA (1, 2, and 13) elements—for example, RLi, R_2Be, and R_3Al—are electron-deficient and can achieve more stable electronic configurations by forming multi-center bonds (Figure 22.3). Some of the synthetic methods currently used

Figure 22.2
Structure of tetrameric methyl lithium $(CH_3Li)_4$.

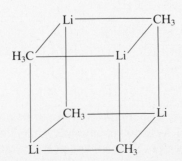

Figure 22.3
Qualitative picture
of a two-electron,
three-center bond
present in many
electron-deficient
compounds such as

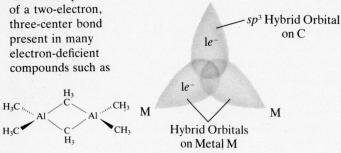

to prepare these types of main-group metal alkyl are as follows:

$$4CH_3Cl + 8Li \xrightarrow{C_6H_6} (CH_3Li)_4 + 4LiCl \qquad (22.3)$$

$$3(C_2H_5)_2Hg + 6Li \longrightarrow (C_2H_5Li)_6 + 3Hg \qquad (22.4)$$

$$xBeCl_2 + 2x(n\text{-}BuLi) \longrightarrow [(n\text{-}Bu)_2Be]_x + 2xLiCl \qquad (22.5)$$

$$x[(CH_3)_2Hg] + xBe \xrightarrow{110\,°C} [(CH_3)_2Be]_x + xHg \qquad (22.6)$$

$$3(CH_3)_2Hg + 2Al \longrightarrow (CH_3)_6Al_2 + 3Hg \qquad (22.7)$$

$$2Al + 3H_2 + 6C_2H_4 \longrightarrow (C_2H_5)_6Al_2 \qquad (22.8)$$

It is also possible to synthesize more complex organometallic compounds using lithium alkyls as starting material:

$$Cp_2Fe + n\text{-}BuLi \rightarrow Cp(\eta^5\text{-}C_5H_4Li)Fe + n\text{-}C_4H_{10} \qquad (22.9)$$

The extremely versatile synthetic reagent sodium cyclopentadienide (Na^+Cp^-) can similarly be used. It is conveniently prepared by reacting cyclopentadiene C_5H_6 with sodium metal dispersed in tetrahydrofuran. Cyclopentadiene is not stable as a monomer, and the $C_{10}H_{12}$ dimer must be cracked thermally immediately prior to use:

$$C_{10}H_{12} \xrightarrow{\Delta} 2C_5H_6 \qquad (22.10)$$

$$2C_5H_6 + 2Na \xrightarrow{THF} 2Na^+Cp^- + H_2 \qquad (22.11)$$

A significant industrial use of Group IA (1) metal organometallics is in the synthesis of ketones:

$$2(CH_3Li)_4 + 4CO_2 \rightarrow 4(CH_3)_2CO + 4Li_2O \qquad (22.12)$$

Lithium alkyls are also used as stereospecific polymerization catalysts in the formation of 1,4-*cis*-poly(isoprene), while aluminum alkyls play an important role in Ziegler–Natta alkene polymerization (see Chapter 23). Organomagnesium compounds such as Grignard reagents, RMgX, are also widely used and best prepared *in situ* by reacting Mg metal with an organic halide RX in either diethyl ether or tetrahydrofuran. The resulting solutions contain a variety of species, some of which are halogen-bridged and solvent-coordinated:

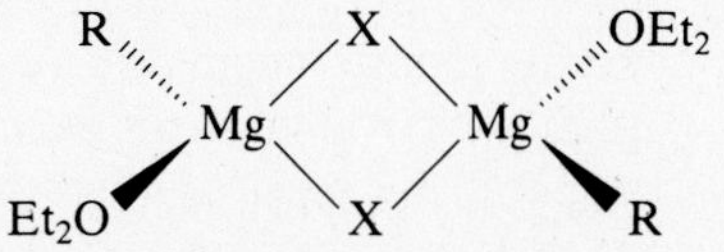

There are numerous organometallic compounds of Groups IIIA (13; Ga, In, Tl) and IVA (14; Si, Ge, Sn, Pb) elements. Some typical synthetic routes to these compounds are indicated below:

$$GaCl_3 + 3C_2H_5MgBr \rightarrow (C_2H_5)_3Ga + 3MgBrCl \qquad (22.13)$$

$$C_5H_6 + TlOH \rightarrow TlCp + H_2O \qquad (22.14)$$

$$SiCl_4 + 2CH_3MgCl \rightarrow (CH_3)_2SiCl_2 + 2MgCl_2 \qquad (22.15)$$

$$4C_2H_5Cl + 4Na/Pb \rightarrow (C_2H_5)_4Pb + 3Pb + 4NaCl \qquad (22.16)$$

These compounds are all highly covalent, soluble in organic solvents, and fairly volatile. Their chemical properties have been exploited both in the research laboratory and industry. For instance, thallium cyclopentadienide (TlCp) is now a recognized laboratory reagent for the introduction of the cyclopentadienyl ligand into transition metal compounds. Tetraethyllead $(C_2H_5)_4Pb$ used to be added to gasoline to increase the combustion efficiency of automobile engines and to eliminate "knocking"; today, however, the attendant pollution hazards and the poisoning effect of lead and its compounds on the automobile exhaust catalysts are leading to the elimination of leaded gasoline.

There is an enormous industrial market for organosilicon compounds such as R_2SiCl_2 because, upon controlled hydrolysis, these compounds afford siloxanes (silicones) $(-R_2Si-O-)_x$ (see page 483).

$$xR_2SiCl_2 \xrightarrow{xH_2O} (-R_2Si-O-)_x + 2xHCl \qquad (22.17)$$

which are used extensively in the manufacture of greases, lubricants, rubbers, and antifoaming and weatherproofing agents. Silicones are usually polymeric (linear, cyclic, or cross-linked) with high thermal stabilities and are resistant to oxidation and chemical attack. Commercially, the polymers usually have $R = CH_3$ or C_6H_5, but a wide range of different materials can be produced by incorporation of small amounts of other organic groups. Commercial development of organosilanes was made possible by the discovery of the so-called direct synthesis of methylsilanes:

$$3Si + 6CH_3Cl \xrightarrow{Cu} CH_3SiCl_3 + (CH_3)_2SiCl_2 + (CH_3)_3SiCl \quad (22.18)$$

The dimethyl compound is the most useful, and the reaction conditions are adjusted to maximize its yield.

There is also a fairly extensive organometallic chemistry of Group VA (15) elements (As, Sb, Bi). Organoarsenic compounds in particular have attracted attention because some of them have proved useful as chemotherapeutic agents in medicine. Group VA (15) organometallics are usually tricovalent—for example, $(C_2H_5)_3As$ and $(C_6H_5)_3Sb$—and can be prepared by treating the corresponding halides with Grignard reagents:

$$MCl_3 + 3RMgCl \rightarrow R_3M + 3MgCl_2 \qquad (22.19)$$

The related $(RO)_3M$ compounds are prepared by the action of alcohols ROH on MCl_3:

$$MCl_3 + 3ROH \rightarrow (RO)_3M + 3HCl \qquad (22.20)$$

Several mixed organohalo-compounds are known, for example,

$$SbCl_3 + 2R_2Hg \rightarrow R_2SbCl + 2RHgCl \qquad (22.21)$$

The R_3M compounds are readily oxidized to the very stable R_3MO compounds, which can also be obtained by reacting oxohalides with Grignard reagents:

$$M(O)Cl_3 + 3RMgCl \rightarrow R_3MO + 3MgCl_2 \qquad (22.22)$$

The R_3M and R_3MO derivatives form complexes with transition metals, for example, *fac*-$Mo(CO)_3(SbPh_3)_3$. Treatment of R_3As or R_3Sb with alkyl or aryl halides (RX) leads to quaternary salts $(R_4M)^+X^-$. The tetraphenylarsonium cation $[Ph_4As]^+$ is an especially large cation that is often used to help in the precipitation of large complex anions. Finally, there are a few cyclic organoarsines known with the general formula $(RAs)_n$ ($n = 4$, 5, ...). These cyclic compounds have single As—As bonds and can be prepared by reacting $RAsCl_2$ with elemental mercury:

$$4CF_3AsCl_2 + 4Hg \rightarrow (CF_3As)_4 + 4HgCl_2 \qquad (22.23)$$

The last main-group elements to be considered in this section are Groups IB (11; Cu, Ag, Au) and IIB (12; Zn, Cd, Hg) elements. In many ways, these elements strongly resemble those of Groups IA (1) and IIA (2), respectively. It is not too surprising, therefore, to find that they also form a variety of organometallic compounds. In fact, as already mentioned, one of the earliest organometallic compounds ever reported is $(C_2H_5)_2Zn$, which can be prepared in the following manner:

$$2Zn + 2EtI \rightarrow 2EtZnI \rightarrow Et_2Zn + ZnI_2 \qquad (22.24)$$

There are relatively few examples of Group IB (11) organometallics, but alkylgold compounds such as linear $[MeAuCl]^-$ and square-planar *cis*-$AuCl(Ph)_2(PPh_3)$ are among the more important. Organogold compounds can be synthesized by reacting Au(I) or Au(III) halides with lithium alkyls or Grignard reagents. In the case of copper, only Cu(I) is found in organometallic compounds—for example, $CpCu(PPh_3)$, which is obtained by reaction of the tetramer $[CuI(PPh_3)]_4$ with TlCp. Solutions of various copper(I) salts absorb acetylenes and CO with the formation of organometallic species such as $CuCl(\eta^2\text{-}C_2H_2)$ and $[Cu(CO)_4]^+$. The latter complex is only stable in very strong acid solution (for example, 98% H_2SO_4) under a pressure of CO gas. There are also examples of metal carbonyl complexes of the other Group IB (11) elements—for example, $Au(CO)Cl$—so that in this respect these elements exhibit strong similarities to the transition metals.

All organosilver compounds contain Ag(I), and those with simple alkyl and aryl groups are not particularly stable, decomposing to elemental silver and hydrocarbons:

$$(CH_3)_4Pb + 4AgNO_3 \xrightarrow[-30\,°C]{C_2H_5OH} 4CH_3Ag + 4PbNO_3 \qquad (22.25)$$

$$2CH_3Ag \rightarrow 2Ag + C_2H_6 \qquad (22.26)$$

The introduction of fluorine groups does, however, lead to greater stability, as in $(CF_3)_2CFAg$.

Dimethylmercury(II) $(CH_3)_2Hg$ is an extremely poisonous liquid (boiling point 92 °C); the danger of this compound in natural waters is now widely recognized. Elemental mercury and its simple inorganic compounds are industrial by-products, which sometimes find their way into neighboring bodies of water (rivers, lakes, and so forth). These mercury-containing materials are converted into $[CH_3Hg]^+$ by certain biochemical processes that occur in aqueous solution. This cation is water-soluble and thus able to easily enter into the food chain. Evidence has been presented that brain damage occurs in persons who eat large quantities of mercury-contaminated fish. However, mercury also exists widely in the environment, and this "natural" mercury may contribute significantly to the pollution problem.

Because of its extraordinary toxicity, $(CH_3)_2Hg$ is best prepared by the following reactions:

$$2Hg + 2CH_3Br \rightarrow (CH_3)_2Hg + HgBr_2 \qquad (22.27)$$

$$HgCl_2 + 2CH_3MgBr \rightarrow (CH_3)_2Hg + 2MgBrCl \qquad (22.28)$$

Other Group IIB (12) organometallics can be synthesized in the following ways:

$$ZnCl_2 + 2RLi \rightarrow R_2Zn + 2LiCl \qquad (22.29)$$

$$CdCl_2 + 2RMgBr \rightarrow R_2Cd + 2MgBrCl \qquad (22.30)$$

These materials are also toxic, low boiling point liquids, for example, $(CH_3)_2Zn$ (melting point $-29.2\,°$; boiling point 44 °C) and $(CH_3)_2Cd$ (melting point $-2.4\,°$; boiling point 105.5 °C).

Diethylzinc(II), $(C_2H_5)_2Zn$, has recently been shown to be useful in the conservation of paper, especially rare books, manuscripts, and other old documents. This is an important discovery because it has been estimated that most of the books from the first half of the twentieth century will not be in a usable condition in the twenty-first century unless some conservation treatment is applied. The organometallic vapor permeates the fibrous structure of the paper, and on exposure to air or moisture, the $(C_2H_5)_2Zn$ spontaneously decomposes to afford finely divided ZnO, thereby providing an alkaline medium to neutralize the acidic content of the paper. Any excess $(C_2H_5)_2Zn$ vapor is easily pumped off. The chemical reactions involved are

$$(C_2H_5)_2Zn + 7O_2 \rightarrow ZnO + 4CO_2 + 5H_2O \tag{22.31}$$

$$(C_2H_5)_2Zn + H_2O \rightarrow ZnO + 2C_2H_6 \tag{22.32}$$

$$ZnO + 2H^+ \rightarrow Zn^{2+} + H_2O \tag{22.33}$$

22.3 *Sigma-Bonded Transition Metal Complexes*

Transition Metal Alkyls

The two major routes to transitional metal alkyls are reaction of a transition metallate anion with an organic halide or reaction of a Group IA (1) or IIA (2) metal alkyl with a transition metal halide. In the first method, various carbonylate anions, $[M(CO)_x]^{n-}$, are now known that can readily be formed *in situ* and then treated with compounds such as CH_3I to directly afford metal–carbon σ bonds. The sodium metal used to reduce the parent metal carbonyls is often in the form of a sodium amalgam (about 5% Na in Hg) because this is less hazardous to work with than the free alkali metal itself:

$$Mn_2(CO)_{10} \xrightarrow[\text{THF}]{\text{Na}} Na[Mn(CO)_5] \tag{22.34}$$

$$\Big\downarrow {\scriptstyle CH_3I}$$

$$CH_3Mn(CO)_5 + NaI$$

$$[CpM(CO)_3]_2 \xrightarrow[\text{THF}]{\text{Na}} Na[CpM(CO)_3] \tag{22.35}$$

$$\Big\downarrow {\scriptstyle C_2H_5I}$$

$$CpM(CO)_3(C_2H_5) + NaI$$

where M = Mo, W.

Two typical examples of the alternative synthetic approach, metal–halide alkylation, are

$$Cp_2TiCl_2 + 2CH_3Li \rightarrow Cp_2Ti(CH_3)_2 + 2LiCl \tag{22.36}$$

$$Pt(PPh_3)_2Cl_2 + 2CH_3MgCl \rightarrow Pt(PPh_3)_2(CH_3)_2 + 2MgCl_2 \tag{22.37}$$

The synthesis of transition metal alkyls is often prevented by the so-called β-hydride elimination reaction (see below; the other ligands have been omitted for the sake of clarity):

$$
\begin{array}{cc}
\text{H} & \text{H} \\
| & | \\
\text{M}-\text{C}^{\alpha}-\text{C}^{\beta}-\text{H} \\
| & | \\
\text{H} & \text{H}
\end{array}
\longrightarrow
\begin{array}{c}
\text{CH}_2 \\
\text{M}-\| \\
| \quad \text{CH}_2 \\
\text{H}
\end{array}
\longrightarrow \text{Products} \tag{22.38}
$$

This β-hydride elimination reaction tends to make the higher alkyl compounds unstable, and the vast majority of known simple σ alkyls contain hydrocarbyl ligands that do not carry a labile β hydrogen such as CH_3, $CH_2C_6H_5$, $CH_2C(CH_3)_3$, and $CH_2Si(CH_3)_3$. It was believed for many years that the failure to synthesize stable alkyl–transition metal compounds was due to thermodynamic instability of the M—C bond; now it is recognized that blockage of the β-hydride elimination reaction can lead to remarkably stable metal alkyls, for example, $(CH_3)_6W$, $Ti(CH_2C_6H_5)_4$, and $Cr[CH_2Si(CH_3)_3]_4$.

Transition Metal Hydrides

Certain organotransition metal compounds contain direct linkages between the transition metal and a hydride (H^-) ligand. The polarities of M—H bonds are similar to those of M—alkyl bonds, and the reactions of metal alkyls with H_2 often produce metal–hydride complexes as intermediates.

Transitional metal carbonyl hydrides are usually synthesized by protonation of carbonylate anions:

$$Mn_2(CO)_{10} \xrightarrow[\text{THF}]{\text{Na}} Na[Mn(CO)_5] \tag{22.39}$$

$$\swarrow \text{H}_3\text{PO}_4$$

$$HMn(CO)_5 + Na(H_2PO_4)$$

$$Na[(CpW(CO)_3] + CH_3COOH \rightarrow CpW(CO)_3H + CH_3COONa \tag{22.40}$$

Another route to transition metal hydrides involves treatment of neutral metal–carbonyl complexes with hydroxyl or borohydride ions:

$$Fe(CO)_5 \xrightarrow{\text{NaOH}} Na[HFe(CO)_4] + Na_2CO_3 + H_2O \qquad \textbf{(22.41)}$$

$$Ru_3(CO)_{12} \xrightarrow{\text{NaBH}_4} Na[Ru_3(CO)_{11}H] \qquad \textbf{(22.42)}$$

All the preparative reactions shown thus far yield complexes with terminal M—H groups. However, complexes with bridging hydride (μ—H) ligands can also be produced:

$$W(CO)_6 \xrightarrow{\text{NaBH}_4} Na[(OC)_5W-H-W(CO)_5] \qquad \textbf{(22.43)}$$

Bridging hydrides are sometimes formed during protonation reactions:

$$Ru_3(CO)_8(dppm)_2 + CF_3COOH \rightarrow$$

$$[Ru_3(CO)_8(dppm)_2(\mu\text{-}H)]^+[CF_3COO]^- \quad \textbf{(22.44)}$$

where dppm = $Ph_2PCH_2PPh_2$ (see page 569).

The metal–carbonyl hydrides are either liquids or low melting point, volatile solids—for example, $HMn(CO)_5$ (colorless liquid, melting point -25 °C), $CpW(CO)_3H$ (yellow crystal solid, melting point 69 °C). The cobalt complex $HCo(CO)_4$ is explosive in its pure liquid form. Hydride groups used to be difficult to locate by X-ray diffraction, and neutron-diffraction techniques were usually neccessary because of the larger cross-section of H atoms to neutrons than to X-rays. However, with modern computing techniques and statistics, it is often possible to find the H atoms with reasonable certainty from X-ray structural determinations. Metal hydrides also exhibit characteristic downfield resonances in their ^{1}H-NMR spectra, and this is one of the most definitive ways of establishing the presence of a metal hydride. In their IR spectra, ν(M—H) bands are often observed in the neighborhood of 1900 cm^{-1}, but care must be exercised because these peaks can sometimes be confused with ν(CO) vibrations.

Transition Metal Carbene (or Alkylidene) Complexes

Carbon monoxide acts as a two-electron σ-donor ligand via the C atom :C≡O in forming a wide variety of transition metal carbonyl complexes. Consequently, it is reasonable to expect that it might be possible to stabilize other two-electron σ-donor ligands such as carbenes (or alkylidenes) :CR$_2$ by coordination to transition metals. Carbenes have been shown to exist in the free state at low temperatures, and like CO (as you will see shortly), they should also be able to function as π-acceptor ligands.

Reported synthetic routes to transition metal carbenes are as follows:

$$Cr(CO)_6 + MeLi \longrightarrow Li^+[Cr(CO)_5C(O)Me]^- \xrightarrow{[Et_3O]^+} (OC)_5Cr{=}C\begin{matrix} \nearrow OEt \\ \searrow Me \end{matrix}$$

$$\Big\downarrow EtNH_2$$

$$(OC)_5Cr{=}C\begin{matrix} \nearrow NHEt \\ \searrow Me \end{matrix} + MeOH \qquad \textbf{(22.45)}$$

$$W(CO)_6 \xrightarrow[\text{(b) } [Me_3O][BF_4]]{\text{(a) LiPh}} Li^+\{W(CO)_5[C(OMe)Ph]\}^-$$

$$\Big\downarrow \begin{matrix} \text{(a) LiPh} \\ \text{(b) HCl, } -78\,°C \end{matrix}$$

$$(OC)_5W{\equiv}CPh_2 + MeOH + LiCl \quad \textbf{(22.46)}$$

Complexes of the parent carbene species $:CH_2$ have also been isolated. One method of formation is by deprotonation of an appropriate alkyl compound:

$$[Cp_2TaMe_2]^+[BF_4]^- \xrightarrow{\text{NaOMe}} Cp_2Ta \begin{matrix} \overset{225\ pm}{\nearrow} CH_3 \\ \underset{203\ pm}{\|} CH_2 \end{matrix} + MeOH + NaBF_4$$

$$\textbf{(22.47)}$$

The significantly shorter Ta—C distance observed for the Ta—CH_2 bond provides strong evidence for Ta—C double bonding in the tantalum(V) complex.

The interest in metal–carbene complexes is more than just academic—it has been shown that they are important intermediates in certain organic reactions catalyzed by transition metal complexes, for example, alkene metathesis:

$$EtCH{=}CH_2 + EtCH{=}CH_2 \xrightarrow{WCl_5/AlEt_3} EtCH{=}CHEt + CH_2{=}CH_2$$

$$\textbf{(22.48)}$$

Transition Metal Carbyne (or Alkylidyne) Complexes

Once it was realized that M=CR_2 bonds existed, it was not too long before the first examples of complexes containing metal–carbyne linkages (M≡CR) were prepared. One method involves the treatment of a metal–carbene complex with HCl:

$$W(CO)_5[C(OMe)Ph] + HCl \xrightarrow{-78\,°C} [(OC)_5W{\equiv}CPh]Cl + MeOH$$

$$\textbf{(22.49)}$$

In another method, a metal–carbene complex is reacted with BF_3:

$$Cr(CO)_4(PMe_3)[C(OMe)Me] + BF_3 \rightarrow$$

$$[MeC\equiv Cr(CO)_3(PMe_3)]F + CO + BF_2(OMe) \quad \textbf{(22.50)}$$

Deprotonation of metal–carbene complexes with BuLi can afford the corresponding metal–carbyne derivative:

$$(Me_3CCH_2)_3Ta=CH(CMe_3) + BuLi \rightarrow$$

$$Li[(Me_3CCH_2)_3Ta\equiv CCMe_3] + C_4H_{10} \quad \textbf{(22.51)}$$

22.4   *Transition Metal Carbonyls*

Metal carbonyls comprise a large area of transition metal organometallic chemistry and occupy a position of particular historical importance. In 1888 tetracarbonylnickel(0) $Ni(CO)_4$ and pentacarbonyliron(0) $Fe(CO)_5$ were the first examples of metal carbonyls to be discovered. The synthesis of $Ni(CO)_4$ is still industrially important because the reaction is reversible—this is the basis of the *Mond process* for the purification of metallic nickel:

$$Ni + 4CO \xrightarrow[\text{101 kPa (1 atm)}]{\text{25 °C}} Ni(CO)_4 \xrightarrow{\text{Distil}} Ni + 4CO \quad \textbf{(22.52)}$$

Impure $\qquad\qquad\qquad\qquad\qquad\qquad$ Pure

In fact, CO is so reactive toward Ni that it will etch the metallic surface in a few minutes.

The simplest metal carbonyls are the neutral, binary $M_x(CO)_y$ compounds; these may be mononuclear ($x = 1$) or polynuclear ($x > 1$). Table 22.3 gives some examples of these binary metal carbonyls. The list is far from complete, especially because new polynuclear complexes are being discovered almost daily. The majority of metal carbonyls are low melting point solids that can be sublimed *in vacuo*; a small number of the compounds are volatile liquids.

There are other mononuclear metal carbonyls, in addition to $Ni(CO)_4$, that can be prepared by direct reaction of the parent metals with gaseous CO at appropriate temperatures and pressures, namely,

$$Fe + 5CO \xrightarrow[\text{10}^4 \text{ kPa (100 atm)}]{\text{200 °C}} Fe(CO)_5 \qquad\qquad \textbf{(22.53)}$$

$$Mo + 6CO \xrightarrow[\text{2.5} \times \text{10}^4 \text{ kPa (250 atm)}]{\text{200 °C}} Mo(CO)_6 \qquad\qquad \textbf{(22.54)}$$

In general, mononuclear and polynuclear metal carbonyls are synthesized by a process known as *reductive–carbonylation* in which a transition metal in a high oxidation state is reduced to zero oxidation state in the presence of CO gas. Often, these reactions are performed at very high pressures in steel bombs (or autoclaves, 100—500 cm^3 capacity). To reduce the possibility of an accident, the bombs are usually operated in explosion-proof rooms located on the top floor of a chemistry building. Some typical metal–carbonyl syntheses employing reductive–carbonylation are shown below:

$$CrCl_3 + Al + 6CO \xrightarrow[\substack{C_6H_6}]{\substack{AlCl_3 \\ Catalyst}} Cr(CO)_6 + AlCl_3 \quad (22.55)$$

$$2CoCO_3 + 2H_2 + 8CO \xrightarrow[\substack{2.5 \times 10^4 \text{ kPa(250 atm)}}]{\substack{About\ 170\ ^{\circ}C}} Co_2(CO)_8 + 2CO_2 + 2H_2O \quad (22.56)$$

$$6Ru(acac)_3 + 9H_2 + 24CO \xrightarrow[\substack{2 \times 10^4 \text{ kPa(200 atm)} \\ MeOH}]{\substack{150\ ^{\circ}C}} 2Ru_3(CO)_{12} + 18acacH \quad (22.57)$$

where acac = acetylacetonate.

$$Re_2O_7 + 17CO \xrightarrow[\substack{3 \times 10^4 \text{ kPa(300 atm)}}]{\substack{300\ ^{\circ}C}} Re_2(CO)_{10} + 7CO_2 \quad (22.58)$$

Other preparative routes originate in the photolysis, pyrolysis, or oxidation of simpler binary metal–carbonyl complexes:

$$2Fe(CO)_5 \xrightarrow[\substack{MeCOOH}]{\substack{h\nu}} Fe_2(CO)_9 + CO$$

$$\Bigg\downarrow \substack{95\ ^{\circ}C. \\ Toluene}$$

$$Fe_3(CO)_{12} \quad (22.59)$$

$$Os_3(CO)_{12} \xrightarrow{\substack{200\ ^{\circ}C}} Os_4(CO)_{13} + Os_5(CO)_{16}$$
$$+ \text{ Higher Polynuclear Complexes} \quad (22.60)$$

$$2Na[V(CO)_6] + 2HCl \longrightarrow 2V(CO)_6 + 2NaCl + H_2 \quad (22.61)$$

The structures and bonding in metal–carbonyl complexes are properties that continue to attract the attention of many inorganic chemists. Carbon monoxide can bond to transition metals not only in a terminal

fashion (M—CO) but also in doubly bridging (M—$\overset{\overset{\displaystyle O}{\|}}{C}$—M) and triply

Table 22.3 *Representative Examples of Binary Metal Carbonyls*[a]

Compound	Physical Form	Molecular Structure (Symmetry)
Ni(CO)₄ structure	Colorless liquid (f.p. −25 °C, b.p. 43 °C)	Tetrahedron (T_d)
Fe(CO)₅ structure (ax, eq)	Yellow liquid (f.p. −20 °C, b.p. 103 °C)	Trigonal-bipyramid (D_{3h})
Os(CO)₅ structure (ax, eq)	Colorless liquid (f.p. −15 °C)	Trigonal-bipyramid (D_{3h})
M(CO)₆ structure (M = Cr, Mo, W)	White crystals (Cr, m.p. 154–155 °C; Mo, dec. 150–151 °C; W, dec. 169–170 °C)	Octahedron (O_h)
V(CO)₆ structure	Blue-green crystals (dec. 60–70 °C)	Slightly distorted octahedron (approximately O_h) (Jahn–Teller effects)
Fe₂(CO)₉ structure	Orange-yellow crystals (dec. 100 °C)	Pseudooctahedron about each Fe atom, with an F—Fe bond and three bridging CO groups (D_{3h})

Table 22.3 (*Continued*)

Compound	Physical Form	Molecular Structure (Symmetry)
	Yellow crystals (m.p. 224 °C)	Trigonal-planar arrangement of three metal–metal bonded Os atoms with octahedral coordination about each Os atom (D_{3h})
	Dark orange crystals (dec. 51–52 °C)	Two $Co(CO)_3$ units joined by two bridging CO groups and a Co—Co bond (C_{2v})
	Black crystals	Three $Co(CO)_2$ units in a plane, joined by three Co—Co bonds and three bridging CO groups, capped by a $Co(CO)_3$ unit (C_{3v})
	Yellow powder (dec. 230 °C)	Four $Ir(CO)_3$ units linked by six Ir—Ir bonds (T_d)
	Yellow crystals (m.p. 152–155 °C)	Two staggered, octahedrally coordinated $Mn(CO)_5$ units joined by a single Mn—Mn bond (D_{4d})

[a] ax = axial and eq = equatorial.

bridging $\left[\begin{array}{c} O \\ \| \\ C \\ M \diagdown \ \diagup M \\ | \\ M \end{array} \right]$ situations:

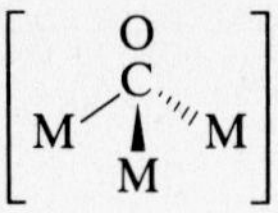

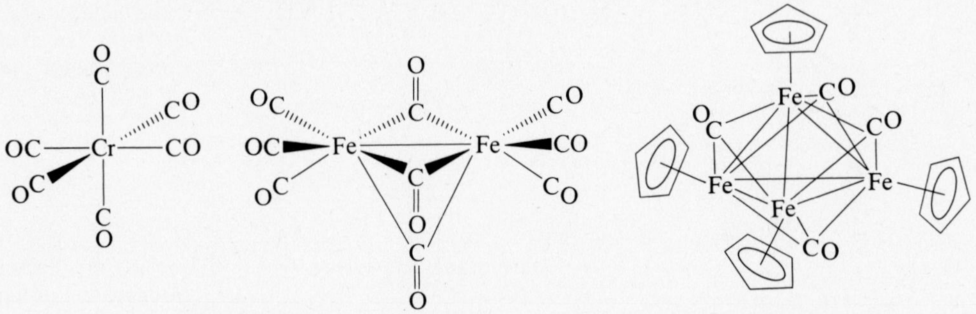

The CO groups are arranged so that there is a minimum amount of interatomic steric repulsion, and symmetric structures are often obtained—for example, tetrahedral $Ni(CO)_4$, trigonal-bipyramidal $Fe(CO)_5$, and octahedral $Cr(CO)_6$. Finally, with the exception of $V(CO)_6$, all the simple binary metal carbonyls are diamagnetic and satisfy the eighteen-electron rule. The larger binary metal–carbonyl clusters such as $Rh_6(CO)_{16}$ tend not to obey the eighteen-electron rule.

The molecular structures of the binary metal carbonyls have been established on the basis of electron-, X ray-, and neutron-diffraction studies. Without the aid of X-ray diffraction, the complicated structures of the higher polynuclear species would never have been appreciated. The existence of doubly and triply bridged CO groups was first shown by X-ray diffraction. Table 22.4 gives some typical metal–CO bond lengths for binary metal carbonyls.

In terminal M—CO linkages, the M—$\hat{\text{C}}$—O angles range between 165–180°, that is, the M—C—O framework is essentially linear. For most metal carbonyls, the C—O bond lengths are approximately 115 pm, that is, slightly longer than that for free CO itself (112.8 pm). From Table 22.4, it can be seen that the M—C distances for terminal CO ligands bonded to first-row transition metals are usually about 187 pm. One important exception is the paramagnetic species $V(CO)_6$; here, the average M—CO bond is almost 15 pm longer. In addition, when CO bridges two metal atoms, as

$$\overset{O}{\underset{\|}{}}$$

in M—C—M, the M—C bond lengths are approximately 20 pm longer. The increased M—C bond lengths in these two situations afford strong evidence of significantly weaker M—C bonding.

Carbon monoxide is not a particularly strong Lewis base—it forms an unstable 1:1 adduct with the Lewis acid BH_3. Why then are there so many thermodynamically stable metal–carbonyl complexes? An analysis of

Table 22.4   *Metal–Carbon and Metal–Metal Bond Lengths (pm) in Some Representative Binary Metal Carbonyls*

Complex	Average Bond Lengths	
	M—C[a]	M—M
$Ni(CO)_4$	183.8	
$Fe(CO)_5$	181.4 (ax)	
	181.2 (eq)	
$V(CO)_6$	200.8	
$Cr(CO)_6$	191.3	
$Mo(CO)_6$	206	
$Mn_2(CO)_{10}$	179 (ax)	292.3
	183 (eq)	
$Fe_2(CO)_9$	183.8 (term)	252.3
	201.6 (br)	
$Co_2(CO)_8$	180 (term)	252.2
	190 (br)	
$Ru_3(CO)_{12}$	194.2 (ax)	285.2
	192.1 (eq)	
$Os_3(CO)_{12}$	194.6 (ax)	287.7
	191.2 (eq)	
$Ir_4(CO)_{12}$	187	269
$Rh_6(CO)_{16}$	186.4 (term)	277.6
	216.8 (tri. br)	

[a] The abbreviations are: ax = axial, eq = equatorial, term = terminal, br = bridging, and tri. br = triply bridging.

metal–CO bonding answers this question. There are two main types of bonding: (a) a σ bond formed by donation of an electron pair from the C atom of CO into a vacant orbital on the metal center; (b) two π bonds formed by *backdonation* of electrons from filled metal $d\pi$ orbitals into the π_2 orbitals of CO. This backbonding leads to an increase in the bond order of the M—C bond and a decrease in the C—O bond order. These σ and π bonds are illustrated pictorially in Figure 22.4, and a simplified MO description of $Cr(CO)_6$ is given in Figure 22.5.

You are now in a position to appreciate why there are differences in the M—CO bond distances for terminal and bridging CO groups. The net σ–π interactions for bridging CO ligands are weaker than those for the terminal ones, and so the resulting M—CO bonds are longer. The bonding differences in M—CO linkages are also reflected in the positions of the ν(CO) peaks in the IR spectra of metal–carbonyl complexes.

Figure 22.4
The σ and π
M—CO bonds in a
typical
metal–carbonyl
compound
containing terminal
M—CO linkages.

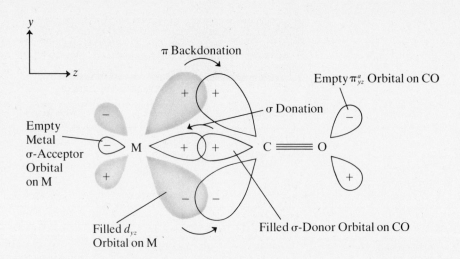

The octahedral complexes $[Mn(CO)_6]^+$, $Cr(CO)_6$, and $[V(CO)_6]^-$ are isoelectronic, and the central metal in each case has a $3d^6$ ground-state electronic configuration. The oxidation states, however, differ: Mn(I), Cr(0), and V(−I). This means that there should be an increase in the extent of π backbonding in going across the series from Mn(I) to V(−I).

Figure 22.5
Qualitative MO
diagram for
octahedral $Cr(CO)_6$.
Only the interaction
with one CO group
is shown. (See also
Chapter 3 for a
description of the
bonding in
heterodiatomics such
as CO.)

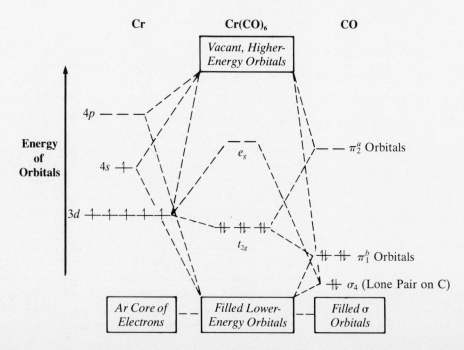

And this, in turn, leads to an increase in M—C bond order and a concomitant decrease in C—O bond order. For coordinated CO groups, the CO bond order and CO force constant (k_{CO}, see Equation 22.62) are closely related so that a decrease in CO bond order results in a decrease in wave number of the t_{1u} CO stretching mode.

$$\nu_{CO} = \frac{1}{2\pi c} \sqrt{\frac{k_{CO}}{\mu_{CO}}} \qquad \textbf{(22.62)}$$

Wave Number of CO–Stretching Vibration   Reduced Mass

This is indeed observed for the octahedral series of complexes: $[Mn(CO)_6]^+$ (2090 cm^{-1}), $Cr(CO)_6$ (1990 cm^{-1}), and $[V(CO)_6]^-$ (1860 cm^{-1}). In general, a decrease in wave number of the $\nu(CO)$ modes for a member of a structurally related series of metal–carbonyl complexes is indicative of a decrease in CO bond order and therefore in the extent of π backbonding.

Terminal and bridging CO groups can usually be distinguished by IR spectroscopy. Some general ranges for the $\nu(CO)$ modes of metal-carbonyl complexes in various bonding situations are shown below. These ranges must be used cautiously in assigning molecular structures, however, because certain anionic complexes display very low-energy $\nu(CO)$ bands, for example, $[Fe(CO)_4]^{2-}$, 1790 cm^{-1}

Type of CO bond	$\nu(CO)$, cm^{-1}
Terminal	2200–1850
Doubly bridging	1850–1750
Triply bridging	1730–1620
Uncoordinated CO	2143

IR spectroscopy is a vital weapon in the organometallic chemist's armory because the number of observed peaks is often an excellent indicator of molecular structure and the subtle shifts in peak positions are directly related to changes in bonding interactions. For example, reaction of $Mn(CO)_5I$ with Et_4NI affords an ionic complex, $Et_4N[Mn(CO)_4I_2]$; its IR spectrum in the $\nu(CO)$ region in $CHCl_3$ solution is shown in Figure 22.6. The group theoretical predictions for the IR-active $\nu(CO)$ modes of octahedral *cis*- and *trans*-$[Mn(CO)_4I_2]^-$ are as follows:

$$\Gamma_{cis}^{C_2v} = 2a_1 + b_1 + b_2 \text{ [four } \nu(CO) \text{ peaks predicted]} \qquad \textbf{(22.63)}$$

$$\Gamma_{trans}^{D_{4h}} = e_u \text{ [one } \nu(CO) \text{ peak predicted]} \qquad \textbf{(22.64)}$$

Figure 22.6
IR spectrum of
$Et_4N[Mn(CO)_4I_2]$ in
the CO–stretching
region ($CHCl_3$
solution).

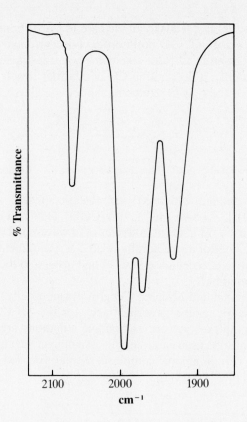

Because the complex exhibits four IR-active $\nu(CO)$ peaks, the structure must be cis. Table 22.5 illustrates the predictions for the number of $\nu(CO)$ modes expected for the most commonly encountered octahedral and trigonal-bipyramidal metal–carbonyl derivatives. Figure 22.7 shows some representative IR spectra.

It should be emphasized, however, that the simple group theoretical spectral predictions are sometimes disobeyed for rational reasons. For instance, $CpMn(CO)_2[S(n\text{-}Bu)_2]$ exhibits four $\nu(CO)$ peaks in n-hexane solution (Figure 22.8), instead of the two expected ($a' + a''$) on group theoretical grounds for the C_s local symmetry of the $Mn(CO)_2[S(n\text{-}Bu)_2]$ moiety. The doubling of the $\nu(CO)$ peaks is due to the presence of the conformational isomers A and B about the Mn—S axis. The observed peak splittings are quite small ($\sim 8 \text{ cm}^{-1}$) and would not have been detected if a polar solvent such as $CHCl_3$ had been used. The dipoles of the polar solvent interact with the CO–bond dipoles, resulting in a broadening of the $\nu(CO)$ peaks.

Table 22.5 *IR-Active $\nu(CO)$ Modes of Some Typical Octahedral and Trigonal-Bipyramidal Metal–Carbonyl Derivatives*[a]

Molecule	Symmetry	Symmetry of $\nu(CO)$ Modes	Molecule	Symmetry	Symmetry of $\nu(CO)$ Modes
(octahedral M with one L, five CO)	C_{4v}	$2a_1 + e$	(TBP M with one L axial, four CO)	C_{3v}	$2a_1 + e$
(octahedral M with two L cis, four CO)	C_{2v}	$2a_1 + b_1 + b_2$	(TBP M with two L, three CO)	C_{2v}	$2a_1 + b_1 + b_2$
(octahedral M with two L trans, four CO)	D_{4h}	e_u	(TBP M with two L axial, three CO)	D_{3h}	e'
(octahedral M with three L, three CO)	C_{3v}	$a_1 + e$	(TBP M with three L, two CO)	C_3	$2a' + a''$
(octahedral M with three L, three CO)	C_{2v}	$2a_1 + b_2$			

[a] L = monodentate ligand.

Figure 22.7
Representative IR
spectra of metal
carbonyl complexes
in the CO–stretching
region. Parts (a)–(d)
reproduced from
D. M. Adams,
*Metal-Ligand and
Related Vibrations*.
London: Arnold,
1967. Part (e)
reproduced from
M. A. El-Sayed and
H. D. Kaesz, *Inorg.
Chem.* 2 (1963):
158.

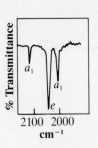

(a) $Mn(CO)_5I$
(C_{4v} Symmetry)
in CCl_4

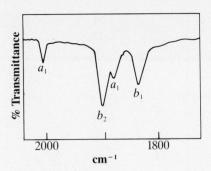

(b) $Mo(CO)_4$ (dipy)
(C_{2v} Symmetry)
in Ethyl Acetate

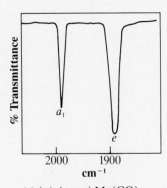

(c) (η^6-Arene) $Mo(CO)_3$
in CH_2Cl_2
Local Symmetry of
$Mo(CO)_3$ Moiety is C_{3v}

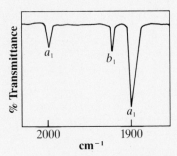

(d) *mer*-$Mo(CO)_3$ $[P(OMe)_3]_3$
(C_{3v} Symmetry) in *n*-Hexane

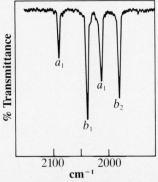

(e) $[Mn(CO)_4I]_2$ in Cyclohexane.
The Local Symmetry of the $Mn(CO)_4I_2$
Fragments is C_{2v}

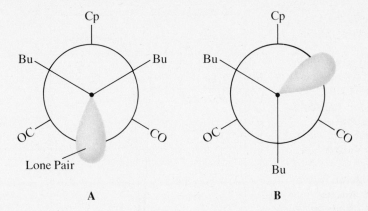

A B

Similarly, *trans*-$Cr(CO)_4(PPh_3)_2$ would be expected to display only one $\nu(CO)$ peak (e_u), yet three are observed in Figure 22.9 because the anticipated D_{4h} molecular symmetry is lowered by the presence of the two bulky PPh_3 ligands. The lowering in symmetry allows the formally IR-inactive a_{1g} and b_{1g} $\nu(CO)$ modes to become weakly IR active.

Figure 22.8
IR spectrum of $CpMn(CO)_2[S(n\text{-}Bu_2]$ in the CO–stretching region (*n*-hexane solution).

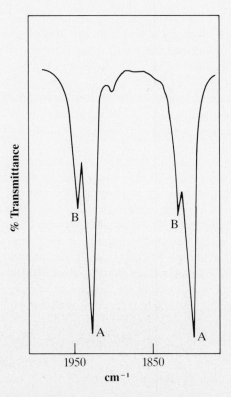

Figure 22.9
IR spectrum of *trans-*$Cr(CO)_4(PPh_3)_2$ in the CO–stretching region ($CHCl_3$ solution). (a_{1g} and b_{1g} are the predicted Raman-active modes.) Reproduced from D. M. Adams, *Metal-Ligand and Related Vibrations*. London: Arnold, 1967.

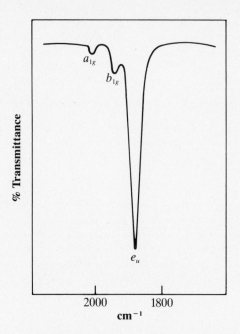

NMR spectroscopy is also critically important in the structural identification of organometallic compounds. The two major nuclei of interest are $^1H(I = \frac{1}{2})$ and $^{13}C(I = \frac{1}{2})$. In addition, low oxidation–state organometallic complexes often contain tertiary phosphorus ligands, and $^{31}P(I = \frac{1}{2})$ is another important NMR nucleus. Moreover, with the advent of high-power multinuclear NMR instrumentation, many relatively insensitive nuclei can also be examined, including metals—for example, $^{195}Pt(I = \frac{1}{2})$.

Most proton chemical shifts for diamagnetic organometallic complexes fall within the 0–10 ppm range, the chief exception being transition metal hydrides. Figure 22.10 illustrates the typical ranges for ^{13}C-NMR chemical shifts of some selected classes of organometallic complexes. Figure 22.11 illustrates the ^{13}C-NMR spectrum of $(\eta^6\text{-}PhCO_2Me)Cr(CO)_3$.

Simple metal carbonyls undergo a wide variety of substitution reduction, and oxidation reactions, as shown for $Mo(CO)_6$ in Figure 22.12.

Carbonyl halides are among the most synthetically versatile derivatives of simple metal carbonyls, and many are prepared by metal–metal bond rupture and subsequent reaction with halogens (X_2):

$$M_2(CO)_{10} + X_2 \longrightarrow 2M(CO)_5X \xrightarrow{120\,°C} [M(CO)_4X]_2 + 2CO \tag{22.65}$$

where M = Mn, Tc, Re; X = Cl, Br, I.

Figure 22.10
Typical ranges for
^{13}C-NMR chemical
shifts of selected
organometallic
complexes.

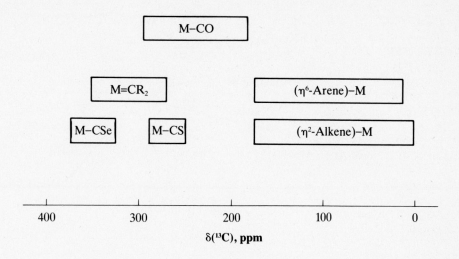

Figure 22.11
^{13}C-NMR spectrum of
$(\eta^6\text{-PhCO}_2\text{Me})\text{Cr(CO)}_3$
in CH_2Cl_2 solution,
together with the
proposed
assignments. Note
that the spectrum is
simplified by proton
decoupling. Adapted
with permission
from G. M. Bodner
and L. J. Todd.
Inorg. Chem. 13
(1974): 360.

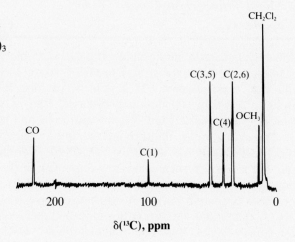

$$[\text{CpM(CO)}_2]_2 + X_2 \rightarrow 2\text{CpM(CO)}_2X \qquad (22.66)$$

where M = Fe, Ru.

$$[\text{CpMo(CO)}_3]_2 + X_2 \rightarrow 2\text{CpMo(CO)}_3X \qquad (22.67)$$

Direct reaction of $Fe(CO)_5$ (which lacks a metal–metal bond) with X_2 produces iron–carbonyl dihalides:

$$Fe(CO)_5 + X_2 \rightarrow \textit{cis-}Fe(CO)_4X_2 + CO \qquad (22.68)$$

Figure 22.12
Typical
reactions of
$Mo(CO)_6$.

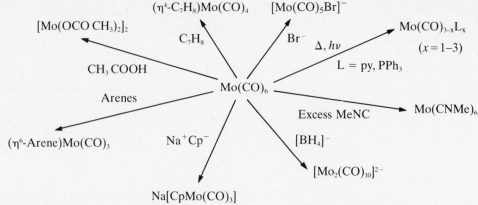

22.5   *Transition Metal Thiocarbonyl and Selenocarbonyl Complexes*

Unlike CO, carbon monosulfide (CS) is unstable above $-160\,°C$ and polymerizes (often explosively) to $(CS)_n$ above this temperature. It is produced *in situ* in about 25% yield by passing an electrical discharge through CS_2 vapor. Carbon monoselenide (CSe) is apparently more unstable because it has eluded isolation, even at liquid-helium temperatures.

Transition metal thiocarbonyl complexes have been known since *trans*-$RhCl(CS)(PPh_3)_2$ and $RhCl_3(CS)(PPh_3)_2$ were reported in 1966, and there are about 200 such complexes containing terminal CS linkages now known. Even the simple derivatives, $M(CO)_5(CS)$ (M = Cr, Mo, W), have been prepared. Analogous selenocarbonyls such as $(\eta^6\text{-}PhCO_2Me)\text{-}Cr(CO)_2(CSe)$ and $Cr(CO)_5(CSe)$ have only been discovered during the past fifteen years, and one tellurocarbonyl, $OsCl_2(CTe)(PPh_3)_3$, has now been reported. X-ray structures of several of these chalcocarbonyl complexes have been determined and the M—C—X (X = S, Se, Te) linkages are linear (175–180°), implying that the bonding is similar to that in M—C—O groups.

There are several different synthetic routes to transition metal thiocarbonyl complexes, among them the following:

1. Cocondensation of CS with a metal vapor. This method has only been successfully applied to the formation of $Ni(CS)_4$ at very low temperatures thus far:

$$Ni + 4CS \xrightarrow{10\ K} Ni(CS)_4 \qquad\qquad (22.69)$$

2. Reaction of CS_2 with a metal complex, often in the presence of a sulfur acceptor such as PPh_3:

$$CpM(CO)_2L + CS_2 + PPh_3 \rightarrow CpM(CO)_2(CS) + PPh_3S + L$$
(22.70)

where M = Mn, Re; L = *cis*-C_8H_{14}, THF, SMe_2, and so on.

$$(\eta^6\text{-Arene})Cr(CO)_2(\textit{cis}\text{-}C_8H_{14}) + CS_2 + PPh_3 \rightarrow$$
$$(\eta^6\text{-Arene})Cr(CO)_2(CS) + PPh_3S + C_8H_{14} \quad (22.71)$$

where Arene = C_6H_6, $PhCO_2Me$, and so on.

The η^2-C,S species shown below are believed to be the intermediates in these reactions:

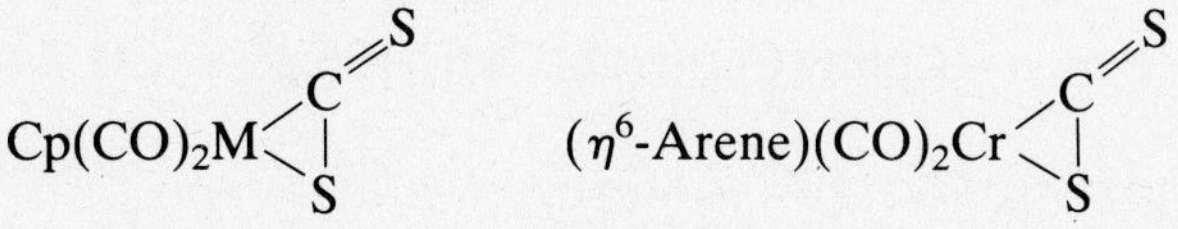

$$Cp(CO)_2M\underset{S}{\overset{C=S}{<|}} \qquad (\eta^6\text{-Arene})(CO)_2Cr\underset{S}{\overset{C=S}{<|}}$$

In the case of the arene–chromium thiocarbonyls, reaction with CO gas under moderate pressure results in arene displacement to yield $Cr(CO)_5(CS)$:

$$(\eta^6\text{-}PhCO_2Me)Cr(CO)_2(CS) + 3CO \xrightarrow[60\,°C]{10^3 \text{ kPa (10 atm)}}$$
$$Cr(CO)_5(CS) + PhCO_2Me \quad (22.72)$$

3. Treatment of a metal complex with CSSe in the presence of PPh_3. In this case, the Se atom is abstracted from the η^2-C,Se intermediate:

$$CpCo(PMe_3)_2 + CSSe + PPh_3 \rightarrow$$
$$CpCo(PMe_3)(CS) + PPh_3Se + PMe_3 \quad (22.73)$$

4. Treatment of a metal carbonyl anion with thiophosgene:

$$2M(CO)_6 \xrightarrow[THF]{Na/Hg} Na_2[M(CO)_5] \xrightarrow{Cl_2CS} M(CO)_5(CS) + 2NaCl$$
(22.74)

where M = Cr, Mo, W.

$$Na_2[Fe(CO)_4] + Cl_2CS \rightarrow Fe(CO)_4(CS) + 2NaCl \quad (22.75)$$

Similar procedures can be employed for the synthesis of the analogous selenocarbonyls:

$$(\eta^6\text{-}PhCO_2Me)Cr(CO)_2(\textit{cis}\text{-}C_8H_{14}) + CSe_2 + PPh_3 \rightarrow$$
$$(\eta^6\text{-}PhCO_2Me)Cr(CO)_2(CSe) + PPh_3Se + C_8H_{14} \quad (22.76)$$

$$(\eta^6\text{-PhCO}_2\text{Me})\text{Cr(CO)}_2(\text{CSe}) + 3\text{CO} \xrightarrow[40\ °C]{10\ atm}$$

$$\text{Cr(CO)}_5(\text{CSe}) + \text{PhCO}_2\text{Me} \quad (22.77)$$

$$\text{CpCo(PMe}_3)_2 \xrightarrow{\text{CSeTe/PPh}_3} \text{CpCo(PMe}_3)(\text{CSe}) + \text{PMe}_3\text{Te} \quad (22.78)$$

The substitution reactions of mixed carbonyl–thiocarbonyl and –selenocarbonyl complexes are interesting in that CO is almost always replaced in preference to CS or CSe. For instance, it has been possible to synthesize CpMn(CO)(CS)_2 and $[\text{CpMn(CO)(NO)(CS)}]^+$ (for a discussion of nitrosyl complexes, see Section 22.6).

$$\text{CpMn(CO)}_2(\text{CS})$$

$$NO^+[PF_6]^- \qquad\qquad h\nu \quad cis\text{-}C_8H_{14}$$

$$[\text{CpMn(CO)(NO)(CS)}]^+\ [\text{PF}_6]^- + \text{CO}$$

$$\text{CpMn(CO)(CS)}(cis\text{-}C_8H_{14}) + \text{CO}$$

$$\qquad\qquad CS_2/PPh_3$$

$$\text{CpMn(CO)(CS)}_2 + \text{PPh}_3\text{S} \quad (22.79)$$

Optically active, chromium–thiocarbonyl complexes have also been isolated, again by making use of the inertness of the CS ligand toward substitution:

$$(22.80)$$

The two chromium–thiocarbonyl diastereomers can be separated by thin-layer chromatography, and subsequent reduction of the CO_2Me function leads to two *o*-xylene complexes that are chiral at the chromium center.

One of the few compounds in which some CS substitution does occur is $[CpFe(CO)_2(CS)]^+$. This complex reacts with tertiary phosphines and related ligands L to give $[CpFe(CO)(CS)L]^+$ and $[CpFe(CO)_2l]^+$; with halide ions (X^-), both $CpFe(CO)(CS)X$ and $CpFe(CO)_2X$ are produced. It appears that a high $\nu(CS)$ value (about 1350 cm^{-1}) is a good indicator of the possibility of CS substitution taking place. Thus, for $CpMn(CO)_2(CS)$ [$\nu(CS)$, 1271 cm^{-1}], only CO substitution has been observed. The lower $\nu(CS)$ values reflect more extensive π backbonding and therefore stronger M—C bonding in the case of the thiocarbonyl complexes.

In addition to terminal M—CS linkages, there is now a variety of bridged–CS complexes, which can usually be distinguished by the position of the $\nu(CS)$ modes in their IR spectra (Table 22.6). Two examples of bridged–CS complexes are

Table 22.6 *Approximate $\nu(CS)$ Ranges for Various Types of CS Bonding*

Type of CS Bonding	Approximate $\nu(CS)$ Range (cm^{-1})
Free CS	1274
Terminal, M—CS	1410–1160
Doubly bridging, M—C(=S)—M	1160–1100
Triply bridging, C(S) over M, M, M	1080–1040
Carbon and sulfur bonded, M—CS—M	1110–1050
Triply bridging and sulfur bonded, M—S—C over M, M, M	~950

The ν(CS) values for terminal M—CS linkages in transition metal thiocarbonyls lie both above and below the value for free CS itself. This is because CS–stretching motions are not absolutely "pure"; there is substantial mixing of the CS–stretching modes at about 1300 cm^{-1} with the MC–stretching modes at about 400 cm^{-1}. In the case of M—CO groups, the CO peaks are due to much purer CO motions because there is comparatively little coupling with the M—C modes. The majority of the CO peaks in metal–carbonyl complexes are below that of free CO (2143 cm^{-1}) due to the increased π backbonding when other ligands are coordinated to the metal center. For the few cases, when ν(CO) is above 2143 cm^{-1}—for example, [Cu(CO)$_4$]$^+$ ($\sim$ 2200 cm^{-1})—there is apparently very little π backbonding, and it is σ donation that explains the increase in CO–bond order.

On the basis of numerous spectroscopic measurements (IR, PES, NMR, and mass) and kinetic data, CS and CSe have been shown to be better σ-donor and π-acceptor ligands than is CO. This accounts for M—CS and M—CSe bonds being generally more difficult to break than are M—CO bonds. Just as with M—CO bonds, the effect of charge on a complex is reflected in the position of the ν(CS) modes. For example, consider the data for the following pair of isoelectronic complexes: CpMn(CO)$_2$(CS), ν(CO) 2010, 1959, ν(CS) 1271; [CpFe(CO)$_2$(CS)]$^+$, ν(CO) 2064, 2039, ν(CS) 1348 cm^{-1}. The positive charge on the iron cation results in less π backbonding to the CO and CS ligands, leading to higher wave numbers being observed for the ν(CO) and ν(CS) modes.

22.6 *Nitric Oxide Complexes*

The diatomic NO molecule has one more electron than does CO, and many complexes are formed that have linear M—N—O linkages. The M—N—O group is similar to the M—C—O group in metal carbonyls, except that NO is acting as three-electron donor. There are also complexes containing NO ligands in which the M—N—O grouping is bent, where NO is considered to be a one-electron donor.

Nitric Oxide as a Three-Electron Donor

For the purposes of the eighteen-electron rule, "linear" NO is regarded as a three-electron donor; the first electron is donated to the metal to give M$^-$, and the resulting NO$^+$ ligand (which is isoelectronic with CO) then bonds to M$^-$. Once again, the presence of extensive π backbonding in the NO$^+$ ligands is demonstrated by the positions of the ν(NO) peaks in the IR spectra. There are a few nitrosonium salts—for example, NO$^+$[HSO$_4$]$^-$ —

that have IR-active NO absorptions in the 2400–2150 cm^{-1} region, while the $\nu(NO)$ modes in metal nitrosyls appear at about 1900–1800 cm^{-1} due to backbonding into the π-antibonding orbitals of the NO$^+$ ligand.

Nitric Oxide as a One-Electron Donor

When the M—N—O linkage is bent, this is reflected in the position of the $\nu(NO)$ mode, for example, [IrCl(CO)(PPh$_3$)$_2$(NO)]$^+$ (M—$\overset{..}{N}$—O = 124°, 1680 cm^{-1}). The $\nu(NO)$ modes for bent M—N—O linkages generally appear in the 1690–1520 cm^{-1} range. In the bent situation, the NO ligand acts as a one-electron donor, thereby leaving a lone pair on the formally sp^2 hybridized N atom:

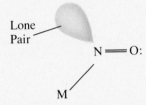

Lone
Pair

N$=\!=$O:

M

Synthesis of Metal Nitrosyl Complexes

There are several synthetic routes to metal nitrosyl complexes:

1. Direct reaction with NO gas:

$$\text{Co}_2(\text{CO})_8 + 2\text{NO} \rightarrow 2\text{Co}(\text{CO})_3(\text{NO}) + 2\text{CO} \qquad \textbf{(22.81)}$$

   This type of reaction, however, is difficult to control, and treatment of Cr(CO)$_6$ with NO at low temperature, for example, results in complete CO substitution:

$$\text{Cr}(\text{CO})_6 + 4\text{NO} \xrightarrow{h\nu} \text{Cr}(\text{NO})_4 + 6\text{CO} \qquad \textbf{(22.82)}$$

2. Reaction with nitrosonium salts:

$$\textit{trans-}\text{Ir}(\text{CO})\text{Cl}(\text{PPh}_3)_2 + \text{NOPF}_6 \rightarrow [\text{Ir}(\text{CO})\text{Cl}(\text{NO})(\text{PPh}_3)_2]\text{PF}_6$$
$$\textbf{(22.83)}$$

3. Reaction with nitrites:

$$\text{Na}[\text{Co}(\text{CO})_4] + \text{NaNO}_2 + 2\text{CH}_3\text{COOH} \rightarrow$$
$$\text{Co}(\text{CO})_3(\text{NO}) + \text{CO} + 2\text{CH}_3\text{COONa} + \text{H}_2\text{O} \quad \textbf{(22.84)}$$

   Like CO, the NO ligand can also act as a bridging ligand between two or three metals. These situations usually can be identified by the position of the $\nu(NO)$ mode:

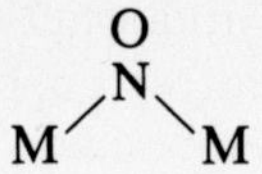

1550–1400 cm^{-1};

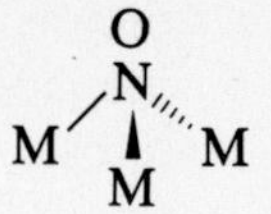

about 1320 cm^{-1}

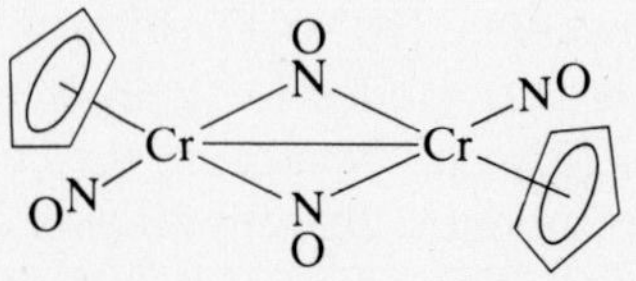

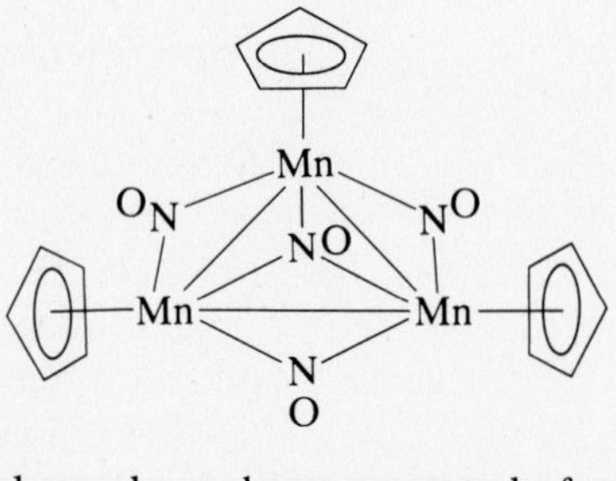

Finally, a few thionitrosyl complexes have been reported, for example,

$$3Na[CpCr(CO)_3] + S_3N_3Cl_3 \xrightarrow{-78\ °C}$$ **(22.85)**

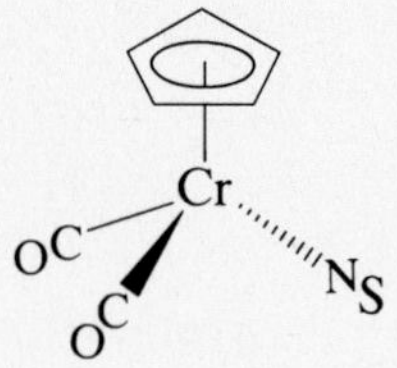

$+ 3CO + 3NaCl$

This complex has a linear Cr—N—S linkage (176.8°), and the $\nu(NS)$ mode appears in the IR spectrum at 1180 cm^{-1}. The NS ligand is apparently more strongly bonded to the Cr atom than is NO in the analogous $CpCr(CO)_2(NO)$ compound.

22.7   *Tertiary Phosphine Complexes*

Stable metal carbonyl complexes often contain tertiary phosphine ligands (PR_3) as well, especially PPh_3. These PR_3 ligands can also participate in the π-backbonding framework because the P atom has vacant $d\pi$ orbitals that can compete with the π_2 orbitals on the CO ligands for the electron density being transferred from the metal center to the surrounding ligands (Figure 22.13).

In phosphine-substituted metal carbonyl complexes, the electronegativity of the R atom affects the position of the $\nu(CO)$ modes. Especially good examples to illustrate this effect are the *fac*-$Mo(CO)_3L_3$ complexes (Table 22.7). These C_{3v} symmetry complexes should exhibit two ($a_1 + e$) $\nu(CO)$ peaks. The effect of the electronegativity of the R group is clearly evident in that the $\nu(CO)$ values increase sharply in wave number in the order Ph < Cl < F. The greater the electron drift toward the R group, the

Figure 22.13
Qualitative description of the bonding between a transition metal M and a tertiary phosphine ligand PR_3.

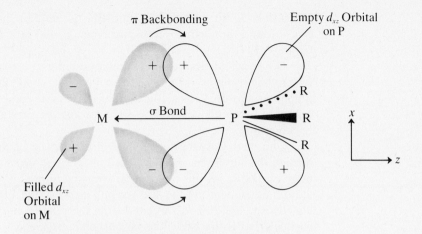

Table 22.7 $\nu(CO)$ Modes for Some Typical fac-$Mo(CO)_3L_3$ Complexes

L in *fac*-$Mo(CO)_3L_3$	$\nu(CO)$ (cm^{-1})	
	a_1	e
PPh_3	1934	1835
$PClPh_2$	1977	1885
PCl_2Ph	2016	1943
PCl_3	2040	1991
PF_3	2090	2055

less π backbonding to the CO groups there is and the higher the energy of the $\nu(CO)$ modes. This is not the only electronic contribution to the position of the $\nu(CO)$ modes, however, because the σ-donor ability of the PR_3 ligands also plays a role, but probably to a lesser degree than does the π-acceptor ability.

The PF_3 ligand is one of the few ligands that leads to complete substitution of the CO groups in metal carbonyls:

$$Ni(CO)_4 + 4PF_3 \rightarrow Ni(PF_3)_4 + 4CO \qquad (22.86)$$

$$Mo(CO)_6 + 6PF_3 \rightarrow Mo(PF_3)_6 + 6CO \qquad (22.87)$$

It is also possible to stabilize complexes such as $Pt(PF_3)_4$ for which no carbonyl analogs exist under normal conditions; $Pt(CO)_4$ has been identified by IR spectroscopy during the cocondensation of Pt metal vapor and

CO gas at about 20 K. Phosphine PH_3 is another ligand that is sometimes found bonded to transition metals, as in $Cr(CO)_4(PH_3)_2$ and $Cr(CO)_3(PH_3)_3$.

Other ligands that exhibit similar bonding to tertiary phosphines are tertiary phosphites—for example, $P(OMe)_3$ and $P(OPh)_3$—and the numerous Group VA (15) ligands such as $AsMe_3$ and $SbPh_3$. All Group VA (15) elements, except N, have vacant $d\pi$ orbitals available for π backbonding.

One of the driving forces in determining the nature of the products formed during organometallic reactions is *steric strain*. For example, the number of PR_3 ligands that can be arranged around a given central metal atom must depend on the size (or bulkiness) of the PR_3 ligands. Tolman has devised a useful way of expressing the relative steric sizes of ligands, particularly PR_3 ligands, in terms of their *cone angles* θ, which are defined (for Ni–PR_3 groups) as shown below:

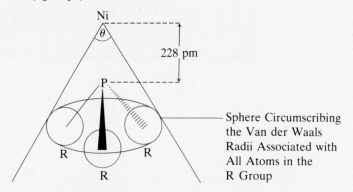

The bulkier the R group, the greater the cone angle θ: the θ values range between $104°$ (PF_3) to $184°$ [$P(C_6F_5)_3$]. The extent of CO substitution by PR_3 ligands in Group VIB (6) metal hexacarbonyls $M(CO)_6$ (M = Cr, Mo, W), for example, critically depends on the steric sizes of the PR_3 ligands. For instance, with the bulky tris(isopropyl)phosphine ligand $P(i\text{-}Pr)_3$ ($\theta = 160°$), complexes such as $Mo(CO)_5[P(i\text{-}Pr)_3]$ and *trans*-$Mo(CO)_4[P(i\text{-}Pr)_3]_2$ are formed. With the less bulky PMe_3 ligand ($\theta = 118°$), however, it is possible to prepare *cis*-$Mo(CO)_4(PMe_3)_2$ and even the tetrasubstituted derivatives $Mo(CO)_2(PMe_3)_4$.

An important tool in the structural identification of organometallic complexes containing phosphorus ligands is ^{31}P-NMR spectroscopy, and the observed spin–spin couplings are often diagnostic of specific geometrical structures. For example, *trans*-PR_3 ligands in square-planar Pt(II) complexes would be expected to have larger [31]P—[31]P coupling constants than would be observed if the two PR_3 ligands were cis to one another. These expectations are borne out by experiment: *trans*-[31]P—[31]P, 550–850 Hz; *cis*-[31]P—[31]P, <100 Hz.

22.8   *Transition Metal Complexes of Unsaturated Organic Ligands*

There is an enormous variety of organotransition metal complexes containing unsaturated organic ligands, such as alkenes, alkynes, cyclopentadienes, and arenes. As you already know, the first documented organometallic complex appears to be the platinum–ethylene complex $[PtCl_3(C_2H_4)]^-$. The major impetus for the development of the field, however, was the discovery in the early 1950s of *sandwich compounds* such as ferrocene Cp_2Fe and dibenzenechromium(0) $(\eta^6\text{-}C_6H_6)_2Cr$.

In 1827 Zeise (a Danish pharmacist) reported the preparation of a neutral compound with the empirical formula $PtCl_2 \cdot C_2H_4$, which he had obtained by passing ethylene gas through an aqueous solution of $[PtCl_4]^{2-}$. It was only in 1953 that the structures of this compound and the water-soluble intermediate $[PtCl_3(C_2H_4)]^-$, which Zeise also isolated from a refluxing alcoholic solution of $K_2[PtCl_4]$, were determined by single-crystal, X-ray diffraction. Both compounds proved to have the ethylene groups lying perpendicular to the molecular plane, and the neutral compound is a chlorine-bridged dimer:

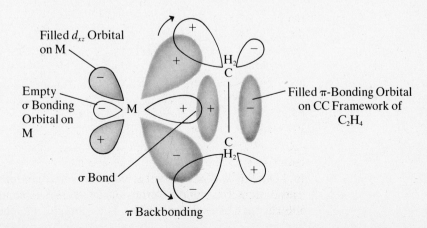

Following the discovery of the structure of Zeise's salt and similar complexes, the nature of the bonding between transition metals and alkenes has been the subject of considerable debate. Figure 22.14 shows the currently accepted description according to the Dewar–Chatt–Duncanson model. Notice that alkene ligands act as π-acid ligands, just like CO, CS, NO, and PR_3.

Figure 22.14
Qualitative description of the bonding between a transition metal M and an alkene ligand such as ethylene C_2H_4.

Filled d_{xz} Orbital on M

Empty σ Bonding Orbital on M

σ Bond

π Backbonding

Filled π-Bonding Orbital on CC Framework of C_2H_4

In these types of complexes, the alkene ligand donates two electrons from its π-bonding MO into a vacant orbital on the metal M. Depending on the type of alkene complex involved, the excess electron density accumulating on M is appropriately reduced by π backbonding from filled metal $d\pi$ orbitals into the π^a orbitals on the alkene. Alkenes attached to metals in higher oxidation states tend to be σ donors; for metals in the zero oxidation state, π backbonding becomes more important:

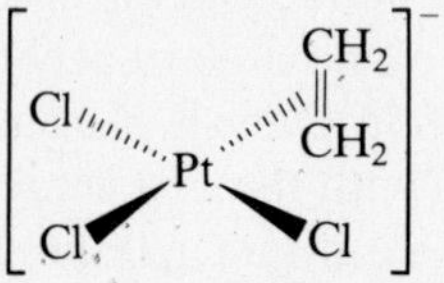

Ethylene Acting Mainly as a σ-donor Ligand; the C=C Bond Is Largely Conserved because Backbonding Is Not Very Important.

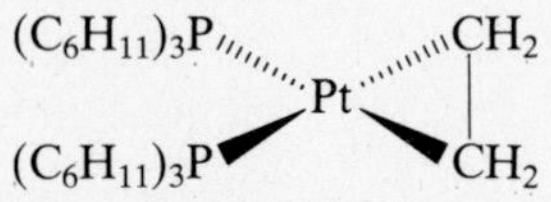

Ethylene Exhibiting σ/π Interaction; the Complex Appears More Like a Metallocyclopropane Complex.

The presence of π backbonding in an alkene complex is demonstrated by about a 100 cm^{-1} decrease in the frequency of the ν(C=C) mode due to the reduction in the bond order of the C=C bond resulting from electrons occupying the π^a orbitals. Molecular structures, determined by X-ray diffraction, also exhibit a range of carbon—carbon bond lengths intermediate between a C—C and a C=C bond (Table 22.8).

There are now numerous examples of organometallic compounds containing not only simple alkenes but also polyalkene ligands such as butadiene, norbornadiene, 1,5-cyclooctadiene, and cycloheptatriene:

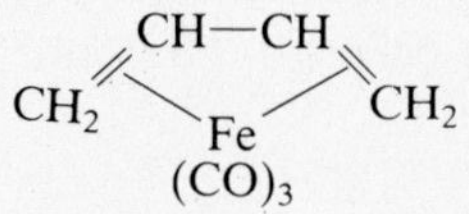

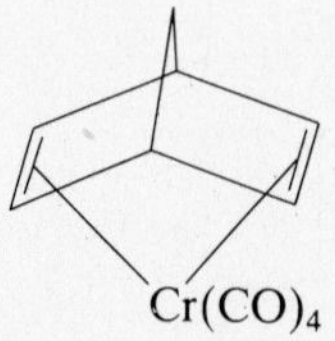

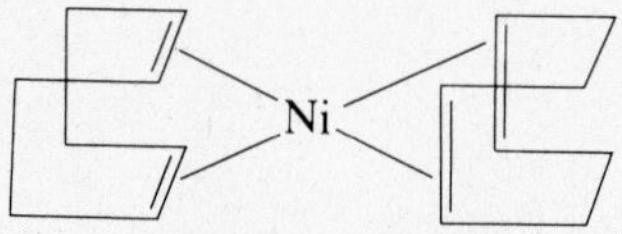

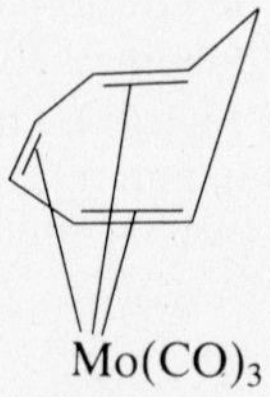

Once it was realized that alkenes would coordinate to a variety of transition metals, an obvious extension was to examine the reactions with alkynes such as acetylene C_2H_2. These organic molecules also form orga-

Table 22.8   *C—C Bond Distances in Some Selected Alkene Complexes*[a]

Compound	C—C Bond Distance (pm)
C—C$_{avg}$ (saturated hydrocarbons)	154
C=C$_{avg}$ (alkenes)	134
C≡C$_{avg}$ (alkynes)	120
C=C (bonded)	140.4
C=C (free)	131.3
C—C$_{avg}$	147
C=C (bonded)	139.8
C=C$_{avg}$ (free)	134

[a] avg = average.

nometallic complexes. However, because the two π bonds of the acetylenes are at right angles to one another, acetylenes can form bridges between two metals with each π bond acting as a two-electron donor in an analogous fashion to the σ–π bonding in metal–alkene complexes. Some examples are

There are several compounds known in which the alkyne is bonded solely to one metal:

$$CpMn(CO)_3 + PhC{\equiv}CPh \xrightarrow{h\nu} CpMn(CO)_2(PhC{\equiv}CPh) + CO$$

$$(22.88)$$

Again, there is X-ray diffraction and IR spectral evidence for π backbonding. The C≡C bond is increased by almost 8 pm upon coordination, while ν(C≡C) decreases from approximately 2200 cm^{-1} in free acetylenes to about 1850 cm^{-1} in the metal complexes. The M—C distances in metal–alkyne complexes are about 7 pm shorter than the corresponding M—C distances in metal–alkene complexes, suggesting stronger bonding interactions in the former.

While work was being reported on metal–alkene and metal–alkyne complexes, synthetic chemists were busy preparing transition metal complexes of a wide range of unsaturated hydrocarbon ligands. The first major investigations in the early 1950s culminated in the identification of sandwich compounds. Two main types of complex were studied: cyclopentadienyl derivatives (Wilkinson, U.K.) and bis(arene) derivatives (Fischer, West Germany).

The extremely air-sensitive Na$^+$[C$_5$H$_5$]$^-$(NaCp) reagent, together with the thallium analog TlCp, are the primary sources of the important Cp group in organometallic chemistry. Recently, it has been established that cyclopentadienyl complexes are often more stable if the pentamethyl ligand [C$_5$Me$_5$]$^-$ (Cp*) is used instead of [C$_5$H$_5$]$^-$. The necessary parent pentamethylcyclopentadiene C$_5$Me$_5$H is available commercially.

Bis (η^5-cyclopentadienyl)iron(II) (or ferrocene Cp$_2$Fe) was discovered in 1951, and its famous sandwich structure shown below was recognized one year later.

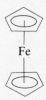

Initially, the major interest in this type of compound derived from its remarkable thermal stability (greater than 500 °C) and its aromatic character. The usual synthetic routes to neutral metallocenes Cp$_2$M are:

1. Treatment of metal halides with NaCp (or TlCp) in THF solution:

    $$FeCl_2 + 2NaCp \rightarrow Cp_2Fe + 2NaCl \qquad \textbf{(22.89)}$$

2. Reacting metal halides with cyclopentadiene in the presence of excess diethylamine:

    $$CrCl_2 + 2C_5H_6 + 2Et_2NH \rightarrow Cp_2Cr + 2[Et_2NH_2]^+Cl^- \qquad \textbf{(22.90)}$$

All neutral metallocenes are soluble in organic solvents and can be purified by sublimation in vacuum. There are examples of neutral Cp$_2$M molecules for all first-row transition elements, and apart from ferrocene, they are all air-sensitive, paramagnetic, and violate the eighteen-electron rule. Nickelocene Cp_2Ni can be safely made in an undergraduate laboratory, but special care should be taken with chromacene Cp_2Cr because it is pyrophoric. Titanocene Cp_2Ti is highly reactive, and attempts to synthesize it lead to complicated reaction products; Cp_2^*Ti has been prepared and is stable. Neutral metallocenes exhibit a wide range of colors: Cp_2V (purple), Cp_2Fe (orange), Cp_2Ni (green). All Cp_2M complexes can be oxidized electrochemically to give $[Cp_2M]^+$ species. The Cp_2Fe—$[Cp_2Fe]^+$ couple ($E^0 = -0.3$ V) is used as a reference in electrochemical work, and there is considerable interest in coating electrodes with organometallic complexes such as Cp_2Fe in an effort to produce more efficient electrochemical cells.

X-ray diffraction has shown that the metallocenes adopt sandwich structures with either eclipsed or staggered Cp rings. The energy difference between the two conformations is 8 kJ mol^{-1}.

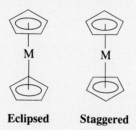

Eclipsed   **Staggered**

The structure of ferrocene (and Cp_2Os) at low temperature is eclipsed, whereas that for Cp_2^*Fe is staggered and the Me groups are pointed away from the central Fe(II) in order to minimize the steric repulsions.

The bonding in ferrocene is complicated, but there is clearly major interaction between π-bonding orbitals of the Cp rings with the $3d_{xz}$ and $3d_{yz}$ AOs of iron. In addition, there is some interaction between the $4p$ AOs of iron with π^a orbitals on the Cp rings.

Cyclopentadienyl rings are usually planar, symmetric pentagons, and early workers noted that the Cp rings exhibit typical aromatic behavior with respect to acetylation, sulfonation, and so on. For example, ferrocene undergoes a Friedel–Craft reaction with CH_3COCl:

$$Cp_2Fe + 2CH_3COCl \xrightarrow{AlCl_3} (C_5H_4COCH_3)_2Fe + 2HCl \qquad (22.91)$$

The great reactivity of titanocene, referred to above, is also characteristic of a number of other metallocenes of the earlier transition groups, for example, Cp_2Mo and Cp_2Nb. Consequently, the simple metallocene chemistry of these elements is poorly developed. Compounds of the type Cp_2MX_2, however, where X is a halogen, halogenoid, or a neutral ligand, are generally stable and have a rich and varied chemistry. Such compounds have pseudotetrahedral coordination, with the centers of the Cp rings occupying two of the tetrahedral positions, as shown below for Cp_2TiCl_2.

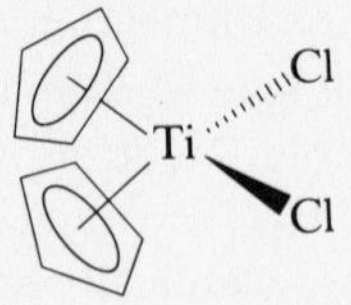

Many Cp_2MX_2 compounds are easily prepared by reaction of MX_4 with NaCp:

$$MCl_4 + 2NaCp \rightarrow Cp_2MCl_2 + 2NaCl \qquad (22.92)$$

Some reactions of the Cp_2MX_2 compounds are shown in the following schemes:

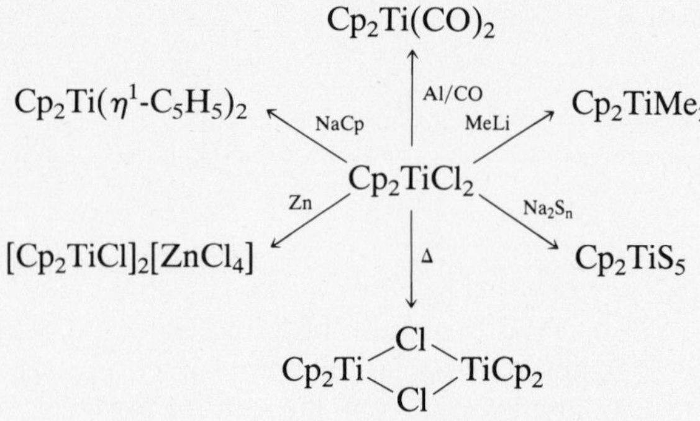

$$(22.93)$$

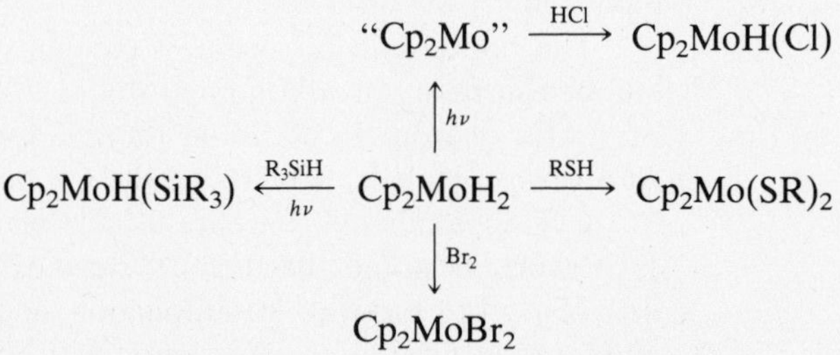

$$(22.94)$$

Once the existence of two Cp rings attached to a transition metal had been established, several complexes containing only one Cp ring were also reported, for example, $CpMn(CO)_3$ and $CpV(CO)_4$. Again, the H atoms of the Cp rings can be replaced by Me groups: $Cp^*Mn(CO)_3$ and $(\eta^5\text{-}C_5H_4Me)Mn(CO)_3$. The latter manganese derivative is a liquid that shows useful antiknock properties, but it cannot be used in internal-combustion engines because it fouls up the exhaust system and the catalytic converter.

Cyclopentadienylmetal–carbonyl complexes can be prepared in several different ways:

1. Direct reaction of a metallocene with CO gas (usually under high pressure):

$$Cp_2V \xrightarrow{\text{CO, } \sim 6 \times 10^3 \text{ kPa (60 atm)}} \text{[complex]} + \text{Polymeric Material} \tag{22.95}$$

These direct reactions afford low yields because of contamination with polymeric material formed from the Cp radical. In the case of nickel, $Ni(CO)_4$ is a convenient *in situ* source of CO:

$$Cp_2Ni + Ni(CO)_4 \longrightarrow \text{[complex]} + 2CO \tag{22.96}$$

2. Direct reaction of a binary metal carbonyl with a Cp source ($C_{10}H_{12}$ or C_5H_6):

$$2Fe(CO)_5 + C_{10}H_{12} \longrightarrow \text{[complex]} + H_2 + 6CO \tag{22.97}$$

The cis product is also known.

3. Reduction of metal–halide complexes by electropositive metals (Na, Mg, and so on) in the presence of a Cp source and CO:

$$2MnCl_2(py)_2 + Mg + 2C_5H_6 + 6CO \xrightarrow{\sim 3 \times 10^4 \text{ kPa (300 atm)}} \text{[complex]} + MgCl_2 + 2py + 2[pyH]^+Cl^- \tag{22.98}$$

Occasionally, new Cp-containing complexes are formed by thermal decomposition of existing cyclopentadienyl derivatives:

$$(22.99)$$

$$2[CpFe(CO)_2]_2 \xrightarrow{\Delta} [CpFe(CO)]_4 + 4CO \qquad (22.100)$$

The chromium product is particularly interesting as it has a metal–metal triple bond Cr≡Cr, as required to satisfy the eighteen-electron rule.

There are numerous reactions of cyclopentadienylmetal–carbonyl compounds. A few of these are indicated for $[CpFe(CO)_2]_2$ in Figure 22.15.

The first six-membered, aromatic-ring sandwich compounds were synthesized by Hein in 1919, but their real structures were only recognized some 35 years later as bis(diphenyl)metal complexes. Hein was investigating the reaction of $CrCl_3$ with the Grignard reagent PhMgBr. The structures of two of the compounds he apparently isolated are indicated below. The neutral species is less stable than ferrocene despite the fact that it satisfies the eighteen-electron rule. Neither compound survives the conditions necessary for aromatic substitution.

A more controlled synthesis of $(\eta^6\text{-}C_6H_6)_2Cr$ and other bis(η^6-Arene)metal(0) complexes involves the reduction of a metal halide in the presence of the appropriate arene ligand:

$$3CrCl_3 + 2Al + AlCl_3 + 6C_6H_6 \xrightarrow{180\,°C} 3[(\eta^6\text{-}C_6H_6)_2Cr]^+[AlCl_4]^-$$

$$\downarrow \text{NaBPh}_4$$

$$(\eta^6\text{-}C_6H_6)_2Cr + H_2O + [SO_3]^{2-} \xleftarrow{Na_2S_2O_4/OH^-} [(\eta^6\text{-}C_6H_6)_2Cr]^+[BPh_4]^-$$

$$(22.101)$$

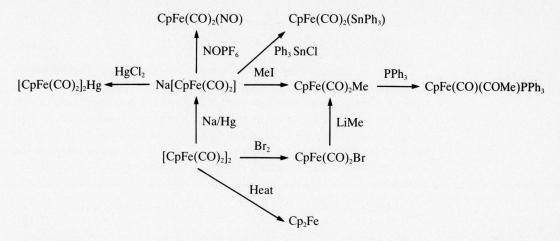

Figure 22.15
Some typical reactions of $[CpFe(CO)_2]_2$.

Cationic complexes of other metals can be prepared in a somewhat similar manner:

$$CoCl_2 + 2AlCl_3 + 2C_6Me_6 \xrightarrow{\text{About 100 °C}} [(\eta^6\text{-}C_6Me_6)_2Co][AlCl_4]_2$$

(22.102)

A radically new method for the synthesis of bis(arene)metal complexes, especially for substituted arenes such as C_6H_4XY (X and Y = F, Cl, CF_3, CO_2Me, and so on), has been devised recently. This cryogenic method involves cocondensation of a metal vapor and the ligand at liquid-nitrogen temperatures:

$$2C_6H_4XY(g) + M(g) \rightarrow (\eta^6\text{-}C_6H_4XY)_2M$$

(22.103)

The yields of the reaction products are normally less than 50%, but the purities are very high. Cryogenic synthesis may become more important in the future because it obviates the need for a solvent, thereby reducing the risk of product loss during subsequent solvent removal. Also, it is sometimes possible to prepare compounds that are not accessible by any other route, for example, the mixed sandwich compound $(\eta^6\text{-}C_6H_6)(\eta^6\text{-}C_6F_6)Cr$.

Bis(benzene)chromium(0) is a black solid (melting point 284 °C). The benzene rings are arranged in an eclipsed fashion, and the bonding is analogous to that in the metallocenes with σ donation from the filled π-bonding MO on the benzene rings, accompanied by π backbonding from the filled $3d$ orbitals on the Cr(0) into the π^a orbitals on the benzene rings.

Substituted π arene–metal complexes are widespread, especially for Group VIB (6) elements (Cr, Mo, W), namely, $(\eta^6\text{-Arene})M(CO)_3$. The complexes can be prepared by heating the parent metal hexacarbonyl with the arene:

$$M(CO)_6 + \text{Arene} \xrightarrow{\text{THF}} (\eta^6\text{-Arene})M(CO)_3 + 3CO \qquad \textbf{(22.104)}$$

These reactions often afford low product yields because of sublimation of the metal hexacarbonyl out of the reaction flask. This difficulty can be minimized by adding a small amount of Et_2O to the reaction mixture. The Et_2O refluxes at a lower temperature than does the solvent and so continually washes the $M(CO)_6$ back from the water condenser into the reaction mixture. Even so, higher product yields are obtained when the arene replaces relatively weakly coordinated ligands already on the parent complex:

$$(MeCN)_3Cr(CO)_3 + C_6H_5Cl \rightarrow (\eta^6\text{-}C_6H_5Cl)Cr(CO)_3 + 3MeCN \qquad \textbf{(22.105)}$$

Molecules with tripods of ligands like $(\eta^6\text{-Arene})Cr(CO)_3$ and $CpMn(CO)_3$ are often referred to as *piano-stool molecules*. In the case of the chromium compounds, the $Cr(CO)_3$ tripods could be either eclipsed or staggered with respect to the plane formed by the arene ring carbon atoms.

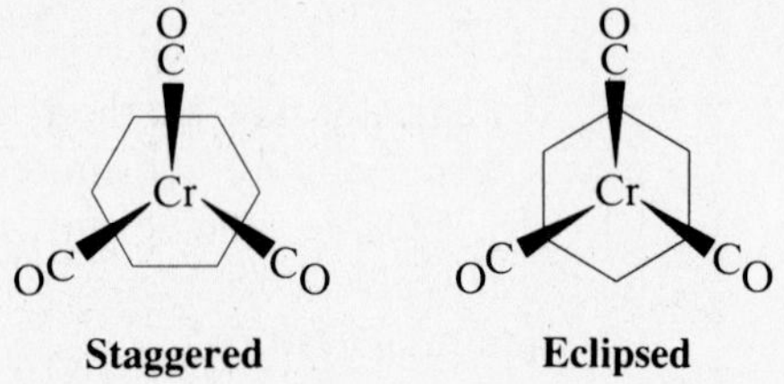

Staggered　　　　**Eclipsed**

At low temperatures, $(\eta^6\text{-}C_6H_6)Cr(CO)_3$ adopts a staggered conformation. The rotation barrier is quite low. Which conformation is adopted in a given situation critically depends on the interplay of steric and electronic effects.

Usually, it is difficult although not impossible to replace the arene ring in $(\eta^6\text{-Arene})M(CO)_3$ complexes by another arene under thermal conditions:

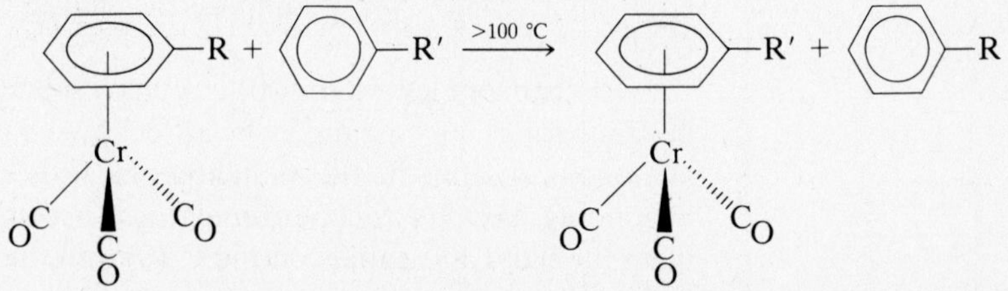

The mechanisms of these types of reactions have been thoroughly investigated, and it is considered likely that bis(carbonyl)-bridged dimers such as that shown below are involved as intermediates in the arene exchange reactions:

Note that the arene ring can act as an η^4 ligand as well as an η^6 ligand. When it is an η^4 ligand, it is necessary to invoke a Cr=Cr bond in order to satisfy the eighteen-electron rule.

Arene substitution in monothiocarbonyl and selenocarbonyl complexes is much easier than in the analogous tricarbonyls:

$$\text{+ 3L} \xrightarrow{<70\ °C} \text{Cr(CO)}_2\text{(CX)L}_3 \text{ + Arene} \tag{22.106}$$

where X = S, Se; L = CO, P(OMe)$_3$, and so on.

While the majority of unsaturated organic systems bonded to transition metals involve Cp, η^6-arene, -alkene, or -alkyne ligands, there are still a few other aromatic ones that should be mentioned, namely, cyclobutadiene ($[C_4H_4]^{2-}$), tropylium ($[C_7H_7]^+$), and cyclopropenyl ($[C_3H_3]^{3-}$). In addition, there are η^1- and η^3-allyl (CH_2=CH—CH_2—) complexes.

Organic chemists had long sought cyclobutadiene before it was eventually isolated in a matrix at about 15 K. There are now several examples of (η^4-cyclobutadiene)metal(0)–carbonyl complexes known, and it is possible to perform organic chemistry on the coordinated $[C_4H_4]^{2-}$ ring. The ring–metal bonds can then be cleaved, liberating an organic molecule that is difficult to obtain by any other synthetic route. The iron derivative (η^4-C_4H_4)Fe(CO)$_3$ is a yellow oil (boiling point 47 °C), while the analogous Group VIB (6) complexes (η^4-C_4H_4)M(CO)$_4$ are low melting point, colored solids (Cr, yellow; Mo and W, orange). These compounds are prepared by direct reaction of dichlorocyclobutene and a binary metal carbonyl:

$$\tag{22.107}$$

where M = Mo, W.

X-ray diffraction studies on several substituted cyclobutadiene complexes have shown that the C_4H_4 ring is essentially square-planar. This has been confirmed for the $[C_4H_4]^{2-}$ ring itself in a low-temperature study of $(\eta^4\text{-}C_4H_4)Fe(CO)_3$; C—C(avg.) = 142.5(7) pm, C—C—C(avg.) = 90.0(3)°.

Surprisingly, the route to seven-membered aromatic tropylium complexes is not through tropylium salts such as $[C_7H_7]^+Br^-$ because these salts only result in oxidation of organometallic compounds. The source of the $[C_7H_7]^+$ species is cycloheptatriene (C_7H_8) complexed to transition metals, as in $(\eta^6\text{-}C_7H_8)Cr(CO)_3$. The requisite hydride loss is achieved by treatment with $[Ph_3C]^+[BF_4]^-$ to yield the orange, piano-stool molecule $[(\eta^7\text{-}C_7H_7)Cr(CO)_3]BF_4$ in quantitative yield:

$$\text{(22.108)}$$

Mixed sandwich compounds involving tropylium are also known:

$$2CpV(CO)_4 + 2C_7H_8 \rightarrow 2CpV(\eta^7\text{-}C_7H_7) + 8CO + H_2 \qquad \text{(22.109)}$$

The purple vanadium complex is paramagnetic (one unpaired electron). All known tropylium complexes have planar $[C_7H_7]^+$ rings in which the C—C—C angles are ~ 128.6°, as expected for a planar heptagon.

There are two other important types of complexes that contain ligands based only on three carbon atoms: cyclopropenyl ($[C_3H_3]^{3-}$) and allyl complexes. Cyclopropenyl complexes have three carbon atoms in a ring—such complexes are usually prepared from $[Ph_3C_3]^+$ salts:

$$3Co_2(CO)_8 + 4[Ph_3C_3]BF_4 \rightarrow$$

$$4(\eta^3\text{-}C_3Ph_3)Co(CO)_3 + 12CO + 2Co(BF_4)_2 \quad \text{(22.110)}$$

$$Ni(CO)_4 + [Ph_3C_3]Cl + 2py \rightarrow (\eta^3\text{-}C_3Ph_3)Ni(py)_2Cl + 4CO$$
$$\text{(22.111)}$$

X-ray diffraction studies on the yellow cobalt complex above and the orange, mixed sandwich compound $Cp(\eta^3\text{-}C_3Ph_3)Ni$ have shown that the Ph substituents on the cyclopropenyl rings are bent out of the plane of the three carbon atoms by about 20°, due probably to a combination of both steric and electronic effects. No complexes of the parent $[C_3H_3]^{3-}$ ligand have yet been reported.

Allylic compounds such as $CH_2=CH—CH_2Cl$ lead to an interesting array of complexes with transition metals. These allyl groups can function as $\eta^1(\sigma$-allyl), η^2(alkene-type), and η^3 donors (π-allyl), with the latter being the most common. Some typical preparative reactions follow:

1. Direct reaction of an allyl Grignard reagent with a metal halide:

$$2C_3H_5MgCl + NiCl_2 \rightarrow (\eta^3\text{-}C_3H_5)_2Ni + 2MgCl_2 \qquad \textbf{(22.112)}$$

2. Treatment of a metal carbonyl anion with an allyl halide:

$$Na[CpFe(CO)_2] + C_3H_5Cl \longrightarrow CpFe(CO)_2(\eta^1\text{-}C_3H_5) + NaCl$$

$$\Big\downarrow h\nu \qquad \textbf{(22.113)}$$

$$CpFe(CO)(\eta^3\text{-}C_3H_5) + CO$$

3. Reaction of an anionic complex with an allyl halide:

$$2[PdCl_4]^{2-} + 2C_3H_5Cl \rightarrow [(\eta^3\text{-}C_3H_5)PdCl]_2 + 4Cl^- + 2Cl_2$$
$$\textbf{(22.114)}$$

4. Direct reaction of a complex containing a labile two-electron ligand with allyl alcohol:

$$CpMn(CO)_2(THF) + C_3H_5OH \longrightarrow \qquad + THF \qquad \textbf{(22.115)}$$

When the allyl ligand acts as a η^3 ligand, the C—C distances are usually approximately equal (140 pm), and the C—C—C angle is $\sim 120°$. An unusual π-allyl type coordination is found when η^7-tropylium complexes undergo reaction with NaCp:

$$(\eta^7\text{-}C_7H_7)Mo(CO)_2I + NaCp \longrightarrow \qquad + NaI \qquad \textbf{(22.116)}$$

This bonding situation arises because the eighteen-electron rule would be violated if the C_7H_7 ligand donated any more than three electrons.

22.9 Organometallics of f-Block Elements

Some inroads have been made into the organometallic chemistry of the lanthanides and actinides. Because of the large size of these elements, the ligands used to stabilize organometallic complexes containing metals in low oxidation states must also be quite large. Useful ligand precursors in this context are cyclooctatetraene (C_8H_8) and its derivatives; these molecules were among the first investigated for lanthanides and actinides. The actual ligands employed are normally dianions such as $[C_8H_8]^{2-}$ (ten-electron donor), which can be generated by treating the parent ligand with an alkali metal at low temperature in THF:

$$C_8H_8 + 2K \xrightarrow[-30\ °C]{THF} K_2[C_8H_8]$$

X-ray diffraction studies on the $[C_8H_8]^{2-}$ ion indicates a planar, octagonal structure with equal C—C distances (~ 141 pm) around the ring. Lanthanide complexes are readily formed by the following types of reactions:

$$2CeCl_3 + 3K_2[C_8H_8] + C_8H_8 \rightarrow 2K[(\eta^8\text{-}C_8H_8)_2Ce] + 4KCl + 2Cl^- \tag{22.117}$$

$$CpHoCl_2 + K_2[C_8H_8] \rightarrow CpHo(\eta^8\text{-}C_8H_8) + 2KCl \tag{22.118}$$

Uranocene, $(\eta^8\text{-}C_8H_8)_2U$ was discovered in 1968 using the following synthesis, which has since been extended to other actinides (Th, Pa, Np, Pu):

$$UCl_4 + 2K_2[C_8H_8] \longrightarrow \quad U \quad + 4KCl \tag{22.119}$$

As with more conventional metallocenes, uranocene and its analogs may have the two C_8H_8 rings staggered or eclipsed with respect to each other. In uranocene, the two rings are eclipsed.

The chemistry of the lanthanides almost entirely involves the III oxidation state. Compounds of the type Cp_2MX are known, but their insolubility and sensitivity to moisture and air have inhibited the investigation of their chemistry. The compounds where $X = CH_3$ are methyl-bridged dimers similar to $Al_2(CH_3)_6$. The structure of $(Cp_2YbCH_3)_2$ is shown below:

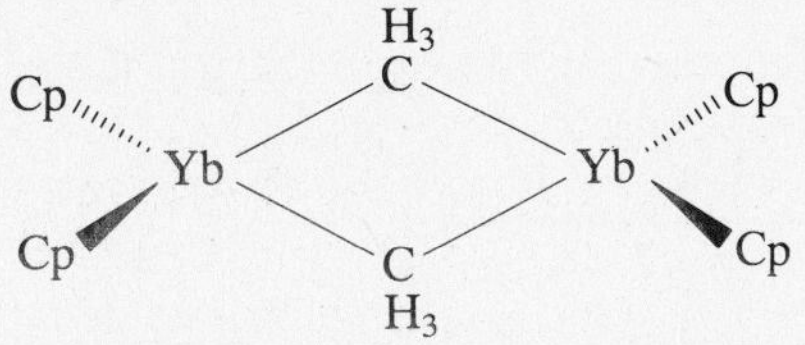

There are strong indications that organolanthanides can exhibit high catalytic activity for hydrogenation and C—H activation. This chemistry is currently under investigation, particularly using Cp* in place of Cp.

The organometallic chemistry of the oxidation state IV of uranium and thorium has recently undergone a rapid development. Compounds of the type Cp_3MX have been known for some time, but they are chemically inert. Attempts to prepare Cp_2MX_2 have always resulted in failure; however, the analogous $Cp_2^*MX_2$ compounds have been made. For example, $Cp_2^*MCl_2$ compounds are readily synthesized by reaction of Cp^*MgCl with MCl_4:

$$2Cp^*MgCl + MCl_4 \rightarrow Cp_2^*MCl_2 + 2MgCl_2 \qquad (22.120)$$

The thorium compound is diamagnetic ($5f^0$ ground-state electronic configuration), whereas the uranium compound is paramagnetic ($5f^1$ ground-state electronic configuration). Although these compounds are thermally very stable, they are easily oxidized in air and extremely susceptible to hydrolysis. This is a reflection of the greater ionic character of the Cp—M bonds for organoactinides, compared to that for most transition metal cyclopentadienyl compounds.

The $Cp_2^*MCl_2$ compounds can be alkylated with methyl lithium to give $Cp_2^*M(CH_3)_2$. These are simple molecular compounds that show great chemical reactivity. For example, they undergo reaction with C—H bonds:

$$Cp_2^*U(CH_3)_2 + 2C_6H_6 \rightarrow Cp_2^*UPh_2 + 2CH_4 \qquad (22.121)$$

Activation of C—H bonds is an important area in organometallic chemistry because alkanes are widespread in the petroleum industry and there is considerable interest in activating C—H bonds under homogeneous catalysis conditions. There are now several ways in which C—H bonds can be activated. It is appropriate to mention two of these at this juncture.

1.   *Ortho*-metallation. This type of reaction is an intramolecular abstraction reaction and is particularly well known for metal–alkyl com-

pounds containing PPh_3 ligands:

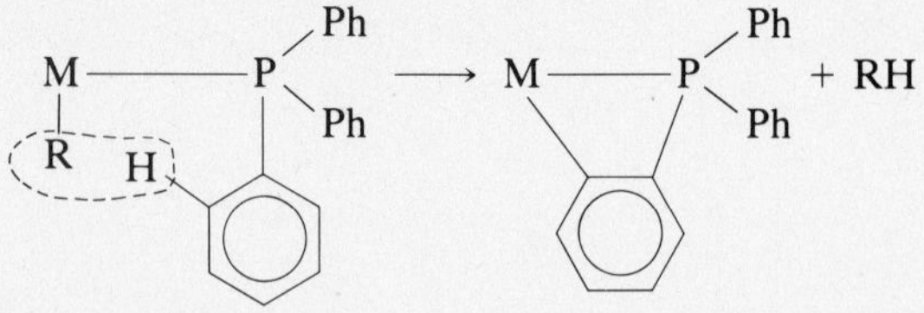
M
P
Ph
Ph
R
H
M
P
Ph
Ph
+ RH

(22.122)

2. Direct reaction of C—H bonds (including those of alkanes) with a coordinated metal center:

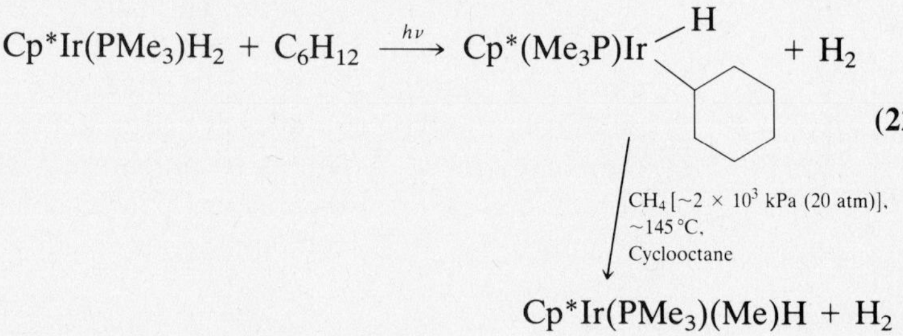
Cp*Ir(PMe3)H2 + C6H12
hν
Cp*(Me3P)Ir
H
+ H2
CH4 [~2 × 10³ kPa (20 atm)],
~145 °C,
Cyclooctane
Cp*Ir(PMe3)(Me)H + H2

(22.123)

22.10 Concluding Remarks

Before concluding this chapter, something should be said about the current interest in the use of organometallic compounds as reagents and catalysts in organic synthesis.

Organometallic fragments such as $Fe(CO)_3$ have found application as protecting agents in synthetic reactions, for example, in the protection of dienes. The diene group of the compound shown below can be protected from hydrogenation by coordination of $Fe(CO)_3$. Once the remaining C=C bond has been hydrogenated, the $Fe(CO)_3$ moiety can be decomplexed by treatment with a variety of agents, including Ce^{4+} and $FeCl_3$.

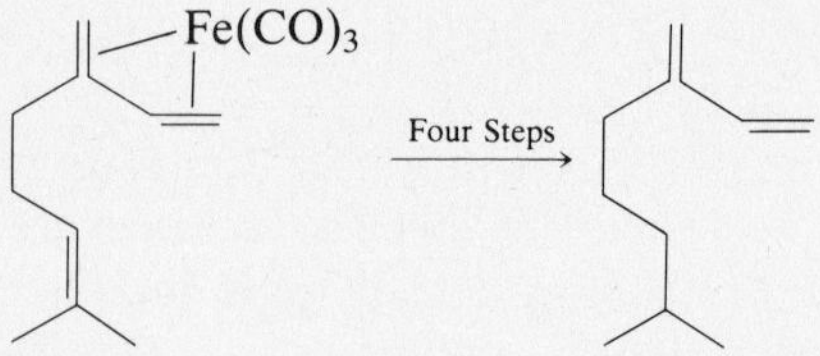
Fe(CO)3
Four Steps

(22.124)

The $Fe(CO)_3$ moiety is also useful in promoting diene isomerization:

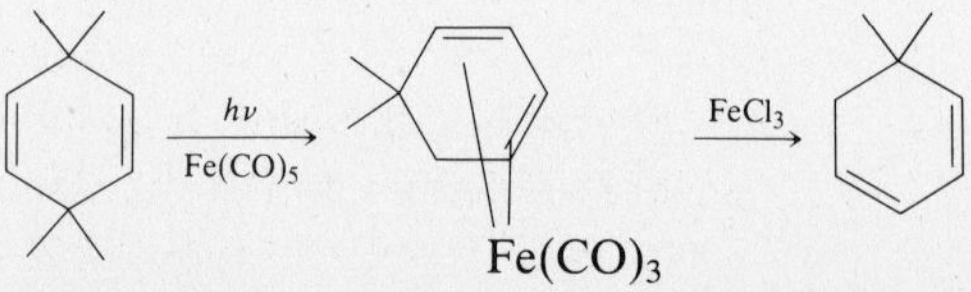
hν
Fe(CO)5
Fe(CO)3
FeCl3

(22.125)

A particularly useful synthetic reagent is $Na_2[Fe(CO)_4]$. This carbonylferrate(-II) dianion converts alkyl or acyl halides into a wide range of products including alkanes, ketones, aldehydes, and carboxylic acids:

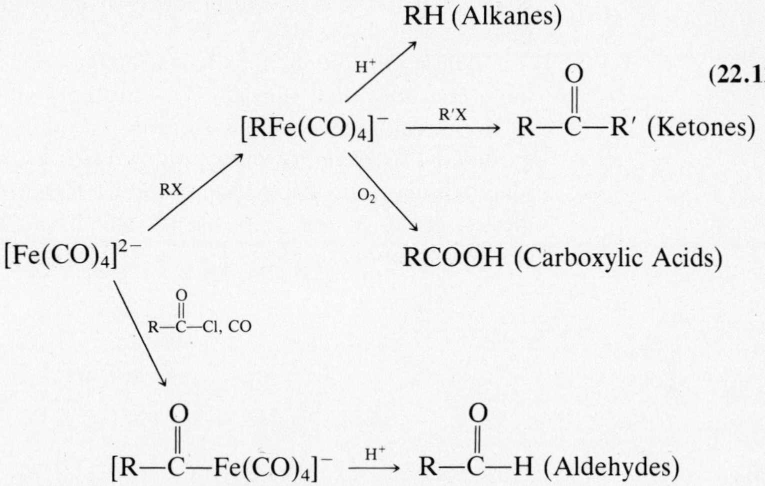

$$RH \text{ (Alkanes)}$$

$$[Fe(CO)_4]^{2-} \xrightarrow{RX} [RFe(CO)_4]^{-} \quad \begin{array}{c} \xrightarrow{H^+} \\ \xrightarrow{R'X} \\ \xrightarrow{O_2} \end{array}$$

RH (Alkanes)

$$R\overset{\overset{\textstyle O}{\|}}{-}C-R' \text{ (Ketones)}$$

RCOOH (Carboxylic Acids)

(22.126)

$$[Fe(CO)_4]^{2-} \xrightarrow{R-\overset{\overset{\textstyle O}{\|}}{C}-Cl,\, CO} [R-\overset{\overset{\textstyle O}{\|}}{C}-Fe(CO)_4]^{-} \xrightarrow{H^+} R-\overset{\overset{\textstyle O}{\|}}{C}-H \text{ (Aldehydes)}$$

Another useful organometallic-mediated reaction is ring closure through intramolecular attack of a carbanion on a metal-complexed arene ring to give fused bicyclic rings:

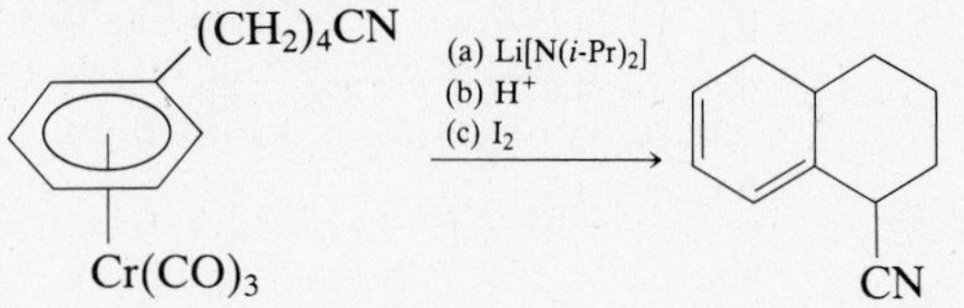

(22.127)

Note that in this case decomplexation of the $Cr(CO)_3$ fragment is achieved by I_2 oxidation.

Cobalt organometallic complexes such as $CpCo(CO)_2$ catalyze the formation of arene rings from suitable alkyne precursors:

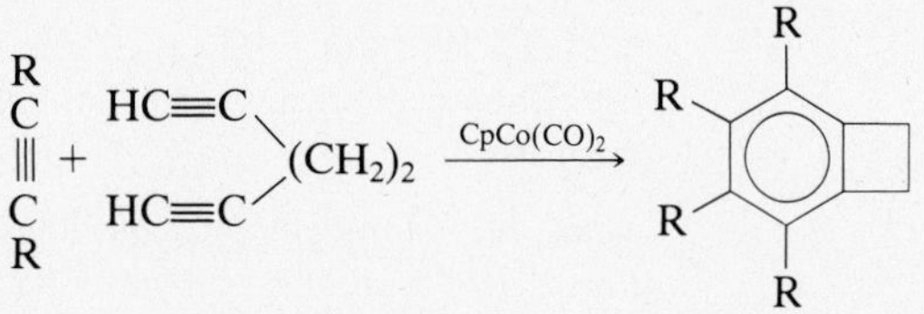

(22.128)

As an illustration of a final application of organometallic compounds, there has been some progress recently in the nonradioisotopic labeling of biochemically important molecules by tagging them with metal carbonyl fragments such as $Cr(CO)_3$, $Mn(CO)_3$, and $Co_2(CO)_6$ and then using FT–IR spectroscopy to detect the $\nu(CO)$ modes of the labels. The $\nu(CO)$

region is devoid of any features due to the steroidal moiety, namely, $\nu(CH)$, $\nu(C=C)$, ring-bending modes, and so on; thus, $\nu(CO)$ modes can be detected fairly easily, although large numbers of scans may have to be coadded if the concentration levels of the organometallic labels are very low.

Among the biochemically interesting molecules investigated thus far is the arene hormonal steroid, 17β-estradiol, which is of crucial importance in the development of breast cancer. Complexation of $Cr(CO)_3$ to the modified 17β-estradiol shown below leads to two diastereomers (α and β) [depending on the disposition of the $Cr(CO)_3$ fragment with respect to the methyl group at the 17-position], which can be separated by thin-layer chromatography.

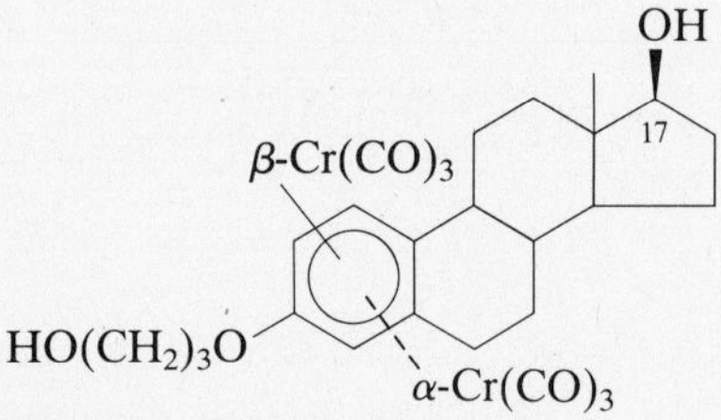

The trans (α) isomer is the more useful for the biochemical work because it has a comparable binding affinity to that of 17β-estradiol for the hormone-receptor sites in protein. This new approach of bioorganometallic labeling holds considerable promise for future biological assay work, especially where conventional radiochemical procedures have proven difficult or hazardous.

BIBLIOGRAPHY

Suggested Reading

Abel, E. W., and F. G. A. Stone, eds. *Organometallic Chemistry*. London: The Chemical Society. An annual survey of all aspects of organometallic chemistry.

Alper, H. "Homogeneous and Phase Transfer Catalyzed Carbonylation Reactions." *J. Organomet. Chem.* 300 (1985): 1.

Collman, J. P., and L. S. Hegedus. *Principles and Applications of Organometallic Chemistry*. Mill Valley, Calif.: University Science Books, 1980.

Cotton, F. A., and G. Wilkinson. *Advanced Inorganic Chemistry*, 5th ed. New York: Wiley, 1988.

Davies, S. G. *Organotransition Metal Chemistry Applications to Organic Synthesis*. Oxford, England: Pergamon Press, 1982.

Heck, R. F. *Organotransition Metal Chemistry: A Mechanistic Approach*. New York: Academic Press, 1974.

Lukehart, C. M. *Fundamental Transition Metal Organometallic Chemistry*. Pacific Grove, Calif.: Brooks/Cole, 1985.

Tsuji, J. *Organic Synthesis by Means of Transition Metal Complexes*. New York: Springer-Verlag, 1975.

Vessières, A., S. Top, A. A. Ismail, I. S. Butler, M. Louer, and G. Jaouen. "Organometallic Estrogen. Synthesis, Interaction with Lamb Uterine Estrogen Receptor, and Detection by Infrared Spectroscopy." *Biochemistry* 27 (1988): 6659.

Wilkinson, G., F. G. A. Stone, and E. W. Abel, eds. *Comprehensive Organometallic Chemistry*. Oxford, England: Pergamon Press, 1982.

Yamamoto, A. *Organotransition Metal Chemistry: Fundamental Concepts and Applications*. New York: Wiley, 1986.

PROBLEMS

22.1   Which of the following compounds lack any M—C bonding?
   a. $[Re(CO)_6]^+$
   b. $[Cr(CO)_5I]^-$
   c. $CpFe(CO)_2Cl$
   d. Ph_2GeCl_2
   e. C_3O_2
   f. Me_2Zn
   g. SnI_4
   h. MeK
   i. Cp_2^*Co

22.2   In which of the following compounds do the central metal atoms obey the eighteen-electron rule?
   a. $CpMn(CO)_2(CS)$
   b. $(\eta^4\text{-}C_4H_4)Fe(CO)_3$
   c. $CpRu(CO)_2(COMe)$
   d. Ph_4Sn
   e. $Co_2(CO)_8$
   f. *trans*-$RhCl(CO)(PPh_3)_2$
   g. $CpTi(\eta^8\text{-}C_8H_8)$
   h. Me_6W
   i. $Cp_2Ti(CO)_2(\eta^2\text{-}C_2Ph_2)$
   j. $Cp(\eta^1\text{-}C_5H_5)Fe(CO)_2$

22.3   Sketch the structures of the following compounds, given that the central metal atoms obey the eighteen-electron rule.
   a. Decacarbonyldirhenium(0)
   b. *trans*-Bis $\{(\mu\text{-carbonyl})carbonyl[\eta^5\text{-cyclopentadienyl}]iron(0)\}$
   c. $(\eta^3\text{-Allyl})$tetracarbonylmanganese(0)
   d. Carbonylchlorobis(triphenylphosphine)(η^2-tetracyanoethylene)-iridium(I)
   e. Potassium tricarbonyl(thiocarbonyl)ferrate(II)
   f. *trans*-Bis[tetracarbonyl(triphenylphosphine)manganese(0)]
   g. Tricarbonyl(1,3,5-cycloheptatrienyl)chromium(0)
   h. Dichlorobis(η^5-pentamethylcyclopentadienyl)zirconium(IV)
   i. Dicarbonyl(η^5-cyclopentadienyl)(thionitrosyl)chromium(0)
   j. Decacarbonylmanganese(0)rhenium(0)
   k. Dicarbonylnitrosyl(η^5-cyclopentadienyl)rhenium(I) cation

22.4 Draw the structures and assign the point-group symmetries for the following complexes:
  a. Benzene(tricarbonyl)molybdenum(0)
  b. (Cyclooctadiene)rhodium(I)-μ-dichloro(cyclooctadiene)rhodium(I)
  c. *Trans*-octacarbonylbis(triphenylphosphine)dirhenium(0)
  d. Tetra(thiocarbonyl)nickel(0)
  e. Pentacarbonyl(selenocarbonyl)chromium(0)
  f. Tricarbonyl(cyclobutadienyl)iron(0)

22.5 Compare the bonding in $W(CO)_6$ and WF_6 from a simple MO standpoint.

22.6 An organometallic compound may be defined as a compound containing carbon-to-metal bonds. Discuss the kinds of iron compounds that may be classified as organometallics on the basis of this definition.

22.7 Trifluorophosphine has π-acceptor properties very similar to those of CO. What would you expect to be the structure of the simplest, binary, zerovalent trifluorophosphine complexes of Cr, Mn, Fe, Co, and Ni? Illustrate schematically the bonding between phosphorus and the metal atom in such complexes.

22.8 An important factor in determining the structures of binary metal carbonyls is the deduction of the molecular point-group symmetry from IR or X-ray diffraction data. Assign a point group and discuss the geometry of the simplest, binary carbonyls of V, W, Tc, Os, Co, and Ni.

22.9 Discuss the bonding in bis(cyclooctadiene)nickel(0) and bis(allyl)nickel(II).

22.10 Which, if any, of the following complexes are likely to be thermodynamically unstable? Explain your answers.
  a. Bis(η^5-cyclopentadienyl)hydridoiron(IV) tetrafluoroborate
  b. η^5-Cyclopentadienyl(cyclooctadiene)cobalt(I)
  c. Tetrahydridotris(triphenylphosphine)ruthenium(IV)
  d. Bis(η^5-cyclopentadienyl)nitrosyltungsten(III) perchlorate

22.11 Write an essay describing the transition metal coordination chemistry of unsaturated organic ligands.

22.12 For what particular aspects of organometallic chemistry are the following chemists particularly known?
  a. Frankland
  b. Zeise
  c. Tolman
  d. Hein
  e. Mond
  f. Wilkinson
  g. Fischer
  h. Dewar–Chatt–Duncanson

22.13 **a.** Which third-row transition element would give the most thermodynamically stable compound of the type $(\eta^5$-cyclopentadienyl$)_2MH_2$, $[(\eta^6\text{-}C_6H_6)M(CO)_3]^+$ and $(\eta^5$-cyclopentadienyl$)M(NO)$?
  b. Describe the coordination geometry of each compound.

22.14 A paramagnetic metal iodide A was treated with a strong reducing agent in the presence of 1,5-cyclooctadiene. The reaction mixture yielded a yellow, air-sensitive, diamagnetic complex B (Anal. C, 70.0; H, 8.7; M, 21.3%).

Compound B reacted with allyl iodide at $-20\ °C$ to give a red, diamagnetic solid C (Anal. C, 15.8; H, 2.2; I, 5.6%; mol wt, 450 ± 10). Reaction of C with iodine produced compound A and allyl iodide. Account for all of the above observations and discuss briefly the structure and bonding of compounds B and C.

22.15 With the aid of suitable examples from your knowledge of transition metal chemistry, how would you justify the following statement in a textbook: "IR spectroscopy is an extremely useful spectroscopic method to demonstrate the coordination of π-acceptor ligands."

22.16 Using group theoretical procedures, predict the IR and Raman spectra of *trans*-$Fe(CO)_3(PPh_3)_2$ and $Fe(CO)_2(NO)_2$ in the $\nu(CO)$ and $\nu(NO)$ regions.

22.17 Account for the following observations:

a. Reaction of $Ru(CO)_5$ with PF_3 produces a series of complexes of which all but one exhibit IR bands in the $2150–1800\ cm^{-1}$ region.

b. Cobalt–alkyne complexes are significantly more stable than are gallium–alkyne complexes.

c. The $\nu(CO)$ band in the IR spectrum of $[Fe(CO)_4]^{2-}$ is at about $1790\ cm^{-1}$, whereas that for $Ni(CO)_4$ is at about $2060\ cm^{-1}$.

d. The C—C bond distance for the C_2H_4 ligand in $[\eta^2\text{-}C_2H_4)PtCl_3]^-$ is significantly longer than is that for the uncoordinated C_2H_4 molecule.

22.18 Decacarbonyldirhenium(0) reacts with chlorine to give product A, which upon heating at elevated temperatures releases CO to form compound B. Treatment of compound B with an equimolar amount of tetraethylammonium chloride (Et_4NCl) affords compound C for which $\mu_D > 0$. Reaction of C with 1 mole of PPh_3 results in the formation of Et_4NCl and compound D, which has the same stereochemistry as compound C. Both compounds B and C exhibit four (CO) bands in their IR spectra.

22.19 An organometallic complex A reacts with sodium metal to afford complex B. Treatment of B with methyl iodide yields a new complex C, which upon reaction with CO affords product D. Given the following data, identify the four complexes A–D and explain what occurred in the reactions:

a. *Complex A*. Mass spectrum: P^+ (m/e 354). IR spectrum: terminal CO groups and Cp rings. ^1H-NMR spectrum: Cp rings. ^{13}C-NMR spectrum: two types of carbons in a 5:2 ratio.

b. *Complex B*. Conductivity measurements: 1:1 electrolyte. IR spectrum: terminal CO groups and Cp rings. ^1H-NMR spectrum: Cp rings. ^{13}C-NMR spectrum: two types of carbons in a 5:2 ratio.

c. *Complex C*. Mass spectrum: P^+ (m/e 190). IR spectrum: terminal CO groups and Cp rings. ^1H-NMR spectrum: Cp rings and CH_3 groups in a 5:3 ratio. ^{13}C-NMR spectrum: three types of carbons in a 5:2:1 ratio.

d. *Complex D*. IR spectrum: terminal and ketonic CO groups and Cp rings. ^1H-NMR spectrum: Cp rings and $COCH_3$ group in a 5:3 ratio. ^{13}C-NMR spectrum: four types of carbon atoms in a 5:2:1:1 ratio.

23

Catalysis by Coordination and Organometallic Complexes

One of the major reasons for the rapid development of organometallic chemistry has been the prospect of using organometallic complexes as homogeneous catalysts in a wide variety of reactions. In fact, there are now several industrial processes that employ organometallic complexes in this way. This chapter describes some recent applications of organometallic complexes in homogeneous catalysis and the proposed mechanisms. Much of the work has been focused on converting by-products of the petroleum industry (for example, alkenes) into useful chemicals for other industries. However, we begin with the heterogeneous catalysis involved in the polymerization of ethylene and propylene, discovered by Ziegler and Natta, because this work has played such an important role in the development of organometallic chemistry.

23.1 *Ziegler–Natta Alkene Polymerization*

Prior to 1950, a number of reactions, some of industrial importance, were discovered in which a transition metal compound catalyzed an organic transformation. These reactions were discovered accidentally, and the mechanisms by which they proceeded remained mysterious. The discovery by Ziegler (West Germany) in 1953 that mixtures of transition metal compounds with metal alkyls of Groups IA (1), IIA (2), or IIIA (13) could catalyze the polymerization of ethylene and higher alkenes to high molecular weight polymers was a watershed in the history of organometallic chemistry. The impact of Ziegler's discovery was greatly amplified by the further discoveries of Natta (Italy): Ziegler-type catalysts exhibit a remarkable stereocontrol over the structure of the polymer chains in both polyalkenes and polydienes.

In a polyalkene such as polypropylene every other backbone carbon atom is chiral (a stereocenter). In the absence of any stereoregulating influence, a random sequence of *d* and *l* centers would be expected. Such a sequence is called an *atactic sequence*. If all of the stereocenters of a given chain have the same chirality, the chain is said to be *isotactic*; if the stereocenters alternate *d* and *l*, it is *syndiotactic*. Natta was able to show that polymers of each kind can be produced by an appropriate choice of catalyst. He found that the structure of polydienes can also be controlled. Polymerization of isoprene (2-methylbuta-1,3-diene) can give rise to a number of different geometric and stereoisomeric units in the chain.

cis-1,4   trans-1,4

Using appropriate catalyst combinations, Natta succeeded in synthesizing several of these polymers in high isomeric purity, in particular, the *cis*- and *trans*-1,4 polymers, which had only been available previously from natural sources (rubber trees). Over the past thirty years or so, Ziegler–Natta catalysis has provided a multimillion-ton-per-year polymer industry, a multimillion-dollar-per-year livelihood for patent lawyers, and a never-ending supply of new and exciting discoveries in organometallic chemistry.

The classic Ziegler–Natta catalyst system for ethylene–propylene polymerization is a mixture of $TiCl_4$ (or $TiCl_3$) and $(C_2H_5)_3Al$. This mixture contains both soluble and insoluble fractions and is extremely chemically complex. The generally accepted mechanism for this catalytic system involves generation of coordinatively unsaturated, five-coordinate Ti atoms on the surface of the catalyst, which can coordinate C_2H_5 groups from $(C_2H_5)_3Al$. Ethylene (or propylene) can then coordinate to the Ti atom in the usual η^2 fashion. Ethyl migration on to the coordinated alkene can take place through a four-center transition state. This migration leaves a vacant coordination site, which can be quickly occupied by another ethylene molecule to which the alkyl chain can migrate. Repetition of this process leads to an alkyl chain of increasing length. A number of termination steps have been identified, one of which is β-hydride elimination to give a vinyl-terminated polymer.

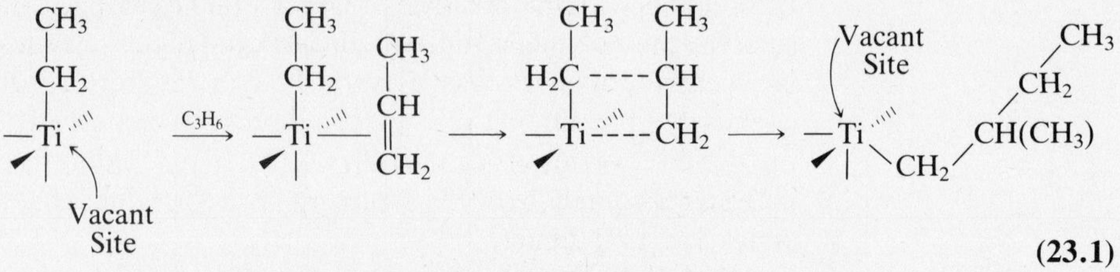

$$(23.1)$$

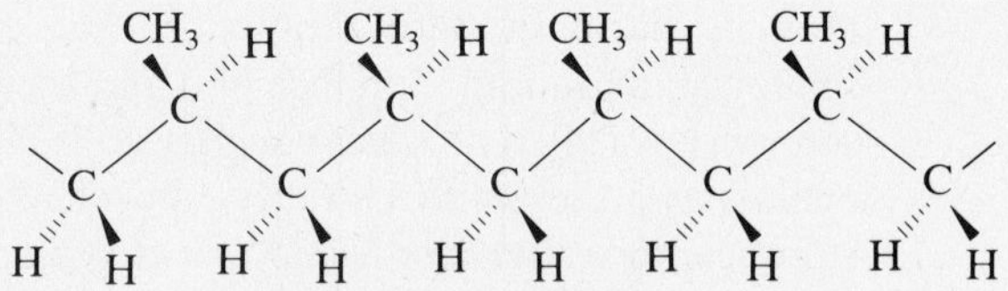

Isotactic Polypropylene

The origin of stereoregulation in heterogeneous catalysis is attributed to the intrinsic chirality of the coordinatively unsaturated Ti atoms. Because the Ti at the surface of a $TiCl_3$ crystal is a chiral center, the two faces of the alkene do not bind to give identical products (the two products are diastereomers), and the product of lower energy is formed preferentially.

Recently, homogeneous catalysts have been prepared using bis(indenyl)ethylene as a ligand to impose chirality at the metal center:

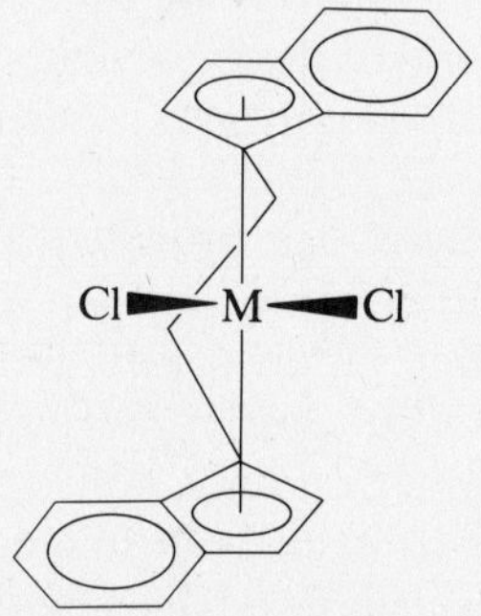

Reaction of the zirconium (or hafnium) complex with alkyl alumoxanes— for example, $[Al(CH_3)—O]_n$—gives a homogeneous catalyst that yields polypropylene with greater than 90% isotacticity.

23.2 Activation of Hydrogen, Oxygen, and Other Small Molecules

Following Vaska's discovery in the early 1960s that four-coordinate, square-planar *trans*-IrCl(CO)(PPh$_3$)$_2$ would bind molecular oxygen reversibly, synthetic chemists became interested in the reactivity of a variety of small molecules such as H$_2$, O$_2$, Cl$_2$, HCl, and CH$_3$I toward organometallic complexes. Reactions involving these simple molecules are often described as *oxidative–addition reactions* because the oxidation state of the central metal is increased, as is the coordination number (see page 629):

$$
trans\text{-IrCl(CO)(PPh}_3)_2 + O_2 \rightleftharpoons
\begin{array}{c}
\text{PPh}_3 \\
| \quad O \\
\text{Cl}\!-\!\!-\!\text{Ir}\!-\!\!-\!\text{O} \\
\diagup \quad | \\
\text{O}^{C} \quad \text{PPh}_3
\end{array}
\tag{23.2}
$$

Ir(I), CN = 4   Ir(III), CN = 6

$$
\text{RhCl(PPh}_3)_3 + H_2 \longrightarrow \text{RhCl(H)(H)(PPh}_3)_2 + \text{PPh}_3
\tag{23.3}
$$

Rh(I), CN = 4   Rh(III), CN = 5

Note that in the latter reaction, a PPh$_3$ ligand is displaced as well.

The reverse of oxidative addition is termed *reductive–elimination*. The interest in the reversible addition to O$_2$ to Vaska's compound arose because the reaction appeared to be analogous to the oxygen uptake of hemoglobin and other naturally occurring metalloproteins. Other transition metal complexes, although not organometallics, have been found to coordinate molecular oxygen. The metals involved are Ru(0), Rh(I), and Pt(0):

$$
\text{Pt(PPh}_3)_3 + O_2 \longrightarrow (\text{Ph}_3\text{P})_2\text{Pt}\!\begin{array}{c} O \\ | \\ O \end{array} + \text{PPh}_3
\tag{23.4}
$$

These types of adducts oxidize CO to CO$_2$ and PPh$_3$ to Ph$_3$PO.

Strictly, for oxidative–addition reactions to take place, the central metal must have oxidation states separated by two units and must accommodate one or two additional ligands in its first coordination sphere. The metals most thoroughly investigated are those that have d^8 or d^{10} ground-state electronic configurations: d^8—Fe(0), Ru(0), Os(0), Rh(I), Ir(I), Pd(II), Pt(II); d^{10}—Ni(0), Pd(0), Pt(0). In some cases, like H$_2$, the atoms usually cease to be bonded to each other after they are coordinated. In others, like O$_2$, the two atoms remain bonded to one another while they are coordinated to the central metal.

A number of dihydrogen complexes have been discovered recently in which the two hydrogen atoms remain substantially bonded to each other. The tungsten complex shown below was the first such complex reported. In such compounds, if HD is used instead of H_2, a strong spin–spin coupling is observed between the H and D atoms in the ^{1}H-NMR spectra; this coupling does not occur for true metal–dihydride complexes.

$$
\begin{array}{c}
\text{H} \\
\diagdown \\
| \quad {}^{\text{\tiny H}} \\
(C_6H_{11})_3P \text{---} W \text{---} P(C_6H_{11})_3 \\
(C_6H_{11})_3P \diagup \quad | \\
\text{C} \\
\text{O}
\end{array}
$$

Another important test for residual H—H bonding between the H atoms is the relaxation time (T_1) of the proton spin in a ^{1}H-NMR experiment. The T_1 value is appreciably shorter (by about an order of magnitude) for the H atoms in a true $M\genfrac{}{}{0pt}{}{\diagup \text{H}}{\diagdown \text{H}}$ complex than it is for the H atoms in a dihydride $M\genfrac{}{}{0pt}{}{\diagup \text{H}}{\diagdown \text{H}}$.

Oxidative additions of H_2 to square-planar complexes usually generate octahedral products in which the two H atoms are cis to one another, whereas halogens (X_2) or alkyl halides (RX) normally afford products in which the X groups or the R and X groups are trans to one another. As an illustration, the oxidative–addition of CH_3X to derivatives of Vaska's iridium complex is shown below:

$$
\begin{array}{cc}
& \quad\quad\quad\quad\quad\quad\quad\quad\quad \text{CH}_3 \\
\text{CO} & \quad\quad\quad\quad\quad\quad\quad | \quad {}^{\text{\tiny CO}} \\
\text{L---Ir---L} + CH_3X \longrightarrow & \text{L---Ir---L} \\
\diagup & \diagup \quad | \\
\text{Y} & \text{Y} \quad \text{X} \quad trans \text{ Addition}
\end{array} \tag{23.5}
$$

L = PPh$_3$, PEt$_3$, PEt$_2$Ph, P(OPh)$_3$
X = Cl, Br, I
Y = F, Cl, Br, I

$$
\Big\Updownarrow CH_3OH
$$

$$
\begin{array}{c}
\text{CH}_3 \\
| \quad {}^{\text{\tiny CO}} \\
\text{L---Ir---L} \\
\diagup \quad | \\
\text{X} \quad \text{Y} \quad cis \text{ Addition}
\end{array}
$$

Kinetic data for the formation of the trans product obey a second-order rate law:

$$
\text{Rate} = k[\text{IrY(CO)L}_2][\text{CH}_3\text{X}] \tag{23.6}
$$

The following associative mechanism has been proposed to explain the observed kinetic data. However, this may not be the complete story because the reactions with other alkyl halides are more complicated and radicals are implicated.

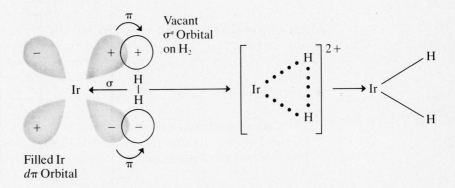

$$(23.7)$$

In the case of oxidative–addition of the hydrogen halides (HX), mixtures of cis and trans products are formed.

Kinetic studies on the oxidative addition of H_2 to square-planar Ir(I) complexes are in accord with a mechanism involving a three-center concerted process, which eventually leads to homolytic cleavage of the H—H bond (see below). The formal similarity between this type of H_2 interaction with a metal center and the σ–π bonding of an alkene should be noted (see also Section 22.3).

A number of other covalent hydrides add to metal centers in an analogous manner to that for H_2. Notable among these are hydrides of Group IVA (14) elements, particularly silicon. For example, addition of the Si—H bond of a R_3SiH molecule to a transition metal complex usually leads to the formation of a silyl–hydride complex:

$$L_nM\begin{array}{c} \diagup SiR_3 \\ \diagdown H \end{array}$$

This process is termed *hydrosilylation*. There are, however, several compounds now known in which complete breakage of the Si—H bond does not occur; strong evidence for this is because the bridging Si—H bonds are significantly longer than are the terminal Si—H bonds:

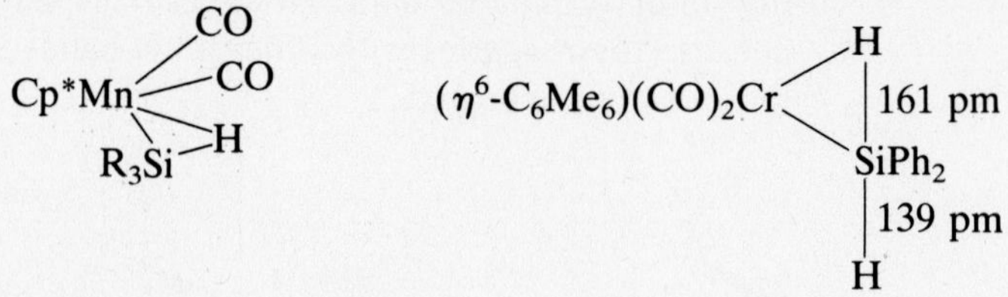

where $(Cp^* = \eta^5{-}C_5(CH_3)_5$.

There are many other examples of the activation of small molecules by organometallic complexes. For our purposes, however, we consider only CS_2, CO_2, and S_n.

Carbon Disulfide

Adducts involving CS_2 are often intermediates in the formation of thiocarbonyl complexes:

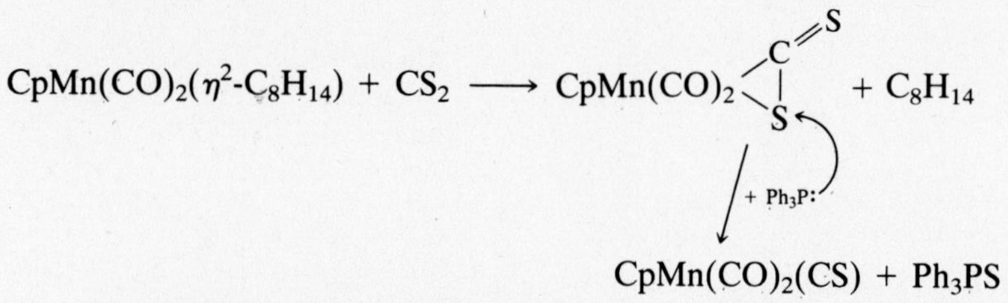

$$\mathrm{CpMn(CO)_2(\eta^2\text{-}C_8H_{14})} + CS_2 \longrightarrow \mathrm{CpMn(CO)_2}\!\!\left\langle\begin{array}{c}C\!\!=\!\!S\\ | \\ S\end{array}\right. + C_8H_{14}$$

(23.8)

+ Ph$_3$P:

$$\mathrm{CpMn(CO)_2(CS)} + Ph_3PS$$

Carbon diselenide undergoes a similar reaction to afford the selenocarbonyl analog. Besides the η^2-C,S coordination above, CS_2 can bridge between two metal atoms and insert into metal–alkyl bonds:

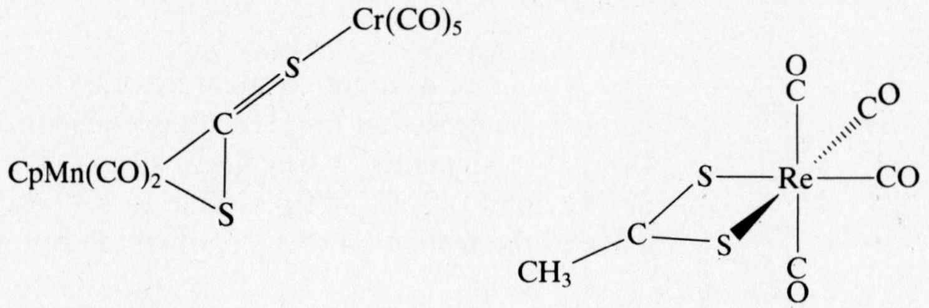

Carbon Dioxide

It has been shown recently that CO_2 also has a rich coordination chemistry. For instance, η^2-C,O bonding, analogous to the η^2-C,S coordination above, has been discovered:

$$[(C_6H_{11})_3P]_3Ni + CO_2 \longrightarrow [(C_6H_{11})_3P]_2Ni\!\!\underset{O}{\overset{C\overset{O}{\diagdown}}{\diagup}}\!\!| \quad + (C_6H_{11})_3P \quad \textbf{(23.9)}$$

In addition, η^1-C coordination of CO_2 has been reported, for example, in the reductive disproportionation of CO_2 by $M_2[M'(CO)_5]$ (M = Li, Na, M' = W; M = K, M' = Cr, Mo, W) to give the corresponding Group VIB (6) metal hexacarbonyl and M_2CO_3:

$$M_2[M'(CO)_5] \xrightarrow{CO_2} \qquad\qquad \textbf{(23.10)}$$

Similar reductive disproportionations leading to coordinated CO occur for other dianionic carbonylmetallates, for example, $Na_2[M(CO)_4]$ (M = Fe, Ru, Os) and $Na_2[CpV(CO)_3]$:

$$Na_2[M(CO)_4] + 2CO_2 \rightarrow M(CO)_5 + Na_2CO_3 \qquad \textbf{(23.11)}$$

$$Na_2[CpV(CO)_3] + 2CO_2 \rightarrow CpV(CO)_4 + Na_2CO_3 \qquad \textbf{(23.12)}$$

S_n Complexes

There has been growing interest in the formation of organometallic complexes containing catenated sulfur ligands (S_n) because it is possible that such complexes may act as *in situ* S_2 generators for use in organic synthesis. For instance, both S_5 and S_3 units can be stabilized by coordination to transition metals; for example, reaction of Cp_2MCl_2 or $Cp_2^*MCl_2$ (M = Ti, Zr, Hf; $Cp^* = C_5Me_5$) with lithium polysulfides (Li_2S_5, Li_2S_3) affords the products shown below:

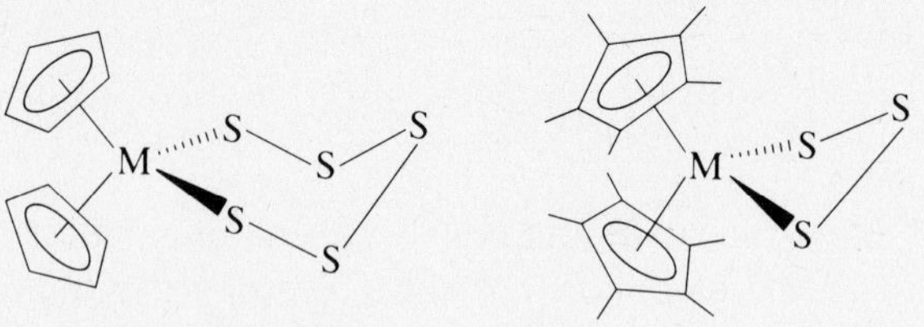

S_2 units can also be coordinated:

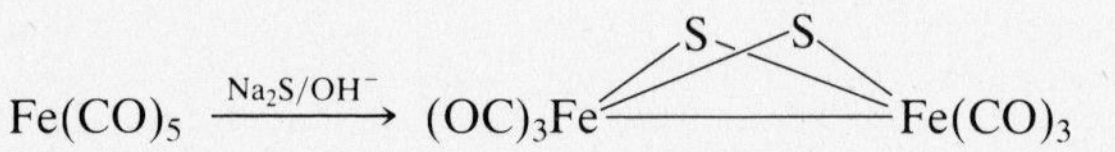

$$Fe(CO)_5 \xrightarrow{Na_2S/OH^-} (OC)_3Fe\!\!\!=\!\!\!=\!\!\!=\!\!\!=\!\!\!Fe(CO)_3 \qquad (23.13)$$

Without the aid of X-ray diffraction, it is extremely difficult to discern the size of a particular MS_n ring. However, it does appear that in complexes of the type Cp_2MS_n, the ring sizes can be distinguished on the basis of the positions of the $\nu(SS)$ modes, which appear in the IR and Raman spectra at about 500 cm^{-1}. In general, the smaller the MS_n ring size, the lower the frequencies of the $\nu(SS)$ vibrations.

Additions to Multiple Bonds

For many years, there has been considerable industrial interest in the homogeneous reduction of unsaturated materials such as alkenes and alkynes under mild conditions. One of the major breakthroughs was the development of *Wilkinson's catalyst*, the square-planar complex $RhCl(PPh_3)_3$. This complex in benzene solution is an extremely selective catalyst for double-bond reduction, for example, in steroidal fragments such as

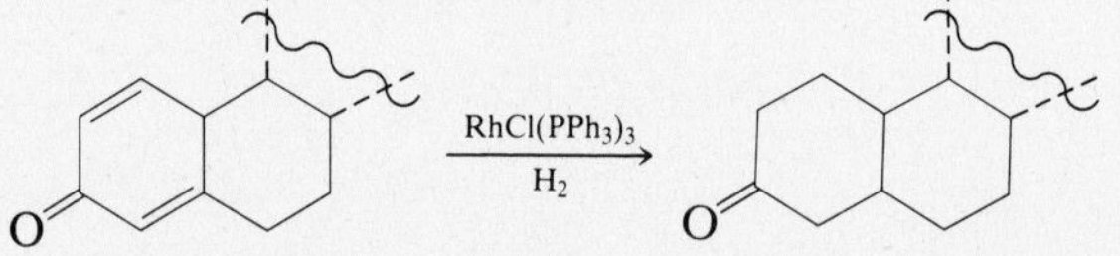

$$(23.14)$$

The mechanism of alkene hydrogenation by Wilkinson's and related catalysts is complicated, and there are numerous side-reactions depending on the nature of the alkene and phosphine ligand concerned. Nevertheless, the principal steps of the catalytic cycle are believed to be those shown in the clock diagram in Figure 23.1. Initially, there is phosphine dissociation from the original catalyst, followed immediately by solvent coordination. It is this intermediate that acts as the actual catalyst. The steps involved are

Figure 23.1
Principal steps in the
catalytic cycle for
the hydrogenation of
an alkene using
Wilkinson's catalyst.
For further details,
see J. Halpern,
T. Okamoto, and
A. Zakhariev.
J. Mol. Cat. 2
(1976): 65.

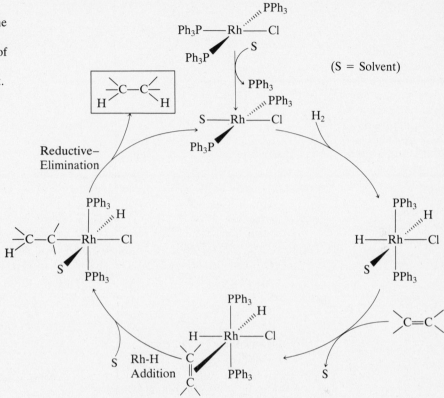

oxidative–addition of H_2, alkene coordination, Rh—H bond addition to
the coordinated alkene accompanied by solvent coordination, and alkane
formation together with regeneration of the original catalyst.

Addition of a Rh—H bond can proceed in either a Markovnikoff or an
anti-Markovnikoff fashion:

$$\longrightarrow RhCH(R)CH_3$$
Markovnikoff

$$\longrightarrow RhCH_2CH_2R$$
Anti-Markovnikoff

Figure 23.2
Catalytic cycle
proposed for the
hydrogenation of
1-alkenes by
RhH(CO)(PPh$_3$)$_3$ at
room temperature
and ~ 100 kPa
(1 atm) pressure.
Adapted from
F. A. Cotton and
G. Wilkinson,
*Advanced Inorganic
Chemistry*, 3rd. ed.
New York:
Wiley-Interscience,
1972.

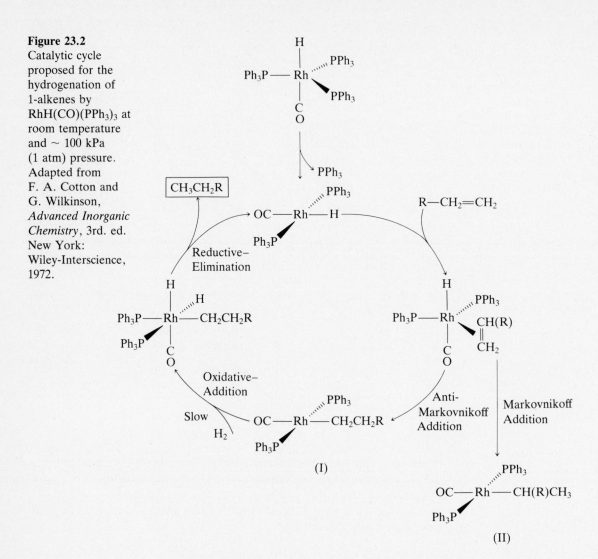

Following this, a second hydrogen transfer together with rupture of
the Rh—C bond yields the alkane CH$_3$—CH$_2$R:

$$\text{Rh}\diagdown\!\!\!\diagup\!\!\!\!\begin{array}{c}\text{H}\\ \vdots \\ \text{CH(R)CH}_3\end{array} \qquad \text{or} \qquad \text{Rh}\diagdown\!\!\!\diagup\!\!\!\!\begin{array}{c}\text{H}\\ \vdots \\ \text{CH}_2\text{CH}_2\text{R}\end{array}$$

Other complexes that are useful for the selective reduction of 1-
alkenes are RhH(CO)(PPh$_3$)$_3$, RuHCl(PPh$_3$)$_3$, and the cationic complexes
[P$_2$MS$_2$]$^+$ (M = Rh, Ir; P$_2$ = chelating diphosphine; S = solvent such as
acetone). The mechanism proposed for the highly selective hydrogenation

of 1-alkenes by $RhH(CO)(PPh_3)_3$ at room temperature and ~100 kPa (1 atm) pressure is shown in the clock diagram in Figure 23.2. It appears that because of steric effects the primary alkyl group ($—CH_2CH_2R$) in the square-planar intermediate I, produced by anti-Markovnikoff Rh—H addition to the 1-alkene, is more stable than the secondary alkyl group [$—CH(R)CH_3$] in intermediate II that is generated by the alternative Markovnikoff addition. Intermediate II is apparently not stable long enough for the slow oxidative–addition of H_2 to take place.

The most important commercial application of homogeneous hydrogenation is the stereoselective reduction of substituted alkenes to give specific stereoisomers in high optical yield. For example, catalysts incorporating chiral phosphines coordinated to Rh(I) have proved highly effective in the synthesis of *l*-DOPA (3,4-dihydroxyphenylalanine), an important drug used in the treatment of Parkinson's disease.

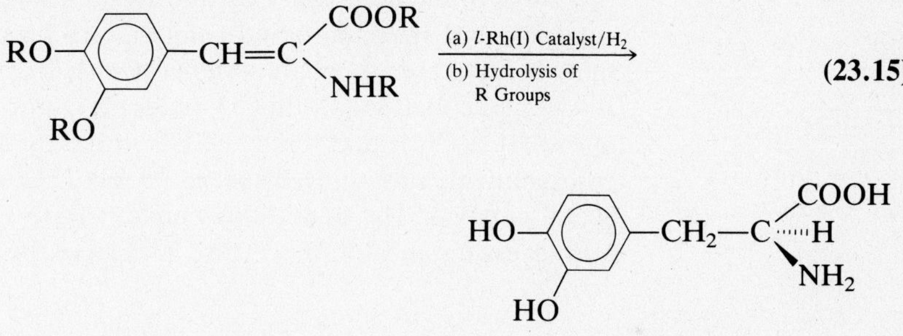

$$(23.15)$$

l-DOPA

The active catalysts are generated by treating Rh(I) complexes such as $[(\eta^4\text{-COD})Rh(DIOP)]^+[BF_4]^-$ and $[(\eta^4\text{-COD})Rh(P^*RR'R'')_2]^+[BF_4]^-$ (COD = 1,5-cyclooctadiene) with molecular hydrogen. The structures of the optically active phosphines DIOP and $P^*RR'R''$ are shown below:

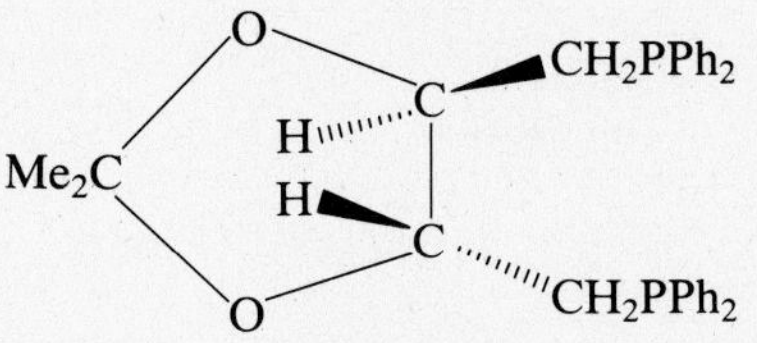

DIOP

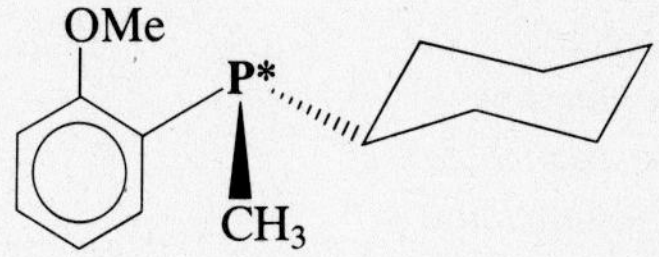

Another important process catalyzed by organometallic carbonyl complexes, including $RhH(CO)(PPh_3)_3$, is the *hydroformylation* of alkenes. As the name implies, a hydrogen atom (H) and a formyl group (HCO) are added to an alkene (most commonly a 1-alkene) to form an aldehyde and ultimately an alcohol. The process actually involves addition of H_2 and

CO, usually in the form of *syn gas*, which can be readily obtained from the high-temperature steam reforming of methane or from coke:

$$CH_4 + H_2O \rightarrow CO + 3H_2 \tag{23.16}$$

$$C + H_2O \rightarrow CO + H_2 \tag{23.17}$$

Butyraldehyde ("butanal") is produced in billion-kilogram quantities per year throughout the world by the hydroformylation of propylene. The butanal is then further hydrogenated to butanol for commercial use:

$$CH_3CH{=}CH_2 + H_2 + CO \longrightarrow CH_3CH_2CH_2CHO \xrightarrow{H_2}$$
$$CH_3CH_2CH_2CH_2OH \tag{23.18}$$

Most of the industrial hydroformylations make use of cobalt–carbonyl catalysts, but rhodium-based catalysts are becoming increasingly important. In the case of cobalt systems, $Co_2(CO)_8$ is formed *in situ* from the reaction of cobalt salts with CO under pressure. This then reacts with H_2 to afford $HCo(CO)_4$ that loses one CO group by dissociation to produce the coordinatively unsaturated species $HCo(CO)_3$, which is believed to be the actual catalyst. The conditions employed are typically 3×10^4 kPa (300 atm) pressure and 150 °C. Figure 23.3 gives the main steps in the catalytic

Figure 23.3
Principal steps proposed for the hydroformylation of a 1-alkene by $HCo(CO)_4$. Adapted from C. M. Lukehart, *Fundamental Transition Metal Organometallic Chemistry*. Pacific Grove, Monterey, Calif.: Brooks/Cole, 1985.

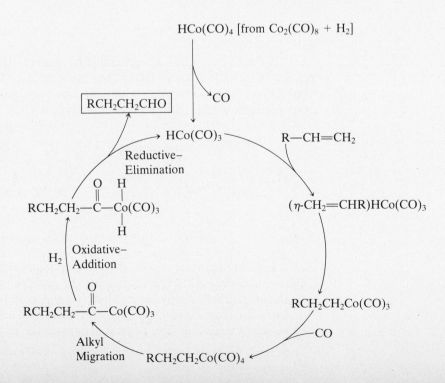

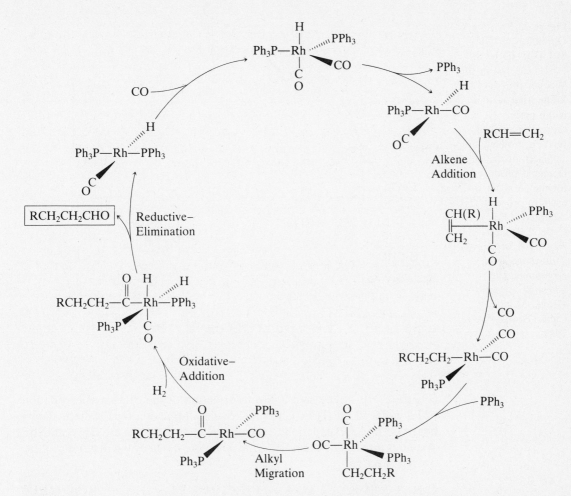

Figure 23.4
Dissociative
mechanism proposed
for hydroformylation
of a 1-alkene by
$RhH(CO)_2(PPh_3)_2$.
At high PPh_3
concentrations, an
associative
mechanism is also
believed to come
into play. [See
D. A. Evans,
J. A. Osborn, and
G. Wilkinson.
J. Chem. Soc. A.
(1968): 3133.]

cycle. The reaction yield of aldehyde product is about 80%, with most of it being linear. Notice that some of the steps involve alkene addition to the Co—H bond and alkyl group migration.

Figure 23.4 shows the dissociative mechanism proposed for hydroformylation using the rhodium catalyst $RhH(CO)(PPh_3)_3$. You should now recognize all the steps involved in this catalytic cycle: tertiary phosphine dissociation, Rh—H bond addition to an alkene, alkyl-group migration, oxidative addition of H_2, and reductive–elimination of aldehyde. This catalyst produces higher yields of the linear aldehyde with fewer by-products than does the cobalt–carbonyl system.

There is a homogeneous palladium-catalyzed process for the production of acetaldehyde from ethylene still used industrially. This process was commercialized in the late 1950s and is known as the *Wacker process*. In this case, unlike hydroformylation, the number of carbon atoms is not increased, but oxidation does occur. The catalytic cycle involves the formation of a palladium–ethylene complex analogous to Zeise's salt

Figure 23.5
Oxidation of C_2H_4
to CH_3CHO by a
Pd(II) catalyst in the
Wacker process.
* The *trans* isomer
is the more
thermodynamically
favored product, but
sufficient of this cis
isomer is produced
for the catalytic
cycle to proceed.
Adapted from
C. M. Lukehart,
*Fundamental
Transition Metal
Organometallic
Chemistry*. Pacific
Grove, Monterey,
Calif.: Brooks/Cole,
1985.

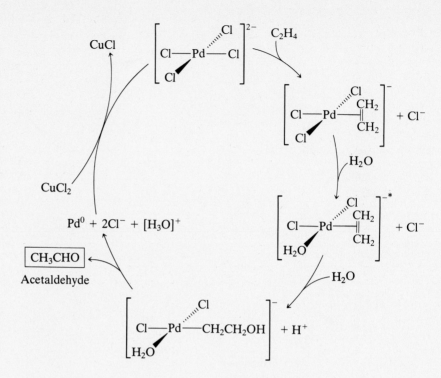

(Figure 23.5). A key step is the reformation of the catalyst via oxidation of Pd metal to Pd(II) by $CuCl_2$. The overall reaction for the steps shown below is ethylene oxidation to acetaldehyde by O_2 in the air. This catalytic system is an elegant example of the complete usage of the materials in the overall reaction.

$$C_2H_4 + K_2[PdCl_4] + H_2O \rightarrow CH_3CHO + Pd + 2HCl + 2KCl \quad \textbf{(23.19)}$$

$$Pd + 2CuCl_2 + 2KCl \rightarrow K_2[PdCl_4] + 2CuCl \quad \textbf{(23.20)}$$

$$2CuCl + 2HCl + \tfrac{1}{2}O_2 \rightarrow 2CuCl_2 + H_2O \quad \textbf{(23.21)}$$

Add _____

$$C_2H_4 + \tfrac{1}{2}O_2 \rightarrow CH_3CHO \quad \textbf{(23.22)}$$

The major use of acetaldehyde is in the preparation of acetic acid. Another important route to acetic acid is the catalyzed carbonylation of methanol:

$$CH_3OH + CO \rightarrow CH_3COOH \quad \textbf{(23.23)}$$

Rhodium-based catalysts are the best available to date because of their great selectivity and the mild reaction conditions necessary (180 °C, 3.5×10^3 kPa (35 atm)). Although there are several catalytic precursors

Figure 23.6
Catalytic cycle proposed for the carbonylation of CH_3OH in the presence of *cis*-$[Rh(CO)_2I_2]^-$. Adapted with permission from D. Forster, *Adv. Organometal. Chem.* 17 (1979): 255.

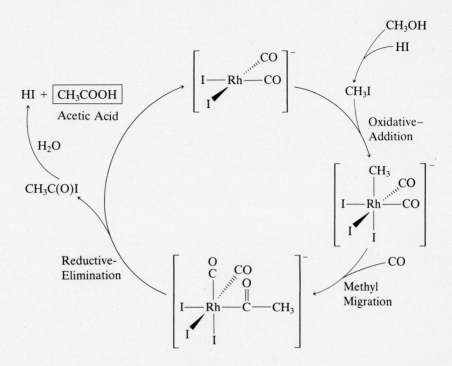

used, the active species is always *cis*-$[Rh(CO)_2I_2]^-$ with the source of iodide ion being CH_3I, I^- in aqueous solution, or I_2 itself. Methanol reacts with hydriodic acid to give CH_3I and water. The CH_3I initiates the cycle by oxidatively adding to the anionic rhodium catalyst (Figure 23.6). The usual steps are followed: oxidative–addition, methyl migration, and reductive–elimination.

Hydrosilylation of alkenes is another important commercial process. Two industrially useful reactions occur, depending on the catalyst employed:

$$RCH{=}CH_2 + HSiR_3 \xrightarrow{H_2PtCl_6} RCH_2CH_2SiR_3 \qquad (23.24)$$

$$2RCH{=}CH_2 + HSiR_3 \xrightarrow{Fe(CO)_5} RCH{=}CHSiR_3 + RCH_2CH_3 \quad (23.25)$$

These reactions are catalyzed homogeneously by several other different catalysts, for example, $Co_2(CO)_8$, $RhCl(PPh_3)_3$, and $(\eta^6\text{-}C_6H_6)Cr(CO)_3$. Much of the kinetic work has focused on the addition of silanes to Ir(I) complexes such as *trans*-$IrH(CO)(PPh_3)_2$ and $[Ir(diphos)_2]^+$ (diphos = $Ph_2PCH_2CH_2PPh_2$) because the reactions almost certainly proceed initially by the oxidative–addition of $HSiR_3$ to the metal species:

$$\textit{trans-}IrH(CO)(PPh_3)_2 + HSiR_3 \rightarrow IrH_2(SiR_3)(CO)(PPh_3)_2 \quad (23.26)$$

The SiR_3 group is a good *trans*-labilizing group and so may activate a coordination site in the oxidative adduct to allow incorporation of the alkene. Insertion of the alkene into the M—H bond followed by reductive–elimination of alkyl and silyl ligands gives the alkylsilane. Insertion of the alkene into the M—Si bond followed by β-hydride elimination yields vinylsilane and H_2. The latter can be consumed in the hydrogenation of alkene.

Another important example of homogeneous catalysis is alkene epoxidation by hydroperoxides. One of the major applications of this process involves the production of propylene oxide by oxygen transfer from an alkyl hydroperoxide to propylene in the presence of Mo(0)—Mo(VI) compounds:

$$\text{MeCH=CH}_2 + t\text{-BuOOH} \longrightarrow \underset{\underset{O}{\diagdown\ \diagup}}{\text{MeCH—CH}_2} + t\text{-BuOH} \qquad (23.27)$$

Propene oxide (or methyloxirane as it is often known commercially) is a vital intermediate in the production of vast quantities of propylene glycol, glycerine, and various polyethers. Yields of propylene oxide approaching 95% can be obtained. Similar high yields are given when other alkenes are used, for example, 1-octene → 1,2-epoxyoctane (92%). Molybdenum(0) compounds such as $Mo(CO)_6$ have the same catalytic activity as Mo(VI) compounds such as MoO_3. Figure 23.7 shows the most probable mechanism for catalytic epoxidation using an alkyl molybdate(VI). The intermediate is presumed to involve oxygen transfer from

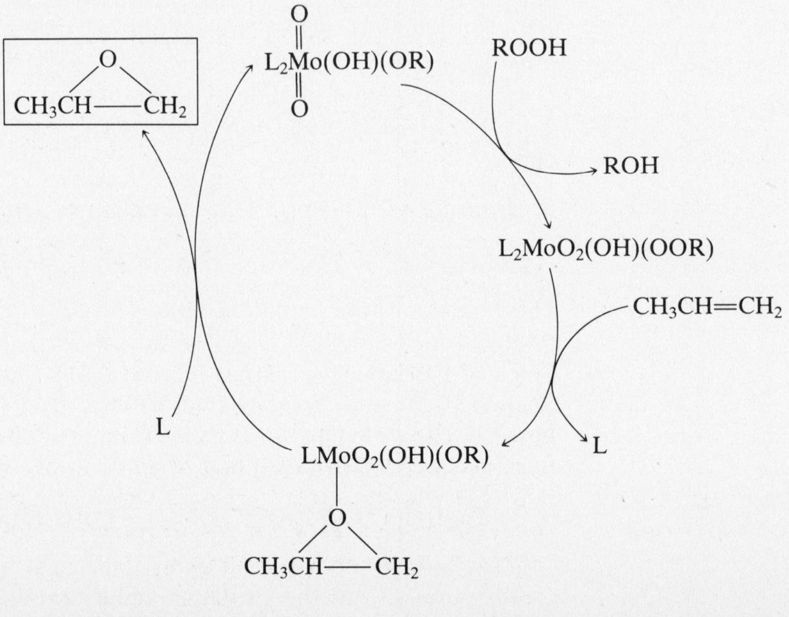

Figure 23.7
Probable steps in the epoxidation of propylene using an alkyl molybdate(VI) as catalyst. L = hexamethylphosphoramide, $(Me_2N)_3P$=O. Adapted from G. W. Parshall, *Homogenous Catalysis: The Applications and Chemistry of Catalysis by Soluble Transition Metal Complexes.* New York: Wiley Interscience, 1980.

the peroxy group bonded to the central Mo atom to the C=C bond of propylene:

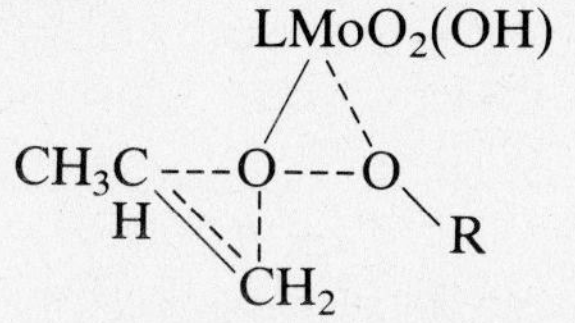

Although molybdenum salts are the usually preferred epoxidation catalysts, a highly selective reaction for the epoxidation of allylic alcohols by *t*-BuOOH occurs in the presence of VO(acac)$_2$ (acac = acetylacetone). For example, only the allylic double bond in 1,5-cyclooctadienol is oxidized:

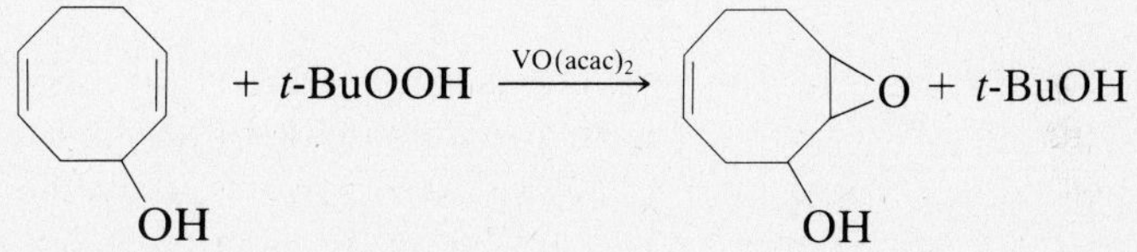

A new metal-catalyzed, asymmetric epoxidation process has recently been described that yields products with greater than 90% optical purity. The process involves treatment of substrates such as allylic alcohols with the appropriate *d*- or *l*-diethyl tartrate and titanium tetraisopropoxide [Ti(O-*i*-C$_3$H$_7$)$_4$] and *t*-BuOOH:

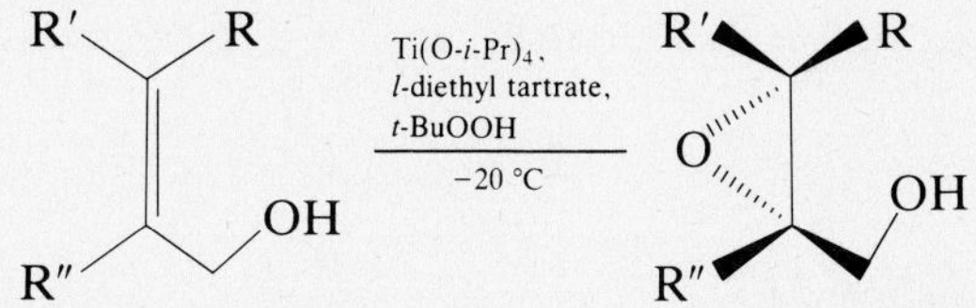

The final example of a homogeneous catalysis reaction involving multiple-bond activation is *alkene metathesis*. This process results in the interchange of CR$_2$ fragments between two alkenes, and there are already important industrial applications. For instance, neohexene is produced on a commercial scale for use in the perfume industry:

$$(Me_3C)CH=CMe_2 + CH_2=CH_2 \rightarrow (Me_3C)CH=CH_2 + CH_2=CMe_2$$
$$\text{Neohexene}$$

The catalysts for this type of reaction are formed from a mixture of WCl$_6$, EtOH, and an alkyl aluminum compound such as EtAlCl$_2$. A metal–carbene complex is initially believed to be produced as the active intermediate, followed by formation of a metallacyclobutane derivative that subsequently splits to yield a new alkene and a new metal-carbene complex. Figure 23.8 summarizes these steps for the formation of neohexene in the reaction scheme.

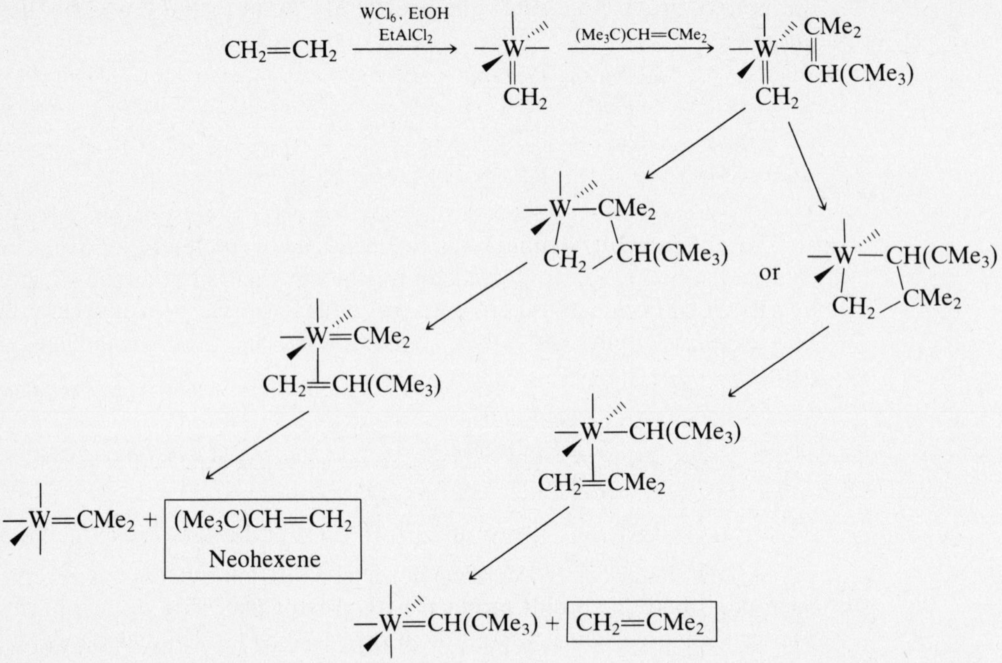

$$CH_2{=}CH_2 \xrightarrow[\text{EtAlCl}_2]{\text{WCl}_6,\ \text{EtOH}} {-}\overset{|}{\underset{\overset{||}{CH_2}}{W}}{-} \xrightarrow{(Me_3C)CH{=}CMe_2} {-}\overset{|}{\underset{\overset{||}{CH_2}}{W}}{-}\overset{CMe_2}{\underset{CH(CMe_3)}{||}}$$

$${-}\overset{|}{\underset{CH_2\diagdown CH(CMe_3)}{W}}{-}CMe_2 \qquad \text{or} \qquad {-}\overset{|}{\underset{CH_2\diagdown CMe_2}{W}}{-}CH(CMe_3)$$

$${-}\overset{|}{\underset{CH_2{=}CH(CMe_3)}{W}}{=}CMe_2$$

$${-}\overset{|}{\underset{CH_2{=}CMe_2}{W}}{-}CH(CMe_3)$$

$${-}\overset{|}{W}{=}CMe_2 + \boxed{(Me_3C)CH{=}CH_2}$$
$$\text{Neohexene}$$

$${-}\overset{|}{W}{=}CH(CMe_3) + \boxed{CH_2{=}CMe_2}$$

Figure 23.8 Principal steps in the alkene metathesis of 2,4,4-trimethyl-2-pentene with ethylene using WCl_6, EtOH, and $EtAlCl_2$ as the catalyst precursors. The new carbene complexes formed presumably react further through similar steps to give neohexene and $CH_2{=}CMe_2$.

BIBLIOGRAPHY

Suggested Reading

Chalk, A. J. "Catalysis and the Organometallic Chemistry of the Transition Metals." *The Spex Speaker* (28 March 1983). (Available from Spex Industries, Inc., Brainy Borosta, Metuchen, N.J. 08840.)

Collman, J. P., and L. S. Hegedus. *Principles and Applications of Organotransition Metal Chemistry*. Mill Valley, Calif.: University Science Books, 1980.

Cotton, F. A., and G. Wilkinson. *Advanced Inorganic Chemistry*, 5th ed. New York: Wiley-Interscience, 1988.

Halpern, J. "Mechanism and Stereoselectivity of Asymmetric Hydrogenation." *Science* 217 (1982): 401.

Lukehart, C. M. *Fundamental Transition Metal Organometallic Chemistry*. Pacific Grove, Calif.: Brooks/Cole, 1985.

Masters, C. *Homogeneous Transition-Metal Catalysis.* New York: Chapman & Hall, 1981.

Parshall, G. W. *Homogeneous Catalysis. The Applications and Chemistry of Catalysis by Soluble Transition Metal Complexes.* New York: Wiley, 1980.

"Symposium on Industrial Applications of Organometallic Chemistry and Catalysis." *J. Chem. Ed.* 63 (1986): 189–225.

Dombek, B. D. "Direct Routes from Synthesis Gas to Ethylene Glycol," 210–212.

Forster, D. W., and T. W. Dekleva. "Catalysis of the Carbonylation of Alcohols to Carboxylic Acids Including Acetic Acid Synthesis from Methanol," 204–206.

Goddall, B. L. "The History and Current State of the Art of Propylene Polymerization Catalysts," 191–195.

Grasselli, R. K. "Selective Oxidation and Ammoxidation of Olefins by Heterogeneous Catalysis," 216–221.

Knowles, W. S. "Application of Organometallic Catalysis to the Commercial Production of *l*-DOPA," 222–225.

Lutz, E. F. "Shell Higher Olefins Process," 202–203.

Murchison, C. B., R. L. Weiss, and R. A. Stowe. "The Production and Recovery of C_2-C_4 Olefins from Syngas," 213–215.

Parshall, G. W., and R. E. Putscher. "Organometallic Chemistry and Catalysis in Industry," 189–191.

Polichnowski, S. W. "Transition-Metal-Catalyzed Carbonylation of Methyl Acetate," 206–209.

Pruett, R. L. "Hydroformylation," 196–198.

Tolman, C. A. "Steric and Electronic Effects in Olefin Hydrocyanation at Du Pont," 199–201.

Yamamoto, A. *Organotransition Metal Chemistry: Fundamental Concepts and Applications.* New York: Wiley, 1986.

PROBLEMS

23.1 Give two examples in each case of the following types of reactions that are involved in catalytic processes:
a. Oxidative–addition
b. Reductive–elimination
c. Metallacyclization
d. Alkyl migration
e. Alkene addition

23.2 Write a mechanism for the hydroformylation of butene (CH_3CH_2-$CH{=}CH_2$) using $HCo(CO)_4$.

23.3   Suggest a route for the synthesis of $(\eta^6\text{-}C_6H_6)Cr(CO)_2(CS)$ from the analogous tricarbonyl complex. Discuss the mechanistic steps involved.

23.4   Predict the products of Rh—H addition to the double bond in propylene.

23.5   Suggest a mechanism for the synthesis of propionic acid from ethanol using a rhodium–carbonyl catalyst.

23.6   Write down the steps involved in the alkene metathesis of propylene to form butene and ethylene.

23.7   Give two criteria for distinguishing between a dihydrogen and dihydride complex.

23.8   Give two examples of coordination catalysis that lead to chiral products.

23.9   Suggest explanations for the following observations:
   a. The complex $[(\eta^2\text{-}C_2H_4)_2RhCl]_2$ in dilute aqueous solution catalyzes the dimerization of ethylene but does not polymerize it.
   b. At high pressure, the rate of $HCo(CO)_4$-catalyzed alkene hydroformylation drops with increasing CO pressure.
   c. Although nearly all of the stereocenters of a given isotactic polypropylene molecule have the same chirality, a sample of isotactic polypropylene is not optically active.
   d. Carbene complexes of the type $(CO)_5W{=}C\begin{smallmatrix}\nearrow OR\\ \searrow R'\end{smallmatrix}$ are poor catalysts for alkene metathesis.

23.10 Bubbling ethylene into a solution of palladium acetate in glacial acetic acid results in the precipitation of palladium metal, and 1 mole of vinyl acetate is produced for each gram-atom of palladium produced. Suggest a mechanism for this reaction.

23.11 Which of the following statements are correct?
   a. If you can do a crystal structure on it, it is not likely to be a catalytic intermediate.
   b. Catalytic intermediates tend to be highly unstable.
   c. A catalyst must bind ligands very tightly to be active.
   d. Methane cannot coordinate because it has no lone pairs.

24

Bioinorganic Chemistry

24.1   What Is Bioinorganic Chemistry?

We tend to think of the living world as being made of organic molecules. Even if we ignore the ubiquitous water, there is a large inventory of inorganic materials that are integral to living systems. For example, in higher animals, skeletal and integumental (outer covering) materials are often highly mineralized, and the lifetime consumption of oxygen by a living entity is a major fraction of its total material needs.

Water, oxygen, calcium phosphate, and calcium carbonate are the most important inorganic constituents of living matter in terms of quantity. Bioinorganic chemistry, however, is more concerned with the uses and management in living systems of more specialized compounds containing elements other than those usually considered to be proprietary to organic chemistry (C, H, N, O, and to a lesser degree P, S, and Cl)) such as the lighter elements of Groups IA and IIA (1 and 2), the transition elements,

721

Table 24.1 *Abundances of Elements in Earth's Crust and in Oceans Compared With Some Abundances in Typical Mammals*[a]

Atomic Number	Symbol	Earth's Crust (ppm)	Ocean (ppm)	Typical Mammal (ppm)
1	H	1400	10^5	10^5
2	He	0.01	10^{-6}	—
3	Li	20	0.2	0.03
4	Be	2	10^{-7}	—
5	B	7	5.0	0.2
6	C	200	28	10^5
7	N	20	16	10^5
8	O	5×10^5	—	5×10^5
9	F	460	1.3	0.1
10	Ne	10^{-4}	10^{-4}	—
11	Na	10^4	10^4	10^3
12	Mg	10^4	10^3	500
13	Al	10^5	10^{-3}	0.5
14	Si	3×10^5	2.9	50
15	F	10^3	10^{-2}	10^4
16	S	300	10^3	10^4
17	Cl	200	10^4	10^3
18	Ar	3	0.5	—
19	K	10^4	10^2	10^3
20	Ca	10^4	10^2	10^4
21	Sc	22	10^{-11}	—
22	Ti	10^4	10^{-3}	—
23	V	10^2	10^{-3}	0.03
24	Cr	10^2	10^{-4}	0.03
25	Mn	10^3	10^{-3}	1.0
26	Fe	10^5	10^{-3}	50
27	Co	28	10^{-4}	0.03
28	Ni	72	10^{-4}	0.03
29	Cu	58	10^{-3}	4.0
30	Zn	82	10^{-3}	25
31	Ga	17	10^{-5}	—
32	Ge	1	10^{-5}	—
33	As	2	10^{-3}	0.05
34	Se	10^{-2}	10^{-4}	0.05
35	Br	4	70	2

[a] The absence of an entry (−) signifies that reliable data are not available. Data given as 10^n are order of magnitude estimates.

Table 24.1 (*Continued*)

Atomic Number	Symbol	Earth's Crust (ppm)	Ocean (ppm)	Typical Mammal (ppm)
36	Kr	10^{-4}	10^{-3}	—
37	Rb	70	0.1	9
38	Sr	450	8	5
39	Y	35	10^{-5}	—
40	Zr	140	10^{-5}	—
41	Nb	20	10^{-5}	—
42	Mo	1	10^{-4}	0.2
44	Ru	10^{-3}	10^{-6}	—
45	Rh	10^{-3}	—	—
46	Pd	10^{-3}	—	—
47	Ag	10^{-2}	10^{-4}	—
48	Cd	0.2	10^{-4}	2
49	In	0.2	—	—
50	Sn	2	10^{-2}	2
51	Sb	0.2	—	—
52	Te	10^{-3}	—	—
53	I	0.5	0.06	1
54	Xe	10^{-5}	10^{-4}	—
55	Cs	2	10^{-3}	—
56	Ba	380	10^{-2}	0.5
57–71	La–Lu	225 (tot)	10^{-5}(tot)	—
72	Hf	4	—	—
73	Ta	2	—	—
74	W	1	—	—
75	Re	10^{-4}	—	—
76	Os	10^{-4}	—	—
77	Ir	10^{-4}	—	—
78	Pt	10^{-2}	—	—
79	Au	10^{-3}	10^{-2}	—
80	Hg	10^{-2}	0.1	—
81	Tl	0.5	10^{-3}	—
82	Pb	10	10^{-2}	0.5
83	Bi	10^{-3}	10^{-2}	—
90	Th	6	10^{-4}	—
92	U	2	3	—

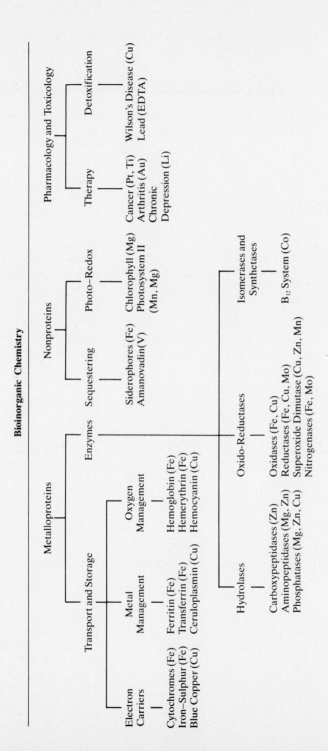

Figure 24.1 Classification scheme for bioinorganic chemistry.

and the heavier congeners of the *p* groups. In almost every case of interest, the "inorganic" species is associated with an organic entity in some way or other.

Table 24.1 lists the abundances of a number of elements in the earth's crust and in the oceans. With the striking exceptions of Al, Si, and Ti, which are among the most abundant of the crustal elements, nature has tended to use the elements according to their availability. Thus, iron is widely used in specialized metalloproteins. Copper and zinc are also commonly used for similar purposes, but the total amount is much less than iron. Finally, there are elements like cobalt, chromium, arsenic, and selenium that are essential in certain living systems, but their uses are highly specialized and they are only needed in tiny amounts.

Because of the competition for trace elements, living systems have developed interesting ways of sequestering them from their environment or stealing them from their neighbors! Highly specific mechanisms for the absorption and transport of these elements have also been developed. Many elements that are not used (and even those that are essential in trace amounts) can be highly toxic if ingested by the cell in amounts above a threshold level. On the other hand, sublethal doses may have beneficial pharmacological effects. For example, certain platinum compounds have found important applications in cancer therapy, and lithium salts are used to control chronic depression.

Figure 24.1 shows a classification scheme for some of the more important areas of bioinorganic chemistry. In most cases, the subject concerns the association of a metallic element or a nonmetallic element other than C, H, N, O, P, or S with an organic ligand of natural origin. Developing a full understanding of a particular system may depend on the synthesis of model compounds in which synthetic ligands closely mimic the natural ones in their chemical and spectroscopic properties or may focus on direct studies of the natural system.

24.2 *Sequestering and Storing Essential Elements*

Iron

One of the simpler examples of a class of bioinorganic sequestering agents are the *siderophores*. These molecules are used by living organisms, particularly microorganisms, for the acquisition and intracellular transport of iron. Unlike higher animals, microbes have relatively little capacity for storage and must constantly acquire essential nutrients from their environment. This may be difficult due to a natural lack of the required element or because the environment (in the form of a living host) may actively depress

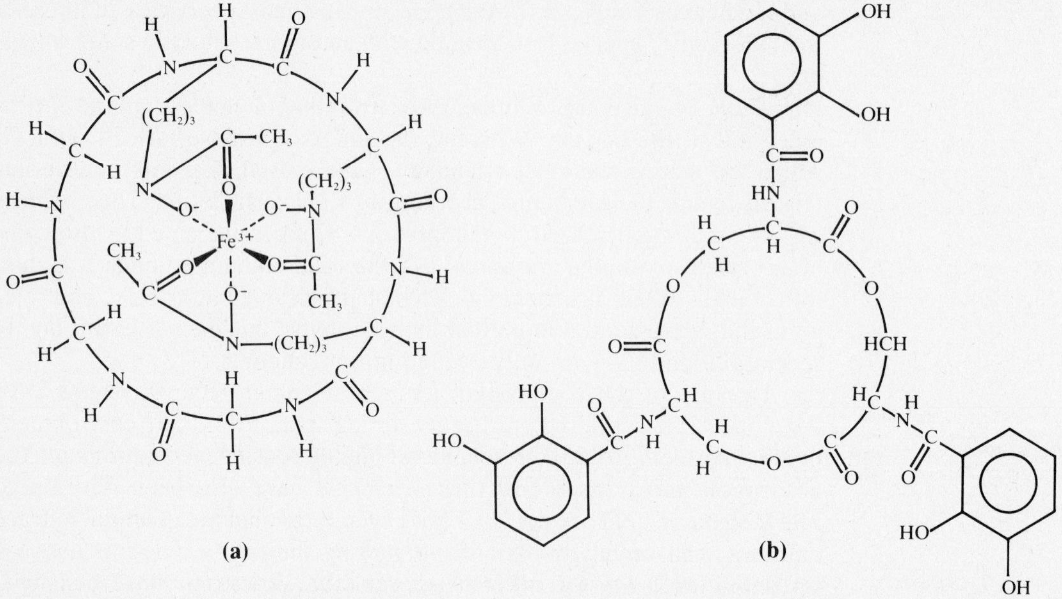

Figure 24.2
Structures of
(a) ferrichrome and
(b) enterobactin.
Reprinted with
permission from
K. N. Raymond,
*ACS. Adv. Chem.
Ser.* no. 162.
R. F. Gould, ed.,
Washington, D.C.:
ACS, 1980.

the concentration of the nutrient in order to discourage invasion by the microbe.

The siderophores are commonly based on either the *o*-catecholate or the hydroxamic acid function:

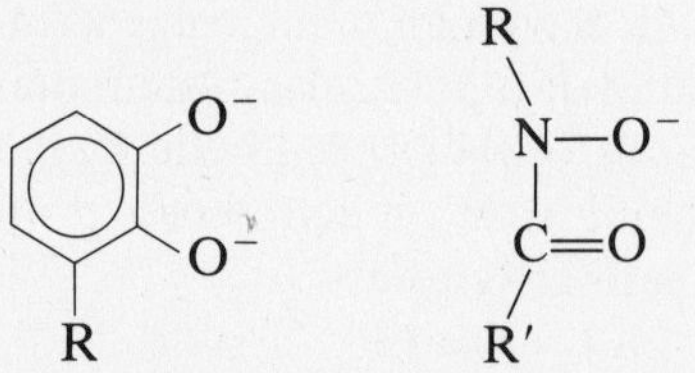

o-Catecholate   Hydroxamate Anion

The natural siderophores contain three such bidentate ligands attached to a short oligopeptide chain, arranged in such a way as to permit highly effective tris-chelation to Fe(III) (Figure 24.2). The binding constants for Fe(III) to such molecules are extremely high. For example, the binding constant for Fe(III) to ethylenediaminetetraacetate (EDTA), one of the strongest binding simple chelates, is about 10^{25}, whereas the binding constants for siderophores can be as high as 10^{29}–10^{33}. Examples of the two major types of siderophore are shown in Figure 24.2.

The bacterium sends the siderophore out into its environment where it picks up Fe(III). The complex is then picked up by a high-affinity receptor on the cell wall and is transported into the cell's interior. In the cell, Fe(III)

Figure 24.3
Structure of
$Fe_{11}O_6(OH_2)_6$-
$[OOCC_6H_5]_{15}$.
(a) The
$Fe_{11}O_6(OH)_6$ core
and **(b)** the whole
molecule except for
the *o*-, *m*-, and
p-carbon atoms of
the phenyl groups.
Reprinted with
permission from
S. M. Gorum
and S. J. Lippard,
Nature, 319
(1986): 666.

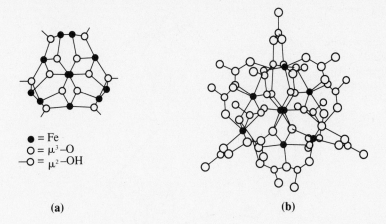

● = Fe
○ = μ^3–O
–○ = μ^2–OH

(a)   **(b)**

is reduced to Fe(II), which has a much lower binding constant to the siderophore, and the iron is then available for use.

The whites of certain birds' eggs provides an interesting example of the struggle for iron. One mechanism that protects the developing embryo against attack by microbes is the severe depletion of free iron in the egg white. This is achieved by the presence of the powerful iron-sequestering protein *conalbumin*, which makes the white an iron-free "desert." Only microbes with a siderophore whose binding constant is greater than conalbumin's can cross. A similar protein, *transferrin*, transports iron in the mammalian circulatory system. The structures of these proteins have not been completely resolved, but the iron is coordinated to the phenoxide functions of tyrosine residues and carbonate.

Complex higher animals have evolved ways of storing iron, enabling them to survive nutritional deficiencies for a certain time. The mammalian storage system *ferritin* consists of an Fe(III) hydrated oxide–phosphate core containing several thousand iron atoms, surrounded by a protein sheath. The intimate structure of ferritin is still not known, but a model complex containing eleven core iron atoms sheathed in benzoate ligands, of formula $Fe_{11}O_6(OH_2)_6[OOCC_6H_5]_{15}$, has recently been prepared and characterized by crystallography (Figure 24.3). It is most likely that the iron is incorporated into and removed from ferritin by an Fe(II) to Fe(III) redox cycle.

Calcium

This element plays a central role in the functioning of nerve and muscle tissues. The functioning of both of these tissues requires rapid and large

Figure 24.4
Schematic representation of the calcium-binding site in parvalbumin. The coordination of the calcium ion is approximately tetrahedral.

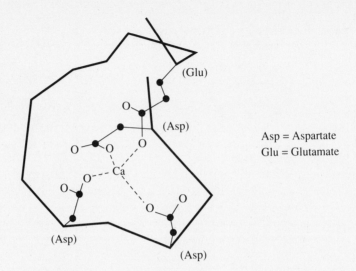

Asp = Aspartate
Glu = Glutamate

changes in calcium concentration. To effect these changes, a family of proteins has evolved for the management of calcium. Some of these proteins are used for transport and storage, while others can either deliver or sequester calcium rapidly in response to a specific biochemical stimulus. Calcium(II), like iron(III), is a hard, unpolarizable cation and has a similar affinity for oxo-type ligands. The calcium-binding proteins are rich in amino acids bearing carboxylate functions, such as aspartic and glutamic acids. Figure 24.4 shows the schematic structure of a calcium-binding molecule from fish muscle, *parvalbumin*.

Vanadium

Certain mushrooms of the Amanita family contain unusually high levels of vanadium. This element is present as the $[VO]^{2+}$ ion complexed to two molecules of the unusual amino acid ligand N-(L-1-carboxyethyl)-N-hydroxy-L-alanine

$$\underset{\substack{| \\ CH_3}}{HOOC-CH}-\overset{\substack{OH \\ |}}{N}-\underset{\substack{| \\ CH_3}}{CH-COOH}$$

which has a very high binding constant for the vanadyl ion (several orders of magnitude higher than EDTA). It is not known whether the vanadium is adventitious or if it has a function in the biochemistry of the mushroom.

24.3 *Heme Proteins and Managing Oxygen*

General Considerations

Heme proteins are the most extensively studied of all metalloproteins. In large part, this is due to their intense color and the ease with which they can be seen, particularly in the case of the oxygen carriers like hemoglobin and myoglobin.

All heme proteins contain iron coordinated to protoporphyrin IX. The porphyrin nucleus is a cyclic tetramer of pyrrole with each pyrrole bridged to its neighbors by a methine bridge. There is a system of nomenclature for porphyrins, based on substitution patterns (which we will not describe here), but protoporphyrin IX has the substitution shown in Figure 24.5. In the neutral, metal-free molecule, protons are attached to two of the nitrogens. These protons are ionized on formation of metal complexes. The dianion is remarkable in that it is totally conjugated around its periphery and the porphyrins and their derivatives are highly colored. Heme is formed by the insertion of an Fe(II) into the square-planar coordination environment provided by the four equivalent nitrogens, as shown in Figure 24.5.

The heme unit is charge-balanced, but if the iron is oxidized to Fe(III), the resulting species (known as "hemin") carries an excess positive charge and must be associated with a charge-balancing anion.

In the low-spin state, both Fe(II) and Fe(III) are usually situated symmetrically within the coordination cavity, but high-spin Fe(II) and Fe(III) are displaced out of the cavity toward one face of the molecular plane. The displacement is accompanied by a distortion of the porphyrin away

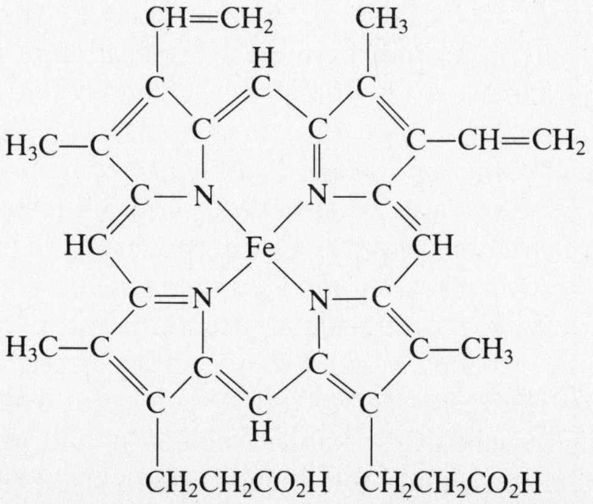

Figure 24.5
Structure of heme.

from planarity (doming). These structural changes are attributed to the larger radii of the high-spin relative to the low-spin ions (see Chapter 13).

The four coordination provided by the porphyrin ring is not sufficient to satisfy the coordination requirements of the iron, and there is a strong tendency for the iron to increase its coordination number. One way that this can be achieved is by reaction with oxygen, and the "naked" Fe(II) porphyrin complexes are extremely reactive with oxygen in the presence of water to give oxo-bridged Fe(III) dimers. The oxidized form of heme is called *hemin*.

As might be expected from their name, heme proteins consist of a protein that is attached to one or more heme units. The attachment can occur through physical forces such as exchange forces, hydrophobic bonding, or hydrogen bonding. It can also occur through covalent binding through the ring periphery—for example, by formation of an ester between a ring carboxylate and an OH group on the protein or by the addition of a protein X—H bond across a ring vinyl group. Finally, the protein normally furnishes one or two coordinating groups to occupy the axial coordination positions of Fe. The common coordinating functions are the imidazole nitrogen of histidine, the carboxylate groups of aspartic and glutamic acids, the phenoxide group of tyrosine, and the sulfide functions of methionine and cysteine. Figure 24.6 shows the structures of common amino acids that possess substituents with a potential for coordination. The variability of protein structure, the selection of available binding modes, and the variability of oxidation and spin states of iron provide for a very large number of different possible systems.

Heme is a near ideal molecular probe because it can be investigated with such a wide range of physicochemical methods. The intense charge-transfer bands (see Chapter 13) arising from metal–porphyrin interactions allow the application of UV/visible spectroscopy and resonance Raman spectroscopy (see Chapter 6). The presence of unpaired electrons on Fe(II) and Fe(III) allow the use of ESR, and iron is an excellent Mössbauer probe (see Chapter 7). The unusually low-field resonances of some of the porphyrin protons permit the application of proton NMR to heme-protein problems, even though the heme-proton concentration is small compared to that of the protein protons. The shifting of heme-proton resonances by the paramagnetic influence of high-spin Fe(II) and Fe(III) can also provide valuable structural data. Ultimately, however, the elucidation of structure requires a successful crystallographic study, and many heme proteins have been obtained in crystalline form. It must be remembered, however, that molecular structure, and particularly protein structure, can be modified by solid-state packing forces. Proteins of the kind we are discussing are not crystalline in their natural state, and their function can be highly sensitive to subtle changes in conformation.

Before describing some specific examples of heme proteins, it should be pointed out that all functional proteins are subject to interspecies

Figure 24.6
Structures of some common amino acids with coordinating substituents.

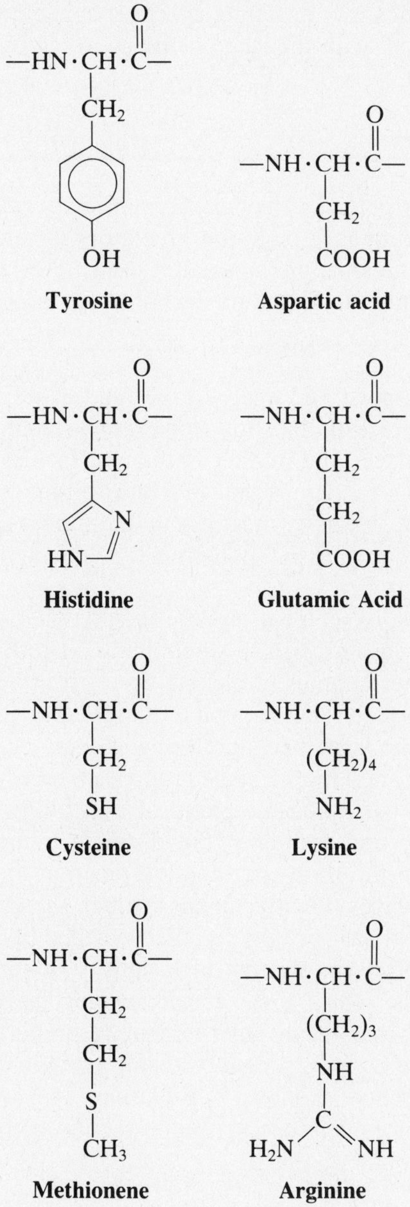

Tyrosine

Aspartic acid

Histidine

Glutamic Acid

Cysteine

Lysine

Methionene

Arginine

variability and may also be subject to intraspecies redundancy and variability. Thus, pig hemoglobin is slightly different from cat hemoglobin; fetal hemoglobin is slightly different from adult hemoglobin; Japanese radish peroxidase can be separated into seven different heme-containing fractions (isozymes), each with a slightly different protein composition and different physical properties (for example, spectroscopic parameters and oxidation potentials), but all with the same heme-coordination site.

Oxygen Carriers

Oxygen carriers perform the functions of acquiring O_2 from the atmosphere, transporting it through the circulatory system of the organism, and delivering it to the locations where it is used. They all bind O_2 rapidly and reversibly according to Equation 24.1:

$$\text{Protein} + O_2 \rightleftharpoons [\text{Protein } O_2] \tag{24.1}$$

The equilibrium must lie largely to the right at the oxygen concentration prevalent at the pick-up location (for example, lung–air interface, muscle cell–blood interface, and mother–fetus interface at the placenta). A key feature of oxygen carriers is that they must interact strongly enough with oxygen to bind it, but they must not undergo irreversible oxidation in the process. This characteristic is extremely difficult to mimic with model complexes of iron.

The simplest oxygen carriers are the myoglobins (Mb), which consist of a single protein unit with a single heme. These compounds serve the function in higher animals of picking up O_2 from the circulatory protein, hemoglobin (Hb), and delivering it to muscle. The red color of mammalian muscle tissue is in large part due to myoglobin.

In both Mb and Hb, the heme group is attached to the protein through coordination to the imidazole group of a histidine residue (the proximal histidine). A second histidine (the distal histidine) occupies space immediately above the sixth-coordination position of the iron, but it is not coordinated. Besides the attachment through the coordinate link, heme is also weakly held in position by the so-called hydrophobic interaction between the porphyrin and the high density of surrounding alkyl amino acid substituents, which form a *hydrophobic pocket*. This hydrophobic pocket places heme in an environment resembling a nonpolar organic solvent. Because O_2 is a stronger oxidizing agent in the presence of protons than in their absence, as shown in Equations 24.2 and 24.3, the hydrophobic environment contributes to the protection of the Fe(II) against irreversible oxidation:

$$O_2 + e^- \rightleftharpoons O_2^- \qquad E^0 = -450 \text{ mV} \tag{24.2}$$

$$O_2 + H^+ + e^- \rightleftharpoons HO_2^{\cdot} \qquad E^0 = -320 \text{ mV} \tag{24.3}$$

Figure 24.7 shows the complete structure of sperm whale myoglobin, one of the early triumphs of metalloprotein crystallography. The Nobel Prize in Chemistry was awarded to Kendrew and Perutz in 1962 for their

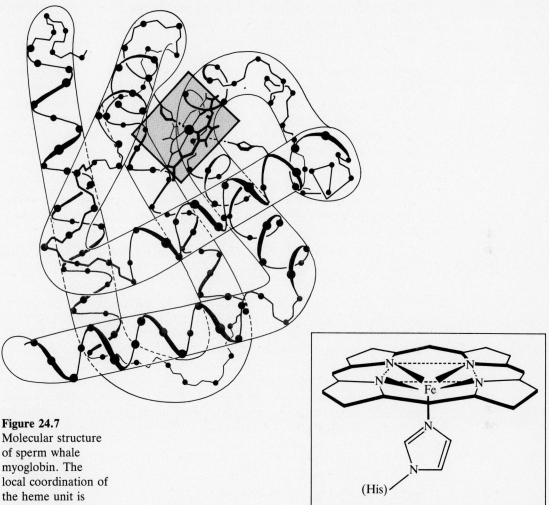

Figure 24.7
Molecular structure of sperm whale myoglobin. The local coordination of the heme unit is shown in the inset. Reprinted with permission from R. E. Dickerson in *The Proteins*, 2nd ed., Vol. 2, p. 103, N. Neurath, ed., New York: Academic Press, 1964.

work on the elucidation of the structures of globular proteins, including myoglobins. The protein structure consists of a number of relatively rigid α-helix runs linked through flexible nonhelical hinges.

In the deoxy form, the iron is high-spin Fe(II) and is displaced from the plane of the porphyrin ring. Oxygenation involves the coordination of the O_2 molecule at the vacant sixth-coordination position, between the iron and the distal histidine. The resulting dioxygen complex is diamagnetic, and it is reasonably certain that the iron moves farther into the plane of the porphyrin on coordination of the oxygen. Crystal structures of

MbO_2 or HbO_2 complexes have not been elucidated. Consequently, knowledge of the structural details come from secondary physical evidence and from model compound studies.

Modeling or mimicking the active sites of metalloproteins with simpler metal-coordination complexes is an important part of the methodology for gaining understanding of their structure and function. However, the design and preparation of models for $HemeO_2$ compounds has been fraught with almost as much difficulty as the structural studies on the actual proteins. Most model hemes react irreversibly with oxygen to produce the μ-oxo-bis(hemin). Two key elements of understanding eventually led to the design of successful models. First, irreversible oxidation can result from the attack of a second heme molecule on the $hemeO_2$ complex; second, protonation of coordinated dioxygen induces irreversible electron transfer from the Fe(II) to O_2. These processes are shown in Equations 24.4 and 24.5:

$$Fe^{II}-O-O + Fe^{II} \rightarrow Fe^{III}-O-O-Fe^{III} \tag{24.4}$$

$$Fe^{II}-O-O + H^+ \rightarrow Fe^{III} + {}^{\bullet}O-O-H \tag{24.5}$$

As written, Equations 24.4 and 24.5 depict the reduction of dioxygen. In the first case, the two iron atoms furnish two electrons to reduce the O_2 to peroxide. In the second case, protonation of the coordinated O_2 promotes electron transfer from the Fe(II) to give superoxide. By sterically blocking one side of the heme plane with bulky ligands to prevent attack on coordinated O_2 on that side and by working in an aprotic system, the remarkable molecule shown in Figure 24.8 was prepared by Collman and his collaborators. This so-called *picket-fence* heme reacted reversibly with O_2 and gave Mössbauer and other spectroscopic parameters very similar to those of the natural oxygen carriers.

Most importantly, Collman's oxygenated picket-fence heme yielded usable crystals. Even though the crystal-structure determination was slightly impaired by a disordering of the dioxygen about the Fe—O bond, the structure did confirm that the O_2 is bonded end-on and nonlinearly, as was predicted by Pauling many years earlier. The Fe—O_2 bonding is now generally accepted to be shown below:

$$\begin{array}{c} \qquad \nearrow O \\ O \\ | \\ Fe \end{array}$$

In heme proteins, it is now clear that the protein serves the functions of preventing attack on the coordinated dioxygen by a second heme and of providing an aprotic medium around the coordination site. Nevertheless, the lifetime of Mb and Hb is not unlimited, and eventually oxidation of Fe(II) to Fe(III) may occur. The irreversible oxidation can lead to extensive damage to the protein, and oxygen carriers must be replaced con-

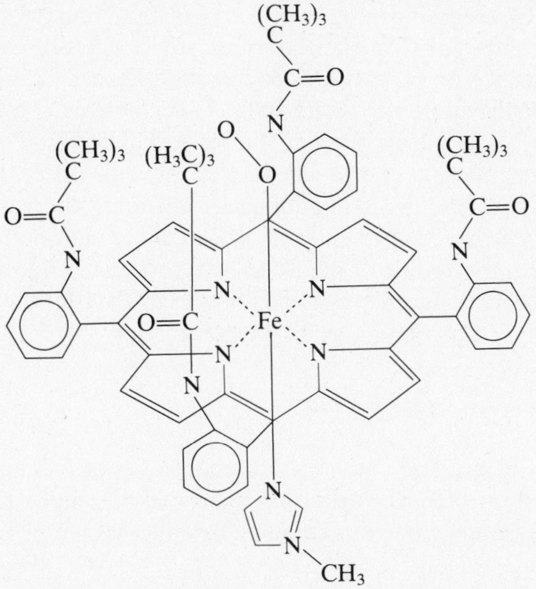

Figure 24.8
Collman's
picket-fence heme.
A model for natural
heme oxygen
carriers.

tinually. Mb and Hb can be chemically oxidized to Fe(III) proteins. Hemin proteins are usually named as for the heme protein with a "met" prefix, as in "metmyoglobin."

Oxygen carriers and other heme proteins can also bind other small ligands quite strongly. Of particular importance are CO, CN^-, and H_2S, which are highly toxic as a result of their ability to bind to heme proteins more strongly than do their natural substrates. In the case of CO, it is the high binding constant to Fe(II) that is important; whereas for CN^- and H_2S, it is the high binding constant for Fe(III). One treatment for CO poisoning is the inhalation of pure oxygen to remove the bound CO by increasing blood–oxygen concentration.

The local details of O_2 binding sites for Hb are almost identical to those of Mb. However, unlike Mb, Hb consists of a tetramer of Mb-like units. The four units consist of two identical pairs, the α- and β-Hb units. The more complicated quaternary structure of hemoglobin gives it the remarkable ability to contravene the law of mass action; that is, over much of the saturation range, the affinity of Hb for O_2 increases with increasing saturation. The details of how this occurs go beyond the scope of this text and are not discussed here. It suffices to say that the most favored model for explaining the functioning of hemoglobin involves a motion of the iron atom into the porphyrin plane as it goes from high spin to low spin on binding oxygen. This motion of the metal atom is translated into a conformational change in the protein, which in turn causes a change in the binding constant of an adjacent subunit.

Changes in amino acid composition remote from the coordination sites of the heme pocket can cause subtle differences in the oxygen-binding properties of O_2 carriers. Such subtle changes can lead to fine tuning to adapt the molecule to its species-specific role. More extensive modifications, particularly those of the proximal or distal histidines, can lead to serious malfunction of the carrier. Several dozen hemoglobin mutations have been identified in humans with varying degrees of pathological symptoms. Sickle cell anemia is caused by a mutational replacement of the proximal histidine of the α-Hb by tyrosine. Coordination of tyrosine to the iron effects its irreversible oxidation to Fe(III), thereby rendering it incapable of functioning as an oxygen carrier.

Peroxidases and Catalases

Peroxidases and catalases are enzymes whose function is to catalyze the oxidation of substrates using hydrogen peroxide. The reactions catalyzed by these enzymes can be generalized as shown in Equation 24.6

$$H_2O_2 + 2AH \rightarrow 2A + 2H_2O \tag{24.6}$$

where AH is a generalized hydrogen-donor substrate.

The difference between the catalases and the peroxidases is that, for the former, 2AH is H_2O_2. That is, the catalases serve the specific function of destroying hydrogen peroxide by disproportionation, thus:

$$2H_2O_2 \xrightarrow{\text{Catalase}} 2H_2O + O_2 \tag{24.7}$$

The peroxidases act on a wider variety of substrates, particularly phenols (for example, the amino acid tyrosine) and indoles (for example, the amino acid tryptophan). Both amino acid substrates are coupled under the influence of peroxidase–hydrogen peroxide to the animal pigment melanine. One important factor in determining the color of melanine is the ratio of tyrosine to tryptophan in the material.

Another important peroxidase is cytochrome c peroxidase. This protein has the special function of oxidizing cytochrome c (see below) by a one-electron transfer process.

Peroxidases and catalases are generally more susceptible to denaturation than are oxygen carriers and therefore more difficult to handle. A crystal structure of cytochrome c peroxidase has been carried out, but so far there is no crystal structure of a plant peroxidase.

A large battery of spectroscopic and chemical techniques has been applied to the study of peroxidase structure, and the consensus is that the iron coordination at the heme site in cytochrome c peroxidase and that in the plant peroxidases is the same. This coordination is also the same as that in the oxygen carriers, namely, the four nitrogens of the porphyrin and a nitrogen from an imidazole residue on the protein. The differences in

Table 24.2 *Heme-Protein–Reduction Potentials*

Protein	$E°$ (mV)
Hemoglobin	-50
Myoglobin	-50
Hb (Saskatoon)	$> +500$ (mutant with distal tyrosine)
Cytochrome c	$+254$
Japanese radish peroxidase	$+120$ to $+220$ (seven isozymes)
Catalase	$> +500$
Rat liver P-450	$+335$

behavior of these proteins reside in the more subtle interactions of the protein with the active site (for example, presence or absence of hydrogen bonding to the proximal imidazole N—H or blocking the access to regions of the active site).

Despite the structural similarity between oxygen carriers and peroxidases, there is enough difference in the fine tuning to make the reduction potentials of peroxidases much higher than those of oxygen carriers; in the resting state of the enzyme, the iron is in the Fe(III) state. Table 24.2 lists the reduction potentials for a selection of heme proteins. The larger the value for the reduction potential, the more easily is Fe(II) oxidized to Fe(III).

Unraveling the mechanism of peroxidase function has been difficult, and there is still a measure of uncertainty about some of the details. One reason for the difficulty is that the catalytic cycle involves Fe(IV) species and such species are virtually unknown in other chemistry of iron. Figure 24.9 shows a scheme for the peroxidase reaction with H_2O_2. In the

Figure 24.9 Intermediates in the reaction of peroxidase with H_2O_2.

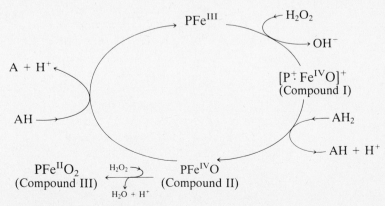

P = Protoporphyrin IX Dianion

scheme, the level of protonation of compound II remains controversial. The main physical evidence for the presence of Fe(IV) in compounds I and II is their unusual Mössbauer isomer shifts (see Chapter 7). Evidence for the porphyrin-cation radical in compound I comes from its abnormal optical spectrum and from ESR measurements. Compound III is the analog of oxymyoglobin. It is produced when peroxidase is treated with a large excess of H_2O_2, but it can also be produced by chemical reduction of peroxidase in the presence of O_2. Like the conventional heme oxygen carriers, oxygen can be displaced from compound III with CO with the formation of the heme carbonyl.

The chemistry of catalase parallels that of peroxidase to a large degree. A number of physical measurements, including the very high reduction potential, indicates that the coordination of the iron is different in catalase. This is confirmed by an X-ray crystal structure that shows that the proximal histidine is replaced by a carboxylate from an aspartate residue. Because of its high oxidation potential, catalase cannot be reduced to Fe(II); therefore, there is no stable analog of compound III for the catalase series.

Mixed-Function Oxidases: Cytochromes P-450

Cytochromes P-450 are a family of heme protein enzymes that use molecular oxygen to oxidize organic molecules according to Equation 24.8:

$$RH + AH_2 + O_2 \xrightarrow{\text{P-450}} ROH + H_2O + A \qquad (24.8)$$

In this reaction, AH_2 is a biochemical reductant such as NADH or a flavin. The half-cell reactions for these reductants are shown in Equations 24.9 and 24.10:

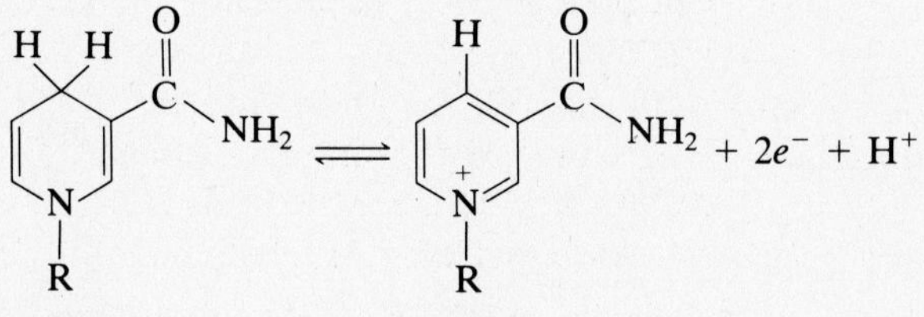

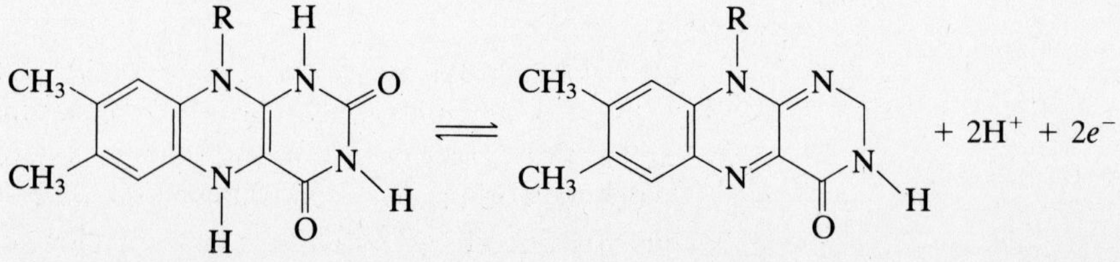

(24.9)

(24.10)

All living organisms use this chemistry for two purposes: the synthesis of specific molecules for use by the organism and the detoxification of cell membranes. An example of selective synthesis is shown in Equation 24.11:

$$\text{11-Deoxycorticosterone} \quad \xrightarrow[-\text{NAD}^+/\text{H}_2\text{O}]{+\text{NADH}/\text{O}_2} \quad \text{Corticosterone} \qquad (24.11)$$

It should be noted that the oxygenation takes place at a secondary aliphatic CH bond. Because they are particularly difficult to attack in a controlled and selective manner with conventional oxidants, these bonds have attracted much interest from chemists.

Hydrophobic molecules (for example, aromatic hydrocarbons and chlorinated hydrocarbons) dissolve in the lipid (oily) part of the cell membrane where they are no longer accessible to the aqueous systems of the cell interior. The membrane-detoxification role of the P-450 enzymes is necessary because the accumulation of large amounts of foreign hydrophobic material in the membrane would eventually lead to its dysfunction or disruption. Selective introduction of hydroxyl groups into the hydrophobic contaminants by P-450–assisted oxidation renders them more hydrophilic and encourages their departure from the cell membrane.

P-450 enzymes are usually stabilized in the cell by being bound to the cell membrane. The presence of the enzyme in rat liver extracts was originally found because the enzyme, in its reduced form, gives a carbonyl complex with a porphyrin band in the optical spectrum at an unusually long wavelength (450 nm; hence, the name of the enzyme). Detachment of the enzyme from the membrane leaves it very susceptible to denaturation, and this has severely limited the application of crystallography to the solution of the structure. The crystal structure of a bacterial P-450, however, has been worked out, and spectroscopic studies on other P-450 enzymes leave little doubt that the iron coordination is the same as in the bacterial protein.

The most obvious structural difference between P-450 proteins and other heme proteins mentioned earlier is the occupation of the fifth coordination position by a thiolato ligand from a cysteine residue, rather than by a histidine residue. The coordination of the thiolato ligand had originally been deduced by studying the properties of metmyoglobin in the presence of sulfur-donor ligands and of model heme compounds with

Figure 24.10
Molecular structure
of a model for the
active site of a
cytochrome P-450.
The porphyrin in
this model is the
dimethylester of
protoporphyrin IX.
The apical ligand is
4-nitrothiophenolate.
The bond distances
given are in
picometers.
Reprinted with
permission from
S. Koch, S. C. Tang,
and R. H. Holm,
J. Am. Chem. Soc.
97 (1975): 916.

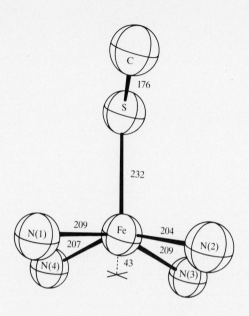

thiolato ligands. These models mimicked to some degree the unusual spectroscopic behavior of the natural enzymes (for example, positions of bands in electronic spectra, high rhombicity in ESR spectra, and characteristic similarities in EXAFS spectra). Figure 24.10 shows the structure of a compound designed to model the heme site of a P-450 enzyme.

In the resting state, P-450 is in the high-spin Fe(III) state. The catalytic cycle begins with the binding of substrate, which among other things leads to a change from the high-spin to low-spin state. Only after substrate binding does the enzyme become activated for reduction to Fe(II) by an electron-transfer protein. Once in the Fe(II) state, the enzyme binds O_2. The intimate details of how this dioxygen complex completes the oxidation of the substrate is still the subject of vigorous investigation, but it is virtually certain that oxygen transfer takes place from an iron species analogous to peroxidase compound II. Figure 24.11 illustrates a cycle showing how this might occur.

A comparison of peroxidase and P-450 illustrates the remarkable degree to which the topography of the protein in relation to the heme site can dramatically alter the function of the enzyme. Both structural studies and chemical studies show that physical access to the heme site is very different in the two enzymes. In the peroxidases, heme is well covered on both sides of the porphyrin plane by protein residues that allow access of hydrogen peroxide to the iron but that restrict access of larger organic

Figure 24.11
A proposed cycle
for the P-450
reaction.

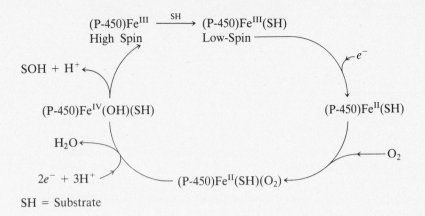

SH = Substrate

molecules. The organic substrate only has access to an edge of the porphy-rin ring, and the electron transfer from substrate to the heme occurs from this edge.

The heme site in P-450 is also buried deep in the protein, but a channel through the protein allows access to one face of the heme by the substrate. Thus, the substrate can encounter the Fe=O group of the oxidized P-450; oxygen transfer, rather than electron transfer, can occur. Figure 24.12 shows schematic representations of the active sites in a peroxidase and a P-450.

As with peroxidase and catalase, the difficulty of elucidating the chemistry of P-450 is compounded by the near absence of Fe(IV) chemistry in other systems than in the heme proteins.

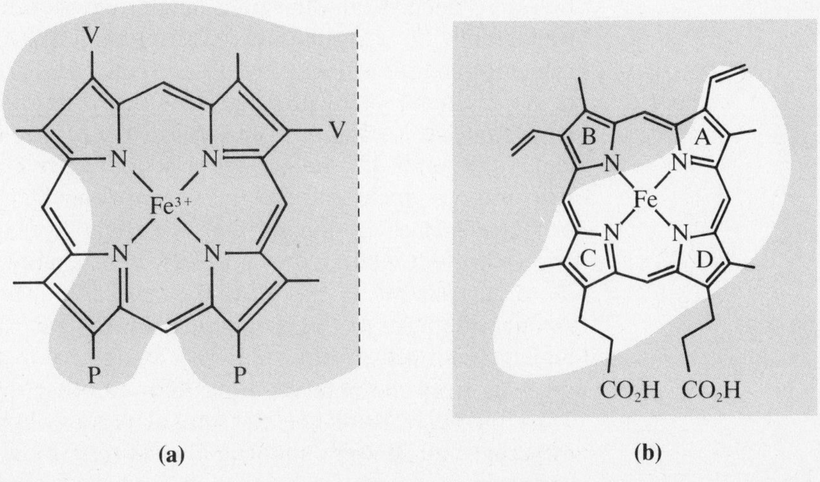

Figure 24.12
Schematic
representations of
the active sites in
(a) a peroxidase and
(b) a P-450. The
shaded areas are the
region of the heme
that are not
accessible to organic
substrate. Reprinted
with permission
from P. R. Ortiz de
Montellano, *Acc.
Chem. Res.* 20
(1987): 289.

(a)   (b)

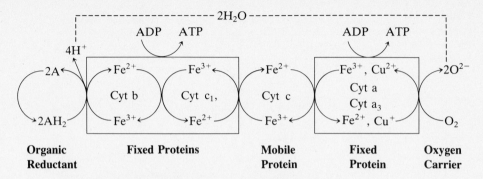

Figure 24.13
A simplified scheme
for the cytochrome
chain in the
mitochondrion.

Electron-Transport Enzymes: Cytochrome c

Cytochromes (literally: "cell color") are highly colored materials associated with the cellular apparatus that transfers carbohydrate-derived electrons to oxygen and uses the chemical potential of the reaction to phosphorylate ADP to energy-storing ATP. This apparatus, the *mitochondrion*, is a highly organized and complex structure that consists of many different components. The electron flow occurs through a series of metalloproteins, which are held immobile, and electrons are carried between them by a set of mobile metalloprotein oxidases and reductases, which bind specifically to the surfaces of the immobile proteins. The binding sites for the oxidases and the reductases are generally different. Figure 24.13 shows a simplified scheme for the cytochrome-containing part of the mitochondrion. An analogous chain of cytochromes is used for transport in the photosynthetic apparatus of green plants.

Cytochrome c, which is situated in the middle of the cytochrome chain of higher animals, can be fairly easily isolated and crystallized. It has therefore been studied in great detail, and its structure resolved by crystallography to a resolution of better than 100 pm. The heme unit of cytochrome c is attached covalently to the protein by the addition of two cysteine —SH groups across the vinyl substituents of the protoporphyrin IX. In addition, the iron is coordinated to a histidine residue and to the sulfur of a methionine residue. Figure 24.14 shows the complete structure of cytochrome c derived from horse heart muscle.

The surface of the protein is rich in hydrophilic groups, but the interior is hydrophobic. It is generally believed that electron transfer from the reductase and to the oxidase occurs over a fairly large distance and through the edge of the porphyrin ring. Because the heme unit is tightly bound and fully coordinated, the metal is not accessible to attacking reagents from the exterior other than by outer-sphere electron transfer. Thus, the Fe(II) form does not form an O_2 or CO complex, and the Fe(III) form does not complex simple anions as they do in the O_2 carriers and the peroxidases.

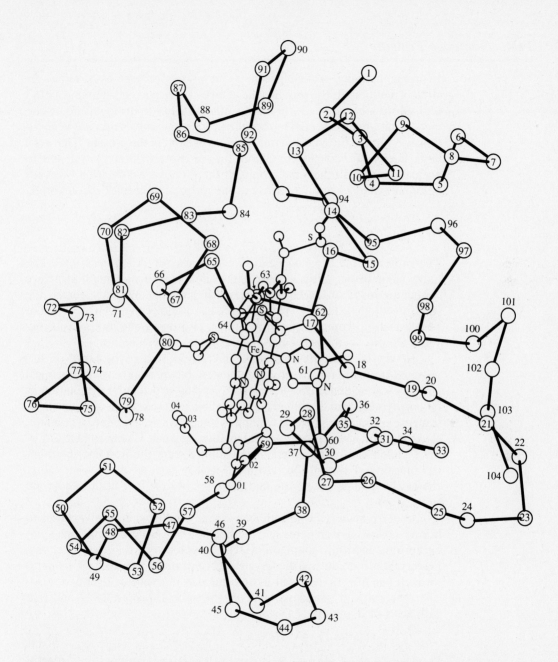

Figure 24.14 Structure of oxidized horse heart cytochrome *c*. In this schematic drawing, each amino acid is represented by a numbered circle. Reprinted with permission from R. E. Dickerson in *The Proteins*, 2nd ed., Vol 2, p. 603, N. Neurath, ed., New York: Academic Press, 1964.

24.4   Nonheme Proteins

In iron-containing proteins, iron is not always present as a heme. As already mentioned, the proteins themselves are rich in coordinating amino acid substituents. Indeed, the task is often difficult to distinguish between metals that are an essential part of the protein and metals that are adventitiously bound to the coordinating residues present in the protein. There is a large family of enzymes containing one or more iron atoms bonded to normal or special functional groups of the protein. We describe two very different examples to give a sense of the range of molecules encountered.

Hemerythrin

This protein is the oxygen carrier of certain classes of marine worms. The functioning hemerythrin system has a complicated polymeric structure containing many of the basic hemerythrin subunits. We only discuss the structure of this subunit. This subunit has a molecular weight of about 14,000 Dalton, contains two iron atoms, and binds one molecule of oxygen. Despite its name, hemerythrin contains no heme unit.

No crystal structure of an oxyhemerythrin or deoxyhemerythrin has been determined, and structural details are based on a range of chemical and physical evidence coupled with the crystal-structure determination of the methemerythrin described below. Amino acid sequences are known for some cases, and the proteins contain cysteine, lysine, tyrosine, glutamate, aspartate, and histidine as potential ligating residues. Chemical blocking of the cysteine and lysine functions does not disrupt the iron coordination, but blocking of two of the tyrosines leads to release of the iron. Other chemical techniques conclude that there are at least four histidines bound to the irons.

In deoxyhemerythrin, both irons are in the high-spin Fe(II) state and in very similar environments. On binding O_2, both irons become high-spin, antiferromagnetically coupled Fe(III). Oxygen can be removed from oxyhemerythrin by reducing the oxygen concentration, to give back hemerythrin. It can also be released by treatment with oxidizing agents or acid. The reaction with oxidizing agents is shown in Equation 24.12 and that with acids in Equation 24.13:

$$HemFe_2O_2 \rightarrow HemFe_2^{III} + O_2 + 2e^- \qquad \textbf{(24.12)}$$

$$HemFe_2O_2 + HX \rightarrow [HemFe_2^{III}X]^+ + [HO_2]^- \qquad \textbf{(24.13)}$$

The resulting protein in both cases contains two Fe(III) atoms. This evidence shows that oxyhemerythrin is best described as a peroxo-complex of Fe(III). Figure 24.15 shows the various states of hemerythrin.

The protein products obtained in Equations 24.12 and 24.13 are called methemerythrin, and the crystal structure of the azido-complex of such a

Figure 24.15
The relationship between the different states of hemerythrin.

[Fe(II), Fe(II)]

Deoxyhemerythrin

$$-O_2 \Big\Updownarrow +O_2$$

[Fe(III), $Fe^{III}(O_2^{2-})$]

Oxyhemerythrin

$$\Big\downarrow 2H^+$$

$[Fe(III), Fe(III)]^{2+} + H_2O_2$

Methemerythrin

Figure 24.16
The iron-coordination site of methemerythrin.

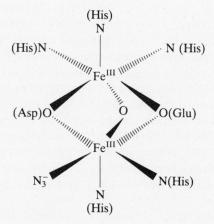

methemerythrin has been determined. In this compound, each iron is octahedrally coordinated, and the two octahedra share a common face. The ligands of this shared face are two carboxylates from aspartine and glutamic acid residues and a μ-oxo bridge. One iron has an additional three histidine ligands, while the other has two histidines and a coordination site occupied by the azide ion (Figure 24.16). It is assumed that in oxyhemerythrin the O^{2-} occupies the same coordination site as does the azide ion of the azidomethemerythrin.

Many other nonheme iron proteins have been identified. These proteins tend to parallel the heme proteins in function, and many oxidative enzymes are known that have nonheme iron-active sites.

Iron–sulfur Proteins

The second class of non-heme iron proteins that we discuss are the so-called iron–sulfur proteins. Like the proteins discussed in the previous section, they contain iron, but they are not coordinated to a porphyrin. In these relatively small proteins (about 100 amino acids), the iron is coordinated uniquely to sulfur. They are found in all living organisms, but due to their generally greater accessibility, those from bacteria have received the most attention. A second factor that has drawn attention to the bacterial proteins is their intimate involvement in nitrogen fixation by anaerobic bacteria (see below).

Several different classes of iron–sulfur proteins are known, differing in the number of iron atoms and the amount of inorganic sulfide in the active site. Rubredoxins have a single iron coordinated to an approximate tetrahedron of cysteine sulfides. Plant ferredoxins have a binuclear iron site consisting of two tetrahedra sharing a common edge. This shared edge contains two bridging inorganic sulfide ligands while the remaining coordination sites are filled by cysteine sulfides. Bacterial ferredoxins and the so-called high-potential iron protein (HIPIP) contain tetrameric iron units. In this site, the irons occupy alternate corners of a cube while triply bridging inorganic sulfides occupy the other corners. The fourth coordination site of the pseudotetrahedral iron is occupied by a cysteine sulfide ligand. Figure 24.17 illustrates the three types of structure.

Independent of the number of iron atoms, all groups shown in Figure 24.17 are one-electron acceptor–donors. The Fe_2S_2 species was shown by Mössbauer spectroscopy to contain two high-spin Fe(III) atoms in the oxidized state. The absence of any paramagnetism or ESR signal indicates strong antiferromagnetic coupling. The reduced form gives ESR and Mössbauer parameters consistent with a coupled high-spin Fe(II)/high-spin Fe(III) pair.

The reduced state of bacterial ferredoxins $[Fe_4S_4]^+$ has one Fe(III) and three Fe(II), and the cluster is paramagnetic due to the odd number of

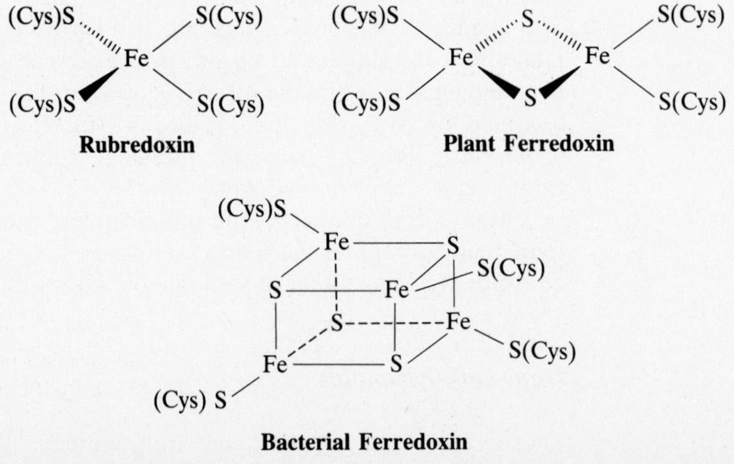

Figure 24.17 Idealized representations of three types of site found in iron–sulfur proteins.

Rubredoxin

Plant Ferredoxin

Bacterial Ferredoxin

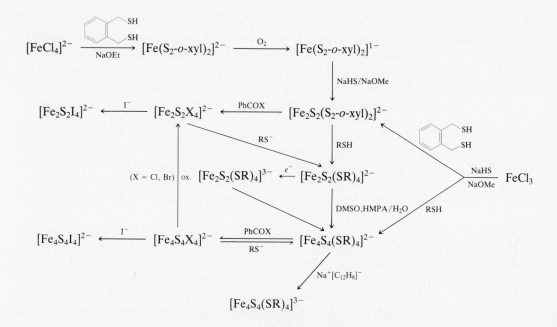

Figure 24.18 Syntheses and reactions of Fe_1, Fe_2, and Fe_4 analogs of the active sites of iron–sulfur proteins. Reprinted with permission from R. H. Holm *Acc. Chem. Res.* 10 (1977): 427.

electrons. The oxidized $[Fe_4S_4]^{2+}$ cluster is magnetically silent due to antiferromagnetic coupling between the even number of electrons.

HIPIP caused some confusion in the early days of iron–sulfur protein research because its reduction potential was anomalously low. Subsequent investigations have shown that the reduced form of HIPIP is isoelectronic with the oxidized form of ferredoxin $[Fe_4S_4]^{2+}$, and the oxidized form of HIPIP is $[Fe_4S_4]^{3+}$. Although the cause of the great difference in the stabilities of the oxidation states of the two kinds of protein is still not fully understood, it is thought that the HOMO of $[Fe_4S_4]^{2+}$ is antibonding in character, weakly in ferredoxins, and more strongly in HIPIP. The weakly antibonding orbital of ferredoxin has little inclination to lose an electron, and it is fairly easy to add an electron. A more strongly antibonding orbital has a greater tendency to lose its electron and very little tendency to acquire an additional one. The cluster of HIPIP undergoes significant dimensional changes on passing between the oxidized and reduced forms; the changes of the ferredoxin-cube dimensions are much smaller. This behavior is in accord with the ferredoxin HOMO being relatively nonbonding in character.

The preparation of synthetic analogs has played a very important part in establishing the structures and chemistries of the iron–sulfur proteins. Much of the work in this area was done by Holm and co-workers; Figure 24.18 summarizes the syntheses and chemical reactions carried out by Holm's group.

24.5   Copper Proteins

Many functions of iron-containing proteins take advantage of the facile one-electron redox process of the Fe(II)/Fe(III) couple. The Cu(I)/Cu(II) couple has similar properties, and copper is widely available in the environment. Evolution has resulted in incorporating a wide range of copper-containing proteins into living systems. To a large measure, these proteins fulfill the same functions as do the iron proteins, namely, oxygen transport, electron transport, and catalysis of redox reactions. In sharp contrast to iron, there is no specific prosthetic group, like heme, through which copper is bound to proteins; the binding takes place entirely through the ligating substituents of the amino acid residues, as it does in the nonheme iron proteins.

Copper occurs in proteins in three different forms. First is an isolated copper ion bound to ligands to give spectral parameters that conform to the normal range of behavior of model copper complexes. In this type of site, copper is bound to oxygen- and nitrogen-ligating atoms; the structural problem is to determine, usually by the study of optical and ESR spectra and by chemical blocking experiments, the number, identity, and geometrical disposition of the ligands.

The second type of copper is known as "blue copper." This may seem strange because many common copper(II) compounds are blue, but in blue copper proteins the intensity of the blue color is an order of magnitude more intense than had ever been observed in model compounds until very recently. All these proteins have an intense absorption band (molar absorptivity ~ 4000) at about 600 nm. In addition, the ESR spectrum of Cu(II) ions in these blue sites exhibit unusually large, hyperfine coupling to the nuclear spin of nitrogen ligands.

The third type of copper is usually referred to as "ESR silent." In this case, the active site contains two copper atoms in close proximity (compare with the two Fe ions in hemerythrin) so that when both are Cu(II) they are antiferromagnetically coupled and the normal ESR signal of ^2D copper(II) is not observed.

Some proteins may contain only a single type of copper, but others contain all three types in the same molecule. Table 24.3 lists some examples of copper proteins.

The Blue copper proteins are particularly interesting from the chemical viewpoint, and we consider one example of such a protein in detail. The fascination with the problem lay in the long history of failure to produce model copper compounds that mimic the intense absorption at 600 nm, which is so characteristic of the blue proteins. For a long time, it was suspected that the 600-nm band was a charge-transfer band associated with an easily oxidizable ligand. Candidates for this ligand were a cysteine sulfide or a tyrosine phenoxide. Synthesis of model compounds containing

Table 24.3 *Examples of Copper-Containing Proteins*

Protein	Source	Function	Type of Copper
Azurin	Bacteria	Electron transport	Blue
Stellacyanin	Plant	Electron transport	Blue
Laccase	Plant	Phenol oxidation	All three
Amine oxidase	Mammals	Amine catabolism	Normal
Tyrosinase	Fungi	Phenol oxidation	ESR silent
Hemocyanin	Mollusks	Oxygen transport	ESR silent
Ceruloplasmin	Mammals	Oxidation of Fe(II); Cu storage	All three

these ligands was thwarted by the ease of irreversible oxidation of the phenolates and thiolates in the presence of Cu(II) and particularly in the presence of oxygen. Even today there are no perfect mimics of the blue copper coordination site. Fortunately, the crystal structures of some blue copper proteins have been reported at a sufficient resolution to confirm that at least in some cases the cysteinate coordination hypothesis is correct. Figure 24.19 shows two such structures. One reason for the continuing difficulty in simulating the spectral characteristics exactly is the difficulty of building in the severely distorted coordination geometry that is imposed on the copper in the blue proteins by the relatively rigid protein matrix.

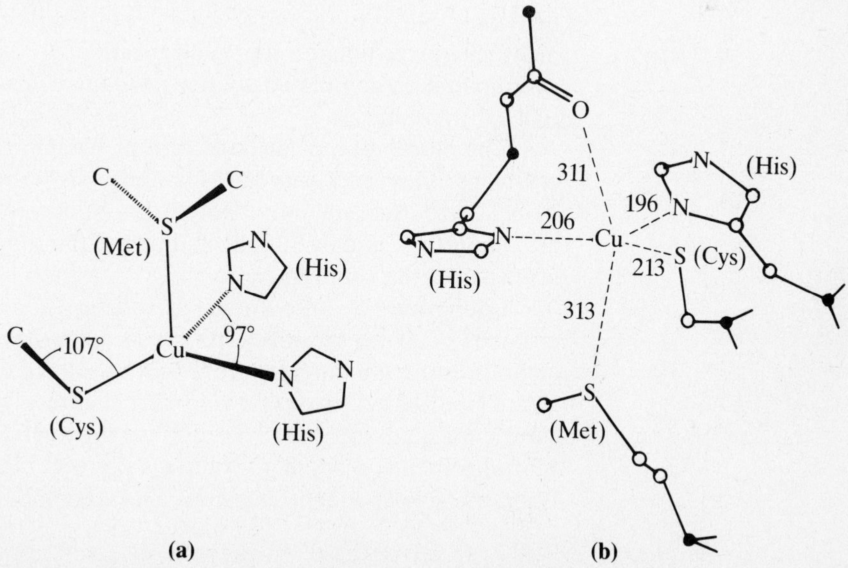

Figure 24.19 Crystallographically determined structures of the active sites of (a) a plastocyanin and (b) an azurin. Reprinted with permission from G. E. Norris, B. F. Anderson, and E. N. Baker, *J. Am. Chem. Soc.* 108 (1986): 2784.

(a)

(b)

The protein of blue copper electron carriers has some functional similarities to that of cytochrome c in that the active site is buried in a hydrophobic cleft, which makes it inaccessible to direct chemical attack. For example, the reduced Cu(I) form of the blue electron carriers does not usually react directly with molecular O_2—a very untypical behavior for a copper(I) complex.

24.6   *Coenzyme B$_{12}$: Nature's Unique Organometallic Reagent*

Coenzyme B_{12} is extraordinary in several respects. It is the only organometallic compound known to play a well-defined and natural role in biochemistry. Although certain bacteria can affect the alkylation of heavy-metal compounds in the environment, such as those of mercury and arsenic, the resulting compounds play no essential role in the biochemistry of organisms and indeed are usually highly toxic. Relatively high concentrations of phenylmyoglobin have been detected in rat tissues, but it is unlikely that this compound performs any biochemical function.

Coenzyme B_{12} is also the only well-defined cobalt-containing biochemical, and it is ligated by a remarkable cyclopolypyrrole ligand, *corrin*. The cobalt corrins were first isolated as the cyanide complex, known and sold as vitamin B_{12} (Figure 24.20a). The coenzyme itself, which is the biochemically active form, has an adenosyl ligand (Figure 24.20b).

The corrin ring system bears some similarity to porphyrin. The absence of one of the methine bridges reduces the size of the ligating cavity and disrupts the closed conjugation. Like porphyrin, corrin has two ionizable protons, and in B_{12} both are lost. The alkyl (or cyanide) ligand completes the neutralization of the Co(III), and the six-coordination is completed by the pendant benzimidazole residue on the ring substitutent.

The elucidation of the structure of vitamin B_{12} was a masterpiece of crystallography and earned its author, Dorothy Crowfoot-Hodgkin, a Nobel Prize. Subsequently, Robert B. Woodward, another Nobel Laureate, undertook and eventually completed the titanic challenge of the total synthesis of the coenzyme.

Coenzyme B_{12}, also known as *cobalamin*, acts as a cofactor to many enzymes in living systems. The kinds of reactions catalyzed by the B_{12} complex are quite different from those we have described above for iron- and copper-based enzymes; one that has received a lot of attention is shown in Equation 24.14. This rearrangement is representative of a class of rearrangements, involving formal alkyl-group migrations, which are catalyzed by B_{12}-dependent enzymes. This particular reaction is catalyzed by the enzyme glutamate mutase:

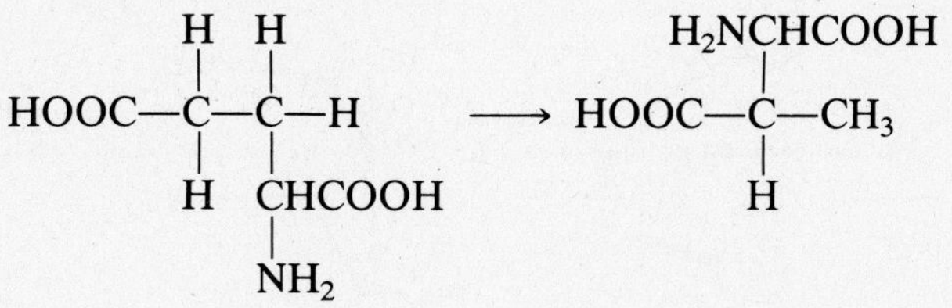

$$\underset{\substack{\displaystyle H \\ \displaystyle CHCOOH \\ \displaystyle NH_2}}{\overset{\displaystyle H}{HOOC-C-}}\overset{\displaystyle H}{\underset{}{C}-H} \longrightarrow \underset{\substack{\displaystyle H}}{\overset{\displaystyle H_2NCHCOOH}{HOOC-C-CH_3}} \qquad (24.14)$$

Because cobalamin is extremely difficult to synthesize and because it can only be obtained from tissue in minute quantities, some good model compounds fortunately have allowed a careful evaluation of the properties of this type of compound. The most widely studied models are derivatives of bis(dialkylglyoximato)Co. Figure 24.21 shows the structure of bis-(dimethylglyoximato)-O-methyl(pyridine)Co(III). This compound has many chemical properties in common with the cobalamins and has been called a *cobaloxime* by analogy. Figure 24.22 shows some of the chemical reactions that both cobalamins and cobaloximes undergo.

Figure 24.20
(a) Vitamin B$_{12}$ and
(b) coenzyme B$_{12}$.

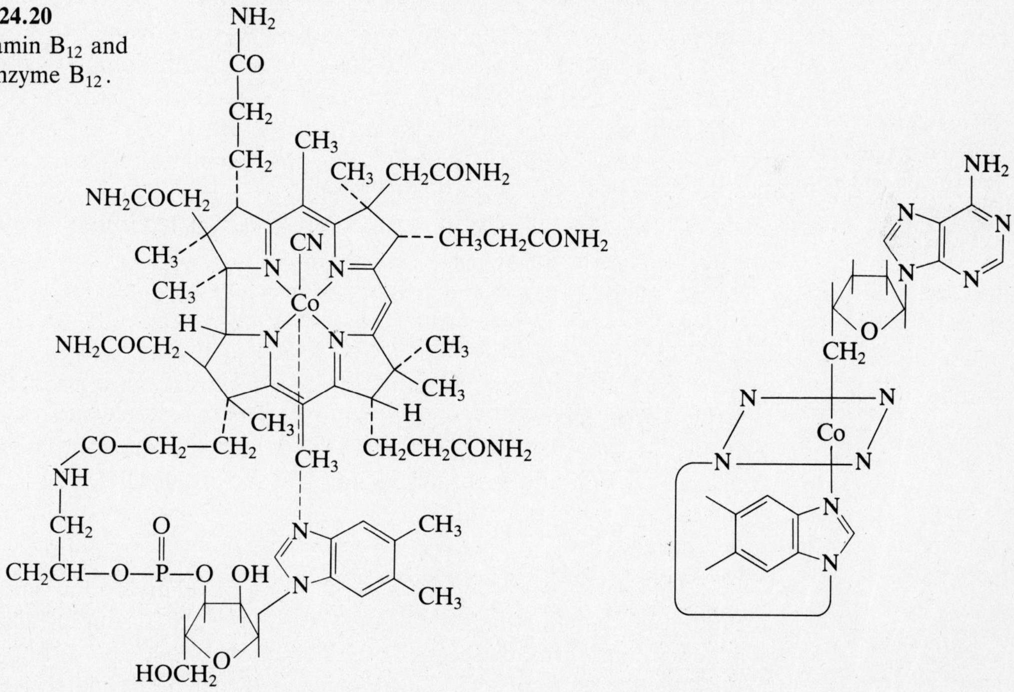

(a)

(b)

Figure 24.21
The structure of a synthetic analog of vitamin B_{12}. *o*-Methyl(Co-C) carboxymethyl [(bisdimethylglyoximato)pyridine]cobalt. Reprinted with permission from G. Lenhert, *J. Chem. Soc. Chem. Commun.*, 980 (1967).

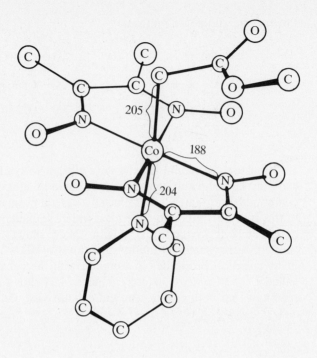

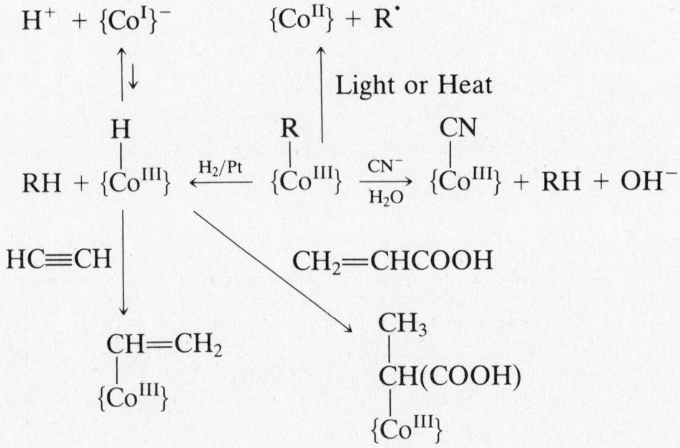

Figure 24.22
Some reactions of the cobalt center in cobalamin and cobaloxime.

$$H^+ + \{Co^I\}^- \qquad \{Co^{II}\} + R^{\bullet}$$

$$\updownarrow \qquad\qquad \uparrow \text{ Light or Heat}$$

$$\begin{array}{ccc} H & R & CN \\ | & | & | \\ RH + \{Co^{III}\} \xleftarrow{H_2/Pt} \{Co^{III}\} \xrightarrow[H_2O]{CN^-} \{Co^{III}\} + RH + OH^- \end{array}$$

$$HC{\equiv}CH \Big| \qquad\qquad CH_2{=}CHCOOH$$

$$\begin{array}{cc} CH{=}CH_2 & CH_3 \\ | & | \\ \{Co^{III}\} & CH(COOH) \\ & | \\ & \{Co^{III}\} \end{array}$$

The bond energy of the Co—C bond is relatively low and can be altered by changing the axial ligand trans to the alkyl group. Studies of the kinetics and thermodynamics of the homolysis of the Co—C bond and its dependence on trans ligand have concluded that the primary biochemical role of cobalamin is to be a source of alkyl radicals. Table 24.4 lists the bond energies for some model compounds and for the adenosyl—Co bond in B_{12}.

Table 24.4 *Bond Energies of Co—C Bonds in Model Compounds and in Adenosyl B_{12}*[a]

Complex	Trans Ligand (L)	Bond Energy $(kJmol^{-1})$
$L(DMG)_2Co(CHMePh)$	Pyridine	82
$L(DMG)_2Co(CHMePh)$	4-Cyanopyridine	75
$L(DMG)_2Co(CHMePh)$	Imidazole	87
$L(saloph)CoC_2H_5$	Pyridine	105
$L(saloph)CoCHMe_2$	Pyridine	84
B_{12} adenosyl	5,6-Dimethyl-3-ribosylbenzimidazole	109

[a] DMG = dimethylglyoximato; saloph = N,N′-disalicylidene-*o*-phenylenediamine.
Source: J. Halpern, *Acc. Chem. Res.*

24.7 Nitrogen Fixation

The Cellular Nitrogen-Fixation Apparatus

The fixation of atmospheric dinitrogen is essential to all forms of life and is carried out in both the soil and oceans on a massive scale by microorganisms. However, only a relatively small number of species, which have evolved the highly specialized cellular apparatus necessary for achieving this difficult reaction, can carry it out. The most successful biological scheme for N_2 reduction involves the formation of ammonia:

$$N_2 + 6e^- + 6H^+ \rightarrow 2NH_3 \tag{24.15}$$

Nitrogen fixation is performed by bacteria and by the so-called *blue-green algae*. The process must be coupled into photosynthesis, either carried out by the organism itself (the algae) or by a symbiotic relation with higher plants. In this partnership, the plant provides the bacterium with its fuel supply in the form of photosynthetic carbohydrate while the bacterium synthesizes ammonia and amino acids from its nitrogen-fixing activity. Figure 24.23 shows a schematic illustration of a typical nitrogen-fixing system. The heart of the system is the *nitrogenase enzyme complex*, which consists of two units—an iron-containing protein and an Fe—Mo protein. The Fe protein contains two subunits and the Fe—Mo protein four subunits. Once N_2 binds to the enzyme complex, it is reduced by a series of six electron transfers. These electrons originate from the degradation of photosynthetic carbohydrate and are carried to the nitrogenase enzyme complex by a reduced ferredoxin. Protons for the formation of ammonia

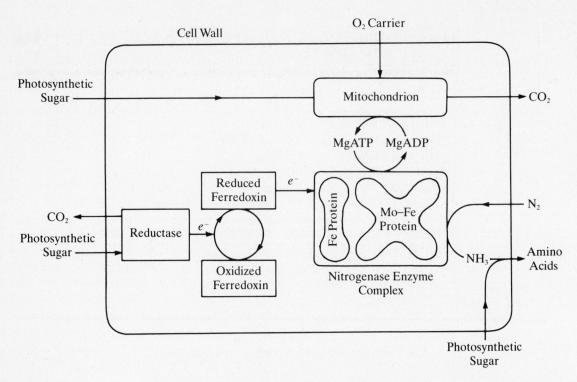

Figure 24.23
Schematic
representation of a
typical bacterial
nitrogen-fixing
system.

are generated through hydrolysis of MgATP to give MgADP and H_3PO_4 (see Equation 16.18). MgATP is regenerated at a mitrochondrion where the oxidation of carbohydrate by oxygen is carried out (see Section 24.3). An important feature of the nitrogenase enzyme complex is its extreme sensitivity to oxygen, a factor that makes isolation and study of the complex extremely difficult. This necessitates that the cell environment remain completely anoxic to protect the enzyme. To achieve this, while allowing for oxygen need for mitochondrial activity, these bacteria use oxygen carriers (leghemoglobins), which have high binding constants and which are specifically bound to the terminal oxidase of the respiratory system.

The nitrogen-fixation process is one of the most highly organized biochemical systems, and its complexity has retarded thorough understanding. The outline given above hides the complexity of the system, and the model shown in Figure 24.23 is schematic and probably not applicable to all nitrogen-fixation systems.

Proteins of the Nitrogenase Enzyme Complex

Isolating the individual components of the nitrogenase complex is very difficult; when they are isolated, establishing the exact metal composition is

Table 24.5	*Compositional Data for Proteins of the Nitrogenase Complex*						
	Mo—Fe Protein				Fe Protein		
Species	Mo	Fe	S^{2-}	RS^-	Fe	S^{2-}	RS^-
Azobacter v.	2	32–38	26–28	41	—	—	—
Clostridium p.	1	14	11–16	21–23	4	4	11
Klebsiella p.	1	17	17	17	4	4	11

difficult. Table 24.5 lists some data on the composition of the Mo—Fe and Fe proteins of several different microorganisms. It is evident from these data that all metalloprotein constituents of the complex are sulfur-rich and of a similar type to the electron-carrying iron–sulfur proteins described in Section 24.4. The Mo—Fe protein of *azobacter vinlandii* has been shown to contain four distinguishable kinds of iron by means of ESR and Mössbauer spectroscopy. The four kinds have the following distinguishing properties:

1. An ESR–active cluster with *g* values of 4.3, 3.7, and 2.01, and S $= \frac{3}{2}$.

2. ESR–active Fe_4S_4 units.

3. An ESR–silent species with Mössbauer parameters indicative of a high-spin Fe(II).

4. An ESR-silent species with no interpretable Mössbauer signal.

After many years of frustrating effort, the active site of the Fe—Mo enzyme has recently been removed successfully from the protein, and its physical and chemical properties carefully compared with those of model compounds. Only two model compounds seem to satisfy all the requirements as good models (Figure 24.24). It is not surprising when one sees these extraordinary molecules that the complete description of the nitrogen-fixation process is such a long and arduous task.

Nitrogen Fixation Through Metal Complexes

Although dinitrogen is isoelectronic with CO, the filled σ_4 and empty π_2 orbitals (see Figure 3.8), which are concentrated on C in CO and are largely responsible for its ability to bond to transition metals, are equally spread on both atoms in N_2. This makes N_2 both a weaker σ donor and a weaker π acceptor than is CO. It is not surprising therefore that the first

Figure 24.24
Two model compounds that have physico-chemical properties very close to those of the Fe—Mo cofactor of nitrogenase. Reprinted with permission. (a) from G. Christon and D. C. Garner, *J. Chem. Soc. Dalton Trans.*, (1980) 2354, (b) from T. E Wolff, J.M. Berg, and R. H. Holm *Inorg. Chem.* 20 (1981) 175.

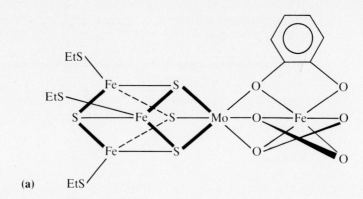

(a)

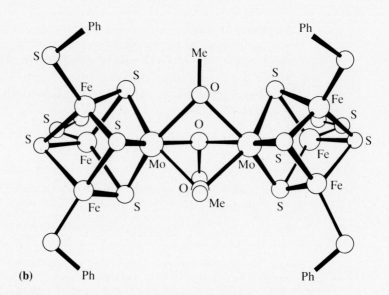

(b)

dinitrogen complex was isolated some seventy years after the first metal carbonyl [Mond made $Ni(CO)_4$ in 1897]. Allen and Senoff, two Canadian scientists, isolated $[Ru(N_2)(NH_3)_5]^{2+}$ salts from the reduction of aqueous ruthenium(III) chloride solutions with hydrazine. It was later shown by Taube that dinitrogen can reversibly displace water from $[Ru(H_2O)-(NH_3)_5]^{2+}$. These discoveries were followed by a number of reports of dinitrogen complexes. In most cases, however, the coordinated nitrogen was no more reactive than molecular N_2, and its study contributed little to the further understanding of nitrogen fixation.

The first authentic demonstration of reduction of a coordinated N_2 ligand to produce ammonia involved the reaction of transition metal halides with Grignard reagents under nitrogen. These systems are extremely difficult to characterize and are now of mainly historical interest. A number of better characterized transition metal compounds have subsequently been shown to react with and to reduce molecular nitrogen. These compounds are so energy-rich, however, that they are of little relevance to enzymatic nitrogen fixation. The most significant result of this line of modeling work remains the demonstration that nitrogen is capable of coordinating to a metal center.

BIBLIOGRAPHY

Suggested Reading

Bezkorovainy, A. *The Biochemistry of Non-Heme Iron*. New York: Plenum, 1980.

Dolphin, D. et al., eds. "Biomimetic Chemistry." *ACS Adv. Chem. Ser.* no. 191. Washington D.C.: ACS, 1980.

Hanzlik, R. P. *Inorganic Aspects of Biological and Organic Chemistry*. New York: Academic Press, 1976.

Ochiai, E-I. *Bioinorganic Chemistry*. Boston: Allyn and Bacon, 1977.

Ortiz de Montellano, P. R. "Control of the Catalytic Activity of Prosthetic Heme by the Structure of Hemoproteins." *Acc. Chem. Res.* 20(8) (1987): 289.

PROBLEMS

24.1 Examination of Table 24.1 reveals that titanium is the second most abundant transition element in the earth's crust. Using the chemical information on titanium given in Chapter 11, suggest reasons why titanium appears to have found no use in living systems. Do the same reasons apply to silicon? (See Chapter 10.)

24.2 Suggest some modifications to ferrichrome (Figure 24.2) that would make it more selective for Ru(III) than for Fe(III).

24.3 Give three factors that have contributed to making synthetic molecules with extremely high binding constants for Fe an evolutionary advantage.

24.4 What features of the porphyrin ring make its effects on a central metal ion distinctly different from the effects that can be achieved by coordinating the same ion to four amino acid nitrogen-base functions?

24.5   Evaluate the validity of the following assertions:
   a. An X-ray crystal structure is the only definitive way to establish the structure of the active site of an enzyme.
   b. Hemoglobin's tetrameric structure is the key to its ability to bind oxygen reversibly.
   c. The special properties of the porphyrin ligand are the key to the ability of hemoglobin to reversibly bind oxygen.
   d. The unique redox properties of the Fe(II)/Fe(III) couple are the key to the reversible oxygen binding of hemoglobin.

24.6   Give four special properties of protein molecules that make them particularly suited to the task of providing a ligand environment for biologically functional transition metal ions.

24.7   The bombardier beetle bombards its enemies with a steam spray of highly toxic quinones. The artillery operates by a rapid mixing of hydrogen peroxide and hydroquinones to give a highly exothermic reaction. What kind of enzyme do you think is responsible for this explosive reaction?

24.8   What are the three main coordination modes encountered in the iron–sulfur proteins? Why is iron persistently tetrahedrally coordinated in the iron–sulfur proteins whereas it tends to be five- or six-coordinated in the heme proteins, hemerythrins, and siderophores?

24.9   Why is it so difficult to make model compounds for the active sites of blue copper proteins?

24.10   Assuming that Co did not exist, suggest a replacement molecule for cobalamin. Give reasons for each design feature of your molecule.

24.11   The coordination geometry of Fe(III) in nonheme proteins is highly variable. Would you expect Co(III) to behave similarly? (See Chapter 13.)

24.12   Construct qualitative MO diagrams to show the difference between a linear Mo—N—N molecule and a linear Mo—O—O molecule. Why are nitrogenases so sensitive to oxygen deactivation?

Appendices

Appendix I

Some Important Character Tables[1]

C_s	E	σ_h		
A'	1	1	x, y, R_z	$x^2, y^2,$ z^2, xy
A''	1	-1	z, R_x, R_y	yz, xz

C_i	E	i		
A_g	1	1	R_x, R_y, R_z	x^2, y^2, z^2 xy, xz, yz
A_u	1	-1	x, y, z	

---

[1] A more complete set of character tables may be found in F. A. Cotton. *Chemical Applications of Group Theory*, 2nd ed. New York: Wiley-Interscience, 1971.

C_2	E	$C_2(z)$		
A	1	1	z, R_z	x^2, y^2, z^2, xy
B	1	-1	x, y, R_x, R_y	yz, xz

D_3	E	$2C_3(z)$	$3C_2$		
A_1	1	1	1		$x^2 + y^2, z^2$
A_2	1	1	-1	z, R_z	
E	2	-1	0	$(x, y)(R_x, R_y)$	$(x^2 - y^2, xy)(xz, yz)$

C_{2v}	E	$C_2(z)$	$\sigma_v(xz)$	$\sigma_v'(yz)$		
A_1	1	1	1	1	z	x^2, y^2, z^2
A_2	1	1	-1	-1	R_z	xy
B_1	1	-1	1	-1	x, R_y	xz
B_2	1	-1	-1	1	y, R_x	yz

C_{3v}	E	$2C_3(z)$	$3\sigma_v$		
A_1	1	1	1	z	$x^2 + y^2, z^2$
A_2	1	1	-1	R_z	
E	2	-1	0	$(x, y)(R_x, R_y)$	$(x^2 - y^2, xy)(xz, yz)$

C_{4v}	E	$2C_4(z)$	$C_2(z)$	$2\sigma_v$	$2\sigma_d$		
A_1	1	1	1	1	1	z	$x^2 + y^2, z^2$
A_2	1	1	1	-1	-1	R_z	
B_1	1	-1	1	1	-1		$x^2 - y^2$
B_2	1	-1	1	-1	1		xy
E	2	0	-2	0	0	$(x, y)(R_z, R_y)$	(xz, yz)

C_{2h}	E	$C_2(z)$	i	σ_h		
A_g	1	1	1	1	R_z	x^2, y^2, z^2, xy
B_g	1	-1	1	-1	R_x, R_y	xz, yz
A_u	1	1	-1	-1	z	
B_u	1	-1	-1	1	x, y	

D_{2h}	E	$C_2(z)$	$C_2(y)$	$C_2(x)$	i	$\sigma(xy)$	$\sigma(xz)$	$\sigma(yz)$		
A_g	1	1	1	1	1	1	1	1		x^2, y^2, z^2
B_{1g}	1	1	-1	-1	1	1	-1	-1	R_z	xy
B_{2g}	1	-1	1	-1	1	-1	1	-1	R_y	xz
B_{3g}	1	-1	-1	1	1	-1	-1	1	R_x	yz
A_u	1	1	1	1	-1	-1	-1	-1		
B_{1u}	1	1	-1	-1	-1	-1	1	1	z	
B_{2u}	1	-1	1	-1	-1	1	-1	1	y	
B_{3u}	1	-1	-1	1	-1	1	1	-1	x	

D_{3h}	E	$2C_3(z)$	$3C_2$	σ_h	$2S_3(z)$	$3\sigma_v$		
A_1'	1	1	1	1	1	1		$x^2 + y^2, z^2$
A_2'	1	1	-1	1	1	-1	R_z	
E'	2	-1	0	2	-1	0	(x, y)	$(x^2 - y^2, xy)$
A_1''	1	1	1	-1	-1	-1		
A_2''	1	1	-1	-1	-1	1	z	
E''	2	-1	0	-2	1	0	(R_x, R_y)	(xz, yz)

D_{4h}	E	$2C_4(z)$	$C_2(z)$	$2C_2'$	$2C_2''$	i	$2S_4(z)$	σ_h	$2\sigma_v$	$2\sigma_d$		
A_{1g}	1	1	1	1	1	1	1	1	1	1		$x^2 + y^2, z^2$
A_{2g}	1	1	1	-1	-1	1	1	1	-1	-1	R_z	
B_{1g}	1	-1	1	1	-1	1	-1	1	1	-1		$x^2 - y^2$
B_{2g}	1	-1	1	-1	1	1	-1	1	-1	1		xy
E_g	2	0	-2	0	0	2	0	-2	0	0	(R_x, R_y)	(xz, yz)
A_{1u}	1	1	1	1	1	-1	-1	-1	-1	-1		
A_{2u}	1	1	1	-1	-1	-1	-1	-1	1	1	z	
B_{1u}	1	-1	1	1	-1	-1	1	-1	-1	1		
B_{2u}	1	-1	1	-1	1	-1	1	-1	1	-1		
E_u	2	0	-2	0	0	-2	0	2	0	0	(x, y)	

D_{5h}	E	$2C_5(z)$	$2C_5^2(z)$	$5C_2$	σ_h	$2S_5(z)$	$2S_5^3(z)$	$5\sigma_v$		
A_1'	1	1	1	1	1	1	1	1		$x^2 + y^2,\ z^2$
A_2'	1	1	1	-1	1	1	1	-1	R_z	
E_1'	2	$2\cos 72°$	$2\cos 144°$	0	2	$2\cos 72°$	$2\cos 144°$	0	(x, y)	
E_2'	2	$2\cos 144°$	$2\cos 72°$	0	2	$2\cos 144°$	$2\cos 72°$	0		$(x^2 - y^2, xy)$
A_1''	1	1	1	1	-1	-1	-1	-1		
A_2''	1	1	1	-1	-1	-1	-1	1	z	
E_1''	2	$2\cos 72°$	$2\cos 144°$	0	-2	$-2\cos 72°$	$-2\cos 144°$	0	(R_x, R_y)	(xz, yz)
E_2''	2	$2\cos 144°$	$2\cos 72°$	0	-2	$-2\cos 144°$	$-2\cos 72°$	0		

D_{6h}	E	$2C_6(z)$	$2C_3(z)$	$C_2(z)$	$3C_2'$	$3C_2''$	i	$2S_3(z)$	$2S_6(z)$	σ_h	$3\sigma_d$	$3\sigma_v$		
A_{1g}	1	1	1	1	1	1	1	1	1	1	1	1		$x^2 + y^2,\ z^2$
A_{2g}	1	1	1	1	-1	-1	1	1	1	1	-1	-1	R_z	
B_{1g}	1	-1	1	-1	1	-1	1	-1	1	-1	1	-1		
B_{2g}	1	-1	1	-1	-1	1	1	-1	1	-1	-1	1		
E_{1g}	2	1	-1	-2	0	0	2	1	-1	-2	0	0	(R_x, R_y)	(xz, yz)
E_{2g}	2	-1	-1	2	0	0	2	-1	-1	2	0	0		$(x^2 - y^2, xy)$
A_{1u}	1	1	1	1	1	1	-1	-1	-1	-1	-1	-1		
A_{2u}	1	1	1	1	-1	-1	-1	-1	-1	-1	1	1	z	
B_{1u}	1	-1	1	-1	1	-1	-1	1	-1	1	-1	1		
B_{2u}	1	-1	1	-1	-1	1	-1	1	-1	1	1	-1		
E_{1u}	2	1	-1	-2	0	0	-2	-1	1	2	0	0	(x, y)	
E_{2u}	2	-1	-1	2	0	0	-2	1	1	-2	0	0		

D_{2d}	E	$2S_4(z)$	C_2	$2C_2'$	$2\sigma_d$		
A_1	1	1	1	1	1		$x^2 + y^2,\ z^2$
A_2	1	1	1	-1	-1	R_z	
B_1	1	-1	1	1	-1		$x^2 - y^2$
B_2	1	-1	1	-1	1	z	xy
E	2	0	-2	0	0	$(x, y);$ (R_x, R_y)	(xz, yz)

D_{3d}	E	$2C_3(z)$	$3C_2$	i	$2S_6(z)$	$3\sigma_d$		
A_{1g}	1	1	1	1	1	1		x^2+y^2, z^2
A_{2g}	1	1	-1	1	1	-1	R_z	
E_g	2	-1	0	2	-1	0	(R_x, R_y)	(x^2-y^2, xy), (xz, yz)
A_{1u}	1	1	1	-1	-1	-1		
A_{2u}	1	1	-1	-1	-1	1	z	
E_u	2	-1	0	-2	1	0	(x, y)	

D_{4d}	E	$2S_8(z)$	$2C_4(z)$	$2S_8^3(z)$	C_2	$4C_2'$	$4\sigma_d$		
A_1	1	1	1	1	1	1	1		x^2+y^2, z^2
A_2	1	1	1	1	1	-1	-1	R_z	
B_1	1	-1	1	-1	1	1	-1		
B_2	1	-1	1	-1	1	-1	1	z	
E_1	2	$\sqrt{2}$	0	$-\sqrt{2}$	-2	0	0	(x, y)	
E_2	2	0	-2	0	2	0	0		(x^2-y^2, xy)
E_3	2	$-\sqrt{2}$	0	$\sqrt{2}$	-2	0	0	(R_x, R_y)	(xz, yz)

D_{5d}	E	$2C_5(z)$	$2C_5^2(z)$	$5C_2$	i	$2S_{10}^3(z)$	$2S_{10}(z)$	$5\sigma_d$		
A_{1g}	1	1	1	1	1	1	1	1		x^2+y^2, z^2
A_{2g}	1	1	1	-1	1	1	1	-1	R_z	
E_{1g}	2	$2\cos 72°$	$2\cos 144°$	0	2	$2\cos 72°$	$2\cos 144°$	0	(R_x, R_y)	(xz, yz)
E_{2g}	2	$2\cos 144°$	$2\cos 72°$	0	2	$2\cos 144°$	$2\cos 72°$	0		(x^2-y^2, xy)
A_{1u}	1	1	1	1	-1	-1	-1	-1		
A_{2u}	1	1	1	-1	-1	-1	-1	1	z	
E_{1u}	2	$2\cos 72°$	$2\cos 144°$	0	-2	$-2\cos 72°$	$-2\cos 144°$	0	(x, y)	
E_{2u}	2	$2\cos 144°$	$2\cos 72°$	0	-2	$-2\cos 144°$	$-2\cos 72°$	0		

T_d	E	$8C_3$	$3C_2$	$6S_4$	$6\sigma_d$		
A_1	1	1	1	1	1		$x^2 + y^2 + z^2$
A_2	1	1	1	-1	-1		
E	2	-1	2	0	0		$(2z^2 - x^2 - y^2, x^2 - y^2)$
T_1	3	0	-1	1	-1	(R_x, R_y, R_z)	
T_2	3	0	-1	-1	1	(x, y, z)	(xy, xz, yz)

O_h	E	$8C_3$	$6C_2$	$6C_4$	$3C_2(=C_4^2)$	i	$6S_4$	$8S_6$	$3\sigma_h$	$6\sigma_d$		
A_{1g}	1	1	1	1	1	1	1	1	1	1		$x^2 + y^2 + z^2$
A_{2g}	1	1	-1	-1	1	1	-1	1	1	-1		
E_g	2	-1	0	0	2	2	0	-1	2	0		$(2z^2 - x^2 - y^2, x^2 - y^2)$
T_{1g}	3	0	-1	1	-1	3	1	0	-1	-1	(R_x, R_y, R_z)	
T_{2g}	3	0	1	-1	-1	3	-1	0	-1	1		(xz, yz, xy)
A_{1u}	1	1	1	1	1	-1	-1	-1	-1	-1		
A_{2u}	1	1	-1	-1	1	-1	1	-1	-1	1		
E_u	2	-1	0	0	2	-2	0	1	-2	0		
T_{1u}	3	0	-1	1	-1	-3	-1	0	1	1	(x, y, z)	
T_{2u}	3	0	1	-1	-1	-3	1	0	1	-1		

I_h	E	$12C_5(z)$	$12C_5^2(z)$	$20C_3$	$15C_2$	i	$12S_{10}(z)$	$12S_{10}^3(z)$	$20S_6$	15σ		
A_g	1	1	1	1	1	1	1	1	1	1		$x^2 + y^2 + z^2$
T_{1g}	3	$\frac{1}{2}(1+\sqrt{5})$	$\frac{1}{2}(1-\sqrt{5})$	0	-1	3	$\frac{1}{2}(1-\sqrt{5})$	$\frac{1}{2}(1+\sqrt{5})$	0	-1	(R_x, R_y, R_z)	
T_{2g}	3	$\frac{1}{2}(1-\sqrt{5})$	$\frac{1}{2}(1+\sqrt{5})$	0	-1	3	$\frac{1}{2}(1+\sqrt{5})$	$\frac{1}{2}(1-\sqrt{5})$	0	-1		
G_g	4	-1	-1	1	0	4	-1	-1	1	0		
H_g	5	0	0	-1	1	5	0	0	-1	1		$(2z^2 - x^2 - y^2,$ $x^2 - y^2,$ $xy, yz, zx)$
A_u	1	1	1	1	1	-1	-1	-1	-1	-1		
T_{1u}	3	$\frac{1}{2}(1+\sqrt{5})$	$\frac{1}{2}(1-\sqrt{5})$	0	-1	-3	$-\frac{1}{2}(1-\sqrt{5})$	$-\frac{1}{2}(1+\sqrt{5})$	0	1	(x, y, z)	
T_{2u}	3	$\frac{1}{2}(1-\sqrt{5})$	$\frac{1}{2}(1+\sqrt{5})$	0	-1	-3	$-\frac{1}{2}(1+\sqrt{5})$	$-\frac{1}{2}(1-\sqrt{5})$	0	1		
G_u	4	-1	-1	1	0	-4	1	1	-1	0		
H_u	5	0	0	-1	1	-5	0	0	1	-1		

Appendix II

Some Common Units and Conversion Factors

There is almost a worldwide acceptance by the scientific community of Système Internationale, or SI, units. These units are constructed from a set of base units and metric prefixes. The relations between some commonly used conventional units and their SI equivalents are tabulated below.

Base Units		
Quantity	Name	Symbol
Length	meter	m
Mass	kilogram	kg
Time	second	s
Temperature	kelvin	K
Amount of substance	mole	mol
Electric current	ampere	A

Metric Prefixes

atto (a)	10^{-18}	deka (da)	10
fempto (f)	10^{-15}	hecto (h)	10^2
pico (p)	10^{-12}	kilo (k)	10^3
nano (n)	10^{-9}	mega (M)	10^6
micro (μ)	10^{-6}	giga (G)	10^9
milli (m)	10^{-3}	tera (T)	10^{12}
centi (c)	10^{-2}	peta (P)	10^{15}
deci (d)	10^{-1}	exa (E)	10^{18}

Abbreviations for Commonly Used Units

Quantity	Conventional	SI
Length	Å	nm, pm
Mass	g	kg
Volume	ml, cm^3, liter	cm^3, dm^3, m^3
Density (D)	$g\ ml^{-1}$, $g\ cm^{-3}$, $g\ liter^{-1}$	$g\ cm^{-3}$, $kg\ dm^{-3}$
Amount of substance	mole	mol
Concentration	$mole\ liter^{-1}$	$mol\ dm^{-3}$
Temperature	°K	K
Pressure	atm, torr	Pa
Time	sec, min, h	s, min, h
Speed	$cm\ sec^{-1}$	$m\ s^{-1}$
Frequency (v)	sec^{-1}	Hz
Energy	cal, erg, eV, cm^{-1}	J, eV, cm^{-1}
Ionization energy	$eV\ atom^{-1}$	$kJ\ mol^{-1}$
Electron affinity	$kcal\ mole^{-1}$	$kJ\ mol^{-1}$
Electric charge	esu, coulomb	C
Electric current	amp	A
Electric potential difference	V	V
Force	dyne	N
Force constant (k)	$mdyne\ Å^{-1}$	$N\ m^{-1}$
Dipole moment (μ_D)	D	C m
Molar absorptivity (ϵ)	$liter\ mole^{-1}\ cm^{-1}$	$dm^3\ mol^{-1}\ cm^{-1}$
Radioactivity	Ci	Bq

Physical Constants (in Both Unit Systems)

Speed of light $(c) = 2.998 \times 10^{10}$ cm sec^{-1} = 2.998×10^{8} m s^{-1}

Planck's constant $(h) = 6.626 \times 10^{-27}$ erg sec = 6.626×10^{-34} J s

Boltzmann's constant $(k) = 1.381 \times 10^{-16}$ erg °K^{-1} = 1.381×10^{-23} J K^{-1}

Rydberg constant $(R_{\mathrm{H}}) = 1.097 \times 10^{5}$ cm^{-1}

Gas constant $(R) = 1.987$ cal mole^{-1} °K^{-1} = 0.08205 liter atm mole^{-1} K^{-1}
$\qquad\qquad = 82.05$ cm^{3} atm mole^{-1} °K^{-1} = 8.314 J mol^{-1} K^{-1}
$\qquad\qquad = 8.314$ dm^{3} kPa mol^{-1} K^{-1}

Avogadro's number $(N) = 6.023 \times 10^{23}$ mole^{-1} = 6.023×10^{23} mol^{-1}

Conversion Factors

1 Å = 10^{-8} cm = 10^{-10} m = 0.1 nm = 100 pm

1 liter = 10^{3} ml = 10^{3} cm^{3} = 1 dm^{3}

1 erg = 10^{-7} J

1 cal = 4.184 J

1 eV = 23.06 kcal mole^{-1} = 96.49 kJ mol^{-1}

1 cm^{-1} = 0.011962 kJ mol^{-1}

1 atm = 1.013×10^{5} Pa

1 debye (D) = 3.336×10^{-30} C m

1 amu = 931.4 MeV

Appendix III

Atomic Masses of Elements Referred to $^{12}C = 12$ (Exactly)

Name	Symbol	Atomic Number	Atomic Weight
Actinium	Ac	89	(227)
Aluminum	Al	13	26.9815[a]
Americium	Am	95	(243)
Antimony	Sb	51	121.75[b]
Argon	Ar	18	39.948[b]
Arsenic	As	33	74.9216[a]
Astatine	At	85	~210
Barium	Ba	56	137.34[b]
Berkelium	Bk	97	(247)
Beryllium	Be	4	9.01218[a]
Bismuth	Bi	83	208.9806[a]
Boron	B	5	10.81[a]
Bromine	Br	35	79.904[a]

[a] Value reliable to ±1 in the last digit.
[b] Value reliable to ±3 in the last digit.

Name	Symbol	Atomic Number	Atomic Weight
Cadmium	Cd	48	112.40
Calcium	Ca	20	40.08
Californium	Cf	98	(251)
Carbon	C	6	12.011[a]
Cerium	Ce	58	140.12
Cesium	Cs	55	132.9055[a]
Chlorine	Cl	17	35.453[a]
Chromium	Cr	24	51.996[a]
Cobalt	Co	27	58.9332[a]
Copper	Cu	29	63.546[a]
Curium	Cm	96	(247)
Dysprosium	Dy	66	162.50
Einsteinium	Es	99	(254)
Erbium	Er	68	167.26
Europium	Eu	63	151.96
Fermium	Fm	100	(257)
Fluorine	F	9	18.9984[a]
Francium	Fr	87	(223)
Gadolinium	Gd	64	157.25[b]
Gallium	Ga	31	69.72
Germanium	Ge	32	72.59[b]
Gold	Au	79	196.9665[a]
Hafnium	Hf	72	178.49[b]
Helium	He	2	4.00260[a]
Holmium	Ho	67	164.9303[a]
Hydrogen	H	1	1.0080[a]
Indium	In	49	114.82
Iodine	I	53	126.9045[a]
Iridium	Ir	77	192.22[b]
Iron	Fe	26	55.847[b]
Krypton	Kr	36	83.80
Lanthanum	La	57	138.9055[b]
Lawrencium	Lr	103	(257)
Lead	Pb	82	207.2[a]
Lithium	Li	3	6.941[a]
Lutetium	Lu	71	174.97
Magnesium	Mg	12	24.305[a]
Manganese	Mn	25	54.9380[a]

[a] Value reliable to ± 1 in the last digit.
[b] Value reliable to ± 3 in the last digit.

Name	Symbol	Atomic Number	Atomic Weight
Mendelevium	Md	101	(256)
Mercury	Hg	80	200.59[b]
Molybdenum	Mo	42	95.94[b]
Neodymium	Nd	60	144.24[b]
Neon	Ne	10	20.179[b]
Neptunium	Np	93	237.0482[a]
Nickel	Ni	28	58.71[b]
Niobium	Nb	41	92.9064[a]
Nitrogen	N	7	14.0067[a]
Nobelium	No	102	(254)
Osmium	Os	76	190.2
Oxygen	O	8	15.9994[b]
Palladium	Pd	46	106.4
Phosphorus	P	15	30.9738[a]
Platinum	Pt	78	195.09[b]
Plutonium	Pu	94	(244)
Polonium	Po	84	(~210)
Potassium	K	19	39.102[b]
Praeseodymium	Pr	59	140.9077[b]
Promethium	Pm	61	(145)
Protactinium	Pa	91	231.0359[a]
Radium	Ra	88	226.0254[a]
Radon	Rn	86	(~222)
Rhenium	Re	75	186.2
Rhodium	Rh	45	102.9055[a]
Rubidium	Rb	37	85.4678[b]
Ruthenium	Ru	44	101.07[b]
Samarium	Sm	62	150.4
Scandium	Sc	21	44.9559[a]
Selenium	Se	34	78.96[b]
Silicon	Si	14	28.086[b]
Silver	Ag	47	107.868[a]
Sodium	Na	11	22.9898[a]
Strontium	Sr	38	87.62[a]
Sulfur	S	16	32.06[a]
Tantalum	Ta	73	180.9479[b]
Technetium	Tc	43	98.9062[a]
Tellurium	Te	52	127.60[b]
Terbium	Tb	65	158.9254[a]
Thallium	Tl	81	204.37[b]

[a] Value reliable to ±1 in the last digit.
[b] Value reliable to ±3 in the last digit.

Name	Symbol	Atomic Number	Atomic Weight
Thorium	Th	90	232.0381[a]
Thulium	Tm	69	168.9342[a]
Tin	Sn	50	118.69[b]
Titanium	Ti	22	47.90[b]
Tungsten	W	74	183.85[b]
Uranium	U	92	238.029[a]
Vanadium	V	23	50.9414[b]
Xenon	Xe	54	131.30
Ytterbium	Yb	70	173.04[b]
Yttrium	Y	39	88.9059[a]
Zinc	Zn	30	65.37[b]
Zirconium	Zr	40	91.22

[a] Value reliable to ±1 in the last digit.
[b] Value reliable to ±3 in the last digit.
SOURCE: *Handbook of Chemistry and Physics*, 52nd ed. Cleveland: The Chemical Rubber Co.

Index